Lecture Notes in Artificial Intelligence **13606**

Subseries of Lecture Notes in Computer Science

More information about this subseries at https://link.springer.com/bookseries/1244

Lu Fang · Daniel Povey · Guangtao Zhai ·
Tao Mei · Ruiping Wang (Eds.)

Artificial Intelligence

Second CAAI International Conference, CICAI 2022
Beijing, China, August 27–28, 2022
Revised Selected Papers, Part III

Springer

Editors
Lu Fang
Tsinghua University
Beijing, China

Daniel Povey
Xiaomi Inc.
Beijing, China

Guangtao Zhai
Shanghai Jiao Tong University
Shanghai, China

Tao Mei
JD Explore Academy
Beijing, China

Ruiping Wang
Chinese Academy of Sciences
Beijing, China

ISSN 0302-9743 ISSN 1611-3349 (electronic)
Lecture Notes in Artificial Intelligence
ISBN 978-3-031-20502-6 ISBN 978-3-031-20503-3 (eBook)
https://doi.org/10.1007/978-3-031-20503-3

LNCS Sublibrary: SL7 – Artificial Intelligence

This Springer imprint is published by the registered company Springer Nature Switzerland AG
The registered company address is: Gewerbestrasse 11, 6330 Cham, Switzerland

Preface

The present book includes extended and revised versions of papers selected from the second CAAI International Conference on Artificial Intelligence (CICAI 2022), held in Beijing, China, during August 27–28, 2022.

CICAI is a summit forum in the field of artificial intelligence and the 2022 forum was hosted by Chinese Association for Artificial Intelligence (CAAI). CICAI aims to establish a global platform for international academic exchange, promote advanced research in AI and its affiliated disciplines, and promote scientific exchanges among researchers, practitioners, scientists, students, and engineers in AI and its affiliated disciplines in order to provide interdisciplinary and regional opportunities for researchers around the world, enhance the depth and breadth of academic and industrial exchanges, inspire new ideas, cultivate new forces, implement new ideas, integrate into the new landscape, and join the new era. The conference program included invited talks delivered by two distinguished speakers, Qiang Yang and Dacheng Tao, as well as five theme tutorials with five talks for each theme, followed by an oral session of 18 papers, a poster session of 127 papers, and a demo exhibition of 19 papers. Those papers were selected from 521 submissions using a double-blind review process, and on average each submission received 3.2 reviews. The topics covered by these selected high-quality papers span the fields of machine learning, computer vision, natural language processing, and data mining, amongst others.

This three-volume series contains 164 papers selected and revised from the proceedings of CICAI 2022. We would like to thank the authors for contributing their novel ideas and visions that are recorded in this book.

The proceeding editors also wish to thank all reviewers for their contributions and Springer for their trust and for publishing the proceedings of CICAI 2022.

September 2022

Lu Fang
Daniel Povey
Guangtao Zhai
Tao Mei
Ruiping Wang

Organization

General Chairs

Lu Fang	Tsinghua University, China
Daniel Povey	Xiaomi, China
Guangtao Zhai	Shanghai Jiao Tong University, China

Honorary Program Chair

Lina J. Karam	Lebanese American University, Lebanon

Program Chairs

Tao Mei	JD Explore Academy, China
Ruiping Wang	Chinese Academy of Sciences, China

Publication Chairs

Hui Qiao	Tsinghua University, China
Adriana Tapus	Institut Polytechnique de Paris, France

Presentation Chairs

Mengqi Ji	Beihang University, Singapore
Zhou Zhao	Zhejiang University, China
Shan Luo	King's College London, UK

Demo Chairs

Kun Li	Tianjin University, China
Fu Zhang	University of Hong Kong, China

International Liaison Chair

Feng Yang	Google Research, USA

Advisory Committee

C. L. Philip Chen	University of Macau, China
Xilin Chen	Institute of Computing Technology, Chinese Academy of Sciences, China
Yike Guo	Imperial College London, UK
Ping Ji	The City University of New York, USA
Licheng Jiao	Xidian University, China
Ming Li	University of Waterloo, Canada
Chenglin Liu	Institute of Automation, Chinese Academy of Sciences, China
Derong Liu	University of Illinois at Chicago, USA
Hong Liu	Peking University, China
Hengtao Shen	University of Electronic Science and Technology of China, China
Yuanchun Shi	Tsinghua University, China
Yongduan Song	Chongqing University, China
Fuchun Sun	Tsinghua University, China
Jianhua Tao	Institute of Automation, Chinese Academy of Sciences, China
Guoyin Wang	Chongqing University of Posts and Telecommunications, China
Weining Wang	Beijing University of Posts and Telecommunications, China
Xiaokang Yang	Shanghai Jiao Tong University, China
Changshui Zhang	Tsinghua University, China
Lihua Zhang	Fudan University, China
Song-Chun Zhu	Peking University, China
Wenwu Zhu	Tsinghua University, China
Yueting Zhuang	Zhejiang University, China

Program Committee

Abdul Rehman	Bournemouth University, UK
Biao Jie	Anhui Normal University, China
Bing Cao	Tianjin University, China
Bo Xue	University of Science and Technology of China, China
Bo Wang	Dalian University of Technology, China
Bochen Guan	OPPO US Research Center, USA
Boyun Li	Sichuan University, China
Chang Yao	East China Normal University, China
Chao Bian	Nanjing University, China

Huan Yin	Hong Kong University of Science and Technology, China
Huanjing Yue	Tianjin University, China
Hui Chen	Tsinghua University, China
Huiyu Duan	Shanghai Jiao Tong University, China
Jiajun Deng	University of Science and Technology of China, China
Jian Zhao	Institute of North Electronic Equipment, China
Jianguo Sun	Harbin Engineering University, China
Jianhui Chang	Peking University, China
Jianing Sun	Dalian University of Technology, China
Jia-Wei Chen	Xidian University, China
Jimin Pi	Baidu, China
Jing Chen	Beijing Research Institute of Precise Mechatronics and Controls, China
Jingwen Guo	Peking University Shenzhen Graduate School, China
Jingyu Yang	Tianjin University, China
Jinjian Wu	Xidian University, China
Jinsong Zhang	Tianjin University, China
Jinyu Tian	University of Macau, China
Jinyuan Liu	Dalian University of Technology, China
Jun Wang	Shanghai University, China
Jupo Ma	Xidian University, China
Kai Hu	Xiangtan University, China
Kaiqin Hu	Xi'an Jiaotong University, China
Kan Guo	Beihang University, China
Ke Xue	Nanjing University, China
Keyang Wang	Chongqing University, China
Keyu Li	Xidian University, China
Kun Cheng	Xidian University, China
Kun Zhang	Hefei University of Technology, China
Le Wang	Xi'an Jiaotong University, China
Le Wu	Hefei University of Technology, China
Lei Wang	University of Wollongong, Australia
Lei Shi	Zhengzhou University, China
Leida Li	Xidian University, China
Liansheng Zhuang	University of Science and Technology of China, China
Liguo Zhang	Harbin Engineering University, China
Likang Wu	University of Science and Technology of China, China

Lili Zhao	University of Science and Technology of China, China
Lizhi Wang	Beijing Institute of Technology, China
Longguang Wang	National University of Defense Technology, China
Meiyu Huang	China Academy of Space Technology, China
Meng Wang	Hefei University of Technology, China
Mengting Xu	Nanjing University of Aeronautics and Astronautics, China
Mengxi Jia	Peking University Shenzhen Graduate School, China
Min Wang	Hefei Comprehensive National Science Center, China
Mingkui Tan	South China University of Technology, China
Mingrui Zhu	Xidian University, China
Min-Ling Zhang	Southeast University, China
Mouxing Yang	Sichuan University, China
Ningyu Zhang	Zhejiang University, China
Peijie Sun	Tsinghua University, China
Pengfei Zhang	National University of Defense Technology, China
Pengyang Shao	Hefei University of Technology, China
Pingping Zhang	Dalian University of Technology, China
Qi Liu	University of Science and Technology of China, China
Qian Ning	Xidian University, China
Qian Xu	CILAB
Qiao Feng	Tianjin University, China
Qing Li	Beijing Normal University, China
Qingbo Wu	University of Electronic Science and Technology of China, China
Qinglin Wang	National University of Defense Technology, China
Qun Liu	Chongqing University of Posts and Telecommunications, China
Richang Hong	Hefei University of Technology, China
Rongjun Ge	Southeast University, China
Ruiping Wang	Institute of Computing Technology, Chinese Academy of Sciences, China
Ruqi Huang	Tsinghua University, China
Sheng Shen	Tianjin University, China
Shishuai Hu	Northwestern Polytechnical University, China

Shuaifeng Zhi	National University of Defense Technology, China
Shuang Yang	Institute of Computing Technology, Chinese Academy of Sciences, China
Shuang Li	The Chinese University of Hong Kong, Shenzhen, China
Shulan Ruan	University of Science and Technology of China, China
Si Liu	Beihang University, China
Sida Peng	Zhejiang University, China
Sinuo Deng	Beijing University of Technology, China
Tan Guo	Chongqing University of Posts and Telecommunications, China
Tao Huang	Xidian University, China
Tao Yue	Nanjing University, China
Tao Zhang	Tsinghua University, China
Tao He	Tsinghua University, China
Ting Shu	Shenzhen Institute of Meteorological Innovation, China
Waikeung Wong	Hong Kong Polytechnic University, China
Wangmeng Zuo	Harbin Institute of Technology, China
Wei Cao	University of Science and Technology of China, China
Wei Jia	Heifei University of Technology, China
Wei Shen	Shanghai Jiao Tong University, China
Wei Sun	Shanghai Jiao Tong University, China
Wei Pang	Beijing Information Science and Technology University, China
Wei Hu	Peking University, China
Weicheng Xie	Shenzhen University, China
Weishi Zheng	Sun Yat-sen University, China
Wenbin Wang	Institute of Computing Technology, Chinese Academy of Sciences, China
Xi Li	Zhejiang University, China
Xia Wu	Beijing Normal University, China
Xiang Ao	Institute of Computing Technology, Chinese Academy of Sciences, China
Xiang Chen	Zhejiang University and AZFT Joint Lab for Knowledge Engine, China
Xiang Gao	Jihua Laboratory, China
Xiang Bai	Huazhong University of Science and Technology, China
Xiangjun Yin	Tianjin University, China

Xiangwei Kong	Zhejiang University, China
Xianwei Zheng	Foshan University, China
Xiaodan Liang	Sun Yat-sen University, China
Xiaopeng Hong	Xi'an Jiaotong University, China
Xiaoyan Luo	Beihang University, China
Xiaoyong Lu	Northwest Normal University, China
Xiaoyun Yuan	Tsinghua University, China
Xin Yang	Dalian University of Technology, China
Xin Yuan	Westlake University, China
Xin Geng	Southeast University, China
Xin Xu	JD.com, China
Xinggang Wang	Huazhong University of Science and Technology, China
Xinglong Zhang	National University of Defense Technology, China
Xinpeng Ding	The Hong Kong University of Science and Technology, China
Xiongkuo Min	Shanghai Jiao Tong University, China
Xiongzheng Li	Tianjin University, China
Xiushan Nie	Shandong Jianzhu University, China
Xiu-Shen Wei	Nanjing University of Science and Technology, China
Xueyuan Xu	Beijing Normal University, China
Xun Chen	University of Science and Technology of China, China
Xuran Pan	Tsinghua University, China
Yang Li	Tsinghua-Berkeley Shenzhen Institute, Tsinghua University, China
Yangang Wang	Southeast University, China
Yaping Zhao	The University of Hong Kong, China
Ye Tian	Harbin Engineering University, China
Yebin Liu	Tsinghua University, China
Yi Hao	Xidian University, China
Yi Zhang	Xidian University, China
Yicheng Wu	Monash University, Australia
Yifan Zhang	National University of Defense Technology, China
Yijie Lin	Sichuan University, China
Ying Fu	Beijing Institute of Technology, China
Yingqian Wang	National University of Defense Technology, China
Yiwen Ye	Northwestern Polytechnical University, Canada
Yonghui Yang	Hefei University of Technology, China

Yu Liu	Dalian University of Technology, China
Yu Wang	Beijing Technology and Business University, China
Yuanbiao Gou	Sichuan University, China
Yuanfang Guo	Beihang University, China
Yuanman Li	Shenzhen University, China
Yuchao Dai	Northwestern Polytechnical University, China
Yucheng Zhu	Shanghai Jiao Tong University, China
Yufei Gao	Zhengzhou University, China
Yulin Cai	Tsinghua University, China
Yulun Zhang	ETH Zurich, Switzerland
Yun Tie	Zhengzhou University, China
Yunfan Li	Sichuan University, China
Zhanxiang Feng	Sun Yat-sen University, China
Zhaobo Qi	University of Chinese Academy of Sciences, China
Zhaoxiang Zhang	Chinese Academy of Sciences, China
Zhaoxin Liu	Xidian University, China
Zheng-Jun Zha	University of Science and Technology of China, China
Zhengming Zhang	Southeast University, China
Zhengyi Wang	Tsinghua University, China
Zhenya Huang	University of Science and Technology of China, China
Zhenyu Huang	Sichuan University, China
Zhenzhen Hu	National University of Defense Technology, China
Zhibo Wang	Tsinghua University, China
Zhiheng Fu	University of Western Australia, Australia
Zhiying Jiang	Dalian University of Technology, China
Zhiyuan Zhu	Western University, Canada
Zhu Liu	Dalian University of Technology, China
Ziwei Zheng	Xi’an Jiaotong University, China
Zizhao Zhang	Tsinghua University, China
Zongzhang Zhang	Nanjing University, China
Zunlei Feng	Zhejiang University, China

Contents – Part III

Intelligent Multilingual Information Processing

A New Method for Assigning Hesitation Based on an IFS Distance Metric 3
Changlin Xu and Yaqing Wen

Adaptive Combination of Filtered-X NLMS and Affine Projection Algorithms for Active Noise Control 15
Xichao Wang, Shifeng Ou, and Yunhe Pang

Linguistic Interval-Valued Spherical Fuzzy Sets and Related Properties 26
Yanzhe Liu, Ye Zhang, Xiaosong Cui, and Li Zou

Knowledge Representation and Reasoning

A Genetic Algorithm for Causal Discovery Based on Structural Causal Model 39
Zhengyin Chen, Kun Liu, and Wenpin Jiao

Stochastic and Dual Adversarial GAN-Boosted Zero-Shot Knowledge Graph 55
Xuejiao Liu, Yaowei Guo, Meiyu Huang, and Xueshuang Xiang

Machine Learning

LS-YOLO: Lightweight SAR Ship Targets Detection Based on Improved YOLOv5 71
Yaqi He, Zi-Xin Li, and Yu-Long Wang

Dictionary Learning-Based Reinforcement Learning with Non-convex Sparsity Regularizer 81
Haoli Zhao, Junkui Wang, Xingming Huang, Zhenini Li, and Shengli Xie

Deep Twin Support Vector Networks 94
Mingchen Li and Zhiji Yang

Region-Based Dense Adversarial Generation for Medical Image Segmentation 107
Ao Shen, Liang Sun, Mengting Xu, and Daoqiang Zhang

Dynamic Clustering Federated Learning for Non-IID Data 119
Ming Chen, Jinze Wu, Yu Yin, Zhenya Huang, Qi Liu, and Enhong Chen

Dynamic Network Embedding by Using Sparse Deep Autoencoder 132
Huimei Tang, Zengyang Shao, Yutao Zhang, Lijia Ma, and Qiuzhen Lin

Deep Graph Convolutional Networks Based on Contrastive Learning: Alleviating Over-smoothing Phenomenon 144
Rui Jin, Yibing Zhan, and Rong Zhang

Clustering-based Curriculum Construction for Sample-Balanced Federated Learning 155
Zhuang Qi, Yuqing Wang, Zitan Chen, Ran Wang, Xiangxu Meng, and Lei Meng

A Novel Nonlinear Dictionary Learning Algorithm Based on Nonlinear-KSVD and Nonlinear-MOD 167
Xiaoju Chen, Yujie Li, Shuxue Ding, Benying Tan, and Yuqi Jiang

Tooth Defect Segmentation in 3D Mesh Scans Using Deep Learning 180
Hao Chen, Yuhao Ge, Jiahao Wei, Huimin Xiong, and Zuozhu Liu

Multi-agent Systems

Crowd-Oriented Behavior Simulation:Reinforcement Learning Framework Embedded with Emotion Model 195
Zhiwei Liang, Lei Li, and Lei Wang

Deep Skill Chaining with Diversity for Multi-agent Systems* 208
Zaipeng Xie, Cheng Ji, and Yufeng Zhang

Natural Language Processing

Story Generation Based on Multi-granularity Constraints 223
Zhenpeng Guo, Jiaqiang Wan, Hongan Tang, and Yunhua Lu

Chinese Word Sense Embedding with SememeWSD and Synonym Set 236
Yangxi Zhou, Junping Du, Zhe Xue, Ang Li, and Zeli Guan

Nested Named Entity Recognition from Medical Texts: An Adaptive Shared Network Architecture with Attentive CRF 248
Junzhe Jiang, Mingyue Cheng, Qi Liu, Zhi Li, and Enhong Chen

CycleResume: A Cycle Learning Framework with Hybrid Attention for Fine-Grained Talent-Job Fit 260
Zichen Zhang, Yong Luo, Yonggang Wen, and Xinwen Zhang

Detecting Alzheimer's Disease Based on Acoustic Features Extracted from Pre-trained Models 272
Kangdi Mei, Zhiqiang Guo, Zhaoci Liu, Lijuan Liu, Xin Li, and Zhenhua Ling

Similar Case Based Prison Term Prediction 284
Siying Zhou, Yifei Liu, Yiquan Wu, Kun Kuang, Chunyan Zheng, and Fei Wu

Interactive Fusion Network with Recurrent Attention for Multimodal Aspect-based Sentiment Analysis 298
Jun Wang, Qianlong Wang, Zhiyuan Wen, Xingwei Liang, and Ruifeng Xu

Developing Relationships: A Heterogeneous Graph Network with Learnable Edge Representation for Emotion Identification in Conversations 310
Zhenyu Li, Geng Tu, Xingwei Liang, and Ruifeng Xu

Optimization

Adaptive Differential Evolution Algorithm with Multiple Gaussian Learning Models 325
Genghui Li, Qingyan Li, and Zhenkun Wang

Online Taxi Dispatching Algorithm Based on Quantum Annealing 337
Chao Wang, Tongyu Ji, and Suming Wang

A Snapshot Gradient Tracking for Distributed Optimization over Digraphs 348
Keqin Che and Shaofu Yang

Differential Evolution Constrained Optimization for Peak Reduction 361
Min Wang, Ting Wen, Xiaoyu Jiang, and Anan Zhang

Evolutionary Multitasking Optimization for Multiobjective Hyperspectral Band Selection 374
Pu Xiong, Xiangming Jiang, Runyu Wang, Hao Li, Yue Wu, and Maoguo Gong

Robotics

Research on Control Sensor Allocation of Industrial Robots 389
Bin Yang, Zhouzhou Huang, and Wenyu Yang

Event-Based Obstacle Sensing and Avoidance for an UAV Through Deep Reinforcement Learning 402
Xinyu Hu, Zhihong Liu, Xiangke Wang, Lingjie Yang, and Guanzheng Wang

Suspension Control of Maglev Train Based on Extended Kalman Filter and Linear Quadratic Optimization 414
Fengxing Li, Yougang Sun, Hao Xu, Guobin Lin, and Zhenyu He

EO-SLAM: Evolutionary Object Slam in Perceptual Constrained Scene 426
Chen Jiahao and Li Xiuzhi

ADRC Based Multi-task Priority Tracking Control for Collaborative Robots ... 439
Kun Fan, Yanhong Liu, Kuan Zhang, Guibin Bian, and Hongnian Yu

Model Predictive Tracking Control for USV with Model Error Learning 451
Siyu Chen, Huiping Li, and Fei Li

Other AI Related Topics

A Hybrid Pattern Knowledge Graph-Based API Recommendation Approach 465
Guan Wang, Weidong Wang, and Dian Li

Intelligence Quotient Scores Prediction in rs-fMRI via Graph Convolutional Regression Network 477
Hao Zhang, Ran Song, Dawei Wang, Liping Wang, and Wei Zhang

A Novel In-Sensor Computing Architecture Based on Single Photon Avalanche Diode and Dynamic Memristor 489
Jiyuan Zheng, Shaoliang Yu, Jiamin Wu, Yuyan Wang, Chenchen Deng, and Zhu Lin

Lane Change Decision-Making of Autonomous Driving Based on Interpretable Soft Actor-Critic Algorithm with Safety Awareness 501
Di Yu, Kang Tian, Yuhui Liu, and Manchen Xu

Demo

GLRNet: Gas Leak Recognition via Temporal Difference in Infrared Video 515
Erqi Huang, Linsen Chen, Tao Lv, and Xun Cao

ATC-WSA: Working State Analysis for Air Traffic Controllers 521
Bo Liu, Xuanqian Wang, Jingjin Dong, Di Li, and Feng Lu

3D-Producer: A Hybrid and User-Friendly 3D Reconstruction System 526
Jingwen Chen, Yiheng Zhang, Zhongwei Zhang, Yingwei Pan, and Ting Yao

Contactless Cardiogram Reconstruction Based on the Wavelet Transform via Continuous-Wave Radar 532
Shuqin Dong, Changzhan Gu, and Xiaokang Yang

Intelligent Data Extraction System for RNFL Examination Reports 537
Chunjun Hua, Yiqiao Shi, Menghan Hu, and Yue Wu

NeRFingXR: An Interactive XR Tool Based on NeRFs 543
Shunli Luo, Hanxing Li, Haijing Cheng, Shi Pan, and Shuangpeng Sun

A Brain-Controlled Mahjong Game with Artificial Intelligence Augmentation 548
Xiaodi Wu, Yu Qi, Xinyun Zhu, Kedi Xu, Junming Zhu, Jianmin Zhang, and Yueming Wang

VAFA: A Visually-Aware Food Analysis System for Socially-Engaged Diet Management 554
Hang Wu, Xi Chen, Xuelong Li, Haokai Ma, Yuze Zheng, Xiangxian Li, Xiangxu Meng, and Lei Meng

XIVA: An Intelligent Voice Assistant with Scalable Capabilities for Educational Metaverse 559
Jun Lin, Yonghui Xu, Wei Guo, Lizhen Cui, and Chunyan Miao

Weld Defect Detection and Recognition System Based on Static Point Cloud Technology 564
Changzhi Zhou, Siming Liu, Fei Huang, Qian Huang, Anbang Yan, and Guocui Luo

3D Human Pose Estimation Based on Multi-feature Extraction 570
Senlin Ge, Huan Yu, Yuanming Zhang, Huitao Shi, and Hao Gao

Visual Localization Through Virtual Views 582
Zhenbo Song, Xi Sun, Zhou Xue, Dong Xie, and Chao Wen

A Synchronized Multi-view System for Real-Time 3D Hand Pose Estimation ... 588
Zhipeng Yu and Yangang Wang

Monocular Real-Time Human Geometry Reconstruction ... 594
Qiao Feng, Yebin Liu, Yu-Kun Lai, Jingyu Yang, and Kun Li

EasyPainter: Customizing Your Own Paintings ... 599
Yuqi Zhang, Qian Zheng, and Gang Pan

Gesture Interaction for Gaming Control Based on an Interferometric Radar ... 605
Yuchen Li, Jingyun Lu, Changzhan Gu, and Xiaokang Yang

Accelerating Allen Brain Institute's Large-Scale Computational Model of Mice Primary Visual Cortex ... 610
Zefan Wang, Kuiyu Wang, and Xiaolin Hu

Artistic Portrait Applet, Robot, and Printer ... 615
Jingjie Zhu, Lingna Dai, Chenghao Xia, Chenyang Jiang, Weiyu Weng, Yiyuan Zhang, Jinglin Zhou, Fei Gao, Peng Li, Mingrui Zhu, and Nannan Wang

Sim-to-Real Hierarchical Planning and Control System for Six-Legged Robot ... 621
Yue Gao, Yangqing Fu, and Ming Sun

Author Index ... 627

Intelligent Multilingual Information Processing

A New Method for Assigning Hesitation Based on an IFS Distance Metric

Changlin Xu(✉) and Yaqing Wen

School of Mathematics and Information Science, North Minzu University, Yinchuan 750021, China
xuchlin@163.com

Abstract. Firstly, we analyze the shortcomings of the existing intuitionistic fuzzy set distance measure, and then propose a method for assigning the hesitation degree of intuitionistic fuzzy sets by considering the fuzzy meanings expressed by the membership degree, non-membership degree, and hesitation degree, and prove through theoretical derivation that the intuitionistic fuzzy set distance measure under the new method satisfies the conditions of the distance measure, and illustrate the rationality of the method with the analysis and comparison of numerical examples. Finally, the method is applied to practical medical diagnosis. The experimental results show that the new method has certain validity and feasibility.

Keywords: Intuitionistic fuzzy sets · Hesitation · Distance

1 Introduction

The world has not only a precise aspect but also a vague one. The world is precise in terms of the unity, regularity, and certainty of the matter in it. In 1965, Zadeh [1], an American computer and cybernetics expert, introduced the concept of fuzzy sets, which laid the foundation for better characterizing the nature of fuzziness between things. 1986, the Bulgarian scholar Atanassov [2] proposed intuitionistic fuzzy sets based on fuzzy set theory. Currently, intuitionistic fuzzy set theory has received a lot of attention and its research results have been widely used in multi-attribute decision making [3,4], medical diagnosis [5], pattern recognition [6], and cluster analysis [7], etc.

The hotspots of research on intuitionistic fuzzy set metrics among existing methods are divided into two main areas. (1)Intuitionistic fuzzy set distance measure: Atanassov [8] proposed an intuitionistic fuzzy set Hamming distance and Euclidean distance based on fuzzy set distance, and achieved the distinction between fuzzy concepts by establishing a distance formula between membership and non-membership. szmdit [9] et al. improved on Atanassov's method by introducing hesitation into the distance measure. 2018 D. ke [10] et al. summarize the

This work is supported by the National Natural Science Foundation of China under Grant 62066001 and the Natural Science Foundation of Ningxia Province under Grant 2022AAC03238.

L. Fang et al. (Eds.): CICAI 2022, LNAI 13606, pp. 3–14, 2022.
https://doi.org/10.1007/978-3-031-20503-3_1

experience of previous studies and define a new distance measure for intuitionistic fuzzy sets based on the interval-valued distance measure and the transformation of intuitionistic fuzzy sets to interval-valued fuzzy sets. Later literatures [11–14] have made new generalization and research on distance measurement of intuitionistic fuzzy sets. (2)Similarity measure for intuitionistic fuzzy sets: Li and Cheng [15] proposed an axiomatic property for the similarity measure of intuitionistic fuzzy sets. They also proposed a similarity measure for intuitionistic fuzzy sets and applied it to pattern recognition. The literature [16] modified it and developed a more general similarity metric. Later, literature [17–19] proposed a new similarity measurement method of intuitionistic fuzzy sets from different perspectives.

Among the above similarity and distance metrics, the distance metric based on the definition of general distance metric has a weak differentiation ability. Therefore, it is necessary to propose a more reasonable and effective distance metric for intuitionistic fuzzy sets to overcome the shortcomings of existing methods. In this paper, we propose a new intuitionistic fuzzy set distance measure based on the potential relationship between the membership, non-membership and hesitation degrees in intuitionistic fuzzy sets. Through theoretical derivation, it is shown that the method satisfies the conditions related to distance and intuitionistic fuzzy set distance. A comparative analysis between the existing method and the new method is carried out by using numerical arithmetic examples to justify its soundness. Finally, the new method is applied to an example to illustrate its validity and feasibility.

2 Preparatory Knowledge

2.1 Intuitionistic Fuzzy Sets

Definition 1. *[2] Let the theoretical domain X be a non-empty set, then the intuitionistic fuzzy set A on X can be expressed as $A = \{< x, \mu_A(x), \nu_A(x) > | x \in X\}$, where $\mu_A(x)$ and $\nu_A(x)$ are the subordination and non-subordination of the element x in the thesis domain X belonging to A, respectively, i.e. $\mu_A(x) : X \to [0,1], x \in X \to \mu_A(x) \in [0,1]$, $\nu_A(x) : X \to [0,1], x \in X \to \nu_A(x) \in [0,1]$ satisfying both $x \in X, 0 \le \mu_A + \nu_A \le 1$, then we have $\pi_A(x) = 1 - \mu_A - \nu_A$. X The elements in x belonging to A are hesitations or uncertainties. For any x , there is $0 \le \pi_A(x) \le 1$. Intuitionistic fuzzy sets are referred to as IFSs.*

2.2 Intuitionistic Fuzzy Set Distance Metric

Definition 2. *Let X be any non-empty set and for any two points x, y in X there is a real number $d(x,y)$ corresponding to it and satisfying*

(1) Nonnegativity, $d(x,y) \ge 0$, and $d(x,y) = 0$ when and only when $x = y$;
(2) Symmetry, $d(x,y) = d(y,x)$;
(3) Triangle inequality, $d(x,z) + d(z,y) \ge d(x,y)$ Call $d(x,y)$ a distance in X;

(4) if $A \subseteq B \subseteq C$, i.e. the corresponding subordination and non-subordination degrees respectively satisfy.

Definition 3. *[20] calls the mapping $D : IFSs(X) \times IFSs(X) \to [0,1]$ a distance measure between intuitionistic fuzzy sets on the theoretical domain X if for any intuitionistic fuzzy set $A, B, C \in IFSs(X)$, the mapping D satisfies*

D1FS1: $0 \leq D(A,B) \leq 1, D(A,B) = 0$ *when and only when* $A = B$;
D2FS2: $D(A,B) = D(B,A)$;
D3FS3: $D(A,C) \leq D(A,B) + D(B,C)$;
D4FS4: if $A \subseteq B \subseteq C$, *i.e. the corresponding subordination and non-subordination degrees respectively satisfy* $\mu_A(x_i) \leq \mu_B(x_i) \leq \mu_C(x_i) \nu_A(x_i) \geq \nu_B(x_i) \geq \nu_C(x_i)$. *Then* $D(A,C) \geq D(A,B)$, $D(A,C) \geq D(B,C)$.

3 Existing Intuitionistic Fuzzy Set Distance Metrics and Their Analysis

3.1 Existing Intuitionistic Fuzzy Set Distance Metric

This paper draws on the notation of the literature [20], so that $\Delta_\mu(i) = \Delta_\mu^{AB}(i) = \mu_A(x_i) - \mu_B(x_i)$, $\Delta_\nu(i) = \Delta_\nu^{AB}(i) = \nu_A(x_i) - \nu_B(x_i)$, $\Delta_\pi(i) = \Delta_\pi^{AB}(i) = \pi_A(x_i) - \pi_B(x_i)$.
(1) Atanassov [8] on the representation of the hamming distance $D_H^{Ata}(A,B)$ for intuitionistic fuzzy sets, with the Euclidean distance $D_E^{Ata} = (A,B)$.

$$\mathrm{D}_H^{Ata}(A,B) = \frac{1}{2}\sum_{i=1}^{n}[|\Delta_\mu(i)| + |\Delta_\nu(i)|], \tag{1}$$

$$D_E^{Ata} = (A,B) = \sqrt{\frac{1}{2}\sum_{i=1}^{n}\left[(\Delta_\mu(i))^2 + (\Delta_\nu(i))^2\right]}. \tag{2}$$

In Eqs. (1) and (2) only the membership and non-membership degrees are considered, and the effect of the hesitation degree on the distance metric is not considered.
(2) Yang [21] et al. introduced hesitation into the intuitionistic fuzzy set distance metric based on the above equation to obtain a new distance metric calculation method.

$$\mathrm{D}_H^{YA}(A,B) = \sum_{i=1}^{n}\max\{|\Delta_\mu(i)|, |\Delta_\nu(i)|, |\Delta_\pi(i)|\}, \tag{3}$$

$$\mathrm{D}_E^{YA}(A,B) = \sqrt{\sum_{i=1}^{n}\max\left\{(\Delta_\mu(i))^2, (\Delta_\nu(i))^2, (\Delta_\pi(i))^2\right\}}. \tag{4}$$

The analysis in the literature [20] shows that although hesitation is introduced in the distance metric in Eqs. (3) (4), the new distance metric formula does

not necessarily satisfy the DIFS4 condition in Definition 3. Therefore the above metric still has certain shortcomings.
(3) An intuitive fuzzy set distance metric calculation method based on hesitation redistribution proposed by Xu [20] and other scholars.

$$\mathrm{MD}(A,B)=\sqrt{\frac{1}{4}\left[\left(\Delta_{\mu}^{AB}\right)^{2}+\left(\Delta_{\nu}^{AB}\right)^{2}+\left(\Delta_{\pi\rightarrow\mu}^{AB}\right)^{2}+\left(\Delta_{\pi\rightarrow\nu}^{AB}\right)^{2}\right]}, \tag{5}$$

$$\mathrm{MD}^{P}(A,B)=\left(\frac{1}{4}\left[\left|\Delta_{\mu}^{AB}\right|^{P}+\left|\Delta_{\nu}^{AB}\right|^{P}+\left|\Delta_{\pi\rightarrow\mu}^{AB}\right|^{P}+\left|\Delta_{\pi\rightarrow\nu}^{AB}\right|^{P}\right]\right)^{\frac{1}{P}}, \tag{6}$$

of which

$$\Delta_{\pi\rightarrow\mu}^{AB}=\frac{1}{2}\left[\Delta_{\pi}^{AB}+2\Delta_{\mu}^{AB}\right],\Delta_{\pi\rightarrow\nu}^{AB}=\frac{1}{2}\left[\Delta_{\pi}^{AB}+2\Delta_{\nu}^{AB}\right]. \tag{7}$$

The intuitionistic fuzzy set metric proposed in the literature [20] integrates the case of membership, non-membership and hesitation degrees, and also satisfies the condition of distance and intuitionistic fuzzy set distance. However, in the allocation formula of hesitation degree(7), the allocation of hesitation degree to both membership and non-membership is $\frac{1}{2}$, and this allocation is not flexible enough.So there is still room for improvement and extension of the method, and this paper is based on Eq. (7).
(4) Juthika [14] et al. propose an intuitive fuzzy set metric.

$$\mathrm{D}^{J}(A,B)=\frac{1}{n}\sum_{i=1}^{n}\frac{\left|\mu_{A}(x_{i})-\mu_{B}(x_{i})\right|+\left|\nu_{A}(x_{i})-\nu_{B}(x_{i})\right|}{\mu_{A}(x_{i})+\mu_{B}(x_{i})+\nu_{A}(x_{i})+\nu_{B}(x_{i})}. \tag{8}$$

The distance metric proposed by Juthika scholars also does not take into account the effect of hesitation on distance.
(5) Chen [12] et al. propose an intuitionistic fuzzy set distance metric.

$$D^{Hh}(A,B)=\frac{1}{2n}\sum_{i=1}^{n}\begin{pmatrix}\left(\left|\mu_{A}(x_{i})-\mu_{B}(x_{i})\right|+\left|\nu_{A}(x_{i})-\nu_{B}(x_{i})\right|\right)\\ \times\left(1-\frac{1}{2}\left|\pi_{A}(x_{i})-\pi_{B}(x_{i})\right|\right)\end{pmatrix}. \tag{9}$$

Although the distance method proposed in the literature [12] takes into account the influence of hesitation on distance generation and also satisfies the definition of distance associated with intuitionistic fuzzy sets, it fails to make the correct distinction for some special intuitionistic fuzzy sets, the specific examples of which are detailed in Table 1, so the method also has some shortcomings.

3.2 Existing Intuitionistic Fuzzy Set Distance Analysis

When calculating the distance between two intuitionistic fuzzy sets, the fuzzy semantics expressed by membership, non-membership and hesitation should be considered together. First of all the above intuitionistic fuzzy set distance measures (1), (2) and (8) do not take into account the effect of hesitancy on the

Table 1. Comparison of distances between the intuitive fuzzy sets A and B.

	1	2	3	4	5	6
A	$< x, 0.3, 0.3 >$	$< x, 0.3, 0.4 >$	$< x, 1.0, 0.0 >$	$< x, 0.5, 0.5 >$	$< x, 0.4, 0.2 >$	$< x, 0.4, 0.2 >$
B	$< x, 0.4, 0.4 >$	$< x, 0.4, 0.3 >$	$< x, 0.0, 0.0 >$	$< x, 0.0, 0.0 >$	$< x, 0.5, 0.3 >$	$< x, 0.5, 0.2 >$
$\mathrm{D}_H^{Ata}(A, B)$	**0.1000**	**0.1000**	**0.5000**	**0.5000**	0.1000	0.0500
$\mathrm{D}_E^{Ata}(A, B)$	**0.1000**	**0.1000**	0.7071	0.5000	0.1000	0.0707
$\mathrm{D}_H^{YA}(A, B)$	**0.2000**	**0.1000**	**1.0000**	**1.0000**	0.2000	0.1000
$\mathrm{D}_E^{YA}(A, B)$	**0.2000**	**0.1000**	**1.0000**	**1.0000**	0.2000	0.1000
$\mathrm{MD}^1(A, B)$	0.0500	0.1000	0.5000	0.2500	0.0500	0.0500
$\mathrm{MD}^2(A, B)$	0.0707	0.1000	0.6124	0.3536	0.0707	0.0612
$\mathrm{D}^J(A, B)$	**0.1429**	**0.1429**	**1.0000**	**1.0000**	0.1429	0.0769
$D^{Hh}(A, B)$	0.0900	0.1000	**0.2500**	**0.2500**	0.0900	0.4750

distance.Secondly, the paper analyses the above intuitionistic fuzzy set distance formulae using the six intuitionistic fuzzy sets used in the literature [22], and the results are shown in Table 1.

(1) From group 5 intuitionistic fuzzy sets and group 6 intuitionistic fuzzy sets we can see that The contribution should contain no more than four levels of $\langle x, 0.4, 0.2\rangle = \langle x, 0.4, 0.2\rangle, \langle x, 0.5, 0.3\rangle \subset \langle x, 0.5, 0.2\rangle$. It is therefore reasonable to assume that there is a difference between the distances of the intuitionistic fuzzy set in group 5 and the intuitionistic fuzzy set in group 6, but the metric MD^1 in Table 1 does not distinguish between these two sets of distances.

(2) From group 1 intuitionistic fuzzy set and group 2 intuitionistic fuzzy set we can see that $\langle x, 0.3, 0.4\rangle \subset \langle x, 0.3, 0.3\rangle, \langle x, 0.4, 0.4\rangle \subset \langle x, 0.4, 0.3\rangle$. According to the definition of distance and the definition of intuitionistic fuzzy set distance, the group 1 intuitionistic fuzzy set distance should be smaller than the group 2 intuitionistic fuzzy set distance. However, from Table 1, only theMD^1andMD^2metrics correctly distinguish between these two sets of intuitionistic fuzzy sets.

(3) It can also be obtained from Table 1 that two sets of intuitionistic fuzzy sets with different meanings actually get the same distance. For example, the 3rd Group intuitionistic fuzzy sets and group 4 intuitionistic fuzzy sets. In terms of the fuzzy meanings expressed in the intuition fuzzy sets, group 3 can be seen as the distance between fully agreeing and not taking a position, while group 4 is the distance between half agreeing and half disagreeing and not taking a position. As the two attitudes are very different, it is reasonable to assume that the distance between group 3 and group 4 should be different. However, the distances calculated in Table 1 for D_H^{Ata}, D_H^{YA}, D_E^{YA}, D^J and D^{Hh} are the same for both sets of intuitionistic fuzzy sets.

In summary, the existing intuitionistic fuzzy set distance measure has certain unreasonableness, in order to improve this problem, this paper proposes a new intuitionistic fuzzy set distance measure.

4 A New Intuitionistic Fuzzy Set Distance Metric

In order to solve the unreasonable problems of the existing intuitionistic fuzzy set distance metric, this paper will improve on the literature [20], which divides the hesitation degree equally into subordinate and non-subordinate degrees. By adding a parameter to the hesitation allocation formula, the allocation of hesitation is made more reasonable .Firstly, the new hesitation allocation formula is given.

Definition 4. *Let $A = \{x_i, \mu_A(x_i), \nu_A(x_i) | x_i \in X\}$ be an intuitionistic fuzzy set on X and the assignment of hesitancy $\pi_A(x_i)$ to membership $\mu_A(x_i)$, and non-membership $\nu_A(x_i)$ are defined as*

$$\begin{aligned} \Delta^A_{\pi\to\mu} &= \frac{1}{h}\left(\pi_A(x_i) + h\mu_A(x_i)\right) \\ \Delta^A_{\pi\to\nu} &= \left(1-\frac{1}{h}\right)\left(\pi_A(x_i) + \frac{1}{\left(1-\frac{1}{h}\right)}\nu_A(x_i)\right)(h \geq 2). \end{aligned} \tag{10}$$

The significance of the parameter in Eq. (10): the degree of hesitation reflects in the intuitionistic fuzzy set a degree of unknown that can be interpreted as an uncertain attitude between affirmation and negation of an unknown object. This uncertain attitude may have an affirmative or a negative component, so this is described in Eq. (10) by introducing the parameter h. The degree of hesitation is divided into subordinate and non-subordinate degrees according to a certain ratio. The variable parameter h is used here since the proportion of hesitation degrees assigned may vary in different cases. The value of should be greater than or equal to 2. When h=2 is used, the formula degenerates to the hesitation allocation in the literature [20]. And since is variable, it makes the calculation of the distance metric also more flexible and more relevant.

The following distance metric formulas are given under the new hesitation assignment method using the intuitionistic fuzzy set distance metric formula (5) from the literature [20].

Definition 5. *Let $A = \{\langle x, \mu_A(x), \nu_A(x)\rangle\}$, $B = \{\langle x, \mu_B(x), \nu_B(x)\rangle\}$ be two intuitionistic fuzzy sets on the domain $X\,(x \in X)$ .Then the distance metric betweenAandBis defined as*

$$\begin{aligned} MD_h(A,B) &= \left(\frac{1}{4}\left[\left(\Delta^{AB}_{\mu}\right)^2 + \left(\Delta^{AB}_{\nu}\right)^2 + \left(\Delta^{AB}_{\pi\to\mu}\right)^2 + \left(\Delta^{AB}_{\pi\to\nu}\right)^2\right]\right)^{\frac{1}{2}} \\ &= \left(\frac{1}{4}\left[\begin{array}{l}\left(\Delta^{AB}_{\mu}\right)^2 + \left(\Delta^{AB}_{\nu}\right)^2 + \left(\frac{1}{h}\left(\pi_A(x_i) + h\mu_A(x_i)\right)\right)^2 \\ + \left(\left(1-\frac{1}{h}\right)\left(\pi_A(x_i) + \frac{1}{\left(1-\frac{1}{h}\right)}\nu_A(x_i)\right)\right)^2\end{array}\right]\right)^{\frac{1}{2}}, \end{aligned} \tag{11}$$

where $\Delta^{AB}_{\pi\to\mu} = \frac{1}{h}\left[\Delta^{AB}_{\pi} + h\Delta^{AB}_{\mu}\right]$, $\Delta^{AB}_{\pi\to\nu} = \left(1-\frac{1}{h}\right)\left[\Delta^{AB}_{\pi} + \frac{1}{\left(1-\frac{1}{h}\right)}\Delta^{AB}_{\nu}\right]$.

Verify that the new intuitionistic fuzzy set distance formula (11) satisfies the definition of distance and the definition of the intuitionistic fuzzy set distance metric. The proof leads to the following conclusions.

Theorem 1. *Let* $A = \{\langle x, \mu_A(x), \nu_A(x)\rangle\}, B = \{\langle x, \mu_B(x), \nu_B(x)\rangle\}$ *be two arbitrary intuitionistic fuzzy sets on the theoretical domain* $X(x \in X)$*, then* $MD_h(A, B)$ *is an intuitionistic fuzzy set distance measure satisfying the distance conditions of Definition 2 and Definition 3.*

Proof. (a) From equation (11) it is clear that $\mathrm{MD}(A, B) \geq 0$ holds. When $A = B$ has $\Delta_{\mu}^{AB} = \Delta_{\nu}^{AB} = \Delta_{\pi\to\mu}^{AB} = \Delta_{\pi\to\nu}^{AB}$. So $\mathrm{MD}(A, B) = 0$. Conversely, if $\mathrm{MD}(A, B) = 0$ .then $\Delta_{\mu}^{AB} = \Delta_{\nu}^{AB} = 0, \mu_A(x) = \mu_B(x), \nu_A(x) = \nu_B(x)$ and thus $A = B$. In addition, since $(\Delta_{\mu}^{AB})^2 \leq 1, (\Delta_{\nu}^{AB})^2 \leq 1$

$$(\Delta_{\pi\to\mu}^{AB})^2 = \left(\frac{\Delta_{\pi}^{AB} + h\Delta_{\mu}^{AB}}{h}\right)^2 = \left(\frac{(h-1)(\mu_A - \mu_B) - (\nu_A - \nu_B)}{h}\right)^2$$

$$= \left(\frac{(h-1)\Delta_{\mu}^{AB} - \Delta_{\nu}^{AB}}{h}\right)^2 \leq \left(\frac{(h-1)\left|\Delta_{\mu}^{AB}\right| + \left|\Delta_{\nu}^{AB}\right|}{h}\right)^2 \leq 1$$

$$(\Delta_{\pi\to\nu}^{AB})^2 = \left(\frac{\Delta_{\pi}^{AB} + \frac{1}{1-\frac{1}{h}}\Delta_{\nu}^{AB}}{\frac{1}{1-\frac{1}{h}}}\right)^2 = \left(\frac{-(\mu_A - \mu_B) + \frac{1}{h-1}(\nu_A - \nu_B)}{\frac{h}{h-1}}\right)^2$$

$$= \left(\frac{-\Delta_{\mu}^{AB} + \frac{1}{h-1}\Delta_{\nu}^{AB}}{\frac{h}{h-1}}\right)^2 \leq \left(\frac{\left|\Delta_{\mu}^{AB}\right| + \frac{1}{h-1}\left|\Delta_{\nu}^{AB}\right|}{\frac{h}{h-1}}\right)^2 \leq 1$$

Therefore we have $(\Delta_{\pi\to\mu}^{AB})^2 \leq 1, (\Delta_{\pi\to\nu}^{AB})^2 \leq 1$. It follows from equation (11) that $0 \leq MD(A, B) \leq 1$ holds. Thus, $MD(A, B)$ satisfies condition (1) of Definition 2 and condition DIFS1 of Definition 3.

(b)For $MD(A, B)$, $MD(A, B) = MD(B, A)$ is constant, so $MD(A, B)$ satisfies the definition of condition (2) with condition DIFS of definition 3 2.

(c)Thinking of $MD(A, B)$ as a Euclidean distance between two points in four dimensions divided by 2 constitutes the Euclidean distance satisfied by the triangle inequality is equally satisfied for this distance. So $MD(A, B)$ satisfies condition (3) of definition 2 and condition DIFS3 of definition 3.

(d)If any three intuitionistic fuzzy setsA, B, C on the theoretical domain $X(x \in X)$ satisfy $A \subset B \subset C$, by the intuitionistic fuzzy set the properties of $0 \leq \mu_A(x) \leq \mu_B(x) \leq \mu_C(x) \leq 1$, and $1 \geq \nu_A(x) \geq \nu_B(x) \geq \nu_C(x) \geq 0$ from which we have$\Delta_{\mu}^{AC} \leq \Delta_{\mu}^{AB} \leq 0, \Delta_{\nu}^{AC} \geq \Delta_{\nu}^{AB} \geq 0$. So, $(\Delta_{\mu}^{AB})^2 \leq (\Delta_{\mu}^{AC})^2$, $(\Delta_{\nu}^{AB})^2 \leq (\Delta_{\nu}^{AC})^2$. Thus we have

$$(\Delta_{\pi\to\mu}^{AB})^2 = \left(\frac{(h-1)\Delta_{\mu}^{AB} - \Delta_{\nu}^{AB}}{h}\right)^2 \leq \left(\frac{(h-1)\Delta_{\mu}^{AC} - \Delta_{\nu}^{AC}}{h}\right)^2 = (\Delta_{\pi\to\mu}^{AC})^2$$

$$(\Delta_{\pi\to\nu}^{AB})^2 = \left(\frac{\Delta_{\pi}^{AB} + \frac{1}{1-\frac{1}{h}}\Delta_{\nu}^{AB}}{\frac{1}{1-\frac{1}{h}}}\right)^2 = \left(\frac{-\Delta_{\mu}^{AB} + \frac{1}{h-1}\Delta_{\nu}^{AB}}{\frac{h}{h-1}}\right)^2$$

$$\leq \left(\frac{-\Delta_{\mu}^{AC} + \frac{1}{h-1}\Delta_{\nu}^{AC}}{\frac{h}{h-1}}\right)^2 = \left(\frac{\Delta_{\pi}^{AC} + \frac{1}{1-\frac{1}{h}}\Delta_{\nu}^{AC}}{\frac{1}{1-\frac{1}{h}}}\right)^2 = (\Delta_{\pi\to\nu}^{AC})^2.$$

This leads to $(\Delta_{\pi\to\mu}^{AB})^2 \leq (\Delta_{\pi\to\mu}^{AC})^2, (\Delta_{\pi\to\nu}^{AB})^2 \leq (\Delta_{\pi\to\nu}^{AC})^2$. Using the formula from MD (A, B) we can find $MD(A,B) \leq MD(A,C)$. The same can be proved for $MD(B,C) \leq MD(A,C)$, so that the Definition 3 condition DIFS4 is satisfied.

In summary, the distance metric (11) given in Definition 5 satisfies all the conditions in Definition 2 and Definition 3, so that in the case of a new assignment of hesitation (11) is an intuitionistic fuzzy set distance metric.

The definition of the Mink-owski distance metric under the new hesitation assignment formula was then obtained based on the definition of the Minkowski distance metric in the literature [20].

Definition 6. *Let A and B be two intuitionistic fuzzy sets on the theoretical domain $X = \{x_1, x_2, \cdots, x_n\}$,then the Minkowski distance metric between and under the new hesitation assignment formula can be defined as*

$$MD_h^p(A,B) = \left(\frac{1}{4}\left[\left|\Delta_\mu^{AB}\right|^P + \left|\Delta_\nu^{AB}\right|^P + \left|\Delta_{\pi\to\mu}^{AB}\right|^P + \left|\Delta_{\pi\to\nu}^{AB}\right|^P\right]\right)^{\frac{1}{P}}$$

$$= \left(\frac{1}{4}\left[\begin{array}{l}\left|\Delta_\mu^{AB}\right|^P + \left|\Delta_\nu^{AB}\right|^P + \left|\frac{1}{h}\left(\pi_A(x_i) + h\mu_A(x_i)\right)\right|^P \\ + \left|\left(1-\frac{1}{h}\right)\left(\pi_A(x_i) + \frac{1}{\left(1-\frac{1}{h}\right)}\nu_A(x_i)\right)\right|^P\end{array}\right]\right)^{\frac{1}{P}}. \tag{12}$$

Similar to Theorem 1, applying the new hesitation allocation formula to Eq. (12) $MD_h^P(A,B)$ leads to the following conclusion.

Theorem 2. *Let A and B be two intuitionistic fuzzy sets on the theoretical domain $X = \{x_1, x_2, \cdots, x_n\}$, then $MD^P(A,B)$ only does not satisfy the condition $MD^P(A,B) \leq 1$ under the new hesitation assignment formula, the rest of the distance conditions in Definition 2 and Definition 3 are satisfied.*

Proof. Based on the literature [20] and Theorem 1 it is clear that the conclusion holds.

The reasonableness of the distance metric under the new hesitation assignment is illustrated below by calculating the six sets of intuitionistic fuzzy set distances in Table 1. The results of the calculations are shown in Table 2.

In Table 2, the distances between the fifth and sixth intuitionistic fuzzy sets are effectively distinguished using the new hesitation allocation formula. The distance between the first and second intuitionistic fuzzy sets also satisfies the distance value of the first set is less than the distance value of the second set, so that the shortcomings of the previous distance measure are properly addressed, and the new method also gives a clear distinction between the third and fourth intuitionistic fuzzy sets, which have very different meanings.

From the above data analysis, it can be seen that the distance measure under the new method can distinguish some special combinations of intuitionistic fuzzy sets, which overcomes the shortcomings of the existing method, has a strong differentiation ability and is more realistic.

Table 2. Comparison of the distance between the intuitionistic fuzzy set A and B for the new hesitation assignment

h		1	2	3	4	5	6
	A	$< x, 0.3, 0.3 >$	$< x, 0.3, 0.4 >$	$< x, 1.0, 0.0 >$	$< x, 0.5, 0.5 >$	$< x, 0.4, 0.2 >$	$< x, 0.4, 0.2 >$
	B	$< x, 0.4, 0.4 >$	$< x, 0.4, 0.3 >$	$< x, 0.0, 0.0 >$	$< x, 0.0, 0.0 >$	$< x, 0.5, 0.3 >$	$< x, 0.5, 0.2 >$
$h = 3$	$\mathrm{MD}^1(A, B)$	0.0667	0.1000	0.5833	0.3333	0.0667	0.0583
	$\mathrm{MD}^2(A, B)$	0.0745	0.1000	0.6871	0.3727	0.0745	0.0687
$h = 3.5$	$\mathrm{MD}^1(A, B)$	0.0714	0.1000	0.6071	0.3571	0.0714	0.0607
	$\mathrm{MD}^2(A, B)$	0.0769	0.1000	0.7107	0.3847	0.0769	0.0711
$h = 4$	$\mathrm{MD}^1(A, B)$	0.0750	0.1000	0.6250	0.3750	0.0750	0.0625
	$\mathrm{MD}^2(A, B)$	0.0791	0.1000	0.7287	0.3953	0.0791	0.0729
$h = 5$	$\mathrm{MD}^1(A, B)$	0.0800	0.1000	0.6500	0.4000	0.0800	0.0650
	$\mathrm{MD}^2(A, B)$	0.0825	0.1000	0.7550	0.4123	0.0825	0.0755
$h = 6$	$\mathrm{MD}^1(A, B)$	0.0833	0.1000	0.6667	0.4167	0.0833	0.0667
	$\mathrm{MD}^2(A, B)$	0.0850	0.1000	0.7728	0.4249	0.0850	0.0773

5 Example Analysis

The validity of the new distance measure proposed in this paper was tested using data provided by the colon cancer diagnostic model in the literature [23,24]. Colon cancer is one of the most common cancers and an unconquered medical challenge for mankind. Patients with colon cancer generally fall into four categories, metastasis, recurrence, progression and recovery. Predicting the movement of the tumour in the patient's treatment and then developing a reasonable prediction plan can increase the patient's chances of survival.

Let A be the set of attributes for a particular patient, where A_1, A_2, A_3, A_4 represents the set of sample attributes for metastasis, recurrence, deterioration, and recovery, respectively. a_1, a_2, a_3, a_4, a_5 denote the five attributes: stool characteristics, abdominal pain, acute bowel obstruction, chronic bowel obstruction, and anaemia, respectively. $A_1 = \{\langle a_1, 0.4, 0.4\rangle, \langle a_2, 0.3, 0.3\rangle, \langle a_3, 0.5, 0.1\rangle, \langle a_4, 0.5, 0.2\rangle, \langle a_5, 0.6, 0.2\rangle\}$, $A_2 = \{\langle a_1, 0.2, 0.6\rangle, \langle a_2, 0.3, 0.5\rangle, \langle a_3, 0.2, 0.3\rangle, \langle a_4, 0.7, 0.1\rangle, \langle a_5, 0.8, 0.0\rangle\}$, $A_3 = \{\langle a_1, 0.1, 0.9\rangle, \langle a_2, 0.0, 1.0\rangle, \langle a_3, 0.2, 0.7\rangle, \langle a_4, 0.1, 0.8\rangle, \langle a_5, 0.2, 0.8\rangle\}$, $A_4 = \{\langle a_1, 0.8, 0.2\rangle, \langle a_2, 0.9, 0.0\rangle, \langle a_3, 1.0, 0.0\rangle, \langle a_4, 0.7, 0.2\rangle, \langle a_5, 0.6, 0.4\rangle\}$. Suppose a colon cancer patient has an attribute value of

$$A = \{\ \langle a_1, 0.3, 0.5\rangle,\ \langle a_2, 0.4, 0.4\rangle,\ \langle a_3, 0.6, 0.2\rangle,\ \langle a_4, 0.5, 0.1\rangle,\ \langle a_5, 0.9, 0.0\rangle\}.$$

According to the $MD_h(A_j, A)$ $(j = 1, 2, 3, 4)$ intuitionistic fuzzy set new distance metric (11) given in this paper, the distance between different cancer development directions A_j and the patient A under the corresponding attribute a_i $(i = 1, 2, 3, 4, 5)$ is calculated and finally the future development direction of the patient's cancer is determined by calculating the combined attribute distance $Z_h^a(A_j)$ $(j = 1, 2, 3, 4)$.$Z_h^a(A_j)$ The smallest one in the equation corresponds to the future direction of cancer development of the patient.

$$Z_h^a(A_j) = \min \sum_i^5 MD_h(A_j, A). \tag{13}$$

In this paper, we choose to calculate the two cases of $h = 3$ and $h = 3.5$, and the results are shown in the table.

Table 3. Distance between patients A and the direction of cancer development A_j under the attribute a_i

	A	$\langle a_1, 0.3, 0.5\rangle$	$\langle a_2, 0.4, 0.4\rangle$	$\langle a_3, 0.6, 0.2\rangle$	$\langle a_4, 0.5, 0.1\rangle$	$\langle a_5, 0.9, 0.0\rangle$
$MD_3(A, A_j)$	A_1	0.1000	0.0745	0.0745	0.0553	0.2609
	A_2	0.1000	0.1000	0.2958	0.1374	0.0687
	A_3	0.2925	0.4888	0.4432	0.5362	0.7426
	A_4	0.4230	0.4598	0.3249	0.1322	0.3436
$MD_{3,5}(A, A_j)$	A_1	0.1000	0.0769	0.0769	0.0539	0.2633
	A_2	0.1000	0.1000	0.3031	0.1421	0.0711
	A_3	0.2882	0.4842	0.4408	0.5296	0.7402
	A_4	0.4279	0.4622	0.3298	0.1380	0.3413

According to Eq. (13) calculate each integrated attribute distance , the results are obtained as

$$Z_3^a(A_1) = 0.5652, Z_3^a(A_2) = 0.7020, Z_3^a(A_3) = 2.5032, Z_3^a(A_4) = 1.6835$$

$$Z_{3.5}^a(A_1) = 0.5711, Z_{3.5}^a(A_2) = 0.7162, Z_{3.5}^a(A_3) = 2.4831, Z_{3.5}^a(A_4) = 1.6991.$$

Due to the $Z_3^a(A_1) < Z_3^a(A_2) < Z_3^a(A_4) < Z_3^a(A_3)$, $Z_{3.5}^a(A_1) < Z_{3.5}^a(A_2) < Z_{3.5}^a(A_4) < Z_{3.5}^a(A_3)$ The most consistent movement of the condition for patients with A was selected as A_1 (metastasis) according to the least principle of $Z_h^a(A_j)$. The conclusions obtained are consistent with those in the literature [23–24]. Based on these results, it can be seen that the results obtained from are correct in practice, regardless of whether is a whole number or a fractional number, demonstrating the flexibility of. The final conclusion is consistent with the literature [23–24] and shows that the new distance measure proposed in this paper has some feasibility and practical significance.

6 Concluding Remarks

This paper investigates the problem of hesitation degree assignment based on distance metrics of intuitionistic fuzzy sets and analyses the shortcomings of existing distance metrics of intuitionistic fuzzy sets. Using the characteristics of intuitionistic fuzzy sets to express the meaning of uncertain objects and the relationship between membership, non-membership and hesitation degrees, a new hesitation degree assignment method in intuitionistic fuzzy set metrics is proposed. It is also shown that the new distance metric formula satisfies the distance-related condition. The new intuitionistic fuzzy set distance measure is verified by numerical comparison to overcome the shortcomings of the existing measures. Finally, examples from medical judgements are used to check that the method is feasible and valid. It is found that the parameter h has different effects on different groups of fuzzy sets. In view of the limitation on the number of pages in the paper, we can further consider the value of the parameter h in the subsequent work and study what value is best for the decision making in practical applications.

References

1. Zadeh, L.A.: Fuzzy sets. Inf. Control **8**(3), 338–353 (1965)
2. Atanassov, K.T.: Intuitionistic fuzzy sets. FuzzySets Syst. **20**(1), 87–96 (1986)
3. Zhang, S.: Grey relational analysis method based on cumulative prospect theory for intuitionistic fuzzy multi-attribute group decision making. J. Intell. Fuzzy Syst. **41**, 3783–3795 (2021)
4. Hua, R., Li, B., Li, Y.: Intuitionistic fuzzy PRI-AND and PRI-or aggregation operators based on the priority degrees and their applications in multi-attribute decision making. IEEE Access **9**, 155472–155490 (2021)
5. Gandhimathi, T.: An application of intuitionistic fuzzy soft matrix in medical diagnosis. J. Comput. Theor. Nanosci. **15**(3), 781–784 (2018)
6. Umar, A., Saraswat, R.N.: New generalized intuitionistic fuzzy divergence measure with applications to multi-attribute decision making and pattern recognition. Recent Adv. Comput. Sci. Commun. **14**(7), 2247–2266 (2021)
7. Mújica-Vargas, D.: Superpixels extraction by an intuitionistic fuzzy clustering algorithm. J. Appl. Res. Technol. **19**(2), 141 (2021)
8. Atanassov, K. T.: Intuitionistic fuzzy sets. 2nd edn. Intuitionistic Fuzzy Sets (1999). https://doi.org/10.1007/978-3-031-09176-6
9. Szmidt, E., Kacprzyk, J.: Distances between intuitionistic fuzzy sets. Fuzzy Sets Syst. **114**(3), 505–518 (2000)
10. Ke, D., Song, Y., Quan, W.: New distance measure for Atanassovs intuitionistic fuzzy sets and its application in decision making. Symmetry **10**(10), 429 (2019)
11. Joshi, R., Kumar, S.: A dissimilarity measure based on Jensen Shannon divergence measure. Int. J. Gen Syst **48**(3–4), 280–301 (2019)
12. Chen, C., Deng, X.: Several new results based on the study of distance measures of intuitionistic fuzzy sets. Iran. J. Fuzzy Syst. **17**(2), 147–163 (2020)
13. Wen, S.: Subtraction and division operations on intuitionistic fuzzy sets derived from the hamming distance. Inf. Sci. **571**, 206–224 (2021)
14. Mahanta, J., Panda, S.: A novel distance measure for intuitionistic fuzzy sets with diverse.applications. Int. J. Intell. Syst. **36**, 615–627 (2021)
15. Li, D., Cheng, C.: New similarity measures of intuitionistic fuzzy sets and application to pattern recognitions. Pattern Recogn. Lett. **23**(1/3), 221–225 (2002)
16. Mitchell, H.B.: On the Dengfeng-Chuntian similarity measure and its application to pattern recognition. Pattern Recogn. Lett. **24**(16), 3101–3104 (2003)
17. Garg, H., Rani, D.: Novel similarity measure based on the transformed right-angled triangles between intuitionistic fuzzy sets and its applications. Cogn. Comput. **13**(2), 447–465 (2021)
18. Xuan, T., Chou, S.: Novel similarity measures, entropy of intuitionistic fuzzy sets and their application in software quality evaluation. Soft. Comput. **26**(4), 2009–2020 (2021)

19. Chen, Z., Liu, P.: Intuitionistic fuzzy value similarity measures for intuitionistic fuzzy sets. Intuitionis. Fuzzy Value Simil. Measur. Intuitionis. Fuzzy Sets **41**(1), 1–20 (2022)
20. Xu, C., Shen, J.: A new intuitionistic fuzzy set distance and its application in decision making. Comput. Appl. Res. **37**(12), 3627–3634 (2020)
21. Yang, Y., Francisco, C.: Consistency of 2D and 3D distances of intuitionistic fuzzy sets. Expert Syst. Appl. **39**(10), 8665–8670 (2012)
22. Li, Y., David, L., Qin, O.Z.: Similarity measures between intuitionistic fuzzy (vague) sets: a comparative analysis. Pattern Recogn. Lett. **28**(2), 278–285 (2007)
23. Iancu, I.: Intuitionistic fuzzy similarity measures based on frank t-norms family. Pattern Recogn. Lett. **42**, 128–136 (2014)
24. Gohain, B., Chutia, R., Dutta, P.: Distance measure on intuitionistic fuzzy sets and its application in decision-making, pattern recognition, and clustering problems. Int. J. Intell. Syst. **37**(3), 2458–2501 (2022)

Adaptive Combination of Filtered-X NLMS and Affine Projection Algorithms for Active Noise Control

Xichao Wang, Shifeng Ou(✉), and Yunhe Pang

School of Physics and Electronic Information, Yantai University, Yantai, China
ousfeng@126.com

Abstract. The filtered-x algorithm is the most popular technology used in active noise control (ANC) system to update the controllers. This paper proposes an adaptive combined normalized filtered-x least mean square (FxNLMS) algorithm and filtered-x affine projection (FxAP) algorithm to balance the convergence speed and noise reduction performance in ANC. The new algorithm requires only single ANC system, which uses a sigmoid function to adaptively combine FxNLMS algorithm and FxAP algorithm, and a coupling factor designed by gradient descent is used to update the filter weights. The simulation experiment results in stationary and nonstationary scenarios demonstrate the better performance of the proposed algorithm as compared with the conventional algorithms.

Keywords: Active noise control · Gradient descent · Adaptive combination

1 Introduction

With the continuous improvement of the level of industrialization, the power of most industrial equipment is increasing. Noise pollution has become a concern of people. Conventional noise control methods mainly use some sound-absorbing or sound-insulating materials to attenuate the propagation of noise. But these methods are not very effective in controlling low-frequency noise. Active noise control (ANC) technology based on the principle of acoustic interference phase elimination effectively fills the gap in the field of low frequency noise control [1]. With the development of adaptive filtering technology, the research of active noise control algorithm has been gradually improved. Now, ANC has very mature applications in noise cancelling headphones and engine noise suppression.

The filtered-x least mean square (FxLMS) algorithm is widely used in the field of ANC because of its low computational complexity [2]. However, FxLMS performs poorly in terms of convergence speed and noise reduction performance. In order to solve this problem, normalized filtered-x least mean square (FxNLMS) algorithm uses the squared Euclidean parametrization of the tapped input vector to normalize the tapped weight adjustment volume, which is more stable based on the fast convergence [3]. The advantage of large steps in the FxLMS algorithm is fast convergence but poor steady-state error, while the opposite is true for small steps. The variable step size filtered-x least

L. Fang et al. (Eds.): CICAI 2022, LNAI 13606, pp. 15–25, 2022.
https://doi.org/10.1007/978-3-031-20503-3_2

mean square (VSS-FxLMS) algorithm uses a variable step size to optimize the algorithm, and the step size adopts a value that varies from large to small, which can achieve fast convergence speed and low steady-state error [4]. The momentum filtered-x least mean square (MFxLMS) algorithm is also derived from the FxLMS algorithm, by adding a momentum term as an estimate of the previous gradient [5]. Its result is smoother, and the weight vector can achieve faster convergence. This improvement is only at the expense of the additional storage requirements of the past weight vectors and an additional scalar multiplication. Among these algorithms, the filtered-x affine projection (FxAP) algorithm has a higher computational burden [6]. But it can achieve better steady-state error than the others, which is to achieve better noise reduction performance. In the feedforward single-channel active noise control system and feedforward multi-channel active noise control system, FxAP algorithm has a relatively good performance [7]. The FxAP algorithm converges faster than FxNLMS, but the steady-state error and computational complexity are higher than FxNLMS, which is why this paper does not use a combination of two FxAP algorithms. In recent years, improved algorithms based on the FxAP algorithm have also been continuously proposed. Such as variable step size filtered-x affine projection algorithm (VSS-FxAP) and the filtered-x affine projection algorithm with evolving order (EFxAP) have also achieved good results in active noise control system [8].

These algorithms mentioned above have more or less their drawbacks, so scholars began to consider the research of combination algorithm. The convex combination of filtered-x affine projection (CFxAP) algorithm is a combination of two ANC systems with different step sizes [9]. The CFxAP algorithm can greatly improve the noise reduction performance and convergence speed of the ANC system. To better improve the performance, a combined FxAP algorithm and FxNLMS algorithm using mean-square deviation derivation is proposed (MSDFxAP-NLMS) [10], and the convergence speed and noise reduction performance of this algorithm are excellent. Since the combination algorithms introduces two adaptive filtering systems, it is a big burden for the computational complexity [11]. To improve this problem, the combined step size filtered-x affine projection (CSSFxAP) algorithm was proposed. The CSSFxAP algorithm requires only one adaptive filtering system for the combination of steps, which can greatly reduce the computational complexity while improving the noise reduction performance and convergence speed. Obviously, the CSSFxAP algorithm also has the inevitable disadvantage of not being stable enough [12].

In this paper, we propose a new combination algorithm. This algorithm uses the sigmoid function to combine the FxNLMS algorithm and the FxAP algorithm, and the combination factor is updated by the gradient descent method. The proposed algorithm is updated using FxAP algorithm in order to achieve fast convergence in the first stage. When the ANC system reaches steady state, it uses the FxNLMS algorithm to obtain a better noise reduction performance. The algorithm proposed in this paper only needs to use an adaptive filtering system for updating, which greatly reduces the computational complexity. The simulation experiment results in stationary and nonstationary scenarios demonstrate the better performance of the proposed algorithm as compared with the conventional algorithm.

This paper is organized as five sections. At Sect. 2, the FxNLMS and FxAP algorithm will be described, In the Sect. 3, a new combination algorithm will be proposed. The simulation of this algorithm will be given in Sect. 4. The conclusion of this paper will be given in the Sect. 5.

2 FxNLMS and FxAP Algorithms in ANC System

2.1 Framework of ANC

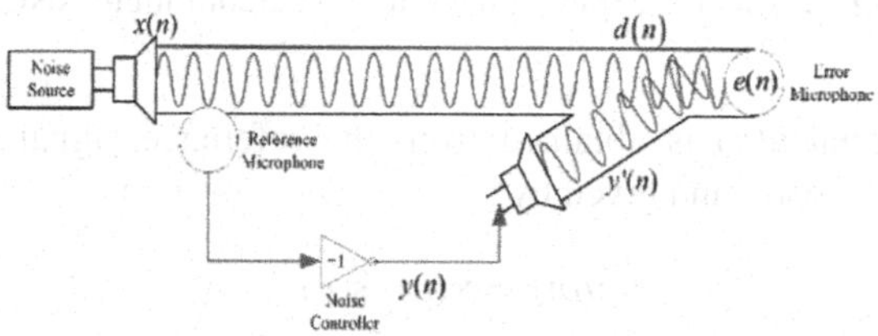

Fig. 1. Framework of ANC.

The framework of ANC is shown in Fig. 1 [13]. The ANC system consists of noise signal, primary channel, secondary channel, reference microphone, noise controller and error microphone. The noise source emits noise that is passed through the main channel to the error microphone to generate the noise to be eliminated $d(n)$. The reference microphone collects the noise signal $x(n)$ into the noise controller, which generates the output signal $y(n)$ through the control algorithm and then passes the reverse signal to the error microphone through the secondary path. At this point the reverse signal is superimposed with the signal $d(n)$ at the error microphone. This results in the elimination of the main channel noise. The sum of the two is recorded as the error signal $e(n)$. The error signal $e(n)$ is fed back to the noise controller to adjust the weight of the noise controller, and the system is continuously circulated until the error signal $e(n)$ reaches the minimum value, which is regarded as the system stable. This is called an adaptive process. It is like the processor gets smarter and can continuously adjust itself according to the noise reduction effect to achieve the best noise reduction performance. With this in mind, adopting a suitable adaptive algorithm is a very critical aspect.

2.2 FxNLMS and FxAP Algorithms

The block diagram of the filtered-x algorithm in ANC is shown in Fig. 2 [14]. Where x(n) is obtained by the reference sensor. Control filter to generate secondary signal $y(n)$. The scalar value $v(n)$ is the background noise, which is generally assumed to be white Gaussian. The error signal e(n) is the difference between the noise signal $d(n)$ obtained at the error sensor and the cancellation signal $y'(n)$. $\hat{S}(z)$ is the available estimate of the secondary channel S(z). In the time domain, it can be modeled as an M-order finite impulse response $\hat{s}(n) = [\hat{s}_0(n), \hat{s}_1(n), ..., \hat{s}_{M-1}(n)]^T$.

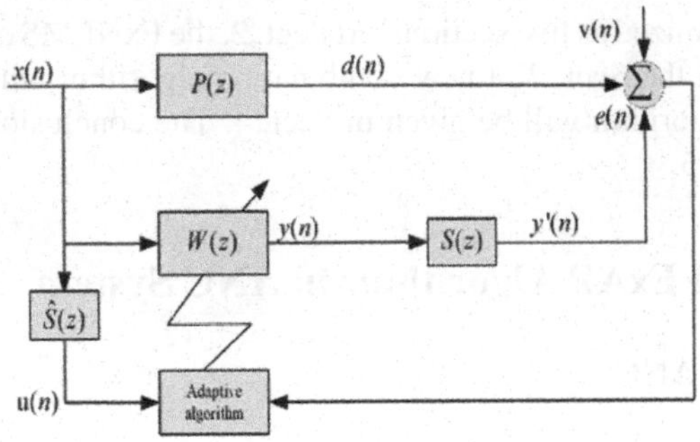

Fig. 2. Block diagram of the filtered-x algorithm in ANC

The filtered signal $u(n)$ is obtained from the reference signal x(n) through the secondary estimation path, and given by

$$u(n) = \mathbf{x}(n)^T \hat{\mathbf{s}}(n) \tag{1}$$

Reference signal $\mathbf{x}(n) = [x(n), x(n-1), \ldots, x(n-M+1)]^T$.

The noise to be cancelled $d(n)$ is generated by the noise source $x(n)$ through the main channel and defined as

$$\mathrm{d(n)} = \mathrm{x(n)} * \mathrm{p(n)} \tag{2}$$

The adaptive filter output signal $y(n)$ is given by

$$y(n) = \mathbf{w}^T(n)\mathbf{x}(n) \tag{3}$$

The expression for the residual error $e(n)$ is computed using

$$e(n) = d(n) - y'(n) \tag{4}$$

where $y'(n) = y(n) * \hat{s}(n)$ is secondary cancelling signal, $\hat{s}(n)$ is the impulse response of $\hat{\mathrm{S}}(\mathrm{z})$, * denotes the discrete convolution operator.

The update equation of the adaptive filter algorithms is expressed as

$$\mathbf{w}(n) = \mathbf{w}(n-1) + \mathbf{G}(n) \tag{5}$$

where G(n) is the adaptation term of the filter at iteration n. The FNxLMS algorithm is widely used to adaptive filter problems because of its simplicity and ease of implementation. The update equation of the FxNLMS algorithm is

$$\mathbf{w}(n) = \mathbf{w}(n-1) + \mu\mathbf{u}(n)(\mathbf{u}^T(n)\mathbf{u}(n) + \delta)^{-1}e(n) \tag{6}$$

where μ is the step-size and δ is the regularization factor, which prevents the input signal from having too small a value and being under-ranked. The input signal $u(n)$ is defined as $\mathbf{u}(n) = [u(n)u(n-1)\ldots u(n-M+1)]^T$, where $u(n)$ is the input signal at time instant n. The adaptation term of the FxNLMS algorithm is given by $\mathbf{G}(n) =$

$\mu\mathbf{u}(n)(\mathbf{u}^T(n)\mathbf{u}(n))^{-1}e(n)$. For noisy signals with high correlation, the convergence speed of FxNLMS algorithm is slow. The misalignment of FxAP algorithm is higher, but it can overcome the disadvantage of the FxNLMS algorithm. The update equation of the FxAP algorithm is

$$\mathbf{w}(n) = \mathbf{w}(n-1) + \mu\mathbf{U}(n)(\mathbf{U}^T(n)\mathbf{U}(n) + \delta\mathbf{I}_\mathrm{k})^{-1}\mathbf{e}(n) \tag{7}$$

where K is the projection orders and $\delta\mathbf{I}_\mathrm{k}$ is regularization matrix. $\mathbf{e}(n) = [e(n)e(n-1)\ldots e(n-K+1)]^T = \mathbf{d}(n) - \mathbf{U}^T(n)\mathbf{w}(n-1)$ represents the error matrix, $\mathbf{U}(n) = [\mathbf{u}(n)\mathbf{u}(n-1)\ldots\mathbf{u}(n-K+1)]$, $\mathbf{d}(n) = [d(n)d(n-1)\ldots d(n-K+1)]$, The FxAP algorithm adaptation term is given by $\mathbf{G}(n) = \mu\mathbf{U}(n)(\mathbf{U}^T(n)\mathbf{U}(n) + \delta\mathbf{I}_\mathrm{k})^{-1}\mathbf{e}(n)$.

3 Combined FxNLMS and FxAP Algorithm

Aiming at the problem that the single-system FxNLMS algorithm and single-system FxAP algorithm cannot take into account the convergence speed and noise reduction performance, this paper proposes an adaptive algorithm with good performance to obtain fast convergence rate and small steady-state error simultaneously. Unlike traditional combinatorial algorithms, the algorithm proposed in this paper requires only one filter for updating. This greatly reduces the computational complexity of the algorithm Contrary to conventional combination algorithms, the algorithm proposed in this paper requires only one filter for updating. This greatly reduces the computational complexity of the Proposed algorithm. The block diagram of the Proposed algorithm is shown in Fig. 3.

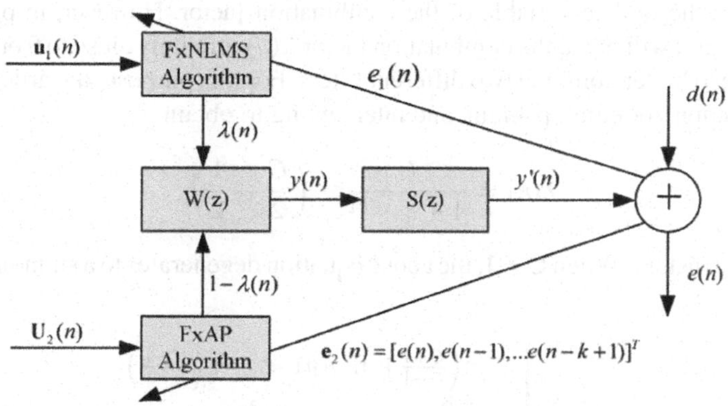

Fig. 3. Block diagram of the adaptive combined FxNLMS and FxAP algorithm

In this paper, FXNLMS and FXAP algorithms are combined by combining factors. And the combined factors are updated using minimized mean squared error and stochastic gradient descent, respectively. Both proposed algorithms can adaptively adjust the weights of different parameters in the algorithm to balance the conflict between system convergence speed and noise reduction performance. For computational convenience, the regularization factors and matrices of the two algorithms are ignored in the derivation. The weight update formulas for the two algorithms are

$$\mathbf{w}_1(n) = \mathbf{w}(n-1) + \mu\mathbf{u}_1(n)(\mathbf{u}_1^T(n)\mathbf{u}_1(n))^{-1}e(n) \tag{8}$$

$$\mathbf{w}_2(n) = \mathbf{w}(n-1) + \mu\mathbf{U}_2(n)(\mathbf{U}_2^T(n)\mathbf{U}_2(n))^{-1}\mathbf{e}(n) \tag{9}$$

where $\mathbf{u}_1(n)$ and $\mathbf{U}_2(n)$ are the inputs to the FxNLMS algorithm and the FxAP algorithm. K is the order of the FxAP algorithm. The two weight coefficient vectors $\mathbf{w}_1(n)$ and $\mathbf{w}_2(n)$ are combined by using the combination factor $\lambda(n)$. The values of the combination factor range between 0 and 1, the combined weight coefficient vectors are

$$\mathbf{w}(n) = \lambda(n)\mathbf{w}_1(n) + (1 - \lambda(n))\mathbf{w}_2(n) \tag{10}$$

The formula for updating the weight coefficients of the algorithm proposed in this paper by substituting (8) and (9) into (10) can be obtained as

$$\begin{aligned} \mathbf{w}(n) = \mathbf{w}(n-1) + \mu[\lambda(n)\mathbf{u}_1(n)(\mathbf{u}_1^T(n)\mathbf{u}_1(n) + \\ (1-\lambda(n))\mathbf{U}_2(n)(\mathbf{U}_2(n)^T\mathbf{U}_2(n))^{-1}e_2(n)] \end{aligned} \tag{11}$$

The combination factor $\lambda(n)$ of this algorithm needs to be updated by stochastic gradient descent. The combination factor is set to

$$\lambda(n) = \frac{1}{1+e^{-a(n)}} \tag{12}$$

where $a(n)$ is the update variable of the combination factor. However, in practice the sigmoid function will make the combination factor $\lambda(n)$ infinitely close to 0 or 1, making it impossible to better combine two different filters. For this purpose, the article does the sigmoid function zooming, panning and intercepting to obtain

$$\lambda(n) = \frac{C}{1+e^{-a(n)}} - \left(\frac{C}{2} - \frac{1}{2}\right) \tag{13}$$

where C is a constant. When $C = 1$, the above equation degenerates to a sigmoid function. $a(n)$ is denoted as

$$a(n) = \begin{cases} -\ln\left(\frac{C+1}{C-1}\right) & \text{if } a(n) < -\ln\left(\frac{C+1}{C-1}\right) \\ 0 & \text{else} \\ \ln\left(\frac{C+1}{C-1}\right) & \text{if } a(n) > \ln\left(\frac{C+1}{C-1}\right) \end{cases} \tag{14}$$

Derivation of the combination factor $\lambda(n)$ yields

$$\frac{\partial\lambda(n)}{\partial a(n)} = C\frac{e^{-a(n)}}{\partial(1+e^{-a(n)})} \tag{15}$$

It can be seen that the value of C determines the speed of change of the combination factor [15]. The larger the value of C, the faster the change of the combination factor, and vice versa, the slower the change of the combination factor. The value of C also affects the value of *a*(*n*). When C = 1, the update formula of the combination factor $\lambda(n)$ degenerates to a sigmoid function, and the range of *a*(*n*) is [-4, 4]. When C > 1, the range of values of *a*(*n*) becomes

$$a(n) = \left[\max\left(-4, \ln\left(\frac{C+1}{C-1}\right)\right), \min\left(-\ln\left(\frac{C+1}{C-1}\right), 4\right)\right] \tag{16}$$

In order to make *a*(*n*) adaptively update the combination factor to minimize the instantaneous squared error of the filter. The article updates *a*(*n*) by stochastic gradient descent method. Define the cost function as

$$J(n) = e^2(n) \tag{17}$$

The update formula for *a*(*n*) is

$$\begin{aligned} a(n) &= a(n-1) - \frac{\mu_a}{2}\nabla J(n) \\ &= a(n-1) + \mu_a \mu \mathbf{u}^T(n)[\frac{\partial\lambda(n-1}{\partial a(n-1)}\mathbf{u}_1^T(n-1) \\ &(\mathbf{u}_1^T(n-1)\mathbf{u}_1(n-1))^{-1}e_1(n-1) - \frac{\partial\lambda(n-1)}{\partial a(n-1)}\mathbf{U}_1^T(n-1) \\ &(\mathbf{U}_1^T(n-1)\mathbf{U}_1(n-1))^{-1}\mathbf{e}_2(n-1)]\mathrm{e}(n) \end{aligned} \tag{18}$$

From formulas (17) and (18), the *a*(*n*) is expressed as

$$\begin{aligned} a(n) = a(n-1) &+ \mu_a \mu \lambda(n)(1-\lambda(n))\mathbf{u}^T(n)[\mathbf{u}_1^T(n-1) \\ &(\mathbf{u}_1^T(n-1)\mathbf{u}_1(n-1))^{-1}e_1(n-1) - \mathbf{U}_1^T(n-1) \\ &(\mathbf{U}_1^T(n-1)\mathbf{U}_1(n-1))^{-1}\mathbf{e}_2(n-1)]\mathrm{e}(n) \end{aligned} \tag{19}$$

where the μ_a is the step size of the stochastic gradient descent method. However, when $\lambda(n) = 0$ or $\lambda(n) = 1$ in the update formula, the update of *a*(*n*) will stop. To prevent the above phenomenon, it is usually necessary to add a small constant ε to avoid this phenomenon. The update formula for *a*(*n*) is modified as follows

$$\begin{aligned} a(n) = a(n-1) &+ \mu_a \mu[\lambda(n)(1-\lambda(n)) + \varepsilon]\mathbf{u}^T(n)[\mathbf{u}_1^T(n-1) \\ &(\mathbf{u}_1^T(n-1)\mathbf{u}_1(n-1))^{-1}e_1(n-1) - \mathbf{U}_1^T(n-1) \\ &(\mathbf{U}_1^T(n-1)\mathbf{U}_1(n-1))^{-1}\mathbf{e}_2(n-1)]\mathrm{e}(n) \end{aligned} \tag{20}$$

Compared with the conventional combination algorithm, the new combined FxAP and FxNLMS algorithm proposed in this paper uses only one filter for updating the coefficients, which greatly reduces the computational complexity burden of the ANC system. This is one of the major advantages of the algorithm proposed in this paper.

The computational complexity of the commonly used active noise control algorithms and the algorithm proposed in this paper are analyzed in Table 1. Since the results of multiplication operations are the main concern when calculating the complexity of the

Table 1. Summary of the computational complexity of the different algorithms in this paper

Algorithm	Multiplication per iterate
FxNLMS	$3L + M + 1$
FxAP	$(K^2 + 2K)L + K^3 + K^2 + KM$
CFxNLMS-AP	$(K^2 + 2K + 5)L + K^3 + K + (K + 1)M + 11$
EFxAP	$(K^2 + 2K)L + K^3 + K^2 + KM + 1$
MSDFxNLMS-AP	$(K^2 + 2K + 4)L + K^3 + K + (K + 1)M + 22$
Proposed	$(K^2 + 2K + 4)L + K^3 + K + (K + 1)M + 8$

algorithm, Table 1 lists the multiplication operations required for each algorithm in this paper. Where K is the order of the AP algorithm, L is the length of the filter, and the length of the secondary channel is M. The analysis shows that the algorithm proposed in this paper requires less computational degree to iterate compared to the MSDFxNLMS-AP algorithm. When the filter of order 32 is used in this paper, the computational degree of the proposed algorithm is also better than that of the CFxNLMS-AP algorithm.

4 Simulation Results

Establish a system model of feedforward ANC system in MATLAB. All the noise sources used in the simulation are the real engine noise obtained by sampling. The main channel is a 32-order system. The secondary channel is modeled as a 10-order system, and the transfer function is [0 0 0 0 0.1 0.3 −0.1 0.05 0.2 0.1].

The simulation result of Proposed algorithm is shown in Fig. 4. In this paper, we compare both the CFxAP-NLMS algorithm and the EFxAP algorithm, and all the algorithm steps are 0.5 and the μ_a is 200. The evaluation criterion used in this paper is mean-square deviation (MSD). All simulations were repeated 50 times using Monte

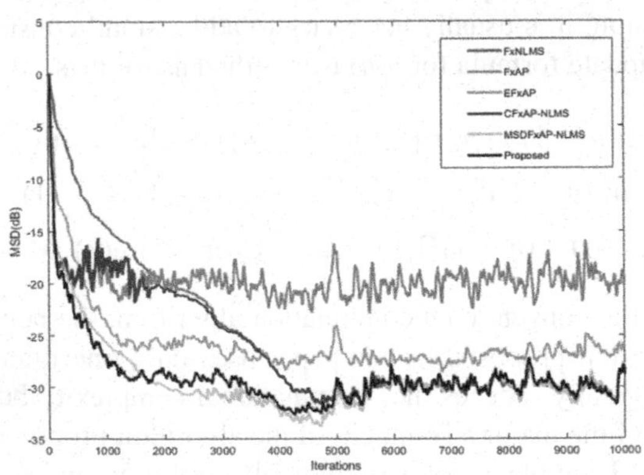

Fig. 4. Comparison of the proposed algorithm with CFxAP algorithm and EFxAP algorithm in stationary environment

Carlo experiments. The FxAP algorithm used in this paper is the order of 8, so the comparison algorithm EFxAP selects the order change from 8 to 1. Figure 4 shows that the algorithm proposed in this paper outperforms the basic FxAP algorithm, FxNLMS algorithm, CFxAP-NLMS algorithm and EFxAP algorithm in terms of convergence speed and noise reduction performance. And it can get about the same performance as the MSDFxAP-NLMS algorithm with less computational complexity.

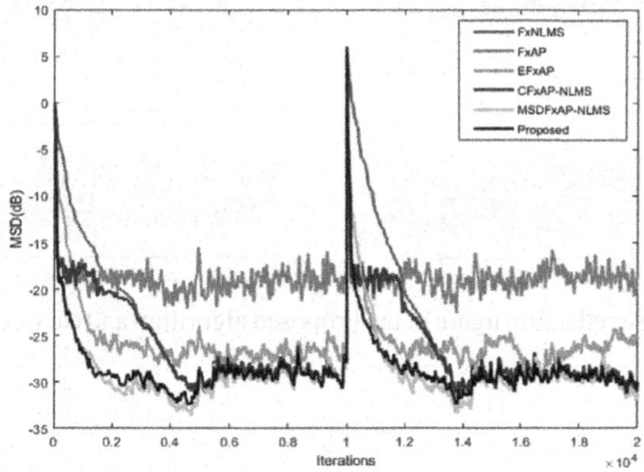

Fig. 5. Comparison of the proposed algorithm with CFxAP algorithm and EFxAP algorithm in non-stationary environment

Figure 5 is the algorithm convergence curve when the environment has a sudden change. As shown in the figure, the external environment has a sudden change at 6000 points. At this time, the algorithm can also quickly converge and achieve a better noise reduction performance. Experiments have proved that the Proposed algorithm is also excellent under non-stationary conditions.

Figure 6 shows the actual noise control effect simulation of the algorithm. The red part is the amplitude of the noise, the green one is the residual of the noise. Through the comparison, it can be concluded that the Proposed algorithm can obtain a smaller residual noise quickly compared with other algorithms, which also reflects the effectiveness of the proposed algorithm.

The simulation shows that the proposed algorithm performs better than the conventional combination algorithm and the evolutionary order algorithm in terms of convergence speed and noise reduction performance. The algorithm proposed in this paper can obtain the same performance as the MSDFxAP-NLMS algorithm with less computational complexity.

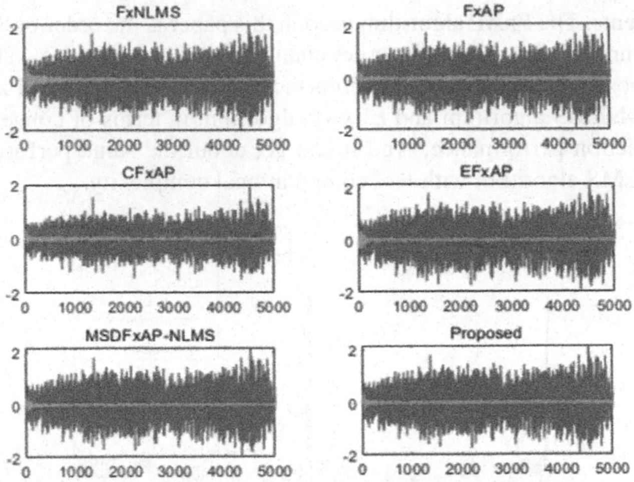

Fig. 6. The actual noise reduction figure of the Proposed algorithm and each comparison algorithm in this paper.

5 Conclusion

This paper Proposes a new combined FxAP and FxNLMS algorithm used for ANC system. It solves the problem that the conventional algorithms cannot balance the convergence speed and noise reduction performance. Simulation shows that the algorithm proposed in this paper is effective and has excellent performance in both stationary and nonstationary environments than the conventional combination algorithm and the evolutionary order algorithm. It has a computational complexity better than the MSDFxAP-NLMS algorithm, which performs similarly. As the algorithm in this paper does not perform stably enough in a non-stationary environment, subsequent work will focus on how to improve the robustness of the algorithm, so as to obtain a better ANC algorithm.

References

1. Kuo, S.M., Morgan, D.R.: Active Noise Control Systems: Algorithms and DSP Implementations. Wiley, New York (1996)
2. Hinamoto, Y., Sakai, H.: A filtered-X LMS algorithm for sinusoidal reference signals—effects of frequency mismatch. IEEE Signal Process. Lett. **14**(4), 259–262 (2007)
3. Kuo, S.M., Mitra, S., Gan, W.S.: Active noise control system for headphone applications. IEEE Trans. Control Syst. Technol. **14**(2), 331–335 (2006)
4. Carmona, J.C., Alvarado, V.M.: Active noise control of a duct using robust control theory. IEEE Trans. Control Syst. Technol. **8**(6), 930–938 (2000)
5. Song, P., Zhao, H.: Filtered-x least mean square/fourth (FXLMS/F) algorithm for active noise control. Mech. Syst. Signal Process. **120**, 69–82 (2019)
6. Guo, J.F., Yang, F.R., Yang, J.: Convergence analysis of the conventional filtered-x affine projection algorithm for active noise control. Signal Process. **170** (2020)
7. Kim, S.E., Kong, S.J., Song, W.J.: An affine projection algorithm with evolving order. IEEE Signal Process. Lett. **16**(11), 937–940 (2009)

8. Kim, D. W., Lee, M., Park, P.: A robust active noise control system with step size scaler in impulsive noise environments. In: 2019 Chinese Control Conference, pp. 3358–3362 (2019)
9. ArenasGarcia, J., AzpicuetaRuiz, L.A., Silva, M.T.M., Nascimento, V.H., Sayed, A.H.: Combinations of adaptive filters: performance and convergence properties. IEEE Signal Process. Mag. **33**(1), 120–140 (2016)
10. Choi, J.H., Kim, S.H., Kim, S.W.: Adaptive combination of affine projection and NLMS algorithms". Signal Process. **100**, 64–70 (2014)
11. Wang, H., Sun, H., Sun, Y., Wu, M., Yang, J.: A narrowband active noise control system with a frequency estimation algorithm based on parallel adaptive notch filter. Signal Process. **154**, 108–119 (2019)
12. Choi, J.H., Kim, S.H., Kim, S.W.: Adaptive combination of affine projection and NLMS algorithms. Signal Process. **100**, 64–70 (2014)
13. Yang, F.R., Guo, J.F., Yang, J.: Stochastic Analysis of the Filtered-x LMS Algorithm for Active Noise Control. IEEE/ACM Trans. Audio Speech Lang. Process. **28**, 2252–2266 (2020)
14. Bouchard, M.: Multichannel affine and fast affine projection algorithms for active noise control and acoustic equalization systems. IEEE Trans. Audio Speech Lang. Process. **11**(1), 54–60 (2003)
15. Huang, F., Zhang, J., Zhang, S.: Combined-step-size affine projection sign algorithm for robust adaptive filtering in impulsive interference environments. IEEE Trans. Circuits Syst. II Express Briefs **63**(5), 493–497 (2016)

Linguistic Interval-Valued Spherical Fuzzy Sets and Related Properties

Yanzhe Liu[1], Ye Zhang[1], Xiaosong Cui[2](✉), and Li Zou[3]

[1] School of Mathematics, Liaoning Normal University, Dalian 116029, People's Republic of China

[2] School of Computer and Information Technology, Liaoning Normal University, Dalian 116081, People's Republic of China
cuixiaosong2004@163.com

[3] School of Computer and Technology, Shandong Jianzhu University, Jinan 250102, People's Republic of China

Abstract. Although traditional spherical fuzzy sets can handle the fuzzy information in the quantitative environment well, it cannot deal with the more realistic qualitative information. The linguistic term set attracts attention due to their ability to handle the qualitative information. Therefore, to provide the more freedom to the decision-makers, in this paper, we propose the linguistic interval-valued spherical fuzzy set and the linguistic interval-valued spherical fuzzy number, whose the membership, non-membership, hesitancy and waiver degree are represented by the interval-valued linguistic terms, for better dealing with the imprecise and uncertain information during the decision-making process. In this paper, we give the concept of the linguistic interval-valued spherical fuzzy set; then we discuss the basic operational laws, some important properties and their related proofs. Subsequently, we give the concept of the linguistic interval-valued spherical fuzzy number, and various operational laws, then the measure formula, score and accuracy functions of the linguistic interval-valued spherical fuzzy number are defined with a brief study of their related properties. At last, an admissible order between the linguistic interval-valued spherical fuzzy numbers using score and accuracy functions is introduced.

Keywords: Linguistic interval-valued spherical fuzzy sets · Linguistic term sets · Spherical fuzzy sets

1 Introduction

Zadeh [1] proposed the fuzzy set that uses the quantitative information to evaluate accessing the alternatives. But only the membership information is considered in the fuzzy set. Therefore, in order to solve this problem, Atanassov [2] proposed the intuitionistic fuzzy set that is the powerful extension of the fuzzy set to deal with the uncertainties by considering the membership, non-membership and hesitation degree at the same time into the analysis. In 1989, Atanassov [3] proposed the interval-valued intuitionistic fuzzy set that uses interval numbers to represent the membership, non-membership and hesitation

L. Fang et al. (Eds.): CICAI 2022, LNAI 13606, pp. 26–36, 2022.
https://doi.org/10.1007/978-3-031-20503-3_3

degree. Atanassov [4] proposed the intuitionistic fuzzy set of second type that is the extension of the intuitionistic fuzzy set. Because in the real decision-making process, the sum of the membership degree and non-membership degree may be larger than 1 but their square sum is equal to or less than 1. This concept does not require decision makers to change their preference information but provides a larger preference area for decision makers.

In 2019, Kutlu Gündoğdu et al. [5] proposed the spherical fuzzy set (SFS), which is an emerging fuzzy set represented by the membership, non-membership, hesitation and waiver degree. SFS has been widely used in research such as decision-making [5–9],medical diagnosis problems [10] and pattern recognition [11, 12]. Subsequently, some researchers proposed the interval-valued spherical fuzzy set (IVSFS) [13, 14] on the basis of the SFS, which improves the ability of dealing with the imprecise and uncertain information. However, in day-to-day life, it is difficult for a decision maker to give his assessments towards the alternative through quantitative information due to complexity and uncertainty of practical problems. Therefore, in order to solve this thorny problem, Zadeh [15–17] proposed the linguistic variables, which enable decision makers to solve practical problems through qualitative evaluation. Jia [18] proposed the linguistic spherical fuzzy set (LSFS) that is the extension of the SFS to deal with the difficulty of obtaining quantitative evaluation in decision-making.

To achieve the preferences of decision makers with linguistic interval-valued, and to provide the more freedom to the decision-makers, in this paper, we have extended the LSFS to the linguistic interval-valued spherical fuzzy set (LIVSFS) in which the membership, non-membership, hesitation and waiver degree are represented by interval-valued linguistic terms. Also, we have discussed various operational laws and their proofs of properties, the measure formula, score and accuracy functions based on it.

The structure of this paper is organized as follows: In Sect. 2, some basic concepts related to the SFS, IVSFS, linguistic term set (LTS) and LSFS are briefly reviewed. In Sect. 3, concept of the LIVSFS and the linguistic interval-valued spherical fuzzy number (LIVSFN) with various operational laws, properties and related properties are proposed. Also, an order between the LIVSFNs using score and accuracy functions is introduced. In Sect. 2 concludes the paper and looks forward to future work.

2 Preliminaries

In this section, we introduce some preliminary notions in order to fix notation. Also, some basic concepts and related properties about the SFS, IVSFS, LTS and LSFS are introduced.

2.1 Spherical Fuzzy Sets

Definition 2.1.1 [5] Let $X \neq \varphi$ be a finite universe. A SFS A in X is defined as

$$A = \{\langle x, (u_A(x), \pi_A(x), \nu_A(x))\rangle | x \in X\} \quad (1)$$

where the function $u_A : X \to [0, 1]$, the function $\pi_A : X \to [0, 1]$ and the function $\nu_A : X \to [0, 1]$ [0, 1] represent the membership degree, the hesitancy degree and the non-membership degree of x to A respectively such that for each $x \in X$,

$$0 \leq u_A^2(x) + \pi_A^2(x) + v_A^2(x) \leq 1 \tag{2}$$

then $r_A(x) = \sqrt{1 - \left(u_A^2(x) + \pi_A^2(x) + v_A^2(x)\right)}$ is the waiver degree of x to A.

2.2 Interval-Valued Spherical Fuzzy Sets

Definition 2.2.1 [14] Let $X \neq \varphi$ be a finite universe. An IVSFS A in X is defined as

$$A = \{\langle x, (\tilde{u}_A(x), \tilde{\pi}_A(x), \tilde{v}_A(x))\rangle | x \in X\} \tag{3}$$

where $\tilde{u}_A(x) = [u_A^L(x), u_A^U(x)]$, $\tilde{\pi}_A(x) = [\pi_A^L(x), \pi_A^U(x)]$ 和 $\tilde{v}_A(x) = [v_A^L(x), v_A^U(x)]$ are subsets of [0, 1], and represent the membership degree, the hesitancy degree and the non-membership degree of x to A respectively such that for all $x \in X$,

$$0 \leq \left(\tilde{u}_A^U(x)\right)^2 + \left(\tilde{\pi}_A^U(x)\right)^2 + \left(\tilde{v}_A^U(x)\right)^2 \leq 1 \tag{4}$$

then $\tilde{r}_A(x) = \left[r_A^L(x), r_A^U(x)\right]$ is subsets of [0, 1], and is the waiver degree of x to A, where $r_A^L(x) = \sqrt{1 - \left(\left(u_A^L(x)\right)^2 + \left(\pi_A^L(x)\right)^2 + \left(v_A^L(x)\right)^2\right)}$, $r_A^U(x) = \sqrt{1 - \left(\left(u_A^U(x)\right)^2 + \left(\pi_A^U(x)\right)^2 + \left(v_A^U(x)\right)^2\right)}$.

2.3 Linguistic Term Sets

Definition 2.3.1 [19] Let $S = \{s_0, s_1, ..., s_g\}$ be a finite linguistic term set (LTS) consisting of $g + 1$ linguistic terms and g is the positive integer. Any LTS must have the following characteristics:

(1) There is a negation operator: $neg(s_i) = s_j j = g - i$;
(2) The set is ordered: $s_i \geq s_j \Leftrightarrow i \geq j$;
(3) There is a max operator: $s_i \geq s_j \Leftrightarrow \max(s_i, s_j) = s_i$;
(4) There is a min operator: $s_i \leq s_j \Leftrightarrow \min(s_i, s_j) = s_i$.

For example, when $g = 6$, a LTS S can be taken as:$S = \{s_0$: None, s_1 : very low, s_2 : low, s_3 : medium, s_4 : high, s_5 : very high, s_6 : perfect$\}$.

Definition 2.3.2 [20] Let $S = \{s_0, s_1, ..., s_g\}$ be a LTS, Xu extended the discrete term set S to a continuous term set $\overline{S} = \{s_z | s_0 \leq s_z \leq s_h, z \in [0, h]\}$, where $h(h > g)$ is a sufficiently large positive integer. For two linguistic terms s_k, $s_l \in \overline{S}$ and λ, λ_1, $\lambda_2 \in [0, 1]$, the operation rules are as follows:

(1) $s_k \oplus s_l = s_l \oplus s_k = s_{k+l}$;
(2) $\lambda s_k = s_{\lambda k}$;
(3) $(\lambda_1 + \lambda_2)s_k = \lambda_1 s_k \oplus \lambda_2 s_k$;
(4) $\lambda(s_k \oplus s_l) = \lambda s_k \oplus \lambda s_l$.

2.4 Linguistic Spherical Fuzzy Sets

Definition 2.4.1 [18] Let $X \neq \varphi$ be a finite universe. A LSFS A in X is defined as

$$A = \{\langle x, (s_{uA}(x), s_{\pi A}(x), s_{\nu A}(x))\rangle | x \in X\} \tag{5}$$

where the function $s_{uA} : X \to S_{[0,h]}$, the function $s_{\pi A} : X \to S_{[0,h]}$ and the function $s_{\pi A} : X \to S_{[0,h]}$ represent the membership degree, the hesitancy degree and the non-membership degree of x to A respectively such that for each $x \in X$,

$$0 \leq uA^2 + \pi A^2 + \nu A^2 \leq h^2 \tag{6}$$

then $s_{rA}(x) = s_{\sqrt{h^2-(u_A^2+\pi_A^2+\nu_A^2)}}$ is the waiver degree of x to A.

3 Linguistic Interval-Valued Spherical Fuzzy Sets

In this section, we have proposed LIVSFS, LIVSFN and their basic operations and properties are also discussed. Finally, the comparison method of LIVSFN is introduced.

3.1 Concepts of LIVSFS and LIVSFN

Definition 3.1.1 Let $X \neq \varphi$ be a finite universe and $\overline{S} = \{s_z | s_0 \leq s_z \leq s_h, z \in [0, h]\}$ be a continuous LTS. A LIVSFS A in X is defined as

$$A = \{\langle x, (\tilde{s}_{uA}(x), \tilde{s}_{\pi A}(x), \tilde{s}_{\nu A}(x))\rangle | x \in X\} \tag{7}$$

where $\tilde{s}_{uA}(x) = [s_{uA}^L(x), s_{uA}^U(x)]$,$\tilde{s}_{\pi A}(x) = [s_{\pi A}^L(x), s_{\pi A}^U(x)]$ and $\tilde{s}_{\nu A}(x) = [s_{\nu A}^L(x), s_{\nu A}^U(x)]$ represent the membership degree, the hesitancy degree and the non-membership degree of x to A respectively such that for each $x \in X$,

$$0 \leq \left(\tilde{s}_{uA}^U(x)\right)^2 + \left(\tilde{s}_{\pi A}^U(x)\right)^2 + \left(\tilde{s}_{\nu A}^U(x)\right)^2 \leq h^2 \tag{8}$$

then $\tilde{s}_{rA}(x) = [r_A^L(x), r_A^U(x)]$ is the waiver degree of x to A, where $r_A^L(x) = s_{\sqrt{h^2-\left((uA^L(x))^2+(\pi A^L(x))^2+(\nu A^L(x))^2\right)}}$,$r_A^U(x) = s_{\sqrt{h^2-\left((uA^U(x))^2+(\pi A^U(x))^2+(\nu A^U(x))^2\right)}}$. $LIVSF(X)$ is called the set of all LIVSFS in X. Usually,$\alpha_A = \left([s_{uA}^L, s_{uA}^U], [s_{\pi A}^L, s_{\pi A}^U], [s_{\nu A}^L, s_{\nu A}^U]\right)$ is called a LIVSFN.

For the convenience, we denote the LIVSFN as $\alpha = \left([s_a, s_b], [s_c, s_d], [s_e, s_f]\right)$, where $[s_a, s_b] \subseteq [s_0, s_h]$, $[s_c, s_d] \subseteq [s_0, s_h]$, $[s_e, s_f] \subseteq [s_0, s_h]$, $0 \leq b^2 + d^2 + f^2 \leq h^2$ and also $s_a, s_b, s_c, s_d, s_e, s_f \in \overline{S}$ hold.

3.2 Basic Operations and Properties of LIVSFS

By means of the definition of LIVSFS, this section will propose its basic operations, properties and related proofs.

Definition 3.2.1 Let $M = \{\langle x, (\tilde{s}_{uM}(x), \tilde{s}_{\pi M}(x), \tilde{s}_{\nu M}(x))\rangle | x \in X\}$ and $N = \{\langle x, (\tilde{s}_{uN}(x), \tilde{s}_{\pi N}(x), \tilde{s}_{\nu N}(x))\rangle | x \in X\}$ be two LIVSFSs in X, where $\tilde{s}_{uA}(x) = [s^L_{uA}(x), s^U_{uA}(x)], \tilde{s}_{\pi A}(x) = [s^L_{\pi A}(x), s^U_{\pi A}(x)]$ and $\tilde{s}_{\nu A}(x) = [s^L_{\nu A}(x), s^U_{\nu A}(x)]$, then

(1) $M \subseteq N$ iff $s^L_{uM}(x) \le s^L_{uN}(x), s^U_{uM}(x) \le s^U_{uN}(x), s^L_{\pi M}(x) \le s^L_{\pi N}(x), s^U_{\pi M}(x) \le s^U_{\pi N}(x)$ and $s^L_{\nu M}(x) \ge s^L_{\nu N}(x), s^U_{\nu M}(x) \ge s^U_{\nu N}(x)$;
(2) $M = N$ iff $M \subseteq N$ and $N \subseteq M$;
(3) $M \cup N =$

$$\{\langle x, [\max\{s_{uM}{}^L(x), s_{uN}{}^L(x)\}, \max\{s_{uM}{}^U(x), s_{uN}{}^U(x)\}], [\min\{s_{\pi M}{}^L(x), s_{\pi N}{}^L(x)\}, \min\{s_{\pi M}{}^U(x), s_{\pi N}{}^U(x)\}]$$
$$[\min\{s_{\nu M}{}^L(x), s_{\nu N}{}^L(x)\}, \min\{s_{\nu M}{}^U(x), s_{\nu N}{}^U(x)\}]\rangle | x \in X\};$$

(4) $M \cap N =$

$$\{\langle x, [\min\{s_{uM}{}^L(x), s_{uN}{}^L(x)\}, \min\{s_{uM}{}^U(x), s_{uN}{}^U(x)\}], [\min\{s_{\pi M}{}^L(x), s_{\pi N}{}^L(x)\}, \min\{s_{\pi M}{}^U(x), s_{\pi N}{}^U(x)\}]$$
$$[\max\{s_{\nu M}{}^L(x), s_{\nu N}{}^L(x)\}, \max\{s_{\nu M}{}^U(x), s_{\nu N}{}^U(x)\}]\rangle | x \in X\};$$

(5) $M^C = \{\langle x, (\tilde{s}_{\nu M}(x), \tilde{s}_{\pi M}(x), \tilde{s}_{uM}(x))\rangle | x \in X\}$.

Theorem 3.2.1 Let
$M = \{\langle x, (\tilde{s}_{uM}(x), \tilde{s}_{\pi M}(x), \tilde{s}_{\nu M}(x))\rangle | x \in X\}$ and $N = \{\langle x, (\tilde{s}_{uN}(x), \tilde{s}_{\pi N}(x), \tilde{s}_{\nu N}(x))\rangle | x \in X\}$ and $K = \{\langle x, (\tilde{s}_{uK}(x), \tilde{s}_{\pi K}(x), \tilde{s}_{\nu K}(x))\rangle | x \in X\}$ be three LIVSFSs in X, then they satisfy the following properties:

(1) Transitivity: If $M \subseteq N$ and $N \subseteq K$,then $M \subseteq K$;
(2) Commutative law: $M \cup N = N \cup M$ and $M \cap N = N \cap M$;
(3) Associative law: $M \cup (N \cup K) = (M \cup N) \cup K$ and $M \cap (N \cap K) = (M \cap N) \cap K$;
(4) Distributive law:$M \cup (N \cap K) = (M \cup N) \cap (M \cup K)$ and $M \cap (N \cup K) = (M \cap N) \cup (M \cap K)$;
(5) De Morgan law: $(M \cup N)^C = M^C \cap N^C$ and $(M \cap N)^C = M^C \cup N^C$;
(6) Idempotent law: $M \cup M = M$ and $M \cap M = M$;
(7) Absorption law: $(M \cup N) \cap M = M$ and $(M \cap N) \cup M = M$.

Proof. (1) Transitivity: By means of Definition 3.2.1,$\because M \subseteq N$,$\therefore$ for $\forall x \in X$, we have $s^L_{uM}(x) \le s^L_{uN}(x), s^U_{uM}(x) \le s^U_{uN}(x); s^L_{\pi M}(x) \le s^L_{\pi N}(x), s^U_{\pi M}(x) \le s^U_{\pi N}(x)$ and $s^L_{\nu M}(x) \ge s^L_{\nu N}(x), s^U_{\nu M}(x) \ge s^U_{\nu N}(x)$.And $\because N \subseteq K$,$\therefore$ for $\forall x \in X$, we also have $s^L_{uN}(x) \le s^L_{uK}(x), s^U_{uN}(x) \le s^U_{uK}(x); s^L_{\pi N}(x) \le s^L_{\pi K}(x), s^U_{\pi N}(x) \le s^U_{\pi K}(x)$ and $s^L_{\nu N}(x) \ge s^L_{\nu K}(x), s^U_{\nu N}(x) \ge s^U_{\nu K}(x)$. So, for $\forall x \in X$, we can get $s^L_{uM}(x) \le s^L_{uK}(x), s^U_{uM}(x) \le s^U_{uK}(x); s^L_{\pi M}(x) \le s^L_{\pi K}(x), s^U_{\pi M}(x) \le s^U_{\pi K}(x)$ and $s^L_{\nu M}(x) \ge s^L_{\nu K}(x), s^U_{\nu M}(x) \ge s^U_{\nu K}(x)$, then $M \subseteq K$.

(2) Commutative law: By means of Definition 3.2.1:
$M \cup N =$

$$\{\langle x,[\max\{s_{uM}{}^L(x),s_{uN}{}^L(x)\},\max\{s_{uM}{}^U(x),s_{uN}{}^U(x)\}],[\min\{s_{\pi M}{}^L(x),s_{\pi N}{}^L(x)\},\min\{s_{\pi M}{}^U(x),s_{\pi N}{}^U(x)\}]$$
$$[\min\{s_{vM}{}^L(x),s_{vN}{}^L(x)\},\min\{s_{vM}{}^U(x),s_{vN}{}^U(x)\}]\rangle|x\in X\}.$$

$N \cup M =$
$$\{\langle x,[\max\{s_{uM}{}^L(x),s_{uN}{}^L(x)\},\max\{s_{uN}{}^U(x),s_{uM}{}^U(x)\}],[\min\{s_{\pi N}{}^L(x),s_{\pi M}{}^L(x)\},\min\{s_{\pi N}{}^U(x),s_{\pi M}{}^U(x)\}]$$
$$[\min\{s_{vN}{}^L(x),s_{vM}{}^L(x)\},\min\{s_{vN}{}^U(x),s_{vM}{}^U(x)\}]\rangle|x\in X\}=M\cup N\text{, then } M\cup N=N\cup M.$$

Similarly, we can prove that $M \cap N = N \cap M$.
(3) Associative law: By means of Definition 3.2.1:

$$M\cup(N\cup K)=\{\langle x,(\widetilde{s}_{uM}(x),\widetilde{s}_{\pi M}(x),\widetilde{s}_{vM}(x))\rangle|x\in X\}\cup$$
$$\{\langle x,[\max\{s_{uN}{}^L(x),s_{uK}{}^L(x)\},\max\{s_{uN}{}^U(x),s_{uK}{}^U(x)\}],[\min\{s_{\pi N}{}^L(x),s_{\pi K}{}^L(x)\},\min\{s_{\pi N}{}^U(x),s_{\pi K}{}^U(x)\}],$$
$$[\min\{s_{vN}{}^L(x),s_{vK}{}^L(x)\},\min\{s_{vN}{}^U(x),s_{vK}{}^U(x)\}]\rangle|x\in X\}$$
$$=\{\langle x,[\max\{s_{uM}{}^L(x),s_{uN}{}^L(x),s_{uK}{}^L(x)\},\max\{s_{uM}{}^U(x),s_{uN}{}^U(x),s_{uK}{}^U(x)\}],$$
$$[\min\{s_{\pi M}{}^L(x),s_{\pi N}{}^L(x),s_{\pi K}{}^L(x)\},\min\{s_{\pi M}{}^U(x),s_{\pi N}{}^U(x),s_{\pi K}{}^U(x)\}],$$
$$[\min\{s_{vM}{}^L(x),s_{vN}{}^L(x),s_{vK}{}^L(x)\},\min\{s_{vM}{}^U(x),s_{vN}{}^U(x),s_{vK}{}^U(x)\}]\rangle|x\in X\}.$$

$(M\cup N)\cup K=$
$$\{\langle x,[\max\{s_{uM}{}^L(x),s_{uN}{}^L(x)\},\max\{s_{uN}{}^U(x),s_{uM}{}^U(x)\}],[\min\{s_{\pi N}{}^L(x),s_{\pi M}{}^L(x)\},\min\{s_{\pi N}{}^U(x),s_{\pi M}{}^U(x)\}]$$
$$[\min\{s_{vN}{}^L(x),s_{vM}{}^L(x)\},\min\{s_{vN}{}^U(x),s_{vM}{}^U(x)\}]\rangle|x\in X\}\cup\{\langle x,(\widetilde{s}_{uK}(x),\widetilde{s}_{\pi K}(x),\widetilde{s}_{vK}(x))\rangle|x\in X\}$$
$$=\{\langle x,[\max\{s_{uM}{}^L(x),s_{uN}{}^L(x),s_{uK}{}^L(x)\},\max\{s_{uM}{}^U(x),s_{uN}{}^U(x),s_{uK}{}^U(x)\}],$$
$$[\min\{s_{\pi M}{}^L(x),s_{\pi N}{}^L(x),s_{\pi K}{}^L(x)\},\min\{s_{\pi M}{}^U(x),s_{\pi N}{}^U(x),s_{\pi K}{}^U(x)\}],$$
$$[\min\{s_{vM}{}^L(x),s_{vN}{}^L(x),s_{vK}{}^L(x)\},\min\{s_{vM}{}^U(x),s_{vN}{}^U(x),s_{vK}{}^U(x)\}]\rangle|x\in X\}=M\cup(N\cup K),$$

then $M \cup (N \cup K) = (M \cup N) \cup K$.
Similarly, we can prove that $M \cap (N \cap K) = (M \cap N) \cap K$.
By means of Definition 3.2.1, the same proof properties (4), (5), (6) and (7).

3.3 Basic Operations and Properties of LIVSFN

Based on the definition of LIVSFN, this section will propose its basic operations, properties and related proofs.

Definition 3.3.1 Let $\alpha_1=([s_{a1},s_{b1}],[s_{c1},s_{d1}],[s_{e1},s_{f1}])$ and $\alpha_2=([s_{a2},s_{b2}],[s_{c2},s_{d2}],[s_{e2},s_{f2}])$ be two LIVSFNs, then

(1) $\alpha_1 = \alpha_2$ iff $a_1 = a_2$, $b_1 = b_2$, $c_1 = c_2$, $d_1 = d_2$, $e_1 = e_2$ and $f_1 = f_2$;
(2) $\alpha_1 \le \alpha_2$ iff $a_1 \le a_2$, $b_1 \le b_2$, $c_1 \le c_2$, $d_1 \le d_2$, $e_1 \ge e_2$ and $f_1 \ge f_2$;
(3) $\alpha_1^c = ([s_{e_1}, s_{f_1}], [s_{c_1}, s_{d_1}], [s_{a_1}, s_{b_1}])$;
(4) $\alpha_1 \cup \alpha_2 =([\max\{s_{a_1}, s_{a_2}\}, \max\{s_{b_1}, s_{b_2}\}], [\min\{s_{c_1}, s_{c_2}\}, \min\{s_{d_1}, s_{d_2}\}], [\min\{s_{e_1}, s_{e_2}\}, \min\{s_{f_1}, s_{f_2}\}])$;
(5) $\alpha_1 \cap \alpha_2 =([\min\{s_{a_1}, s_{a_2}\}, \min\{s_{b_1}, s_{b_2}\}], [\min\{s_{c_1}, s_{c_2}\}, \min\{s_{d_1}, s_{d_2}\}], [\max\{s_{e_1}, s_{e_2}\}, \max\{s_{f_1}, s_{f_2}\}])$.

Definition 3.3.2 Let $\alpha_1 = ([s_{a1}, s_{b1}], [s_{c1}, s_{d1}], [s_{e1}, s_{f1}])$ and $\alpha_2 = ([s_{a2}, s_{b2}], [s_{c2}, s_{d2}], [s_{e2}, s_{f2}])$ be two LIVSFNs, λ is any positive integer, then

$$(1)\ \alpha_1 \oplus \alpha_2 = \left(\left[s_{\sqrt{a_1^2+a_2^2-\frac{a_1^2a_2^2}{h^2}}}, s_{\sqrt{b_1^2+b_2^2-\frac{b_1^2b_2^2}{h^2}}}\right], \left[s_{\frac{c_1c_2}{h}}, s_{\frac{d_1d_2}{h}}\right], \left[s_{\frac{e_1e_2}{h}}, s_{\frac{f_1f_2}{h}}\right]\right);$$

$$(2)\ \alpha_1 \otimes \alpha_2 = \left(\left[s_{\frac{a_1a_2}{h}}, s_{\frac{b_1b_2}{h}}\right], \left[s_{\sqrt{c_1^2+c_2^2-\frac{c_1^2c_2^2}{h^2}}}, s_{\sqrt{d_1^2+d_2^2-\frac{d_1^2d_2^2}{h^2}}}\right], \left[s_{\sqrt{e_1^2+e_2^2-\frac{e_1^2e_2^2}{h^2}}}, s_{\sqrt{f_1^2+f_2^2-\frac{f_1^2f_2^2}{h^2}}}\right]\right);$$

$$(3)\ \lambda\alpha_1 = \left(\left[s_{\sqrt{h^2-h^2\left(1-\frac{a_1^2}{h^2}\right)^\lambda}}, s_{\sqrt{h^2-h^2\left(1-\frac{b_1^2}{h^2}\right)^\lambda}}\right], \left[s_{h\left(\frac{c_1}{h}\right)^\lambda}, s_{h\left(\frac{d_1}{h}\right)^\lambda}\right], \left[s_{h\left(\frac{e_1}{h}\right)^\lambda}, s_{h\left(\frac{f_1}{h}\right)^\lambda}\right]\right);$$

$$(4)\ \alpha_1^\lambda = \left(\left[s_{h\left(\frac{a_1}{h}\right)^\lambda}, s_{h\left(\frac{b_1}{h}\right)^\lambda}\right], \left[s_{\sqrt{h^2-h^2\left(1-\frac{c_1^2}{h^2}\right)^\lambda}}, s_{\sqrt{h^2-h^2\left(1-\frac{d_1^2}{h^2}\right)^\lambda}}\right], \left[s_{\sqrt{h^2-h^2\left(1-\frac{e_1^2}{h^2}\right)^\lambda}}, s_{\sqrt{h^2-h^2\left(1-\frac{f_1^2}{h^2}\right)^\lambda}}\right]\right).$$

Theorem 3.3.1 Let $\alpha = ([s_a, s_b], [s_c, s_d], [s_e, s_f])$, $\alpha_1 = ([s_{a1}, s_{b1}], [s_{c1}, s_{d1}], [s_{e1}, s_{f1}])$ and $\alpha_2 = ([s_{a2}, s_{b2}], [s_{c2}, s_{d2}], [s_{e2}, s_{f2}])$ be three LIVSFNs, λ, λ_1 and λ_2 be any positive integer, then

(1) $\alpha_1 \oplus \alpha_2 = \alpha_2 \oplus \alpha_1$;
(2) $\alpha_1 \otimes \alpha_2 = \alpha_2 \otimes \alpha_1$;
(3) $(\alpha_1 \oplus \alpha_2) \oplus \alpha = \alpha_1 \oplus (\alpha_2 \oplus \alpha)$;
(4) $(\alpha_1 \otimes \alpha_2) \otimes \alpha = \alpha_1 \otimes (\alpha_2 \otimes \alpha)$;
(5) $\lambda(\alpha_1 \oplus \alpha_2) = \lambda\alpha_1 \oplus \lambda\alpha_2$;
(6) $(\lambda_1 + \lambda_2)\alpha = \lambda_1\alpha \oplus \lambda_2\alpha$;
(7) $(\alpha_1 \otimes \alpha_2)^\lambda = \alpha_1^\lambda \otimes \alpha_2^\lambda$;
(8) $\alpha^{\lambda_1} \otimes \alpha^{\lambda_2} = \alpha^{(\lambda_1+\lambda_2)}$.

Proof. (1) By means of Definition 3.3.1:

$$\alpha_1 \oplus \alpha_2 = \left(\left[s_{\sqrt{a_1^2+a_2^2-\frac{a_1^2a_2^2}{h^2}}}, s_{\sqrt{b_1^2+b_2^2-\frac{b_1^2b_2^2}{h^2}}}\right], \left[s_{\frac{c_1c_2}{h}}, s_{\frac{d_1d_2}{h}}\right], \left[s_{\frac{e_1e_2}{h}}, s_{\frac{f_1f_2}{h}}\right]\right),$$

$$\alpha_2 \oplus \alpha_1 = \left(\left[s_{\sqrt{a_2^2+a_1^2-\frac{a_2^2a_1^2}{h^2}}}, s_{\sqrt{b_2^2+b_1^2-\frac{b_2^2b_1^2}{h^2}}}\right], \left[s_{\frac{c_2c_1}{h}}, s_{\frac{d_2d_1}{h}}\right], \left[s_{\frac{e_2e_1}{h}}, s_{\frac{f_2f_1}{h}}\right]\right),$$

then $\alpha_1 \oplus \alpha_2 = \alpha_2 \oplus \alpha_1$.

(2) According to (1), easy to proof that (2) holds.

(3) By means of Definition 3.3.2:

$$(\alpha_1 \oplus \alpha_2) \oplus \alpha = \left(\left[s_{\sqrt{a_1^2+a_2^2-\frac{a_1^2a_2^2}{h^2}}}, s_{\sqrt{b_1^2+b_2^2-\frac{b_1^2b_2^2}{h^2}}}\right], \left[s_{\frac{c_1c_2}{h}}, s_{\frac{d_1d_2}{h}}\right], \left[s_{\frac{e_1e_2}{h}}, s_{\frac{f_1f_2}{h}}\right]\right) \oplus \left([s_a, s_b], [s_c, s_d], [s_e, s_f]\right)$$

$$= \left(\left[s_{\sqrt{a^2+a_1^2+a_2^2-\frac{a_1^2a_2^2}{h^2}-\frac{a^2}{h^2}\left(a_1^2+a_2^2-\frac{a_1^2a_2^2}{h^2}\right)}}, s_{\sqrt{b^2+b_1^2+b_2^2-\frac{b_1^2b_2^2}{h^2}-\frac{b^2}{h^2}\left(b_1^2+b_2^2-\frac{b_1^2b_2^2}{h^2}\right)}}\right], \left[s_{\frac{c_1c_2c}{h}}, s_{\frac{d_1d_2d}{h}}\right], \left[s_{\frac{e_1e_2e}{h}}, s_{\frac{f_1f_2f}{h}}\right]\right),$$

$$\alpha_1 \oplus (\alpha_2 \oplus \alpha) = \left([s_{a_1}, s_{b_1}], [s_{c_1}, s_{d_1}], [s_{e_1}, s_{f_1}]\right) \oplus \left(\left[s_{\sqrt{a_2^2+a^2-\frac{a_2^2a^2}{h^2}}}, s_{\sqrt{b_2^2+b^2-\frac{b_2^2b^2}{h^2}}}\right], \left[s_{\frac{c_2c}{h}}, s_{\frac{d_2d}{h}}\right], \left[s_{\frac{e_2e}{h}}, s_{\frac{f_2f}{h}}\right]\right)$$

$$= \left(\left[s_{\sqrt{a^2+a_1^2+a_2^2-\frac{a_1^2a_2^2}{h^2}-\frac{a^2}{h^2}\left(a_1^2+a_2^2-\frac{a_1^2a_2^2}{h^2}\right)}}, s_{\sqrt{b^2+b_1^2+b_2^2-\frac{b_1^2b_2^2}{h^2}-\frac{b^2}{h^2}\left(b_1^2+b_2^2-\frac{b_1^2b_2^2}{h^2}\right)}}\right], \left[s_{\frac{c_1c_2c}{h}}, s_{\frac{d_1d_2d}{h}}\right], \left[s_{\frac{e_1e_2e}{h}}, s_{\frac{f_1f_2f}{h}}\right]\right),$$

then $(\alpha_1 \oplus \alpha_2) \oplus \alpha = \alpha_1 \oplus (\alpha_2 \oplus \alpha)$.

(4) According to (3), easy to proof that (4) holds.

(5) By means of Definition 3.3.2:

$$\lambda(\alpha_1 \oplus \alpha_2) = \lambda\left(\left[s_{\sqrt{a_1^2+a_2^2-\frac{a_1^2a_2^2}{h^2}}}, s_{\sqrt{b_1^2+b_2^2-\frac{b_1^2b_2^2}{h^2}}}\right], \left[s_{\frac{c_1c_2}{h}}, s_{\frac{d_1d_2}{h}}\right], \left[s_{\frac{e_1e_2}{h}}, s_{\frac{f_1f_2}{h}}\right]\right)$$

$$= \left(\left[s_{\sqrt{h^2-h^2\left(1-\frac{a_1^2a_2^2-\frac{a_1^2a_2^2}{h^2}}{h^2}\right)^{\lambda}}}, s_{\sqrt{h^2-h^2\left(1-\frac{b_1^2b_2^2-\frac{b_1^2b_2^2}{h^2}}{h^2}\right)^{\lambda}}}\right], \left[s_{h\left(\frac{c_1c_2}{h}\right)^{\lambda}}, s_{h\left(\frac{d_1d_2}{h}\right)^{\lambda}}\right], \left[s_{h\left(\frac{e_1e_2}{h}\right)^{\lambda}}, s_{h\left(\frac{f_1f_2}{h}\right)^{\lambda}}\right]\right),$$

$$\lambda\alpha_1 \oplus \lambda\alpha_2 = \left(\left[s_{\sqrt{h^2-h^2\left(1-\frac{a_1^2}{h^2}\right)^{\lambda}}}, s_{\sqrt{h^2-h^2\left(1-\frac{b_1^2}{h^2}\right)^{\lambda}}}\right], \left[s_{h\left(\frac{c_1}{h}\right)^{\lambda}}, s_{h\left(\frac{d_1}{h}\right)^{\lambda}}\right], \left[s_{h\left(\frac{e_1}{h}\right)^{\lambda}}, s_{h\left(\frac{f_1}{h}\right)^{\lambda}}\right]\right)$$

$$\oplus \left(\left[s_{\sqrt{h^2-h^2\left(1-\frac{a_2^2}{h^2}\right)^{\lambda}}}, s_{\sqrt{h^2-h^2\left(1-\frac{b_2^2}{h^2}\right)^{\lambda}}}\right], \left[s_{h\left(\frac{c_2}{h}\right)^{\lambda}}, s_{h\left(\frac{d_2}{h}\right)^{\lambda}}\right], \left[s_{h\left(\frac{e_2}{h}\right)^{\lambda}}, s_{h\left(\frac{f_2}{h}\right)^{\lambda}}\right]\right)$$

$$= \left(\left[s_{\sqrt{h^2-h^2\left(1-\frac{a_1^2a_2^2-\frac{a_1^2a_2^2}{h^2}}{h^2}\right)^{\lambda}}}, s_{\sqrt{h^2-h^2\left(1-\frac{b_1^2b_2^2-\frac{b_1^2b_2^2}{h^2}}{h^2}\right)^{\lambda}}}\right], \left[s_{h\left(\frac{c_1c_2}{h}\right)^{\lambda}}, s_{h\left(\frac{d_1d_2}{h}\right)^{\lambda}}\right], \left[s_{h\left(\frac{e_1e_2}{h}\right)^{\lambda}}, s_{h\left(\frac{f_1f_2}{h}\right)^{\lambda}}\right]\right),$$

then $\lambda(\alpha_1 \oplus \alpha_2) = \lambda\alpha_1 \oplus \lambda\alpha_2$.

(6) According to (5), easy to proof that (6) holds.

(7) By means of Definition 3.3.2:

$$(\alpha_1 \otimes \alpha_2)^\lambda = \left(\left[S_{\frac{a_1 a_2}{h}}, S_{\frac{b_1 b_2}{h}}\right], \left[S_{\sqrt{c_1^2+c_2^2-\frac{c_1^2 c_2^2}{h^2}}}, S_{\sqrt{d_1^2+d_2^2-\frac{d_1^2 d_2^2}{h^2}}}\right], \left[S_{\sqrt{e_1^2+e_2^2-\frac{e_1^2 e_2^2}{h^2}}}, S_{\sqrt{f_1^2+f_2^2-\frac{f_1^2 f_2^2}{h^2}}}\right]\right)^\lambda$$

$$= \left(\left[S_{h\left(\frac{a_1 a_2}{h^2}\right)^\lambda}, S_{h\left(\frac{b_1 b_2}{h^2}\right)^\lambda}\right], \left[S_{\sqrt{h^2-h^2\left(1-\frac{c_1^2+c_2^2-\frac{c_1^2 c_2^2}{h^2}}{h^2}\right)^\lambda}}, S_{\sqrt{h^2-h^2\left(1-\frac{d_1^2+d_2^2-\frac{d_1^2 d_2^2}{h^2}}{h^2}\right)^\lambda}}\right] \left[S_{\sqrt{h^2-h^2\left(1-\frac{e_1^2+e_2^2-\frac{e_1^2 e_2^2}{h^2}}{h^2}\right)^\lambda}}, S_{\sqrt{h^2-h^2\left(1-\frac{f_1^2+f_2^2-\frac{f_1^2 f_2^2}{h^2}}{h^2}\right)^\lambda}}\right]\right)$$

$$\alpha_1^\lambda \otimes \alpha_2^\lambda = \left(\left[S_{h\left(\frac{a_1}{h}\right)^\lambda}, S_{h\left(\frac{b_1}{h}\right)^\lambda}\right], \left[S_{\sqrt{h^2-h^2\left(1-\frac{c_1^2}{h^2}\right)^\lambda}}, S_{\sqrt{h^2-h^2\left(1-\frac{d_1^2}{h^2}\right)^\lambda}}\right], \left[S_{\sqrt{h^2-h^2\left(1-\frac{e_1^2}{h^2}\right)^\lambda}}, S_{\sqrt{h^2-h^2\left(1-\frac{f_1^2}{h^2}\right)^\lambda}}\right]\right)$$

$$\otimes \left(\left[S_{h\left(\frac{a_2}{h}\right)^\lambda}, S_{h\left(\frac{b_2}{h}\right)^\lambda}\right], \left[S_{\sqrt{h^2-h^2\left(1-\frac{c_2^2}{h^2}\right)^\lambda}}, S_{\sqrt{h^2-h^2\left(1-\frac{d_2^2}{h^2}\right)^\lambda}}\right], \left[S_{\sqrt{h^2-h^2\left(1-\frac{e_2^2}{h^2}\right)^\lambda}}, S_{\sqrt{h^2-h^2\left(1-\frac{f_2^2}{h^2}\right)^\lambda}}\right]\right)$$

$$= \left(\begin{array}{l} \left[S_{\frac{h\left(\frac{a_1}{h}\right)^\lambda h\left(\frac{a_2}{h}\right)^\lambda}{h}}, S_{\frac{h\left(\frac{b_1}{h}\right)^\lambda h\left(\frac{b_2}{h}\right)^\lambda}{h}}\right], \left[S_{\sqrt{h^2-h^2\left(1-\frac{c_1^2}{h^2}\right)^\lambda + h^2-h^2\left(1-\frac{c_2^2}{h^2}\right)^\lambda - \frac{\left(h^2-h^2\left(1-\frac{c_1^2}{h^2}\right)^\lambda\right)\left(h^2-h^2\left(1-\frac{c_2^2}{h^2}\right)^\lambda\right)}{h^2}}}, S_{\sqrt{h^2-h^2\left(1-\frac{d_1^2}{h^2}\right)^\lambda + h^2-h^2\left(1-\frac{d_2^2}{h^2}\right)^\lambda - \frac{\left(h^2-h^2\left(1-\frac{d_1^2}{h^2}\right)^\lambda\right)\left(h^2-h^2\left(1-\frac{d_2^2}{h^2}\right)^\lambda\right)}{h^2}}}\right], \\ \left[S_{\sqrt{h^2-h^2\left(1-\frac{e_1^2}{h^2}\right)^\lambda + h^2-h^2\left(1-\frac{e_2^2}{h^2}\right)^\lambda - \frac{\left(h^2-h^2\left(1-\frac{e_1^2}{h^2}\right)^\lambda\right)\left(h^2-h^2\left(1-\frac{e_2^2}{h^2}\right)^\lambda\right)}{h^2}}}, S_{\sqrt{h^2-h^2\left(1-\frac{f_1^2}{h^2}\right)^\lambda + h^2-h^2\left(1-\frac{f_2^2}{h^2}\right)^\lambda - \frac{\left(h^2-h^2\left(1-\frac{f_1^2}{h^2}\right)^\lambda\right)\left(h^2-h^2\left(1-\frac{f_2^2}{h^2}\right)^\lambda\right)}{h^2}}}\right] \end{array}\right)$$

$$= \left(\left[S_{h\left(\frac{a_1 a_2}{h^2}\right)^\lambda}, S_{h\left(\frac{b_1 b_2}{h^2}\right)^\lambda}\right], \left[S_{\sqrt{h^2-h^2\left(1-\frac{c_1^2+c_2^2-\frac{c_1^2 c_2^2}{h^2}}{h^2}\right)^\lambda}}, S_{\sqrt{h^2-h^2\left(1-\frac{d_1^2+d_2^2-\frac{d_1^2 d_2^2}{h^2}}{h^2}\right)^\lambda}}\right], \left[S_{\sqrt{h^2-h^2\left(1-\frac{e_1^2+e_2^2-\frac{e_1^2 e_2^2}{h^2}}{h^2}\right)^\lambda}}, S_{\sqrt{h^2-h^2\left(1-\frac{f_1^2+f_2^2-\frac{f_1^2 f_2^2}{h^2}}{h^2}\right)^\lambda}}\right]\right),$$

then $(\alpha_1 \otimes \alpha_2)^\lambda = \alpha_1^\lambda \otimes \alpha_2^\lambda$.
(8) According to (7), easy to proof that (8) holds.

3.4 Comparing Methods and Measurement Formulas of LIVSFNs

To rank LIVSFNs, this section presents a comparison low of LIVSFNs based on score function and accuracy function.

Definition 3.4.1 Let $\alpha = \left([s_a, s_b], [s_c, s_d], [s_e, s_f]\right)$ be a LIVSFN. The score function and the accuracy function of α are respectively defined as

$$S(\alpha) = s_{\sqrt{\frac{2h^2+a^2+b^2-e^2-f^2-\left(\frac{c}{2}\right)^2-\left(\frac{d}{2}\right)^2}{4}}} \tag{9}$$

$$H(\alpha) = s_{\sqrt{\frac{a^2+b^2+e^2+f^2+c^2+d^2}{2}}} \tag{10}$$

Then, the comparison low based on the score function and the accuracy function between two different LIVSFNs will be defined as follows.

Theorem 3.4.1 Let $\alpha_1 = \left([s_{a1}, s_{b1}], [s_{c1}, s_{d1}], [s_{e1}, s_{f1}]\right)$ and $\alpha_2 = \left([s_{a2}, s_{b2}], [s_{c2}, s_{d2}], [s_{e2}, s_{f2}]\right)$ be two different LIVSFNs, the comparison lows be defined as follows:

(1) If $S(\alpha_1) > S(\alpha_2)$, then $\alpha_1 > \alpha_2$;
(2) If $S(\alpha_1) < S(\alpha_2)$, then $\alpha_1 < \alpha_2$;
(3) If $S(\alpha_1) = S(\alpha_2)$, then

(a) If $H(\alpha_1) > H(\alpha_2)$, then $\alpha_1 > \alpha_2$;
(b) If $H(\alpha_1) < H(\alpha_2)$, then $\alpha_1 < \alpha_2$;
(c) If $H(\alpha_1) = H(\alpha_2)$, then $\alpha_1 = \alpha_2$.

Definition 3.4.2 Let $\alpha_1 = \left([s_{a1}, s_{b1}], [s_{c1}, s_{d1}], [s_{e1}, s_{f1}]\right)$ and $\alpha_2 = \left([s_{a2}, s_{b2}], [s_{c2}, s_{d2}], [s_{e2}, s_{f2}]\right)$ be two different LIVSFNs, the measure formula $d(\alpha_1, \alpha_2)$ be defined as follows:

$$d(\alpha_1, \alpha_2) = \frac{1}{4h^2}\Big(\left|a_1^{\,2} - a_2^{\,2}\right| + \left|b_1^{\,2} - b_2^{\,2}\right| + \left|c_1^{\,2} - c_2^{\,2}\right| + \left|d_1^{\,2} - d_2^{\,2}\right| + \left|e_1^{\,2} - e_2^{\,2}\right| + \left|f_1^{\,2} - f_2^{\,2}\right|$$
$$+ \left|r_1^{L^2} - r_2^{L^2}\right| + \left|r_1^{U^2} - r_2^{U^2}\right|\Big) \tag{11}$$

where $r_i^L = \sqrt{h^2 - b_i^2 - d_i^2 - f_i^2}$, $r_i^U = \sqrt{h^2 - a_i^2 - c_i^2 - e_i^2}$; $s_{r_i} = [s_{r_i^L}, s_{r_i^U}](i = 1, 2)$ is the waiver degree of α_i, and satisfy the following properties:

(1) $d(\alpha_1, \alpha_2) \in [0, 1]$;
(2) $d(\alpha_1, \alpha_2) = d(\alpha_2, \alpha_1)$;
(3) $d(\alpha_1, \alpha_2) = 0$ iff $\alpha_1 = \alpha_2$.

4 Conclusion

The traditional IVSF theory is a more effective mathematical tool for dealing with the imprecise and uncertain data under the quantitative environment, the LSFS theory expresses the uncertainty in qualitative aspect. It is of great theoretical value and practical basis to extend the traditional spherical fuzzy set from quantitative environment to qualitative environment by combining the linguistic term set theory. In view of this, in this paper, we have proposed the LIVSFS in which membership, non-membership and hesitancy degree represented by the interval-valued linguistic terms. Also, we have given the concept of the LIVSFN, and have discussed various operational laws, related properties and their proofs, the measure formula, score and accuracy functions based on it. In the future, we will develop the application in other disciplines.

References

1. Zadeh, L.A.: Fuzzy sets. Inf. Control **8**(3), 338–353 (1965)
2. Atanassov, K.T.: Intuitionistic fuzzy sets. Fuzzy Sets Syst. **20**(1), 87–96 (1986)
3. Atanassov, K.T., Gargov, G.: Interval-valued intuitionistic fuzzy sets. Fuzzy Sets Syst. **31**(3), 343–349 (1989)
4. Atanassov, K.T.: Geometrical interpretation of the elements of the intuitionistic fuzzy objects. Int. J. Bioautom. **20**(S1), S27–S42 (2016)
5. Kutlu, G.F., Kahraman, C.: Spherical fuzzy sets and spherical fuzzy TOPSIS method. J. Intell. Fuzzy Syst. **36**(1), 337–352 (2019)
6. Ashraf, S., Abdullah, S., Mahmood, T., et al.: Spherical fuzzy sets and their applications in multi-attribute decision making problems. J. Intell. Fuzzy Syst. **36**(3), 2829–2844 (2019)
7. Guleria, A., Bajaj, R.K.: T-spherical fuzzy graphs: operations and applications in various selection processes. Arab. J. Sci. Eng. **45**(3), 2177–2193 (2020)
8. Guleria, A., Bajaj, R.K.: Eigen spherical fuzzy set and its application to decision-making problem. Scientia Iranica **28**(1), 516–531 (2021)
9. Ullah, K., Garg, H., Mahmood, T., Jan, N., Ali, Z.: Correlation coefficients for T-spherical fuzzy sets and their applications in clustering and multi-attribute decision making. Soft. Comput. **24**(3), 1647–1659 (2019). https://doi.org/10.1007/s00500-019-03993-6
10. Mahmood, T., Ullah, K., Khan, Q., Jan, N.: An approach toward decision-making and medical diagnosis problems using the concept of spherical fuzzy sets. Neural Comput. Appl. **31**(11), 7041–7053 (2018). https://doi.org/10.1007/s00521-018-3521-2
11. Wu, M.Q., Chen, T.Y., Fan, J.P.: Divergence measure of T-spherical fuzzy sets and its applications in pattern recognition. IEEE Access **8**, 10208–10221 (2019)
12. Ullah, K., Mahmood, T., Jan, N.: Similarity measures for T-spherical fuzzy sets with applications in pattern recognition. Symmetry **10**(6), 193 (2018)
13. Gul, M., Ak, M.F.: A modified failure modes and effects analysis using interval-valued spherical fuzzy extension of TOPSIS method: case study in a marble manufacturing facility. Soft. Comput. **25**(8), 6157–6178 (2021). https://doi.org/10.1007/s00500-021-05605-8
14. Duleba, S., Kutlu, G.F., Moslem, S.: Interval-valued spherical fuzzy analytic hierarchy process method to evaluate public transportation development. Informatica **32**(4), 661–686 (2021)
15. Zadeh, L.A.: The concept of a linguistic variable and its application to approximate reasoning-I. Inf. Sci. **8**(3), 199–249 (1975)
16. Zadeh, L.A.: The concept of a linguistic variable and its application to approximate reasoning—II. Inf. Sci. **8**(4), 301–357 (1975)
17. Zadeh, L.A.: The concept of a linguistic variable and its application to approximate reasoning-III. Inf. Sci. **9**(1), 43–80 (1975)
18. Jia, R.: Linguistic spherical fuzzy set and related properties. Oper. Res. Fuzziol. **11**(3), 282–228 (2021)
19. Herrera F., Martínez L.: A model based on linguistic 2-tuples for dealing with multigranular hierarchical linguistic contexts in multi-expert decision-making. IEEE Transactions on Systems, Man, and Cybernetics, Part B (Cybernetics) 31(2), 227–234 (2001)
20. Liao, H., Xu, Z., Zeng, X.J.: Distance and similarity measures for hesitant fuzzy linguistic term sets and their application in multi-criteria decision making. Inf. Sci. **271**, 125–142 (2014)

Knowledge Representation and Reasoning

A Genetic Algorithm for Causal Discovery Based on Structural Causal Model

Zhengyin Chen[1,2], Kun Liu[1,2], and Wenpin Jiao[1,2](✉)

[1] Institute of Software, School of Computer Science, Peking University, Beijing 100871, China
{chenzy512,kunl,jwp}@pku.edu.cn

[2] Key Laboratory of High Confidence Software Technology, (Peking University), MOE, China

Abstract. With a large amount of data accumulated in many fields, causal discovery based on observational data is gradually emerging, which is considered to be the basis for realizing strong artificial intelligence. However, the existing main causal discovery methods, including constraint-based methods, structural causal model based methods, and scoring-based methods, cannot find real causal relations accurately and quickly. In this paper, we propose a causal discovery method based on genetic algorithm, which combines structural causal model, scoring method, and genetic search algorithm. The core of our method is to divide the causal relation discovery process into the evaluation phase based on the features of structural causal model and the search phase based on the genetic algorithm. In the evaluation phase, the causal graph is evaluated from three aspects: model deviation, noise independence, and causal graph cyclicity, which effectively ensures the accuracy of causal discovery. In the search phase, an efficient random search is designed based on genetic algorithm, which greatly improves the causal discovery efficiency. This paper implements the corresponding algorithm, namely SCM-GA (Structural Causal Model based Genetic Algorithm), and conducts experiments on several simulated datasets and one widely used real-scene dataset. The experiments compare five classic baseline algorithms, and the results show that SCM-GA has achieved great improvement in accuracy, applicability, and efficiency. Especially on the real scene dataset, SCM-GA achieves better results than the state-of-the-art algorithm, with similar SHD (Structure Hamming Distance) value, 40% higher recall rate, and 83.3% shorter running time.

Keywords: Causal discovery · Genetic algorithm · Structural causal model

Supported by National Science and Technology Major Project (Grant No. 2020AAA 0109401).

L. Fang et al. (Eds.): CICAI 2022, LNAI 13606, pp. 39–54, 2022.
https://doi.org/10.1007/978-3-031-20503-3_4

1 Introduction

Causal relation is the objective relationship between the "cause" event and the "effect" event, and the "cause" event is the reason why the "effect" event occurs [2]. Compared with correlation, causal relation strictly distinguishes the cause and the effect and can better reflect the nature of the interaction between events, so as to understand the laws behind the changes of events, and then guide people on how to predict, make decisions, intervene and control. Causal relation is an important relationship in modern scientific research and is needed in almost all disciplines of scientific research.

A traditional method to identify causal relations is *Randomized Controlled Experiments*, which is acknowledged as the golden rule for determining causal relations [6]. Randomized Controlled Experiments refer to experiments that explore whether the target variable has changed under the condition that the irrelevant variables are kept the same and the experimental variables are set different. However, in real life, Randomized Controlled Experiments are often faced with the constraints of cost and ethics and are difficult to implement.

In recent years, a large amount of data accumulated in various disciplines such as computer science, life science, economics, and sociology [15]. Causal discovery based on observational data[1] has drawn more and more researchers' attention. The observational data can be temporal series or non-temporal series. Since temporal series data contains time information, data collection is difficult and quality requirements are high. Besides, temporal series data contains more causal clues. The research and application of causal discovery based on temporal series data are greatly limited, so this paper mainly focuses on causal discovery based on non-temporal series data.

The causal discovery based on non-temporal series observational data is essentially to analyze and process the data according to a certain method or principle, and gradually find the hidden causal relations from it, so as to identify the complete causal graph structure. Existing research can mainly be divided into three categories [1,13]: constraint-based methods (e.g. [20–22]), structural causal model based methods (e.g. [12,19,25]), and scoring-based methods (e.g. [3,7,14,27]). The basic idea of these methods is to construct a causal graph first, and then remove the edges between those factors (i.e. nodes) that obviously do not have causal relation. The core is to construct and search for a valid causal graph and to measure and evaluate the effectiveness of the causal graph, which are related to the efficiency and accuracy of causal discovery, respectively.

However, existing methods may not ensure accuracy and efficiency at the same time. For example, in scoring-based methods, the scoring functions are mostly based on fitting causal mechanisms, which are not reliable and generalizable, and difficult to guarantee accuracy [1,11]. To improve the accuracy, it is necessary to consider the features of structural causal model such as noise independence [17], but this will lead to complex scoring functions and low com-

[1] It is commonly referred to as *causal discovery*, this abbreviation is also used in related literature, so in this article we do the same.

putational efficiency. In addition, the search for causal graphs is NP-hard [3,5]. To improve the efficiency of search, existing methods generally use heuristic methods ([3]) or gradient-based methods ([14,27,29]), but most of them only consider the fitting degree of causal mechanisms and the DAG (Directed Acyclic Graph) constraints of the causal graph, the accuracy cannot be guaranteed.

To overcome the trade-off between efficiency and accuracy, we propose a causal discovery method based on genetic algorithm, which combines structural causal model, scoring method, and genetic search algorithm. The discovery process is divided into an evaluation phase based on the features of structural causal model and a search phase based on genetic algorithm. In the search phase, several candidate causal graphs are generated through random search of the genetic algorithm. In the evaluation phase, the candidate causal graphs are scored and evaluated by the metrics based on structural causal model.

The main contributions of this paper include: Firstly, a general causal graph evaluation metric is designed which comprehensively considers three features of structural causal model – causal mechanism fitting degree, noise independence, and causal graph cyclicity. It can effectively deal with the heterogeneous noise in the real environment and improve the accuracy of causal discovery. Secondly, we implement SCM-GA (Structural Causal Model based Genetic Algorithm). SCM-GA borrows ideas from genetic algorithm, it has high implicit parallel ability and excellent global optimization ability. Finally, we conduct experiments on multiple simulated datasets and a real dataset to verify the effectiveness and efficiency of our method compared with five baseline algorithms.

The remainder of this paper is structured as follows. In Sect. 2 we introduce related works. In Sect. 3 we describe the method we propose and the corresponding algorithm implementation. We conduct some experiments to evaluate our method in Sect. 4. Finally, we summarize our work in Sect. 5.

2 Related Work

Since the 1980 s,s, pioneer researchers in the field of causal discovery have explored how to discover causal relations from non-temporal series data from a statistical point of view. They use causal graph as the basic tool, which is derived from a probabilistic graphical model based on Bayesian network, adding a causal relation explanation. Over the past few decades, the field of causal discovery has attracted an increasing number of researchers, along with a series of advances in causal discovery algorithms. Existing research methods can mainly be divided into three categories [1,13]: constraint-based methods, structural causal model based methods, and scoring-based methods.

Constraint-based methods mainly exploit conditional independence between variables to recover causal structure, usually requiring the use of conditional independence tests based on statistics or information theory. Typical algorithms include Inductive Causation (IC) algorithm [22], SGS algorithm [20], and PC algorithm [21]. However, these methods may have undirected causal edges due to the existence of Markov equivalence classes with the same conditional independence [10].

Methods based on structural causal model exploit the asymmetry of causal mechanisms to discover causal direction based on specific assumptions about the mechanism of data generation. The classic methods are Linear Non-Gaussian Acyclic Model (LiNGAM) [19], Additive Noise Model (ANM) [12], and Post-Nonlinear Causal Model (PNL) [25]. These methods effectively improve the accuracy of causal discovery, but this improvement requires that the data generation satisfies structural causal model hypothesis. However, the causal mechanisms and noise mechanisms in the real world are often complicated, so it is difficult to apply such methods in real life.

Scoring-based methods regard the learning of causal graphs as a search problem and support flexible search strategies and scoring functions. Even when there exists Markov equivalence class, selecting an appropriate scoring function can direct all causal edges. The classic methods include Greedy Equivalence Search (GES) [3] based on heuristic search strategy, NOTES [27], GraN-DAG [14], and CGNN (Causal Generative Neural Networks) model [7] based on gradient search. There are two challenges in these methods: one is that the complexity of the algorithms is considerably high as the exponential explosion problem in high-dimensional data; the other is that the accuracy of the algorithm may decrease when facing the problem of heterogeneous noise in real scenarios.

One of the challenges in the accuracy of causal discovery is the problem of noise heterogeneity, that is, in real-world variables are often generated by heterogeneous noise of mixed types or mixed variances. For example, RL-BIC [29] assumes that the noise of all observational variables follows a uniform mechanism. It may erroneously absorb noise into the fitting process, resulting in dependencies between the residuals (i.e. noise) of the predicted variables which is contrary to structural causal model theory. In real scenarios, this overfitting problem often leads to spurious edges and missing edges in the generated structures [11], reducing the accuracy of causal discovery. Independence-based scoring functions [17] and DARING [11] consider the features of structural causal model such as noise independence in addition to the causal mechanism fitting degree. But this leads to more complex fitness functions or lower computational inefficiency.

For the efficiency challenge of causal discovery, current researchers mainly adopt two schemes, the heuristic method, and the gradient-based method. The most representative of the heuristic methods is the GES algorithm [3]. The scoring function of this algorithm is based on the Markov equivalence class so the search space is reduced from the original DAG space to a Markov equivalence class. More recently, NOTES [27] gave a smooth characterization of the DAG constraint, thereby formulating the problem as a continuous optimization problem. Subsequent GraN-DAG [14] and RL-BIC [29] also adopted negative log-likelihood and least-squares loss as loss functions respectively. But these works rarely consider the problem of noise heterogeneity so the accuracy is often poor.

3 Causal Discovery Based on Genetic Algorithm

In this section, we introduce our causal discovery method based on genetic algorithm by synthesizing structural causal model and scoring method. It includes three modules: data pre-processing, causal relation pre-discovery, and causal graph pruning. The data pre-processing module normalizes the observational data to obtain normalized data with a mean of zero and a unit variance. The causal relation pre-discovery module iteratively performs the search and evaluation of the causal graph, that is, firstly searches for a good enough causal graph, and then evaluates whether the found causal graph satisfies the termination condition, if so, outputs the current best causal graph, otherwise continues the next round of search. Since the causal graph output by the causal relation pre-discovery module often has spurious edges, a pruning operation is required to obtain the final causal graph. Figure 1 presents an overview of our method.

3.1 Data Pre-processing

In our method, we use three metrics to evaluate the causal graph. So a data standardization process is carried out in advance to remove the influence of the magnitude of the data. Z-score normalization is a commonly used data normalization method. It can uniformly convert raw data of different magnitudes into data with a mean of 0 and a variance of 1, see Eq. 1:

$$x = \frac{x' - v}{\sigma} \tag{1}$$

where v is the mean of the original data, and σ is the standard deviation of the original data.

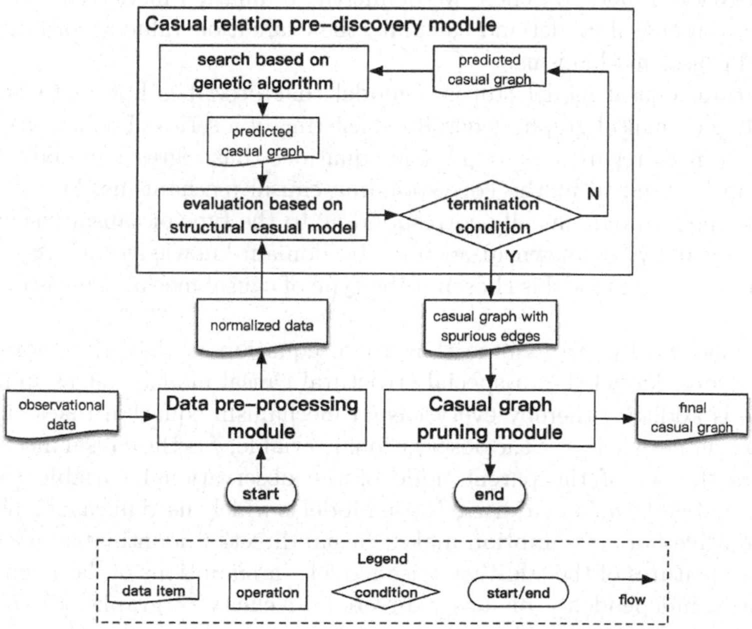

Fig. 1. Causal discovery based on genetic algorithm

We do not remove noise in the original data because the noise variable contains important causal direction information [16,24,26]. For example, the independence of noise is an important basis for judging the causal direction [11].

3.2 Causal Relation Pre-discovering

The process of causal relation pre-discovery is as follows (see Algorithm 1): First, an initial population $\mathcal{M}$ is generated (line 1), which includes a set of candidate causal graphs (individuals). Second, it enters the evaluation phase (line 4–6), calculates the evaluation value s_i of each individual m_i through the metrics based on structural causal model, and then judges whether the termination condition is met (line 7–9); if so, the algorithm ends and returns the individual with the best evaluation value and its corresponding causal graph; otherwise, it enters the search phase (line 11–23), performs evolutionary operations (selection, crossover, and mutation) to generate a new population, and then goes back to the second step.

Evaluation Based on Structural Causal Model. The purpose of the evaluation phase is to accurately measure the difference between the predicted causal graph and the real causal graph, which is essentially an evaluation problem. In our method, the core of the evaluation phase is to fit the causal mechanisms and define the evaluation metrics. That is, firstly, under the given observational data and predicted causal graph, the predicted causal mechanisms are obtained by the regression method. Then, we define the evaluation metrics by synthesizing structural causal model and calculate the evaluation value according to the predicted causal mechanisms.

Structural causal model [16] is a model that describes how data is generated through a causal graph, generally consisting of a series of equations. These equations can be regarded as regression equations thus regression analysis techniques can be used to fit the corresponding causal mechanisms. We select the regression method dynamically corresponding to the type of causal mechanisms which can usually be known in advance by domain knowledge or pre-analysis. Gaussian regression model is chosen if the type of causal mechanisms is unknown in advance.

More specifically, we assume that these equations satisfy the (non-linear) Additive Noise Model [17], a special structural causal model that assumes that the noise is additive, thereby every casual mechanism equation takes the form $x_i = f_i(pa_i)+n_i$, where x is the observational variable, f is the causal mechanism, and pa is the set of the parent node of the observational variable (possible "reasons"), n is the noise. Additive Noise Model is widely used in casual discovery for its additive noise assumption makes casual direction identifiable. Therefore, we use the features of the additive noise model – assumptions of data generation mechanism, independence of noise, directed acyclicity of graphs, to construct evaluation metrics for causal graphs.

Algorithm 1: Causal relation pre-discovering

Input: X, observational data; K, number of population; γ, mutation rate; c_{elite}, proportion of the preserved elites; W, initial weights of evaluation metrics

Output: G, predicted casual graph

$\mathcal{M}$ = createInitialPopulation(X,K);
T = 0;
while *True* **do**
  foreach $m_i \in \mathcal{M}$ **do** /* evaluation phase */
    s_i = calculateScore(X,m_i,W) ;
  end
  $\mathcal{I} = \{< m_i, s_i >| 1 \leq i \leq K\}$;
  if *any of the terminating creteria* **then**
    break;
  end
  $\mathcal{M}' \leftarrow \emptyset$; /* search phase */
  $\mathcal{I} = \{< m_i, calculateFitness(s_i) >| 1 \leq i \leq K\}$;
  while $|M'| < K$ **do**
    m_0', m_1' = selectionRet2($\mathcal{I}$);
    m_2', m_3' = crossover(m_0', m_1');
    $\mathcal{M}' \leftarrow \mathcal{M}' \cup \{m_2', m_3'\}$;
  end
  for i = *1 to* $K \cdot \gamma$ **do**
    m_0' = selection($\mathcal{I}$);
    m_1' = mutation(m_0');
    $\mathcal{M}' \leftarrow \{\mathcal{M}' - m_0'\} \cup \{m_1'\}$;
  end
  $\mathcal{M} = bestIndividuals(\mathcal{M}', \mathcal{M}, c_{elite})$;
  T += 1;
end
Let $\mathcal{M}_{min}$ be the minimum score in $\mathcal{I}$;
Return G corresponding to $\mathcal{M}_{min}$

For the predicted causal mechanism $\hat{F}$, the evaluation metric $S(X, \hat{G}, W)$ is calculated as,

$$S(X, \hat{G}, W) = w_0 * S_{lik}(X, \hat{F}) + w_1 * S_{dep}(X, \hat{F}) + w_2 * S_{DAG}(\hat{G}) \quad (2)$$

where, X is the observational data; $\hat{G}$ is the predicted causal graph; $W = \{w_0, w_1, w_2\}$ is the weight parameter; $\hat{F}$ is the predicted causal mechanisms, which are obtained by $\hat{G}$ and using the regression model; $S_{lik}(X, \hat{F})$ is the model deviation degree of the causal mechanism; $S_{ind}(X, \hat{F})$ is the noise independence degree; $S_{DAG}(\hat{G})$ is the cyclicity of the causal graph $\hat{G}$.

Model Deviation degree $S_{lik}(X, \hat{F})$ measures the ability of the causal mechanisms to recover the data X. In this paper we use the likelihood-based Bayesian Information Criterion (BIC) score because it not only satisfies the overall consistency

[9], but its decomposition formula also satisfies local consistency [3]. Therefore, the model deviation based on the BIC score is defined as follows:

$$S_{lik}(X, \hat{F}) = -2\log p(X; \hat{F}) + c_f \log M \tag{3}$$

where, c_f is a hyper-parameter that represents the complexity of the model; M is thc sample size of the observational data.

As we have $\hat{x}_i = \hat{f}_i(pa_i)$ represents the predicted value of the ith variable, and x_i^k represents the value of the ith variable in the kth sample, Eq. 3 can be further defined as:

$$S_{lik}(X, \hat{F}) = \sum_{i=1}^{D}(M \log(\sum_{k=1}^{M}(\hat{x}_i^k - x_i^k)^2/M)) + |G| \log M \tag{4}$$

The first term of Eq. 4 represents the likelihood objective, which is also equivalent to the log-likelihood objective in RL-BIC [29] and GraN-DAG [14]. The second term represents a penalty term for the number of edges in the causal graph, whose coefficient increases as the sample size increases. It can be seen from this equation that the smaller the model deviation degree, the better the predicted causal mechanisms.

Noise Independence degree $S_{ind}(X, \hat{F})$ measures the independence between noises. The smaller the value, the stronger the independence degree between noises, and the more consistent with the noise independence assumption of structural casual model. Here the noise is expressed as $\hat{n}_i$, which refers to the residual value between the predicted value and the true value.

$$\hat{n}_i = |x_i - \hat{x}_i| = |x_i - \hat{f}_i(pa_i)| \tag{5}$$

To measure the independence between noises, we use HSIC (Hilbert-Schmidt independence criterion) [8], which, like covariance, can be used to measure the independence between two random variables. In addition, studies have shown [24] that the independence between all noise (residual value) variables is equivalent to the independence between the noise variable and all parent nodes of this random variable. Therefore, in this paper, we define the noise independence as follows:

$$S_{ind}(X, \hat{F}) = \sum_{i=1}^{D} \sum_{x_j \in pa_i} \frac{HSIC(\hat{n}_i, x_j)}{|pa_i|} \tag{6}$$

where $HSIC(\hat{n}_i, x_j)$ represents the independence degree between two variables.

The Cyclicity of the causal graph $S_{DAG}(\hat{G})$ measures the degree to which the causal graph G conforms to the directed acyclic graph. The smaller the value, the better the causal graph G conforms to the directed acyclic graph. According to NOTEARS [27], the cyclicity of the causal graph satisfies:

Theorem 1. *Casual graph* $G \in \mathbb{R}^{D \times D}$ *is a DAG, if and only if* $\operatorname{tr}\left(e^{G \circ G}\right) - D = 0$

where,

- $e^M = \sum_{k=0}^{\infty} \frac{M^k}{k!}$ represents the matrix exponential;
- $\circ$ represents Hadamard product;
- tr represents the trace of the matrix.

Studies have shown that the value of the formula $\mathrm{tr}\left(\mathrm{e}^{G \circ G}\right) - D$ is non-negative, and the larger the value, the more cycles in the causal graph [29]. Therefore, we define the cyclicity of the causal graph as follows:

$$S_{DAG}(\hat{G}) = \mathrm{tr}\left(\mathrm{e}^{\hat{G} \circ \hat{G}}\right) - D \tag{7}$$

Search Based on Genetic Algorithm. The goal of the search phase is to find a sufficiently "good" causal graph, which is essentially a causal graph generation problem. Sufficiently "good" means that the generated causal graph should be the closest to the real causal graph, so the causal graph generation problem in the search phase can be regarded as an optimization problem. In this paper, genetic algorithm is applied to the search phase to quickly search for an optimal or near-optimal causal graph in the causal graph space. The solution space of this problem is all possible causal graphs, and the optimization objective adopts the evaluation metric based on structural causal model as we mentioned above.

In this paper, the D-variable causal graph is represented by a $D \times D$ adjacency matrix, where the matrix elements take the value 1 or 0, representing the presence or absence of a causal edge. Through binary coding, the individual is defined as $D \times D$ genes, which ensures the one-to-one correspondence between the individual search space and the solution space of the problem.

In the evaluation phase, we calculate the evaluation value s_i of each individual m_i in the population using the metric that we mentioned above. So we have a population with evaluation value $\mathcal{I} = \{< m_i, s_i >| \, 1 \le i \le K\}$. Based on the evaluation value, we can calculate the individual fitness, which is realized by reciprocating and normalizing the evaluation value of the individuals. Specifically, the calculation is as follows:

$$fit_i = \frac{1}{s_i + 1} \tag{8}$$

where fit_i is the fitness of the ith individual; s_i is the evaluation value of the ith individual.

The search process of the genetic algorithm mainly includes three genetic operators, namely selection, crossover, and mutation. Generally speaking, it is necessary to select existing genetic operators or design new operators according to the problem scenarios. In our method, the roulette selection, uniform distribution crossover, and binary mutation are used. In addition, to deal with the problem of "population degradation", we use "elite preservation strategy" in the selection process, which is the function *bestIndividuals* in line 23 of Algorithm 1. When the next generation is generated, the individuals with the best fitness in a certain proportion (c_{elite}) of the parent population are selected and directly added to the new population.

3.3 Causal Graph Pruning

In causal discovery, the predicted causal graph may contain spurious edges (i.e., edges that do not exist in the real causal graph). This spurious edge problem is very common and has been encountered in related works such as ICA-LiNGAM [19] and NOTEARS [27], so further processing of the output causal graph is required. ICA-LiNGAM adopts the method of conditional independence test to judge whether there are redundant edges. NOTEARS adopts a threshold pruning method based on least squares regression, thereby reducing the number of spurious edges.

After the causal relation pre-discovery module finishes, the algorithm outputs a causal graph with the best score. Then, based on the weights of the causal edges, the spurious edges will be pruned, thereby reducing false discovery. In regression problems, post-processing the coefficient estimates by a fixed threshold can effectively reduce the number of false discoveries [23,28]. Inspired by these works, we set an appropriate threshold (0.3) to prune the causal graph, that is, delete causal edges whose weights are less than the threshold.

4 Experiments

In this section, we conduct several experiments on both simulated datasets and a real-life dataset. The simulated datasets include different sample sizes, node sizes, noise mechanisms, and causal mechanisms. The real-life dataset is Sachs dataset which is widely used in causal discovery [18].

4.1 Research Questions

We study the following three research questions to evaluate our method:

- RQ1: Accuracy. What is the accuracy of our algorithm compared with the baseline algorithms?
- RQ2: Applicability. How does our algorithm perform in the real scenarios compared with the baseline algorithms?
- RQ3: Efficiency. What is the time efficiency of our algorithm compared with the same type of algorithms?

we conduct experiments on simulated datasets for RQ1, on real-life dataset for RQ2, and on both simulated datasets and real-life dataset for RQ3.

4.2 Experiment Setting

Combined with previous studies, we use Precision Rate, Recall Rate, and Structural Hamming Distance (SHD) three metrics in this paper. We select five classic algorithms as baseline algorithms, covering three types of causal discovery methods, including constraint-based method PC algorithm [21], structural causal model based method ICA-LiNGAM [19], and scoring-based methods GES [4], NOTEARS [27] and GraN-DAG [14].

Our method is based on structural causal model and genetic algorithm, referred to as *SCM-GA*. Its parameters mainly include the parameters of genetic algorithm and the parameters of the evaluation function. The parameter settings in the experiments are shown in Table 1.

Table 1. Algorithm parameters setting

Description	Symbol	Value
Number of population	K	nodeNum*50
Maximum iterations	maxGeneration	nodeNum*100
Mutation rate	mutationRate	0.1
Proportion of the preserved elites	eliteRate	0.05
Iterations when the best score remains stable	plateauSize	20
Initial weight for evaluation function	W	0.8,0.1,0.1

4.3 Results and Analysis

In this section, we show the results of the experiments and analyze the results.

RQ1(Accuracy). To more comprehensively verify the accuracy of the algorithm, we conduct four set experiments, which the simulated datasets vary in sample size, node size, causal mechanism, and noise mechanism respectively. The specific descriptions are as follows:

- Sample size, that is, how many samples are generated. We conduct the experiments under three sample sizes of 1000, 2000, and 4000;
- Node size, that is, how many variables are studied in the causal graph. We choose 10, 20, and 40 node scales in the experiments;
- Causal mechanism, indicates how the effect variable is generated by other causal variables. We consider three different causal mechanisms, linear, polynomial, and Gaussian process. Our method follows the additive noise model assumption, so the noise is additive;
- Noise mechanism, indicates what kind of distribution the noise obeys. We use three different distributions in our experiments – mixed-type noise with equal variance, Gaussian noise with mixed variance, and non-Gaussian noise with mixed variance.

The results are shown in Fig. 2. Each row represents a dataset setting, followed by different sample sizes, node sizes, causal mechanisms, and noise mechanisms. We can see from the results that GraN-DAG algorithm performs the best in precision rate and SCM-GA also ranks at the forefront. As for recall rate, SCM-GA performs the best, followed by the PC algorithm. When we look into the comprehensive metric SHD, SCM-GA always outperforms the others, which shows that SCM-GA has shown good accuracy on different simulated datasets.

RQ2(Applicability). To explore the performance of our algorithm in the real scene, we select the protein network dataset that is widely used in causal discovery, namely the Sachs dataset [18]. The real causal graph of this dataset contains 11 protein nodes and 17 edges. This dataset contains both observational and interventional data. Since our method belongs to causal discovery based on observational data, only observational data of 853 samples are considered, which is the same as the case of GraN-DAG [14] and RL-BIC [29].

Table 2 shows the results on the Sachs dataset. The table shows the total number of predicted edges, the number of correct edges, precision rate, recall rate, and SHD, respectively. It is worth noting that since the true causal graph is very sparse, the SHD of an empty graph is only 17. The total number of predicted edges of the traditional methods PC and GES algorithms both exceed 25, but the number of correct edges is less than 10, so the performance is not good, the SHD values exceed 20 and recall rates are below 0.5. The recently popular NOTEARS, ICA-LiNGAM, and GraN-DAG algorithms have SHD values between 13–18, and their performances are significantly better than traditional methods because they consider more features of structural causal model. Among them, the GraN-DAG performs the best, with an SHD of only 13. Meanwhile, our algorithm also takes into account the features of structural causal model and benefits from the search

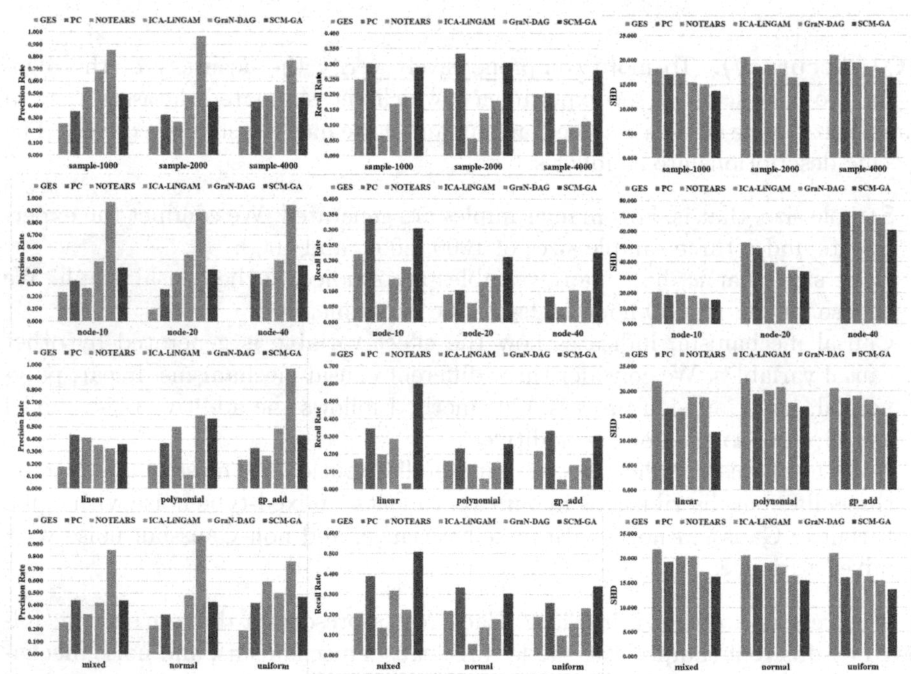

(a) **Precision rate** comparison results (b) **Recall rate** comparison results (c) **SHD** comparison results

Fig. 2. Experiment results of different sample sizes, node sizes, causal mechanisms, and noise mechanisms ability of the genetic algorithm, so it also achieves good results, the SHD is also 13, and the precision rate, recall rate, and discovery of correct edges are the best. Compared with the GraN-DAG algorithm, the recall rate is increased by 40%, and the precision rate is increased by 7.6%.

Through the experiment results in real scenes, we can find that the accuracy of SCM-GA is significantly improved compared with the baseline algorithms, and it has better results in real scenes.

Table 2. Result on Sachs dataset

Method	Number of all predicted edges	Number of correct edges	Precision rate	Recall rate	SHD
GES	36	5	0.139	0.294	32
PC	27	4	0.296	0.235	24
NOTEARS	20	6	0.300	0.353	19
ICA-LiNGAM	8	4	0.500	0.235	14
GraN-DAG	10	5	0.500	0.294	**13**
SCM-GA	13	**7**	**0.538**	**0.412**	**13**

RQ3(Efficiency). In order to explore the efficiency of SCM-GA, we compare the running time of the same type of algorithms (the scoring-based methods GES, NOTES, and GraN-DAG). In terms of dataset selection, since the running time is mainly related to the node size and sample size of the dataset, we select the real scene Sachs dataset, three simulated datasets with different node scales and three simulated datasets with different sample sizes. Table 3 shows the experiment results, where the unit of time is second.

It can be seen from Table 3 that on all datasets, the running time of SCM-GA is the shortest. The running time of the GES algorithm is the longest because the algorithm first adds edges according to the greedy principle, and then deletes edges. The search space for both operations increases exponentially as the node size increases. On the Sachs dataset with 853 samples of 12 nodes, the GES algorithm runs up to 852.364 s, and the running time of SCM-GA is only 1.4% of it. With the increase of node size and sample size, the running time of all algorithms increases. Among them, the GES algorithm has the fastest growth rate and it fails to obtain the results in 5 d (the maximum task duration specified by the platform) in the 40 nodes case. When the node size increases from 10 to 20, the running time of the GES algorithm increases by 4.3 times, while SCM-GA only increases by 2.0 times. SCM-GA only takes 4 more seconds when the sample size increases from 2000 to 4000, while the running time of the GES algorithm increases by 829 s.

In general, SCM-GA we proposed is efficient on all the datasets. The accuracy of the predicted causal graph obtained by SCM-GA is also better than the five classical baseline algorithms.

Table 3. Experiment results of the running time of different algorithms

Dataset	GES	NOTEARS	GraN-DAG	SCM-GA
Sachs	852.364	65.549	76.046	**12.676**
Simul-Node-10[a]	981.026	40.343	76.163	**21.845**
Simul-Node-20	4254.214	174.176	102.794	**43.842**
Simul-Node-40	–	304.385	191.124	**69.256**
Simul-Sample-1000	457.461	31.664	65.106	**17.683**
Simul-Sample-2000	981.026	40.324	76.163	**21.845**
Simul-Sample-4000	1810.374	56.746	89.707	**25.060**

[a]Since the default node size is 10 and the default sample size is 2000, the simulation dataset of 10 nodes is the same as that of 2000 samples

5 Conclusion

In recent years, causal discovery has received more and more attention, and related algorithms have emerged one after another. However, they still have certain limitations, either the algorithm accuracy suffers from the heterogeneous noise problem, or the algorithm efficiency is limited to the exponential explosion of node size. To better take into account the requirements of accuracy and efficiency, we combine the advantages of scoring-based and structural causal model based methods to explore a genetic algorithm based causal discovery method, which includes data pre-processing, causal relation pre-discovery, and causal graph pruning three modules. In the causal relation pre-discovery module, firstly, we deeply explore the data generation mechanism. Based on the features of structural causal model, the evaluation metrics of model deviation degree, noise independence degree, and causal graph cyclicity degree are designed, which can accurately evaluate the predicted causal graph, thereby improving the accuracy of the algorithm. Secondly, based on the idea of genetic algorithm, we design an efficient search process to ensure the efficiency of the algorithm. Since the genetic algorithm supports a flexible fitness function, SCM-GA can well combine the evaluation metrics we design, so as to efficiently and accurately search for the correct causal graph. Finally, three sets of experiments have shown that SCM-GA has achieved great improvements in algorithm accuracy, applicability, and efficiency. To further analyze our method, we may conduct ablation studies to explore the effectiveness of each component in the future.

The method in this paper is currently only applicable to causal discovery from the observational data of continuous random variables, but cannot deal with the case of discrete random variables. In real scenarios, many random variables are discrete. In the future work, we will consider how to support discrete variables in the regression method, and how to calculate the scoring metrics in the evaluation phase. In addition, our method uses the Gaussian process regression method to predict the causal mechanism which is not accurate enough for nonlinear data. Further research will be conducted on more accurate regression methods to obtain better predicted causal mechanisms. Besides, genetic algorithm that we use for searching causal graphs may not perform well in all scenarios, we will consider other search or optimization methods in expectation of higher effectiveness or efficiency.

References

1. Cai, R.C., Chen, W., Zhang, K., Hao, Z.F.: A survey on non-temporal series observational data based causal discovery (in Chinese). Chin. J. Comput. **40**(6), 1470–1490 (2017)
2. Cai, R.C., Hao, Z.F.: Casual discovery in big data (in Chinese). Science Press (2018)
3. Chickering, D.M.: Learning Bayesian networks is NP-complete. In: Fisher, D., Lenz, H.J. (eds.) Learning from Data. Lecture Notes in Statistics, vol. 112, pp. 121–130. Springer, New York, NY (1996). https://doi.org/10.1007/978-1-4612-2404-4_12
4. Chickering, D.M.: Optimal structure identification with greedy search. J. Mach. Learn. Res. **3**(3), 507–554 (2002). https://doi.org/10.1162/153244303321897717
5. Chickering, M., Heckerman, D., Meek, C.: Large-sample learning of Bayesian networks is NP-hard. J. Mach. Learn. Res. **5**, 1287–1330 (2004)
6. Glymour, C., Zhang, K., Spirtes, P.: Review of causal discovery methods based on graphical models. Front. Genet. **10**, 524 (2019). https://doi.org/10.3389/fgene.2019.00524
7. Goudet, O., Kalainathan, D., Caillou, P., Guyon, I., Lopez-Paz, D., Sebag, M.: Learning functional causal models with generative neural networks. In: Escalante, H.J., et al. (eds.) Explainable and Interpretable Models in Computer Vision and Machine Learning. TSSCML, pp. 39–80. Springer, Cham (2018). https://doi.org/10.1007/978-3-319-98131-4_3
8. Gretton, A., Herbrich, R., Smola, A., Bousquet, O., Schölkopf, B., et al.: Kernel methods for measuring independence. J. Mach. Learn. Res. **6**, 2075–2129 (2005). https://doi.org/10.1007/s10846-005-9001-9
9. Haughton, D.M.: On the choice of a model to fit data from an exponential family. Ann. Stat. **16**, pp. 342–355 (1988). https://doi.org/10.1214/aos/1176350709
10. He, Y.-B., Geng, Z., Liang, X.: Learning causal structures based on Markov equivalence class. In: Jain, S., Simon, H.U., Tomita, E. (eds.) ALT 2005. LNCS (LNAI), vol. 3734, pp. 92–106. Springer, Heidelberg (2005). https://doi.org/10.1007/11564089_9
11. He, Y., Cui, P., Shen, Z., Xu, R., Liu, F., Jiang, Y.: Daring: differentiable causal discovery with residual independence. In: Proceedings of the 27th ACM SIGKDD Conference on Knowledge Discovery Data Mining, pp. 596–605 (2021). https://doi.org/10.1145/3447548.3467439
12. Hoyer, P., Janzing, D., Mooij, J.M., Peters, J., Schölkopf, B.: Nonlinear causal discovery with additive noise models. In: Advances in Neural Information Processing Systems 21 (2008)
13. Kitson, N.K., Constantinou, A.C., Guo, Z., Liu, Y., Chobtham, K.: A survey of Bayesian network structure learning. arXiv preprint arXiv:2109.11415 (2021)
14. Lachapelle, S., Brouillard, P., Deleu, T., Lacoste-Julien, S.: Gradient-based neural DAG learning. In: International Conference on Learning Representations (2020)
15. Mattmann, C.A.: A vision for data science. Nature **493**(7433), 473–475 (2013). https://doi.org/10.1038/493473a

16. Peters, J., Janzing, D., Schölkopf, B.: Elements of causal inference: foundations and learning algorithms. The MIT Press (2017)
17. Peters, J., Mooij, J.M., Janzing, D., Schölkopf, B.: Causal discovery with continuous additive noise models. J. Mach. Learn. Res. **15**(58), 2009–2053 (2014). http://hdl.handle.net/2066/130001
18. Sachs, K., Perez, O., Pe'er, D., Lauffenburger, D.A., Nolan, G.P.: Causal protein-signaling networks derived from multiparameter single-cell data. Science **308**(5721), 523–529 (2005). https://doi.org/10.1126/science.1105809
19. Shimizu, S., Hoyer, P.O., Hyvärinen, A., Kerminen, A., Jordan, M.: A linear non-gaussian acyclic model for causal discovery. J. Mach. Learn. Res. **7**(10), 2003–2030 (2006). https://doi.org/10.1007/s10883-006-0005-y
20. Spirtes, P., Glymour, C., Scheines, R.: Causality from probability (1989)
21. Spirtes, P., Glymour, C.N., Scheines, R., Heckerman, D.: Causation, prediction, and search. MIT press (2000)
22. Verma, T.S., Pearl, J.: Equivalence and synthesis of causal models. In: Probabilistic and Causal Inference: the works of Judea Pearl, pp. 221–236 (2022). https://doi.org/10.1145/3501714.3501732
23. Wang, X., Dunson, D., Leng, C.: No penalty no tears: least squares in high-dimensional linear models. In: International Conference on Machine Learning, pp. 1814–1822. PMLR (2016)
24. Zhang, K., Hyvärinen, A.: Causality discovery with additive disturbances: an information-theoretical perspective. In: Buntine, W., Grobelnik, M., Mladenić, D., Shawe-Taylor, J. (eds.) ECML PKDD 2009. LNCS (LNAI), vol. 5782, pp. 570–585. Springer, Heidelberg (2009). https://doi.org/10.1007/978-3-642-04174-7_37
25. Zhang, K., Hyvärinen, A.: Distinguishing causes from effects using nonlinear acyclic causal models. In: Causality: Objectives and Assessment, pp. 157–164. PMLR (2010)
26. Zhang, K., Hyvarinen, A.: On the identifiability of the post-nonlinear causal model. arXiv preprint arXiv:1205.2599 (2012)
27. Zheng, X., Aragam, B., Ravikumar, P.K., Xing, E.P.: DAGs with no tears: continuous optimization for structure learning. In: Advances in Neural Information Processing Systems 31 (2018)
28. Zhou, S.: Thresholding procedures for high dimensional variable selection and statistical estimation. In: Advances in Neural Information Processing Systems 22 (2009)
29. Zhu, S., Ng, I., Chen, Z.: Causal discovery with reinforcement learning. In: International Conference on Learning Representations (2020)

Stochastic and Dual Adversarial GAN-Boosted Zero-Shot Knowledge Graph

Xuejiao Liu, Yaowei Guo, Meiyu Huang, and Xueshuang Xiang(✉)

Qian Xuesen Laboratory of Space Technology, China Academy of Space Technology, Beijing, China

huangmeiyu@qxslab.cn, xiangxs2@163.com

Abstract. Zero-shot knowledge graph (KG) has gained much research attention in recent years. Due to its excellent performance in approximating data distribution, generative adversarial network (GAN) has been used in zero-shot learning for KG completion. However, existing works on GAN-based zero-shot KG completion all use traditional simple architecture without randomness in generator, which greatly limits the ability of GAN mining knowledge on complex datasets. Moreover, the discriminator not only needs to distinguish true data from generated data but also needs to classify generated data correctly at the same time, which affects the optimization process of the generator. In this work, we propose a novel zero-shot KG framework based on stochastic and dual adversarial GAN (SDA) to better mine the association between semantic information and extracted features. Specifically, we introduce a stochastic generator and an additional classifier to improve the model's ability of approximating features and classifying unseen tasks. The experiments on NELL-ZS and Wiki-ZS datasets show that the proposed SDA outperforms the classic methods in zero-shot KG completion task. In particular, the proposed SDA receives a 0.6% and 0.7% increase on mean reciprocal ranking (MRR) for NELL-ZS and Wiki-ZS datasets, respectively.

Keywords: Zero-shot learning · Knowledge graph completion · Generative adversarial network

1 Introduction

Deep learning has achieved great success in supervised and semi-supervised learning tasks [2,10,30]. Knowledge graph completion, which is a typical supervised learning task, has been studied by many researchers [3,16,32,37] these

Supported by the National Natural Science Foundation of China under Grant 12001525, and the Beijing Nova Program of Science and Technology under Grant Z191100001119129.

X. Liu and Y. Guo—These authors contributed equally.

L. Fang et al. (Eds.): CICAI 2022, LNAI 13606, pp. 55–67, 2022.
https://doi.org/10.1007/978-3-031-20503-3_5

years. However, knowledge graphs (KG) in real world are always encountering new knowledge, which the above mentioned works cannot handle efficiently. Zero-shot learning (ZSL) is an ideal approach to deal with this newly emerging knowledge. Generative adversarial network (GAN) [11] has attracted much attention because its strong generative ability and is widely used in the field of computer vision to generate realistic images [4,7,12,15,19,21]. In recent years, GAN has been applied to zero-shot KG completion tasks, that is, a generator is used to obtain categorical features through the auxiliary information of corresponding categories, and a discriminator is used to distinguish and classify the generated data. The architecture of GAN models used in present works [8,24] is very simple, which generally based on a single hidden layer neural network, therefore limits the approximation ability of the generator. Moreover, the discriminator needs to take into account the two tasks of distinguishing true data from generated data and classifying generated data to correct class, which leads to the inconsistent optimization objective of the training process and is not conducive to obtain the optimal generator.

In this work, we mainly focus on improving the association mining ability between semantic information and features of GAN in zero-shot KG completion. The contributions of this work are as followings:

- We propose a novel zero-shot knowledge graph framework based on stochastic and dual adversarial GAN (SDA) which successfully improves the mapping between semantic information and features.
- In this framework, a stochastic generator and an additional classifier are introduced to improve the approximation ability and optimization process.
- We conduct experiments on NELL-ZS and Wiki-ZS datasets which show that the proposed SDA framework outperforms classic models for zero-shot knowledge graph completion.

2 Related Work

The way of learning embeddings for KG has been well-studied. Effective models such as TransE [3], DistMult [37], ComplEx [32], and SimplE [16] are able to generate reasonable embeddings for KG. However, one major shortage of above mentioned models is that they can only handle cases where all entities and relations are seen during training stage. If some new relations are added into KG during testing, the only way for these models to learn embeddings for those relations is re-training the entire knowledge graph.

Many of works focus on learning the embeddings of unseen entities. [13] applies graph neuron network to calculate embeddings of unseen entities from averaging their neighbors' embeddings during test time. [34] uses aggregation methods to embed unseen entities by their existing neighbors. [14] utilizes attributes of entities to generate embeddings of unseen entities. Other works such as [31] aim to learn the behind graph structure and relational information which represent the semantic meaning of triple and then use semantic information to represent unseen entities. Differ from above mentioned works that rely on computing embeddings of unseen entities through constructing connections between unseen and seen entities, some works follow another track that utilizes additional auxiliary information to learn embeddings. [27,35] introduce textual description corpus to generate word embeddings for unseen entities. [28] uses content masking technique to select key words from entity's textual description and learns a network to denote embeddings of new entities from those words. [6] combines structural embeddings learned from graph and textual embeddings learned from description to embed unseen entities. [23,36] import other types of auxiliary information such as images and attributes to model the embeddings.

There are also a few works concentrate on learning embeddings of unseen relations. [24] takes use of GAN to generate embeddings of unseen relations from their text-descriptions. [8] exploits ontological knowledge base information as auxiliary information which captures the correlations between different types of relations. However, the simple GAN structure in both [8,24] cannot fully utilize the extracted features from either textual description or ontology graph. We notice that at the same time there are other works [9,18,29] that also focus on unseen relation. [9,29] both concentrate on learning representations for ontological information in which [9] learns disentangled embedding and [29] leverages graph convolution network. [18] studies the embeddings of unseen relation by character-level n-grams graph. Our proposed SDA, instead, leverages a stochastic generator and dual adversarial training process to improve the association mining ability of GAN in zero-shot KG framework.

3 Methods

3.1 Background

In this section, we first define the task of KG completion. Knowledge graph is consisted of a great amount of RDF triples (h, r, t) where h, t and r denote head entity, tail entity and relation between entities respectively. KG completion aims to predict the missing value in a triple when the other two values are presented. For consistence, we suppose all the missing values are tail entities in this work. Formally, for a given query tuple $(h,\ r)$, we want to predict the correct $t_{predict}$ from a set of candidate tail entities $\{t_c\}$ which makes $(h,\ r,\ t_{predict})$ hold true. For each $t \in t_c$, KG completion model can compute a score to represent the correctness of $(h,\ r,\ t)$. Therefore, $t_{predict}$ is selected as the one with highest score that outperforms all others in $\{t_c\}$.

In above mentioned KG completion task, all entities and relations in KG are available (seen) during training process. However, in ZSL setting, some entities or relations do not appear (unseen) during training and are only seen by the model during testing. This requires the zero-shot KG completion model being able to deal with unseen entities or relations without re-training. In this work, we only take into account unseen relations. Formally, relations in KG are divided into seen set R_s and unseen set R_u without overlapping. The model is trained by triples (h, r_s, t) where $r_s \in R_s$. During testing stage, the model is asked to predict the missing tail entity given query tuple (h, r_u) where $r_u \in R_u$ is never seen by the model before.

3.2 Motivation

GAN, as an outstanding method for approximating data distribution, has been applied in several zero-shot KG completion models. However, existing zero-shot learning methods based on GAN [8,24] usually adopt traditional and relatively simple network architecture without randomness in the generator, which will greatly limit the ability of GAN to mine knowledge [19]. In addition, the discriminator needs to undertake two different tasks at the same time, namely, distinguishing true data from generated data and classification. However, the optimization objectives of the two tasks are inconsistent. On one hand, the discriminator is required to judge the generated data as false; on the other hand, the discriminator is required to correctly classify the generated data. This will affect the optimization process of the generator. Therefore, we propose a zero-shot KG framework by introducing a stochastic generator and a dual adversarial training with an independent classifier.

3.3 Framework of SDA

As shown in Fig. 1, the framework of SDA contains a feature encoder, an auxiliary information encoder, a stochastic generator and dual adversarial networks: discriminator and classifier. The feature encoder is used to capture meaningful features of seen samples, and the auxiliary information encoder is used to extract semantic information from text description or ontology graph. The stochastic generator is applied to generate embedding of seen/unseen relations from noise vectors and semantic information, while the discriminator and classifier are used to distinguish generated features from true features produced by feature encoder and perform classification, respectively. We will further introduce each module in detail in this section.

Feature Encoder. Similar to the work of [24], a neighbor encoder which takes one-hop structure as input and an entity encoder with a feed-forward layer fully connected network are adopted to learn the cluster-structure distribution. For the entity e, its one-hop structure T_e is consisted of all relations and entities that satisfied (e, r, t) in the triple set T. Formally, $T_e = \{(r, t)|(e, r, t) \in T\}$. Then the neighbor encoder computes the embedding u_e of entity e as

$$u_e = \tanh(E_{(r,t)\in T_e} f_1(v_r, v_t)), f_1(v_r, v_t) = W_1([v_r, v_t]) + b_1, \tag{1}$$

where v_r and v_t are the relation embedding and tail entity embedding vectors obtained by traditional KG representation learning method with same embedding dimension d. W_1 and b_1 represent the weight and bias of the fully connected network f_1, respectively. $[.,.]$ is the concatenation operation.

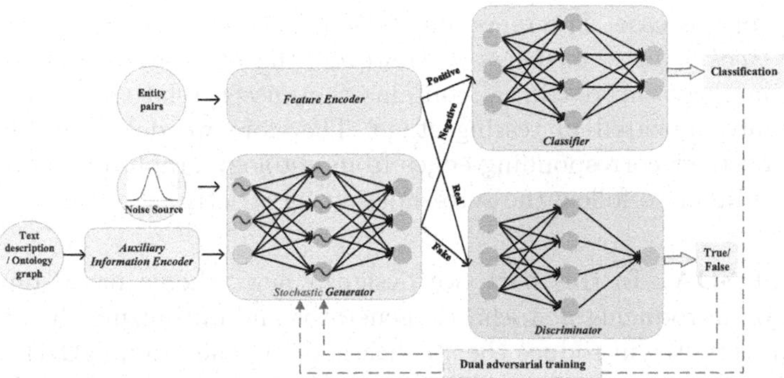

Fig. 1. Overview of SDA framework contains a feature encoder, an auxiliary information encoder, a stochastic generator and dual adversarial networks: a discriminator and a classifier. For training stage, the semantic information of seen relation is obtained from the text description or ontology graph through the auxiliary information encoder, and the semantic information is used as the input of the stochastic generator together with the noise vector to generate the corresponding feature of this seen relation. Then, the generated feature will be input into the discriminator and classifier together with the real feature corresponding to the entity pair to distinguish and classify. For testing stage, the auxiliary information of unseen relation is used to generate the corresponding feature for prediction. The solid line represents the data flow and the dotted line represents the dual adversarial training process.

For entity encoder, the embedding u_{ht} of entity pair (h, t) is computed as

$$u_{ht} = \tanh([f_2(v_h), f_2(v_t)]), f_2(v) = W_2(v) + b_2, \tag{2}$$

where W_2 and b_2 represent the weight and bias of the fully connected network f_2. v_h is head entity embedding vectors.

Then, the feature $u_{(h,t)}$ of entity pairs (h, t) for the seen relation r can be represented as

$$u_{(h,t)} = [u_h, u_{ht}, u_t]. \tag{3}$$

Auxiliary Information Encoder. The auxiliary information in our work includes textual information and ontology graph information. For textual information, we follow the same setting as [24], i.e., words are vectorized by pre-trained word embedding model Word2Vec [20]. Furthermore, in order to reduce the influence of unimportant words, TF-IDF [26] method is adopted.

The ontology graph used in [8] is essentially a newly-constructed knowledge graph. The nodes in ontology graph are relations and entity classes in original knowledge graph, while edges in ontology graph describe the relationships between those nodes. In addition, each node in ontology graph has a specific edge that connect to its textual description extracted from dataset. For instance, node *concept*:$languageOfUniversity$ in ontology graph have three edges. Two of them connect to classes of head and tail entities, which is *language* and *university* in this case. The remaining edge connects to its textual description *Language spoken at a university*. However, the ontology graph which that inputs to ontology encoder in [8] contains the unseen relations which are supposed to only appeared in testing stage. Therefore we delete all the unseen relations and their corresponding edges from ontology graph and leave the rest as encoder's input to follow the zero-shot setting strictly.

Scheme of SDA. In this part, we design a new scheme based on WGAN-GP [1,12] by introducing a stochastic generator and optimizing the adversarial training process. We introduce the stochastic data generation (SDG) [19] and a classification loss into generator. SDG imposes the randomness into the generator by leveraging Gaussian distribution assumption between two layers of neural network. As discussed in [19], SDG improves the approximation ability of generator. In order to facilitate SDG algorithm, reparameterization method [17,25] is adopted to learn the mean and variance of Gaussian distribution. Therefore, in SDA, the target of the stochastic generator $L(G(z, \epsilon, a_r; \theta))$ is

$$\begin{aligned}\theta^* = \arg\min_{\theta}\{&\mathbb{E}_{z\sim p_z,\epsilon\sim\mathcal{N}(0,I)}(-D(G(z,\epsilon,a_r;\theta))\\&+\max(0,\gamma+score(C(G(z,\epsilon,a_r;\theta)))-score(C(x_n))))\},\end{aligned} \tag{4}$$

where θ is the trained parameter of the stochastic generator, $z \sim p_z$ is input noisy, $\epsilon \sim \mathcal{N}(0, I)$ is the parameter of reparameterization, a_r refers to the semantic information output by the auxiliary information encoder and γ is the margin parameter. The first term requires that the generated data is identified as true samples by the discriminator, and the second term is the margin loss which regulates that the generated data should be positive samples and are correctly classified. x_n represents the negative triple.

The discriminator D is adopted as WGAN-GP. The target of the discriminator $L(D(x; \eta))$ is

$$\begin{aligned}\eta^* = \arg\min_{\eta}\{&-\mathbb{E}_{x\sim p(x)}D(x;\eta)+\mathbb{E}_{z\sim p_z,\epsilon\sim\mathcal{N}(0,I)}D(G(z,\epsilon,a_r;\theta);\eta)\\&+\lambda\mathbb{E}_{\hat{x}\sim p(\hat{x})}(||\nabla_{\hat{x}}D(\hat{x})||_2-1)^2\},\end{aligned} \tag{5}$$

where η is the trained parameter of the discriminator. $p(\hat{x})$ is the interpolation distribution with $\hat{x} = \alpha x + (1 - \alpha)x'$ where $\alpha \in (0, 1)$, x follows the real data distribution $p(x)$ and x' follows the generated data distribution $p_g(x')$. The first two terms represent the critic loss which requests discriminator to view real data x and generated data x' be as true and fake samples, respectively. The last term denotes the gradient penalty.

Inspired by [7], an additional exclusive classifier is added for determining the category of real data and generated data. i.e., the target of the classifier $L(C(x; \phi))$ is

$$\begin{aligned}\phi^* = \arg\min_{\phi}\{&\mathbb{E}_{z\sim p_z,\epsilon\sim\mathcal{N}(0,I)}\max(0,\gamma+score(C(x,\phi))-score(C(x_g,\phi)))\\&+\mathbb{E}_{x\sim p(x)}\max(0,\gamma+score(C(x,\phi))-score(C(x_n,\phi)))\},\end{aligned} \tag{6}$$

where ϕ is the trained parameter of the classifier and $x_g = G(z, \epsilon, a_r; \theta)$. To adversarial training with generator, the classifier will recognize generated data as negative samples.

Predicting Tasks Based on Unseen Relation. After above dual adversarial training, the generator is able to output reasonable embeddings for unseen relation from their auxiliary information. During testing stage, given a query tuple (h, r_u) where r_u represents unseen relation and a set of candidate tail entities $\{t_c\}$, generator will produce embedding u_{r_u} for unseen relation r_u. Classifier will then calculate cosine similarity score between $u_{(h,t)}$ and u_{r_u} where t is candidate tail entity from $\{t_c\}$. In order to eliminate the effect of different random sampled z, for each relation we sample 20 times and calculate their average cosine similarity score as final score used for predicting. The predicted tail entity will be the entity which has the highest average cosine similarity score.

4 Experiments

In this section, we conduct experiments of our proposed SDA framework on zero-shot KG completion task on two different datasets NELL-ZS and Wiki-ZS. We record the results and compare them with previous zero-shot knowledge graph embedding models. We then analyze our results and study the effects of different modules in our model.

4.1 Datasets Description

The datasets we use for the experiments are NELL-ZS and Wiki-ZS proposed by [24]. These datasets are constructed from the open source large-scale knowledge graph NELL by [22] and Wiki by [33] respectively. The candidate triples in NELL-ZS and Wiki-ZS are separated into training, validation, and testing where the relations in testing are guaranteed not to appear in both training and validation stages. The textual descriptions in two datasets contain different components. The description part in NELL-ZS is composed by relation descriptions and entity class description. For instance, the entity class description of relation *concept:languageOfUniversity* are sentences that explain *language*(*Language is a system...*) and *university*(*A university is a...*) respectively, which serve as constraint information that describes the domain and range of relation. In Wiki-ZS, relation descriptions include attributes *P*31(*subject*), *P*1629(*relationship*), and *P*185(*example*) as supplementary information to describe the constraints. It is also worth to note that the useful word length of relation descriptions in both datasets are relatively short. The detailed statistics of datasets are shown in Table 1.

4.2 Baselines and Implementation Details

The baselines that we are going to compare with contain four different models: two well-studied classic KG embedding models TransE [3] and DistMult [37], and

Table 1. Statisitcs of NELL-ZS and Wiki-ZS datasets. # Triple and # Entity denote the number of triple and entity in each dataset. # Relation(Tr/V/Te) denotes the number of relation in training/validation/testing respectively.

Datasets	# Triple	# Entity	# Relation(Tr/V/Te)
NELL-ZS	188,392	65,567	181(139/10/32)
Wiki-ZS	724,967	605,812	537(469/20/48)

two zero-shot KG embeddings models ZSGAN [24] and OntoZSL [8] which both leverage the generative ability of GAN to learn embeddings for unseen relations from auxiliary information as input. As mentioned above, TransE and DistMult are not capable of learning embeddings for unseen relations, thus we apply the results of ZS-TransE and ZS-DistMult from paper [24]. For OntoZSL, [8] feeds entire ontology graph to ontology encoder so that causes the unseen relation being seen by the network before testing stage, we delete the part of unseen relation from the original ontology graph to follow close to the line of zero-shot setting and leave the rest as input to ontology encoder. We also select TransE and DistMult as pre-trained KG embedding models in our feature extractor to maintain consistency with baselines. It is noteworthy that we use a variant version of entity class constraint to generate candidate triples for testing when some classes contains only a little entities. For instance, candidate tail entities of relation *concept*:$languageOfUniversity$ should be generated from class $university$; however, if number of entities under class *univerisity* is smaller then 10, we may select candidate tail entities from the similar class such as *building* or *locaion* to maintain the candidate's quality. We apply two widely used metrics in KG completion task which are hits at K(HITS@K) and mean reciprocal ranking (MRR) [3], where K is often 10, 5, and 1.

Now we give the implementation details. For dimension setting, the embedding dimension for NELL-ZS and Wiki-ZS are 100 and 50, respectively. The dimension of word embedding is fixed to 300 (ZSGAN) and 600 (OntoZSL) on both datasets and the noise dimension is set to 15. For feature extractor, we delete all triples that contain unseen relation from input. Learning rate is set to $5e^{-4}$. Max number of neighbor entity is 50. Training batch size is set to 64 for NELL-ZS and 128 for Wiki-ZS since Wiki-ZS is much larger. The reference number in each epoch we use are 30. The margin γ that we use to calculate entity pair embedding is 10 in both datasets. For the stochastic generator, we use two parallel feed-forward layers with the same neurons as set in ZSGAN or OntoZSL in the first hidden layer to learn the mean and variance of the Gaussian distribution, and the Gaussian resampled variables are propagated to the next layer. For adversarial training part, we design classifier to use same network architecture with discriminator except for the last fully connected layer. The default word embedding model we adopt is Word2Vec [20]; however, we also apply other transformer-based model such as BERT [5] which we will further discuss in later section.

Table 2. Results (%) for zero-shot KG completion with unseen relation. From top to bottom are classic embedding models ZS-TransE, ZS-DistMult, two zero-shot embedding models ZSGAN, OntoZSL and our proposed model SDA. SDA_{Des} and SDA_{Onto} represent SDA using textual description or ontology graph as auxiliary information respectively. In the table, bold numbers represent the best score for different embedding method (TransE and DistMult) using different auxiliary information, underlined results represent the highest score among all models.

Embedding Models	NELL-ZS				Wiki-ZS			
	HITS@10	HITS@5	HITS@1	MRR	HITS@10	HITS@5	HITS@1	MRR
ZS-TransE	20.3	14.7	4.3	9.7	11.9	8.1	1.8	5.3
ZSGAN(TransE)	34.8	29.2	15.6	22.5	26.6	21.2	14.0	18.6
SDA_{Des}(TransE)	**35.8**	**30.3**	**15.9**	**23.0**	**27.6**	**22.1**	**14.7**	**19.3**
OntoZSL(TransE)	40.6	33.7	17.8	25.5	30.3	25.3	16.6	21.6
SDA_{Onto}(TransE)	<u>**40.8**</u>	<u>**34.1**</u>	**17.9**	**25.9**	**31.4**	**26.1**	**16.9**	**22.1**
ZS-DistMult	32.6	28.4	18.5	23.5	23.6	21.0	16.1	18.9
ZSGAN(DistMult)	**35.3**	**29.4**	18.0	24.2	**29.2**	24.1	16.8	21.2
SDA_{Des}(DistMult)	34.9	29.2	**18.5**	**24.5**	**29.2**	**24.3**	**16.9**	**21.4**
OntoZSL(DistMult)	38.8	32.6	20.0	26.6	31.6	25.8	18.3	23.0
SDA_{Onto}(DistMult)	**40.2**	**33.9**	<u>**20.2**</u>	<u>**27.2**</u>	<u>**31.7**</u>	<u>**26.4**</u>	<u>**18.7**</u>	<u>**23.4**</u>

4.3 Result

We compare our SDA model with baselines on KG completion task in zero-shot setting. As shown in Table 2, it is clearly to see that SDA achieves stable increase on MRR for both embedding methods (TransE and DistMult) and datasets (NELL-ZS and Wiki-ZS). The results illustrate that our proposed SDA method is capable of generating more proper embeddings for unseen relation which demonstrates its relatively powerful extraction and generation ability. In addition, we also note that our proposed model obtains a little higher increase on the larger scale Wiki-ZS dataset than NELL-ZS dataset. For instance, our proposed SDA model receives up to 0.7% increase for Wiki-ZS and up to 0.6% increase for NELL-ZS on MRR.

4.4 Ablation Study

We study the effect of different parts contributed to final results by removing SDG (-SDG) or classifier (-C) from SDA and analyze the corresponding performance drop. The detail results are shown in Table 3. We observe that in most cases removing SDG from SDA results cause an obvious decline in performance (as shown in the column of MRR), which demonstrates that SDG module plays a relatively important role in improving generative ability of GAN. Moreover, by comparing the results using SDG (i.e., removing classifier) and the baseline in Table 2, we found that MRR results achieved stable improvement, further implying that stochastic generator can improve knowledge mining ability. When SDA uses DistMult as embedding model and takes ontology graph as input auxiliary information, we notice that removing classifier module actually leads to much larger performance drop. We think this may be because our proposed classifier can deal with complex input information more effectively by treating the generated embeddings as negative samples. The performance drop of removing SDG or classifier indicates that both module are indispensable for SDA.

Table 3. Results (%) of zero-shot KG completion on two datasets when SDG (-SDG) or classifier (-C) is removed from SDA. SDA_{Des} and SDA_{Onto} represent SDA using textual description or ontology graph as auxiliary information respectively.

Embedding	NELL-ZS				Wiki-ZS			
Models	HITS@10	HITS@5	HITS@1	MRR	HITS@10	HITS@5	HITS@1	MRR
-SDG (TransE)	34.5	28.8	15.7	22.3	26.9	21.7	14.1	18.7
-C(TransE)	35.5	29.7	15.8	22.7	27.6	22.0	14.2	19.0
SDA_{Des}(TransE)	35.8	30.3	15.9	23.0	27.6	22.1	14.7	19.3
-SDG (TransE)	40.7	33.6	18.0	25.7	30.9	25.6	16.7	21.8
-C(TransE)	40.5	33.6	18.2	25.8	30.7	25.6	16.9	21.9
SDA_{Onto}(TransE)	40.8	34.1	17.9	25.9	31.4	26.1	16.9	22.1
-SDG (DistMult)	35.0	29.0	18.3	24.2	29.1	24.0	16.8	21.2
-C(DistMult)	35.7	29.2	18.5	24.4	29.6	24.2	17.0	21.4
SDA_{Des}(DistMult)	34.9	29.2	18.5	24.5	29.2	24.3	16.9	21.4
-SDG (DistMult)	40.1	33.7	19.6	26.7	31.1	26.0	18.6	23.2
-C(DistMult)	40.6	33.7	20.0	26.9	31.4	26.3	18.4	23.1
SDA_{Onto}(DistMult)	40.2	33.9	20.2	27.2	31.7	26.4	18.7	23.4

4.5 Influence of Word Embedding Method

In this section, we inspect the influence of different word embedding methods toward KG completion results. As shown in Table 4, we can see that the transformer-based model BERT reaches relatively lower results compered with Word2Vec. The reason is that embeddings learned by BERT-base contains higher dimension (768) compared with that of Word2Vec, which makes generator difficult to extract meaningful features from word embeddings. However, SDA outperforms ZSGAN on a larger scale (0.8% mean increase on MRR) using BERT-model than that of using Word2Vec (0.4% mean increase on MRR), which again demonstrates the superior generative ability of SDA.

Table 4. Results (%) of zero-shot KG completion using Word2Vec or BERT (BERT-base) to embed textual description on NELL-ZS dataset.

Word series Embedding	Models	NELL-ZS			
		HITS@10	HITS@5	HITS@1	MRR
Word2Vec	ZSGAN(TransE)	34.8	29.2	15.6	22.5
	SDA_{Des}(TransE)	**35.8**	**30.3**	15.9	23.0
	ZSGAN(DistMult)	35.3	29.4	18.0	24.2
	SDA_{Des}(DistMult)	34.9	29.2	**18.5**	**24.5**
BERT	ZSGAN(TransE)	34.2	28.1	14.7	21.3
	SDA_{Des}(TransE)	34.7	28.5	15.0	21.8
	ZSGAN(DistMult)	35.1	28.6	16.7	23.0
	SDA_{Des}(DistMult)	34.9	28.7	18.1	24.0

5 Conclusions

This paper has proposed a novel zero-shot knowledge graph framework by introducing a stochastic generator and dual adversarial training. Under this framework, the proposed SDA is auxiliary-information free method which can be generalized to other zero-shot tasks. Compared with previous zero-shot KG models ZSGAN and OntoZSL, SDA achieves higher performance under the same conditions. This work is in its early stage. The theoretical effectiveness of the SDA method needs to be discussed in future work.

References

1. Arjovsky, M., Chintala, S., Bottou, L.: Wasserstein generative adversarial networks. In: International Conference on Machine Learning, pp. 214–223 (2017)
2. Berthelot, D., Carlini, N., Goodfellow, I., Papernot, N., Oliver, A., Raffel, C.A.: MixMatch: a holistic approach to semi-supervised learning. In: Neural Information Processing Systems, pp. 5049–5059 (2019)
3. Bordes, A., Usunier, N., Garcia-Duran, A., Weston, J., Yakhnenko, O.: Translating embeddings for modeling multi-relational data. In: Advances in Neural Information Processing Systems 26 (2013)
4. Brock, A., Donahue, J., Simonyan, K.: Large scale GAN training for high fidelity natural image synthesis (2019)
5. Devlin, J., Chang, M.W., Lee, K., Toutanova, K.: BERT: pre-training of deep bidirectional transformers for language understanding. arXiv preprint arXiv:1810.04805 (2018)
6. Ding, J., Ma, S., Jia, W., Guo, M.: Jointly modeling structural and textual representation for knowledge graph completion in zero-shot scenario. In: Cai, Y., Ishikawa, Y., Xu, J. (eds.) APWeb-WAIM 2018. LNCS, vol. 10987, pp. 369–384. Springer, Cham (2018). https://doi.org/10.1007/978-3-319-96890-2_31

7. Dong, J., Lin, T.: MarginGAN: adversarial training in semi-supervised learning. In: Neural Information Processing Systems, pp. 10440–10449 (2019)
8. Geng, Y., et al.: Ontozsl: ontology-enhanced zero-shot learning. In: Proceedings of the Web Conference 2021 (2021)
9. Geng, Y., et al.: Disentangled ontology embedding for zero-shot learning. arXiv preprint arXiv:2206.03739 (2022)
10. Gong, C., Wang, D., Liu, Q.: Alphamatch: improving consistency for semi-supervised learning with alpha-divergence. In: Computer Vision and Pattern Recognition, pp. 13683–13692 (2021)
11. Goodfellow, I., et al.: Generative adversarial nets. In: Neural Information Processing Systems, pp. 2672–2680 (2014)
12. Gulrajani, I., Ahmed, F., Arjovsky, M., Dumoulin, V., Courville, A.: Improved training of Wasserstein GANs. In: Neural Information Processing Systems, pp. 5769–5779 (2017)
13. Hamaguchi, T., Oiwa, H., Shimbo, M., Matsumoto, Y.: Knowledge transfer for out-of-knowledge-base entities: a graph neural network approach. arXiv preprint arXiv:1706.05674 (2017)
14. Hamilton, W., Ying, Z., Leskovec, J.: Inductive representation learning on large graphs. In: Advances in Neural Information Processing Systems 30 (2017)
15. Karras, T., Laine, S., Aila, T.: A style-based generator architecture for generative adversarial networks. arXiv preprint arXiv:1812.04948 (2018)
16. Kazemi, S.M., Poole, D.: Simple embedding for link prediction in knowledge graphs. In: Advances in Neural Information Processing Systems 31 (2018)
17. Kingma, D.P., Welling, M.: Auto-encoding variational Bayes. In: International Conference on Learning Representations (2014)
18. Li, M., Chen, J., Mensah, S., Aletras, N., Yang, X., Ye, Y.: A hierarchical n-gram framework for zero-shot link prediction. arXiv preprint arXiv:2204.10293 (2022)
19. Liu, X., Xu, Y., Xiang, X.: Towards GANs' approximation ability. In: 2021 IEEE International Conference on Multimedia and Expo (ICME), pp. 1–6 (2021). https://doi.org/10.1109/ICME51207.2021.9428197
20. Mikolov, T., Sutskever, I., Chen, K., Corrado, G.S., Dean, J.: Distributed representations of words and phrases and their compositionality. In: Advances in Neural Information Processing Systems 26 (2013)
21. Mirza, M., Osindero, S.: Conditional generative adversarial nets. In: Computer Science, pp. 2672–2680 (2014)
22. Mitchell, T., et al.: Never-ending learning. Commun. ACM **61**(5), 103–115 (2018)
23. Pezeshkpour, P., Chen, L., Singh, S.: Embedding multimodal relational data for knowledge base completion. arXiv preprint arXiv:1809.01341 (2018)
24. Qin, P., Wang, X.E., Chen, W., Zhang, C., Xu, W., Wang, W.Y.: Generative adversarial zero-shot relational learning for knowledge graphs. In: AAAI (2020)
25. Rezende, D.J., Mohamed, S., Wierstra, D.: Stochastic backpropagation and approximate inference in deep generative models. In: International Conference on Machine Learning (2014)
26. Salton, G., Buckley, C.: Term-weighting approaches in automatic text retrieval. Inf. Process. Manage. **24**(5), 513–523 (1988)
27. Shah, H., Villmow, J., Ulges, A., Schwanecke, U., Shafait, F.: An open-world extension to knowledge graph completion models. In: Proceedings of the AAAI Conference on Artificial Intelligence, vol. 33, pp. 3044–3051 (2019)
28. Shi, B., Weninger, T.: Open-world knowledge graph completion. In: Proceedings of the AAAI Conference on Artificial Intelligence, vol. 32 (2018)

29. Song, R., et al.: Ontology-guided and text-enhanced representation for knowledge graph zero-shot relational learning. In: ICLR 2022 Workshop on Deep Learning on Graphs for Natural Language Processing (2022)
30. Springenberg, J.T.: Unsupervised and semi-supervised learning with categorical generative adversarial networks. In: International Conference on Learning Representations (2016)
31. Teru, K., Denis, E., Hamilton, W.: Inductive relation prediction by subgraph reasoning. In: International Conference on Machine Learning, pp. 9448–9457. PMLR (2020)
32. Trouillon, T., Welbl, J., Riedel, S., Gaussier, É., Bouchard, G.: Complex embeddings for simple link prediction. In: International Conference on Machine Learning, pp. 2071–2080. PMLR (2016)
33. Vrandečić, D., Krötzsch, M.: Wikidata: a free collaborative knowledgebase. Commun. ACM **57**(10), 78–85 (2014)
34. Wang, P., Han, J., Li, C., Pan, R.: Logic attention based neighborhood aggregation for inductive knowledge graph embedding. In: Proceedings of the AAAI Conference on Artificial Intelligence, vol. 33, pp. 7152–7159 (2019)
35. Xie, R., Liu, Z., Jia, J., Luan, H., Sun, M.: Representation learning of knowledge graphs with entity descriptions. In: Proceedings of the AAAI Conference on Artificial Intelligence, vol. 30 (2016)
36. Xie, R., Liu, Z., Luan, H., Sun, M.: Image-embodied knowledge representation learning. arXiv preprint arXiv:1609.07028 (2016)
37. Yang, B., Yih, W.T., He, X., Gao, J., Deng, L.: Embedding entities and relations for learning and inference in knowledge bases. arXiv preprint arXiv:1412.6575 (2014)

Machine Learning

LS-YOLO: Lightweight SAR Ship Targets Detection Based on Improved YOLOv5

Yaqi He, Zi-Xin Li, and Yu-Long Wang(✉)

School of Mechatronic Engineering and Automation, Shanghai University, Shanghai 200444, China
{zixinli,yulongwang}@shu.edu.cn

Abstract. At present, two main problems, which are the multi-scale of ship targets and the lightweight of detection models, restrict the real-time and on-orbit detection of ship targets on SAR images. To solve two problems, we propose a lightweight ship detection network (LS-YOLO) based on YOLOv5 model for SAR images. In the proposed network, we propose two modules, namely, Feature Refinement Module (FRM) and DCSP. The FRM module is designed to solve the multi-scale problem of ship targets in SAR images. This structure can effectively expand the receptive field of the model and improve the detection ability of small target ships. DCSP is lightweight module based on YOLOv5 CSP. This module effectively reduces model parameters and computation while keeping feature extraction ability as much as possible. The LS-YOLO detection speed is up to 1.2 ms, the accuracy (AP) is 96.6%, and the model size is only 3.8 MB. It can balance detection accuracy and detection speed, and provide reference for the construction of real-time detection network.

Keywords: Real-time ship detection · SAR images · YOLO

1 Introduction

In the field of remote sensing, synthetic aperture radar (SAR) provides all-weather ground observation capability, which promotes the rapid development of SAR technology in recent years [1,2]. Many countries have developed their own space-borne technology, such as Germany's TerraSAR-X, China's Gaofen-3 and Canada's RADARSAT-2. Meanwhile, SAR technology has been widely applied in target detection [3], disaster detection [4], military operations [5], and resource exploration [6]. Among them, SAR technology plays an important role in ocean surveillance, so it is of great significance to achieve the real-time detection of ship targets in spaceborne SAR images [7].

At present, ship target detection methods in SAR images are mainly divided into two categories: traditional detection methods and target detection methods based on deep learning. Before there is a major breakthrough in deep learning, ship target detection techniques in SAR images include Constant False Alarm Rate(CFAR) [8,9], Graph Cuts segmentation algorithms based on Graph Theory

L. Fang et al. (Eds.): CICAI 2022, LNAI 13606, pp. 71–80, 2022.
https://doi.org/10.1007/978-3-031-20503-3_6

[10], etc. With the continuous development of deep learning technology in natural image object detection [11–16], deep learning algorithm has become the mainstream algorithm for ship target recognition in SAR images [17,18]. For example, R. Zhang et al. [19] proposed a new ship detection method based on CNN. C. Fu et al. [20] proposed a single-stage ship target detection algorithm, the deconvolution module was used to realize up-sampling of deep features. Although the above method can deal with multi-scale ship targets and achieve high-precision detection of ship targets, its complex model and huge amount of calculation make it difficult to be carried on SAR satellites with weak computing ability. Therefore, the research of lightweight SAR ship target detection algorithm is of paramount significance. This is the research motivation of this paper.

Based on the above analysis, this paper proposes a lightweight detection method LS-YOLO, which can effectively solve multi-scale ship target detection and complex model on satellite SAR images. To be specific, to improve the detection ability of YOLYv5 for multi-scale ship targets, the feature refinement module is designed. Moreover, DCSP is improved to reduce model parameters and computation. The main contributions of our work are given as follows:

- In the task of ship target detection based on SAR images, LS-YOLO detection algorithm with less parameters and less computation is proposed. LS-YOLO algorithm can achieve high recognition accuracy that is comparable to YOLOv5 algorithm.
- In order to solve the multi-scale problem of ship targets, we propose two module, FRM and DCSP, which can be embedded in the backbone of the network to ensure the feature extraction ability of the module and improve the receptive field of the network.

The rest of this article is organized as follows: Sect. 2 describes in detail the structure of our target detection network and the implementation details of our two proposed modules. Section 3 describes the dataset and experiments, i.e., we conduct a series of comparative experiments on the dataset and compare the results. The conclusions are summarized in Sect. 4.

2 Related Work

In this section, the YOLOv5 algorithm and CSP module related to this work are elaborated in a nutshell.

2.1 YOLOv5 Algorithm

As a one-stage target recognition network, YOLO takes the target detection problem as a regression problem. Compared with the two-stage algorithm led by Faster R-CNN, YOLO is faster and more suitable for algorithmic lightweight. YOLOv5 is one of the latest and most popular improved algorithms in the YOLO family. The network structure of YOLOv5 is shown in Fig. 1.

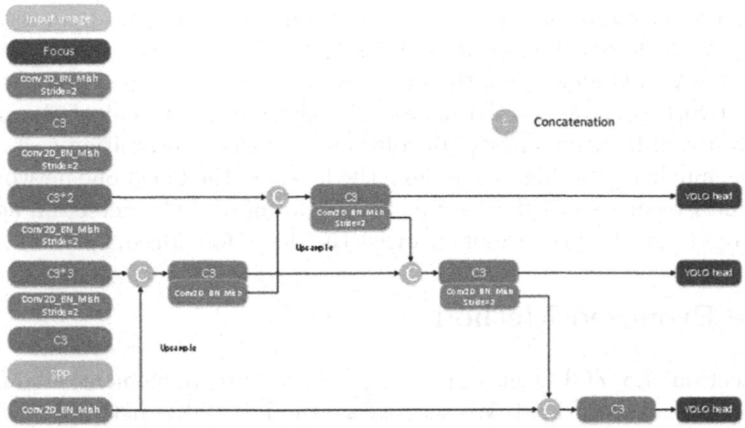

Fig. 1. The network structure of YOLOv5

2.2 CSP Module

The CSP module is one of the main components of the YOLOv5 network. The CSP module sends the input information to two branches. One branch outputs the information through Conv module with kernel size of 3×3 and a series of RES units. The other branch aims to change the number of channels through a simple convolution operation with kernel size of 1×1. The two branches are concatenated together, and the output feature layers go through BN and activation function Relu successively. The output information is obtained by a convolution with kernel size of 1×1 that is used to change the number of output channels. The CSP module can deepen the network depth and improve the feature extraction ability of the network. Specifically, the depth of the CSP module can be changed by adjusting the number of RES units. Although the feature extraction ability of the CSP module is very strong, the number of parameters and layers of the module still need to be further optimized to realize real-time SAR image ship detection.

Compared with other algorithms, although YOLOv5 algorithm has shown its advantages in terms of detection accuracy and detection speed, it is not specially designed for target detection based on SAR images. That is, the YOLOv5 algorithm can not achieve high-precision recognition of ship targets with different scales in SAR images. To solve this problem, Tang et al. [22] utilized the convolutional block attention module (CBAM) to reduce the loss of feature information in YOLOv5 network. It is worth noting that the embedding of CBAM increases the computation cost. In other words, algorithms may not support real-time detection. Zhou et al. [23] proposed CSPMRes2 modules. The CSPMRes2 module introduces a coordinate attention module (CAM) for multi-scale feature extraction in scale dimension. However, the CSPMRes2 module is also not suitable for the construction of lightweight network.

In this paper, we construct an end-to-end convolutional neural network based on YOLOv5 for detecting ships in SAR images. Different from above-mentioned methods, LS-YOLO focuses on the fusion of multi-scale features and the reduction of network parameters. To be specific, the receptive field of the network can be improved by using empty convolution, and the computation costs can be reduced as much as possible by reducing the layers of the backbone network. The experimental results show that our method can improve the detection accuracy of ship targets on the premise of ensuring the detection speed.

3 The Proposed Method

In this section, LS-YOLO network structure, feature refinement module, and DCSP module are described. We summarize the LS-YOLO network in Fig. 2.

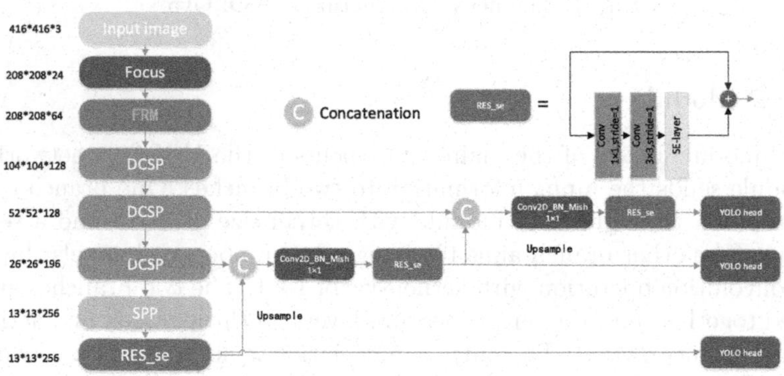

Fig. 2. The network structure of LS-YOLO

3.1 LS-YOLO Network Structure

In order to realize real-time SAR image detection on satellite, there are two main difficulties in algorithm application: (1) multi-scale problem of SAR ship target. Networks designed for natural images are difficult to solve this problem, which results in poor target detection. (2) Due to the limited satellite computation ability, the parameters and complexity of our model are also limited, and a more lightweight network structure is needed to solve the problem of real-time detection. To sum up, we propose a new LS-YOLO network to solve the problems of small targets and algorithm lightweight in SAR images.

In the construction of LS-YOLO network, we design two new modules FRM and DCSP in backbone of the proposed network according to the characteristics of SAR image ship targets. In addition, we replace PANet [21] with an improved FPN-SE structure in the head part of the network to meet the lightweight requirements of the ship detection algorithm. For multi-scale ship target detection, the feature extraction ability of the proposed network outperforms the original network, and the model is lighter than the original model. The detailed network structure is shown in Fig. 2.

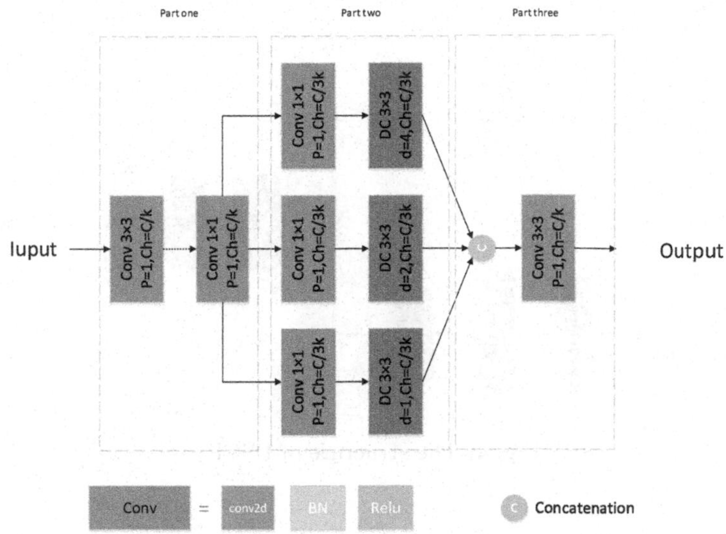

Fig. 3. The structure of FRM

Feature Refinement Module (FRM). There are a lot of small target ships in the existing SAR images. A constant receptive field can not effectively detect small target ships. To improve the generalization performance on the small targets, we design the feature refinement module (FRM). Different from existing methods using ordinary convolution module to introduce the context information, the key idea is to use network hole convolution to extract multi-scale context information. Figure 3 shows the concrete structure of FRM, which consists of three parts.

In the first part of the module, in order to improve channel information representation ability, two Conv modules with kernel sizes of 3x3 and 1x1 are introduced. From the analysis on FRM structure, the first 1x1 Conv operation is stable in order to make the model of information flow. The subsequent 3x3 Conv is used for further feature extraction of the input information.

In the second part of the module, three parallel convolution layers with kernel size of 1x1 are used to change the number of feature channels from the original C/k to C/3k. At the same time, different size expansion rates of empty convolution are used to get more contextual information for small target detection. Among them, P is the fill of the convolution operation, Ch indicates the number of output layers, DC indicates dilated convolutions, d indicates cavity expansion rate of convolution, k is a super parameters and is set to be 2.

At the end of the module, the features of these three branches are fused by concatenation operation. Ordinary convolutions with kernel size of 3×3 are introduced to alleviate the splicing feature. The aliasing effect effectively ensures the stability of features and the relative abundance of information. Therefore, the feature layer output derived by FRM module contains rich multi-scale context information.

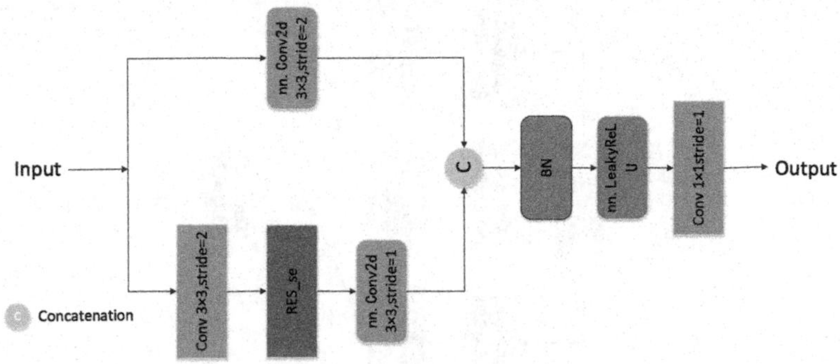

Fig. 4. The structure of DSCP

DCSP Module. DCSP is a backbone feature extraction module, which is designed based on CSP module for lightweight network model. It further reduces the complexity of the network by integrating down-sampling operations. The concrete structure of the module is shown in Fig. 4.

Firstly, the module passes the input data through two different branches. One branch consists of several residual network modules with SE and a Conv module with the kernel size of 3×3 and the stride of 2. The other branch completes the down-sampling operation by a 3×3 convolution. Then, the two branches are joined together by a concatenation operation. BN normalization and ReLU activation function are used to improve the nonlinear capability of the model. Finally, a 1×1 Conv module is used to adjust the number of output layers. DSCP module can reduce the parameters of the backbone network, improve the operation speed, and ensure the depth of the backbone network.

4 Experiment and Results

In this section,the experimental configurations and dataset, evaluation indicators, experimental results and analysis are mainly described. To verify the effectiveness of LS-YOLO detection method, comparative experiments are carried out under the same experimental conditions.

4.1 Experimental Configurations and Dataset

The experimental software and hardware configurations are shown in Table 1. In the experiment, all comparison methods are conducted on the same platform.

The experiment uses the SSDD dataset, which is commonly used in SAR ship target detection. The SSDD dataset contains 1160 SAR images with 2456 ship targets, there are 2.12 ships on average per image. The dataset includes ships occupying different pixel sizes, ranging from small target ships of 7×7 pixels to large target ships of 211×298 pixels. To ensure the reliability of the experimental data, the SSDD dataset is randomly divided into a training set, a validation set, and a test set according to the ratio of 7:2:1.

In the training process, the mosaic data enhancement method is used, the training epochs are set to be 300, the training batch size is taken as 8, the optimizer selects SGD, the initial learning rate is set to be 0.01, and then the learning rate is adjusted 0.5 times after every 50 rounds. Cosine annealing learning rate is used for training weights, the momentum factor is 0.9, and the image input size is 416×416.

Table 1. Experimental software and hardware configuration.

Configuration	Parameter
CPU	Intel(R) Core(TM) i9-10980HK
RAM	32.0 GB
GPU	NVIDIA GeForce RTX2080 supe
Operating system	Windows10
Development tools	Pytorch,CUDA10.1

4.2 Evaluation Indicators

Average Precision (AP), which is the most common metric that is used to evaluate all types of models, involves Precision, Recall, and Intersection over Union (IoU). True positive (TP) represents that a sample is considered a positive sample by the classifier, and it is indeed a positive sample. False negative (FP) represents that a sample is considered a positive sample by the classifier, but it is not actually a positive sample. FN represents that a sample is considered a negative sample by the classifier, but it is not a negative sample. IoU is calculated as the ratio of the overlap between the prediction frame and the real frame. The calculation equations are expressed as follows:

$$Preceision = \frac{TP}{TP + FP}, \quad Recall = \frac{TP}{TP + FN}, \quad AP = \int_0^1 P(R)dR. \tag{1}$$

4.3 Experimental Results and Analysis

Experiments are conducted to demonstrate the superiority of our method against the following five CNN-based methods: Faster RCNN, SSD, YOLOv3, YOLOv3-tiny, and YOLOv5s. The comparison results are shown in Table 2. According to the final detection results, the precision of LS-YOLO is only 1.4% lower than that of YOLOv3 algorithm, but its model size is only 3.08% of that of YOLOv3 algorithm, its detection speed is 5.4 times higher than that of YOLOv3 algorithm, and its AP is also 1.6% higher than that of YOLOv3 algorithm. Moreover, the AP of LS-YOLO is also 3.04% higher than that of lightweight algorithm yolov3-tiny. Compared with the original algorithm YOLOv5, using 26.77% of algorithm parameters and 54.88% of computation, LS-YOLO can achieve an AP that is comparable to YOLOv5 algorithm. In brief, compared with other comparison algorithms, LS-YOLO performs well.

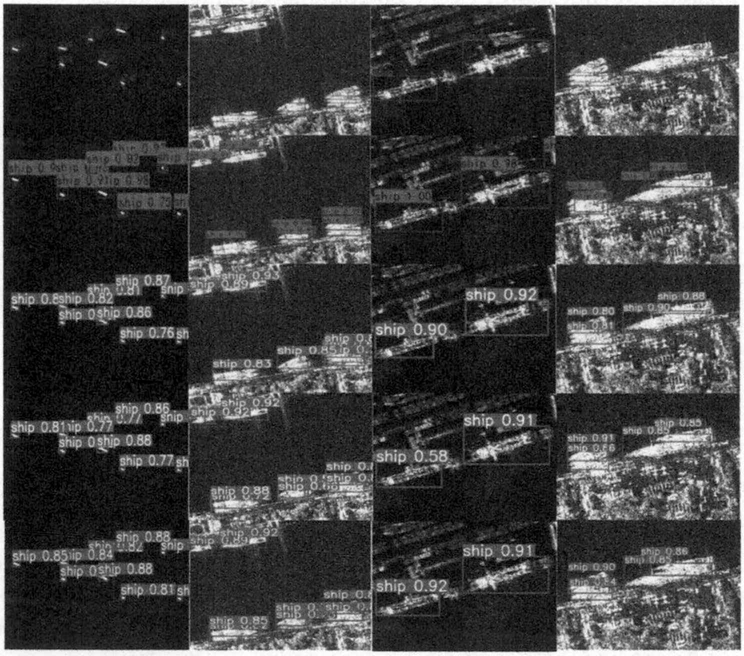

Fig. 5. The detection results obtained by different algorithms in different scenarios. From top to bottom: original image, Faster R-CNN, SSD, YOLOv5s, LS-YOLO

Table 2. Experimental results

Model	Model Size (MB)	Precision(%)	Recall (%)	AP (%)	Time (ms)
Faster RCNN	108.12	59.65	89.65	82.02	23.5
SSD	90.6	75.30	91.21	85.62	9.2
YOLOv3	123.4	**82.72**	94.60	95.06	6.5
YOLOv3-tiny	17.4	81.51	93.74	93.62	1.7
YOLOv5s	14.4	79.20	96.20	96.50	2.0
LS-YOLO	**3.8**	81.32	**97.42**	**96.66**	**1.2**

In order to judge the effects of the new algorithm more intuitively, we select four images with different scenes and different sizes of ship targets as test cases outside the training dataset. See Fig. 5 for details.

As one can see from Table 2 and Fig. 5, LS-YOLO not only achieves the fastest detection speed, but also achieves favorable performance in terms of both experimental performance metrics and actual detection results.

5 Conclusions

In this paper, a lightweight ship detection network LS-YOLO has been proposed. Specifically, FRM has been proposed to improve the detection ability for multi-scale targets. In addition, we have improved DCSP to reduce model parameters and computation. The effectiveness of the proposed lightweight network has been verified by conducting comparative experiments on SAR dataset. The experiment results have shown that the proposed ship detection method, which uses only 26.77% of algorithm parameters and 54.88% of computation, can achieve the AP that is comparable to YOLOv5 algorithm. In short, LS-YOLO can effectively improve the target detection accuracy, and maintain real-time detection performance.

Acknowledgments. This work was supported in part by the National Science Foundation of China (Grant Nos. 61873335, 61833011); the Project of Science and Technology Commission of Shanghai Municipality, China (Grant Nos. 20ZR1420200, 21SQBS01600, 22JC1401400, 19510750300, 21190780300); and the 111 Project, China under Grant No. D18003.

References

1. Li, D., Liang, Q., Liu, H., Liu, Q., Liu, H., Liao, G.: A novel multidimensional domain deep learning network for SAR ship detection. IEEE Trans. Geosci. Remote Sens. **60**, 1–13 (2021)
2. Wu, Z., Hou, B., Jiao, L.: Multiscale CNN with autoencoder regularization joint contextual attention network for SAR image classification. IEEE Trans. Geosci. Remote Sens. **59**(2), 1200–1213 (2020)
3. Lan, D.U., Wang, Z.C., Wang, Y., Wei, D., Lu, L.I.: Survey of research progress on target detection and discrimination of single-channel SAR images for complex scenes. Radars **9**(1), 34–54 (2020)
4. Saepuloh, A., Bakker, E., Suminar, W.: The significance of SAR remote sensing in volcano-geology for hazard and resource potential mapping. In: Proceedings of the AIP Conference Proceedings, article no. 070005 (2017)
5. Schumacher, R., Schiller, J.: Non-cooperative target identification of battlefield targets-classification results based on SAR images. In: Proceedings of the IEEE International Radar Conference, pp. 167–172 (2005)
6. Xu, X., Zhang, X., Zhang, T.: Multi-scale SAR ship classification with convolutional neural network. In: Proceedings of the International Geoscience and Remote Sensing Symposium, pp. 4284–4287 (2021)

7. Yang, Y., Liao, Y., Ni, S., Lin, C.: Study of algorithm for aerial target detection based on lightweight neural network. In: Proceedings of the International Conference on Consumer Electronics and Computer Engineering, pp. 422–426 (2021)
8. Wang, C., Bi, F., Zhang, W., Chen, L.: An intensity-space domain CFAR method for ship detection in HR SAR images. IEEE Geosci. Remote Sens. Lett. **14**(4), 529–533 (2017)
9. Pappas, O., Achim, A., Bull, D.: Superpixel-level CFAR detectors for ship detection in SAR imagery. IEEE Geosci. Remote Sens. Lett. **15**(9), 1397–1401 (2018)
10. Shi, Z., Yu, X., Jiang, Z., Li, B.: Ship detection in high-resolution optical imagery based on anomaly detector and local shape feature. IEEE Trans. Geosci. Remote Sens. **52**(8), 4511–4523 (2013)
11. Redmon, J., Divvala, S., Girshick, R., Farhadi, A.: You only look once: unified, real-time object detection. In: Proceedings of the IEEE Conference on Computer Vision and Pattern Recognition, pp. 779–788 (2016)
12. Redmon, J., Farhadi, A.: YOLO9000: better, faster, stronger. In: Proceedings of the IEEE Conference on Computer Vision and Pattern Recognition, pp. 7263–7271 (2017)
13. Redmon, J., Farhadi, A.: Yolov3: an incremental improvement. arXiv preprint arXiv: 1804.02767 (2018)
14. Liu, W., et al.: SSD: single shot multibox detector. In: Leibe, B., Matas, J., Sebe, N., Welling, M. (eds.) Computer Vision – ECCV 2016. ECCV 2016. Lecture Notes in Computer Science(), vol. 9905, pp. 21–37. Springer, Cham (2016). https://doi.org/10.1007/978-3-319-46448-0_2
15. Girshick, R., Donahue, J., Darrell, T., Malik, J.: Rich feature hierarchies for accurate object detection and semantic segmentation. In: Proceedings of the IEEE Conference on Computer Vision and Pattern Recognition, pp. 580–587 (2014)
16. Girshick, R.: Fast R-CNN. In: Proceedings of the IEEE International Conference on Computer Vision, pp. 1440–1448 (2015)
17. Chen, C., He, C., Hu, C., Pei, H., Jiao, L.: A deep neural network based on an attention mechanism for SAR ship detection in multiscale and complex scenarios. IEEE Access **7**, 104848–104863 (2019)
18. Cui, Z., Li, Q., Cao, Z., Liu, N.: Dense attention pyramid networks for multi-scale ship detection in SAR images. IEEE Trans. Geosci. Remote Sens. **57**(11), 8983–8997 (2019)
19. Zhang, R., Yao, J., Zhang, K., Feng, C., Zhang, J.: S-CNN-based ship detection from high-resolution remote sensing images. In: International Archives of the Photogrammetry, Remote Sensing Spatial Information Sciences 41 (2016)
20. Fu, C. Y., Liu, W., Ranga, A., Tyagi, A., Berg, A.C.: DSSD: deconvolutional single shot detector. arXiv preprint arXiv:1701.06659 (2017)
21. Wang, K., Liew, J. H., Zou, Y., Zhou, D., Feng, J.: Panet: few-shot image semantic segmentation with prototype alignment. In: Proceedings of the IEEE/CVF International Conference on Computer Vision, pp. 9197–9206 (2019)
22. Tang, G., Zhuge, Y., Claramunt, C., Men, S.: N-Yolo: a SAR ship detection using noise-classifying and complete-target extraction. Remote Sens. **13**(5), 871 (2021)
23. Zhou, K., Zhang, M., Wang, H., Tan, J.: Ship detection in SAR images based on multi-scale feature extraction and adaptive feature fusion. Remote Sens. **14**(3), 755 (2022)

Dictionary Learning-Based Reinforcement Learning with Non-convex Sparsity Regularizer

Haoli Zhao[1,2], Junkui Wang[1,2], Xingming Huang[3], Zhenini Li[4,5(✉)], and Shengli Xie[6,7]

[1] School of Automation, Guangdong University of Technology, Guangzhou 510006, China
1521860803@aliyun.com

[2] 111 Center for Intelligent Batch Manufacturing Based on IoT Technology (GDUT), Guangzhou 510006, China

[3] School of Computer Science and Engineering, Sun Yat-sen University, Guangzhou 510006, China
huangxm66@mail2.sysu.edu.cn

[4] Guangdong Provincial Key Laboratory of IoT Information Technology (GDUT), Guangzhou 510006, China

[5] Guangdong-HongKong-Macao Joint Laboratory for Smart Discrete Manufacturing (GDUT), Guangzhou 510006, China
lizhenni2012@gmail.com

[6] Key Laboratory of Intelligent Detection and The Internet of Things in Manufacturing (GDUT), Ministry of Education, Guangzhou 510006, China
shlxie@gdut.edu.cn

[7] Key Laboratory of Intelligent Information Processing and System Integration of IoT (GDUT), Ministry of Education, Guangzhou 510006, China

Abstract. Spare representations can help improve value prediction and control performances in Reinforcement Learning (RL), by capturing most essential features from states and ignoring unnecessary ones to avoid interference. However, existing sparse coding-based RL methods for control problems are optimized in the neural network methodology, which can not guarantee convergence. To this end, we propose a dictionary learning-based RL with the non-convex sparsity regularizer for RL control. To avoid the black-box optimization with the SGD, we employ the dictionary learning model in RL control, guaranteeing efficient convergence in control experiments. To obtain accurate representations in RL, we employ the non-convex ℓ_p norm ($0 < p < 1$) beyond the convex ℓ_1 norm as the sparsity regularizer in dictionary learning-based RL, for capturing more essential features from states. To obtain solutions efficiently, we employ the proximal splitting method to update the multivariate optimization problem. Hence, the non-convex sparsity regularized dictionary learning-based RL is developed and validated in different benchmark

The work described in this paper was supported by the National Natural Science Foundation of China (62203122,62273106,61973087,U1911401).

L. Fang et al. (Eds.): CICAI 2022, LNAI 13606, pp. 81–93, 2022.
https://doi.org/10.1007/978-3-031-20503-3_7

RL environments. The proposed algorithm can obtain the best control performances among compared sparse coding-based RL methods with around 10% increases in reward. Moreover, the proposed method can obtain higher sparsity in representations in different environments.

Keywords: Reinforcement learning · Dictionary learning · Non-convex sparsity regularizer

1 Introduction

Reinforcement Learning (RL) is playing an important role in the artificial intelligence systems [1]. Effective representations can obtain more essential features from states, and thus achieve more accurate value estimation and prediction in RL control tasks with large state and action spaces [7,21]. However, suffering from the locality and high correlation of continuous space states in RL models, the learned representations are usually redundant and unstable, resulting in catastrophic interference between nonlocal representations [2,10]. Consequently, inaccurate representations and state-values prediction make RL methods difficult to train and obtain good control performance [9,15].

Recently, studies have shown that sparse coding-based representation learning can obtain efficient representations in various RL tasks [3,6,8,19]. The sparse coding model considers describing states in RL environments with a possibly small number of key features, i.e., sparse representations, leading to high independence between representations of non-local states. Thus, effective sparse representations can help enhancing action value prediction and control performances.

Existing sparse coding-based RL methods mainly consist of two kinds. One kind is to employ sparsity regularizers in Neural Network (NN) to learn sparse representations [3,19]. These methods are optimized with Stochastic Gradient Descent (SGD), which may easily converge to a local minimum because of the batch stochasticity. Another one is the Dictionary Learning-based RL (DLRL) [6], which is directly optimized based on sparse coding theory. However, the existing work uses the convex ℓ_1 norm as the sparsity regularizer, which can hardly obtain sufficiently high sparsity and may lead to bias in representations during the optimization [20]. Furthermore, this work only employs the dictionary learning model in the value prediction problem for RL.

In this paper, we concentrate on the control problem based on the DLRL model. To avoid the black-box optimization with SGD, we employ the sparse coding-based dictionary learning model for the RL control problem. To efficiently obtain representations with the dictionary learning model, we propose to use the non-convex ℓ_p norm $(0 < p < 1)$ as the sparsity regularizer in the DLRL model. Hence, a Non-Convex sparsity regularized Dictionary Learning-based Reinforcement Learning (NCDLRL) is developed, which can ensure the optimization convergence with the sparse representation theory, and obtain representations with high sparsity based on the non-convex sparsity.

This is the first work to employ the dictionary learning model for control tasks based on reinforcement learning, main contributions are summarized as follows,

1. To avoid the black-box optimization with the SGD, we employ the sparse coding based dictionary learning model in RL control problems, guaranteeing efficient convergence in control experiments based on sparse representation theories.
2. To obtain accurate representations in RL, we employ the non-convex ℓ_p norm $(0 < p < 1)$ as the sparsity regularizer in dictionary learning-based RL, by capturing more essential features from states beyond the convex ℓ_1 norm.
3. To optimize the proposed problem efficiently, we use the proximal alternating iteration method to update the multivariate optimization problem, ensuring sufficient decreases to a critical point.

In Sect. 2, we review related works of RL using sparse models. In Sect. 3, we detail the NCDLRL model and further present the optimization process of the proposed model. Then, we present experimental validation on three benchmark RL environments in Sect. 4. Finally, conclusions are given in Sect. 5.

2 Related Work

In RL, the target is to control an agent which interacts discretely with a certain environment. The agent receives observations of states and selects actions to maximize the reward from the environment. Modeled by a Markov Decision Process (MDP), The environment is consisted of $(S, A, \Pr, R, \gamma)$, where $S \in \{\mathbf{s}_t, t = 1, ..., T+1\}$ is the state set describing conditions of the environment and the agent, $A \in \{a_t, t = 1, ..., T\}$ is an action set of the agent, $\Pr : S \times A \times S \rightarrow [0, 1]$ is the transition probability, and $\gamma \in [0, 1]$ is the discount rate. The environment is called Markov if $\Pr(\mathbf{s}_{t+1}|\mathbf{s}_t, a_t) = \Pr(\mathbf{s}_{t+1}|\mathbf{s}_t, a_t, ..., \mathbf{s}_1, a_1)$.

The goal of RL is to find an optimal policy $\pi : S \times A \rightarrow [0, 1]$, which can obtain a maximal expected return from each state. The policy is either directly learned, as in policy gradient methods [12,13] or inferred by acting near-greedily according to the learned action values, as in the Q-learning [17]. State-Action-Reward- State-Action (SARSA) is one on-policy control example which defines the action-value function as:

$$Q^{\pi}(\mathbf{s}, a) = \mathbb{E}[\sum_{i=t}^{T} \gamma^{i-t} r_{i+1}|\mathbf{s}_t = \mathbf{s}, a_t = a] = \mathbb{E}[g_{i+1}|\mathbf{s}_t = \mathbf{s}, a_t = a], \tag{1}$$

where Q^{π} maps the state $\mathbf{s}$ and action a based on policy π to the cumulative reward, and g_{t+1} is the accumulated discount reward with a discount rate γ.

The RL models consist of highly correlated data, resulting in significant interference in training. Therefore, the estimated action values are usually unstable, even states change slightly. One idea for alleviating the interference is the target networks, and there are successive evidences that it can improve the stability of training [2,14]. However, this idea prevents the DNN training from the most recent knowledge from the environment, which can reduce the efficiency of learning [5,8]. Another idea suggests obtaining action values accurately with efficient representations. In [16], Wang et al. utilized the mixture regularization to improve the generalization of representations from mixed observations from different environments. [8] proposed to use a sparsity distribution as the regularizer in NN, which can encode representations to a desired sparse distribution, thus improving control performances. [3] proposed to utilize the ℓ_1 norm and ℓ_2 norm regularizers on neurons to obtain sparse representations. Furthermore, [6] proposed to use the ℓ_1 norm regularized dictionary learning algorithm for the RL model, and validated that it can outperform conventional sparse models like Tile Coding (TC) in predicting more accurate action values [12], nevertheless, the sparsity induced by the ℓ_1 norm is biased, thus can not pursue accurate sparse representations efficiently.

3 Dictionary Learning-Based Reinforcement Learning with Non-convex Sparsity Regularizer

In this section, we first detail the NCDLRL problems, where we use the synthetic model to obtain accurate sparse representations for RL. Then, we detail how to solve the proposed problem with an efficient algorithm.

3.1 Problem Formulation

Considering that we have a trajectory of states, actions, and rewards from a control environment, i.e., $\{(\mathbf{s}_t, a_t, r_{t+1}, \mathbf{s}_{t+1})\}_{t=1}^{T}$, the target is to estimate accurate action values for the input state $\mathbf{s}_t \in \mathbb{R}^m$ via the synthetic dictionary learning. Denoting the parameter set as $\boldsymbol{\theta}$, which contains all parameters used in the problem, the targeted policy π for the action-value function can be approximated with representations from the synthetic dictionary learning,

$$Q^{\pi}(\mathbf{s}_t, a) \approx Q_{\boldsymbol{\theta}}(\mathbf{s}_t, a) := \mathbf{w}_a^{\mathrm{T}} \mathbf{z}_t, \quad t = 1, ..., T+1, \tag{2}$$

where $\mathbf{w}_a \in \mathbb{R}^n$ denotes the corresponding action-value parameter for different actions a in control, $\mathbf{z}_t \in \mathbb{R}^n$ is the sparse representation for $\mathbf{s}_t$ from the synthetic dictionary learning.

To alleviate interference in RL, we use the synthetic dictionary learning model to obtain adaptive sparse representations from states. Denoting the synthetic dictionary as $\mathbf{D} \in \mathbb{R}^{m \times n}$ and all states as $\mathbf{S} = [\mathbf{s}_1, ..., \mathbf{s}_{T+1}]$, we obtain the sparse representation in the following linear model, $\mathbf{S} = \mathbf{DZ} + \mathbf{N}$, where $\mathbf{Z} = [\mathbf{z}_1, ..., \mathbf{z}_{T+1}] \in \mathbb{R}^{n \times (T+1)}$ is the sparse representation matrix for all states, and $\mathbf{N}$ denotes noises in the states. the "synthetic" means that the state $\mathbf{s}_t$ is composed of the few important atoms from the dictionary $\mathbf{D}$, in which each column $\mathbf{d}_i \in \mathbb{R}^m$ is an atom. To ensure the state to be reconstructed accurately, the state approximation error f is defined by the Frobenius norm as follows,

$$f(\mathbf{D}, \mathbf{Z}) = \frac{1}{2} \|\mathbf{S} - \mathbf{DZ}\|_F^2. \tag{3}$$

Since the dictionary $\mathbf{D}$ is overcomplete with $n > m$ and full-rank, this is an underdetermined system with infinite solutions. To obtain the accurate and possibly sparse representation, we need to enforce sparsity in representations with the ℓ_p norm ($0 < p < 1$) as sparsity regularizer g_z, defined as follows,

$$g_z(\mathbf{Z}) = \|\mathbf{Z}\|_p^p = \sum_t^{T+1} \sum_{i=1}^n |z_{t,i}|^p. \tag{4}$$

Although the ℓ_p norm ($0 < p < 1$) is non-convex and hard to optimize, it can obtain sparser and more accurate representations than the popular ℓ_1 norm, thus leading to more accurate action-value prediction in this problem. Moreover, with a smaller p closer to 0, ℓ_p norm ($0 < p < 1$) is more concave and can induce higher sparsity, but also making it harder to find global optimal solutions.

To estimate accurate action values for states in the environment, the Mean-Squared Return Error (MSRE) [6] is employed, which considers the accumulated expected return reward in Eq. (1). Denoting $\mathbf{g} = [g_1, ..., g_{T+1}]$, the reward estimation error h defined as follows,

$$h(\mathbf{Z}, \mathbf{w}_a) = \frac{1}{2}\|\mathbf{g} - \mathbf{w}_a^{\mathrm{T}}\mathbf{Z}\|_2^2. \tag{5}$$

Consequently, we integrate the state approximation error f, the sparsity regularizer g_z, and the reward estimation error h to form the loss function of the NCDLRL model with the parameter set $\boldsymbol{\theta} = \{\mathbf{D}, \mathbf{w}_a\}$, detailed as follows,

$$\begin{aligned}\min_{\boldsymbol{\theta},\mathbf{Z}} L(\boldsymbol{\theta}, \mathbf{Z}) = f(\mathbf{D}, \mathbf{Z}) + \lambda g_z(\mathbf{Z}) + \beta h(\mathbf{Z}, \mathbf{w}_a) = \frac{1}{2}\|\mathbf{S} - \mathbf{D}\mathbf{Z}\|_F^2 + \lambda\|\mathbf{Z}\|_p^p + \frac{\beta}{2}\|\mathbf{g} - \mathbf{w}_a^{\mathrm{T}}\mathbf{Z}\|_2^2 \\ s.t. \quad 0 < p < 1, \ \|\mathbf{d}_i\|_2 = m, \ \|\mathbf{w}_a\|_2 \le c, \ \forall i, \forall a,\end{aligned} \tag{6}$$

where λ is the parameter to control the sparsity of representations pursued by the regularizer g_z, and β is the parameter to balance the trade-off between state approximation and reward estimation. The value scales of dictionary atoms are set to m in ℓ_2 norm to make all atoms function similarly. Moreover, the maximum of $\mathbf{w}_a$ is restricted to avoid value explosion.

3.2 Optimization and Algorithm

To efficiently solve the proposed NCDLRL model in Eq. (6), the proximal alternating iteration method is employed to obtain solutions in the non-convex multivariate optimization problem. The updating processes consist of mainly three phases, i.e., the sparse representation updating for $\mathbf{Z}$, the synthetic dictionary learning for $\mathbf{D}$, and the action-value weight updating for $\mathbf{w}_a$.

1) The sparse representation updating: Fixing the synthetic dictionary $\mathbf{D}$ and the action-value weight $\mathbf{w}_a$, we update the sparse representations $\mathbf{Z}$ for all input states in the following subproblem:

$$\mathbf{Z} = \arg\min_{\mathbf{Z}} L(\mathbf{Z}) = \frac{1}{2}\|\mathbf{DS} - \mathbf{Z}\|_F^2 + \lambda\|\mathbf{Z}\|_p^p + \frac{\beta}{2}\|\mathbf{g} - \mathbf{w}_a^{\mathrm{T}}\mathbf{Z}\|_2^2. \tag{7}$$

Based on the proximal splitting method, the sparse representations in iteration k for the input state set $\mathbf{S}$ can be obtained as follows,

$$\mathbf{Z}^{(k+1)} = \arg\min_{\mathbf{Z}} \hat{L}_z(\mathbf{Z}) = \frac{1}{2}\|\mathbf{U}_z^{(k)} - \mathbf{Z}\|_2^2 + \lambda\|\mathbf{Z}\|_p^p = Prox_p(\mathbf{U}_z^{(k)}, \lambda), \tag{8}$$

where $\mathbf{U}_z^{(k)} = \mathbf{Z}^{(k)} - \frac{1}{\alpha_z}(\mathbf{D}^{\mathrm{T}}(\mathbf{D}\mathbf{Z}^{(k)} - \mathbf{S}) + \beta\mathbf{w}_a(\mathbf{w}_a^{\mathrm{T}}\mathbf{Z}^{(k)} - g_{t+1}))$ is obtained by the proximal gradient. The adaptive step size α_z is defined by the Lipschitz constant to ensure the convergence of this subproblem, which should be larger than the largest eigenvalue of $(\mathbf{D}^{\mathrm{T}}\mathbf{D} + \beta\mathbf{w}_a\mathbf{w}_a^{\mathrm{T}})$. With different p value, the solution can be obtained with different proximal operator $Prox_p$. For example, $p = \frac{1}{2}$ with corresponding explicit close-form solutions for proximal operators can be employed to obtain the non-convex sparse solutions in Eq. (8). For a vector $\mathbf{z} \in \mathbb{R}^n$, the half proximal operator $Prox_{\frac{1}{2}}$ [18] is defined as,

$$Prox_{\frac{1}{2}}(\mathbf{z}, \lambda) := \begin{cases} \frac{2}{3}z_i(1 + \cos(\frac{2}{3}\arccos(-\frac{3^{\frac{3}{2}}}{4}\lambda|z_i|^{-\frac{3}{2}}))) & , |z_i| \geq \frac{3}{2}\lambda^{\frac{2}{3}} \\ 0 & , |z_i| < \frac{3}{2}\lambda^{\frac{2}{3}} \end{cases}, \tag{9}$$

2) The synthetic dictionary learning: We then fix the sparse coefficient $\mathbf{Z}$ and the action-value weight $\mathbf{w}_a$, and update the synthetic dictionary $\mathbf{D}$ with the following constrained subproblem:

$$\mathbf{D} = \arg\min_{\mathbf{D}} L(\mathbf{D}) = \frac{1}{2}\|\mathbf{S} - \mathbf{D}\mathbf{Z}\|_F^2 \quad s.t. \quad \|\mathbf{d}_i\|_2 = m, \ \forall i, \tag{10}$$

where the normalization constraint force the learned synthetic dictionary atoms to a set ℓ_2 norm valued m. Denoting J_d as an indicator function for all dictionary atoms, $J_d(\mathbf{D}) = \begin{cases} 0 & , \|\mathbf{d}_i\|_2 = m, \forall i \\ +\infty & , otherwise \end{cases}$. Therefore, for updating the synthetic dictionary $\mathbf{D}$ in iteration k, the optimization subproblem is formed as follows,

$$\mathbf{D}^{(k+1)} = \arg\min_{\mathbf{D}} \hat{L}_d(\mathbf{D}) = \frac{1}{2}\|\mathbf{D} - \mathbf{U}_d^{(k)}\|_F^2 + J_d(\mathbf{D}) = Prox_d(\mathbf{U}_d^{(k)}), \tag{11}$$

where $\mathbf{U}_d^{(k)} = \mathbf{D}^{(k)} - \frac{1}{\alpha_d}(\mathbf{D}^{(k)}\mathbf{Z} - \mathbf{S})\mathbf{Z}^{\mathrm{T}}$ is obtained based on the proximal gradient. Then, α_d is the adaptive step size defined by the Lipschitz constant to ensure the convergence of this subproblem, being larger than the largest eigenvalue of $\mathbf{Z}\mathbf{Z}^{\mathrm{T}}$. Moreover, $Prox_d$ decides to normalize which column in $\mathbf{U}_d^{(k)}$.

3) The action-value weight updating: Fixing the sparse representation $\mathbf{Z}$ and the synthetic dictionary $\mathbf{D}$, the action-value weight $\mathbf{w}_a$ is updated in the following constrained subproblem:

$$\mathbf{w}_a = \arg\min_{\mathbf{w}_a} L(\mathbf{w}_a) = \frac{\beta}{2}\|\mathbf{g} - \mathbf{w}_a^{\mathrm{T}}\mathbf{Z}\|_2^2 \quad s.t. \ \|\mathbf{w}_a\|_2 \leq c, \ \forall a, \tag{12}$$

where the value scales of $\mathbf{w}_a$ are restricted by c in terms of the ℓ_2 norm. Similarly, by denoting J_w as the indicator function for the normalization constraint as follows,

$$J_w(\mathbf{w}_a) = \begin{cases} 0 & , \|\mathbf{w}_a\|_2 \leq c \\ +\infty & , otherwise \end{cases}. \tag{13}$$

Therefore, we can update the action weights in iteration k with the following optimization problem,

$$\mathbf{w}_a^{(k+1)} = \arg\min_{\mathbf{w}_a} \hat{L}_w(\mathbf{w}_a) = \frac{1}{2}\|\mathbf{w}_a - \mathbf{u}_a^{(k)}\|_2^2 + J_w(\mathbf{w}_a) = Prox_w(\mathbf{u}_a^{(k)}), \tag{14}$$

where $\mathbf{u}_a^{(k)} = \mathbf{w}_a^{(k)} - \frac{1}{\alpha_a}\mathbf{Z}(\mathbf{Z}^{\mathrm{T}}\mathbf{w}_a^{(k)} - \mathbf{g}^{\mathrm{T}})$ is obtained by the proximal gradient. Again, α_a is the adaptive step size defined by the Lipschitz constant to ensure the convergence of this subproblem, which should be larger than the largest eigenvalue of $\mathbf{ZZ}^{\mathrm{T}}$. Moreover, $Prox_w$ is to decide whether $\mathbf{w}_a$ need normalization.

In the end, we summarize the proposed NCDLRL in Algorithm 1.

Algorithm 1: The proposed NCDLRL algorithm

Input:
Dataset $\{(\mathbf{s}_i, a_i, r_{i+1}, \mathbf{s}_{i+1})\}_{i=1}^{N}$, λ, β, and p.
Initialization:
$\mathbf{D} \in \mathbb{R}^{m \times n}$, $\mathbf{Z} = \mathbf{0} \in \mathbb{R}^{n \times N}$, $\mathbf{w}_a \in \mathbb{R}^n$, and $k = 0$.
Main iterations:
1: **Repeat**
2: $k = k + 1$;
3: Adjust α_ϕ based on $\mathbf{D}$ and $\mathbf{w}_a$;
4: Fix $\mathbf{D}$ and $\mathbf{w}_a$, update the sparse representation $\mathbf{Z}$ based on Eq. (8);
5: Adjust α_d based on $\mathbf{Z}$;
6: Fix $\mathbf{Z}$ and $\mathbf{w}_a$, and update the synthetic dictionary $\mathbf{D}$ based on Eq. (11);
7: Adjust α_a based on $\mathbf{Z}$;
8: Fix $\mathbf{Z}$ and $\mathbf{D}$, and update the action-value weights $\mathbf{w}_a$ based on Eq. (14);
9: **Until** stopping criterion is met.
Output: $\theta = \{\mathbf{w}_a, \mathbf{D}\}$.

4 Experiments and Discussions

In this section, the performances of the proposed NCDLRL will be validated in various RL environments. Firstly, the experiment details and algorithms for comparison are detailed. After that, we show how parameter selection may affect the performances of NCDLRL. Then, we will present detailed performance comparison for different RL algorithms with corresponding optimal parameters.

4.1 Experiment Details

We validate performances of the proposed algorithm in three benchmark RL environments: Acrobot, Cart Pole, and Puddle World. All experiments of the proposed algorithm were performed via pytorch 1.8, and programs were run on a CPU with 2.5 GHz Intel cores and 60G RAM.

Initialization. All algorithms compared in this experiment are trained and then control in two separated phases. In the training phase, we train the parameter set of each algorithm to learn appropriate representations for an environment from trajectory data generated by a certain policy. In Acrobot[1], training data are generated with a random policy, which selects actions from the action space

[1] https://gym.openai.com/envs/Acrobot-v1/.

randomly with the same probability. In Cart Pole[2], a random policy that the agent selects to push the cart to left or right randomly with the same probability is employ for training data generation. In Puddle World, where the agent aims to reach the goal without touching two adjunct puddles in the field, training data are obtained with the policy that selects to go north or east with the same probability.

After the training with the generated data converges, the parameter set $\boldsymbol{\theta}$ of each algorithm is then fixed and used in the control phase where only the action-value weights $\mathbf{w}_a$ that respond to different actions are fine-tuned by an agent.

Algorithms. To validate the performance of NCDLRL, we compared it with different sparse coding-based RL algorithms, consisting of mainly two types of methods:
(1) DQN structured algorithm: These algorithms use the DNN to predict the action values in the DQN architecture. Firstly, A vanilla forwarding Neuron Network (NN) is used as the baseline, consisting of 2 dense hidden layers of the size $[32, 256]$ with ReLU activation and a linear output layer without bias, and the following DNN based algorithms all share the same structures. Secondly, Dropout [11], which randomly dilutes a ratio of neurons to zeros, is compared to show the effects of random sparsity for RL. Then, L_{1w}-NN [3], which applies the convex ℓ_1 norm regularizer on the weights of all layers in the DNN, is compared to show the performances of sparsity regularized DNN. In the training phase, a dataset with $N = 10000$ is used as the training buffer of all DNN based algorithms with the batch size set to 64. In the control phase, we use SARSA with MSTDE forming the RL algorithm, and update $\mathbf{w}_a$ through the basic SGD by searching the learning rate in $[0.001, 0.1]$ for obtaining appropriate output weights $\mathbf{w}_a$. Target network and replay buffer are not used in the control phase.
(2) Dictionary Learning-based algorithm: These algorithms use the synthetic dictionary learning models to obtain sparse representations from the state, and then predict action values to control agents in different RL environments. The same sized dictionaries with $n = 64$ atoms and a training dataset sized $N = 5000$ are employed to obtain representations for all dictionary learning-based algorithm. The synthetic dictionary learning-based reinforcement learning with the convex ℓ_1 norm sparsity regularizer (DLRL-L_1) is compared to show the effectiveness of the non-convex sparsity in RL. Different from the original work [6], which employs the ℓ_1 norm for only value prediction, we have expanded the work in control by using the same control procedures as the proposed NCDLRL. For the proposed NCDLRL, $p = 0.5$ is validated in the following experiments, i.e., NCDLRL-$\mathrm{L}_{1/2}$.

4.2 Parameter Selection

In this subsection, we analyze how parameters affect the performances of the proposed NCDLRL-$\mathrm{L}_{1/2}$. λ for adjusting the effect of the $\ell_{1/2}$ norm sparsity

[2] https://gym.openai.com/envs/CartPole-v1/.

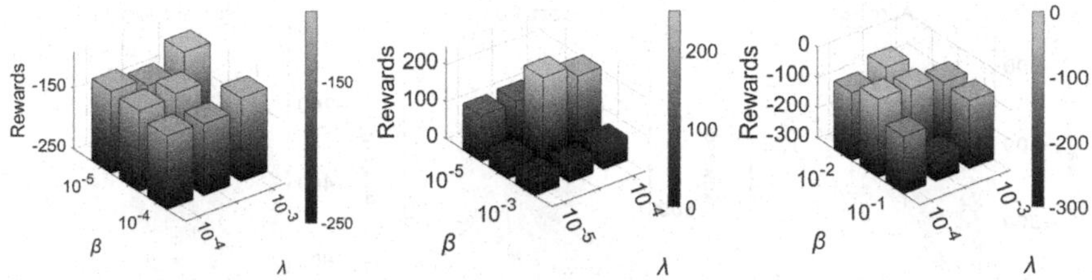

(a) Acrobot (b) Cart Pole (c) Puddle World

Fig. 1. Performances of the proposed NCDLRL with a range of λ and β in different RL environments

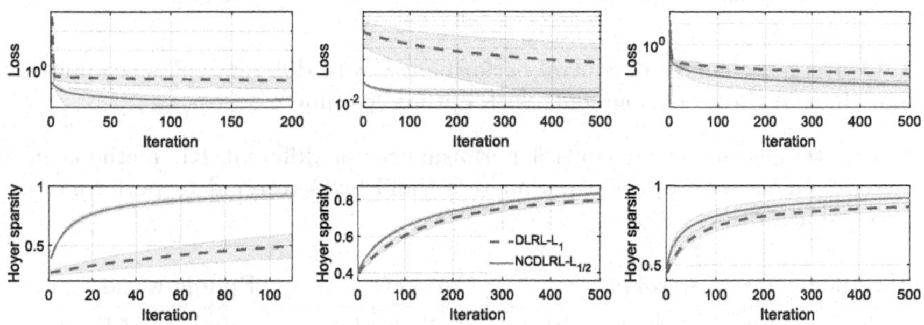

(a) Acrobot (b) Cart Pole (c) Puddle World

Fig. 2. Convergence curves of loss value and Hoyer sparsity of dictionary learning-based RL algorithms in three environments with optimal parameters

regularizer is searched in $[10^{-5}, 10^{-3}]$. β for balancing between the state approximation error and the action-value estimation error is searched in $[10^{-5}, 10^{-1}]$. The control performances of NCDLRL-$L_{1/2}$ with different λ and β are shown in Fig. 1. Overall, NCDLRL-$L_{1/2}$ can obtain optimal performances in different environments when λ is around 10^{-4}. Since the value scale of representations is restricted in a range because of the normalization of input states and dictionary atoms, the similar λ indicate similar optimal sparsity of representations in different environments. On the other hand, the optimal β is varied largely in different environments, indicating that balancing state approximation and action-value estimation can help obtain accurate action-value prediction.

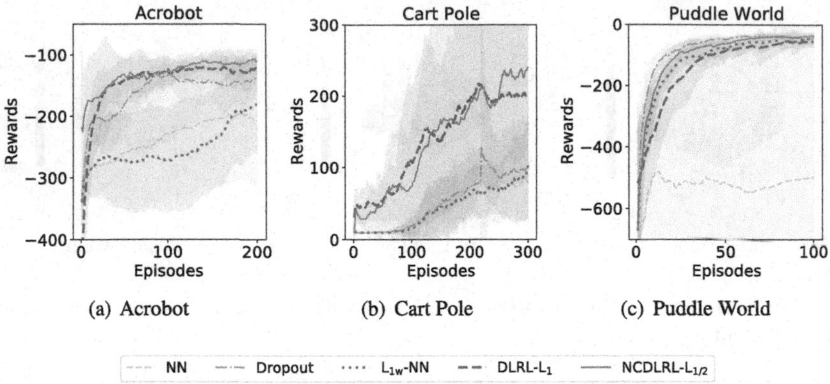

(a) Acrobot (b) Cart Pole (c) Puddle World

Fig. 3. Convergence curves of control performances with different sparse coding-based RL algorithms in three environments with optimal parameters

Table 1. Average converged control performances of different RL methods in the three environments with optimal parameters (bold texts respond to performances of the proposed algorithm)

Method	Acrobot	Cart pole	Puddle world
NN	–209.59 ± 103.69	79.25 ± 53.21	–498.49 ± 547.04
Dropout [11]	–137.45 ± 28.65	98.05 ± 66.04	–44.29 ± 20.38
L_{1w}-NN [3]	–182.15 ± 79.50	102.71 ± 73.77	–44.61 ± 17.37
DLRL-L_1 [6]	–122.20 ± 23.84	197.74 ± 119.04	–58.62 ± 19.53
NCDLRL-$L_{1/2}$	**–110.28 ± 18.81**	**239.61 ± 110.40**	**–36.33 ± 4.62**

4.3 Performances Comparison

In this subsection, we present detailed comparisons between our proposed method and different sparse coding-based RL algorithms, results of all algorithms are averages of 20 repeat runs. We first show the convergence performances of two dictionary learning-based RL methods in the training phase, in terms of the loss function value and the Hoyer sparsity measure [4]. The Hoyer sparsity measure maps a vector $\mathbf{z} \in \mathbb{R}^n$ to a fix range $[0, 1]$, defined as follows,

$$\text{Hoyer sparsity}(\mathbf{z}) = \frac{\sqrt{n} - \sum_i |\mathbf{z}_i| / \sqrt{\sum_i \mathbf{z}_i^2}}{\sqrt{n} - 1}. \tag{15}$$

The sparsity measure reaches 1 when $\mathbf{z}$ is the sparsest with only one nonzero element, and decreases with increase of nonzero numbers. Shown in Fig. 2, the proposed NCDLRL-$L_{1/2}$ can converge to lower loss than the convex DLRL-L_1 in different environments. One reason is that the non-convex sparsity can lead to higher sparsity in representations, as shown in the Hoyer sparsity convergence diagrams.

Figure 3 further presents the convergence analysis in the control phase. Overall, the proposed NCDLRL-$L_{1/2}$ can converge after interacting with different

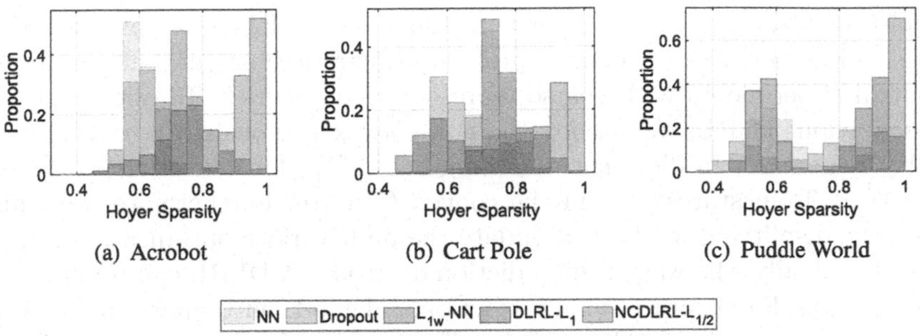

(a) Acrobot (b) Cart Pole (c) Puddle World

Fig. 4. Comparison of the Hoyer sparsity in representations of different algorithms in different RL environments (distributions with higher Hoyer values closer to 1 indicate higher sparsity in representations)

environments in reasonable numbers of episodes. The convergence speeds of NCDLRL-$L_{1/2}$ are as fast as DLRL-L_1 and Dropout, and faster than the others in these environments.

The converged detail control performances of different RL algorithms are shown in Table 1. Overall, the proposed NCDLRL-$L_{1/2}$ can obtain the best converged control performances among these algorithms in different environments, supporting the effectiveness of dictionary learning-based RL in exploiting accurate and sparse representations in control tasks. The vanilla NN holds the worst performances in different environments, suggesting that it is heavily affected by interference in RL environments. Moreover, both L_{1w}-NN and DLRL-L_1 employ the ℓ_1 norm as sparsity regularizer, but L_{1w}-NN is optimized with SGD-based ADAM, and DLRL-L_1 is optimized with the dictionary learning methodology. The advantages of DLRL-L_1 over L_{1w}-NN can also support the effectiveness of dictionary learning-based RL. Furthermore, NCDLRL-$L_{1/2}$ can outperform DLRL-L_1 in all these environments, supporting the effectiveness of non-convex sparsity in obtaining effective representations than the convex ℓ_1 norm.

To further analysis the differences of representations learned by different algorithms in these environments, we present the Hoyer sparsity distributions of different algorithms in Fig. 4. Overall, the proposed NCDLRL-$L_{1/2}$ can obtain higher sparsity in representations than the others in different environments. The sparsity of the vanilla NN is generally smaller than the others in different environments. The sparsity distribution of Dropout is close to NN in Acrobot and Cart Pole, suggesting that the random sparsity is not efficient in finding sparse and efficient representation. Furthermore, NCDLRL-$L_{1/2}$ can obtain higher sparsity than DLRL-L_1, supporting the effectiveness of non-convex sparsity in RL control tasks.

5 Conclusion

In this paper, we propose an efficient dictionary learning-based RL method with non-convex sparsity regularizer for control problems, i.e., the NCDLRL model.

To avoid the black-box optimization with the SGD in DRL, we employ the dictionary learning model in RL control problems, guaranteeing efficient convergence in control experiments with sparse representation theories. To obtain accurate representations in RL, we employ the non-convex ℓ_p norm $(0 < p < 1)$ as the sparsity regularizer in dictionary learning-based RL, for capturing more essential features from states beyond the convex ℓ_1 norm. Furthermore, we employ the proximal splitting method to update the multivariate optimization problem efficiently. Finally, the well-trained dictionary from NCDLRL can obtain sparse representations for accurate value prediction and good control performances. The control experiments have validated the effectiveness of the proposed method. In three benchmark RL environments, the proposed NCDLRL-$L_{1/2}$ can all obtain the best control performances with around 10% increases in rewards. Moreover, NCDLRL-$L_{1/2}$ can always obtain higher sparsity in representations because of the non-convex sparsity.

References

1. Bengio, Y., Courville, A., Vincent, P.: Representation learning: a review and new perspectives. IEEE Trans. Pattern Anal. Mach. Intell. **35**(8), 1798–1828 (2013)
2. Fan, J., Wang, Z., Xie, Y., Yang, Z.: A theoretical analysis of deep q-learning. In: Proceedings of the 2nd Conference on Learning for Dynamics and Control, pp. 486–489. PMLR (2020)
3. Hernandez-Garcia, J.F., Sutton, R.S.: Learning sparse representations incrementally in deep reinforcement learning. arXiv preprint arXiv:1912.04002 (2019)
4. Hoyer, P.O.: Non-negative matrix factorization with sparseness constraints. J. Mach. Learn. Res. **5**(9) (2004)
5. Kim, S., Asadi, K., Littman, M., Konidaris, G.: Deepmellow: removing the need for a target network in deep q-learning. In: Proceedings of the 28th International Joint Conference on Artificial Intelligence (2019)
6. Le, L., Kumaraswamy, R., White, M.: Learning sparse representations in reinforcement learning with sparse coding. In: Proceedings of the 26th International Joint Conference on Artificial Intelligence, pp. 2067–2073 (2017)
7. Li, Z., Xu, M., Nie, J., Kang, J., Chen, W., Xie, S.: Noma-enabled cooperative computation offloading for blockchain-empowered internet of things: a learning approach. IEEE Internet Things J. **8**, 2364–2378(2020)
8. Liu, V., Kumaraswamy, R., Le, L., White, M.: The utility of sparse representations for control in reinforcement learning. In: Proceedings of the AAAI Conference on Artificial Intelligence, vol. 33, pp. 4384–4391 (2019)
9. Luo, X., Meng, Q., Di He, W.C., Wang, Y.: I4r: promoting deep reinforcement learning by the indicator for expressive representations. In: Proceedings of the 29th International Joint Conference on Artificial Intelligence, pp. 2669–2675 (2020)
10. Sarafian, E., Tamar, A., Kraus, S.: Constrained policy improvement for efficient reinforcement learning. In: Proceedings of the 29th International Joint Conference on Artificial Intelligence (2020)
11. Srivastava, N., Hinton, G., Krizhevsky, A., Sutskever, I., Salakhutdinov, R.: Dropout: a simple way to prevent neural networks from overfitting. J. Mach. Learn. Res. **15**(1), 1929–1958 (2014)

12. Sutton, R.S., Barto, A.G.: Reinforcement Learning: An Introduction. MIT Press (2018)
13. Sutton, R.S., McAllester, D.A., Singh, S.P., Mansour, Y.: Policy gradient methods for reinforcement learning with function approximation. In: Advances in Neural Information Processing Systems, pp. 1057–1063 (2000)
14. Van Hasselt, H., Doron, Y., Strub, F., Hessel, M., Sonnerat, N., Modayil, J.: Deep reinforcement learning and the deadly triad. arXiv preprint arXiv:1812.02648 (2018)
15. Wang, K., Kang, B., Shao, J., Feng, J.: Improving generalization in reinforcement learning with mixture regularization. In: Conference on Neural Information Processing Systems (2020)
16. Wang, K., Kang, B., Shao, J., Feng, J.: Improving generalization in reinforcement learning with mixture regularization. In: Advances in Neural Information Processing Systems (2020)
17. Watkins, C.J., Dayan, P.: Q-learning. Mach. Learn. **8**(3–4), 279–292 (1992)
18. Xu, Z., Chang, X., Xu, F., Zhang, H.: $l_{1/2}$ regularization: a thresholding representation theory and a fast solver. IEEE Trans. Neural Networks Learn. Syst. **23**(7), 1013–1027 (2012)
19. Zhao, H., Wu, J., Li, Z., Chen, W., Zheng, Z.: Double sparse deep reinforcement learning via multilayer sparse coding and nonconvex regularized pruning. IEEE Transactions on Cybernetics (2022)
20. Zhao, H., Zhong, P., Chen, H., Li, Z., Chen, W., Zheng, Z.: Group non-convex sparsity regularized partially shared dictionary learning for multi-view learning. Knowl. Based Syst. **242**, 108364 (2022)
21. Zhou, Q., Kuang, Y., Qiu, Z., Li, H., Wang, J.: Promoting stochasticity for expressive policies via a simple and efficient regularization method. Adv. Neural Inform. Process. Syst. **33** (2020)

Deep Twin Support Vector Networks

Mingchen Li and Zhiji Yang(✉)

Yunnan University of Finance and Economics, Kunming, China
yzhiji@foxmail.com

Abstract. Twin support vector machine (TSVM) is a successful improvement for traditional support vector machine (SVM) for binary classification. However, it is still a shallow model and has many limitations on prediction performance and computational efficiency. In this paper, we propose deep twin support vector networks (DTSVN) which could enhance its performances in all aspects. Specifically, we put forward two version of DTSVN, for binary classification and multi classification, respectively. DTSVN improves the abilities of feature extraction and classification performance with neural networks instead of a manually selected kernel function. Besides, in order to break the bottleneck that the original model cannot directly handle multi classification tasks, multiclass deep twin support vector networks (MDTSVN) is further raised, which could avoid the inefficient one-vs-rest or one-vs-one strategy. In the numerical experiments, our proposed DTSVN and MDTSVN are compared with the other four methods on MNIST, FASHION MNIST and CIFAR10 datasets. The results demonstrate that our DTSVN achieves the best prediction accuracy for the binary problem, and our MDTSVN significantly outperforms other existing shallow and deep methods for the multi classification problem.

Keywords: SVM · TSVM · Deep learning · Neural networks

1 Introduction

Traditional support vector machine (SVM) is a widely used supervised learning model [2,12]. In the spirit of the SVM, a number of variants have been proposed, such as ν-SVM, support vector data description (SVDD) and twin support vector machine (TSVM). ν-SVM sets parameter ν to control an upper bound on the fraction of training errors and a lower bound of the fraction of support vectors [9,23]. SVDD is mainly used for anomaly detection [26]. TSVM has the main advantage of reducing the computational complexity of the model to solve the quadratic programming problem (QPP) [15]. The methods above work well as a shallow model for small sample problems. However, there are some limitations. The computational complexity of solving the QPP is still high, and it is time consuming to select the appropriate kernel function and the corresponding parameters with grid search.

L. Fang et al. (Eds.): CICAI 2022, LNAI 13606, pp. 94–106, 2022.
https://doi.org/10.1007/978-3-031-20503-3_8

When faced with large-scale problems, deep learning models often have stronger advantages. Deep neural networks (DNN) typically use the mini-batch approach to train networks by back propagation with stochastic gradient descent, which reduces the computational complexity compared to solving a large optimization problem at a time [16]. Additionally, several advanced DNN methods have been demonstrated to achieve outstanding performances on the feature extraction and prediction [8,11,24]. Therefore, it is inspired to take advantage of deep learning to improve the traditional shallow methods.

It has been shown that the SVM-type methods can be improved by using neural networks [4,30]. Methods such as multilayer SVM and multilayer TSVM utilize their specific classifiers to replace the original perceptron of neural networks [18,22,27]. However, the limitation of such methods is that it still needs to solve large QPPs. Moreover, these methods still need to manually choose the appropriate kernel function. [7,14,29,31]. The deep multi-sphere SVDD approach explores the combination of SVDD and neural networks [6]. The model is solved in two stages, first using a neural network to accomplish feature extraction and then using the results as input to the SVDD. In [3,19], the proposed deep SVM follows the similar idea above, which improves the two-stage solution into a unified formulation.

Moreover, the deep SVM-type models above mainly focus on anomaly detection and binary classification problems [20,25]. When dealing with multi-class issues, a common way is by one-vs-one or one-vs-rest strategy [10,13].

In this paper, two models are proposed to break the limitations above. Firstly, a deep twin support vector networks (DTSVN) is proposed, which extends the original shallow TSVM to deep learning framework. Secondly, a multiclass deep twin support vector networks (MDTSVN) is put forward, which can handle multi classification problem directly and efficiently.

The main contributions in this paper are as follows.

1) From a model construction of view, it utilizes the neural networks and avoids the manual selection of kernel functions and parameters, which greatly enhances the capability of feature extraction and data representation.
2) From a model optimization point of view, our DTSVN could update the neural network and classification hyperplane parameters simultaneously by back propagation, which replaces the previous inefficient two-stage strategy.
3) From a model application point of view, our MDTSVN can directly handle the multi classification task rather than generating multiple classifiers by one-vs-rest or one-vs-one approach.

2 Related Work

We first briefly review the shallow TSVM in this section. Consider the binary classification problem with dataset $\mathcal{D} = \{(\boldsymbol{x}_1, y_1), \ldots, (\boldsymbol{x}_n, y_n)\}$, where $\boldsymbol{x}_i \in \mathcal{R}^m, \boldsymbol{y}_i \in \{-1, 1\}$. For such a dataset with only positive and negative classes, it can further be abbreviated as $D^{\pm} = \{\boldsymbol{x}_i \mid i \in \mathcal{I}^{\pm}\}$, where $\mathcal{I}^{\pm}$ corresponds to the sample set of label $\boldsymbol{y}_i = \pm 1$, respectively.

The objective of TSVM is to find positive and negative class hyperplanes of the following equation in $\mathcal{R}^m$:

$$\boldsymbol{w}_+^T\boldsymbol{x} + b_+ = 0, \quad \boldsymbol{w}_-^T\boldsymbol{x} + b_- = 0 \tag{1}$$

The main idea of the method is to find two hyperplanes so that they are close to one class of samples and far from the other. The optimization problem is given as follows.

$$\begin{aligned} \min_{\boldsymbol{w},b,\xi} \quad & \frac{1}{2}\sum_{i\in\mathcal{I}^+}\left(\boldsymbol{w}_+^T\boldsymbol{x}_i + b_+\right)^2 + C_1\sum_{j\in\mathcal{I}^-}\xi_j \\ \text{s.t.} \quad & -\boldsymbol{w}_+^T\boldsymbol{x}_j - b_+ \geq 1 - \xi_j \\ & \xi_j \geq 0, j \in \mathcal{I}^- \end{aligned} \tag{2}$$

$$\begin{aligned} \min_{\boldsymbol{w},b,\xi} \quad & \frac{1}{2}\sum_{j\in\mathcal{I}^-}\left(\boldsymbol{w}_-^T\boldsymbol{x}_j + b_-\right)^2 + C_2\sum_{i\in\mathcal{I}^+}\xi_i \\ \text{s.t.} \quad & \boldsymbol{w}_-^T\boldsymbol{x}_i + b_- \geq 1 - \xi_i \\ & \xi_i \geq 0, i \in \mathcal{I}^+ \end{aligned} \tag{3}$$

where $C_1, C_2 > 0$ is the penalty factor given in advance. The first term of the objective function guarantees that the positive (negative) class hyperplane can be closer to the positive (negative) class samples. The constraints restrict the function distance of the positive and negative class samples from the other class hyperplane to be at least 1.

For the test simple $\boldsymbol{x}$, its class is determined based on its distance from the two hyperplanes..

$$k = \arg\min_{y=+1,-1}\left(\frac{|\boldsymbol{w}_+^T\boldsymbol{x} + b_+|}{\|\boldsymbol{w}_+\|}, \frac{|\boldsymbol{w}_-^T\boldsymbol{x} + b_-|}{\|\boldsymbol{w}_-\|}\right) \tag{4}$$

3 Deep Twin Support Vector Networks

As mentioned in the Introduction, the shallow method often encounters some problem. The original TSVM involves the manual selection of kernel functions, solving large-scale QPPs, and matrix inverse operation. Extending the shallow TSVM to deep learning framework can avoid these problems.

3.1 DTSVN for Binary Classification

The original TSVM finds a pair of non-parallel hyperplanes. For the DTSVN, it obtains these two hyperplanes by training two neural networks.

As shown in Fig. 1, the parameters of the output layer of the neural network are replaced with the parameters of the hyperplane of TSVM. Unlike the design of siamese network [1], the input values for two neural networks are the same training data, and they do not share weights.

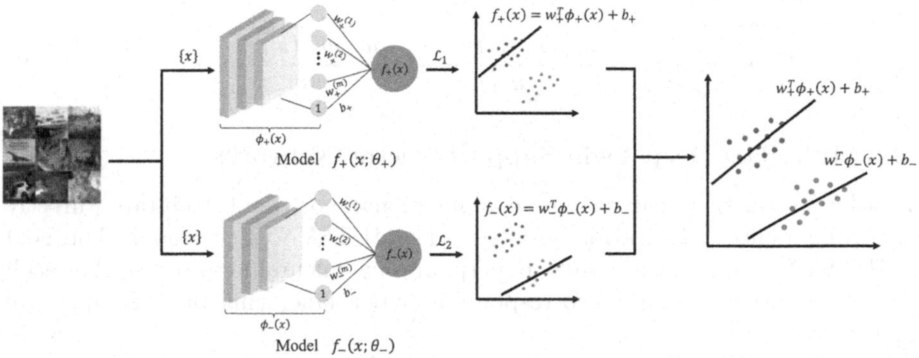

Fig. 1. An illustation of DTSVN. Two neural networks are used to construct positive and negative class hyperplanes, respectively. The hidden layer for feature extraction is denoted by $\phi_+(x)$ and $\phi_-(x)$. The output layer parameters represent hyperplane parameters.

The goal of our DTSVN is to learn the following two models (5).

$$f_+(\boldsymbol{x};\theta_+) = \boldsymbol{w}_+^T\boldsymbol{\phi}_+(\boldsymbol{x}) + b_+, \quad f_-(\boldsymbol{x};\theta_-) = \boldsymbol{w}_-^T\boldsymbol{\phi}_-(\boldsymbol{x}) + b_- \tag{5}$$

Here, we abstract the original data $\boldsymbol{x}$ as $\boldsymbol{\phi}_\pm(\boldsymbol{x})$ through two neural networks. To simplify the representation, we use $\theta_\pm(\boldsymbol{\phi}_\pm, \boldsymbol{w}_\pm, b_\pm)$ to represent all coefficients needed to be solved in DTSVN, including the weights of hidden layers and the coefficients $\boldsymbol{w}_\pm, b_\pm$ for the classification hyperplanes.

To minimize the empirical risk of the training data, the following formulation is given.

$$\min_{\theta\in\Theta}\sum_{i=1}^{n}\mathcal{L}\left(f\left(\mathbf{x}_i;\theta\right), \mathbf{y}_i\right) \tag{6}$$

where $\mathcal{L}$ refers to a certain loss function, e.g., mean squared error or cross-entropy loss. θ refers to the neural network parameters.

Based on the original TSVM which minimizes the objective function (2) and (3), our DTSVN is to minimize the following loss functions.

$$\mathcal{L}_1 = \frac{1}{2}\sum_{i\in\mathcal{I}^+} f_+(\boldsymbol{x}_i;\theta_+)^2 + C\sum_{j\in\mathcal{I}^-}\left[\max\left(1 - y_j f_+(\boldsymbol{x}_j;\theta_+), 0\right)\right] \tag{7}$$

$$\mathcal{L}_2 = \frac{1}{2}\sum_{j\in\mathcal{I}^-} f_-(\boldsymbol{x}_j;\theta_-)^2 + C\sum_{i\in\mathcal{I}^+}\left[\max\left(1 - y_i f_-(\boldsymbol{x}_i;\theta_-), 0\right)\right] \tag{8}$$

During the model $f_\pm(\boldsymbol{x};\theta_\pm)$ training process, the hidden layer $\boldsymbol{\phi}_\pm(\boldsymbol{x})$ parameter and the hyperplane parameter $\boldsymbol{w}_\pm, b_\pm$ of the output layer can be updated simultaneously, without the previous two-stage process.

The class to which the test set belongs is judged according to its distance from the two hyperplanes, with the following discriminant (9).

$$k = \arg\min_{y=+1,-1}\left(\frac{|\boldsymbol{w}_+^T\phi_+(\boldsymbol{x}) + b_+|}{\|\boldsymbol{w}_+\|}, \frac{|\boldsymbol{w}_-^T\phi_-(\boldsymbol{x}) + b_-|}{\|\boldsymbol{w}_-\|}\right) \tag{9}$$

3.2 Multiclass Deep Twin Support Vector Networks

In order to further reduce the computational cost and facilitate directly addressing multi classification problem, the MDTSVN is proposed. The goal of MDTSVN is to construct multi hyperplanes by a neural network, so that each class of samples is close to its corresponding hyperplane, while other samples are far from it.

The above DTSVN uses two different neural networks to generate two class hyperplane. Contrastively, in the MDTSVN, we just train a neural network and output multiple class hyperplanes, which could greatly improve the computational efficiency.

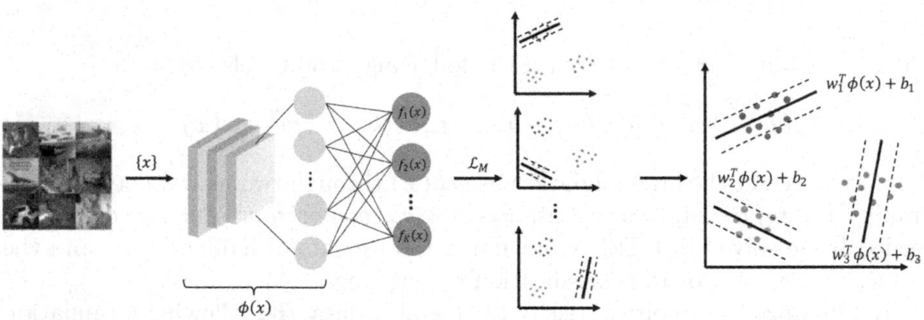

Fig. 2. An illustation of MDTSVN.

Given a dataset containing K classes $D = \{\boldsymbol{x}_i \mid i \in \mathcal{I}^k, k = 1, \cdots, K\}$, where $\mathcal{I}^k$ denotes the set of samples of the k-th class. K represents the total number of categories. The k-th neuron of the output layer corresponds to the k-th class hyperplane, which is given as follows.

$$f_k(\boldsymbol{x}, \theta_k) = \boldsymbol{w}_k^T\boldsymbol{\phi}(\boldsymbol{x}) + b_k \ , \ k = 1, \cdots, K \tag{10}$$

Here, we use $\theta_k(\boldsymbol{\phi}, \boldsymbol{w}_k, b_k)$ to represent all coefficients needed to be solved in MDTSVN, including the weights of the neural network and the coefficients $\boldsymbol{w}_k, b_k$ for the k-th class hyperplanes.

The loss function is given as

$$\mathcal{L}_M = \sum_{k=1}^{K}\left(\frac{1}{2}\sum_{i\in\mathcal{I}^k} f_k(\boldsymbol{x}, \theta_k)^2 + C\sum_{j\notin\mathcal{I}^k} \max\left(1 - |f_k(\boldsymbol{x}, \theta_k)|, 0\right)\right) \tag{11}$$

As shown in Fig. 2, the k-th output layer of MDTSVN corresponds to the k-th class hyperplane $f_k(\boldsymbol{x}, \theta_k)$. The first term of Eq. (11) ensures that each class hyperplane is as close as possible to the corresponding class sample. The second term keeps the hyperplanes as far away as possible from the samples from other classes. If the functional distance between the k-th hyperplane and the samples not in k-th class is less than 1, a penalty will be given. It is shown as the dotted line in the figure.

The desicion function of the MDTSVN is given as follows.

$$k = \arg\min_{y=1,\cdots,K}\left(\frac{|\boldsymbol{w}_1^T\phi(\boldsymbol{x}) + b_1|}{\|\boldsymbol{w}_1\|},\cdots,\frac{|\boldsymbol{w}_K^T\phi(\boldsymbol{x}) + b_K|}{\|\boldsymbol{w}_K\|}\right) \tag{12}$$

To deal with multi classification task, MDTSVN does not need to use the one-vs-one or one-vs-rest strategies. Compared with DTSVN, it just requires to train one neural network to realize the feature extraction, which greatly reduces the computational cost.

3.3 Algorithm

The pseudo code for the training process of MDTSVN is shown in Algorithm 1.

In the forward pass, the input is the original data $\boldsymbol{x}$, and the output layer neurons are each class hyperplane $f_k(\boldsymbol{x}, \theta_k)$ of MDTSVN. The training loss is obtained through Eq. (11).

In the backward pass, the network weights and parameters of the classification hyperplanes are updated simultaneously by gradient descent method to minimize the training loss Eq. (11).

The algorithm of DTSVN is similar, which is omitted here.

Algorithm 1. Training process of MDTSVN

Input: Training data $\mathcal{D} = \{(\boldsymbol{x}_n, y_n), 1 \le n \le N\}$.
Output: Hyperplane $f_k(\boldsymbol{x}, \theta_k), k = 1, \cdots, K$.

```
Random initial weights of neural networks θ_0.
Initialize hyperparameters maxepoch, maxbatch, C and learning rate.
for epoch=1:maxepoch do
    for batch=1:maxbatch do
        # Forward Pass
        f_k(x) = w_k^T φ(x) + b_k, k = 1, ··· , K
        Compute training loss: L_M(f_k(x; θ), y)
        # Backward Pass
        Compute ∇_θ L_M(f_k(x; θ), y)
        Update parameters by gradient descent with learning rate.
    end for
end for
```

3.4 Comparision with Shallow TSVM and Traditional DNN

Compared with the shallow TSVM, firstly, our proposed MDTSVN builds a neural networks to extract features, instead of manually selecting kernel functions.

Secondly, our MDTSVN is trained with back propagation in the mini-batch process, which avoids calculating large matrices or solving large QPPs at a time. It is more suitable for handling large-scale datasets. Thirdly, our MDTSVN also has a significant advantage in multi classification problem, while in shallow multi-class TSVM, the inefficient one-vs-rest or one-vs-one strategy is required.

Our MDTSVN has a similar network structure with the traditional DNN. However, the MDTSVN outputs functional distances of the samples to each classification hyperplane, which enables the hidden layers $\boldsymbol{\phi}(\boldsymbol{x})$ and output layers $\boldsymbol{f}_k(\boldsymbol{x}, \boldsymbol{\theta}_k)$ to have explicit geometric interpretation. The functional distances can be regarded as scores of how close the sample is to a certain class. It also makes data visualization easier, which is further discussed in Sect. 4.2. On the contrary, traditional DNN outputs the probability that the samples belong to each class, but has the narrow model interpretation.

4 Numerical Experiments

To demonstrate the validity of the model, we compare the performance of our DTSVN and MDTSVN with the traditional SVM, TSVM, deep SVM [3] and DNN methods on the MNIST, FASHION MNIST and CIFAR10 dataset [17,28]. The description of data sets are presented in Table 1.

Table 1. Description on three benchmark data sets

Name	Training set	Test set	Class	Dim
MNIST	60000	10000	10	$28 * 28 = 784$
FASHION MNIST	60000	10000	10	$28 * 28 = 785$
CIFAR10	50000	10000	10	$3 * 32 * 32 = 3072$

All algorithms were implemented in python on Windows operating system with an Intel i5-9400F CPU 2.90 GHz with 8.00 GB of RAM. The shallow models are implemented by sklearn packages, and deep models are implemented by pytorch [5,21].

Since the SVM, TSVM and DSVM models are binary classification models, the one-vs-rest method is employed.

In the comparison of deep learning models, the structure of the hidden layer is referenced from LeNet-5 and ResNet18. In addition, the hyperparameters are set consistently. In short, the differences between deep learning models is the output layer and loss function. In the shallow model, grid search approach is used to find the optimal parameters, and the hyperparameters are not consistent with deep model so the time consumption is not listed.

4.1 Experiments on Benchmark Datasets

The accuracy and time results for binary classification are shown in Tables 2, 3, and 4. The classification performance of DTSVN is much more outstanding,

Table 2. Accuracy and cost time (in minutes) on MNIST dataset for binary classification

	SVM	TSVM	DSVM	TIME	DNN	TIME	DTSVN	TIME	MDTSVN	TIME
0	0.9977	0.9607	**0.9992**	9.23	0.9990	8.69	**0.9992**	20.15	0.9990	10.59
1	0.9981	0.9804	0.9990	9.08	0.9990	8.96	**0.9993**	19.46	0.9990	10.56
2	0.9947	0.9490	0.9985	9.06	0.9985	8.87	0.9983	19.26	**0.9986**	10.54
3	0.9953	0.9152	0.9987	9.12	**0.9988**	8.89	0.9984	18.90	**0.9988**	10.54
4	0.9954	0.9397	**0.9991**	9.09	**0.9991**	8.93	0.9987	19.15	0.9987	10.57
5	0.9957	0.9212	0.9987	8.96	0.9984	8.98	**0.9988**	20.27	0.9987	10.60
6	0.9965	0.9687	0.9982	8.95	0.9983	8.83	0.9981	19.70	**0.9985**	10.63
7	0.9929	0.9795	**0.9981**	10.65	0.9979	8.89	**0.9981**	19.09	0.9978	10.67
8	0.9937	0.9339	0.9986	8.85	0.9982	8.92	**0.9987**	19.41	0.9985	10.67
9	0.9913	0.9253	**0.9971**	8.91	0.9970	8.83	**0.9971**	19.40	0.9969	10.72
AVE	0.9951	0.9474	**0.9985**	9.19	0.9984	8.88	**0.9985**	19.48	**0.9985**	10.61

Table 3. Accuracy and cost time (in minutes) on FASHSION MINST dataset for binary classification

	SVM	TSVM	DSVM	TIME	DNN	TIME	DTSVN	TIME	MDTSVN	TIME
0	0.9658	0.9528	0.9712	8.95	0.9732	8.82	**0.9743**	22.33	0.9724	10.67
1	0.9954	0.9852	0.9985	8.73	**0.9987**	8.89	0.9984	21.10	0.9983	10.71
2	0.9599	0.9219	0.9744	8.88	0.9748	8.64	**0.9772**	21.17	0.9763	10.62
3	0.9771	0.9500	0.9830	8.97	0.9839	8.86	**0.9853**	20.00	0.9844	10.78
4	0.9590	0.9200	0.9729	8.88	0.9722	8.74	**0.9755**	22.15	0.9721	11.08
5	0.9917	0.9443	0.9973	8.70	0.9967	8.67	**0.9979**	21.14	0.9978	10.65
6	0.9409	0.9000	0.9526	8.88	0.9523	8.65	**0.9562**	21.89	0.9521	10.70
7	0.9880	0.9638	0.9945	8.94	0.9945	8.69	**0.9952**	21.19	0.9941	10.61
8	0.9936	0.9726	0.9977	8.73	0.9974	8.77	**0.9978**	21.52	0.9977	10.57
9	0.9916	0.9743	0.9950	8.87	0.9946	8.68	**0.9956**	22.43	0.9949	10.62
AVE	0.9763	0.9485	0.9837	8.85	0.9838	8.74	**0.9853**	21.49	0.9840	10.70

especially when dealing with high-dimensional data. However, the shallow models perform less well than the deep models. In particular, on the CIFAR10 data experiments, the shallow model will incorrectly classify all samples into one class because of the unbalanced data distribution, so the experimental results are not given.

As show in Table 5, the results of DTSVN and MDTSVN compared to other methods with WIN/DRAW/LOSS are summarized. Compared with the original model, the DTSVN has gained an overall superiority in classification ability. Compared with other deep models, the outstanding performance of our DTSVN is more pronounced on the CIFAR10 dataset.

Table 4. Accuracy and cost time (in minutes) on CIFAR10 dataset for binary classification

	DSVM	TIME	DNN	TIME	DTSVN	TIME	MDTSVN	TIME
0	0.9535	11.59	0.9553	11.67	**0.9564**	37.60	0.9521	13.52
1	0.9703	11.74	0.9664	11.58	**0.9729**	41.58	0.9706	14.07
2	0.9289	11.83	0.9298	11.69	**0.9378**	41.16	0.9328	13.59
3	**0.9182**	12.38	0.9139	11.54	**0.9182**	41.29	0.9170	13.81
4	0.9363	12.01	0.9336	11.62	**0.9422**	42.15	0.9408	13.13
5	0.9354	12.10	0.9326	11.67	**0.9408**	41.50	0.9388	13.48
6	0.9578	12.80	0.9549	11.52	**0.9617**	40.56	0.9554	14.09
7	0.9573	12.23	0.9563	11.67	**0.9610**	40.72	0.9564	13.70
8	0.9678	11.71	0.9688	11.59	0.9712	41.92	**0.9716**	13.63
9	0.9618	11.58	0.9610	11.53	**0.9692**	42.44	0.9652	13.41
AVE	0.9487	12.00	0.9473	11.61	**0.9531**	41.09	0.9501	13.64

Table 5. WIN/DRAW/LOSS results of DTSVN and MDTSVN for binary classification

	DTSVN				MDTSVN			
	SVM	TSVM	DSVM	DNN	SVM	TSVM	DSVM	DNN
MNIST	11/0/0	11/0/0	3/3/5	7/0/4	11/0/0	11/0/0	3/2/6	5/3/3
FASHION MNIST	11/0/0	11/0/0	10/0/1	10/0/1	11/0/0	11/0/0	5/1/5	6/0/5
CIFAR10	11/0/0	11/0/0	10/1/0	11/0/0	11/0/0	11/0/0	7/0/4	10/0/1

The experiments are also conducted for directly handling multi classification task. The results are shown in Table 6. It is demonstrated that our MDTSVN achieves the best prediction performance.

From these results above, we can conclude that the proposed DTSVN has the significant prediction advantages in binary classification accuracy. Besides, the proposed MDTSVN is superior than other models for multi classification in prediction accuracy with a comparable computational time.

4.2 Discussion on Parameter C

In the TSVM-type models, parameter C is mainly used to control the weight on slack variable ξ. In the constraint term, the role of ξ is to give some penalty to the negative (positive) class samples whose function distance from the positive (negative) class hyperplane is less than 1. Obviously, when C takes a larger value, the model changes the position of the hyperplane so that it is as far away from the other class of sample points as possible. When C takes a small value, the hyperplane is close to the sample points of this class.

We conduct the experiments on a synthetic dataset to demonstrate this properties. The data follow two-dimensional gaussian distributions. As show in

Table 6. Accuracy and cost time (in minutes) for multi-classification on three benchmark datasets

	DNN	TIME	MDTSVN	TIME
MNIST	0.9922	6.93	**0.9934**	13.71
FASHION MNIST	0.9153	6.85	**0.9175**	13.31
CIFAR10	0.8750	14.95	**0.8850**	17.76
AVE	0.9275	9.58	**0.9320**	14.93

Fig. 3, the red and blue points are the positive and negative classes, respectively. $X_+ \sim N(\mu_1, \Sigma_1), X_- \sim N(\mu_1, \Sigma_1)$, where $\mu_1 = [0.5, -3]^T, \mu_2 = [-0.5, 3]^T$, $\Sigma_1 = \Sigma_2 = \begin{pmatrix} 0.2 & 0 \\ 0 & 3 \end{pmatrix}$ and $n = 200$.

It can be seen that the distance between the positive and negative class hyperplanes increases with C, which also affects the location of the decision hyperplane.

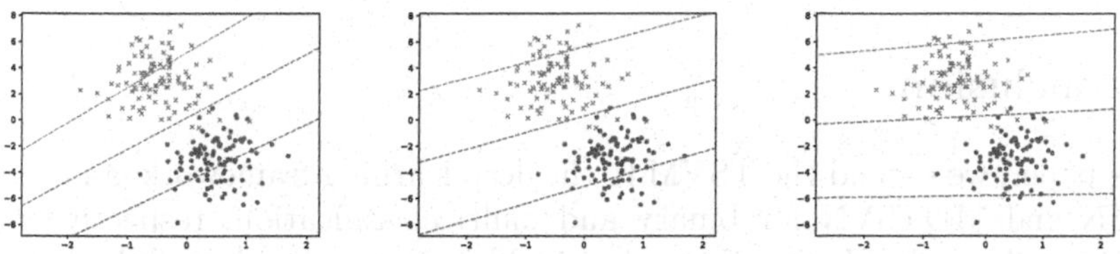

(a) C=1 (b) C=10 (c) C=100

Fig. 3. An illustation of the original TSVM on the synthetic dataset at $C = 1, 10, 100$, respectively. The decision hyperplane (green line) is defined as the line in the middle of the positive and negative hyperplanes. (Color figure online)

The experiment of MDTSVN on this two-dimensional dataset is also conducted. The last layer of the hidden layer is set to two perceptrons in order to show $\phi(\boldsymbol{x})$. The output layer of the neural network also has two perceptrons, corresponding to positive and negative class hyperplanes. The results are plotted in the Fig. 4. When the value of C is small, the hidden layer of the neural network will map the samples to the positive and negative class hyperplanes as much as possible. Anomalies or noise will have less effect on the location of the hyperplane. When C takes gradually increasing values, it gains more penalties on a class of sample points which is too close to another class of hyperplane. The hyperplane is then gradually pushed away from the sample points. It demonstrates that our MDTSVN has the similar property with the shallow TSVM.

Unlike the original TSVM, due to the strong feature extraction ability of the neural network, the samples mapped as $\phi(\boldsymbol{x})$ are located roughly

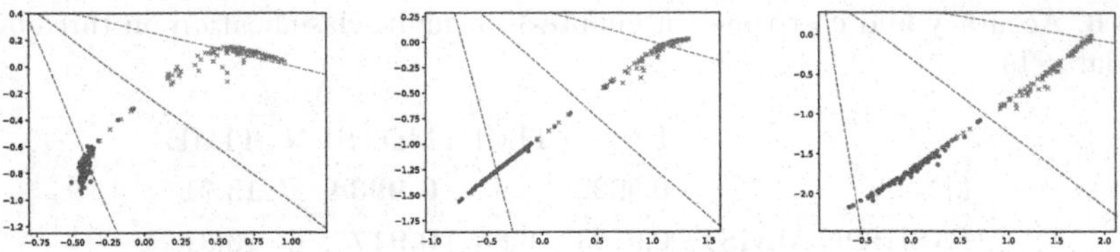

(a) C=0.1 (b) C=1 (c) C=10

Fig. 4. An illustration of MDTSVN on the synthetic dataset at $C = 0.1, 1, 10$, respectively.

perpendicular to the decision hyperplane, which is more benefit to classify the samples according to their distance from the hyperplane.

As can be seen, MDTSVN not only inherits some of the properties of TSVM, the feature extraction capability of the neural network better helps the model to find the hyperplane. In practical applications, C can be chosen with reference to the number of noise and anomalies in the data set, and the value of C can usually be found using a grid search method.

5 Conclusion

In this paper, we extend the TSVM to the deep learning framework, and propose DTSVN and MDTSVN for binary and multi classification, respectively. The experimental results demonstrate the significant superiorities of the proposed DTSVN and MDTSVN, compared with other shallow and deep models. Our proposed methods could give a guidance for extending other shallow SVM-type models to deep learning framework. In addition, the cross-entropy loss function is used in many deep models, and the loss function of MDTSVN provides a new option for these models.

Acknowledgments. This work was supported in part by Graduate innovation fund project of Yunnan university of finance and economics (No. 2021YUFEYC081), Scientific research fund project of Yunnan provincial department of education (No. 2022Y546), Scientific research fund project of Yunnan provincial department of science and technology (No. 202001AU070064) and National Natural Science Foundation of China (No. 62006206).

References

1. Chopra, S., Hadsell, R., LeCun, Y.: Learning a similarity metric discriminatively, with application to face verification. In: 2005 IEEE Computer Society Conference on Computer Vision and Pattern Recognition (CVPR 2005), vol. 1, pp. 539–546. IEEE (2005)
2. Cortes, C., Vapnik, V.: Support-vector networks. Mach. Learn. **20**(3), 273–297 (1995)

3. Diaz-Vico, D., Prada, J., Omari, A., Dorronsoro, J.: Deep support vector neural networks. Integrated Computer-Aided Engineering (Preprint), pp. 1–14 (2020)
4. Díaz-Vico, D., Prada, J., Omari, A., Dorronsoro, J.R.: Deep support vector classification and regression. In: Ferrández Vicente, J., Álvarez-Sánchez, J., de la Paz López, F., Toledo Moreo, J., Adeli, H. (eds.) IWINAC 2019. LNCS, vol. 11487, pp. 33–43. Springer, Cham (2019). https://doi.org/10.1007/978-3-030-19651-6_4
5. Feurer, M., Klein, A., Eggensperger, K., Springenberg, J., Blum, M., Hutter, F.: Efficient and robust automated machine learning. Adv. Neural Inform. Process. Syst. **28** (2015)
6. Ghafoori, Z., Leckie, C.: Deep multi-sphere support vector data description. In: Proceedings of the 2020 SIAM International Conference on Data Mining, pp. 109–117. SIAM (2020). https://doi.org/10.1137/1.9781611976236.13
7. Gönen, M., Alpaydın, E.: Multiple kernel learning algorithms. J. Mach. Learn. Res. **12**, 2211–2268 (2011)
8. Goodfellow, I., et al.: Generative adversarial nets. Adv. Neural Inform. Process. Syst. **27** (2014)
9. Hao, P.Y.: New support vector algorithms with parametric insensitive/margin model. Neural Networks **23**(1), 60–73 (2010). https://doi.org/10.1016/j.neunet.2009.08.001
10. Hastie, T., Tibshirani, R.: Classification by pairwise coupling. Adv. Neural Inform. Process. Syst. **10** (1997)
11. He, K., Zhang, X., Ren, S., Sun, J.: Deep residual learning for image recognition. In: Proceedings of the IEEE Conference on Computer Vision and Pattern Recognition, pp. 770–778 (2016)
12. Hearst, M.A., Dumais, S.T., Osuna, E., Platt, J., Scholkopf, B.: Support vector machines. IEEE Intell. Syst. Appl. **13**(4), 18–28 (1998). https://doi.org/10.1109/5254.708428
13. Hsu, C.W., Lin, C.J.: A comparison of methods for multiclass support vector machines. IEEE Trans. Neural Networks **13**(2), 415–425 (2002). https://doi.org/10.1109/72.991427
14. Hussain, M., Wajid, S.K., Elzaart, A., Berbar, M.: A comparison of SVM kernel functions for breast cancer detection. In: 2011 Eighth International Conference Computer Graphics, Imaging and Visualization, pp. 145–150. IEEE (2011). (10/fc9b7v)
15. Khemchandani, R., Chandra, S.: Twin support vector machines for pattern classification. IEEE Trans. Pattern Anal. Mach. Intell. **29**(5), 905–910 (2007). (10/bkq688)
16. Krizhevsky, A., Sutskever, I., Hinton, G.E.: Imagenet classification with deep convolutional neural networks. Adv. Neural Inform. Process. Syst. **25** (2012)
17. LeCun, Y.: The MNIST database of handwritten digits (1998).http://yann.lecun.com/exdb/mnist/
18. Li, D., Tian, Y., Xu, H.: Deep twin support vector machine. In: 2014 IEEE International Conference on Data Mining Workshop, pp. 65–73. IEEE (2014). (10/gpfwfh)
19. Li, Y., Zhang, T.: Deep neural mapping support vector machines. Neural Networks **93**, 185–194 (2017). (10/gbspkm)
20. Liu, W., Wen, Y., Yu, Z., Yang, M.: Large-margin softmax loss for convolutional neural networks. In: ICML, vol. 2, p. 7 (2016)
21. Paszke, A., et al.: Pytorch: an imperative style, high-performance deep learning library. Adv. Neural Inform. Process. Syst. **32** (2019)
22. Ruff, L., et al.: Deep one-class classification. In: International Conference on Machine Learning, pp. 4393–4402. PMLR (2018)

23. Schölkopf, B., Smola, A.J., Williamson, R.C., Bartlett, P.L.: New support vector algorithms. Neural Comput. **12**(5), 1207–1245 (2000). https://doi.org/10.1162/089976600300015565
24. Simonyan, K., Zisserman, A.: Very deep convolutional networks for large-scale image recognition. arXiv preprint arXiv:1409.1556 (2014)
25. Tang, Y.: Deep learning using support vector machines. CoRR, abs/1306.0239 2, 1 (2013)
26. Tax, D.M., Duin, R.P.: Support vector data description. Mach. Learn. **54**(1), 45–66 (2004). https://doi.org/10.1023/B:MACH.0000008084.60811.49
27. Wiering, M., Schomaker, L.R.: Multi-Layer Support Vector Machines. Chapman & Hall/CRC Press (2014)
28. Xiao, H., Rasul, K., Vollgraf, R.: Fashion-mnist: a novel image dataset for benchmarking machine learning algorithms. arXiv preprint arXiv:1708.07747 (2017)
29. Zareapoor, M., Shamsolmoali, P., Jain, D.K., Wang, H., Yang, J.: Kernelized support vector machine with deep learning: an efficient approach for extreme multiclass dataset. Pattern Recogn. Lett. **115**, 4–13 (2018). https://doi.org/10.1016/j.patrec.2017.09.018
30. Zhang, S.X., Liu, C., Yao, K., Gong, Y.: Deep neural support vector machines for speech recognition. In: 2015 IEEE International Conference on Acoustics, Speech and Signal Processing (ICASSP), pp. 4275–4279. IEEE (2015). https://doi.org/10.1109/ICASSP.2015.7178777
31. Zou, Y., et al.: MK-FSVM-SVDD: a multiple kernel-based fuzzy SVM model for predicting DNA-binding proteins via support vector data description. Current Bioinform **16**(2), 274–283 (2021). https://doi.org/10.2174/1574893615999200607173829

Region-Based Dense Adversarial Generation for Medical Image Segmentation

Ao Shen, Liang Sun, Mengting Xu, and Daoqiang Zhang(✉)

MIIT Key Laboratory of Pattern Analysis and Machine Intelligence, College of Computer Science and Technology, Nanjing University of Aeronautics and Astronautics, Nanjing 211106, China
dqzhang@nuaa.edu.cn

Abstract. Deep neural networks (DNNs) have achieved great success in medical image segmentation. However, the DNNs are generally deceived by the adversarial examples, making robustness a key factor of DNNs when applied in the field of medical research. In this paper, in order to evaluate the robustness of medical image segmentation networks, we propose a novel Region-based Dense Adversary Generation (RDAG) method to generate adversarial examples. Specifically, our method attacks the DNNs on both pixel-level and region-of-interesting (ROI) level. The pixel-level attack makes DNNs mistakenly segment each individual pixel. Meanwhile, the ROI-level attack will generate perturbation based on region information. We evaluate our proposed method for medical image segmentation on DRIVE and CELL datasets. The experimental results show that our proposed method achieves effective attack results on both datasets for medical image segmentation when compared with several state-of-the-art methods.

Keywords: Adversarial examples · Medical image segmentation · Region-based dense adversary generation

1 Introduction

Deep learning methods have achieved great performance in many medical tasks, such as medical image segmentation [18], lesion detection [5], image reconstruction [1], and disease diagnosis [9], etc. To ease the high costs of labor expenses, some ongoing theories are discussing the possibility of replacing manual operations with automated deep learning models. However, in most medical image segmentation tasks, only a limited amount of training data can be used to train the over-parameterized deep learning models, which leads to the memorization of the training data and the low predictability of unseen data. Also, noisy data

This work was supported by the National Natural Science Foundation of China under Grants 62136004, and 62006115.

L. Fang et al. (Eds.): CICAI 2022, LNAI 13606, pp. 107–118, 2022.
https://doi.org/10.1007/978-3-031-20503-3_9

acquired by medical equipment can make deep learning models produce incomprehensible mistakes. Besides the forementioned problems, szegedy et al. [19] find that small perturbations can cause significant performance decline of deep learning methods, and the lack of interpretability of deep learning models [16] [10] makes it extremely difficult to find a valid defense method. Previous studies have shown that these artificial imperceptible perturbations i.e., adversarial examples, can trick deep learning models applied in medical research and clinic operations [15,17], causing irresistible instability.

To avoid potential risks caused by deep learning models, it is highly required to assess the stability of deep learning models [3]. Robustness is a key factor to evaluate this aspect, especially in the field of medical image segmentation. Recently, many adversarial examples generation strategies are proposed for evaluating the robustness of deep learning models for classification tasks. For example, the fast gradient sign method (FGSM) [6] and projected gradient descent (PGD) [12] are two widely used methods to generate adversarial examples. However, FGSM only applies a one-step pixel-level update along with the sign of the gradient, which has limited influence on non-linear models (e.g., deep neural networks). PGD uses iterative steps to perturb the input images, but it still has a rather poor performance on image segmentation tasks. Liu et al. [11] suggest global perturbation might lead to specific types of segmentation errors by employing FGSM method, but fails to investigate more sophisticated adversaries and their impacts on deep learning models. It is highly desired to study more effective adversarial attack methods for evaluating the robustness of the medical image segmentation model and enhancing adversarial defense (Fig. 1).

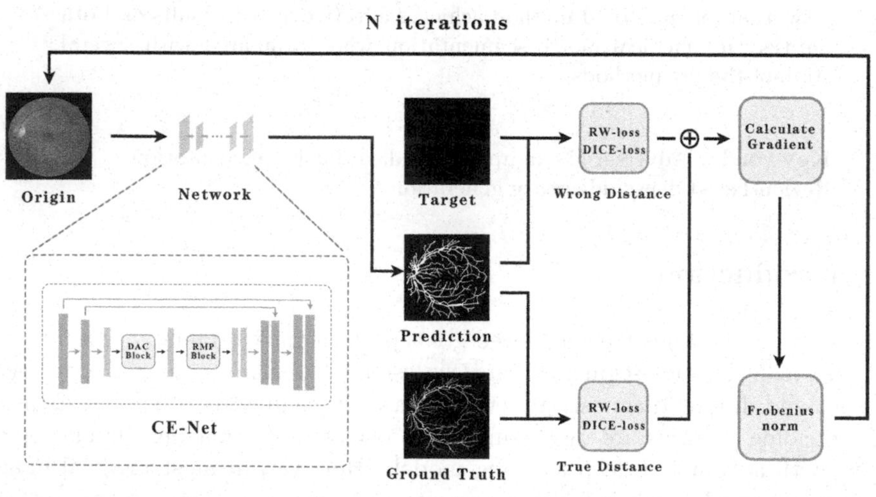

Fig. 1. Overview of the proposed pipeline for region-based dense adversary generation method. Specifically, we use CE-Net to generate prediction segmentation results from the original image. We calculate the region-wise loss and Dice loss between the prediction, the target adversarial image, and the ground truth. Then we calculate gradient of the overall losses, we use the Frobenis norm to restrict the perturbation and add it back to the original image.

In this paper, we focus on generating adversarial examples for medical image segmentation. Different from classification tasks, generating adversarial examples for deceiving medical image segmentation models requires resembling per-pixel. To this end, we propose a Region-based Dense Adversary Generation (RDAG) to generate adversarial examples. Specifically, inspired by Xie et al. [22] and Cheng et al. [2], we employ a widely used pixel-level loss function to add perturbations through iterative gradient back-propagation to make networks mistakenly segment pixels on images. Meanwhile, to simultaneously mistakenly pixels in the segmented mask, we also employ an ROI-level loss function to cheat the network by the adversarial example, i.e., region wise loss [21]. We compare our method with state-of-the-art adversarial example generation methods (i.e., FGSM and PGD) on a widely used network (i.e., CE-Net [7]) on DRIVE and CELL datasets for medical image segmentation. The main contributions of this work are summarized as follows,

1) We propose a region-based dense adversary generation method by combining region-wise loss and pixel-wise loss to generate sophisticated adversarial examples to perturb the image in both pixel-level and ROI-level.
2) We use our methods to attack two widely used neural networks for medical image segmentation tasks on two datasets, we compare our method to the state-of-the-art adversarial generating method, and the experimental results prove our RDAG method is much more effective and less perceptible in medical image segmentation tasks.
3) We use adversarial examples to augment the original datasets and the results show that the deep learning models trained with our adversarial examples have better performance in both accuracy and robustness, when compared with models trained by the clean data.

2 Methods

2.1 Adversarial Examples

Szegedy et al. [19] discovered the blind spots in neural networks and its intriguing weakness in the context of image classification tasks. They show that despite deep neural networks achieve remarkable performance on many high precision computations, these models still present a surprisingly low stability in the face of small perturbations which are maliciously designed to trick models (i.e., adversarial examples). Futhermore, these generated perturbations present a rather high transferability. Perturbations generated by the pretrain model are capable of distorting the prediction of other deep learning models. Based on Szegedy's research, Goodfellow et al. [6] proposed Fast Gradient Sign Method (FGSM) to generate adversarial examples aiming the task of image classification. FGSM calculates the gradient through the loss between pixels of ground truth and predicition image, and moves a single step based on the sign of the gradient by adding the gradient to the original image. Madry et al. [12] extended FGSM by introducing an iterative algorithm to calculate the total perturbations named as

Projected Gradient Descend (PGD). The mathematical notation of adversarial examples generation can be described as follows:

In training stage, adversarial examples aim to add a small perturbation to the input image to make DNNs produce a wrong prediction. We denote a deep neural network classifier as $F(X, \theta)$, where X is the input image, and θ is the trainable parameters. For giving a ground truth Y of the input image, we can generate an adversarial example $X^{'} = X + r$ where $F(X, \theta) = Y$,$F\left(X^{'}, \theta\right) = Y^{'}$,$Y^{'} \neq Y$. By constraining the perturbation r to a minimum value, we can make the perturbation small enough to be imperceptible to humans. In the attack stage, we use gradient descent method to adjust the weights of DNNs to minimize the loss between ground truth and prediction.

2.2 Target Adversary Generation

The aim of adversarial attack for medical image segmentation is to make the trained segmentation model assign a wrong label to each pixel, while ensuring that the generated image is visually imperceptible. To make the adversary example generation method is more suitable to the segmentation tasks, we replace the cross-entropy loss function with Dice coefficient loss function which is commonly used in medical image segmentation task. However, the pixel-level Dice coefficient loss only measure the overlap ratio. Hence, it cannot reflect the region information of these pixels. For example, pixels near the edge of segmentation mask are more ambiguous to distinguish. Hence, background pixels near the edge of mask may increase the probability of being segmented as the mask pixels after applying the pixel-level gradient sign method. Meanwhile, the pixel-level attack methods move the pixels a step towards the sign of the gradient, the step size of each pixel shares the same value. These pixel-level stepping attack methods are easily alleviated by an encoder-decoder based deep neural network for medical image segmentation. Thereby the pixel-level loss makes sub-optimal results in generating adversary examples to disturb the deep neural networks for medical image segmentation.

Hence, dense adversarial generation (DAG) method [22] propose a dense adversary generation (DAG) method to make networks produce wrong segmented labels based on target and non-target generation method. DAG minimizes the loss between the prediction score of each pixel with the ground truth $\mathcal{L} = \{l_0, \cdots, l_c\}$ and prediction score of each pixel with the incorrect label set $\mathcal{L}^{'} = \left\{l_0^{'}, \cdots, l_c^{'}\right\}$, where C is the number of classes. Hence, the loss function $L_p\left(X, \mathcal{T}, \mathcal{L}, \mathcal{L}^{'}\right)$ is defined as follows,

$$L_p = \sum_{c=1}^{C} \left[f_{l_c}(X, t_c) - f_{l_c^{'}}(X, t_c)\right] \tag{1}$$

where $\mathcal{T} = \{t_1, \cdots, t_c\}$ denotes the targets of the input image X, and $f_{l_c}(X, t_c)$ is the prediction score of the target t_n which is classified as the true label l_n.

Meanwhile, an adversarial label $l_c^{'} \in \mathcal{L}^{'}$ is randomly selected from the incorrect label set. $f_{l_c^{'}}(X, t_c)$ is the prediction score of the target t_c which is classified as the wrong label $l_c^{'}$. By minimizing the loss L_p, all pixels within or outside the mask are equally encouraged to be segmented follow the pattern of target images.

2.3 Region-Based Dense Adversary Generation

The segmentation task aims to acquire a precise ROI. To further boost the adversarial example generation ability, and solve the problem of lacking pattern distribution. We leverage both region-wise (RW) loss and pixel-wise loss to learn information of the input image in region-level and pixel-level for adversary example generation, denoted as Region-based Dense Adversarial Generation (RDAG). In the ROI-level, we use the RW loss function to capture region-wise information. RW loss is proposed by Valverde et al. [21] to jointly consider class imbalance and pixel importance. We modify the original RW loss function to make it more adaptable to our region-based dense adversary generation method. RW loss is defined as follows,

$$L_{RW} = \frac{1}{C}\sum_{c=1}^{C}[X \cdot rrwmap] \tag{2}$$

$$rrwmap = \begin{cases} \frac{-\|i-b_{ic}\|_2}{max_{j \in \Omega_c}\|j-b_{ic}\|_2} & \text{if } i \in \Omega_c, \\ 1 & \text{otherwise.} \end{cases} \tag{3}$$

here C is the number of the class and $rrwmap$ is a label map generated from the input image. It produces a matrix with the same shape as the input image, where each element stands for the distance between its corresponding pixel i of the input image and the closet boundary pixel b_{ic} of the region area Ω_c of class c. To ensure optimization stability, $rrwmap$ is rectified by enforcing every element outside the region to be equal for each class. In our case, we set these pixels to be 1. By employing this loss function we take position information into consideration. When we calculating the gradient of each pixel through backpropagation, pixels in the center of mask have larger gradient than pixels in the edge of mask and are more severely perturbed after stepping towards the sign of the gradient. In the pixel-level we use Dice coefficient loss [4].

$$L_{Dice} = \frac{1}{C}\sum_{c=1}^{C}\left[1 - \frac{\left(2\sum_n^N p_{cn}g_{cn}\right)}{\sum_n^N p_{cn}^2 + \sum_n^N g_{cn}^2}\right] \tag{4}$$

Hence, we minimize the Dice coefficient loss and RW loss between the ground truth and the automatic segmentation mask, while maximizing the values between the adversarial labels and the automatic segmentation result. At the m_{th} iteration, the current image (after m-1 times iterations) is denoted as X_m. We calculate the gradient concerning the current image r_m as follows

$$\frac{1}{C}\sum_{c=1}^{C}\left[\nabla X_m[L(G)_{Dice} + L(G)_{RW}] - \nabla X_m[L(G^{'})_{Dice} + L(G^{'})_{RW}]\right] \quad (5)$$

where G is the ground truth and $G^{'}$ is the target image (i.e., a blank label) given by the user. We also apply Frobenius norm on r_m to ensure numerical stability [22]: $r_m^{'} = \frac{\gamma \cdot r_m}{\|r_m\|_F}$ where γ is a hyper-parameter to determine the weight of perturbation in one iteration which is usually set to 1. It is then added to the input image $X_{m+1} = X_m + r_m^{'}$ and proceed to the next iteration. Notice that we clamp the value of each pixel of input to an interval of $[0, 1]$ (the image is normalized from $[0, 255]$ to $[0, 1]$) after every iteration to avoid overflow. After n_{th} iteration, the final adversarial perturbation is obtained as $r = \sum_m r_m^{'}$.

3 RESULT

3.1 Materials and Configurations

We use a widely used network (i.e., CE-Net) in the medical image segmentation task to evaluate our proposed RDAG method. The models were trained with the soft Dice loss, and model optimization is performed with ADAM optimizer with the initial learning rate of 2e–4 and batch size of 8. During the training stage, we apply a dynamic learning rate descent where the learning rate is reduced by half when training loss stops dropping for 10 continuous epochs. The maximum epoch is 300. During attacks, the iteration of PGD, DAG and RDAG are 20.

Two datasets are used in our study. DRIVE [14] is established for comparative studies on the segmentation of blood vessels in retina images. The dataset was originally split into a training set (20 scans) and a testing set (20 scans) and was manually labeled by the trained independent human observer. CELL [20] is a part of NuCLS dataset which contains 30 cell images with labeled nuclei from breast cancer images from TCGA. We divided this dataset into a training set (20 scans) and a testing set (20 scans). Both datasets were resized into 448 × 448 and are preprocessed by random rotation, flipping, shifting, and saturation.

3.2 Evaluation Metrics

To evaluate the performance of each attacking method, we use IOU to measure the similarity of prediction generated by networks from adversarial examples since our method generates perturbations partly based on Dice coefficient and would naturally increase the attack success rate when evaluated by it. Meanwhile, the Structural Similarity Index (SSIM), Root Mean Squared Error (RMSE) and Peak Signal-to-Noise Ratio (PSNR) [8] are used to quantify the total values of adversarial perturbation.

Following [13,19], the perceptibility of the adversarial perturbation r for an image contains K pixels is defined as $p = \left(\frac{1}{K}\sum_{k=1}^{K} \| r_k \|_2^2\right)^{\frac{1}{2}}$, where r_k is the intensity vector (3-dimensional in the RGB color space, $k = 1, 2, 3$) normalized in [0, 1]. Hence, we can not only control the exact perceptibility value of each perturbation but also manage to restrict the perceptibility value among attacking methods in the same order of magnitude over the entire test set. The perceptibility values of each pixel are set to be 1.4e–2 and 2.5e–2 on the DRIVE and CELL datasets, respectively. These values must be set relatively small, which guarantees the imperceptibility of the generated adversarial perturbation.

Table 1. The IOU achieved by CE-Net on the DRIVE (first line) and CELL (second line) datasets attacked by adversarial example generation methods. SSIM, RMSE, PSNR are used to evaluate the similarity of perturbed images and original images.

Data	Attack type	IOU↓	SSIM↑	RMSE↓	PSNR↑
DRIVE	No Attack	0.7675	–	–	–
	FGSM	0.6774	0.9982	0.0015	56.547
	PGD	0.1804	0.8910	0.0129	37.805
	DAG(only pixel)	0.2411	0.9483	0.0088	41.119
	RDAG (pixel+region)	**0.0103**	**0.9673**	**0.0073**	**42.809**
CELL	No Attack	0.7736	–	–	–
	FGSM	0.7372	0.9998	0.0015	56.516
	PGD	0.3681	0.9701	0.0217	33.247
	DAG(only pixel)	0.2006	0.9903	0.0144	36.827
	RDAG (pixel+region)	**0.0945**	**0.9896**	**0.0146**	**36.748**

3.3 Results on DRIVE and CELL Datasets

We perform the segmentation on the DRIVE and CELL datasets attacked by our RDAG and the competing methods. The segmentation results achieved by CE-Net with the attacked data by different methods are shown in Table 1.

It can be observed in Table 1 that CE-Net achieves a high IOU ratio with clean images on DRIVE and CELL datasets for segmentation tasks (0.7675 and 0.7736), respectively. By using CE-Net for image segmentation on DRIVE and CELL datasets, our RDAG method achieves state-of-the-art attack performance. For example, The IOU ratio achieved by CE-Net on DRIVE and CELL datasets with our RDAG are 0.0103 and 0.0945, which is much better than FGSM (0.6774 and 0.7372), PGD (0.1804 and 0.3681) and DAG (0.2411 and 0.2006). These results suggest that using our proposed RDAG method with pixel-level and ROI-level can achieve attack performance for medical image segmentation tasks on DRIVE and CELL datasets when compared with several state-of-the-art methods.

The degree of adversarial perturbation is a key factor for the segmentation performance. To demonstrate the effectiveness of our proposed methods, we further use SSIM, RMSE, and PSNR to evaluate the perturbation. As is shown in

Table 1, PSNR (42.809 and 36.748) and SSIM (0.9673 and 0.9896) suggest that RDAG produces a measurable difference between the original image and perturbed image, while RMSE (0.0073 and 0.0146) are low. RDAG can perturb the image with less perturbation than FGSM, which only applies a single step of gradient descent. It also achieves better results than PGD and original DAG where iterative perturbation strategies are used for adversarial examples generation. The possible reason is that our RDAG methods leverage an additional ROI-level loss for adversarial example generation. Figure 2 shows the original images and attacked images by different methods. Although the image attacked by our RDAG method is imperceptible, the CE-Net achieves very poor segmentation results on DRIVE dataset.

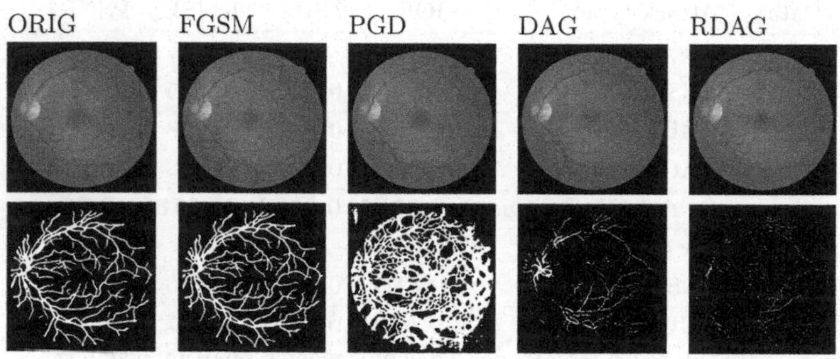

Fig. 2. Predictions generated by CE-Net for segmentation under different types of attack on DRIVE dataset. The first line is the original image and the adversarial images. The second line is the predictions of each adversarial image. Notice that all adversarial perturbations are imperceptible to human observation.

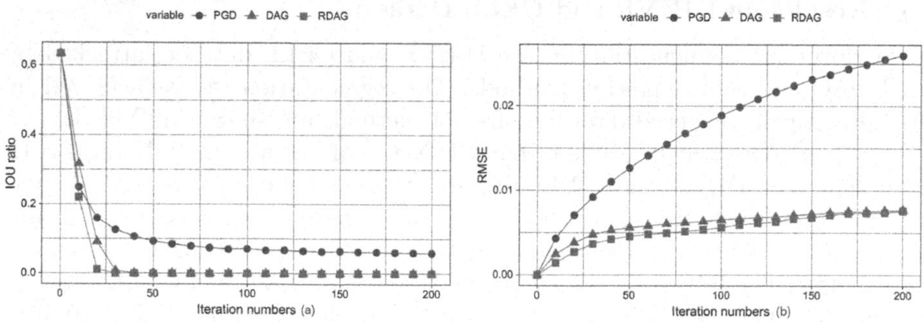

Fig. 3. The convergence of different adversarial generation methods measured by the IOU ratio with respect to the number of iterations (a). And the curve of Root Mean Squared Error (RMSE) respecting to the similarity of perturbed image and original image (b). The experiment is conducted on DRIVE dataset using a pretrain CE-Net model as prediction model.

We further study the convergence of RDAG. We investigate the optimal iteration for generating a desired adversarial examples. We use IOU and RMSE to quantify the attack performance and degree of perturbation, respectively. The results are shown in Fig. 3. As shown in Fig. 3(a), the average numbers of convergence iterations are 110.23, 40.28, and 37.02 for PGD, DAG, and RDAG, respectively. These results suggest that using both pixel-level and ROI-level loss can result in faster convergence in medical image segmentation tasks on DRIVE dataset with CE-Net. Figure 3(b) shows that DAG and RDAG achieve stable values in terms of RMSE when the iteration number rises, while the curve of PGD rises sharply along with the iteration number. This suggests that RDAG is capable of a severely interfering network while restringing the perturbation added to the input images to a low degree. It is worth noting that the adversarial examples generated by PGD are visibly different from the original image after 200 iterations, while differences generated by RDAG remain imperceptible.

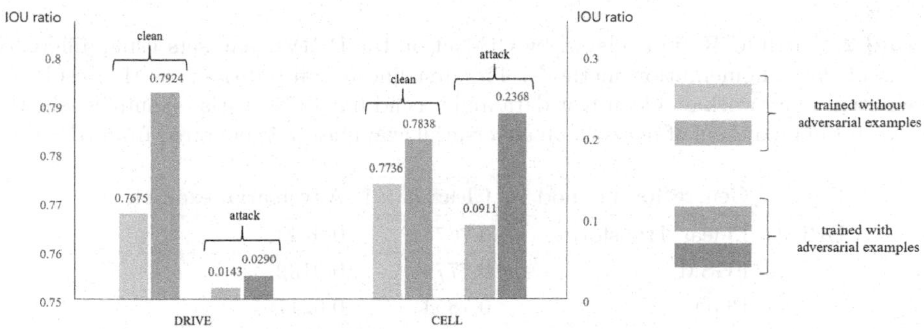

Fig. 4. Experiments of CE-Net on DRIVE and CELL datasets. We compare the performance of models when trained with and without adversarial examples. Both accuracy (yellow bar, the segmentation results of clean data) and robustness (blue bar, the segmentation results when facing adversarial attacks) increase when trained with adversarial examples (Color figure online)

3.4 The Adversarial Examples for Data Augmentation

It is well known that deep learning has a huge demand for training samples, but the medical image datasets generally have a limited number. One important possible application of the adversarial examples is to increase the number of the training sample. Hence, we study the influence of adversarial examples on the medical image segmentation. Specifically, we use the adversarial examples generated by our RDAG method to help train the original deep neural networks (CE-Net) by adding these adversarial examples into the training set. The results are shown in Fig. 4. The DRIVE and CELL train datasets are newly-added 20 perturbed images respectively. The masks of perturbed images are duplicates of masks of original images. As shown in the Fig. 4., the segmentation results of DRIVE and CELL datasets yield 0.0249 improvements and 0.0102 improvements, respectively. These results suggest that the generated adversarial examples can improve the segmentation performance of CE-Net on DRIVE and CELL datasets. The robustness of deep neural networks benefits from our adversarial examples. We train CE-Net with the clean data and the adversarial examples generated by using different methods. Then, we test the adversarial examples generated from RDAG. The results are reported in Table 2. As shown in Table 2, our RDAG achieved the best segmentation results on both clean data and adversarial example. For example, our CE-Net achieve 0.7924 in terms of IOU by using the adversarial example generated from our RDAG for clean data segmentation, which is better than the adversarial example generated from linear transform (0.7675), FGSM (0.7779), PGD (0.7854), and DAG (0.7839). Meanwhile, our RDAG methods also achieve the best results on adversarial example segmentation. These results prove the possibility of using our adversarial examples to assist in training deep neural networks and increase their accuracy and robustness.

Table 2. The IOU Ratio achieved by CE-Net on the DRIVE datasets using different kinds of data augmentation methods. The first line (clean data) shows the result of prediction of network on clean test data and second line (adversarial examples) shows the result of prediction of network on adversarial examples (all generated from RDAG).

Data	Generation method	Clean data↑	Adversarial examples↑
DRIVE	Linear Transform	0.7675	0.0143
	FGSM	0.7779	0.0132
	PGD	0.7854	0.0217
	DAG(only pixel)	0.7839	0.0282
	RDAG (pixel+region)	**0.7924**	**0.0290**

4 Conclusion

In this paper, we focus on the adversarial example in the medical image segmentation tasks. We propose a Region-based Dense Adversary Generation (RDAG) method for attacking medical images. Specifically, the RDAG method optimizes the loss function by increasing the distance between prediction and ground truth while decreasing the distance between prediction and target image on both pixel-level and ROI-level. Experimental results suggest that our proposed RDAG method can achieve superior results than the traditional attacking method and the original DAG method while maintaining a relatively low perceptibility of the adversarial perturbation. It's worth noticing that, although we do not exploit a practicable method to defend this attack, our work still suggests that this proposed method can produce more diverse examples with less perturbation, which is important in the field of medical research where massive data is usually scarce. Our work proves the feasibility of using these adversarial examples to augment training samples and help increase the accuracy and robustness of DNNs, which may also help promote the study of investigating more effective adversarial defense.

References

1. Chen, J., Qian, L., Urakov, T., Gu, W., Liang, L.: Adversarial robustness study of convolutional neural network for lumbar disk shape reconstruction from MR images. In: Medical Imaging 2021: Image Processing, vol. 11596, p. 1159615. International Society for Optics and Photonics (2021)
2. Cheng, B., Schwing, A., Kirillov, A.: Per-pixel classification is not all you need for semantic segmentation. Adv. Neural Inform. Process. Syst. **34** (2021)
3. Daza, L., Pérez, J.C., Arbeláez, P.: Towards robust general medical image segmentation. In: de Bruijne, M., et al. (eds.) MICCAI 2021. LNCS, vol. 12903, pp. 3–13. Springer, Cham (2021). https://doi.org/10.1007/978-3-030-87199-4_1
4. Dice, L.R.: Measures of the amount of ecologic association between species. Ecology **26**(3), 297–302 (1945)
5. Drukker, K., Giger, M.L., Horsch, K., Kupinski, M.A., Vyborny, C.J., Mendelson, E.B.: Computerized lesion detection on breast ultrasound. Med. Phys. **29**(7), 1438–1446 (2002)
6. Goodfellow, I.J., Shlens, J., Szegedy, C.: Explaining and harnessing adversarial examples. arXiv preprint arXiv:1412.6572 (2014)
7. Gu, Z., et al.: CE-Net: context encoder network for 2D medical image segmentation. IEEE Trans. Med. Imaging **38**(10), 2281–2292 (2019)
8. Hore, A., Ziou, D.: Image quality metrics: PSNR vs. SSIM. In: 2010 20th International Conference on Pattern Recognition, pp. 2366–2369. IEEE (2010)
9. Li, R., Zhang, W., Suk, H.I., Wang, L., Li, J., Shen, D., Ji, S.: Deep learning based imaging data completion for improved brain disease diagnosis. In: Golland, P., Hata, N., Barillot, C., Hornegger, J., Howe, R. (eds.) MICCAI 2014. LNCS, vol. 8675, pp. 305–312. Springer, Cham (2014). https://doi.org/10.1007/978-3-319-10443-0_39
10. Liu, Y., Chen, X., Liu, C., Song, D.: Delving into transferable adversarial examples and black-box attacks. arXiv: Learning (2016)
11. Liu, Z., Zhang, J., Jog, V., Loh, P.L., McMillan, A.B.: Robustifying deep networks for image segmentation. arXiv preprint arXiv:1908.00656 (2019)
12. Madry, A., Makelov, A., Schmidt, L., Tsipras, D., Vladu, A.: Towards deep learning models resistant to adversarial attacks. arXiv preprint arXiv:1706.06083 (2017)
13. Moosavi-Dezfooli, S.M., Fawzi, A., Frossard, P.: Deepfool: a simple and accurate method to fool deep neural networks. In: Proceedings of the IEEE Conference on Computer Vision and Pattern Recognition, pp. 2574–2582 (2016)
14. Niemeijer, M., Staal, J., van Ginneken, B., Loog, M., Abramoff, M.D.: Comparative study of retinal vessel segmentation methods on a new publicly available database. In: Medical Imaging 2004: Image Processing, vol. 5370, pp. 648–656. International Society for Optics and Photonics (2004)
15. Ozbulak, U., Van Messem, A., Neve, W.D.: Impact of adversarial examples on deep learning models for biomedical image segmentation. In: Shen, D., et al (eds.) MICCAI 2019. LNCS, vol. 11765, pp. 300–308. Springer, Cham (2019). https://doi.org/10.1007/978-3-030-32245-8_34

16. Papernot, N., McDaniel, P., Goodfellow, I., Jha, S., Celik, Z.B., Swami, A.: Practical black-box attacks against machine learning. In: Computer and Communications Security (2017)
17. Paschali, M., Conjeti, S., Navarro, F., Navab, N.: Generalizability vs. robustness: adversarial examples for medical imaging. arXiv preprint arXiv:1804.00504 (2018)
18. Pham, D.L., Xu, C., Prince, J.L.: Current methods in medical image segmentation. Ann. Rev. Biomed. Eng. **2**(1), 315–337 (2000)
19. Szegedy, C., et al.: Intriguing properties of neural networks. arXiv preprint arXiv:1312.6199 (2013)
20. Tomczak, K., Czerwińska, P., Wiznerowicz, M.: The Cancer Genome Atlas (TCGA): an immeasurable source of knowledge. Contemp. Oncol. **19**(1A), A68 (2015)
21. Valverde, J.M., Tohka, J.: Region-wise loss for biomedical image segmentation. arXiv preprint arXiv:2108.01405 (2021)
22. Xie, C., Wang, J., Zhang, Z., Zhou, Y., Xie, L., Yuille, A.: Adversarial examples for semantic segmentation and object detection. In: Proceedings of the IEEE International Conference on Computer Vision, pp. 1369–1378 (2017)

Dynamic Clustering Federated Learning for Non-IID Data

Ming Chen[1,2(✉)], Jinze Wu[1,2], Yu Yin[1,2], Zhenya Huang[1,2], Qi Liu[1,2], and Enhong Chen[1,2]

[1] Anhui Province Key Laboratory of Big Data Analysis and Application, School of Data Science & School of Computer Science and Technology, University of Science and Technology of China, Hefei, China
{chenming166613,hxwjz,yxonic}@mail.ustc.edu.cn,
{huangzhy,qiliuql,cheneh}@ustc.edu.cn

[2] State Key Laboratory of Cognitive Intelligence, Hefei, China

Abstract. Federated learning (FL) aims to raise a multi-client collaboration approach in the case of local data isolation. In particular, the clients with non-IID data frequently participate in or leave the federated learning training process asynchronously, resulting in dynamic federated learning (DFL) scenario, which attracts more and more attention. Indeed, an effective DFL solution has to address the following two challenges: 1) Statistical Dynamics. The distributions of local data from clients are always non-IID and the global data distribution is dynamic due to the participation or departure of clients. 2) Expiration Dynamics. After clients leave the federated training process, their historical updated models have a certain validity to reuse in subsequent rounds but it is hard to quantify this validity. In this paper, we first consider clustering the clients with similar data distribution to make them much closer to IID and concentrating on the training the models in each cluster. Then we analyze the changing trend of model validity named model quality and define one suitable function to describe expiration dynamics. As a solution, we propose **D**ynamic **C**lustering **F**ederated **L**earning (DCFL) framework to improve federated learning on non-IID data in DFL. Specifically, DCFL follows the client-server architecture as the standard FL. On the client side, the local devices calculate the related information of the local data distribution for client clustering. On the server side, we design two strategies for the challenges above. We propose dynamic clustering aggregation strategy (including a dynamic clustering algorithm and a two-stage aggregation) by dynamically clustering clients and then aggregating the local models to overcome Statistical Dynamics. Besides, we propose expiration memory strategy by reusing the historical models and then adjusting to model quality of historical models as the basis for model aggregation to overcome Expiration Dynamics. Finally, we conduct extensive experiments on public datasets, which demonstrate the effectiveness of the DCFL framework.

Keywords: Federated learning · Dynamics · Clustering · Non-IID data

L. Fang et al. (Eds.): CICAI 2022, LNAI 13606, pp. 119–131, 2022.
https://doi.org/10.1007/978-3-031-20503-3_10

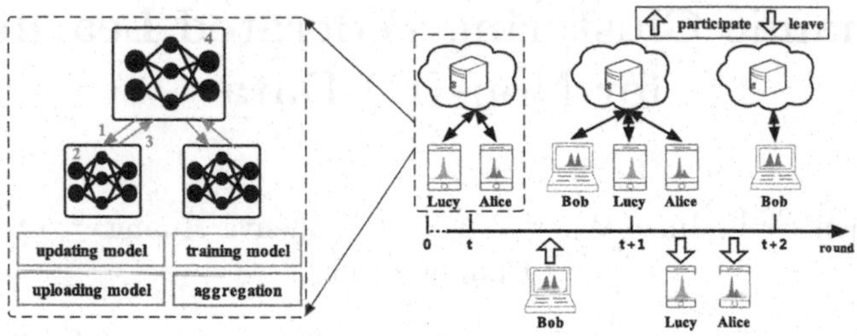

Fig. 1. Dynamic Federated Learning scenario (the screen of client device shows the distribution of their personal data).

1 Introduction

Under the guarantee of protecting data privacy, Federated Learning (FL) has received widespread attention [20,31] and gained incredible achievements in many fields, such as education [16,25], recommendation [15,21], computer vision [33].

In particular, Dynamic Federated Learning (DFL) is general due to less-limited communication [19], incentive mechanism [29], link reliability [12], etc. Concretely, Fig. 1 is a toy example of the DFL scenario. DFL follows original client-server architecture and training framework. Figure 1 shows that in the training round t, local models are initialized/updated with the global model and retrained individually with personal isolated data, and then the server takes a weighted average of all local models to obtain a united global model [19]. However, the situation that the overall distribution changes all the time because of clients with non-IID data participating or leaving frequently and asynchronously badly affects the performance of DFL models. In Fig. 1, client Bob participates in training before round $t+1$ and client Lucy and Alice leave before round $t+2$. Therefore, the distribution at round t is the overall distribution of Lucy and Alice, and the client's distribution changes to the overall distribution of Bob, Lucy and Alice after Bob participates. And then Client Lucy and Alice leave before round $t+2$, causing the distribution to be dominated by Bob's local data. It can be seen that during the DFL training process, the overall data distribution changes frequently, eventually leading to more unstable training and poor performance. However, the above problems of DFL are still underexplored.

Therefore, we conclude these difficulties faced in DFL scenarios as two main challenges: 1) *Statistical Dynamics.* On one hand, in real-world federated learning scenarios, the data distribution of the clients is almost inevitably non-IID [19,32]. On the other hand, since the clients participate in or leave the training process dynamically, the overall data distribution changes dynamically. 2) *Expiration Dynamics.* The expiration of the model in DFL is dynamic [27]. In DFL scenarios, clients leave the training framework, which leads to frequent changes of the number of local models participating in aggregation and affects the stability and reliability of the global model [27]. As Fig. 1 shows, after $T+1$ round, client Lucy and Alice leave the framework and only the local model from Bob is retrained in the process. As a result, such drastic changes of clients affect the stability of subsequent aggregations and updates. Furthermore, their historical updated models have a certain validity to reuse in subsequent rounds [27]. Therefore, we should reasonably assess changes in model validity.

In this paper, we first consider clustering the clients with similar data distribution and concentrate on training the models in each cluster. In this way, the distributions of clients in the same group are almost IID. As clients participate, the dynamic clustering algorithm can classify them into corresponding groups. If there is no correspond group for the new client, the algorithm would create a new cluster for the new client algorithmically. Then we analyze the changing trend of model effectiveness. The validity of the model decreases with the number of rounds, and even the rate of decrease increases [27]. Therefore, we quantify the changing trend into one suitable function. As a solution, we propose a novel *Dynamic Clustering Federated Learning* (DCFL) framework. DCFL framework follows the standard client-server architecture. Specifically, on the client side, besides training and updating local model, it only calculates its data distribution characteristics once for dynamic clustering, which hardly breaks data privacy. On the server side, we design two strategies for the aggregation of server-side models. The first strategy is called *dynamic clustering aggregation strategy*, including a dynamic clustering algorithm and a two-stage aggregation process, to overcome the statistical dynamics. First, the server processes the dynamic clustering algorithm for clients with the local distribution characteristics. Then the server process a two-stage aggregation process to aggregate the local models referring to the clusters. In the first stage, the server first aggregates clients within each cluster. And in the second stage, server aggregates all cluster models into a high-quality global model. The second strategy is *expiration memory strategy* to get over the expiration dynamics challenge. We use the model quality as the basis for model aggregation to preserve the recent available models and enhance the stability of the federated training process. It should be pointed out that the process of expiration memory strtegy happens after the dynamic clustering process and before the two-stage aggregation process.

Finally, we conduct extensive experiments on two public datasets including MNIST and Cifar-10. The results clearly demonstrate that DCFL outperforms the baselines in terms of accuracy performance and clustering division in scenarios with varying degrees of dynamics and non-IID.

2 Related Work

Federated Learning. As people paying more attention to privacy protection [24], FL has received supernumerary attention continuously [11,19,28], and gained great achievements [3,9,17,29]. However, the asynchrony problem and the non-IID problem (none independent and identically distributed) terribly affect the training results [7,14,30]. Fortunately, researchers have got excellent achievements on these problems. For non-IID problem, FedAvg [19] considers the data size as weighting basis for global aggregation. FedProx [13] aims to reduce the distance between local model and global model by a proximal term. Moreover, Zhao et al. [32] create global-shared dataset to make the isolated data more similar. For asynchrony problem, Xie et al. [27] cancel the setting of global epoch. Chai et al. [2] tier clients based on submission speed. However, the above problems in DFL are still underexplored.

Table 1. Notations for our framework.

Notation	Meaning	Notation	Meaning
C	All client models	$\Theta_k^{'t}$	Updated parameters of model k
C^o	Original client models	Θ_k^t	Uploaded parameters of model k
C^*	New client models	λ_k^t	Weight of model k for aggregation
ξ	All cluster models	$\mathcal{QW}(t)_{c_j}$	Model quality weight after t rounds
Φ	The center model	ϱ_k	Point of model k in clustering space
R	Clustering division radius	$\mathcal{DIS}(\varrho_i, \varrho_j)$	Distance between ϱ_i and ϱ_j

Clustering Federated Learning. Clustering algorithm plays an important role in FL with its clustering properties for non-IID problem. Specifically, the clustering algorithm can cluster clients with similar characteristics and makes the distributions of clients in the same group much closer to IID. Extensive studies have been succeeded for this issue. For instance, Shlezinger et al. [23] and Ghosh et al. [8] optimize the operation of client data and solve the problem to a certain extent. CFL [22] uses the cosine similarity of updated parameters as basis for clustering. Moreover, Kim et al. [12] points out that in the traditional FL scenario, the data of each client changes through rounds, which in turn changes the clustering division. However, in DFL scenarios, the number and division of clusters should dynamically change and these methods are not suitable for DFL. In this paper, we make up for the challenges mentioned in DFL.

3 Problem Definition

In DFL scenario, the number of clients are dynamic. So we clarify a dynamic scenario in a certain round t. We assume that there are $|C|$ clients represented as $C = \{c_1, c_2, \cdots\}$, including $|C^o|$ original clients as $C^o = \{c_1^o, c_2^o, \cdots\}$ and $|C^*|$ new clients as $C^* = \{c_1^*, c_2^*, \cdots\}$. For any client $c_i \in C$, the local model $\theta_{c_i}^t$ is trained using only local data. In our problem, our goal is to coordinate all asynchronously participating or exiting training clients to obtain a valid global model θ_Φ^t while guaranteeing the validity of each local model $\theta_{c_i}^t$. Besides, all notations mentioned above and other notations used in this article with their meanings are clearly listed in Table 1.

4 DCFL Framework

4.1 Client Design

In our DCFL framework shown in Fig. 2, the client is mainly responsible for two phases: calculating phase and traditional local training process. As for calculating phase, each new client calculates its distribution characteristic information (In this paper, $\varrho_c =$ (mean, variance, standard deviation) for client c.) to supply the dynamic clustering algorithm. In particular, we assume that the data of client c is represented as $\{(x_1, y_1), \ldots, (x_{\mathcal{L}(c)}, y_{\mathcal{L}(c)})\}$, where y_i is the label of i-th sample and $\mathcal{L}(c)$ is data size of client c. As for mean, we use the formula as $\frac{\sum_{i=1}^{\mathcal{L}(c)} y_i}{\mathcal{L}(c)}$ to gain that. Similarly, we can get other characteristics. As for training process phase shown in Fig. 2, local model first gets initialized or updated from server parameters. Then the client retrains local model on personal data individually. Finally, local device uploads updated model parameters to the server side.

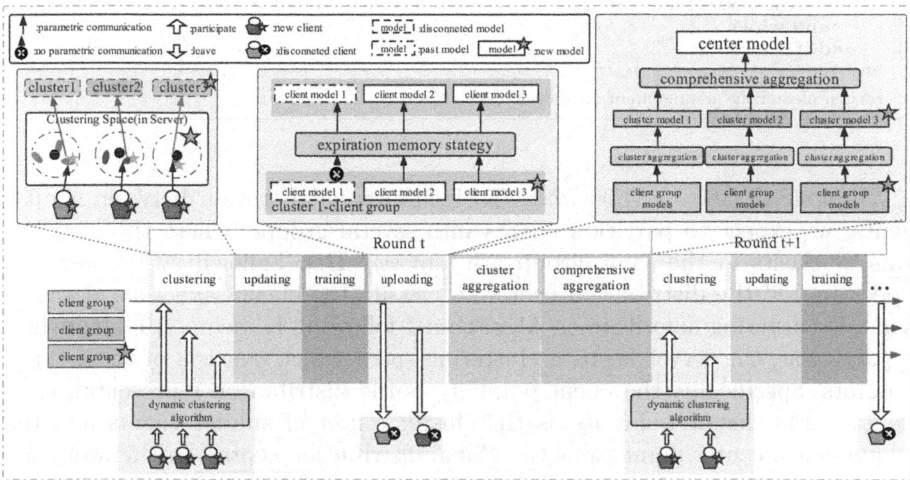

Fig. 2. Dynamic Clustering Federated Learning framework.

4.2 Server Design

The server mainly processes dynamic clustering arrangement of clients and the core task of model aggregation which can expand the available information for clients with two strategies described in detail below.

Dynamic Clustering Aggregation Strategy. In order to resolve statistical dynamics, we proposed the dynamic clustering aggregation strategy including the dynamic clustering algorithm and a two-stage aggregation.

Algorithm 1. Dynamic Clustering Algorithm

Input: C^*: all new clients, i.e. $c_1^*, c_2^*, \cdots$; ϱ_i: the new point composed of distribution features; ϱ_{ξ_i}: all the cluster points, i.e. $i = 1, 2, \cdots$; ϱ_Φ: the center point; R: clustering division radius;
Output: clustering arrangement;

for each new client $c_i^* \in C^*, i = 1, 2, \cdots$ **do**
  compute $\text{minDist}_{c_i^*} = \min(\underbrace{\mathcal{DIS}_1, ..., \mathcal{DIS}_j, ...}_{|\xi_{all}|}, R)$, where $\mathcal{DIS}_j$ is $\mathcal{DIS}(\varrho_{c_i^*}, \varrho_{\xi_j})$.
  if $\text{minDist}_{c_i^*}$ ¡ R **then**
    arrange c_i^* to the cluster where c_i^* gets the $\text{minDist}_{c_i^*}$.
  else
    create a new cluster and copy the center model as its cluster model.
    arrange c_i^* to the new cluster and take $\varrho_{c_i^*}$ as the new cluster point $\varrho_{\xi_{new}}$.
  end if
  add $\varrho_{c_i^*}$ to the cluster space and adjust center/cluster points.
  if the rate of new clients to all clients ≥ 0.2 since the last time when the framework does the following Step 1 to 3 **then**
    take cluster points as the initial cluster centers.
    *Step*1: calculate the distance between each client point and each cluster point, and assign each client point to the cluster center which is closest to it.
    *Step*2: each cluster center is recalculated based on the existing points in the cluster.
    *Step*3 :
    while the cluster centers change **do**
      repeat *Step*1 and *Step*2.
    end while
  end if
end for
return clustering arrangement to server.

Dynamic Clustering Algorithm. To facilitate collaboration between similar clients, we expect to partition clients into several groups, where the distributions of clients in the same group are almostly IID. Following this view, we expect to find the distribution center to classify the clients and then design a dynamic clustering algorithm as Algorithm 1 following K-means [10] algorithm. In particular, the server keeps a clustering space which consists of three types of points. Specifically, the client point ϱ_{c_i} is the distribution representation of client c_i, the cluster point ϱ_{ξ_j} is the cluster center of similar clients and the unique global center point ϱ_Φ is the global distribution center. In this work, we use euclidean distance calculated as $\mathcal{DIS}(\varrho_i, \varrho_j) = \sqrt{|\varrho_i - \varrho_j|^2}$ to represent the distribution distance. The clustering division radius R is used to control the distance between points within the same group shorter and the distance between groups longer and also to create a new cluster when conditions establish. In summary, the division and number of clusters are dynamic in Algorithm 1.

Cluster Aggregation and Comprehensive Aggregation. Based on clustering client, we design a two-stage aggregation including cluster aggregation and comprehensive aggregation instead of the original one-stage aggregation.

As for cluster aggregation, we define the aggregation weight λ_{c_k} as $\frac{\mathcal{L}(c_k)}{\sum_{i=1}^{N} \mathcal{L}(c_i)}$ following FedAvg [19], where $\mathcal{L}(c_k)$ is data size of client c_k. Besides, we get the updated parameters Θ_{ξ_k} of the cluster model by $\Theta_{\xi_k} = \sum_{i=1}^{|\xi_k|} \lambda_{c_i} \times \Theta_{c_i}$, where clients c_i belongs to cluster ξ_k.

As for comprehensive aggregation, we take the distribution map into consideration. If cluster center point ϱ_{ξ_k} is close to the global center point ϱ_{Φ}, the distribution that cluster ξ_k represents is similar to the global distribution, and its weight should be larger. So we calculate the distance between each cluster point and center point by $\mathcal{DIS}(\varrho_{\xi_k}, \varrho_{\Phi}) = \sqrt{|\varrho_{\xi_k} - \varrho_{\Phi}|^2}$, and then obtain the adjusted weight by $\lambda_{\xi_k} = \frac{\mathcal{L}(\xi_k) \times e^{-\mathcal{DIS}(\varrho_{\xi_k}, \varrho_{\Phi})+1}}{\sum_{i=1}^{N} \mathcal{L}(\xi_i) \times e^{-\mathcal{DIS}(\varrho_{\xi_i}, \varrho_{\Phi})+1}}$, where $\mathcal{L}(\xi_k) = \sum_{i=1}^{|\xi_k|} \mathcal{L}(c_i)$ and client c_i belongs to cluster ξ_k. Finally, we can get the updated parameters of center model by $\Theta_{\Phi} = \sum_{i=1}^{N} \lambda_{\xi_i} \times \Theta_{\xi_i}$. We clearly state the logic of this two-stage aggregation in the Fig. 2's cluster and comprehensive aggregation phases.

Algorithm 2. Expiration memory strategy

Input: t: current round; c_k: client c_k; $state_{c_k}$: the statement of client c_k (disconnection/connection); θ_{c_k}: the uploaded parameter of client c_k; $LIBRARY$: the historical parameter library represented as $\{(\theta^*_{c_1}, t'_{c_1}), (\theta^*_{c_2}, t'_{c_2}), ...\}$, where θ^* is the historical parameter and t' is time stamp to mark the uploaded round of θ^*;

Output: λ_{c_k}: weight of for cluster aggregation; θ_{c_k}: parameter for cluster aggregation;

if $state_{c_k}$ is connection **then**
  Replace $(\theta^*_{c_k}, t'_{c_k})$ by (θ_{c_k}, t)
end if
$T = t - t'_{c_k}$
Compute the quality weight $\mathcal{QW}_{c_k}$ by $\mathcal{QW}(T)_{c_j} = \begin{cases} 1 & T = 0, \\ \frac{1}{1+e^{\frac{1}{2}(T-4)}} & 0 < T. \end{cases}$
Compute the weight λ_{c_k} by $\lambda_{c_k} = \frac{\mathcal{L}(c_k) \times \mathcal{QW}(T)_{c_k}}{\sum_{i=1}^{N} \mathcal{L}(c_i) \times \mathcal{QW}(T)_{c_k}}$.
Update θ_{c_k} by $\theta^*_{c_k}$
Return λ_{c_k} and θ_{c_k}

Expiration Memory Strategy. To solve the expiration dynamics, we propose the expiration memory strategy as Algorithm 2. The asynchrony issue reflects the phenomenon of frequent disconnection of the client. Their historical updated models have a certain validity to reuse in subsequent rounds [27], so when some clients leave the framework, we think about reusing their historical retrained models to slow down the change of global distribution and keep training steady. Inspired by Xie et al. [27], we consider the quality of the historical model changes through training rounds. In the first few rounds after a client leaves, its historical model is of high quality. After that, this quality declines sharply, even harmful to the framework. Therefore, we give up using the historical model after several global rounds. To describe the quality changing trend, we define one function used in Algorithm 2 following the sigmoid function.

Updating Phase. After the aggregation, the server should distribute the updated server-side parameters to the clients. Therefore, we design the server to gain these parameters by $\Theta'_{c_k} = \frac{\Theta_{\xi_i} + \Theta_{\Phi}}{2}$, where client c_k belongs to cluster ξ_i. The center model parameters is for expanding the available global information to clients, and the cluster model parameters is to weaken the influence of the client inconsistency and speed up the convergence rate.

5 Experiments

5.1 Experimental Datasets

Datasets. To evaluate our DCFL framework, we conduct extensive experiments with two public datasets including MNIST[1] and Cifar-10[2]. In particular, **MNIST** is a famous dataset used in classification tasks and records 60000/10000 train/test digital images which are 28 * 28 gray-scale images of 0 to 9. **CIFAR-10** is also a classical dataset for classification tasks but differently keeps 60,000 32 * 32 three-channel color images which contain ten kinds of real-world objects.

Data Partition. To simulate the real federated learning scenario, we need to divide the complete dataset into one global testing dataset, $\mathcal{N}$ local testing datasets and $\mathcal{N}$ local training datasets. For MNIST and Cifar-10, $\mathcal{N}$ is 500.

For MNIST, we take the original testing dataset with 10000 samples as global testing dataset. And we take 1,000 samples from each category to obtain a total of 10,000 samples, shuffle these 10,000 samples and then randomly allocate them to $\mathcal{N}$ clients on average as local testing datasets. For Cifar-10, we take 2,000 samples from each category to obtain 20,000 samples and randomly select 10,000 samples from them as global testing dataset. Then we shuffle the rest 10,000 samples and randomly assign $\frac{10{,}000}{\mathcal{N}}$ samples to each of $\mathcal{N}$ clients as its local testing dataset.

As for local training datasets, in order to control the degree of non-IID, we follow the classic method applied in ensemble-FedAvg [1]. Taking MNIST as an example, we assign the sample with label i from the remained training dataset to the i-th group with probability ϖ or to each remaining group with probability $\frac{1-\varpi}{9}$ while the group number is equal to the number of label classes. Then we arrange it to a client in the group uniformly at random. A larger ϖ called degree of non-IID to control local training data distribution reflects a larger degree of non-IID [1]. Similarly, we can get local training datasets of Cifar-10.

DFL Scenario Simulation. In every global round, the framework randomly selects κ of n remaining clients as new clients for training and the number n which represents the number of remaining clients changes to $n - \kappa.(\kappa \in (0, n)$. In the beginning, $n = \mathcal{N}$.) Then the fraction $frac$ of training clients is $\in (0\%, 100\%)$ to simulate the departure of clients.($frac$ equal to 100% means there is no client left.) And we will put the clients who have left back to the remaining set (n changes to $n + (100\% - frac) \times (\mathcal{N} - n)$). If a client has left for a certain global rounds (In this paper, we assume the time is when $\mathcal{QW} < 0.4$.), we give up using its historical model because of harmful low-quality. In summary, the above is a method of sampling with replacement, which means that n is often greater than zero. In this way, FL training process could go along with the client's changes, which eventually simulates the DFL scenario.

5.2 Experimental Settings

Settings. We setup learning rate as 0.01, SGD momentum as 0.5, and loss function as NLL fucntion. The local epoch is 3 and the local batch size is 10.

---

[1] http://yann.lecun.com/exdb/mnist.

[2] http://www.cs.toronto.edu/~kriz/cifar.html.

Other settings for all baselines refer to the original article and we initialize the same parameters of downstream model with all methods. For DCFL, we setup R as 0.4 and initial $|C|$ as 100 for MNIST and Cifar-10.

Baselines.

First, we implement some classical baselines. **FedSGD** [19] is a standard FL method based on stochastic gradient descent. **FedAvg** [19] aggregates multiple-epoch-trained local models to a united global model different from FedSGD. **FedProx** [13]

adds a proximal term to shorten the distance between local and global model. **FedAsync** [27] arranges clients to obtain global model at any time, and global model to aggregate in a single-weight manner with any local model. **CFL** [22] considers the cosine similarity of model parameters as basis of clustering. Then, we choose ablation versions **EMS** as FedAvg with expiration memory strategy and **DCAS** as FedAvg with dynamic clustering aggregation strategy to compare with the baselines. Finally, all methods are implemented by Pytorch, and trained on a Macbook with Apple M1 silicon and 16G memory. All experiments are repeated five times with different random seed and the average results are taken as the final results.

Evaluation Metrics. We evaluate methods at multi-classification tasks and use Prediction Accuracy (ACC) metric to measure the proximity between prediction and ground truth [26].

5.3 Experimental Results

Different Dynamic Scenario Performances. To explore the stability and effectiveness of our framework on scenes with different degrees of dynamic, we carry out two comparative experiments with different dynamic degree settings on MNIST and Cifar-10. The results in Fig. 3 show the following: (1) our proposed DCFL performs better than other baselines in DFL scenario through rounds. (2) DCFL performs more stable than other baselines for different dynamics. Those demonstrate that our DCFL framework with two strategies is appropriate for DFL scenario. (3) our ablation version DCAS as FedAvg with dynamic clustering aggregation strategy performs better than FedAvg and other baselines. It shows our dynamic clustering aggregation strategy aiming to solve Statistical Dynamics keeps effective in scenes of varying degrees of dynamics.

Different Non-IID Scenario Performances. We verify the effectiveness of DCFL in DFL scenario with three degrees of non-IID ($\varpi = 0.1/0.5/0.8$) and use the average of the five highest ACCs during training as the result. Some key observations from Table 2 as follow: (1) the larger the degree of non-IID is, the better DCFL performs than other baselines. It show that DCFL can better deal with the situation where the degree of non-IID is high in DFL. (2) Our ablation version EMS as FedAvg with expiration memory strategy perform better than FedAvg when the client data is non-IID but worse when the client data is IID. Those demonstrate that expiration memory strategy is beneficial to solve the situation where the data is not independent and identically distributed. When the data distribution is similar, the framework can choose to cancel this

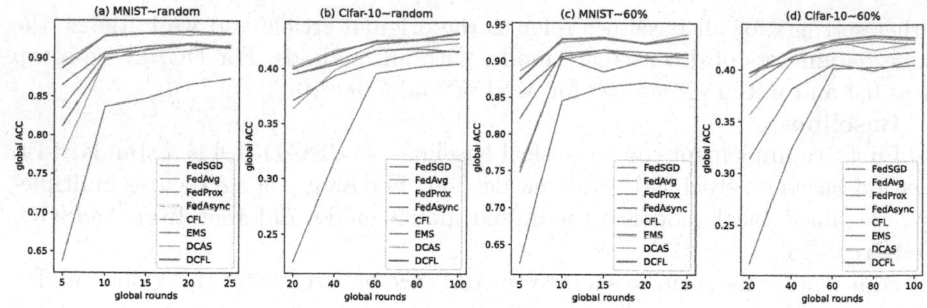

Fig. 3. Accuracy performances through training rounds on both datasets. (a) and (b): κ = random value $\in (0, n)$, $frac$ = random value $\in (0, 1)$, ϖ = random value $\in (0, 1)$; (c) and (d): $\kappa = 60$, $frac = 40\%$, ϖ = random value $\in (0, 1)$.

Table 2. Accuracy performances with three degrees of non-IID on two datasets. ("L-' means local and 'G-' means global. κ = random value $\in (0, n)$, $frac$ = random value $\in (0, 1)$.)

Methods	MNIST			Cifar-10		
	$\varpi = 0.1$(IID)	$\varpi = 0.5$(non-IID)	$\varpi = 0.8$(non-IID)	$\varpi = 0.1$(IID)	$\varpi = 0.5$(non-IID)	$\varpi = 0.8$(non-IID)
	L/G-ACC	L/G-ACC	L/G-ACC	L/G-ACC	L/G-ACC	L/G-ACC
FedSGD	0.815/0.894	0.738/0.878	0.742/0.861	0.335/0.415	0.320/0.408	0.306/0.407
FedAvg	0.832/0.912	0.828/0.907	0.820/0.903	0.357/0.430	0.344/0.425	0.340/0.418
FedProx	0.845/**0.913**	0.825/0.909	0.822/0.906	0.361/0.435	0.355/0.430	0.343/0.423
FedAsync	0.853/0.912	0.835/0.909	0.827/0.904	0.367/**0.437**	0.361/0.424	0.346/0.420
CFL	**0.864**/–	**0.858**/–	0.845/–	**0.375**/–	0.362/–	0.353/–
EMS	0.784/0.874	0.830/0.908	0.824/0.903	0.302/0.397	0.343/0.428	0.345/0.421
DCAS	0.855/0.911	0.847/0.923	0.853/0.913	0.356/0.432	0.360/0.434	0.361/0.429
DCFL	0.796/0.890	0.849/**0.925**	**0.854/0.913**	0.312/0.406	**0.362/0.435**	**0.363/0.430**

strategy at its discretion. (3) CFL outperforms same better as DCFL in several cases. It is because CFL can also handle dynamic clients, but CFL can only perform clustering after the client joins several rounds of training. Our method can complete the clustering as soon as the client joins. From this point of view, our method works earlier than CFL.

Clustering Performances. We adopt some common used indicators to evaluate the clustering effectiveness on MNIST and Cifar-10. One is Davies-Bouldin Index (DBI) [4] used to measure the intracluster compactness. The smaller the DBI, the better the performance. Firstly, we assume that there are k clusters and each one is represented as $\xi = \{\varrho_1, \cdots, \varrho_{|\xi|}\}$. In particular, DBI is defined as $\frac{1}{k}\sum_{i=1}^{k} \max_{i\neq j}(\frac{avg(\xi_i)+avg(\xi_j)}{\mathcal{DIS}(\mu_i,\mu_j)})$, where $\mu_I = \frac{1}{|\xi_I|}\sum_{1\leq i\leq|\xi_I|}\varrho_i$ and $avg(\xi_I) = \frac{2}{|\xi_I|(|\xi_I|-1)}\sum_{1\leq i<j\leq|\xi_I|}\mathcal{DIS}(\varrho_i,\varrho_j)$. The other is Dunn Index (DI) [18] used for measuring the inter-class dispellability. A larger DI means a better clustering performance. Specifically, DI is formulated as $\min_{1\leq i\leq k}\{\min_{i\neq j}(\frac{d_{min}(\xi_i,\xi_j)}{\max_{1\leq l\leq k} diam(\xi_l)})\}$, where $d_{min}(\xi_i,\xi_j) = \min_{\varrho_i\in\xi_i;\varrho_j\in\xi_j}\mathcal{DIS}(\varrho_i,\varrho_j)$ and $diam(\xi_I) = \max_{1\neq i<j\neq|\xi_I|}\mathcal{DIS}(\varrho_i,\varrho_j)$. More-

over, we add three more mainstream clustering algorithms including K-means [10], DBScan [6] and AHC [5] as baselines and we set the number of clusters as $\sqrt[e]{k}$ for K-means and AHC (k is the number of clients). Table 3 reports the results of DBI and DI on two datasets. We can conclude that DCFL performs better than other clustering methods on both DBI and DI, meaning that our clustering method can better cluster clients with similar characteristics in DFL.

Table 3. Results of DBI and DI on MNIST and Cifar-10.

MINST	DCFL	K-means	DBScan	AHC	CFL	Cifar-10	DCFL	K-means	DBScan	AHC	CFL
DBI	**0.517**	1.245	0.651	0.852	0.602	DBI	**0.526**	1.119	0.591	0.915	0.636
DI	**0.532**	0.084	0.043	0.173	0.385	DI	**0.408**	0.121	0.049	0.206	0.341

6 Conclusion

In this paper, we designed a Dynamic Clustering Federated Learning framework. It enabled DFL to gain high-quality models with non-IID data. Specifically, we came up with two strategies named dynamic clustering aggregation strategy and expiration memory strategy for statistical dynamics and expiration dynamics. In server, the framework clustered the clients dynamically and used comprehensive aggregation and cluster aggregation for global model updating, while the framework reused or abandoned retrained local models for stabling training. Finally, our results showed that DCFL outperformed existing federated learning methods, which indicated DCFL is more suitable for DFL scenarios.

Acknowledgements. This research was partially supported by grant from the National Key Research and Development Program of China (No. 2021YFF0901003), the National Natural Science Foundation of China (Grant No. 61922073 and U20A20229), the Fundamental Research Funds for the Central Universities (Grants No. WK2150110021), and the Iflytek joint research program.

References

1. Cao, X., Jia, J., Gong, N.Z.: Provably secure federated learning against malicious clients. In: Thirty-Fifth AAAI Conference on Artificial Intelligence, AAAI 2021, Thirty-Third Conference on Innovative Applications of Artificial Intelligence, IAAI 2021, The Eleventh Symposium on Educational Advances in Artificial Intelligence, EAAI 2021, Virtual Event, 2–9 Feb 2021, pp. 6885–6893 (2021)
2. Chai, Z., Chen, Y., Anwar, A., Zhao, L., Cheng, Y., Rangwala, H.: Fedat: a high-performance and communication-efficient federated learning system with asynchronous tiers. In: SC '21: The International Conference for High Performance Computing, Networking, Storage and Analysis, 14–19 Nov 2021, pp. 60:1–60:16. St. Louis, Missouri, USA (2021)

3. Chen, M., Shlezinger, N., Poor, H.V., Eldar, Y.C., Cui, S.: Communication-efficient federated learning. Proc. Natl. Acad. Sci. **118**(17) (2021)
4. Davies, D.L., Bouldin, D.W.: A cluster separation measure. IEEE Trans. Pattern Anal. Mach. Intell. **2**, 224–227 (1979)
5. Day, W.H., Edelsbrunner, H.: Efficient algorithms for agglomerative hierarchical clustering methods. J. Classif. **1**(1), 7–24 (1984)
6. Ester, M., Kriegel, H.P., Sander, J., Xu, X.: Density-based spatial clustering of applications with noise. In: International Conference Knowledge Discovery and Data Mining, vol. 240, p. 6 (1996)
7. Gao, Y., Zuo, M., Jiang, T., Du, J., Ma, J.: Asynchronous consensus of multiple second-order agents with partial state information. Int. J. Syst. Sci. **44**(5), 966–977 (2013)
8. Ghosh, A., Chung, J., Yin, D., Ramchandran, K.: An efficient framework for clustered federated learning. In: Advances in Neural Information Processing Systems 33: Annual Conference on Neural Information Processing Systems 2020, NeurIPS 2020, 6–12 Dec 2020, virtual (2020)
9. Hamer, J., Mohri, M., Suresh, A.T.: Fedboost: a communication-efficient algorithm for federated learning. In: International Conference on Machine Learning, pp. 3973–3983. PMLR (2020)
10. Hartigan, J.A., Wong, M.A.: Algorithm as 136: a k-means clustering algorithm. J. Royal Statist. Soc. Ser. C (Appl. Statist.) **28**(1), 100–108 (1979)
11. Kairouz, P., et al.: Advances and open problems in federated learning. Found. Trends Mach. Learn. **14**(1–2), 1–210 (2021)
12. Kim, Y., Al Hakim, E., Haraldson, J., Eriksson, H., da Silva, J.M.B., Fischione, C.: Dynamic clustering in federated learning. In: ICC 2021-IEEE International Conference on Communications, pp. 1–6. IEEE (2021)
13. Li, T., Sahu, A.K., Zaheer, M., Sanjabi, M., Talwalkar, A., Smith, V.: Federated optimization in heterogeneous networks. In: Proceedings of Machine Learning and Systems 2020, MLSys 2020, 2–4 March 2020. Austin, TX, USA (2020)
14. Li, X., Huang, K., Yang, W., Wang, S., Zhang, Z.: On the convergence of fedavg on non-iid data. In: 8th International Conference on Learning Representations, ICLR 2020, 26–30 April 2020. Addis Ababa, Ethiopia (2020)
15. Liu, Q., Chen, E., Xiong, H., Ge, Y., Li, Z., Wu, X.: A cocktail approach for travel package recommendation. IEEE Trans. Knowl. Data Eng. **26**(2), 278–293 (2014). https://doi.org/10.1109/TKDE.2012.233
16. Liu, Q., et al.: EKT: exercise-aware knowledge tracing for student performance prediction. IEEE Trans. Knowl. Data Eng. **33**(1), 100–115 (2021). https://doi.org/10.1109/TKDE.2019.2924374
17. Lyu, L., Xu, X., Wang, Q., Yu, H.: Collaborative fairness in federated learning. In: Yang, Q., Fan, L., Yu, H. (eds.) Federated Learning. LNCS (LNAI), vol. 12500, pp. 189–204. Springer, Cham (2020). https://doi.org/10.1007/978-3-030-63076-8_14
18. Ujjwal, M., Sanghamitra, B.: Performance evaluation of some clustering algorithms and validity indices. IEEE Trans. Pattern Anal. Mach. Intell. **24** (2002)
19. McMahan, B., Moore, E., Ramage, D., Hampson, S., y Arcas, B.A.: Communication-efficient learning of deep networks from decentralized data. In: Artificial Intelligence and Statistics, pp. 1273–1282. PMLR (2017)
20. Ouadrhiri, A.E., Abdelhadi, A.: Differential privacy for deep and federated learning: a survey. IEEE Access **10**, 22359–22380 (2022)
21. Qi, T., Wu, F., Wu, C., Huang, Y., Xie, X.: Privacy-preserving news recommendation model training via federated learning (2020)

22. Sattler, F., Müller, K.R., Samek, W.: Clustered federated learning: Model-agnostic distributed multitask optimization under privacy constraints. IEEE Trans. Neural Networks Learn. Syst. **99** (2020)
23. Shlezinger, N., Rini, S., Eldar, Y.C.: The communication-aware clustered federated learning problem. In: 2020 IEEE International Symposium on Information Theory (ISIT), pp. 2610–2615. IEEE (2020)
24. Voigt, P., Von dem Bussche, A.: The EU general data protection regulation (GDPR). A Practical Guide, 1st ed, vol. 10, p. 3152676. Springer, Cham (2017)
25. Wu, J., et al.: Federated deep knowledge tracing. In: Proceedings of the 14th ACM International Conference on Web Search and Data Mining, pp. 662–670 (2021)
26. Wu, J., et al.: Hierarchical personalized federated learning for user modeling. In: Proceedings of the Web Conference 2021, pp. 957–968 (2021)
27. Xie, C., Koyejo, S., Gupta, I.: Asynchronous federated optimization. arXiv preprint arXiv:1903.03934 (2019)
28. Yang, Q., Liu, Y., Chen, T., Tong, Y.: Federated machine learning: concept and applications. ACM Trans. Intell. Syst. Technol. **10**(2), 1–19 (2019)
29. Yu, H., et al.: A fairness-aware incentive scheme for federated learning. In: Proceedings of the AAAI/ACM Conference on AI, Ethics, and Society, pp. 393–399 (2020)
30. Zhang, B., Jia, Y., Du, J., Zhang, J.: Finite-time synchronous control for multiple manipulators with sensor saturations and a constant reference. IEEE Trans. Control. Syst. Technol. **22**(3), 1159–1165 (2014)
31. Zhang, K., Song, X., Zhang, C., Yu, S.: Challenges and future directions of secure federated learning: a survey. Frontiers Comput. Sci. **16**(5), 165817 (2022)
32. Zhao, Y., Li, M., Lai, L., Suda, N., Civin, D., Chandra, V.: Federated learning with non-iid data. arXiv preprint arXiv:1806.00582 (2018)
33. Zhuang, W., et al.: Performance optimization of federated person re-identification via benchmark analysis. In: Proceedings of the 28th ACM International Conference on Multimedia, pp. 955–963 (2020)

Dynamic Network Embedding by Using Sparse Deep Autoencoder

Huimei Tang, Zengyang Shao(✉), Yutao Zhang, Lijia Ma, and Qiuzhen Lin

The College of Computer Science and Software Engineering, Shenzhen University, Shenzhen 518060, China
zyshao0302@gmail.com

Abstract. Learning network representation, which aims to capture various properties of networks in low-dimensional feature space, has recently attracted significant attention. Almost all existing static network embedding and dynamic network embedding methods that employ deep models adopt dense structures. Deep models can ensure that the network embedding achieves a good effect on the task (link prediction, network reconstruction, etc.); however, all works of this kind ignore the high complexity of the deep model training process. In this paper, we propose an embedding method that learns dynamic network embedding by using a sparse deep model. The general idea underpinning our approach involves greatly reducing the number of connections between layers in the deep model. Moreover, the sparse structure of the deep model evolved during the training process to achieve the purpose of fitting the network data. Experimental results on simulated benchmark networks and real-world networks prove that, compared with existing network embedding methods utilizing dense structures, our method is able to greatly reduce the number of training weights, while minimally affecting or sometimes even improving the effect of network tasks.

Keywords: Network embedding · Dynamic networks · Deep model · Sparse structure · Low-dimensional feature space

1 Introduction

In the real world, many systems can be expressed in the form of networks [12, 19, 21]. Discovering the structural characteristics of the network is critical to network tasks. Among the many network-related works, embedding is a highly effective method for mining network structure characteristics [6, 11, 26, 28], which aims to find a low-dimensional vector to express a node in the network.

In recent years, most existing network embedding methods focus primarily on static networks, the structures of which do not change over time. However, many systems are dynamic and evolve over time in the real world. Recently, some methods of dynamic network embedding methods have been proposed [25], which can be divided into two types: (1) Single snapshot: These methods obtain the

L. Fang et al. (Eds.): CICAI 2022, LNAI 13606, pp. 132–143, 2022.
https://doi.org/10.1007/978-3-031-20503-3_11

embedding of the network at a single moment of the dynamic network, simply by using the link structure information of networks [4,30,32]; (2) Multiple snapshots: The embedding of the network is obtained on multiple snapshots of the dynamic network, which utilizes evolutionary information about the structure of the network on the information of the dynamic network derived from multiple moments [3,17].

With the development of deep learning in academia and industry, there has also been a sharp increase in network embedding methods that utilize deep learning. These methods focus primarily on the structural attributes of the network and design the objective function of deep learning by observing the network attributes. In the method of dynamic network embedding, DynGEM [4] minimizes the reconstruction loss, reduces the distance between adjacent nodes in the network, and determines the adaptive deep architecture size. EvolveGCN [18] proposed an approach that uses RNN to evolve the GCN parameters, and which can also capture the dynamism of the graph sequence. However, all previous deep learning-based network embedding methods have ignored the high complexity of the deep learning training process, particularly in the case of a dynamic network containing multiple snapshots. Furthermore, previous works have shown that most weights in the neural network training process are close to zero [5]. In order to solve this problem, some methods have recently been proposed to reduce the training complexity of neural networks by reducing the number of weights [1,7,14]. The ProbMask [31] algorithm uses probability as a global criterion for all layers to measure the weight importance. Yuan et al. [27] defined an auxiliary discrete variable that combines the accuracy and sparsity as an optimization object to reduce the number of weights. Reducing the weight of the deep model is very important for solving the problem of high training complexity in deep model-based dynamic network embedding.

To solve the above problems, we propose a novel dynamic network embedding method (called SPDNE), which considers the high complexity of the depth model in dynamic network embedding and unsupervised learning weight updating. The main idea of SPDNE is to find a sparse topology to learn the structures of the dynamic network. Our main contributions can be summarized as follows.

- We propose a novel embedding framework (called SPDNE) with a deep sparse topology for dynamic networks, which can reduce the complexity of the training process by reducing the number of weights in the dynamic network embedding based on the deep model.
- We design an adaptive algorithm to evolve a deep sparse topology, which solves the problem of weight updating based on unsupervised learning and can achieve a better data fit to ensure the effects of the network tasks.
- We apply the method to three dynamic embedding methods (SDNE [24], DynGEM [4], ElvDNE [13]) based on deep models, which prove that the effect of network tasks (such as network reconstruction [22,29] and link prediction [10,15]) on three simulated networks and four real-world networks was as good as that of the fully connected deep model.

The remainder of this paper is organized as follows. In Sect. 2, we provide the problem definition and formulation of the dynamic network embedding problem in SPDNE. In Sect. 3, we introduce the proposed deep autoencoder algorithm. Experimental results are analyzed and some comparisons are made in Sect. 4. Finally, we provide our concluding remarks and discuss some possible avenues for future work in Sect. 5.

2 Definition and Problem Formulation

2.1 Definition

In this section, the definitions of the dynamic network, autoencoder structure, and dynamic network embedding are provided.

Definition 1. ***Dynamic Network:*** *A dynamic network $\mathcal{G}$ can be represented as a series of snapshots $\{\mathcal{G}^1, \mathcal{G}^2, \ldots, \mathcal{G}^k\}$, where $\mathcal{G}^t = \{\mathcal{V}, \mathcal{E}^t\}$ $(1 \le t \le k)$, k is the number of snapshots, and $\mathcal{V}$, $\mathcal{E}^t$ is the set of nodes and edges at the t-th timestep for the dynamic network, and $\mathcal{V} = \{v_1, v_2, ...v_n\}$, $\mathcal{E}^t = \{e^t_{v_i v_j}\}$ $(1 \le i, j \le n)$. We represent a network snapshot $\mathcal{G}^t$ as an adjacent matrix $\mathrm{A}^t \in R^{n \times n}$.*

Definition 2. ***Autoencoder Structure:*** *An autoencoder structure contains an encoder structure $M = [m^l]$ $(0 \le l \le h)$ and a decoder structure $\hat{M} = [\hat{m}^l]$ $(0 \le l \le h)$. Here, m^0 is the input layer of the autoencoder, $\hat{m}^l$ is the output layer of the autocncoder, and h denotes the number of autoencoder layers in the encoder and decoder. The weights of encoder are represented by $\mathcal{W} = \{\mathrm{W}^l\}$ $(1 \le l \le h)$, while those of the decoder are $\hat{\mathcal{W}} = \{\hat{\mathrm{W}}^l\}$ $(1 \le l \le h)$. In traditional autoencoder structure, the layers are fully connected; thus, the number of weights in its encoder and decoder layers are $\mathrm{W}^l \in \mathrm{R}^{|m^{l-1}| \times |m^l|}$ and $\hat{\mathrm{W}}^l \in \mathrm{R}^{|\hat{m}^{l-1}| \times |\hat{m}^l|}$ respectively. The weight quantity of an autoencoder is $s = \sum_{l=1}^{h}(|m^{l-1}| \times |m^l|) + \sum_{l=1}^{h}(|\hat{m}^{l-1}| \times |\hat{m}^l|)$. Consider a dynamic network $\mathcal{G}$, represented as a series of adjacent matrices $\{\mathrm{A}^1, \mathrm{A}^2, \ldots, \mathrm{A}^t\}$. The low-dimensional representation of the nodes is obtained by the encoder; the calculation method is as follows:*

$$\mathrm{H}_{i.}^{l,t} = \mathrm{S}(\mathrm{W}^{l,t} \cdot \mathrm{H}_{i.}^{l-1,t} + \mathrm{b}^{l-1,t}), \tag{1}$$

where $\mathrm{H}^{l,t}$ is the output of the l-th encoder layer at the t-th timestep, while S is a sigmoid function. Finally, we obtain the output for the last layer of the encoder $\mathrm{H}_{i.}^{h,t}$, which is the embedding of node i at the t-th timestep.

Definition 3. ***Dynamic Network Embedding (DNE):*** *Consider a series of dynamic network snapshots $\mathcal{G} = \{\mathcal{G}^1, \mathcal{G}^2, \ldots, \mathcal{G}^k\}$ and the dimension of latent representation d. For each snapshot $\mathcal{G}^t$, DNE aims to learn a mapping function $\mathbf{F}^t$: $\mathrm{X}^t \to \mathrm{H}^t$, $t = 1, 2, \ldots, k$, where $\mathrm{X}^t \in \mathrm{R}^{n \times n}$ denotes the input data for the mapping function, while $\mathrm{H}^t \in \mathrm{R}^{n \times d}$ is the nodes' latent representation at the t-th timestep. The mapping function is mainly determined by the autoencoder loss function. It can be uniformly expressed as follows:*

$$\mathbf{L}^t = \mathbf{L}_g^t + \alpha \cdot \mathbf{L}_l^t + \beta \cdot \mathbf{L}_s^t, \tag{2}$$

where $\mathbf{L}_g^t$ represents the global loss function at time t, while $\mathbf{L}_l^t$ represents the local loss function at time t and $\mathbf{L}_s^t$ represents the smooth loss function at time t. Here, α and β are the weight of the local structure and smooth part, respectively.

For the loss function of the framework, we use a combination of a global structural loss function, local loss function and smooth loss function. Here, hyperparameter α is range from 0 to 100, while β is range from 0 to 10. $\mathbf{L}_g^t = \sum_{i=1}^{n} \|\hat{\mathrm{X}}_{i.}^t - \mathrm{X}_{i.}^t\|_2^2 = \|\hat{\mathrm{X}}^t - \mathrm{X}^t\|_2^2$ is the global structure loss function part. The local structure loss function is the same as SDNE, $\mathbf{L}_l^t = \sum_{i,j=1}^{n} \mathrm{X}_{ij}^t \|\mathrm{H}_{i.}^t - \mathrm{H}_{j.}^t\|_2^2$, while $\mathrm{H}_{i.}$ and $\mathrm{H}_{j.}$ represent the embedding of node i and j respectively. This means that neighboring nodes should be close in the embedded low-dimensional space. $\mathbf{L}_s^t$ is the smooth function part, which makes use of the similarity between the same node at adjacent moments; this can ensure the stability of dynamic network embedding, and is defined as:

$$\mathbf{L}_s^t = \begin{cases} \sum_{i=1}^{n} \|\mathrm{H}_{i.}^t - \mathrm{H}_{i.}^{t-1}\|_2^2 & \text{if } t > 1 \\ 0 & \text{if } t = 0, \end{cases} \tag{3}$$

2.2 Problem Formulation

Previous work on the autoencoder-based dynamic network embedding method has failed to consider the sparsity of the network structure and the high complexity of the autoencoder training process. If the weight quantity of the autoencoder is reduced to the original r ratio, the formula is: $r \times (\mathrm{W}_{u,v}^l) = r \times (|m^{l-1}| \times |m^l|)$, meaning that the weight quantity of the training process will become $r \cdot s \cdot e \cdot k$. When r is very small, this will greatly reduce the number of weights in the autoencoder. For the dynamic network embedding of unsupervised tasks, if the traditional method of removing near-zero values and retaining non-near-zero values is adopted, too much information may be lost in the dimensionality reduction process. Inspired by the idea of the simulated annealing algorithm [9,23], some near-zero values are retained while others are not, so as to retain some implicit information about the network. The simulated annealing formula is as follows:

$$p(\Delta t) = exp(\frac{-\Delta t}{f}), \tag{4}$$

where $\Delta t < 0$; when the value of Δ is higher, the probability of $p(\Delta t)$ is lower, and when the value of f is lower, the probability of $p(\Delta t)$ is also lower.

3 Our Algorithm: SPDNE

In this section, we present an algorithm governing the setting of the number of sparse structure weights of the autoencoder and the evolution method of the structure. In order to reduce the number of weights in the autoencoder, it is necessary to transform the autoencoder model layer from a fully connected structure into a sparse structure. First, we set the number of weights in the autoencoder, which is done according to the proportion of full connection. Second, the sparse topology of the autoencoder is updated in order to find the optimal structure for fitting data.

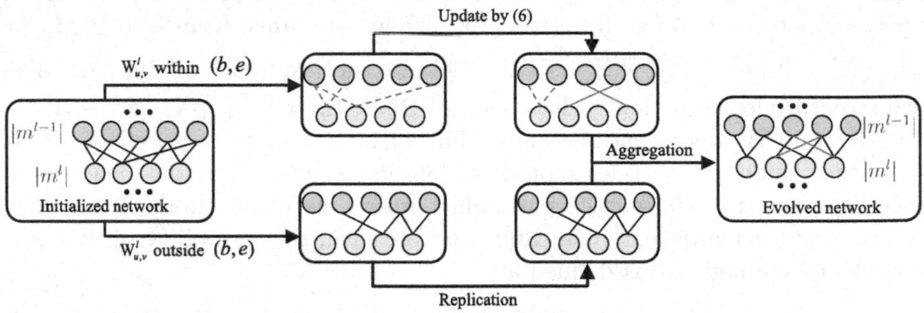

Fig. 1. In the updating process of network layer l, first, $\mathrm{W}^l_{u,v}$ is calculated; if $\mathrm{W}^l_{u,v}$ is inside (b, e), the weight is retained. Else, it will be determined whether the weight should be retained or deleted according to (6). As shown in the figure, two red dotted lines are deleted during the update process. The algorithm will then randomly select two new blue lines, and finally aggregate all the weights to form a new network structure.

Sparse Structure Initialization: Initially, the sparse autoencoder is a random sparse structure. Subsequently, it evolves from this random sparse structure, and retains the number of connections of the sparse structure during the training iterations. u, v are the pair nodes between adjacent layers of the neural network, and the probability of their connection is given by:

$$p(\mathrm{W}^l_{u,v}) = \frac{\ell(|m^{l-1}| + |m^l|)}{|m^{l-1}| \times |m^l|}, \tag{5}$$

where ℓ is the parameter of a sparse neural structure that determines the number of parameters in the neural network. The number of weight parameters between layer $l-1$ and layer l in the sparse network is $\ell(|m^{o-1}|+|m^o|)$. If $\ell \ll |m^{l-1}|$ and $\ell \ll |m^l|$, then there is a linear rather than quadratic increase in the number of connections within the fully connected layers.

Sparse Structure Evolution: In order to learn the optimal structure of the sparse neural network for data, during training, a proportion ξ of the smallest positive weights and the largest negative weights of $\mathrm{W}^l_{u,v}$ are set as thresholds. the smallest positive weights threshold is e and the largest negative weight is b. If the weight of the autoencoder is not in the interval (b, e), this means that these weights are far away from zero and play an important role in the network; thus, these weights need to be maintained. If the weight of the autoencoder is within the interval (b, e), we take inspiration from the idea of simulated annealing, where the probability of the weight being removed is greater the closer it is to zero.

The calculation method is as follows:

$$p(\mathrm{W}_{u,v}^{l}=\mathrm{W'}_{u,v}^{l})=\begin{cases}1-exp(\frac{\mathrm{W}_{u,v}^{l}-e}{f}) & \text{if } \mathrm{W}_{u,v}^{l}>0\\ 1-exp(\frac{b-\mathrm{W}_{u,v}^{l}}{f}) & \text{if } \mathrm{W}_{u,v}^{l}<0.\end{cases} \tag{6}$$

where $\mathrm{W}_{u,v}^{l}$ is the weight in the interval (b, e) of an autoencoder. $\mathrm{W'}_{u,v}^{l}$ is a new random value within the range $(0, 1)$. f can control the probability of the weight being replaced by the new random value. The larger the value of f, the smaller the change of the sparse structure will be.

After each forward process of the autoencoder, the weight value will be obtained, after which the topological structure will be updated according to (6). In the process of updating the autoencoder weight in reverse, we use the stochastic gradient descent method (SGD), which can consistently identify the global optimal solution. The updating process of the network is illustrated in Fig. 1. while the detailed procedure of the algorithm is shown in Algorithm 1.

Algorithm 1. The algorithm framework of SPDNE

```
Input: Dynamic network: 𝒢 = {𝒢^1, 𝒢^2, ..., 𝒢^k}, the parameter: ℓ, f;
Output: Ĥ = {Ĥ^1, ..., Ĥ^t}.
Data preprocessing;
Initialize the parameters of the deep autoencoder by (5);
Initialize the structure of the deep autoencoder;
for each snapshot 𝒢^t, t = 1, 2, ..., k do
    for the weights of the deep autoencoder are not converged do
        Update weights by SGD;
        Calculate the threshold value(b, e) from the encoder according to the parameter ℓ;
        Selective removal of weights of the deep autoencoder by (6);
        Randomly initializes the weight of deep autoencoder by the same number as the removed weights;
    end for
    Initializes the deep autoencoder of the structure for the next time
    Ĥ ← Ĥ^t;
end for
```

4 Experimental Results

In this section, we test SPDNE on three deep autoencoder methods (SDNE, DynGEM, ElvDNE). Each method is implemented on three simulated networks and four real-world networks, and two criteria are used for comprehensive performance evaluation.

4.1 Experimental Settings

The test networks chosen are the GN (SYN-FIX, SYN-VAR) [2], the stochastic block (SBM) [8] benchmark networks, and four real-world dynamic networks (ia-contacts [20], football [16], ia-radoslaw-email [20] and ca-cit-Hep [20]). In order to evaluate the effectiveness of our proposed method, we adopt two different topological modes (Dense, Sparse) of the deep autoencoder. To verify the performance of SPDNE, we use the low-dimension representations of nodes learned by three deep autoencoder frames under three different topologies (Dense, Sparse, SPDNE). We further used network reconstruction and link prediction tasks to evaluate the effect of the experiment.

The SPDNE was tested with the parameter settings for the deep architectures (1.0/5.0/10/0.4/0.1 for five hyperparameters $\alpha/\beta/\zeta/\ell/f$ on SDNE and DynGEM methods, while 1.0/5.0/2.0/10/0.4/0.1 for $\alpha/\beta/\lambda/\zeta/\ell/f$ on the ElvDNE). In order to ensure the fairness of the experimental results, we use the same hyperparameters for all dynamic networks across the different methods. We ran 30 experiments on each method independently to reduce variation.

Network Reconstruction: It reconstructs the network structure by calculating the distance between these node representations. We use mean average reconstruction precision (p_r) to measure the network reconstruction effect, which is computed as follows:

$$p_r = \frac{\sum_{t=1}^{k} \sum_{i=1}^{n} \left(\sum_{j=1}^{n} \mathrm{X}_{ij}^t \cdot \mathrm{A}_{ij}^t \right) / ||\mathrm{A}_{i.}^t||}{t}, \tag{7}$$

where $\mathrm{X}_{ij}^t \in \{0, 1\}$ represents the proximity between the node representations in low-dimensional space. When p_r is larger, the effect of the low-dimensional representations on the nodes of the dynamic network embedding is better.

Link Prediction: The goal of this task is to predict the edges that will appear in the network through the representations of the low-dimensional space of nodes. We use mean average reconstruction precision (p_r) to measure the network reconstruction effect. It is computed as follows:

$$p_p = \frac{\sum_{t=1}^{k-1} \sum_{i=1}^{n} \left(\sum_{j=1}^{n} \mathrm{X}_{ij}^{t+1} \cdot \mathrm{A}_{ij}^{t+1} \right) / ||\mathrm{A}_{i.}^{t+1}||}{t-1}: \tag{8}$$

where $\mathrm{X}_{ij}^t \in \{0, 1\}$ is determined by the nodes' low-dimensional representations at time t. When p_p is larger, there is a better low-dimensional representation effect of the nodes of the dynamic network embedding.

4.2 Experimental Results

Experimental Results on ElvDNE: In order to demonstrate the effectiveness of our method, we use an autoencoder embedding framework that employs a dynamic network evolution structure to apply our method.

As can be seen from Table 1, this framework of fully connected autoencoder works best for the network tasks (network reconstruction and link prediction) in most networks. Due to the addition of the evolutionary dynamic structure in the dynamic network for smoothing purposes, the difference between the results of the three methods is relatively small compared with the previous two frameworks, showing better stability. Overall, SPDNE works better than the other two methods on three simulated networks, although the difference between all methods is less than 4 percent. However, our method has less than 30 percent of the weight in the autoencoder compared to the fully connected method, particularly in large-scale networks, where it has less than 4 percent of the weight.

Table 1. Results on the test networks. $\bar{p}$ and $\boldsymbol{p}$ are the mean and best results respectively on criterion p for each algorithm over 30 independent trials. The best result for each network is marked in boldface.

Networks	Criteria	Dense	Sparse	SPDNE	Ratio
SYN-FIX	$\bar{p}_r$	0.4076	0.3868	**0.5310**	0.2951
	$\boldsymbol{p}_r$	0.4309	0.4136	**0.5415**	
	$\bar{p}_p$	0.4025	0.3889	**0.4699**	
	$\boldsymbol{p}_p$	0.4193	0.4102	**0.4863**	
SYN-VAR	$\bar{p}_r$	0.3983	0.4128	**0.4915**	0.2298
	$\boldsymbol{p}_r$	0.4143	0.4277	**0.5040**	
	$\bar{p}_p$	0.2707	0.2736	**0.2842**	
	$\boldsymbol{p}_p$	0.2785	0.2838	**0.2894**	
SBM	$\bar{p}_r$	0.2550	0.2589	**0.2623**	0.0394
	$\boldsymbol{p}_r$	0.2599	0.2630	**0.2661**	
	$\bar{p}_p$	0.2554	0.2583	**0.2614**	
	$\boldsymbol{p}_p$	0.2606	0.2631	**0.2645**	
ia-contacts	$\bar{p}_r$	0.3128	0.3088	**0.3340**	0.3113
	$\boldsymbol{p}_r$	0.3372	0.3279	**0.3484**	
	$\bar{p}_p$	0.3372	0.3387	**0.3657**	
	$\boldsymbol{p}_p$	0.3659	0.3653	**0.3739**	
Football	$\bar{p}_r$	0.7284	0.7207	**0.8530**	0.3033
	$\boldsymbol{p}_r$	0.7618	0.7620	**0.8690**	
	$\bar{p}_p$	0.7583	0.7544	**0.8599**	
	$\boldsymbol{p}_p$	0.7949	0.7912	**0.8732**	
ia-radoslaw	$\bar{p}_r$	**0.3966**	0.3935	0.3835	0.2767
	$\boldsymbol{p}_r$	**0.4078**	0.4023	0.3928	
	$\bar{p}_p$	**0.4123**	0.4076	0.3966	
	$\boldsymbol{p}_p$	**0.4265**	0.4204	0.4095	
ca-cit-Hep	$\bar{p}_r$	**0.5224**	0.5121	0.4956	0.0340
	$\boldsymbol{p}_r$	**0.5248**	0.5157	0.4983	
	$\bar{p}_p$	**0.5562**	0.5408	0.5217	
	$\boldsymbol{p}_p$	**0.5584**	0.5429	0.5249	
Best/all		8/28	0/28	**20**/28	

In the real-world network, the autoencoder of the full-connection method performs better than the other two methods; however, the effect of our method on the ia-radoslaw and ca-cit-Hep networks is reduced by about 2 and 2 percent respectively compared with the full-connection method. Similarly, compared with the fully connected method, the weights of our method on the real-world dataset can be reduced by about 3 percent to 30 percent.This experiment proves that our method can maintain a good effect compared with the full-connection method by smoothing the evolutionary dynamic network structure and greatly reducing the number of weights. Furthermore, our weighting ratio continues to decrease as the network size increases, and the number of weights of our method is reduced by 80 percent on average when compared to the fully connected network.

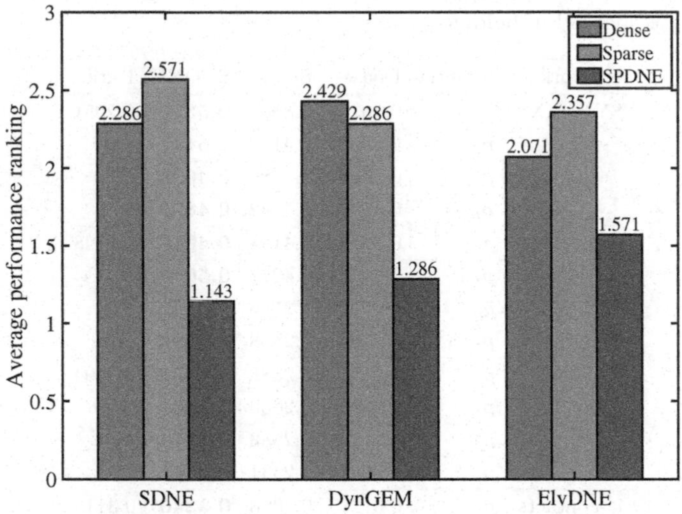

Fig. 2. Average performance ranking of three methods under the three frameworks in terms of p_p and p_r on all networks.

By comparing the three methods under the three frameworks, our method can achieve a greater effect than the fully connected network while greatly reducing the weight. Figure 2 plots the average performance ranks of three methods under the three frameworks in terms of all criteria. For each network in each framework, the three methods are ranked from 1 to 3 based on their performance. The results show that SPDNE achieves an average performance rank (1.143, 2.429 and 2.071) that is much smaller than that of Dense (2.286, 1.286 and 1.571) and Sparse (2.5713, 2.286 and 2.357). This further proves that our method is of great value for achieving dimensionality reduction in unsupervised dynamic networks.

Effects of Parameter Settings: In this section, we analyze the laws of parameters ℓ and ζ on the network reconstruction and link prediction tasks between the ElvDNE model and SPDNE method.

The parameter ℓ represents the sparsity of the weights of the autoencoder model. Figure 3(a) and Fig. 3(b) plot the model performance under different values of the parameter. We can determine that when the value of the parameter is greater than 5 and less than 40, the model performance is very stable. This shows that the performance of the model can be improved by our proposed method under the appropriate level of sparsity. And the parameter ζ controls the number of model weights that are updated during the iteration. Figure 3(c) and Fig. 3(d) reveal that the model performance first increases and then decreases as the value of the parameter increases. This is because updating too few model weights prevent all model parameters from being evaluated; accordingly, excellent sparse topology cannot be obtained, and too many parameter updates can cause the preserved parameters to be lost in the model.

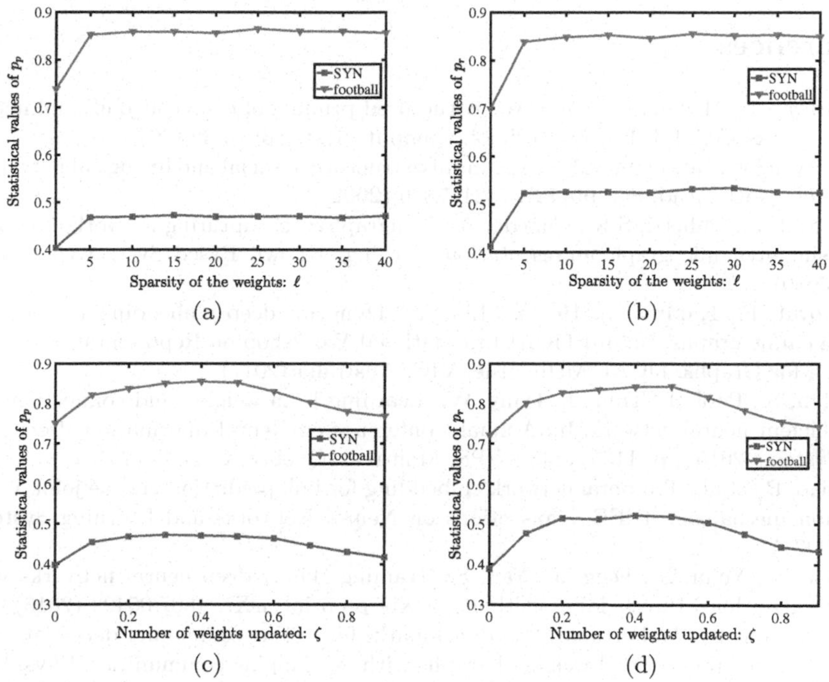

Fig. 3. Statistical values of p_p and p_r by ElvDNE with SPDNE method. (a, b) sparsity of the weight ℓ; (c, d) number of weights updated ζ.

5 Conclusion

In this paper, we propose a sparse topology method that can be applied to the framework of an autoencoder. This framework can not only reduce the number of weights in the autoencoder, but also maintain the performance of the network at a level equivalent to that of a fully connected autoencoder. In this method, the sparse topology of the autoencoder based on the idea of simulated annealing, can fit the dynamic network data and complete the learning of unsupervised node representations. We demonstrate that our method performs well in network tasks (network reconstruction and link prediction) by comparing the other methods in three frameworks.

This work determines the weight quantity of sparse autoencoder in advance for subsequent training purposes. Moreover, it is concluded from the experiment that a network with a complex internal structure and more implicit structure yields a relatively general effect. In future work, we will study how to determine the minimum number of weights in the encoder while maintaining the network embedding effect; in addition, special topological structure exploration can be carried out for networks with more hidden structures to achieve the effect of fitting data.

References

1. Anwar, S., Hwang, K., Sung, W.: Structured pruning of deep convolutional neural networks. ACM J. Emerg. Technol. Comput. Syst. **13**(3), 32 (2017)
2. Girvan, M., Newman, M.E.: Community structure in social and biological networks. Proc. Natl. Acad. Sci. **99**(12), 7821–7826 (2002)
3. Goyal, P., Chhetri, S.R., Canedo, A.: Dyngraph2vec: capturing network dynamics using dynamic graph representation learning. Knowl. Based Syst. **187**, 104816 (2020)
4. Goyal, P., Kamra, N., He, X., Liu, Y.: Dyngem: deep embedding method for dynamic graphs. In: 3rd IJCAI International Workshop on Representation Learning for Graphs. IJCAI. Melbourne, VIC, Australia (2017)
5. Han, S., Pool, J., Tran, J., Dally, W.: Learning both weights and connections for efficient neural network. In: Annual Conference on Neural Information Processing Systems 2015, pp. 1135–1143. NIPS, Montreal, Quebec, Canada (2015)
6. Jiao, P., et al.: Temporal network embedding for link prediction via vae joint attention mechanism. IEEE Transactions on Neural Networks and Learning Systems (2021)
7. Jin, X., Yuan, X., Feng, J., Yan, S.: Training skinny deep neural networks with iterative hard thresholding methods. arXiv preprint arXiv:1607.05423 (2016)
8. Lancichinetti, A., Fortunato, S.: Benchmarks for testing community detection algorithms on directed and weighted graphs with overlapping communities. Phys. Rev. E Statist. Nonlinear Soft Matter Phys. **80**(1), 016118 (2009)
9. Lee, J., Perkins, D.: A simulated annealing algorithm with a dual perturbation method for clustering. Pattern Recogn. **112**, 107713 (2020)
10. Li, T., Zhang, J., Philip, S.Y., Zhang, Y., Yan, Y.: Deep dynamic network embedding for link prediction. IEEE Access **6**, 29219–29230 (2018)
11. Liu, Q., Long, C., Zhang, J., Xu, M., Lv, P.: Triatne: tipartite adversarial training for network embeddings. IEEE Trans. Cybern. **52**, 9634–9645 (2021)
12. Ma, L., Li, J., Lin, Q., Gong, M., Coello, C.A.C., Ming, Z.: Cost-aware robust control of signed networks by using a memetic algorithm. IEEE Trans. Cybern. **50**(10), 4430–4443 (2020)
13. Miikkulainen, R., et al.: Evolving deep neural networks. In: Artificial Intelligence in the Age of Neural Networks and Brain Computing, pp. 293–312. Elsevier (2019)

14. Mocanu, D.C., et al.: Evolutionary training of sparse artificial neural networks: a network science perspective. arXiv preprint arXiv:1707.04780 (2017)
15. Nasiri, E., Berahmand, K., Rostami, M., Dabiri, M.: A novel link prediction algorithm for protein-protein interaction networks by attributed graph embedding. Comput. Biol. Med. **137**, 104772 (2021)
16. Newman, M.E., Girvan, M.: Finding and evaluating community structure in networks. Phys. Rev. E Statist. Nonlinear Soft Matter Phys. **69**(2), 026113 (2004)
17. Nguyen, G.H., Lee, J.B., Rossi, R.A., Ahmed, N.K., Koh, E., Kim, S.: Continuous-time dynamic network embeddings. In: The Web Conference 2018, pp. 969–976. Lyons, FRANCE (2018)
18. Pareja, A., et al: Evolvegcn: evolving graph convolutional networks for dynamic graphs. In: 34th AAAI Conference on Artificial Intelligence, pp. 5363–5370. AAAI, New York, NY, USA (2020)
19. Park, P.S., Blumenstock, J.E., Macy, M.W.: The strength of long-range ties in population-scale social networks. Science **362**(6421), 1410–1413 (2018)
20. Rossi, R.A., Ahmed, N.K.: The network data repository with interactive graph analytics and visualization. In: 29th AAAI Conference on Artificial Intelligence. AAAI, Austin, Texas, USA (2015). http://networkrepository.com
21. Santolini, M., Barabási, A.L.: Predicting perturbation patterns from the topology of biological networks. Proc. Natl. Acad. Sci. **115**(27), E6375–E6383 (2018)
22. Teng, X., Liu, J., Li, L.: A synchronous feature learning method for multiplex network embedding. Inform. Sci. **574**, 176–191 (2021)
23. Tóth, J., Toman, H., Hajdu, A.: Efficient sampling-based energy function evaluation for ensemble optimization using simulated annealing. Pattern Recogn. **107**, 107510 (2020)
24. Wang, D., Cui, P., Zhu, W.: Structural deep network embedding. In: 22nd ACM International Conference on Knowledge Discovery and Data Mining, pp. 1225–1234. ACM, San Francisco, CA, USA (2016)
25. Xue, G., Zhong, M., Li, J., Chen, J., Zhai, C., Kong, R.: Dynamic network embedding survey. Neurocomputing **472**, 212–223 (2022)
26. Yang, M., Zhou, M., Kalander, M., Huang, Z., King, I.: Discrete-time temporal network embedding via implicit hierarchical learning in hyperbolic space. In: Proceedings of the 27th ACM SIGKDD Conference on Knowledge Discovery & Data Mining, pp. 1975–1985 (2021)
27. Yuan, X., Savarese, P., Maire, M.: Growing efficient deep networks by structured continuous sparsification. arXiv preprint arXiv:2007.15353 (2020)
28. Zhang, W., Guo, X., Wang, W., Tian, Q., Pan, L., Jiao, P.: Role-based network embedding via structural features reconstruction with degree-regularized constraint. Knowl. Based Syst. **218**, 106872 (2021)
29. Zhao, Z., Zhou, H., Li, C., Tang, J., Zeng, Q.: Deepemlan: deep embedding learning for attributed networks. Inform. Sci. **543**, 382–397 (2021)
30. Zhou, L.k., Yang, Y., Ren, X., Wu, F., Zhuang, Y.: Dynamic network embedding by modeling triadic closure process. In: 32nd AAAI Conference on Artificial Intelligence. AAAI, New Orleans, Louisiana, USA (2018)
31. Zhou, X., Zhang, W., Xu, H., Zhang, T.: Effective sparsification of neural networks with global sparsity constraint. In: Proceedings of the IEEE/CVF Conference on Computer Vision and Pattern Recognition, pp. 3599–3608 (2021)
32. Zhu, D., Cui, P., Zhang, Z., Pei, J., Zhu, W.: High-order proximity preserved embedding for dynamic networks. IEEE Trans. Knowl. Data Eng. **30**(11), 2134–2144 (2018)

Deep Graph Convolutional Networks Based on Contrastive Learning: Alleviating Over-smoothing Phenomenon

Rui Jin[1], Yibing Zhan[2], and Rong Zhang[1](✉)

[1] Department of Electronic Engineering and Information Science, University of Science and Technology of China (USTC), Hefei, China
zrong@ustc.edu.cn

[2] JD Explore Academy, Hefei, China

Abstract. Graphs are a common and important data structure, and networks such as the Internet and social networks can be represented by graph structures. The proposal of Graph Convolutional Network (GCN) brings graph research into the era of deep learning and has achieved better results than traditional methods on various tasks. For ordinary neural networks, more layers can often achieve better results. However, for GCN, the deepening of the number of layers will cause a catastrophic decline in performance, including the gradual indistinguishability of node features, the disappearance of gradients, and the inability to update weights. This phenomenon is called over-smoothing. The occurrence of over-smoothing makes training deep GCNs a difficult problem. Compared with deep GCNs, shallow GCNs tend to perform better. Therefore, we design a contrastive learning model such that the deep GCN learns the features of the same node (positive samples) of the shallow GCN while alienating the features of other nodes (negative samples) , so that the deep GCN can learn the performance of the shallow GCN. Experiments show that our method can effectively alleviate the over-smoothing phenomenon. At the same time, we apply this model to other over-smoothing methods, and also achieve better results.

Keywords: Graph convolutional network · Graph contrastive learning · Over-smoothing

1 Introduction

Graph Convolutional Networks (GCN) [11] can transfer information between graph nodes and show superior ability in graph representation learning. GCN has shown remarkable breakthroughs in a large number of tasks, including node classification [3,11], object detection [13], traffic prediction [15], and more. However, despite the remarkable success of GCNs, current GCNs still suffer from a serious problem: when the number of layers increases, the performance drops severely.

L. Fang et al. (Eds.): CICAI 2022, LNAI 13606, pp. 144–154, 2022.
https://doi.org/10.1007/978-3-031-20503-3_12

This phenomenon is generally considered to be due to over-smoothing [2,19, 21]. Each convolution operation mixes the features of adjacent nodes through information transfer. Therefore, when stacking convolutional layers, the features of adjacent nodes gradually become indistinguishable. This phenomenon is called over-smoothing in GCN [19] proves that the feature space of nodes in GCN gradually becomes smaller with layer-by-layer convolution. Considering that for most neural networks, the deeper the depth, the better the effect, how to alleviate the over-smoothing problem and develop a deep graph neural network has become a research hotspot.

To alleviate the over-smoothing problem, many models have been proposed [9,16,21,26,28]. DropEdge [21] and DropNode [6], respectively, randomly remove edges and nodes during training to alleviate over-smoothing by making the graph more sparse. SkipNode [16] randomly selects nodes to skip the convolutional layer. PairNorm [28] adds a regularization layer between convolutional layers to keep the feature distances of node pairs in space unchanged. ResGCN, DenseGCN [12]and JKNet [26] choose to add connections between GCN layers. Contrastive learning on graph neural networks is a recently emerging research hotspot. We have tried to alleviate this problem by using contrastive learning and achieved good performance.

We Propose a Contrastive Learning Model that Alleviates the Over-Smoothing Phenomenon. Shallow GCNs have good expressiveness, so we use the target deep GCNs to contrast with shallow GCNs. For the nodes in the deep GCN, we treat the same node in the shallow GCN as a positive sample, and different nodes as a negative sample, and obtain our contrastive model by performing contrastive learning through InfoNCE loss.

We Conduct Semi-Supervised Node Classification Experiments on the Cora, Citeseer, and Pubmed Datasets [7,22]. The experiments are carried out at different depths. The results show that our method effectively alleviates the over-smoothing phenomenon, and even achieves more than 30 percent accuracy at some layers.

We Combine the Contrastive Learning Model in this Paper with Other Frameworks that Alleviate Over-Smoothing. In this paper, we use four models: DropEdge [21], SkipNode [16], ResGCN [12] and JKNet [26]. We combine our contrast network with them and experiments prove that our models can help them achieve better results.

2 Related Work

2.1 GCNs

Originally, [1] first designed a graph spectral theory based on graph convolution operations. However, considering the high computational cost, it is difficult to apply spectral-based models [1,5] to very large graphs. Space-based model types [8,17] exhibit good scalability since their convolution operations can be directly applied to spatial neighborhoods of unseen graphs.

GCN [11] simplifies the previous work and proposes the currently commonly used graph convolution structure, which achieved excellent results. in the node classification task. Based on GCN, many works have been proposed to improve graph representation learning [24,25,25,27] studied the expressive power of different aggregates and proposed a simple model GIN. SGC [24] reduces the complexity of GCN by discarding non-linear functions and weight matrices. Although the performance of GCN has been gradually improved, due to the existence of over-smoothing, deep GCNs are still an unsolved problem.

2.2 Over-smoothing

The features between the connected nodes gradually become indistinguishable when stacking too many graph convolutional operations. This phenomenon is referred to as the over-smoothing problem.

[14] pointed out that Laplacian smoothing is the key to over-smoothing, and proved that the operation of graph convolution is a type of Laplace smoothing. The Laplacian smoothing algorithm computes the new features of a vertex as a weighted average of itself and its neighbors. Since vertices in the same cluster tend to be densely connected, smoothness makes their characteristic, which makes the subsequent classification task easier, resulting in a huge performance improvement, while over-smoothing phenomenon presents a simple proof.

[19] proved that GCNs lose the ability to distinguish nodes exponentially. giving a basis for determining whether over-smoothing occurs. Numerous studies [10,14] also gave proof of over-smoothing convergence. Meanwhile, [19] pointed out that densely connected graphs are very prone to over-smoothing, and the sparsity of graphs in practical applications makes GCNs better used. From the computational results, the occurrence of over-smoothing makes the node representations within the same connected component of the graph will converge to the same value and the nodes are completely homogeneous, thus destroying the functionality of the GCN.

2.3 Graph Contrastive Learning

Contrastive learning on graphs is a novel research field. At present, contrastive learning can be mainly divided into two categories. One is the contrast of graph structure, and the other is the contrast of graph features.

The contrast of graph structure mainly focuses on data enhancement on the graph [20] mainly study the transfer learning of graphs in different scenarios, so after randomly sampling subgraphs, the mask operation is further carried out, while the choice of data augmentation is proposed to be closely related to the domain of the dataset, and the effects of different data augmentation options are explored.

The contrast on graph features mainly includes random masking of features, adding Gaussian noise to feature vectors, etc. [29] proposed a set of methods to calculate the importance of each dimension of the feature vector, and perform the masking operation according to the importance [23] adopted a node-by-node feature contrast method, which inspired the work in this paper: constructing GCNs at two different levels and making the deep GCN learn the performance capability of the shallow GCN by performing a node-by-node contrast model.

3 Method

In contrastive learning, it is very important to construct positive and negative samples. We constructed a contrastive learning model of node features using nodes of deep and shallow GCNs as positive and negative samples.

3.1 Contrast Structure and Data Augmentation

Deep GCNs tend to suffer from over-smoothing and perform poorly in training, while shallow GCNs performs well, if we can use contrastive learning techniques to make the node feature of deep GCN close to shallow GCN, it is possible to alleviate the over-smoothing phenomenon.

In the contrastive learning framework of this paper, positive samples are easy to find, and it is expected in the experiments that the deep GCN can learn the node features of the shallow GCN. Therefore, for a certain node, the positive sample of the deep GCN is the same node in the shallow GCN.

For the other nodes, we cannot determine whether they have similar feature vectors as our target node, so we same treat them as negative samples (see in Fig. 1).

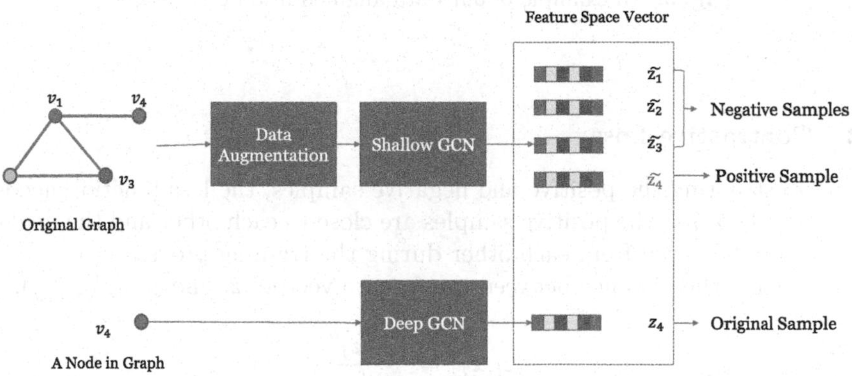

Fig. 1. The positive and negative samples in our contrast structure

To obtain more positive and negative samples for our contrastive learning, we designed two methods to perform data augmentation on the graph, one is to randomly remove edges in a certain proportion, and the other is to randomly mask a portion of the node features in a certain proportion (see in Fig. 2). We tune the probability of random parameters in our experiments to get better results.

As a result, we obtained three groups of shallow GCNs, and we calculated the contrastive losses of these three groups of GCNs respectively and added them together to obtain the final contrastive loss:

$$L_{cont} = L_{shallowGCN} + L_{RemoveEdge} + L_{MaskFeature} \tag{1}$$

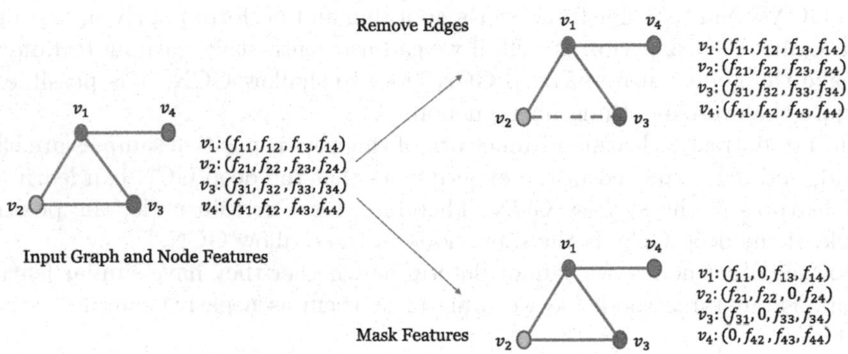

Fig. 2. An example of our data augmentation on a graph

3.2 Contrastive Loss

After constructing the positive and negative samples, the loss function needs to be designed so that the positive samples are close to each other and the negative samples are far away from each other during the training process.

We define the distance between the feature vectors z_i and z_j as (z_i, z_j):

$$(z_i, z_j) = \frac{z_i * {z_j}^T}{|z_i||z_j|} \tag{2}$$

In this paper, we refer to the loss function of simCLR [4], and define z_i as the node feature vector in the target deep network and $\widetilde{z}_i$ as the node feature vector of the same node in the shallow GCN, then the pair contrast infoNCE loss is as follows:

$$L_{pc(z_i,\widetilde{z}_i)} = -log\frac{e^{(z_i,\widetilde{z}_i)/\tau}}{e^{(z_i,\widetilde{z}_i)/\tau} + \sum_{j=1,j\neq i}^{N} e^{(z_i,\widetilde{z}_j)/\tau} + \sum_{j=1,j\neq i}^{N} e^{(z_i,z_j)/\tau}} \tag{3}$$

where N is the total number of classes and τ is the temperature coefficient. The contrast loss function is defined as the average of the total number of nodes after contrasting the shallow and deep GCNs with each other:

$$L_{shallowGCN} = \frac{\sum_{i=1}^{N}(L_{PC(z_i,\tilde{z}_i)} + L_{PC(\tilde{z}_i,z_i)})}{2N} \tag{4}$$

$L_{RemoveEdge}$ and $L_{MaskFeature}$ are also calculated by this method.

3.3 For Node Classification Tasks on GCN

In this paper, we conduct node classification experiments under semi-supervision and hope to optimize the classification results using contrastive learning. Therefore, the feature space vector of the experiment is for the classification results, i.e., the probability vector of the samples in different classifications.

For example, if dataset X contains N classes, then when training the contrast network, our feature node vector is the N-dimensional node classification probability.

In training, we use the softmax function to calculate the classification probability, for class i, the GCN training result is scored as $score_i$, then the probability of being classified in class i is:

$$softmax_i = \frac{e^{score_i/\tau}}{\sum_{j=1}^{M} e^{score_j/\tau}} \tag{5}$$

where M is the total number of classes and τ is the temperature coefficient.

Then our eigenvector z is given as:

$$z = (softmax_1, softmax_2, ..., softmax_M) \tag{6}$$

In the experiment, the deep and shallow GCNs also need to be trained, so the final loss function is the classification loss of the shallow and deep networks plus the contrast loss function, and the classification loss uses the cross-entropy function, as:

$$Lclass = -\frac{\sum_{i=1}^{N}\sum_{c=1}^{M} y_{ic}\log p_{ic}}{N} \tag{7}$$

where N is the number of nodes; M is the number of categories; y_{ic} is a symbolic function that takes 1 if the true category of sample i, and 0 otherwise; p_{ic} is the predicted probability that the observed sample i belongs to category c (the predicted probability uses a softmax function, as in Equation 5).

Let the classification loss of deep and shallow GCNs be $Lclass_{deep}$ and $Lclass_{shallow}$, and the loss function is:

$$L = L_{cont} + \alpha Lclass_{deep} + \beta Lclass_{shallow} \tag{8}$$

where α and β are hyperparameters that need to be adjusted in the experiment. We chose to pre-train the shallow GCNs before the experiments (see Sect. 4.2

for details), avoiding the need to train both GCNs with tuning references during the experiments. Thus, the final loss function is:

$$L = L_{cont} + \alpha Lclass_{deep} \tag{9}$$

4 Experiments

To reveal the effectiveness of our contrastive learning framework, extensive experiments have been conducted on the task of semi-supervised node classification.

4.1 Experimental Details

We use three widely-used citation networks datasets [7,18,22]: Cora, Citeseer and Pubmed to perform semi-supervised node classification experiments. In the citation networks, the nodes represent documents and their links refer to citations between documents, where each node is associated with a bag-of-words feature vector and a ground-truth label. The dataset statistics are summarized in Table 1.

Table 1. Dataset details

Datasets	Nodes	Edges	Features
Cora	2708	5429	1433
Citeseer	3327	4732	3703
Pubmed	19717	44338	500

There are five models used in the experiments: GCN [11], DropEdge [21], SkipNode [16], ResGCN [12] and JKNet [26]. The last four models are all designed to alleviate over-smoothing. DropEdge randomly deletes the edges of the graph, SkipNode randomly selects nodes to skip the convolution operation, ResGCN adds inter-layer residual connections, and JKNet adds adaptive inter-layer aggregation.

To better observe the effect and over-smoothing phenomenon of the model at different depths of GCN, we conduct semi-supervised node classification experiments at six depths of 2, 4, 8, 16, 32, and 64, with a GTX 1080Ti GPU and hidden layer parameters all set to 64.

4.2 Pretraining Shallow GCN

The original loss function is shown in Eq. 7. This loss function represents that we train deep and shallow GCNs while comparing. In experiments, we find that the performance of deep GCN is far inferior to shallow GCN, and the loss of deep GCN will occupy the main part of the loss function, which will interfere with the training of shallow GCN in the experiment. At the same time, it is also difficult to tune loss functions with complex hyperparameters.

So we conducted experiments on GCNs with three depths of 2, 3 and 4 layers (see Table 2), and based on the experimental results, we decided to choose the 2-layer GCN as our pre-trained shallow network.

Table 2. Semi-supervised Classification Accuracy on Shallow GCN(%)

GCN layers	Dataset		
	Cora	Citeseer	Pubmed
2	**81.5**	71.4	**79.0**
3	81.0	**72.1**	78.2
4	79.2	69.1	77.3

Therefore, we decided to pre-train a 2-layer GCN network independently and no longer train simultaneously with the deep GCN in the experiments. For the input graph, the node predictions obtained by the shallow GCN are directly trained by the pre-trained GCN. Therefore, we do not need to add the classification loss of shallow GCN to the experiment, the loss function becomes Eq. 9.

4.3 Node Classification Results

We conducted experiments on many depths and models (see Table 3). This paper first conducts experiments on GCN, and it can be seen that there is an obvious over-smoothing phenomenon on the three datasets: at layers 2, 4 and 8, GCN can maintain a high accuracy, but starting from layer 16, there is a large drop in accuracy, proving the occurrence of over-smoothing.

Using the contrasting network in this paper, and experimenting again under the same conditions, it can be found that at 16 and 32 layers, the network in this paper can still maintain high accuracy, especially at layer 16, the network in this paper can improve the accuracy by more than 20 percentage points, which is far better than GCN. This shows that the model in this paper effectively alleviates the over-smoothing phenomenon and helps to train deep GCNs.

The contrast network in this paper improves the over-smoothing phenomenon by adding a contrast structure, and a shallow GCN is designed separately in addition to the deep GCN for contrast. Therefore, the network in this paper can also be used in combination with existing over-smoothing frameworks. This paper adds some existing frameworks to the contrast network. It turns out that the network in this paper can also improve the ability of these frameworks to the over-smoothing problem.

Table 3. Semi-supervised Classification Accuracy on Different Models(%)

Model	Layers	Dataset					
		Cora		Citeseer		Pubmed	
		Original	contrast network	Original	contrast network	Original	contrast network
GCN	2	81.5	**81.5**	71.4	**71.8**	79.0	**79.9**
	4	79.2	**80.5**	**69.1**	68.4	77.3	**80.0**
	8	**81.4**	81.2	64.5	**67.4**	76.8	**79.5**
	16	41.0	**75.4**	37.2	**64.5**	41.6	**76.3**
	32	24.2	**71.1**	23.9	**34.5**	41.3	**58.1**
	64	21.3	**25.6**	30.6	**34.1**	40.7	**43.7**
DropEdge	2	**81.6**	81.4	**71.4**	69.5	79.5	**79.7**
	4	**81.1**	80.2	**69.1**	68.4	78.4	**79.7**
	8	**81.4**	79.6	65.6	**68.4**	78.0	**79.1**
	16	41.0	**73.1**	39.9	**64.5**	41.6	**79.3**
	32	25.6	**71.1**	24.1	**24.1**	41.3	**58.1**
	64	21.3	**25.6**	30.6	**34.1**	41.3	**43.7**
SkipNode	2	**81.5**	81.4	**71.4**	69.7	79.5	**79.5**
	4	81.1	**81.2**	**72.1**	68.4	78.4	**79.7**
	8	**81.4**	79.6	66.8	**67.4**	77.7	**78.7**
	16	43.5	**77.1**	38.1	**66.3**	56.6	**77.4**
	32	31.1	**71.3**	24.9	**40.1**	42.6	**60.2**
	64	24.2	**24.2**	32.3	**34.1**	43.9	**46.5**
ResGCN	2	**81.5**	81.6	**71.4**	71.5	79.0	**80.0**
	4	78.2	**79.2**	**66.7**	66.9	77.9	**78.8**
	8	**78.7**	77.0	51.9	**61.7**	77.6	**77.1**
	16	64.8	**70.9**	41.0	**56.1**	41.6	**70.6**
	32	24.7	**60.5**	29.4	**36.3**	50.6	**62.8**
	64	16.9	**45.4**	23.0	**34.9**	55.2	**57.3**
JKNet	2	**81.4**	80.9	**71.8**	69.3	79.1	**79.3**
	4	78.9	**80.8**	**66.0**	64.3	77.7	**78.9**
	8	**77.9**	77.6	60.5	**64.9**	77.1	**78.8**
	16	79.7	**79.7**	60.0	**61.7**	76.5	**77.6**
	32	74.2	**77.1**	59.9	**60.9**	79.4	**77.2**
	64	72.9	**75.9**	53.8	**61.9**	75.7	**76.3**

5 Conclusion

Over-smoothing can lead to significant performance degradation of graph convolutional networks. In this paper, a contrastive learning network is proposed to alleviate the over-smoothing problem using contrastive learning, and good results are achieved. On GCN, experiments are conducted on various depths in this paper, and the experimental results show that the model in this paper is very effective.

The method in this paper does not change the structure of the graph but chooses to improve the over-smoothing phenomenon through contrastive learning. Therefore, the method in this paper is independent and can be applied to various models and play a role. We also add this paper's contrast network model to various popular frameworks for solving the over-smoothing problem, and also help them to achieve better results.

The models in this paper can be applied to various structures and networks, and the combination between models usually improves the performance, and we expect to combine the existing excellent models for further training development to better alleviate the over-smoothing problem.

References

1. Bruna, J., Zaremba, W., Szlam, A., LeCun, Y.: Spectral networks and locally connected networks on graphs. arXiv preprint arXiv:1312.6203 (2013)
2. Chen, D., Lin, Y., Li, W., Li, P., Zhou, J., Sun, X.: Measuring and relieving the over-smoothing problem for graph neural networks from the topological view. In: Proceedings of the AAAI Conference on Artificial Intelligence, vol. 34, pp. 3438–3445 (2020)
3. Chen, M., Wei, Z., Huang, Z., Ding, B., Li, Y.: Simple and deep graph convolutional networks. In: International Conference on Machine Learning, pp. 1725–1735. PMLR (2020)
4. Chen, T., Kornblith, S., Norouzi, M., Hinton, G.: A simple framework for contrastive learning of visual representations. In: International Conference on Machine Learning, pp. 1597–1607. PMLR (2020)
5. Defferrard, M., Bresson, X., Vandergheynst, P.: Convolutional neural networks on graphs with fast localized spectral filtering. Adv. Neural Inf. Process. Syst. **29** (2016)
6. Do, T.H., Nguyen, D.M., Bekoulis, G., Munteanu, A., Deligiannis, N.: Graph convolutional neural networks with node transition probability-based message passing and dropnode regularization. Expert Syst. Appl. **174**, 114711 (2021)
7. Getoor, L.: Link-based classification. In: Advanced Methods for Knowledge Discovery from Complex Data. AIKP, pp. 189–207. Springer, London (2005). https://doi.org/10.1007/1-84628-284-5_7
8. Hamilton, W., Ying, Z., Leskovec, J.: Inductive representation learning on large graphs. Adv. Neural Inf. Process. Syst. **30** (2017)
9. He, K., Zhang, X., Ren, S., Sun, J.: Deep residual learning for image recognition. In: Proceedings of the IEEE Conference on Computer Vision and Pattern Recognition, pp. 770–778 (2016)
10. Huang, W., Rong, Y., Xu, T., Sun, F., Huang, J.: Tackling over-smoothing for general graph convolutional networks. arXiv preprint arXiv:2008.09864 (2020)
11. Kipf, T.N., Welling, M.: Semi-supervised classification with graph convolutional networks. arXiv preprint arXiv:1609.02907 (2016)
12. Li, G., Muller, M., Thabet, A., Ghanem, B.: Deepgcns: Can GCNS go as deep as cnns? In: Proceedings of the IEEE/CVF International Conference on Computer Vision, pp. 9267–9276 (2019)
13. Li, H., Miao, S., Feng, R.: DG-FPN: learning dynamic feature fusion based on graph convolution network for object detection. In: 2020 IEEE International Conference on Multimedia and Expo (ICME), pp. 1–6. IEEE (2020)
14. Li, Q., Han, Z., Wu, X.M.: Deeper insights into graph convolutional networks for semi-supervised learning. In: Thirty-Second AAAI Conference on Artificial Intelligence (2018)

15. Li, Z., et al.: A hybrid deep learning approach with GCN and LSTM for traffic flow prediction. In: 2019 IEEE Intelligent Transportation Systems Conference (ITSC), pp. 1929–1933. IEEE (2019)
16. Lu, W., et al.: Skipnode: On alleviating over-smoothing for deep graph convolutional networks. arXiv preprint arXiv:2112.11628 (2021)
17. Monti, F., Boscaini, D., Masci, J., Rodola, E., Svoboda, J., Bronstein, M.M.: Geometric deep learning on graphs and manifolds using mixture model cnns. In: Proceedings of the IEEE Conference on Computer Vision and Pattern Recognition, pp. 5115–5124 (2017)
18. Namata, G., London, B., Getoor, L., Huang, B., Edu, U.: Query-driven active surveying for collective classification. In: 10th International Workshop on Mining and Learning with Graphs. vol. 8, p. 1 (2012)
19. Oono, K., Suzuki, T.: Graph neural networks exponentially lose expressive power for node classification. arXiv preprint arXiv:1905.10947 (2019)
20. Qiu, J., et al.: GCC: graph contrastive coding for graph neural network pre-training. In: Proceedings of the 26th ACM SIGKDD International Conference on Knowledge Discovery & Data Mining, pp. 1150–1160 (2020)
21. Rong, Y., Huang, W., Xu, T., Huang, J.: Dropedge: towards deep graph convolutional networks on node classification. arXiv preprint arXiv:1907.10903 (2019)
22. Sen, P., Namata, G., Bilgic, M., Getoor, L., Galligher, B., Eliassi-Rad, T.: Collective classification in network data. AI Mag. **29**(3), 93–93 (2008)
23. Wan, S., Zhan, Y., Liu, L., Yu, B., Pan, S., Gong, C.: Contrastive graph poisson networks: Semi-supervised learning with extremely limited labels. Adv. Neural Inf. Process. Syst. **34** (2021)
24. Wu, F., Souza, A., Zhang, T., Fifty, C., Yu, T., Weinberger, K.: Simplifying graph convolutional networks. In: International Conference on Machine Learning, pp. 6861–6871. PMLR (2019)
25. Xu, K., Hu, W., Leskovec, J., Jegelka, S.: How powerful are graph neural networks? arXiv preprint arXiv:1810.00826 (2018)
26. Xu, K., Li, C., Tian, Y., Sonobe, T., Kawarabayashi, K.i., Jegelka, S.: Representation learning on graphs with jumping knowledge networks. In: International Conference on Machine Learning, pp. 5453–5462. PMLR (2018)
27. You, J., Ying, R., Leskovec, J.: Position-aware graph neural networks. In: International Conference on Machine Learning, pp. 7134–7143. PMLR (2019)
28. Zhao, L., Akoglu, L.: Pairnorm: Tackling oversmoothing in gnns. arXiv preprint arXiv:1909.12223 (2019)
29. Zhu, Y., Xu, Y., Yu, F., Liu, Q., Wu, S., Wang, L.: Graph contrastive learning with adaptive augmentation. In: Proceedings of the Web Conference 2021, pp. 2069–2080 (2021)

Clustering-based Curriculum Construction for Sample-Balanced Federated Learning

Zhuang Qi, Yuqing Wang, Zitan Chen, Ran Wang, Xiangxu Meng, and Lei Meng(✉)

Shandong University, Jinan, Shandong, China
{z_qi,yuqing_wang,chenzt,wr}@mail.sdu.edu.cn,
{mxx,lmeng}@sdu.edu.cn

Abstract. Federated learning is a distributed machine learning scheme that provides data privacy-preserving solution. A key challenge is data distribution heterogeneity of on different parties in federated learning. Existing methods only focus on the training rule of local model rather than data itself. In this paper, we reveal an fact that improving the performance of the local model can bring performance gain to the global model. Motivated by this finding, this paper proposes a Clustering-based curriculum construction method to rank the complexity of instances, and develops a Federation curriculum learning algorithm (FedAC). Specifically, FedAC assigns different weights to training samples of different complexity, which is able to take full advantage of the valuable learning knowledge from a noisy and uneven-quality data. Experiments were conducted on two datasets in terms of performance comparison, ablation studies, and case studies, and the results verified that FedAC can improve the performance of the state-of-the-art Federated learning methods.

Keywords: Curriculum learning · Federated learning · Neural networks

1 Introduction

Federated learning, as a privacy-preserving distributed machine learning paradigm, has attracted much attention in artificial intelligence [1–3]. It typically uses a central server to coordinate multiple clients for collaborative modeling, and protects the privacy of training data of all parties, aiming to achieve the same or similar performance as data sharing [4]. However, existing studies have verified that the data heterogeneity leads to the poor generalization ability of the fused model [5]. Many studies have focused on alleviating the problem of distribution heterogeneity in federated learning, but there is a lack of solutions to uneven data quality.

Existing studies trying to improve the model performance can be divided into two levels: data-level [7,8] and model-level [9–11]. Data-level methods often use

L. Fang et al. (Eds.): CICAI 2022, LNAI 13606, pp. 155–166, 2022.
https://doi.org/10.1007/978-3-031-20503-3_13

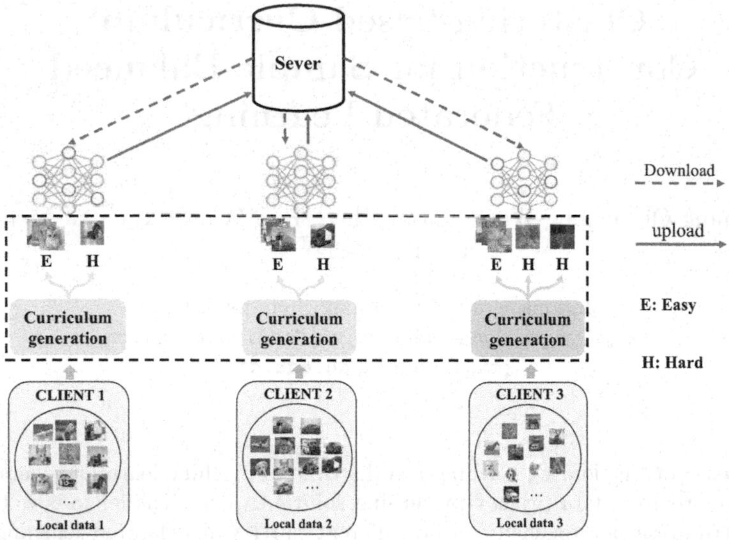

Fig. 1. Illustration to FedAC for image classification in federated learning.

two techniques, including data augmentation and sampling. They alleviate the imbalance by increasing the data during training, but ignore the quality of self-generated data. Many model-level approaches combine knowledge distillation to reduce global and local distribution bias. They add constraints between global and local models (e.g., model parameters [12], feature maps [13], probability distributions of predicted results [14].) to improve the similarity between local and global information. However, in the insufficient model training phase, distillation training is meaningless and even hindering model convergence. As shown in Fig. 2, the accuracy of distillation method in the early stage is lower than baseline. The above methods only focus on the training rule of local model, while the impact of data quality to model performance have not been well analyzed.

To address aforementioned problems, this paper presents a Clustering-based curriculum construction for sample-balanced Federated learning method. As illustrated in Fig. 1, it first ranks the complexity of data instances. Specifically, a clustering method is used to generate the hierarchy of image-class pairs from training set, which learns the different image patterns of same class. The hierarchy of each clients, serving as a personalized knowledge base, is able to filter the noisy data. To rank the complexity of all the data, a rule is adopted, which defines the score based on cluster size. The larger the cluster, the higher the data score it contains. In training phase, a loss weighting mechanism based on score is adopted. Then, an adaptive curriculum construction method is applied in local model training process. As observed, FedAC is able to get performance gain of global model by improving the robustness of local models.

Experiments have been conducted on the CIFAR-10 and CIFAR-100 datasets in terms of performance comparison, ablation study and case studies for the effec-

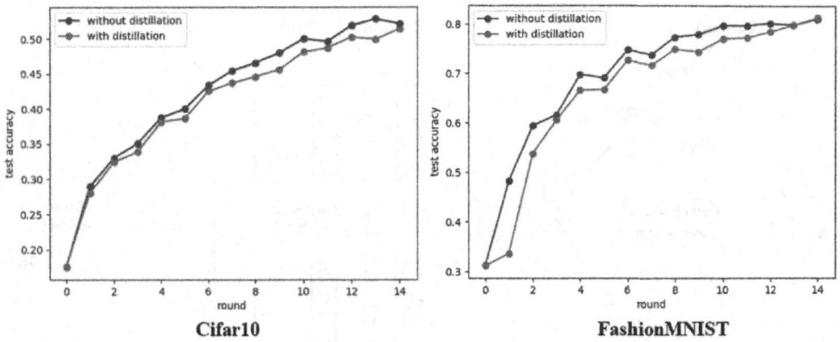

Fig. 2. Illustration the accuracy of with and without distillation methods.

tiveness of FedAC. The results verify Adaptive curriculum learning can improve the performance of local model and bring performance gain to the global model.

To summarize, this paper includes two main contributions:

- An curriculum learning method, termed CG, is proposed, which enables achieve adaptive grouping of data. This can effectively use data to learn knowledge.
- A model-agnostic framework, termed FedAC, is proposed to combine curriculum learning method and can achieve local-global model performance gain.

2 Related Work

2.1 Curriculum Learning

Curriculum learning (CL) imitates the human learning process, which ranks all instances based on complexity, and adopts the knowledge learning method of easy first and then hard [15]. The core problem of the curriculum learning is to get a ranking function, termed as Difficulty Measuring, which gives its learning priority for each piece of data or task. In addition, the training rule is determined by the training scheduler [16]. Therefore, there are many CL methods based on the framework of "difficulty measuring and training scheduler". CL can be divided into two categories based on whether it is automatically designed or not, namely Predefined CL [17] and Automatic CL [18]. Both the difficulty measuring and training scheduler of Predefined CL are designed by human experts using human prior knowledge, while at least one of Automatic CL is automatically designed in a data-driven manner.

2.2 Federated Learning

Many strategies have been proposed to address the data heterogeneity problem in Federated Learning (FL), which are mainly from two perspectives: data-level and model-level. Data level methods usually generate extra data to achieve data

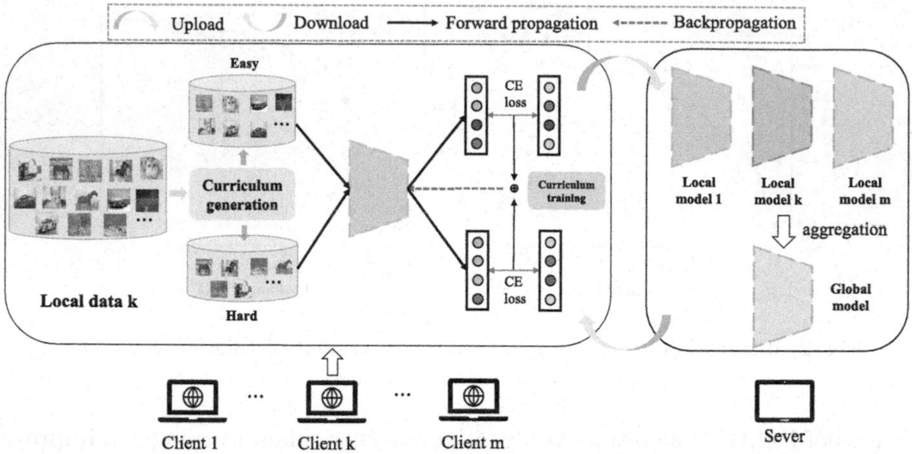

Fig. 3. The framework of FedAC.

balance [6]. For example, Astraea uses data augmentation based on global data distribution to alleviate imbalances [8]. Model-level methods focus on optimization strategies to make the diversity between client local models and global model limited [19]. Fedprox restricts local model parameter updates from being far away from the global model [12]. MOON utilizes the similarity between local and global model representations to correct local training for all parties [5]. FML coordinates all clients to jointly train global models and independently train personalized models, and realize global knowledge transfer through deep mutual learning [9]. For studies in the aggregation phase, FedMA matches and averages weights in a hierarchical manner [20], FedNova normalizes local updates before averaging [21]. Notably, these aggregation ideas can be combined with our study.

3 Problem Formulation

This paper investigates curriculum learning for local training in the image classification task of federated learning. Suppose there are N clients, $(C_1, C_2, ..., C_N)$. And N clients hold heterogeneous data partition $\mathcal{D}^1, \mathcal{D}^2, ..., \mathcal{D}^N$, respectively. The goal is to learning a global model ω over the dataset $\mathcal{D} \triangleq \bigcup_{k\in[N]} \mathcal{D}^k$ with the coordination of the central server without data share. For local training, client k starts with copying the weight vector $w^k \in \mathbb{R}^d$, and the goal is mininzing the local loss function $f_k(w^k)$. The updated weight vector w^k can be obtain gradient decent method, i.e. $w^{k+1} \leftarrow w^k - \eta \nabla f_k\left(w^k\right)$. Finally, the global model can be obtained by aggregating all local models, i.e. $w^{\text{global}} \leftarrow \sum_{k=1}^{N} \frac{|D^k|}{|D|} w^k$.

Beyond conventional settings, our proposed FedAC first introduces a curriculum learning method to learn the image patterns of the same class. This enables the data grouping as a curriculum for all local data, i.e. $\mathrm{D}^k \mapsto \{D_1^{\mathrm{k}}, D_2^{\mathrm{k}}\}$. Subsequently, FedAC uses a image encoder to learn knowledge from D_1^{k} and D_2^{k}, respectively. By weighting the corresponding loss function $L = \alpha L_1 + \beta L_2$ as total loss. This enables encoders learn accurate knowledge.

4 Federation Curriculum Learning

FedAC introduces a curriculum learning framework to learn the complexity of data. As shown in Fig. 3, FedAC contains two main modules in each clients, including Curriculum Generation (CG) Module and Curriculum Training (CT) Module, as illustrated in the following sub-sections.

4.1 Curriculum Generation (CG) Module

FedAC explores the complexity of all data in CG module, with the aim of completing data grouping. As shown in the Fig. 4, CG module divides the raw data into two parts. Given a dataset including images $\mathcal{I} = \{\mathbf{I}_i | i = 1, 2, ..., N\}$ and corresponding labels of J classes $\mathcal{C} = \{c_j | j = 1, 2, ..., J\}$. Specifically, CG module first uses multi-channel clustering method to learn the data pattern of each classes. Notably, the ART algorithm [22] is extended to gather the similar features of the same class into a cluster. The details are shown as follows:

- For a data $p = (\mathbf{x}, O_x)$, O_x is one-hot version of the label, let $I = [\mathbf{x}, 1 - \mathbf{x}]$, the match score can be calculated by formula 1:

$$C_p = \left\{ c_j \mid \min \left\{ \frac{|\mathbf{I} \wedge w_j|}{|\mathbf{I}|}, O_j^T O_x \right\} \geq \rho, j = 1, \ldots, J \right\} \tag{1}$$

  where w_j, O_j are the weight vector and indicator of cluster c_j, O_j^T denotes the transposition of O_j, $p \wedge q = \min\{p, q\}, |p| = \sum_i p_i, \rho \in [0, 1]$ is vigilance parameter. If C_p is a empty set, generate a new cluster, otherwise proceed to the next step.
- For each candidate cluster $c_j \in C_p$, the choice function T_j with a choice parameter α as shown in formula 2:

$$T_j = \frac{|\mathbf{I} \wedge \mathbf{w_j}|}{\alpha + |w_j|} \tag{2}$$

- For the final-match cluster c_j*, using formula 3 to update the corresponding weight vector w_j*, and $\beta \in [0, 1]$

$$\hat{w}_{j*} = \beta \left(I \wedge w_{j*} \right) + (1 - \beta) w_{j*} \tag{3}$$

After the cluster process, many clusters C with the same class are obtained, the overall process can be expressed as formula 4:

$$C = Clustering(\mathcal{D}, \mathcal{C}) \tag{4}$$

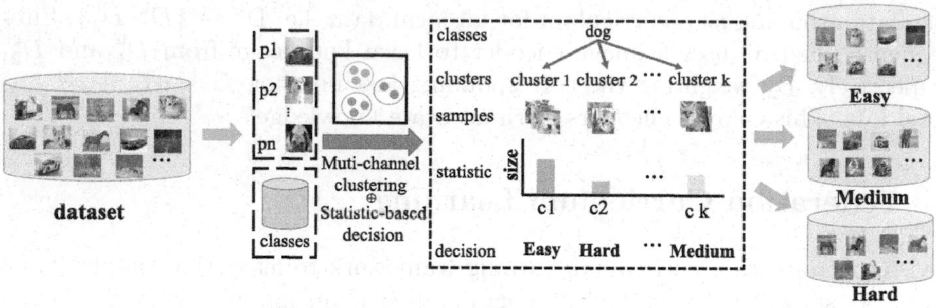

Fig. 4. Illustration of the CG module.

Table 1. Statistics of the datasets used in the experiments.

Datasets	#class	#Training	#Testing
CIFAR-10	10	50000	10000
CIFAR-100	100	50000	10000

where $\mathcal{D}$ denotes all image data, and $\mathcal{C}$ is the corresponding class labels.

Then, the Statistic-based Cluster filtering method $S(.)$ is used to divide all clusters into two parts based on cluster size, i.e.

$$P_1, P_2 = S(C) \tag{5}$$

where $P_1 = \{c_i | \ if \ |c_i| < T\}$ and $P_2 = \{c_i | \ if \ |c_i| \geq T\}$, and T is a threshold.

4.2 Curriculum Training (CT) Module

After data grouping, CT Module trains a image encoder $E(\cdot)$ on P_1 and P_2. This paper proposes a weighting loss approach to reduce the adverse impact of uneven-quality data during training phase. The Eqs. 6 and 7 represent the loss function of easy and hard data, respectively.

$$L_1 = CE(p_1, y_1) \tag{6}$$

$$L_2 = CE(p_2, y_2) \tag{7}$$

where p_i and y_i denote the prediction and ground-truth of data x_i. The total loss is weighted by L_1 and L_2, i.e.

$$L_{total} = \alpha L_1 + \beta L_2 \tag{8}$$

5 Experiments

5.1 Experimental Setup

Datasets. We use two benchmarking datasets CIFAR-10 and CIFAR-100 that are commonly used in federated classification for experiments. Their statistics are showing in Table 1. Like recent studies, the Dirichlet distribution $Dir_N(\cdot)$ is used to generate the non-IID distribution among all parties. Specifically, we use $p_k \sim \text{Dir}_N(\beta)$ to sample and allocate a $p_{k,j}$ proportion of the instances of class k to client j, where β is a concentration parameter. We set 10 clients by default in all experiments. The data partition results of different parameters are shown in Fig. 5.

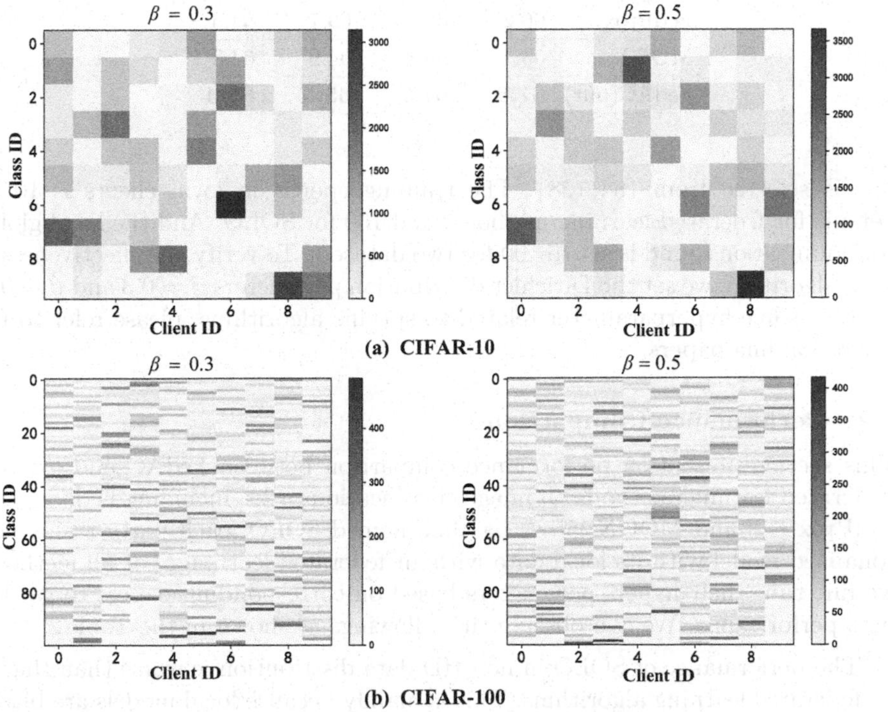

Fig. 5. Data distribution corresponding to different β values on CIFAR-10 and CIFAR-100 datasets.

Evaluation Measures. In the experiment, the Top-1 Accuracy is used to evaluate the performance of all models, the define of Accuracy is:

$$\text{Accuracy} = (TP + TN)/(P + N) \tag{9}$$

where TP, TN, P and N denote True Positives, True Negatives, Positives, and Negatives.

Implementation Details. Following the work of MOON [5], we use Simple-CNN (including 3 Convolution layers and 2 fully connected layers with ReLU activation) model on CIFAR-10. For CIFAR-100, the ResNet-18 [23] model is used as a base encoder. We use Stochastic Gradient Descent (SGD) optimizer with a learning rate 0.01 and weight decay 0.00001 for all methods. The batch

Table 2. The top-1 accuracy of FedAC and the other baselines on CIFAR-10 and CIFAR-100 datasets. For all datasets, dirichlet distribution parameter selection 0.3 and 0.5.

Methods	CIFAR-10		CIFAR-100	
	$\alpha = 0.3$	$\alpha = 0.5$	$\alpha = 0.3$	$\alpha = 0.5$
SOLO	46.8	47.3	22.5	23.4
FedAvg	65.9	66.2	63.8	64.1
FedProx	66.2	66.7	63.7	64.6
MOON	66.4	66.8	64.9	64.3
FedAC(our)	**67.6**	**67.7**	**65.5**	**65.4**

size is selected from $\{64, 128\}$. The training epochs in local clients is 10 by default for federated learning methods, and 100 for SOLO. And the local-global communication round is set to 100 for two datasets. To verify the effectiveness of each algorithm, we set the Dirichlet distribution parameters $\beta = 0.3$ and $\beta = 0.5$. Finally, some hyperparameter related to specific algorithms, please refer to the corresponding papers.

5.2 Performance Comparison

This section presents a performance comparison between FedAC and existing Federated learning methods on image classification tasks, including FedAvg [24], FedProx [12] and MOON [5]. A baseline named SOLO, each party trains personalized model with its local data without federated learning. For all methods, we fine-tune their hyper- parameters based on corresponding papers to get the best performance. We can observe the followings as shown in the Table 2:

- The performance of SOLO in non-IID data distribution is worse than that of federated learning algorithms. This is mainly because local models are biased classifiers with poor performance. And this verifies the benefits of federated learning.
- FedProx and MOON outperformed FedAvg on both datasets, this verifies that global knowledge distillation is conducive to guide local model training. And global features seem more useful than parameters.
- For different dirichlet distribution parameters, all methods achieve better performance at $\alpha = 5$, this is mainly because the data distribution is more dispersed at $\alpha = 3$.
- On both datasets, the proposed FedAC achieves significant performance improvement than existing methods, demonstrating optimize the local training process can bring performance gain to the global model.

5.3 Ablation Study

In this section, we investigate the effect of curriculum learning method for local training. From Table 3, the following observations can be drawn:

- For SOLO method, the FedAC(SOLO) achieves 5% performance improvement on the CIFAR10 at $\alpha = 0.5$. And it is more effective on CIFAR10 than on CIFAR100.
- For FedProx and MOON, FedAC version methods also make progress. This verifies that curriculum learning can bring performance gain for local model and global model.
- For both datasets, MOON-based methods are always better than the FedProx-based approaches, which may be because the feature is more representative of knowledge than the model parameters.

Table 3. The top-1 accuracy of SOLO, FedProx, MOON and their FedAC version.

Methods	CIFAR-10		CIFAR-100	
	$\alpha = 0.3$	$\alpha = 0.5$	$\alpha = 0.3$	$\alpha = 0.5$
SOLO	46.8	47.3	22.5	23.4
FedAC(SOLO)	**49.5**	**50.2**	**24.1**	**23.5**
FedProx	66.2	66.7	63.7	64.6
FedAC(FedProx)	**66.8**	**67.3**	**66.4**	**65.8**
MOON	66.4	66.8	64.9	64.3
FedAC(MOON)	**67.2**	**67.6**	**66.3**	**66.8**

5.4 Case Study

In this section, we will further analyze how FedAC improves local and global model performance in view of hidden vectors distribution. To this end, 100 images of all classes were randomly selected in the test set, which were unknowable in the training phase. In this paper, T-SNE [25] was used to explore the distribution changes of their corresponding representations. We randomly selected three clients and visualized their hidden vectors in the test set.

As shown in Fig. 5, the hidden vectors distribution of FedAvg and FedAC have a large difference. Specifically, the training model in FedAvg algorithm learns poor features, and the feature representation of most classes is even mixed, which can not be distinguished. For FedAC, we can observe that the points with the same class are more divergent in Fig. 6(b) compared with Fig. 6(a) (e.g., see class 0, 1 and 7). Compared with FedAC, FedAvg algorithm learns worse representation in local training phase. This may leads a poor performance of fused model. Therefore, improving the generalization ability of local models may also bring performance gains to global models.

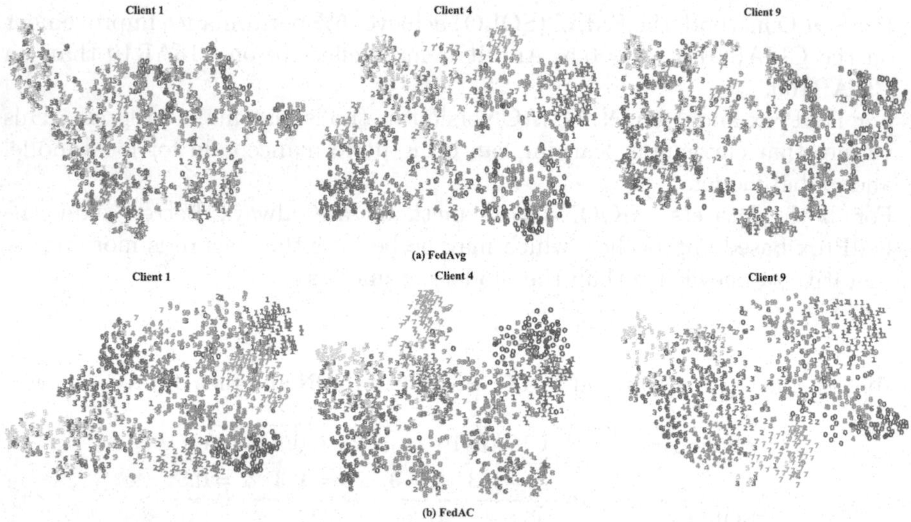

Fig. 6. T-SNE visualizations of hidden vectors on CIFAR-10.

6 Conclusion

As an effective method to solve data silos, federated learning has attracted attention in many fields, such as medicine and finance. To improve the performance of models on non-IID datasets, we proposed Adaptive curriculum Federated learning (FedAC), simple and effective approach for federated learning. To alleviate the negative impact of uneven quality, FedAC introduces a novel curriculum learning method for local training. It utilizes the complexity of the data to design the weighting method. Experiments results show that FedAC achieves significant improvement of local model to obtain the global gain on image classification tasks.

Future work of this study can be further explored in two directions. First, stronger curriculum inference techniques can significantly improve performance in guiding the local model to learn personalized knowledge. Second, FedAC can be applied to other problems, such as Natural Language Processing (NLP) and Recommendation System (RS).

Acknowledgments. This work is supported in part by the Excellent Youth Scholars Program of Shandong Province (Grant no. 2022HWYQ-048) and the Oversea Innovation Team Project of the "20 Regulations for New Universities" funding program of Jinan (Grant no. 2021GXRC073).

References

1. Liu, Y., et al.: Fedvision: an online visual object detection platform powered by federated learning. In: Proceedings of the AAAI Conference on Artificial Intelligence, pp. 13172–13179 (2020)

2. Yang, Q., Liu, Y., Chen, T., Tong, Y.: Federated machine learning: concept and applications. ACM Trans. Intell. Syst. Technol. (TIST), **10**(2), 1–19 (2019)
3. Kairouz, P., et al.: Advances and open problems in federated learning. arXiv preprint arXiv:1912.04977 (2019)
4. Kaissis, G.A., Makowski, M.R., Ruckert, D., Braren., R.F.: Secure, privacy-preserving and federated machine learning in medical imaging. Nat. Mach. Intell. 1–7 (2020)
5. Li, Q., He, B., Song, D.: Model-contrastive federated learning. In: Proceedings of the IEEE/CVF Conference on Computer Vision and Pattern Recognition, pp. 10713–10722 (2021)
6. Hao, W., El-Khamy, M., Lee, J., et al.: Towards fair federated learning with zero-shot data augmentation. In: Proceedings of the IEEE/CVF Conference on Computer Vision and Pattern Recognition, pp. 3310–3319 (2021)
7. Lin, T., Kong, L., Stich, S.U., et al.: Ensemble distillation for robust model fusion in federated learning. Adv. Neural Inf. Process. Syst. 23512363 (2020)
8. Duan, M., Liu, D., Chen, X., et al.: Astraea: self-balancing federated learning for improving classification accuracy of mobile deep learning applications. In: 2019 IEEE 37th International Conference on Computer Design (ICCD). IEEE, pp. 246–254 (2019)
9. Shen, T., Zhang, J., Jia, X., et al.: Federated mutual learning. arXiv preprint arXiv:2006.16765 (2020)
10. Zhu, Z., Hong, J., Zhou, J.: Data-free knowledge distillation for heterogeneous federated learning. In: International Conference on Machine Learning. PMLR, pp. 12878–12889 (2021)
11. Yao, X., Sun, L.: Continual local training for better initialization of federated models. In: 2020 IEEE International Conference on Image Processing (ICIP). IEEE, pp. 1736–1740 (2020)
12. Li, T., Sahu, A.K., Zaheer, M., Sanjabi, M., Talwalkar, A., Smith, V.: Federated optimization in heterogeneous networks. In: Third Conference on Machine Learning and Systems (MLSys), vol. 2, pp. 429–450 (2020)
13. Wu, C., Wu, F., Liu, R., et al.: FedKD: Communication Efficient Federated Learning via Knowledge Dis tillation. ArXiv, abs/2108.13323 (2021)
14. Li, D., Wang, J.: FEDMD: Heterogenous federated learning via model distillation. arXiv preprint arXiv:1910.03581 (2019)
15. Bengio, Y., Louradour, J., Collobert, R., et al.: Curriculum learning. In: Proceedings of the 26th Annual International Conference on Machine Learning, pp. 41–48 (2009)
16. Soviany, P., Ionescu, R.T., Rota, P., et al.: Curriculum learning: a survey. Int. J. Comput. Vis. 1–40 (2022)
17. Jiang, L., Meng, D., Zhao, Q., et al.: Self-paced curriculum learning. In: Twenty-Ninth AAAI Conference on Artificial Intelligence (2015)
18. Braun, S., Neil, D., Liu, S.C.: A curriculum learning method for improved noise robustness in automatic speech recognition. In: 2017 25th European Signal Processing Conference (EUSIPCO). IEEE, pp. 548–552 (2017)
19. Yao, D., Pan, W., Dai, Y., et al.: LocalGlobal Knowledge Distillation in Heterogeneous Federated Learning with NonIID Data. ArXiv, abs/2107.00051 (2021)
20. Wang, H., Yurochkin, M., Sun, Y., Papailiopoulos, D., Khazaeni, Y.: Federated learning with matched averaging. In International Conference on Learning Representations (2020)

21. Wang, J., Liu, Q., Liang, H., Joshi, G., Poor, H.V.: Tackling the objective inconsistency problem in heterogeneous federated optimization. Adv. Neural Inf. Process. Syst. **33** (2020)
22. Meng, L., Tan, A.H., Miao, C.: Salience-aware adaptive resonance theory for large-scale sparse data clustering. Neural Netw. **120**, 143–157 (2019)
23. He, K., Zhang, X., Ren, S., Sun, J.: Deep residual learning for image recognition. In: Proceedings of the IEEE Conference on Computer Vision and Pattern Recognition, pp. 770–778 (2016)
24. McMahan, B., Moore, E., Ramage, D., Hampson, S.Y., Arcas, B.A., et al.: Communication-efficient learning of deep networks from decentralized data. arXiv preprint arXiv:1602.05629 (2016)
25. van der Maaten, L., Hinton, G.: Visualizing data using t-sne. J. Mach. Learn. Res. **9**, 2579–2605 (2008)

A Novel Nonlinear Dictionary Learning Algorithm Based on Nonlinear-KSVD and Nonlinear-MOD

Xiaoju Chen, Yujie Li, Shuxue Ding, Benying Tan(✉), and Yuqi Jiang

School of Artificial Intelligence, Guilin University of Electronic Technology, Guilin 541004, China
{20022201005,20022201018}@mails.guet.edu.cn, {yujieli,sding, by-tan}@guet.edu.cn

Abstract. At present, scholars have proposed numerous linear dictionary learning methods. In the field of dictionary learning, linear dictionary learning is the most commonly applied method, and it is typically utilized to address various signal processing problems. However, linear dictionary learning cannot meet the requirements of nonlinear signal processing, and the nonlinear signals cannot be accurately simulated and processed. In this study, we first construct a nonlinear dictionary learning model. Then we propose two algorithms to solve the optimization problem. In the dictionary update stage, based on the K-SVD and the method of optimal directions (MOD), we design nonlinear-KSVD (NL-KSVD) and nonlinear-MOD (NL-MOD) algorithms to update the dictionary. In the sparse coding stage, the nonlinear orthogonal matching pursuit (NL-OMP) algorithm is designed to update the coefficient. Numerical experiments are used to verify the effectiveness of the proposed nonlinear dictionary learning algorithms.

Keywords: Sparse representation · Dictionary learning · Nonlinear · Nonlinear-KSVD · Nonlinear-MOD

1 Introduction

In the past few decades, there has been an increasing interest in studying dictionary learning and sparse representation, and many novel approaches have been proposed and applied in engineering. Dictionary learning and sparse representation have become popular research directions in machine learning [1, 2]. Learning a dictionary from a set of sample signals is called dictionary learning, which can better match the represented signal [3, 4]. A more accurate signal representation will be obtained by sparse coding.

With the development of dictionary learning, sparse representation using the dictionary learning method has certain advantages. It has been widely used in signal denoising, image denoising, image classification, indoor localization, face recognition, etc. [3–12]. The basic concept of sparse representation is that a signal can be linearly combined with a small number of atoms from an overcomplete dictionary [3, 13]. Good or bad dictionaries affect the performance of sparse representations, so sparse representation using

L. Fang et al. (Eds.): CICAI 2022, LNAI 13606, pp. 167–179, 2022.
https://doi.org/10.1007/978-3-031-20503-3_14

dictionary learning can improve the performance [13]. Usually, we formulate different sparsity constraints depending on the problem to obtain the solution we expect.

Various methods are usually used to perform the calculations for the different sparse constraints formulated. Calculating ℓ_0-norm can achieve strong sparsity, but calculating ℓ_0-norm is an NP-hard problem [3, 14]. The NP-hard problem can be solved by greedy algorithm, such as OMP [6, 15, 16]. Usually ℓ_0-norm is relaxed to ℓ_1-norm. The ℓ_1-norm is a convex optimization problem. This problem always uses the iterative shrinkage thresholding algorithm (ISTA) [17] as a solver. However, compared to ℓ_0-norm, ℓ_1-norm does not achieve strong sparsity and tends to be less effective for some applications.

As an improvement, a nonconvex sparsity constraint is introduced to obtain a more sparse solution. In [18], ℓ_p-norm ($0 < p < 1$) was introduced as a sparse regularization to solve the nonconvex model. The method improves the sparsity of the solution and obtains an almost unbiased solution. The generalized minimax-concave (GMC) was introduced as a sparse regularization in [19] to get a more exact solution while allowing a simple solving by convex optimization methods. The *log*-regularization was introduced in [20], which was solved by the proximal operator method, yielding more accurate estimated solutions.

In recent years, more researchers have proposed many methods for dictionary learning [3, 14, 21–23], and the well-known methods are MOD and K-SVD. The basic idea of the MOD method is to update the dictionary matrix by least-squares [4, 23], and the K-SVD method is to update each atom column by column of the dictionary matrix by SVD decomposition [3, 8, 23]. Both methods use the OMP as a solver for sparse coding. Direct optimization algorithm (Dir), majorization method (MM), GMC-based dictionary learning algorithm and other dictionary learning methods have been proposed by researchers one after another [14, 24–26].

Presently, research on dictionary learning is based on linear dictionary learning. Since nonlinear signals do not have linear sparse representation by dictionary atoms, they are unable to be accurately simulated. The drawback of linear dictionary learning has inspired research on nonlinear dictionary learning. In [27], a nonlinear dictionary learning was proposed and used for image classification. In [28], a nonlinear dictionary learning model was constructed based on deep neural networks and an optimization algorithm was developed to significantly improve the classification performance. In [29], a kernel regularized nonlinear dictionary learning method was developed and extensively experimentally validated to achieve excellent performance. Different with the above methods, in this study, we propose a new nonlinear dictionary learning model, which enforce the nonlinear operation on the dictionary and coefficient.

The contributions of this paper include the following aspects. Firstly, we function a new nonlinear dictionary learning model to meet the requirements of nonlinear signal processing. Secondly, we propose the NL-OMP algorithm to solve the nonlinear sparse coding problem. Thirdly, we propose two algorithms, NL-KSVD and NL-MOD, to solve the nonlinear dictionary learning problem. They are respectively based on the K-SVD and MOD methods for linear dictionary learning.

2 Model and Formulation

In this paper, the proposed nonlinear dictionary learning model is constructed by the following constrained optimized equation:

$$\min_{\mathbf{W},\mathbf{X}} \sum_i \|\mathbf{x}_i\|_0, \text{ s.t.} \|\mathbf{Z} - f(\mathbf{WX})\|_F^2 \leq \varepsilon, \tag{1}$$

where $\mathbf{X} \in \mathbb{R}^{n \times N}$ is the sparse matrix, $\mathbf{x}_i$ is the ith column of the sparse matrix $\mathbf{X}$, $\mathbf{Z} \in \mathbb{R}^{m \times N}$ denotes the sample signals, and $\mathbf{W} \in \mathbb{R}^{m \times n}$ is the overcomplete dictionary with $m < n$. Different with linear dictionary learning, nonlinear dictionary learning adds a nonlinear function f. f denotes a nonlinear and monotonic function, such as Sigmoid, Tanh, Leaky ReLU, etc., and $f(\cdot)$ denotes the mapping through the nonlinear function f for each element of the matrix. It is common to transform the constrained optimization into the unconstrained optimization, as follows:

$$\min_{\mathbf{W},\mathbf{X}} \|\mathbf{Z} - f(\mathbf{WX})\|_F^2 + \lambda \|\mathbf{X}\|_0, \tag{2}$$

where the first term denotes the approximation error, λ denotes the regularization parameter, and $\|\mathbf{X}\|_0 = \sum_i \|\mathbf{x}_i\|_0$ in the second term is used to achieve strong sparsity of $\mathbf{X}$. The corresponding optimization of (2) is as follows:

$$\begin{aligned} (\widehat{\mathbf{W}}, \widehat{\mathbf{X}}) &= \operatorname*{argmin}_{\mathbf{W},\mathbf{X}} \|\mathbf{Z} - f(\mathbf{WX})\|_F^2 + \lambda \|\mathbf{X}\|_0 \\ & s.t. \|\mathbf{w}_i\|_2 = 1, 1 \leq i \leq n. \end{aligned} \tag{3}$$

Here the dictionary matrix is column ℓ_2-norm normalized. And $\mathbf{w}_i$ denotes the *ith* column of the dictionary matrix $\mathbf{W}$. To directly solve (3) directly is difficult because there are two variables. The general solving method is to update the dictionary matrix and sparse coefficient matrix alternatively.

The nonlinear sparse coding step: fix the dictionary matrix $\mathbf{W}$, update the sparse matrix $\mathbf{X}$,

$$\widehat{\mathbf{X}} = \operatorname*{argmin}_{\mathbf{X}} \|\mathbf{Z} - f(\mathbf{WX})\|_F^2 + \lambda \|\mathbf{X}\|_0. \tag{4}$$

We use the equivalent transformation method to solve (4) in this study. Then we use the NL-OMP algorithm to solve the equivalent cost function, which is also solves the NP-hard problem of ℓ_0-norm. We will present the details in Sect. 3.1.

The nonlinear dictionary update step: fix the sparse matrix $\mathbf{X}$, update the dictionary matrix $\mathbf{W}$,

$$\begin{aligned} \widehat{\mathbf{W}} &= \operatorname*{argmin}_{\mathbf{W}} \|\mathbf{Z} - f(\mathbf{WX})\|_F^2 \\ & s.t. \|\mathbf{w}_i\|_2 = 1, 1 \leq i \leq n. \end{aligned} \tag{5}$$

We made an equivalent conversion of (5) and used the NL-KSVD and NL-MOD algorithms to solve it. More details will be described in Sect. 3.2.

3 Algorithm

This section details the process of solving the nonlinear dictionary learning and the nonlinear sparse representation problem through algorithms. We can obtain NL-OMP based on the OMP algorithm to solve the nonlinear sparse representation problem. This algorithm can get the sparse matrix **X**. We propose the nonlinear dictionary learning algorithms, NL-KSVD and NL-MOD, based on K-SVD [3] and MOD [4]. For the two algorithms, we can get the dictionary matrix **W**. A brief overview of the nonlinear dictionary learning algorithm is shown in Fig. 1.

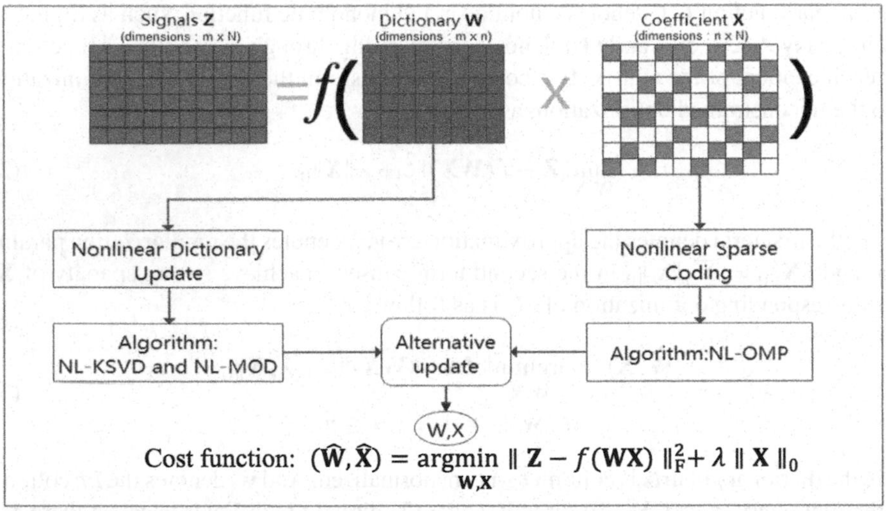

Fig. 1. Overview of the proposed nonlinear dictionary learning method. f is a nonlinear function, NL-KSVD and NL-MOD are used as solvers for updating nonlinear dictionary, and NL-OMP is used as a solver for nonlinear sparse coding.

Algorithm 1: Nonlinear sparse coding by NL-OMP

Input: Data $\mathbf{z} \in \mathbb{R}^m$, dictionary matrix $\mathbf{W} \in \mathbb{R}^{m \times n}$, the error threshold ε_0, sparse value T_0.
Calculate $f^{-1}(\mathbf{z})$; normalize the columns of $\mathbf{W}$ to the ℓ_2-norm.
Initialize: $k = 0, \mathbf{x}^{(0)} = \mathbf{0}, \mathbf{r}^{(0)} = f^{-1}(\mathbf{z}) - \mathbf{W}\mathbf{x}^{(0)} = f^{-1}(\mathbf{z}), S^{(0)} = \text{support}\{\mathbf{x}^{(0)}\} = \emptyset$.
for $k = 1$ to T_0 **do**
  for $j = 1 : n$
    $\mathbf{y}_j^* = \mathbf{w}_j^{\mathrm{T}} \mathbf{r}^{(k-1)}$
    $\varepsilon(j) = \left\| \mathbf{w}_j \mathbf{y}_j^* - \mathbf{r}^{(k-1)} \right\|_2^2$
    $\forall j \notin S^{(k-1)}, \varepsilon(j_0) \le \varepsilon(j), S^{(k)} = S^{(k-1)} \cup \{j_0\}$
    $\mathbf{x}^{(k)} = \underset{\mathbf{x}}{\operatorname{argmin}} \|\mathbf{W}\mathbf{x} - f^{-1}(\mathbf{z})\|_2^2, s.t.\ \text{support}\{\mathbf{x}\} = S^{(k)}$
  end for
  $\mathbf{r}^{(k)} = f^{-1}(\mathbf{z}) - \mathbf{W}\mathbf{x}^{(k)}$
  Stopping rule:stop if $\|\mathbf{r}^{(k)}\|_2 \le \varepsilon_0$
end for
output: $\mathbf{x}^{(k)}$

In this study, the most critical step to solve (4) and (5) lies in transforming the two equations into the following two forms:

$$\widehat{\mathbf{X}} = \underset{\mathbf{X}}{\operatorname{argmin}} \|f^{-1}(\mathbf{Z}) - \mathbf{W}\mathbf{X}\|_{\mathrm{F}}^2 + \lambda \|\mathbf{X}\|_0, \tag{6}$$

$$\widehat{\mathbf{W}} = \underset{\mathbf{W}}{\operatorname{argmin}} \|f^{-1}(\mathbf{Z}) - \mathbf{W}\mathbf{X}\|_{\mathrm{F}}^2, \tag{7}$$

where $f^{-1}(\cdot)$ denotes the inverse function of the monotonic function $f(\cdot)$, and the existence of $f^{-1}(\cdot)$ requires that $f(\cdot)$ is monotonic. By the above equivalent transformation, the sparse matrix $\mathbf{X}$ and the dictionary matrix $\mathbf{W}$ can be derived by alternately updating (6) and (7). The solver for nonlinear sparse coding uses the NL-OMP algorithm, and the solver for updating the nonlinear dictionary uses NL-KSVD and NL-MOD algorithms.

3.1 Nonlinear Sparse Coding Based on NL-OMP

In this study, the ℓ_0-norm regularization formulated in (6) is an NP-hard problem that can be solved by generalizing the OMP [30] algorithm to obtain NL-OMP, using it as a solver for nonlinear sparse coding, as listed in Algorithm 1.

3.2 Nonlinear Dictionary Update

In this study, two methods, K-SVD and MOD, are classical and effective in solving linear dictionary learning. The K-SVD algorithm is to update the whole dictionary by updating the dictionary atoms one by one, as detailed in [3, 8]. The MOD algorithm is to update the entire dictionary directly, as described in [4]. In this section, NL-KSVD and NL-MOD, based on K-SVD and MOD, will be introduced as solvers for updating the nonlinear dictionary.

Nonlinear Dictionary Update Based on NL-KSVD. We proposed NL-KSVD to solve the nonlinear dictionary learning problem. The process is shown below. In (7), we note that $\mathbf{w}_k$ is the kth atom of the dictionary matrix $\mathbf{W}$ to be updated currently, and $\mathbf{x}_\mathrm{T}^k$ is the kth row vector of the sparse matrix $\mathbf{X}$ to be updated currently. Then,

$$\begin{aligned}\mathbf{w}_k &= \arg\min_{\mathbf{w}_k} \left\| f^{-1}(\mathbf{Z}) - \sum\nolimits_{j=1}^{n} \mathbf{w}_j \mathbf{x}_\mathbf{T}^j \right\|_\mathbf{F}^2 \\ &= \arg\min_{\mathbf{w}_k} \left\| \left[f^{-1}(\mathbf{Z}) - \sum\nolimits_{j \neq k} \mathbf{w}_j \mathbf{x}_\mathbf{T}^j \right] - \mathbf{w}_k \mathbf{x}_\mathbf{T}^k \right\|_\mathbf{F}^2 \\ &= \arg\min_{\mathbf{w}_k} \| \mathbf{E}_k - \mathbf{w}_\mathbf{k} \mathbf{x}_\mathbf{T}^k \|_\mathbf{F}^2,\end{aligned} \tag{8}$$

where $\mathbf{E}_k = f^{-1}(\mathbf{Z}) - \sum_{j \neq k} \mathbf{w}_j \mathbf{x}_\mathrm{T}^j$ denotes the residuals.

At this moment, the optimization of (8) can be described as follows:

$$\mathbf{w}_k = \arg\min_{\mathbf{w}_k} \| \mathbf{E}_k - \mathbf{w}_k \mathbf{x}_\mathrm{T}^k \|_\mathrm{F}^2. \tag{9}$$

We need to adjust $\mathbf{w}_k$ and $\mathbf{x}_\mathrm{T}^k$ to make their products as close as possible to $\mathbf{E}_k$. If SVD decomposition of $\mathbf{E}_k$ directly is used to update $\mathbf{w}_k$ and $\mathbf{x}_\mathrm{T}^k$, then it will make $\mathbf{X}$ not sparse. To ensure sparsity, $\mathbf{x}_\mathrm{T}^k$ only retains the non-zero positions, and $\mathbf{E}_k$ only retains the term after the product of non-zero positions in $\mathbf{w}_k$ and $\mathbf{x}_\mathrm{T}^k$.

At this moment, the optimized equation can be described as follows:

$$\mathbf{w}_k = \arg\min_{\mathbf{w}_k} \| \mathbf{E}_k' - \mathbf{w}_k \mathbf{x}_\mathrm{T}^{\prime k} \|_\mathrm{F}^2. \tag{10}$$

We need to adjust $\mathbf{w}_k$ and $\mathbf{x}_\mathrm{T}^{\prime k}$ to make their product as close as possible to $\mathbf{E}_k'$. This can be achieved by adjusting $\mathbf{w}_k$ and $\mathbf{x}_\mathrm{T}^{\prime k}$ based on SVD decomposition of $\mathbf{E}_k'$, i.e., $\mathbf{E}_k' = \mathbf{U} \sum \mathbf{V}^\mathrm{T}$. We take the first column of the left singular matrix $\mathbf{U}$ as $\mathbf{w}_k$, i.e., $\mathbf{w}_k = \mathbf{U}(:, 1)$. We take the product of the first row of the right singular matrix and the first singular value as $\mathbf{x}_\mathrm{T}^{\prime k}$, i.e., $\mathbf{x}_\mathrm{T}^{\prime k} = \sum(1, 1)\mathbf{V}^\mathrm{T}(1, :)$. After getting $\mathbf{x}_\mathrm{T}^{\prime k}$, we update the original $\mathbf{x}_\mathrm{T}^k$ with its counterpart.

To summarize above, we propose the NL-KSVD method for updating the nonlinear dictionary, as listed in Algorithm 2.

Algorithm 2: Nonlinear dictionary update based on NL-KSVD

Input: Data matrix $\mathbf{Z} \in \mathbb{R}^{m \times N}$.
Calculate $f^{-1}(\mathbf{Z})$.
Initialize: Construct $\mathbf{W}_{(0)} \in \mathbb{R}^{m \times n}$, and randomly select m samples in $f^{-1}(\mathbf{Z})$.
Normalize the columns of $\mathbf{W}_{(0)}$ to the ℓ_2-norm.
for $j = 1$ to max iteration **do**
  Using NL-OMP for nonlinear sparse coding: $\mathbf{X}_{(j)} \in \mathbb{R}^{n \times N}$
  for $k = 1 : n$
    Calculate $\mathbf{E}_k^{'}$
    $\mathbf{E}_k^{'} = \mathbf{U}\Sigma\mathbf{V}^{\mathrm{T}}$
    $\mathbf{w}_k = \mathbf{U}(:,1)$
    $\mathbf{x}_{\mathrm{T}}^{'k} = \Sigma(1,1)\mathbf{V}^{\mathrm{T}}(1,:)$
    Corresponding update $\mathbf{x}_{\mathrm{T}}^{k}$
  end for
end for
Output: $\mathbf{W}_{(j)}$, $\mathbf{X}_{(j)}$

Algorithm 3: Nonlinear dictionary update by NL-MOD

Input: Data matrix $\mathbf{Z} \in \mathbb{R}^{m \times N}$.
Calculate $f^{-1}(\mathbf{Z})$.
Initialize: Construct $\mathbf{W}_{(0)} \in \mathbb{R}^{m \times n}$, and randomly select m samples in $f^{-1}(\mathbf{Z})$.
normalize the columns of $\mathbf{W}_{(0)}$ to the ℓ_2-norm.
for $j = 1$ to max iteration **do**
  Using NL-OMP for nonlinear sparse coding: $\mathbf{X}_{(j)} \in \mathbb{R}^{n \times N}$
  $\mathbf{W}_{(j)} = f^{-1}(\mathbf{Z})\mathbf{X}_{(j)}^{\mathrm{T}}(\mathbf{X}_{(j)}\mathbf{X}_{(j)}^{\mathrm{T}})^{-1}$
end for
Output: $\mathbf{W}_{(j)}$

Nonlinear Dictionary Update Based on NL-MOD. For nonlinear dictionary learning, we obtain NL-MOD based on MOD. NL-MOD is used to solve the nonlinear dictionary learning problem. A detailed overview of the specific algorithm is shown in Algorithm 3.

4 Numerical Experiments

In this section, we conduct numerical experiments to verify the algorithm proposed in the previous section. MATLAB is performed simulation experiments, and we discussed the scheme of nonlinear dictionary learning and nonlinear sparse representation. In the experiment, we conducted experiments using synthetic data to verify the effectiveness of the proposed algorithm for the nonlinear dictionary learning model.

We use proposed NL-KSVD and NL-MOD algorithms to solve the nonlinear dictionary learning problem. And we use NL-OMP as a solver for nonlinear sparse coding. Three evaluation indicators are used to evaluate: 1) the convergence performance of the relative error of the two proposed algorithms for nonlinear dictionary learning; 2) the convergence performance of the dictionary recovery ratio of the two proposed algorithms for nonlinear dictionary learning; 3) the convergence performance of the Hoyer sparsity of the two proposed algorithms for nonlinear dictionary learning.

4.1 Experimental Settings

In this section, we generate a dictionary matrix $\mathbf{W} \in \mathbb{R}^{20\times 50}$ whose elements obey a Gaussian distribution and normalize each column by the ℓ_2-norm. $\mathbf{W}$ is called the ground-truth dictionary. We use the dictionary $\mathbf{W}$ to generate the signals. Then, we generate a sparse matrix $\mathbf{X} \in \mathbb{R}^{50\times 1500}$ with three non-zero elements randomly in each column, which are independently located and obey a uniform distribution. The magnitudes of the values of these three elements follow a Gaussian distribution. For nonlinear dictionary learning, we make $\mathbf{Z} = f(\mathbf{WX})$, i.e., by combining the three columns of the dictionary matrix $\mathbf{W}$ linearly, the sample signals are generated by the mapping of the nonlinear function f. In this experiment, Leaky ReLU is chosen as the nonlinear function. The nonlinear dictionary learning model is solved by two algorithms, NL-KSVD and NL-MOD. For the nonlinear sparse representation, NL-OMP is used to solve it. In this paper, three evaluation indicators are used to evaluate the proposed algorithm for nonlinear dictionary learning. Among them, the Leaky ReLU function and its inverse function are defined as follows:

$$f(x) = \begin{cases} x, x \geq 0 \\ \alpha x, x < 0 \end{cases}, \tag{11}$$

$$f^{-1}(x) = \begin{cases} x, x \geq 0 \\ \frac{x}{\alpha}, x < 0 \end{cases}. \tag{12}$$

4.2 Evaluation Indicators

We validate our proposed algorithm's effectiveness by following three indicators, i.e., average relative error, average dictionary recovery ratio, and average Hoyer sparsity.

The expression used to evaluate the relative error are as follows:

$$\text{Relative error} = \frac{\|\mathbf{Z} - f(\widehat{\mathbf{W}}\widehat{\mathbf{X}})\|_F^2}{\|\mathbf{Z}\|_F^2}. \tag{13}$$

The method to calculate the dictionary recovery ratio is shown below. We compare the learned dictionary $\widehat{\mathbf{W}}$ with the ground-truth dictionary $\mathbf{W}$. For each atom $\{\mathbf{w}_i\}_{i=1}^n$ of the ground-truth dictionary $\mathbf{W}$ and each atom $\{\widehat{\mathbf{w}}_j\}_{j=1}^n$ of the learned dictionary $\widehat{\mathbf{W}}$, we calculate the distance between each $\mathbf{w}_i$ and $\widehat{\mathbf{w}}_j$. If $\left|\mathbf{w}_i^T\widehat{\mathbf{w}}_j\right| > 0.99$ or $1 - \left|\mathbf{w}_i^T\widehat{\mathbf{w}}_j\right| < 0.01$ is satisfied, the atom $\mathbf{w}_i$ is considered to be successfully recovered. Then we will obtain the recovery ratio of the learned dictionary by calculating the ratio of successfully recovered atoms to the total atoms.

Hoyer sparsity is used to evaluate the sparsity of a sparse matrix, which is defined as follows:

$$\text{Sparsity}(\mathbf{x}_i) = \frac{\sqrt{n} - \|\mathbf{x}_i\|_1 / \|\mathbf{x}_i\|_2}{\sqrt{n} - 1} \in [0, 1], \tag{14}$$

where $\mathbf{x}_i$ denotes the *ith* column of the sparse matrix $\widehat{\mathbf{X}}$ and n denotes the dimension of the sparse vector $\mathbf{x}_i$. We can calculate the sparsity of each column of the sparse matrix $\widehat{\mathbf{X}}$ by (14). Then, we take the average to get the Hoyer sparsity of the sparse matrix.

4.3 Experimental Results

Firstly, based on the three evaluation indicators in Sect. 4.2, we evaluate different parameters α. Comparing the result, we select α which performs best as the parameters in nonlinear function Leaky ReLU. Then we use the selected α to analyze the proposed NL-KSVD and NL-MOD. The nonlinear sparse coding is solved by NL-OMP. We evaluate the above algorithms based on the three evaluation indicators. All the results are the average obtained by repeating the experiments 50 times.

Experimental Results for Different Parameters. We observe the results of three evaluation indicators corresponding to each different α. For each parameter α, we perform 200 iterations and repeat the experiment 50 times to get the average result. Table 1 and Table 2 show the performance of varying α corresponding to NL-KSVD and NL-MOD on the three indicators, respectively.

As seen in Table 1 and Table 2, NL-KSVD and NL-MOD show the best performance when $\alpha = 0.075$. Therefore, the parameter α selected as 0.075 is the optimal choice for the currently generated synthetic data.

Table 1. Performance of three indicators corresponding to different parameters α on NL-KSVD algorithm

α	0.060	0.065	0.070	**0.075**	0.080	0.085	0.090	0.095
Relative error	0.0131	0.0133	0.0150	**0.0118**	0.0134	0.0122	0.0132	0.0132
Sparsity	0.9183	0.9183	0.9180	**0.9185**	0.9183	0.9185	0.9183	0.9183
Recovery ratio (%)	92.920	92.920	91.120	**93.840**	92.600	93.320	92.640	93.200

Table 2. Performance of three indicators corresponding to different parameters α on NL-MOD algorithm

α	0.060	0.065	0.070	**0.075**	0.080	0.085	0.090	0.095
Relative error	0.0137	0.0139	0.0137	**0.0127**	0.0138	0.0135	0.0132	0.0133
Sparsity	0.9182	0.9182	0.9183	**0.9184**	0.9182	0.9182	0.9183	0.9183
Recovery ratio (%)	91.880	92.120	92.040	**92.840**	92.640	92.240	92.960	92.640

Experimental Results for Optimal Parameters. The parameter $\alpha = 0.075$ is chosen. At the same time, NL-KSVD and NL-MOD are evaluated using three indicators: relative error, sparsity, and recovery ratio. Moreover, the average results are obtained by repeating the experiment 50 times.

Figure 2, Fig. 3, and Fig. 4 show that both algorithms have better performance and verify the effectiveness of the proposed two algorithms for nonlinear dictionary learning. Figure 2 and Fig. 3 show that the NL-KSVD outperforms the NL-MOD algorithm in terms of the average relative error and the average dictionary recovery ratio. Figure 4 shows that the NL-KSVD and the NL-MOD algorithm perform almost the same and both can achieve high sparsity in terms of the average Hoyer sparsity.

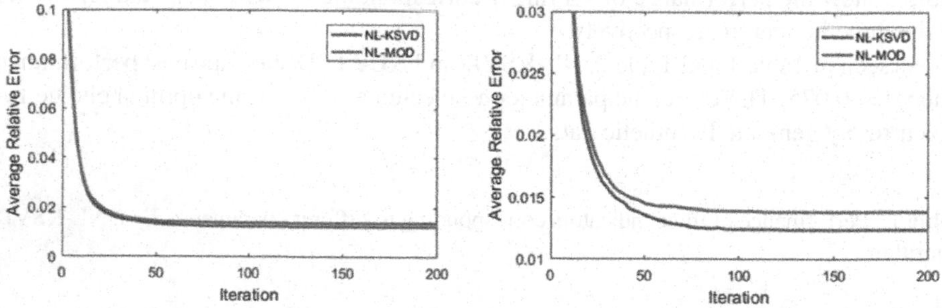

Fig. 2. On the left is the relationship between the number of iterations and the average relative error for the proposed NL-KSVD and NL-MOD algorithms; on the right is the partial enlarged picture.

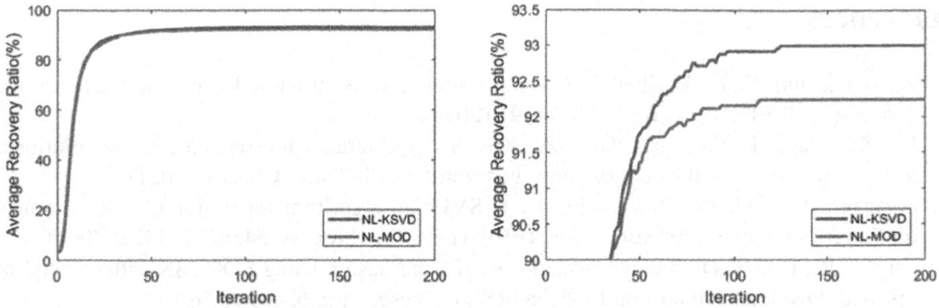

Fig. 3. On the left is the relationship between the number of iterations and the average recovery ratio for the proposed NL-KSVD and NL-MOD algorithms; on the right is the partial enlarged picture.

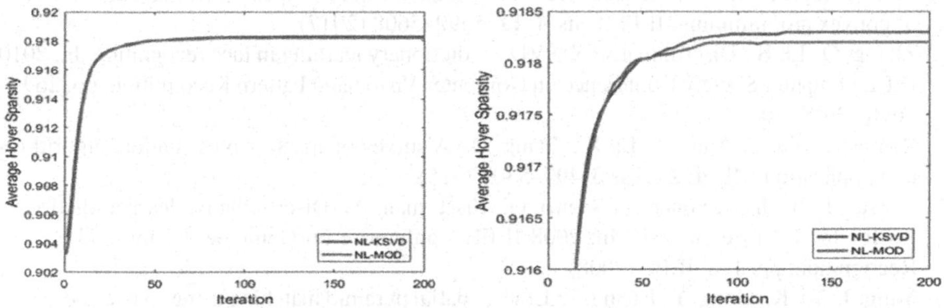

Fig. 4. On the left is the relationship between the number of iterations and the average Hoyer sparsity for the proposed NL-KSVD and NL-MOD algorithms, and on the right is the partial enlarged picture.

5 Conclusions and Discussions

As linear dictionary learning methods can hardly meet the requirements for signal processing, we propose a nonlinear dictionary learning model to meet the requirements of nonlinear signal processing. We propose the NL-KSVD and NL-MOD algorithms in this study. We use the above two algorithms and NL-OMP to update the dictionary and sparse matrix alternately. Experiments with synthetic signals verify the effectiveness of nonlinear dictionary learning. The experiments show that both algorithms have better performance. The nonlinear dictionary learning algorithm achieves better performance on synthetic data. In future work, it is expected that the nonlinear dictionary learning algorithm will be applied to real experiments such as image denoising and neural networks.

Acknowledgements. This research was partially funded by the Guangxi Postdoctoral Special Foundation and the National Natural Science Foundation of China under Grants 61903090 and 62076077, respectively.

References

1. Zhao, H., Ding, S., Li, X., Huang, H.: Deep neural network structured sparse coding for online processing. IEEE Access **6**, 74778–74791 (2018)
2. Li, Z., Zhao, H., Guo, Y., Yang, Z., Xie, S.: Accelerated log-regularized convolutional transform learning and its convergence guarantee. IEEE Trans. Cybern. (2021)
3. Aharon, M., Elad, M., Bruckstein, A.: K-SVD: an algorithm for designing overcomplete dictionaries for sparse representation. IEEE Trans. Sig. Process. **54**, 4311–4322 (2006)
4. Engan, K., Rao, B.D., Kreutz-Delgado, K.: Frame design using FOCUSS with method of optimal directions (MOD), In: Proc. NORSIG, Citeseer, pp. 65–69 (1999)
5. Dumitrescu, B., Irofti, P.: Dictionary Learning Algorithms and Applications. Springer, Cham (2018). https://doi.org/10.1007/978-3-319-78674-2
6. Elad, M.: Sparse and Redundant Representations: From Theory to Applications in Signal and Image Processing. Springer, New York (2010). https://doi.org/10.1007/978-1-4419-7011-4
7. Li, X., Ding, S., Li, Z., Tan, B.: Device-free localization via dictionary learning with difference of convex programming. IEEE Sens. J. **17**, 5599–5608 (2017)
8. Zhang, Q., Li, B.: Discriminative K-SVD for dictionary learning in face recognition. In: 2010 IEEE computer Society Conference on Computer Vision and Pattern Recognition, pp. 2691–2698. IEEE (2010)
9. Zhang, Z., Xu, Y., Yang, J., Li, X., Zhang, D.: A survey of sparse representation: algorithms and applications. IEEE Access **3**, 490–530 (2015)
10. Mairal, J., Bach, F., Ponce, J., Sapiro, G., Zisserman, A.: Discriminative learned dictionaries for local image analysis. In: 2008 IEEE Conference on Computer Vision and Pattern Recognition, pp. 1–8. IEEE (2008)
11. Yang, J., Yu, K., Gong, Y., Huang, T.: Linear spatial pyramid matching using sparse coding for image classification. In: 2009 IEEE Conference on Computer Vision and Pattern Recognition, pp. 1794–1801. IEEE (2009)
12. Zhang, K., Tan, B., Ding, S., Li, Y., Li, G.: Cybernetics, device-free indoor localization based on sparse coding with nonconvex regularization and adaptive relaxation localization criteria. Int. J. Mach. Learn. **I**, 1–15 (2022)
13. Hu, J., Tan, Y.-P.: Nonlinear dictionary learning with application to image classification. Pattern Recogn. **75**, 282–291 (2018)
14. Tan, B., Li, Y., Zhao, H., Li, X., Ding, S.: A novel dictionary learning method for sparse representation with nonconvex regularizations. Neurocomputing **417**, 128–141 (2020)
15. Tropp, J.A.: Greed is good: algorithmic results for sparse approximation. IEEE Trans. Inf. Theory **50**, 2231–2242 (2004)
16. Wang, J., Kwon, S., Shim, B.: Generalized orthogonal matching pursuit. IEEE Trans. Sig. Process. **60**, 6202–6216 (2012)
17. Daubechies, I., Defrise, M., De Mol, C.: An iterative thresholding algorithm for linear inverse problems with a sparsity constraint. Commun. Pure **57**, 1413–1457 (2004)
18. Cui, A., Peng, J., Li, H., Wen, M., Jia, A. J.: Mathematics, Iterative thresholding algorithm based on non-convex method for modified lp-norm regularization minimization, **347**, 173–180 (2019)
19. Selesnick, I.: Sparse regularization via convex analysis. IEEE Trans. Sig. Process. **65**, 4481–4494 (2017)
20. Li, Z., Ding, S., Hayashi, T., Li, Y.: Incoherent dictionary learning with log-regularizer based on proximal operators. Digit. Sig. Process. **63**, 86–99 (2017)
21. Kreutz-Delgado, K., Murray, J.F., Rao, B.D., Engan, K., Lee, T.-W., Sejnowski, T.: Dictionary learning algorithms for sparse representation. Neural Comput. **15**, 349–396 (2003)

22. Van Nguyen, H., Patel, V.M., Nasrabadi, N.M., Chellappa, R.: Kernel dictionary learning. In: 2012 IEEE International Conference on Acoustics, Speech and Signal Processing (ICASSP), pp. 2021–2024. IEEE (2012)
23. Cai, S., Weng, S., Luo, B., Hu, D., Yu, S., Xu, S.: A dictionary-learning algorithm based on method of optimal directions and approximate K-SVD. In: 2016 35th Chinese control conference (CCC), pp. 6957–6961. IEEE (2016)
24. Yaghoobi, M., Blumensath, T., Davies, M.E.: Dictionary learning for sparse approximations with the majorization method. IEEE Trans. Sig. Process. **57**, 2178–2191 (2009)
25. Li, Z., Yang, Z., Zhao, H., Xie, L.S.: Systems, direct-optimization-based dc dictionary learning with the MCP Regularizer (2021)
26. Rakotomamonjy, A.: Direct optimization of the dictionary learning problem. IEEE Trans. Sig. Process. **61**, 5495–5506 (2013)
27. Hu, J., Tan, Y.-P.: Nonlinear dictionary learning with application to image classification, **75**, 282–291 (2018)
28. Zhang, H., Liu, H., Song, R., et al.: Nonlinear dictionary learning based deep neural networks. In: 2016 International Joint Conference on Neural Networks (IJCNN), pp. 3771–3776. IEEE (2016)
29. Liu, H., Liu, H., Sun, F., et al.: Kernel regularized nonlinear dictionary learning for sparse coding. IEEE Trans. Syst. Man Cybern. Syst. **49**(4), 766–775 (2017)
30. Tropp, J.A., Gilbert, A.C.: Signal recovery from random measurements via orthogonal matching pursuit. IEEE Trans. Inf. Theory **53**, 4655–4666 (2007)

Tooth Defect Segmentation in 3D Mesh Scans Using Deep Learning

Hao Chen(✉), Yuhao Ge(✉), Jiahao Wei(✉), Huimin Xiong(✉), and Zuozhu Liu(✉)

ZJU -UIUC Institute, ZJU -Angelalign R&D Center for Intelligent Healthcare, Zhejiang University, Zhejiang, China
{haoc.19,yuhao.19,jiahao.19,huimin.21,zuozhuliu}@intl.zju.edu.cn

Abstract. Computer-aided systems are widely used in digital dentistry to help human experts for accurate and efficient diagnosis and treatment planning. In this paper, we study the problem of tooth defect segmentation in 3-Dimensional (3D) mesh scans, which is a prerequisite task in many dental applications. Existing models usually perform poorly in this task due to the highly imbalanced characteristic of tooth defects. To tackle this issue, we propose a novel triple-stream graph convolutional network named TripleNet to learn multi-scale geometric features from mesh scans for end-to-end tooth defect segmentation. With predefined geometrical features as inputs and a focal loss for training guidance, we achieve state-of-the-art performance on 3D tooth defect segmentation. Our work exhibits the great potential of artificial intelligence for future digital dentistry.

Keywords: Tooth defect segmentation · Deep learning · 3D intraoral scans · Point cloud · Imbalanced data

1 Introduction

Tooth defect delineation in the intraoral scans (IOS) is an import diagnosis step in dentistry, while effective automatic solutions are yet under development. In this paper, we formulate is as a 3-dimensional (3D) semantic segmentation task over the point clouds downsampled from IOSs, where the tooth defect areas such as tooth holes should be segmented correctly and completely for each 3D tooth. The main challenges for this task are as follows: (1) diversity and heterogeneity of teeth. The surface shape and structure vary across different types of teeth and even within the same type of teeth, thus affecting judgment on the defect areas. Meanwhile, it's difficult to distinguish the naturally occurring complex tooth structures and the defect areas. (2) class imbalance. Namely, the area of the normal tooth areas is much larger than the defect areas.

On the one hand, as potential solutions, diverse Deep Learning (DL) based methods have been proposed for 3D object segmentation in point clouds. Previous research can be coarsely cast into two categories. The first line attempts

These authors contributed equally.

L. Fang et al. (Eds.): CICAI 2022, LNAI 13606, pp. 180–191, 2022.
https://doi.org/10.1007/978-3-031-20503-3_15

to handle the inputs in the regular representation domain, where the input shapes are transformed into 3D volume grids [6,9,20] or multi-view 2D images sets [2,12,23], which are subsequently fed to typical Convolution Neural Networks(CNNs) for representation learning. However, geometric information loss and quantization artifacts are inevitably introduced in this process and the voxelization usually brings extra computational and memory costs.

Another stream directly operates on the point clouds. Some pioneering works process each point independently by shared multilayer perceptron (MLP), after which a symmetric aggregation function is applied to capture permutation invariant global features [16,17]. Recently, more advanced architectures are proposed. For example, the graph-based methods perform feature learning and aggregation on the graph constructed by considering each point as a vertex of a graph and generating edges based on its neighbors [21]. The convolution-based methods learn convolutional kernels according to local neighborhoods for aggregating features adaptively [13,26,27]. The transformer-based methods utilize attention mechanisms to model long-range dependencies for better feature aggregation [7,28]. Nevertheless, due to the shape and structure discrepancy between teeth and natural shapes, and the data class unbalance in the tooth dataset, these fashions may not work ideally when being applied to the tooth defect segmentation task. Therefore, it's necessary to devise specialized networks.

On the other hand, many 3D tooth-related tasks receive extensive attention recently and various solutions have been proposed to cope with them, e.g. 3D tooth segmentation [4,8,10,14,24,29], dental landmark localization [11,19,25], dental restoration [1], dental diagnosis [18], teeth margin detection [3], orthodontic treatment process [5,22] etc. However, tooth defect segmentation, which is equally important in clinical practice, has not been clearly investigated yet.

To this end, we propose a triple-stream graph convolutional network called TripleNet to learn abundant geometric information for end-to-end tooth defect segmentation on 3D IOSs. Firstly, by taking the central point of each triangular face, the raw tooth mesh is transformed to a tooth point cloud. Each point is associated with an 18-dimension geometric feature vector extracted from the original mesh face, including 3D Euclidean coordinates, face normals, face shape descriptors, and additional geometric features. Thereafter, in order to capture multi-scale geometric contextual information, our network is designed as three parallel streams, where each is constructed based on EdgeConv [21] operation with different neighborhood ranges to directly consume the tooth point cloud, thus facilitating the capture of multi-scale structural information. Then the features output from triple streams are aggregated to yield the final predictions. Considering the data imbalance over the normal areas and the defect areas, we adopt the focal loss[15] to pay more attention to the hard parts, namely the defect areas for the balance of data processing.

In summary, the main contributions of our paper are three folds:

• We are the first to explore the tooth defect segmentation approaches for 3D dental IOS models and propose a simple yet effective segmentation framework named TripleNet to tackle this challenging task. We hope that our work could inspire more deep learning based methods to flourish in this specific field.

- We devise effective features tailored for 3D tooth data as the input of the network and propose the triple stream graph convolutional neural network to capture the multi-scale geometric contexts for more accurate segmentation of the defect areas. Besides, focal loss is used to handle the data imbalance issue.
- We conduct extensive experiments to verify the effectiveness of our network and achieve great performance improvement on this task compared to the baseline. Experimental results indicate that our TripleNet has achieved state-of-the-art performance on 3D tooth defect segmentation.

2 Related Work

2.1 3D Shape Segmentation

Processing as Grid-Structured Data. This type of approaches transforms the input into regular representation domain, after which standard CNNs are applied to handle it, e.g. the discretization-based methods voxelize the shape surfaces into 3D volume grids to construct 3D CNNs [6,9,20] and the projection-based methods render the surfaces into multi-view 2D image collections then use 2D CNNs to extract features [2,12,23]. Though existing effective implementations, the discretization-based methods still result in expensive computational and memory overhead as the spatial resolution increases. Both operations inevitably lead to information loss and transformation artifacts.

Directly Operating on Point Clouds. Rather than doing any transformation, this type of manners directly deals with raw input surfaces with DL based schemes in the irregular domain. As a pioneering work, PointNet [16] learns the spatial encoding of the input points by successive shared MLP layers and applies a symmetrical max pooling function to obtain global features. However, it fails to capture local dependency effectively. To this end, PointNet++ [17] was proposed to learn local structure information by applying PointNet recursively on a nested partitioning of the input point cloud. Besides, DGCNN [21], which constructs local neighborhood graphs and applies EdgeConv operations on the edges connecting neighboring pairs of points, can recover topology inherently lacking for point clouds. PointCNN [13] learns χ-transformations from input point clouds to simultaneously weight and re-permute the input features for subsequent general convolution. In view of the learned χ-transformations are not permute-invariant, PointConv [26] was proposed to directly learn the corresponding kernel weights for different points combining with the point cloud density at different locations, yet suffering from heavy computation and memory costs. While PAConv [27] constructs convolution kernels by combining basic weight matrices in Weight Bank, thus having higher efficiency. Recently, transformer-based networks [7,28] arose in the point cloud processing area.

Although these methods perform well in general shape segmentation dataset, e.g. ShapeNet or ModelNet, they are not suitable for our tooth defect segmentation task. First, the teeth differ from the natural objects in the standard datasets in terms of shape and structure. They have diversity and heterogeneity, namely though common geometric characteristics exist in human teeth, their shapes vary dramatically across individuals. Second, the data unbalance is ubiquitous in our tooth dataset, in which the area of the defect part is pretty small compared to the non-defective part, hampering the segmentation accuracy significantly.

2.2 3D Teeth-related Tasks

In recent years, 3D teeth-based explorations have been increasingly launched for better clinical practice in digital dentistry. The first is 3D tooth segmentation, whose target is to delineate the gingiva and each tooth on the 3D dental surface models from IOSs. Conventional methods usually separate the dental models relying on predefining geometry criteria such as the face curvatures [10], harmonic fields [29], morphological skeleton [24] etc. The popular deep learning based approaches learn task-oriented geometry representations by CNNs for fully-automated tooth segmentation [4,8,14].

The second is dental landmark localization/digitalization, i.e. localizing anatomical landmarks of dental models automatically, which is essential for analyzing dental models in orthodontic planning or orthognathic surgery yet substantially sensitive to varying shape appearances of teeth. Some deep learning based methods are proposed to solve it by regressing the landmark heatmaps or enabling attribute transfer, e.g. TS-MDL [25], c-SCN [19], . DLLNet [11].

Furthermore, the 3D teeth-related tasks include dental restoration [1], dental diagnosis [18], teeth margin detection [3], orthodontic treatment process [5,22] etc.

Still, as a fundamental and crucial task in the clinical diagnosis and treatment, the tooth defect segmentation solutions are never probed, motivating us to develop a fully automated manner to free the dentists from time-consuming and tedious annotation. Therefore, in this paper, we are the first to study the 3D tooth defect segmentation problem and propose a simple but effective novel framework to solve it.

3 Method

3.1 Data Pre-processing

Input Conversion and Label Extraction. Each individual tooth case contains the raw tooth mesh from IOS which represents the structure of the tooth, and the patched mesh representing the defect area that needs to be filled which is created by the dentists. During pre-processing, by taking the center of each triangular face in mesh as the corresponding point in the point cloud, the raw tooth mesh is transformed into the point cloud form as it has merits like permutation invariance and is easier to handle. Then, for each produced point, we calculate the signed distance between it and the surface of the mesh of the defect area and compare it with a threshold. The point will be labeled as a part of hole if it is very close to or inside the defect area surface.

Feature Extraction. As the point cloud is taken as the input of the network, topological structure information will be lost while we transfer the mesh data into the point cloud data. Meantime, in view of the diversity and heterogeneity of teeth, local geometrical features should be recorded as detailed as possible to help the model learn the common features of defect area among different kinds of teeth and distinguish that with the normal rugged area. Therefore, it is necessary to extract local features from the mesh data and concatenate those onto each point they transfer to in order to maintain those structural information. Two kinds of features are devised in our work, as elaborated as follows.

Coordinate and Vector Information. 15 dimensions of input data is of this kind, including coordinates of the point (3D), normal vector of the triangle face (3D), corner feature of the triangle (9D). As illustrated in Fig. 1, c is the triangle center, v_1, v_2, and v_3 are the vertices of the triangle, and the corner feature consists of those three vector v1-c, v2-c, v3-c.

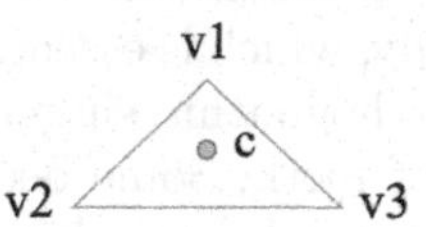

Fig. 1. The corner feature of a triangle face consists of the vector of v1-c,v2-c,v3-c

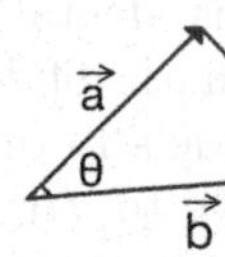

Fig. 2. Computation of the inner angle in a triangle face

Calculated Geometrical Information. As only the coordinate and vector information may not be enough for feature extraction, more geometrical information is needed. We use the cosine value of three triangle inner angles to use as supplement which can be easily calculated by $Cos(\theta) = \frac{\vec{a} \cdot \vec{b}}{|\vec{a}||\vec{b}|}$ where $\vec{a}$ and $\vec{b}$ are two adjacent edge vectors that make up the angle, as shown in Fig. 2. This feature is used to provide the local shape of mesh and enable the model to capture the overall shape of the tooth regions. However, this feature has a relatively small scale and it will be easily ignored while using the baseline method, which will be explained in the experiment section. The TripleNet framework is introduced to solve this problem which will be introduced in the next section.

3.2 Model Architecture

Our model tackles the problem of the diversity and heterogeneity of teeth surface in the segmentation task by utilizing different types of features and combine them while training the model and making predictions. Besides, inspired by DGCNN [21], the Edge-Conv operator is used to learn those local features. We further

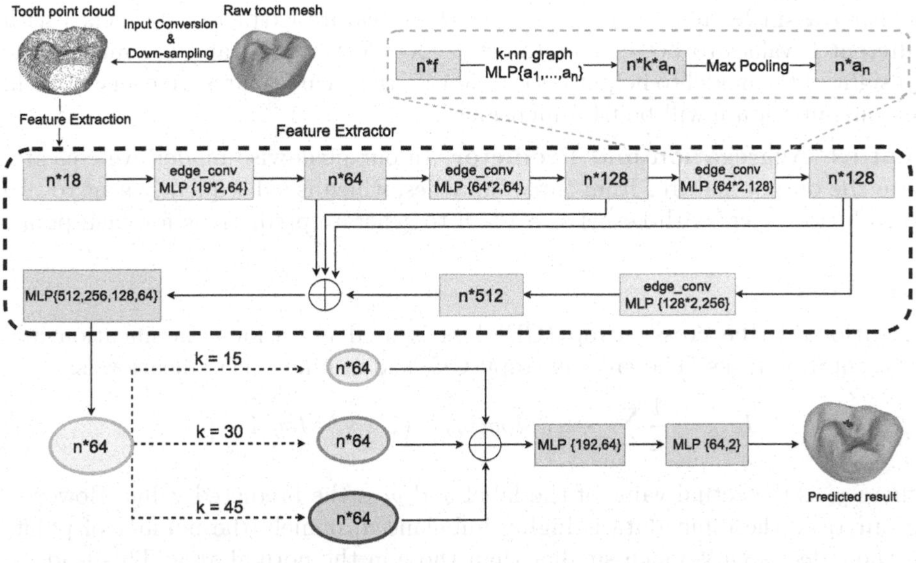

Fig. 3. Model architecture: Our model takes n points sampled from the raw tooth mesh as input, the upper part is the feature extractor, which is inspired from the DGCNN. It uses edge convolution to capture the local feature of each point. In our model, we use three branches of such extractor with different size of k, the number of the nearest neighbor of each point. The output of three extractors will be concatenated and send the feature predictor to make pixel-wise prediction.

adopt a multi-scale design for more comprehensive feature learning. The overall pipeline is illustrated in Fig. 3.

Feature Learning. Our input point cloud has $F = 18$ dimensions as mentioned above, and is denoted as $X = \{x_1, ..., x_n\}$, where $x_i \in R^F$. Due to the complexity of the teeth feature, it will be tough to capture the information with a single scale view. In this case, we propose a multi-scale feature extraction architecture. To be specific, we use edge convolution to capture local features of the top k nearest neighbors of each point x_i. The edge features of each point are the concatenation of its own features and the difference with its nearest k neighbors, namely

$$h_\Theta(x_i, x_j) = h_\Theta(x_i, x_i - x_j) \tag{1}$$

where the x_j is the k nearest neighbors of x_i. And the EdgeConv operator can be written as:

$$e_{ij} = ReLU(\theta \cdot (x_j - x_i) + \phi_i \cdot x_i) \tag{2}$$

where the $e_{ij} \in R^{2F}$ indicates the edge features of edge consisting of x_i and x_j. Among them, the learnable parameters Θ can be formulated as

$$\Theta = (\theta_1, ..., \theta_k, \phi_1, ...\phi_k) \tag{3}$$

Same as the feature extractor in the DGCNN, we apply the convolution layer with kernel size 1×1 after each edge convolution operation. However, instead of using a single value of k, we set up three feature extraction branches with different k values to better maintain the local feature. We aim to enhance the eyesight of the model to help it better identify the local features. In every branch, the output feature will be 64 dimensions.

Feature Aggregation and Predictor. In our pixel-wise model, we concatenate the output features from three branches, which is subsequently sent to two convolution layers with kernel size $1 \times n$ to generate predictions for each point.

3.3 Loss

Traditionally, the cross-entropy(CE) loss is used in various classification and segmentation tasks. The cross-entropy(CE) loss of binary classification is:

$$L_{CE} = \frac{1}{n} \sum_{i} -\left[y_i \cdot log(p_i) + (1 - y_i) \cdot log(1 - p_i)\right] \tag{4}$$

where y_i is the actual value of the label and p_i is the predicted value. However, in our task, the input data is highly imbalanced, namely the number of points in the defect area is much smaller than those in the normal area. This leads to a low weight learning performance and thus a bad prediction result as the loss is small even when predicting all the areas as good. To solve this problem, focal loss [15] is used in our task.

$$L_{Focal} = w_{y_i} \times (1 - p_i)^2 \times \frac{1}{n} \sum_{i} -\left[y_i \cdot log(p_i) + (1 - y_i) \cdot log(1 - p_i)\right] \tag{5}$$

Here, y_i is the actual value of the label and p_i is the predicted value. In our experiment, we set the weight w_y of the class of normal regions to be 0.04 and 0.96 for the class of defect regions.

4 Experiment

4.1 Dataset and Experimental Setup

The input dataset consists of 32 different kinds of teeth with 16 on the maxillary and 16 on the mandible. Although the features between different kinds of teeth are slightly different, one general framework is used to conduct the prediction. In general, there are 2884 teeth samples in total. 70% of the data is used as the train set, 20% of the data is used as the test set, and 10% of the data is used as the validation set. Our experiments run on four 3090 GPUs, and we set the batch size to be four. We use the Adam optimizer. The total iteration of each experiment is 100. For each tooth sample, we select 8000 points randomly for training, while all the points of the test data are used in the testing. As for criterion, we use the intersection over union(IoU) as our main standard while the precision, the accuracy of defect area Pos_{acc} and F1 score are also used as the auxiliary measurements. In our problem, we hope to extract the whole defected area but focus less on the healthy area, so all the IoU and accuracy is only with respect to the defected class. In our data, we identify the defect area of tooth as positive area and the healthy area as negative area.

4.2 The Overall Performance

As our TripleNet model is closely related to DGCNN, we use the DGCNN model as our baseline. The cross-entropy loss is used and the input feature dimension is 15 which only includes the coordinate and vector information. As for the TripleNet, we use 18 dimension features and focal loss as mentioned in the previous chapter to show our final result. We compare the performance of the TripleNet with all methods used and the baseline as shown in Table 1, to represent the overall improvement of our work. The result shows that the baseline have very low accuracy of segmentation. Comparing with the baseline, the TripleNet improves almost 47% IOU and 7.67% precision on this problem which proves the effectiveness of our solution to this task. To make the experiment more comprehensive, PointNet and PointNet++ are also tested in out task which also accept 15-D input feature and use the cross-entropy loss. However, the performance is extremely bad as they even classify all the point to normal area which result in 0 accuracy and IoU, so they are not included in the chat. As highly imbalanced data is one of the most important issue in this task, the use of focal loss contributes a lot for the performance improvement which meet the expectation. Although the introduction of our new local features and the TripleNet framework may not have such a big influence on the IoU as the use of focal loss, the overall precision increase in the visualization is still obvious. The details will be introduced the next two sections.

Table 1. Comparison of the overall performance. "ce" indicates the cross-entropy loss and "focal" indicates the focal loss. The number in the parenthesis indicates the dimension of the features used in the experiment. We compute the IoU with respect to the defect part.

Model	Loss	IoU(%)	Precision(%)	Pos_{acc}(%)	F1(%)
DGCNN (15)	Ce	30.26	76.08	30.59	39.74
Triple (18)	Focal	**77.57**	**83.75**	**90.52**	**86.25**

4.3 Ablation Studies

The Effectiveness of Loss. As for Table 2, we mainly focus on the validation of the improvement from the loss. While keeping the model and the features the same, the performance of different kinds of loss function can be compared. The result clearly shows that focal loss has great improvement on the performance on our imbalanced data compared with the Cross-Entropy loss. In this table we can find that the focal loss improves the performance at a large amount and the mechanism has been discussed in Sect. 3.5. For the PointNet models, it improves by around 70% IoU and for the DGCNN model, it improves by around 42% IoU. According to the experiment results, we conclude that the focal loss brings a great improvement on this problem.

Table 2. Segmentation performance of all models with different losses. The meaning of each symbol is the same as Table 1

Model	Loss	IoU(%)	Precision(%)	Pos_{acc}(%)	F1(%)
PointNet (15)	Focal	73.26	82.71	84.16	81.86
PointNet++ (15)	Focal	75.10	83.16	88.55	84.17
DGCNN (15)	Focal	75.82	82.75	89.44	85.49
TripleNet (18)	Focal	**77.57**	**83.75**	**90.52**	**86.25**
PointNet (15)	Ce	0	0	0	0
PointNet++ (15)	Ce	0	0	0	0
DGCNN (15)	Ce	30.26	76.08	30.59	39.74
TripleNet (18)	Ce	32.17	77.84	31.21	40.15

Table 3. Segmentation performance of models with different feature dimensions and different architectures. "15" denotes the basic 15-dimensional features, and "18" denotes the features after supplementing the three inner angles contained in each triangular face of the mesh to the basic 15-dimensional features.

Model	Dim	IoU(%)	Precision(%)	Pos_{acc}(%)	F1(%)
DGCNN	15	75.82	82.75	89.44	85.49
DGCNN	18	75.70	84.12	87.65	85.10
TripleNet	15	76.50	**84.28**	88.31	85.30
TripleNet	18	**77.57**	83.75	**90.52**	**86.25**

The Effectiveness of Features and Architecture. Table 3 is used to prove the improvement of our new introduced features and innovative model on this problem. Focal loss is used here in all cases as a fixed variable. From the table it can be concluded that our new designed model has improvement compared with the original DGCNN model (0.68% while using 15 dimensions of input feature, 1.87% while using 18 dimensions of input feature). As we said in the previous chapter, the angle is relatively small among all the features. To be specific, the range of the angle feature is $[-1, 1]$ and most of them are below 0.5, while other features like the x,y,z coordinates features range from $[-10, 10]$. so it will be somehow ignored when the convolution operation is applied in the original DGCNN model. However, in our TripleNet, we use a multi-scale method to enhance the model's ability to capture local features and keep the angle information. It is worth noting that the newly designed model improves the effect of adding the angle feature, and the new added angle feature boosts the performance of our new model. With our new feature and model, the results show that our system has gained better performance for all four criterions.

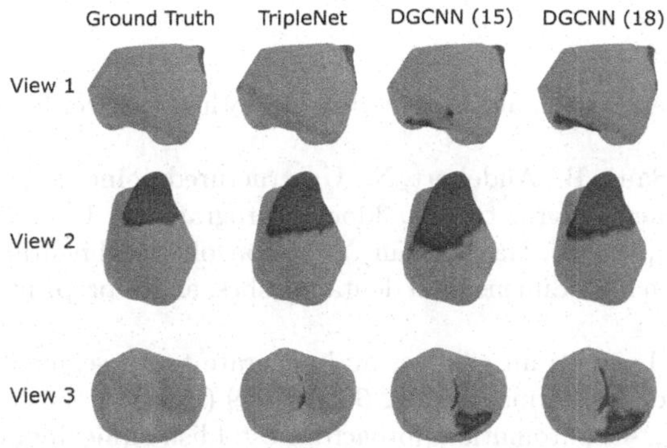

Fig. 4. Visualization of the output from different models.

4.4 Visualization

To better illustrate the result of the segmentation, we present the output of different models with respect to the same tooth and we use three views to fully show the overall output of the prediction. In Fig. 4, The green area indicates the healthy part of the tooth while the red area indicates the fragment. Compared with the ground truth, TripleNet successfully predicts the tooth fragment. The prediction closely matches the ground truth. Compared with the other two models, the TripleNet along with the new introduced local features successfully reduces the excessive predictions of fragment area due to imbalanced data.

5 Conclusion

In this paper, we mainly focus on the application of computer-aid system on the tooth defect segmentation, which has few existing research to the best of our knowledge. We propose a simple yet effective model, TripleNet, for end-to-end accurate and efficient tooth defect segmentation on 3D IOS. Our model employs a multi-scale feature extraction paradigm to learn rich contextual representations from the input point clouds, which is samples from the original tooth mesh with each point associated with predefined geometric features. A tweaked focal loss is employed to better work with imbalance distributions of normal and defect areas in each tooth. Comprehensive experimental results demonstrate that we can achieve impressive performance improvement over conventional 3D point cloud processing networks. Our work exhibit the great potential of deep learning for future digital dentistry.

Acknowledgement. This work was supported by the National Science Foundation of China 62106222 and the ZJU-Angelalign R&D Center for Intelligent Healthcare. We thank Dr. Huikai Wu and Dr.Chenxi Liu for the helpful discussions and prior work.

References

1. Beckett, H.: Preservation and restoration of tooth structure. Br. Dent. J. **198**(11), 727–727 (2005)
2. Boulch, A., Le Saux, B., Audebert, N.: Unstructured point cloud semantic labeling using deep segmentation networks. 3dor@ Eurograph. **3**, 1–8 (2017)
3. Chen, H., Li, H., Hu, B., Ma, K., Sun, Y.: A convolutional neural network for teeth margin detection on 3-dimensional dental meshes. arXiv preprint arXiv:2107.03030 (2021)
4. Cui, Z., et al.: Tsegnet: an efficient and accurate tooth segmentation network on 3D dental model. Med. Image Anal. **69**, 101949 (2021)
5. Deng, H., et al.: An automatic approach to establish clinically desired final dental occlusion for one-piece maxillary orthognathic surgery. Int. J. Comput. Assist. Radiol. Surg. **15**(11), 1763–1773 (2020)
6. Graham, B., Engelcke, M., Van Der Maaten, L.: 3D semantic segmentation with submanifold sparse convolutional networks. In: Proceedings of the IEEE conference on computer vision and pattern recognition, pp. 9224–9232 (2018)
7. Guo, M.H., Cai, J.X., Liu, Z.N., Mu, T.J., Martin, R.R., Hu, S.M.: PCT: point cloud transformer. Comput. Vis. Media **7**(2), 187–199 (2021)
8. Hao, J., et al.: Toward clinically applicable 3-dimensional tooth segmentation via deep learning. J. Dent. Res. **101**(3), 304–311 (2022)
9. Huang, J., You, S.: Point cloud labeling using 3D convolutional neural network. In: 2016 23rd International Conference on Pattern Recognition (ICPR), pp. 2670–2675. IEEE (2016)
10. Kumar, Y., Janardan, R., Larson, B., Moon, J.: Improved segmentation of teeth in dental models. Comput.-Aid. Des. Appl. **8**(2), 211–224 (2011)
11. Lang, et al.: DLLNet: an attention-based deep learning method for dental landmark localization on high-resolution 3D digital dental models. In: de Bruijne, M., et al. (eds.) MICCAI 2021. LNCS, vol. 12904, pp. 478–487. Springer, Cham (2021). https://doi.org/10.1007/978-3-030-87202-1_46
12. Lawin, F.J., Danelljan, M., Tosteberg, P., Bhat, G., Khan, F.S., Felsberg, M.: Deep projective 3D semantic segmentation. In: Felsberg, M., Heyden, A., Krüger, N. (eds.) CAIP 2017. LNCS, vol. 10424, pp. 95–107. Springer, Cham (2017). https://doi.org/10.1007/978-3-319-64689-3_8
13. Li, Y., Bu, R., Sun, M., Wu, W., Di, X., Chen, B.: Pointcnn: convolution on x-transformed points. Adv. Neural Inf. Process. Syst. **31** (2018)
14. Lian, C., et al.: Deep multi-scale mesh feature learning for automated labeling of raw dental surfaces from 3D intraoral scanners. IEEE Trans. Med. Imaging **39**(7), 2440–2450 (2020)
15. Lin, T.Y., Goyal, P., Girshick, R., He, K., Dollár, P.: Focal loss for dense object detection (2017). 10.48550/ARXIV.1708.02002, https://arxiv.org/abs/1708.02002
16. Qi, C.R., Su, H., Mo, K., Guibas, L.J.: Pointnet: deep learning on point sets for 3D classification and segmentation. In: Proceedings of the IEEE Conference on Computer Vision and Pattern Recognition, pp. 652–660 (2017)
17. Qi, C.R., Yi, L., Su, H., Guibas, L.J.: Pointnet++: deep hierarchical feature learning on point sets in a metric space. Adv. Neural Inf. Process. Syst. **30** (2017)
18. Raith, S., et al.: Artificial neural networks as a powerful numerical tool to classify specific features of a tooth based on 3D scan data. Comput. Biol. Med. **80**, 65–76 (2017)

19. Sun, D., et al.: Automatic tooth segmentation and dense correspondence of 3D dental model. In: Martel, A.L., et al. (eds.) MICCAI 2020. LNCS, vol. 12264, pp. 703–712. Springer, Cham (2020). https://doi.org/10.1007/978-3-030-59719-1_68
20. Tchapmi, L., Choy, C., Armeni, I., Gwak, J., Savarese, S.: Segcloud: semantic segmentation of 3D point clouds. In: 2017 International Conference on 3D Vision (3DV), pp. 537–547. IEEE (2017)
21. Wang, Y., Sun, Y., Liu, Z., Sarma, S.E., Bronstein, M.M., Solomon, J.M.: Dynamic graph cnn for learning on point clouds. ACM Trans. Graph. (TOG) (2019)
22. Wei, G., et al.: TANet: towards fully automatic tooth arrangement. In: Vedaldi, A., Bischof, H., Brox, T., Frahm, J.-M. (eds.) ECCV 2020. LNCS, vol. 12360, pp. 481–497. Springer, Cham (2020). https://doi.org/10.1007/978-3-030-58555-6_29
23. Wu, B., Wan, A., Yue, X., Keutzer, K.: Squeezeseg: convolutional neural nets with recurrent CRF for real-time road-object segmentation from 3D lidar point cloud. In: 2018 IEEE International Conference on Robotics and Automation (ICRA), pp. 1887–1893. IEEE (2018)
24. Wu, K., Chen, L., Li, J., Zhou, Y.: Tooth segmentation on dental meshes using morphologic skeleton. Comput. Graph. **38**, 199–211 (2014)
25. Wu, T.H., et al.: Two-stage mesh deep learning for automated tooth segmentation and landmark localization on 3D intraoral scans. IEEE Trans. Med. Imaging (2022)
26. Wu, W., Qi, Z., Fuxin, L.: Pointconv: deep convolutional networks on 3D point clouds. In: Proceedings of the IEEE/CVF Conference on Computer Vision and Pattern Recognition, pp. 9621–9630 (2019)
27. Xu, M., Ding, R., Zhao, H., Qi, X.: Paconv: position adaptive convolution with dynamic kernel assembling on point clouds. In: Proceedings of the IEEE/CVF Conference on Computer Vision and Pattern Recognition, pp. 3173–3182 (2021)
28. Zhao, H., Jiang, L., Jia, J., Torr, P.H., Koltun, V.: Point transformer. In: Proceedings of the IEEE/CVF International Conference on Computer Vision, pp. 16259–16268 (2021)
29. Zou, B.J., Liu, S.J., Liao, S.H., Ding, X., Liang, Y.: Interactive tooth partition of dental mesh base on tooth-target harmonic field. Comput. Biol. Med. **56**, 132–144 (2015)

Multi-agent Systems

Crowd-Oriented Behavior Simulation:Reinforcement Learning Framework Embedded with Emotion Model

Zhiwei Liang, Lei Li, and Lei Wang(✉)

University of Science and Technology of China, Hefei 230000, China
wangl@ustc.edu.cn

Abstract. In the study of group behavior, group emotion is an important factor affecting group behavior. A simplified mechanism model based on psychological theory can be established to better explain crowd behavior, but the model is too rough. Pure machine learning methods can reproduce crowd behavior well, but cannot reasonably explain crowd behavior, let alone predict the ability. Combining the two is a promising research direction. This paper designs a reinforcement learning framework embedded with emotion model. Then, the decision and behavior parts of the framework are implemented according to the emotion model and reinforcement learning algorithm.A variety of simulation application scenarios are built, and the behavior decision-making process of an agent in emergency situations is analyzed by using computer simulation technology. The simulation results show that the proposed framework can simulate the crowd behavior more realistically, and has good explanatory ability and certain predictive ability.

Keywords: Reinforcement learning · Emotional model · Crowd behavior · Agent simulation

1 Introduction

In emergency situations, emotion model is crucial to control the conduct of an agent for different decision choices, such as running, close to, to avoid, for the agent's actions, most of the current study of single modeling method based on rules or learning simulation implementation [1–3], but in the face of complex real scene, these methods have some limitations in the simulation application.Rule-based modeling methods generally tend to make the agent's behavior conform to the intuitive performance through artificial construction of rules [4], so as to achieve the expected simulation effect. But in the face of the micro motion control, this kind of assimilation method is easy to cause the behavior of the agent, and the simulation of complex behaviors such as running, beyond cannot

Supported by High-tech innovation special zone project (20-163-14-LZ-001-004-01).

L. Fang et al. (Eds.): CICAI 2022, LNAI 13606, pp. 195–207, 2022.
https://doi.org/10.1007/978-3-031-20503-3_16

be adaptive to adjust, lost to the motion path of merit and criterion, and the crowd density high, narrow channels such as the abnormal static, quiver when two-way traffic microcosmic distortion.

With the rapid development of machine learning, reinforcement learning algorithm is also concerned in the field of simulation [5]. Through repeated training of empirical data, intelligent agent has the ability to deal with complex environment, can complete some nonlinear and irregular flexible actions, and achieve more fine control effect. This method is universal and adaptable in the process of implementation without relevant video data set and complex manual design [6]. However, the learning method may fail to converge in the training process with more factors to be considered, and the time cost of realizing simple behavior is high.

All the above model methods have their own advantages and disadvantages. Generally speaking, different model methods can be combined to give full play to the advantages of each model method, so as to meet the requirements of simulation in various situations. For example, Neumann et al. [7,8] consider individual emotional infection based on cellular automata model and social force model respectively, and design a mechanism of emotional infection based on improved SIR model.Lv et al. [9] model the behavior of rioters and police under the background of mass violence and terror, the reinforcement learning algorithm is used to realize the action selection process against the crowd. However, these studies did not explore the design of simulation framework, and the coupling between various model methods is strong, which is difficult to expand.

On the basis of the above research and analysis, we attempts to design a comprehensive simulation framework, which combines the emotion model with the dynamics model, and selects the appropriate dynamics model according to the specific situation, so as to effectively avoid the limitations brought by the current single dynamics model and achieve more complex crowd behavior. At the same time, the framework meets certain extensibility, which is convenient for model switching and extension in different scenarios.

2 Related Work

According to the principle of EROS proposed by Jager et al. [10], adding emotional factors to the crowd simulation system can improve the authenticity of simulation results,Turner et al. [11] points out that the formation process of crowd emotion mainly includes four stages: emotional arousal, emotional infection, deindividuation and behavioral imitation. Bourgais [12] summarizes three agent-based emotion models, including emotion representation model, emotion cognitive evaluation model and emotion infection model. Current simulation models of emotional infection mainly include three types, ASCRIBE model [13], Durupinar model [14], Newton model [15]. Boss et al. [13] based on thermodynamic studies in which adjacent substances transfer energy to each other at a rate unique to each substance, the continuous deterministic, interactiving-based ASCRIBE emotion model is introduced. Lungu et al. [15] defines the key concepts of Newton's emotional model. Rincon et al. [16] further extends Newton's emotional model to emotional infection between individuals, using the calculation method of universal gravitation to represent the emotional stress between individuals, expressing emotions as positions in mechanics, Durupinar et al. [14]based on the mechanism of infectious disease model, designed the emotional transmission among individuals, and proposed a Durupinar emotional model based on probabilistic threshold infection. This model is similar to that of infectious diseases. It mainly considers that an individual is in two states: susceptible and infected [17]. When the number of infected people around the individual R is greater, the probability of the individual changing from susceptible to infection to infected is greater.We use Durupinar model to construct corresponding emotional decision nodes, and introduce multiple emotional states and corresponding motivation theories to construct emotional nodes, so as to fully realize agent decision-making and selection.

In some studies, Q-learning algorithm is used to optimize the sequential strategy and realize the corresponding behavior nodes [18]. However, due to its insufficient learning ability, its application scope is limited. Deep reinforcement learning uses the powerful function of neural network to solve the problem of high-dimensional state space, and its learning ability is greatly enhanced. Proximal Policy Optimization(PPO) [19], as the most classical deep reinforcement learning algorithm, has a wide range of applications and is easy to implement and debug. The agent's behavior can be continuous or discrete control, so this article will take the PPO algorithm as the simulation node behaviors in the framework of the implementation, however, in the current crowd simulation research based on reinforcement learning algorithm, the intrinsic reward are often ignored, we will be training in the 3D simulation environment, prone to explore space is larger, the rewards of sparse situation, as a result, it is difficult for an agent to converge within a limited time, so it is necessary to improve the PPO algorithm by designing intrinsic rewards. Pathak et al. [20] in the study proposed a kind of Intrinsic Curiosity Module, the model of individual curiosity is designed to strengthen the intrinsic reward of learning, The purpose is to encourage agents to perform different actions, enable agents to conduct effective space exploration, and facilitate the convergence rate of reinforcement learning. Therefore, we designs PPO algorithm driven by curiosity reward to solve the problem of sparse rewards mentioned above.

3 Methodology

3.1 Pedestrian Simulation Framework

Due to the lack of relevant software tools and the support of mainstream simulation platforms, deep reinforcement learning is difficult to be applied to the field of pedestrian simulation. Naturally, a universal pedestrian simulation framework based on deep reinforcement learning is also lacking. Therefore, we design a universal pedestrian simulation framework based on deep reinforcement learning

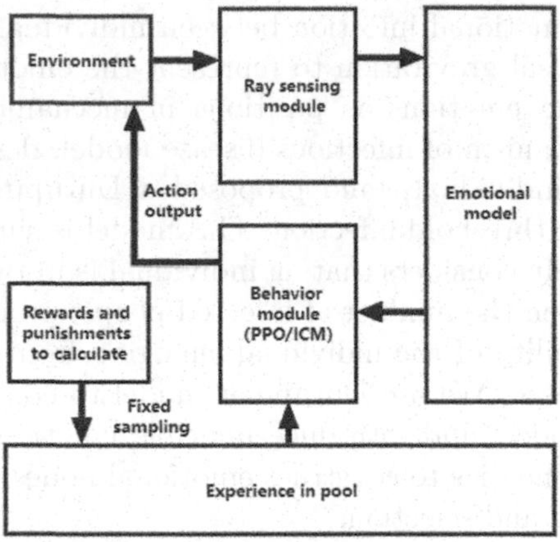

Fig. 1. The simulation framework of emotion model and reinforcement learning algorithm is introduced

based on agent model structure [21–24], See Fig. 1. With the rapid development of reinforcement learning, there are many excellent single-agent reinforcement learning algorithms. PPO algorithm [19] as the decision-making model of pedestrians. This algorithm is the mainstream single-agent deep reinforcement learning algorithm, which can be applied to continuous or discrete control tasks simultaneously.

3.2 Decision Realization Based on Emotion Model

The infectious disease model assumes that a person is in two states: susceptible and infected. When the number of infected people around the individual R is large, the probability of the individual changing from susceptible to infection is large [25].

$$D_R(t) = \sum_{t'=t-K+1}^{t} d_R(t') \tag{1}$$

$$d_R = log\text{-}\mathcal{N}(\mu_{d_R}, \sigma_{d_R}^2) \tag{2}$$

$$q_R(t) = q_R(t-1) - \beta \cdot q_R(t-1) \tag{3}$$

At each time t', the individual randomly selects another individual from the surrounding group. If you take an infected individual, from the selected averages μ_{d_R} and variance for $\sigma_{d_R}^2$ normal distribution generated in the emotional dose $d_R(t')$, length of the individuals to save time for K mood dose of history. If the cumulative emotional value exceeds the infection threshold, the individual changes from a susceptible state to an infected state. β is the emotion attenuation ratio. If the emotion attenuation of the infected individual is 0, it will change from the infected state to the susceptible state.

In the field of sentiment classification, experts believe that people have a common basic emotion, the sentiment driven by physiological factors, were not acquired environment interaction in the learning process [26], has nothing to do with the individual culture, gender, everyone will be unified to express these feelings, and can be others understand perception, the six most representative basic emotions are joy, anger, sadness, fear, disgust and surprise [27]. Cacioppo and others in the study [28,29] a mood bivariate model is put forward, the above basic emotion can be divided into two different types of variables, there are Positive emotions and Negative emotions respectively, thus generating two motivations: Approach and Avoidance. Harmon-Jones et al. [30] further summarizes a motivational dimensional model that takes motivational cues from direction, intensity and type.Based on the above theoretical analysis, the characteristics of individual motivation can be summarized into the following aspects:

$$\begin{array}{ll} M_x =< Val, Dir, FeelingType > & Val \in [0,1] \\ Dir \in \{Approach, Avoidance\} & FeelingType \in \{Positive, Negative\} \end{array} \tag{4}$$

Among them M_x represents the dimension vector of motivation,we usually use a three-dimensional list $< Val, Dir, FeelingType >$ express,among them Val represents motivation size,the range is 0 to 1,and the higher the number,the stronger the motivation,Dir indicates the direction of motivation,$FeelingType$ represents the type of emotion associated with motivation.Table 1 show the basic classification of emotions and the corresponding motivation direction,and the classification and motivation of some emotions are not clear.

Table 1. Basic emotion classification and motivation.

	Joy	Anger	Sad	Fear	Hate	Surprise
Emotional type	Positive	Negative	Negative	Negative	Negative	Unknown
Motive direction	Approach	Approach	Unknown	Avoidance	Avoidance	Unknown

We mainly considers the modeling method of emotion bivariate model,and takes the positive emotion of crowd joy and the negative emotion of crowd fear as an example to express individual emotion with the emotion value of two dimensions:

$$q_R =< q_R^H, q_R^F > \tag{5}$$

q_R^H is the joy emotion value, q_R^F is the fear emotion value,the emotional value of each dimension is [0,1], and the larger the value is, the stronger the corresponding emotion will be, so the emotion state of individual x $FeelingType$ can be represented as $Positive$,$Negative$, the motivation is expressed as:

$$\begin{array}{l} M_x^H =< q_R^H, Approach, Positive > \\ M_x^F =< q_R^F, Avoidance, Negative > \end{array} \tag{6}$$

In the face of various emotional states of the crowd under emergencies, fear was the main emotion and joy was the secondary emotion, and the sequence of emotional nodes was determined. In addition, the emotional state of *Normal* needs to be added to the implementation, which represents Normal individual behavior. When the execution flow Tick arrives, the mood control node needs to update each mood individually. As mentioned above, the renewal process mainly includes cognitive evaluation, affective infection and attenuation.

3.3 Behavior Realization Based on ICM and PPO Algorithm

PPO algorithm is an algorithm based on actor-critic architecture. By introducing a strategy evaluation mechanism, it can solve problems such as high variance and difficulty in convergence. The actor network parameters of the target policy are updated using the gradient rise method, and the loss function is as follows:

$$L(\theta) = \frac{1}{T}\sum_{t=0}^{T} \min\left(\frac{\pi_\theta(a|s_t)}{\pi_{\theta_k}(a|s_t)} A^{\pi_{\theta_k}}(s_t, a), g(\epsilon, A^{\pi_{\theta_k}}(s_t, a))\right) \tag{7}$$

$$g(\epsilon, A) = \begin{cases} (1+\epsilon)A, A \geq 0 \\ (1-\epsilon)A, A<0 \end{cases} \tag{8}$$

Among them π_θ is the target strategy,update network parameters,π_{θ_k} is behavioral strategy, the experience buffer pool will be generated after execution in the environment. Since the network structure of the two policies is consistent, π_θ is periodically copied to π_{θ_k},while $A^{\pi_{\theta_k}}$ is according to the current value function V_{ϕ_k} and reward expected $\hat{R}_t$ calculated advantage function:

$$A^{\pi_{\theta_k}}(s_t, a) = \hat{R}_t - V_{\phi_k}(s_t) \tag{9}$$

The gradient descent method is used for the Critic network parameters, and the regression variance of the above dominant functions is used to update them:

$$L(\phi) = \frac{1}{T}\sum_{t=0}^{T}\left(V_\phi(s_t) - \hat{R}_t\right)^2 \tag{10}$$

We designs a PPO algorithm driven by curiosity reward model [20] to solve the sparse reward problem. Therefore, at time t, agent's remuneration extends to:

$$r_t = r_t^i + r_t^e \tag{11}$$

Among them r_t^i is intrinsic reward for curiosity,r_t^e is the reward that the agent gets from interacting with the environment,the PPO policy network parameter θ_P is updated in the same way as before to maximize the total reward:

$$\max_{\theta_P} E_{\pi_{\theta_P}}[\sum_t \gamma^t r_t] \tag{12}$$

In the training phase, DeepMind's Distributed Proximal Policy Optimization (DPPO) [31] parallel execution method is adopted, which makes full use of computer resources to speed up the training speed of PPO algorithm,support large-scale Agent training.

4 Experiments

4.1 Introduction to The Experimental System

In order to build the simulation framework of the agent and the 3D simulation scene, all experiments in this section will be carried out in the Unity3D environment. For the realization of reinforcement learning algorithm, ML-agents[1] open source toolkit, in the Unity3D environment, the toolkit can train agents through simple and easy to use Python API, and also supports packaging training environment into exe executable file, which is convenient for different parameter configuration and algorithm invocation.

4.2 Behavior Realization Comparative Experiment

To verify the availability of behavior nodes and the superiority of reinforcement learning, two simplified scenarios, the degree of crowd avoidance and the rationality of evacuation route, are designed for testing. The partial hyperparameter Settings of PPO training are shown in Table 2 where Episode is a complete training of the current scene. When each Episode reaches the termination state, the next Episode will be restarted. MaxSteps is the maximum number of steps executed in a single Episode, TrainEpisodeNum is the maximum number of training turns, the hidden layer networks of Actor and Critic are fully connected neural networks, LayersNum is the number of network layers, UnitsNum is the number of neurons, and the activation function is relu. The reward discount rate γ was set to 0.99, the clipping rate ϵ was set to 0.2, and the ICM model had two layers of network, 256 neurons, a learning rate of 0.0003, and a scaling factor of η of 0.02. In addition, the DPPO parallel operation scenario was set to 20.

Table 2. PPO algorithm part of the hyperparameter settings

Parameter	Value	Parameter	Value
TrainEpisodeNum	100000	MaxSteps	5000
ActorLayersNum	3	CriticLayersNum	2
ActorUnitsNum	512	CriticUnitsNum	512
ActorLearningRate	0.0001	CriticLearningRate	0.0003
ActorHiddenActivation	relu	CriticHiddenActivation	relu
ICMUnitsNum	256	ICMLearningRate	0.0003
γ	0.99	ϵ	0.2

Figure 2 is the schematic diagram of the simulation comparison results of scene 1. The individual with yellow halo represents the simulation agent, the green arrow represents the direction of pedestrian movement, and the one without mark represents the agent of the general rule model. Figure (a) shows that the

[1] https://github.com/Unity-Technologies/ml-agents.

(a) Evacuation process based on general rules

(b) Evacuation process based on reinforcement learning

Fig. 2. Result of model comparison in Scene 1. (a) Evacuation process based on general rules. (b) Evacuation process based on reinforcement learning.

agent based on general rules cannot complete the avoidance and transcendence behavior in crowd fleeing; Figure (b) shows that the agent based on reinforcement learning behavior node can perceive the position and action of other individuals, so that it can timely adjust its own direction and behavior in advance and flexibly complete the avoidance and transcendence behavior.

Figure 3 is the comparison of the simulation results of scene two. In the rule model, agent only select adjacent export rules based on distance without considering the evacuation time and crowded degree, result in slow progress in large populations, presents the state of the linear line at the exit, and intelligence based on reinforcement learning node consider red area population density situation, so as to avoid the congestion exports, take the longest but shortest route to avoid other people at the exit.

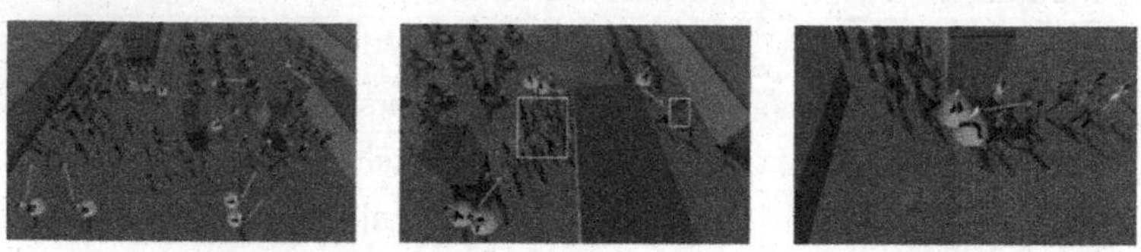

(a) Evacuation process based on general rules

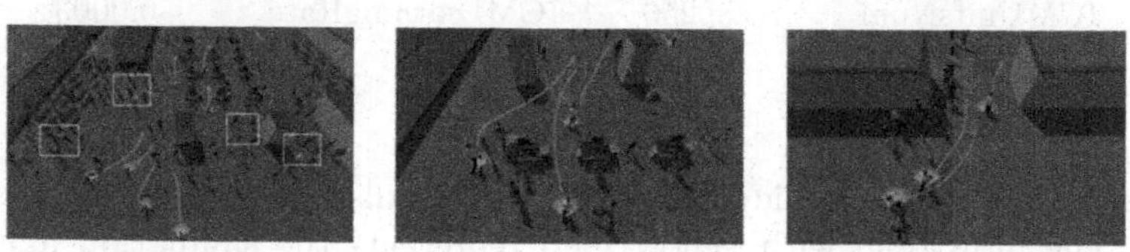

(b) Evacuation process based on reinforcement learning

Fig. 3. Result of model comparison. (a)Evacuation process based on general rules. (b)Evacuation process based on reinforcement learning.

As shown in Fig. 4 the off-line training curves of different scenarios are shown. Scene 1 is trained by a single agent, so the total number of steps is small, while scene 2 is trained by multiple agents in the same scenario at the same time, and the environmental information is relatively complex. So the training steps are more. At the beginning, due to the agent's continuous attempts, the collision frequency is high and the time is long, and the reward is negative. With the increase of the reward in space exploration, the agent plans the best course of action, and the reward value is close to 1. It can be seen from the figure that the PPO algorithm driven by curiosity has more exploration space and therefore converges faster than the PPO algorithm alone, and finally obtains more rewards. The experimental results well reflect the superiority of the ICM+PPO algorithm. Then, the reinforcement learning node of ICM+PPO is introduced into the simulation framework mentioned above, and the training behavior is slightly modified by online means. Next, the framework is applied to each scene to analyze the final performance of agents under different models.

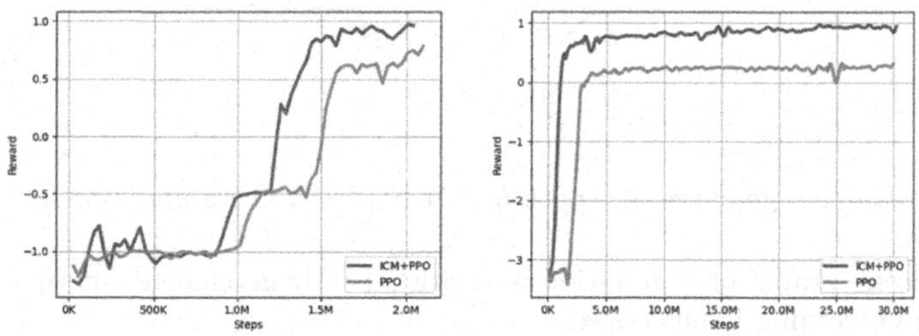

(a) Scene 1 (b) Scene 2

Fig. 4. Training curve of ICM+PPO algorithm in different scenarios (a) Scene 1. (b) Scene 2.

4.3 A Variety of Emotional Simulation Experiments

In this section, the experimental reference real scene Ankara bombing[2]. As the video picture is not stable enough, the occurrence process of the event is described qualitatively. At the beginning, people gather in the square with joyful emotions, and then there is the scene of terror attack and explosion and crowd fleeing. Based on the experiment in the previous section, positive emotion nodes are added to verify the expansibility of the simulation framework. Since the scene is relatively open and the actions in the approaching and walking

[2] https://www.theguardian.com/world/2015/oct/10/turkey-suicide-bomb-killed-in-ankara.

process are not easy to be distorted, the general navigation nodes are selected as the movement control. The reinforcement learning node is still used for the escape behavior under the panic node, which also reflects the flexibility of the simulation framework in building the simulation model.

Figure 5 (a) shows the mean change curve of emotion. It can be found that the emotion value decreases rapidly to 0 at about 15 s due to the explosion event, while the contagion process of panic is fast. It reached its peak at about 15 s–20 s, and began to decline slowly at about 25 s. The two emotions were consistent with the theory of primary and secondary emotions, which was relatively consistent with the real situation. Figure 5 (b) shows the convergence of rewards during training.

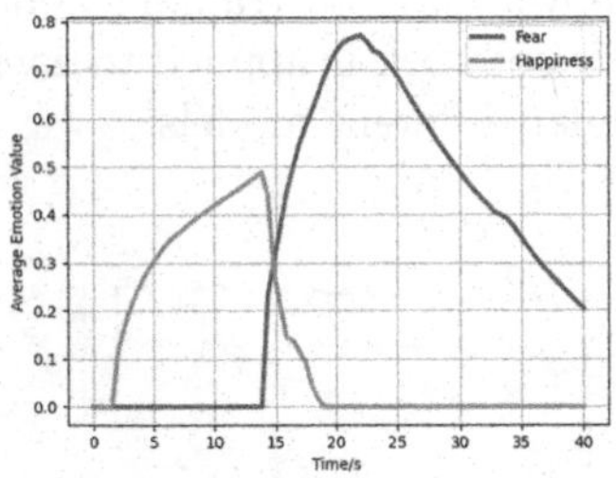

(a) Mean change curve of emotion

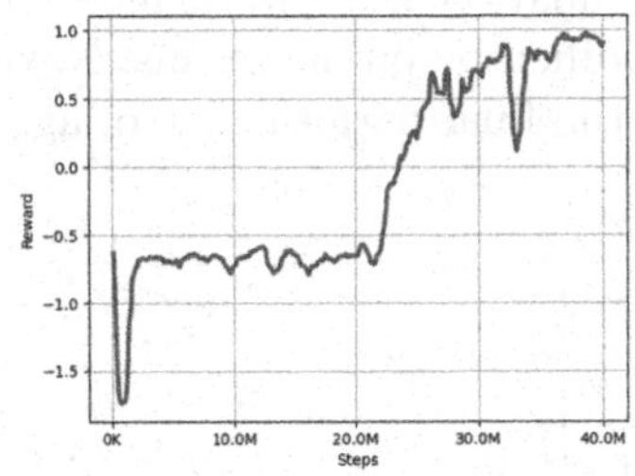

(b) ICM+PPO training bonus curve

Fig. 5. Process graph of simulation scenario. (a)Mean change curve of emotion. (b)ICM+PPO training bonus curve.

Figure 6 is the simulation result in this scenario.The pedestrian without halo represents an emotional state of *Normal*, the yellow halo represents *Happiness*, the red halo represents *Fear*, (a) after the stimulus was given, the nearby individuals showed joyful emotion, while the distant individuals were not disturbed; (b) most of the individual emotional states are infected with the motivation to approach, gradually approaching the target region; (c) when the explosion occurred, the nearby individuals appeared red panic mood, but still most of the individuals remained happy mood; (d) the mood of the whole individual changes to panic and moves away from the stimulus in all directions. Experimental results show that the proposed method can simulate multiple emotions in the same scene, and the usability and expansibility of the simulation framework are verified.

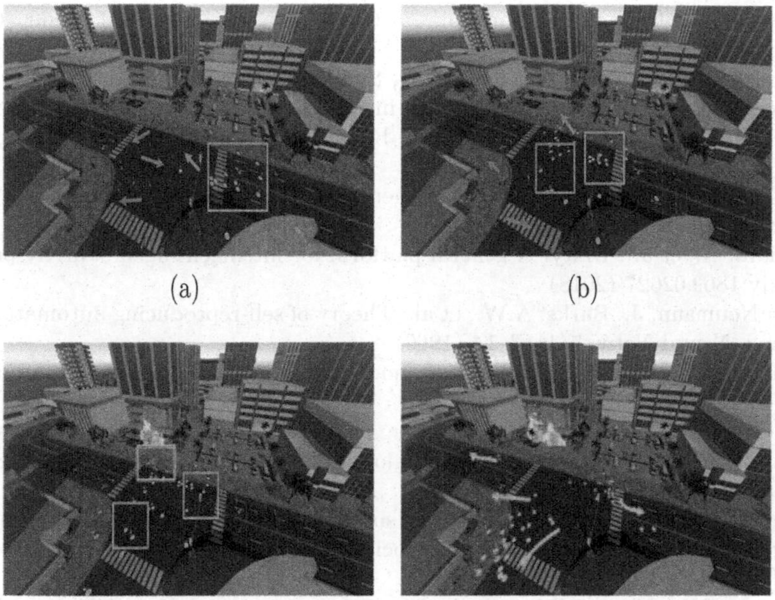

(a) (b) (c) (d)

Fig. 6. A variety of emotional simulation experiments

5 Conclusion

In this paper, we propose a simulation framework based on crowd emotion model. The existing problems of simulation models and simulation frameworks are discussed, and the idea of combining emotion model with multi-dynamic model is introduced. The decision module and behavior module of the framework are implemented by using specific emotion model and reinforcement learning algorithm respectively, and the realization process is extended and improved by using emotion classification theory and ICM model. The advantages and feasibility of the framework are verified through three-dimensional simulation experiments, and the application of the evaluation method of emotion model and layered simulation framework in crowd behavior simulation is demonstrated in combination with real emergencies. The research work mainly focuses on emotional factors under emergencies. In the future, various factors such as role relationship, background culture and gender difference can be further introduced into the design of agent to make the simulation effect richer and more real.

References

1. Henderson, L.: The statistics of crowd fluids. Nature **229**(5284), 381–383 (1971)
2. Hughes, R.L.: A continuum theory for the flow of pedestrians. Transp. Res. Part Methodological **36**(6), 507–535 (2002)

3. Treuille, A., Cooper, S., Popović, Z.: Continuum crowds. ACM Trans. Graph. **25**(3), 1160–1168 (2006)
4. Bera, A., Randhavane, T., Kubin, E., Shaik, H., Gray, K., Manocha, D.: Data-driven modeling of group entitativity in virtual environments. In: Proceedings of the 24th ACM Symposium on Virtual Reality Software and Technology, pp. 1–10 (2018)
5. Sutton, R.S., Barto, A.G.: Reinforcement Learning: An Introduction. MIT press, Cambridge (2018)
6. Juliani, A., et al.: Unity: A general platform for intelligent agents. arXiv preprint arXiv:1809.02627 (2018)
7. Von Neumann, J., Burks, A.W., et al.: Theory of self-reproducing automata. IEEE Trans. Neural Netw. **5**(1), 3–14 (1966)
8. Helbing, D., Molnar, P.: Social force model for pedestrian dynamics. Phys. Rev. E **51**(5), 4282 (1995)
9. Lv, P., Xu, B., Li, C., Yu, Q., Zhou, B., Xu, M.: Antagonistic crowd simulation model integrating emotion contagion and deep reinforcement learning. arXiv preprint arXiv:2105.00854 (2021)
10. Jager, W.: Enhancing the realism of simulation (EROS): On implementing and developing psychological theory in social simulation. J. Artif. Soc. Soc. Simul. **20**(3), 1–15 (2017)
11. Turner, J.H., Turner, P.R., et al.: The Structure of Sociological Theory. Dorsey Press Homewood, IL, Kawana (1978)
12. Bourgais, M., Taillandier, P., Vercouter, L., Adam, C.: Emotion modeling in social simulation: a survey. J. Artif. Soc. Soc. Simul. **21**(2), 1–22 (2018)
13. Bosse, T., Duell, R., Memon, Z.A., Treur, J., Van Der Wal, C.N.: Multi-Agent Model for Mutual Absorption of Emotions. ECMS, pp. 22–37 (2009)
14. Durupınar, F.: From audiences to mobs: Crowd simulation with psychological factors. Ph.D. thesis, Bilkent University, Ankara (2010)
15. Lungu, V.: Newtonian Emotion System, pp. 307–315 (2013)
16. Rincon, J.A., Costa, A., Novais, P., Julian, V., Carrascosa, C.: A dynamic emotional model for agent societies. In: Demazeau, Y., Ito, T., Bajo, J., Escalona, M.J. (eds.) PAAMS 2016. LNCS (LNAI), vol. 9662, pp. 169–182. Springer, Cham (2016). https://doi.org/10.1007/978-3-319-39324-7_15
17. Dodds, P.S., Watts, D.J.: A generalized model of social and biological contagion. J. Theor. Biol. **232**(4), 587–604 (2005)
18. Dey, R., Child, C.: QL-BT: enhancing behaviour tree design and implementation with q-learning. In: 2013 IEEE Conference on Computational Inteligence in Games, pp. 1–8. IEEE (2013)
19. Schulman, J., Wolski, F., Dhariwal, P., Radford, A., Klimov, O.: Proximal policy optimization algorithms. arXiv preprint arXiv:1707.06347 (2017)
20. Pathak, D., Agrawal, P., Efros, A.A., Darrell, T.: Curiosity-driven exploration by self-supervised prediction. In: International Conference on Machine Learning, pp. 2778–2787. PMLR (2017)
21. Fu, Y., Qin, L., Yin, Q.: A reinforcement learning behavior tree framework for game AI. In: 2016 International Conference on Economics, Social Science, Arts, Education and Management Engineering, pp. 573–579. Atlantis Press (2016)
22. Zhang, Q., Sun, L., Jiao, P., Yin, Q.: Combining behavior trees with maxq learning to facilitate CGFS behavior modeling. In: 2017 4th International Conference on Systems and Informatics, pp. 525–531. IEEE (2017)

23. Kartasev, M.: Integrating Reinforcement Learning into Behavior Trees by Hierarchical Composition. Master's thesis, School of Electrical Engineering and Computer Science, Islamabad (2019)
24. Kartašev, M., Saler, J., Ögren, P.: Improving the performance of backward chained behavior trees using reinforcement learning. arXiv preprint arXiv:2112.13744 (2021)
25. Li, C., et al.: ACSEE: Antagonistic crowd simulation model with emotional contagion and evolutionary game theory. arXiv preprint arXiv:1902.00380 (2019)
26. Shi, Y., Zhang, G., Lu, D., Lv, L., Liu, H.: Adaptive intervention for crowd negative emotional contagion. In: 2021 IEEE 24th International Conference on Computer Supported Cooperative Work in Design, pp. 18–23. IEEE (2021)
27. Ekman, P.: An argument for basic emotions. Cogn. Emot. **6**(3–4), 169–200 (1992)
28. Cacioppo, J.T., Berntson, G.G.: Relationship between attitudes and evaluative space: a critical review, with emphasis on the separability of positive and negative substrates. Psychol. Bull. **115**(3), 401 (1994)
29. Cacioppo, J.T., Gardner, W.L., Berntson, G.G.: Beyond bipolar conceptualizations and measures: the case of attitudes and evaluative space. Pers. Soc. Psychol. Rev. **1**(1), 3–25 (1997)
30. Gable, P., Harmon-Jones, E.: The motivational dimensional model of affect: implications for breadth of attention, memory, and cognitive categorisation. Cogn. Emot. **24**(2), 322–337 (2010)
31. Heess, N., et al.: Emergence of locomotion behaviours in rich environments. arXiv preprint arXiv:1707.02286 (2017)

Deep Skill Chaining with Diversity for Multi-agent Systems*

Zaipeng Xie[1,2](✉), Cheng Ji[2], and Yufeng Zhang[2]

[1] State Key Laboratory of Hydrology-Water Resources and Hydraulic Engineering, Hohai University, Nanjing, China

[2] College of Computer and Information, Hohai University, Nanjing, China
{zaipengxie,chengji,yufengzhang}@hhu.edu.cn

Abstract. Multi-agent reinforcement learning requires the reward signals given by the environment to guide the convergence of individual agents' policy networks. However, in a high-dimensional continuous space, the non-stationary environment may provide outdated experiences that lead to the inability to converge. The existing methods can be ineffective in achieving a satisfactory training performance due to the inherent non-stationary property of the multi-agent system. We propose a novel reinforcement learning scheme, MADSC, to generate an optimized cooperative policy. Our scheme utilizes mutual information to evaluate the intrinsic reward function that can generate a cooperative policy based on the option framework. In addition, by linking the learned skills to form a skill chain, the convergence speed of agent learning can be significantly accelerated. Hence, multi-agent systems can benefit from MADSC to achieve strategic advantages by significantly reducing the learning steps. Experiments are performed on the SMAC multi-agent tasks with varying difficulties. Experimental results demonstrate that our proposed scheme can effectively outperform the state-of-the-art methods, including IQL, QMIX, and hDQN, with a single layer of temporal abstraction.

Keywords: Reinforcement learning · Multi-agent systems · Temporal abstraction · Mutual information · Skill discovery

1 Introduction

As one of the most promising technologies to realize general AI, reinforcement learning has become one of the main focuses of multi-agent system research. Unlike the typical supervised or unsupervised learning with data set, reinforcement learning algorithms are learned via interactions. The agent continuously learns knowledge according to the rewards obtained, making it more adaptable to the environment. However, in reality, many complex problems cannot be modeled as a single agent interacting with the environment. Instead, they should be

*Supported by The Belt and Road Special Foundation of the State Key Laboratory of Hydrology-Water Resources and Hydraulic Engineering under Grant 2021490811, and the National Natural Science Foundation of China under Grant No. 61872171.

L. Fang et al. (Eds.): CICAI 2022, LNAI 13606, pp. 208–220, 2022.
https://doi.org/10.1007/978-3-031-20503-3_17

modeled as multi-agent collaboration or competition problems [1]. As a popular research area in distributed AI, a multi-agent system (MAS) can solve many complex real-time dynamic problems, such as network routing collaboration [2], trash recycling problems [3], and urban traffic control [4].

Recently, the state-action space's nonstationary property and its exponential growth have received much attention [5]. Several works have proposed multi-agent reinforcement learning (MARL) algorithms to improve policy effectiveness and convergence speed. However, these methods can be incompetent when a large number of agents learn simultaneously.

The basic paradigm of deep reinforcement learning (DRL) is a two-stage rule: *the Evaluation Stage* and *the Improvement Stage* [6]. On this premise, the agent's policy network needs to continuously improve the accuracy driven by the reward signals from the environment. However, in a collaborative MAS, we consider how each agent's policy network converges to the optimal state while the collaborative policies between different agents are still suboptimal. This is often difficult because the input to the policy networks of the agents may not be a direct global state. All the agents are simultaneously interacting with and learning from the environment. For an individual agent, its fellow agents can also be a part of the environment, leading to constant changes in the environment [7], i.e., one's best policy may change as the other agents' policies update.

We propose a method named Multi-agent Deep Skill Chaining (MADSC) that can effectively enable agents to extend their actions via temporal abstraction. Furthermore, we utilize the intrinsic reward to inspire cooperation and guide the agents' exploration trajectories. A neural network is developed based on the value decomposition method to aggregate the policy functions of each agent. Hence, our proposed algorithm can significantly improve the convergence speed and performance of cooperative policy networks in MAS.

2 Related Works

In a large-scale MAS, it can be infeasible to train each agent separately. Existing multi-agent reinforcement learning approaches [8–10] often employ a centralized training, decentralized execution(CTDE) method, where centralized Q-networks are used for training the global network to stabilize the evaluation process. Meanwhile, a decentralized execution approach can be employed, allowing each agent to use their observations as a critical basis for decision-making during execution.

Rashid et al. [11] proposed an algorithm named QMIX to solve the credit assignment problem. By using the decomposition of the value function, QMIX can improve the efficiency of cooperation. However, evaluating the utility of each agent's behavior in MAS can be difficult as all agents are performing simultaneously. Bacon et al. [12] propose that the abstraction of learning tasks can be based on the option that is essentially a sequence of actions to complete the corresponding sub-tasks in a specific state subspace. The option itself can be considered a unique action that consists of an action set, including the primitive action. Tang et al. [13] propose hDQN, a hierarchical control structure is constructed through the invocations between the upper and lower policy layers. In addition, various optimization algorithms have been proposed to improve the convergence speed. For example, Deep Skill Network (DSN) [14] combines the hierarchical scheme with the deep Q-learning algorithm to improve the reusability of existing trained skills. Hindsight Experience Replay method [15] integrates the experience pool design with the hierarchical approach to improve the experience sampling of the layered learning efficiency.

In the options framework, skill is often modeled as an option [16]. When an agent explores a particular environment state, its observations can directly determine the actions performed in the following multiple time steps at the macro level. Sharma et al. [17] propose a skill discovery method, using the mutual information between state sequences and potential skills to construct internal rewards that encourage agents to explore a collection of skills with diversity. Bagaria et al. [16] propose deep skill chaining that can autonomously discover skills in high-dimensional continuous domains. The algorithm constructs skills that implicitly represent relationships between several related skills, allowing an agent to execute the skill chain sequentially. However, the lack of an efficient exploration method may still lead to a decline in the algorithm's convergence performance.

3 MARL with Skill Discovery

A Markov Decision Process can be formally defined as (N, S, P, R, γ). When multiple agents are involved, MDP is no longer satiable to describe the environment because the agents' actions are strongly correlated with the overall environment state. Thus we can extend the definition of MDP into Markov games [18], also called stochastic games. The definition of a Markov game [18] can be described as $(N, S, \{A^{(i)}\}_{i \in N}, P, \{R^{(i)}\}_{i \in N}, \gamma)$, where N is the total number of agents, S is the overall state distribution, $\{A^{(i)}\}_{(i \in N)}$ represents the set of all agents, and P, R, γ are state transition, reward function, and discount factor correspondingly.

3.1 How Agents Learn their Policies

There are various methods [19–21] to learn a policy in an MDP. One popular method opts to learn an action-value function $Q^{\pi}(s_t, a_t)$, i.e., the sum of the cumulative rewards that the agent may achieve according to the current policy. And the goal for each agent is set to maximize the value of this action-value function give by $\pi(s_t) = \arg\max_a Q(s_t, a_t)$, where a_t and s_t denote the action and state at time t. The dynamic migration process of MAS can be described by $s_{t+1} = f_t(s_t, \boldsymbol{a}_t)$, where f_t is the environment-dependent deterministic function, and $\boldsymbol{a}_t$ denotes the joint-action. The reward function, as the extrinsic reward R_E, can be defined as a linear sum of the rewards of all the agents given by $R_E = \sum_{s_t, \boldsymbol{a}_t \in \tau^{(i)}} R(s_t, \boldsymbol{a}_t)$, where $\tau^{(i)}$ is the state trajectory of the i-th agent.

3.2 Option Framework

The concept of the option framework is derived from the temporal abstraction technique [22] in hierarchical reinforcement learning (HRL). An option can be formally defined as $\omega \in \Omega \triangleq (\mathcal{I}_\omega, \pi_\omega, \beta_\omega)$, where Ω denotes all the available options. All options consist of three parts: the set of initial states $\mathcal{I}_\omega$, the internal policy π_ω, and the set of termination states β_ω. The agent uses the option internal policy π_ω to select actions at low-level time steps to interact with the environment. The option internal policy can map the state s to the low-level action a.

3.3 Problem Formulation

Consider a MAS with a CTDE architecture, and the option framework is employed based on HRL temporal abstraction for learning and updating the behavioral policies of each agent. From each agent's perspective, its option internal policy learns each primitive action $a_t^{(i)}$ on the low-level time scale and updates the policy parameters through a deep policy network. Our goal on the low-level time scale is to maximize the cumulative rewards of all the primitive actions. Meanwhile, on the high-level time scale, due to the use of temporal abstraction, each agent trains their option policy $\pi_\Omega^{(i)}$ to select the applicable option at the high-level temporal sequence.

We assume a greedy approach to measure the sum of rewards of all agents in a collaborative MAS and choose an option that maximizes the high-level Q-function. Our optimization goal can be formally described as:

$$\pi_\Omega^{(i)} \in \arg\max_{\pi_\omega} \mathbb{E}\left[\sum_{i=1}^{N}\sum_{t=1}^{T_{max}} R(s_t, a_t^{(i)})\right], \tag{1}$$

where T_{max} is the time budget available to the agent when an option is applied. If the time budget is reached, the control of the option is reclaimed, and the option policy π_Ω determines the next option candidate.

4 Methodology

4.1 Option Learning in MAS

All the agents can start constructing an option repository. The option repository within every agent is defined by

$$\mathcal{O} = \{\omega_k^{(i)} \mid (\mathcal{I}_{\omega_k^{(i)}}(s_t) = 1) \cap (\beta_{\omega_k^{(i)}}(s_t) = 0)\}, \tag{2}$$

where the $\mathcal{I}_{\omega_k^{(i)}}$ and $\beta_{\omega_k^{(i)}}$ denote the initiation set and termination set, $\omega_k^{(i)}$ represents the i-th agent's k-th option. We denote ω_G as the global option whose $T_{max} = 1$. As the agents explore the environment and exploit their experiences, their option repository can be adaptively expanded. We use a simple binary classifier to train and predict the initiation set. If an agent reaches a specific target state and successfully triggers the $\beta(s_t)$ function of the option for K times, we train it with a sequence of state trajectories of length k as the input to the classifier $p(\omega_k^{(i)} \mid \tau_k^{(i)})$. During the classifier's training, its output determines how each agent may execute the option at an appropriate state trajectory.

In general, the independent Q-learning (IQL) method [23] can be adequate to fulfill the requirements of training policy networks with improved Q value prediction accuracy. However, the IQL method may not be able to promote collaborations among agents and lead to suboptimal results. Therefore, to mitigate the issue, we proposed to set up a Mixer network at the top layer of our proposed learning framework and the network employs a value decomposition algorithm similar to that in the QMIX [11] algorithm.

4.2 Skill Chaining for MARL

The option is considered a skill that can be combined to achieve improved results. Bagaria et al. [16] propose that skills chaining can produce an improved convergence performance in single-agent tasks. Since it is desired to improve training performance in MAS, we aim to build the skill chain for each option. However, establishing a skill chain can be hindered because the agent may not be able to reach the goal state due to the non-stationary problem, resulting in poor exploration efficiency and deteriorated convergence performance.

To solve the abovementioned performance bottleneck, we propose a DRL scheme, named multi-agent deep skill chaining (MADSC), to improve the convergence performance in MAS. The proposed MADSC scheme can optimize the MAS architecture based on the skill chaining technique.

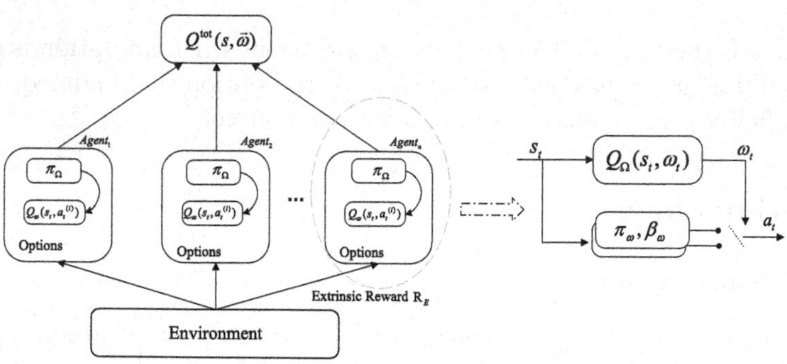

Fig. 1. Diagram of MADSC with the option framework

Figure 1 shows the overall diagram of the proposed MADSC scheme. Each agent has a complete repository of options and is independently selected and trained by an option policy. Meanwhile, agents are coordinated through a Mixer network denoted as Q^{tot}. For each agent, its policy network π_Ω is updated through Q-learning,The update target y_t is given by

$$y_t = \sum_{t'=t}^{\tau} \gamma^{t'-t} r_{t'} + \gamma^{\tau-t} \cdot Q_\Omega(s_{t+\tau}, \arg\max_{\omega^{(i)}} Q(s_{t+\tau}, \omega)), \tag{3}$$

where τ denotes the trajectory of all time steps. Hence, the option policies of each agent is updated at each training step with the update target y_t.

The key to the skill chaining is to make the initiation set of the current option ω_i intersect with the termination set of the next option ω_{i-1}. The agents can naturally execute the next option when the option is executed to its termination set. Hence, it is desired that the learning of the initiation set $\mathcal{I}_{\omega_i}$ is continuously trained until it is accurate during the training process. For each agent, the condition $\beta_{\omega_{i+1}} = \mathcal{I}_{\omega_i}$ needs to be satisfied, so that the current option ω_i can be chained successfully to the next option. Figure 2 illustrates the diagram of the skill chaining process.

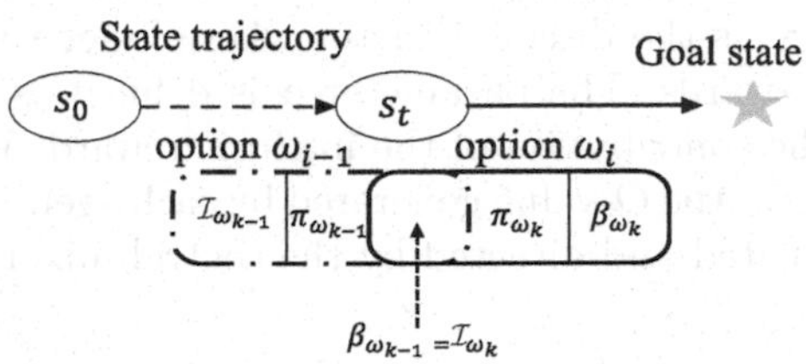

Fig. 2. Diagram of the skill chaining process

4.3 Mutual Information for Space Exploration

The mutual information is a general measure to describe the correlation between two random variables, and it is defined as the KL-diversity [24] between the joint distribution $p(s;\omega)$ and $p(s) \cdot p(\omega)$ given by

$$I(S;\Omega) = \iint p(s,\omega) \log \frac{p(s,\omega)}{p(s) \cdot p(\omega)} ds d\omega, \tag{4}$$

Maximizing mutual information of two random variables in deep learning training has been shown to perform well in many other domains [25–27]. A high mutual information between the state distribution S and the option sample distribution denotes that the uncertainty of the state distribution can be reduced when the option $\omega^{(i)}$ is fixed. Thus, we can effectively increase the diversity of the exploration with a maximized mutual information.

In our proposed MADSC method, the option's internal policy network is driven by the rewards from the environment. Meanwhile, the high-level cooperative policy is updated using a combination of intrinsic and extrinsic rewards constructed through mutual information. We define the high-level value decomposition-based cooperative policy as π^{tot}. Therefore, our goal is to maximize the mutual information between the state trajectory distribution $p(\tau)$ and the sample distribution of option $\omega^{(i)}$. The resulting construction of the intrinsic reward function R_I for the i-th agent can be written as

$$R_I^{(i)} = I(S;\Omega) - \frac{1}{N}\sum_{i=1}^{N} I(S;\omega^{(i)}) = \log p(\Omega \mid S) - \frac{1}{N}\sum_{i=1}^{N} \log p(\omega^{(i)} \mid s_t) \quad (5)$$

The rewards function is implemented to promote efficient exploration in MDP and mitigate the performance bottleneck in the skill chain technique. Therefore, the reward that ultimately update the policy Q^{tot} is a mixture of extrinsic and intrinsic rewards given by

$$R^*(\tau,\omega) = \alpha \cdot \sum_{s_t,\boldsymbol{a}_t \in \tau^{(i)}} \gamma^t R(s_t,\boldsymbol{a}_t) + (1-\alpha)\cdot \sum_{i=1}^{N} R_I^{(i)}, \quad (6)$$

where $\sum_{s_t,\boldsymbol{a}_t \in \tau^{(i)}} \gamma^t R(s_t,\boldsymbol{a}_t)$ denotes the extrinsic rewards R_E and R_I represents the intrinsic reward, γ^t is the discount factor where lower values are placed on immediate extrinsic rewards. The parameter α is defined as a factor that may be used to regulate the contribution of the intrinsic reward. We incorporate the mixer network to process the Q-value generated by each agent so that each policy update can be coordinated and directed by the central mixer network.

Algorithm 1: The MADSC process running on the central server

```
Init : Initialize with env.reset(), set parameters T_max for option framework
       Global option for each agent: ω_G = (I_{ω_G^(i)}, β_{ω_G^(i)}, T_max = 1)
       Initialize Agents' option repository O ← {ω_G}
while time steps t < Max training steps N^t do
    if (t%evaluate cycle==0) then
        forall Agents do
            Evaluate current policies and record win rates
    forall Agents do
        while t < T_max do
            Sample current option ω_t^(i) with π_{ω_k^(i)}(s_t) = arg max_{a_t^(i) ∈ a_t} Q_ω^π(s_t, a_t)
            Execute option ω_t^(i) and save experiences
            Update π_{ω_k^(i)}(s_t) using Q-learning and append it to O
    Generate intrinsic reward R_I using Eq. (5)
    Update Q^tot using R_I and extrinsic rewards R_E sampled from experiences
```

Algorithm 2: The MADSC process running on the individual agent

Init : Get state s_t, observation obs_t, $\beta_{\omega^{(i)}}(s_t) \leftarrow 0$

low-level time scale $t_0 = t$

T_{max} is the option's time budget

while *s_t is not the termination state* **do**

  while *($\beta_{\omega^{(i)}}$ == False) and ($t < T_{max}$)* **do**

    $a_t^{(i)} \leftarrow \pi_{\omega^{(i)}}(obs_t)$

    $r_t, s_{t+1} \leftarrow env.step()$

    $t \leftarrow t + 1$

    $s_t \leftarrow s_{t+1}$

  Train option initiation set $\mathcal{I}_{\omega^{(i)}}$ with binary classifier

Since our MADSC scheme is developed for the CTDE architecture, it has two major processes: the first process is for centralized training, and the second process is for decentralized execution. Algorithm 1 summarizes the process running on the centralized server, where all the information required including the observations and state trajectories are collected and fed into the policy networks. Then the central server uses information to train those networks. Algorithm 2 describes the process running on an individual agent, where each agent utilizes its observation as an input to the policy network and subsequently selects an action and executes it.

5 Experimental Evaluations

In order to evaluate the performance of our proposed MADSC algorithm, we have performed multiple experiments using the StarCraft II (SMAC) [28] as the multi-agent test environment. In the experiments, we vary the difficulty of the SMAC maps and train our agents for three million time steps.

5.1 SMAC Environment Setup

We choose various maps of StarCraft II as the evaluation MAS environment. The main purpose is to train a group of agents for a cooperative goal, i.e., defeating the AI rival. Table 1 summarizes the information of the SMAC maps.

We first evaluate the performance by varying the value of α, and the results show that when $\alpha = 0.7$, we can achieve the best result. In addition, for the Option's time budget $T_{\max}$, we set $T_{\max} = 300$ as commonly used in most state-of-the-art work on SMAC.

Table 1. SMAC maps for our experimental evaluations

Name	Ally units	Enemy units	Map difficulty
2s3z	2 Stalkers, 3 Zealots	2 Stalkers, 3 Zealots	Easy
3m	3 Marines	3 Marines	Easy
1c3s5z	1 Colos,3 Stalkers,5 Zealots	1 Colos,3 Stalkers,5 Zealots	Normal
3s5z	3 Stalkers, 5 Zealots	3 Stalkers, 5 Zealots	Normal
8m vs 9m	8 Marines	9 Marine	Normal
8m	8 Marines	8 Marines	Normal
10m vs 11m	10 Marines	11 Marines	Hard
25m	25 Marines	25 Marines	Hard
Corridor	6 Zealots	24 Zerglings	Hard

5.2 Mutual Information Evaluation

Due to the inherent property of POMDP, the experiences can be outdated when the agents interact with the environment. Decision policies based on the invalidated samples may prevent the agents from obtaining the maximum expected return in the subsequent training process.

Ideally, we expect that agents can effectively converge to a target state using a skill chain. We also desire that the skill chains be constructed backward sequentially until the s_0 state of the MDP is included in the initiation set. However, in a large MAS, the states evolve rapidly as each agent interacts with the environment. The agents may not observe the global state directly. Thus, using skill chains may also lead to the trap of a local optimal due to the ineffective exploration, resulting in poor convergence of the policy networks. We propose to incorporate the mutual information between the state trajectory distribution and the option sampling distribution and the mutual information is used to construct the intrinsic rewards of the high-level strategy network. Figure 3 demonstrates the average win rate of our proposed MADSC with and without mutual information.

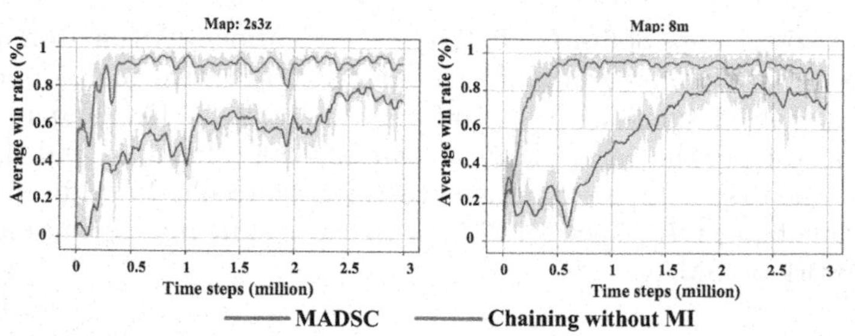

Fig. 3. Comparison of the mutual information in MADSC using intrinsic rewards

The construction of intrinsic rewards is based on the mutual information $I(S;\Omega)$ between the state distribution $p(s)$ and the ω sampling distribution $p(\omega)$. Hence, the convergence of exploration can be effectively secured against falling into the local optimum. We observe that, on the MAS tasks, including *8* m, *2s3z*, our method converges after 0.5 million time steps by utilizing the mutual information reward method. Meanwhile, the vanilla multi-agent skill chaining shows some difficulty reaching convergence even at three million steps.

5.3 Performance Evaluation

To evaluate the efficiency of our proposed MADSC scheme, We also compared the convergence performance of MADSC and some state-of-the-art MARL algorithms, including QMIX, IQL, and hDQN. The evaluation results are demonstrated in Fig 4.

A total of nine SMAC maps are utilized in the evaluation with various difficulties. The experiments are categorized into three difficulty levels, i.e., easy, normal, and hard. The main difference between each map exists in the number of agents and the difficulty variations in achieving the target state.

Our method can generally achieve the best performance among four cooperative MARL algorithms. For the easy-level maps, including *8* m, *2s3z*, because the action-state space is relatively small, both our MADSC method, QMIX, and

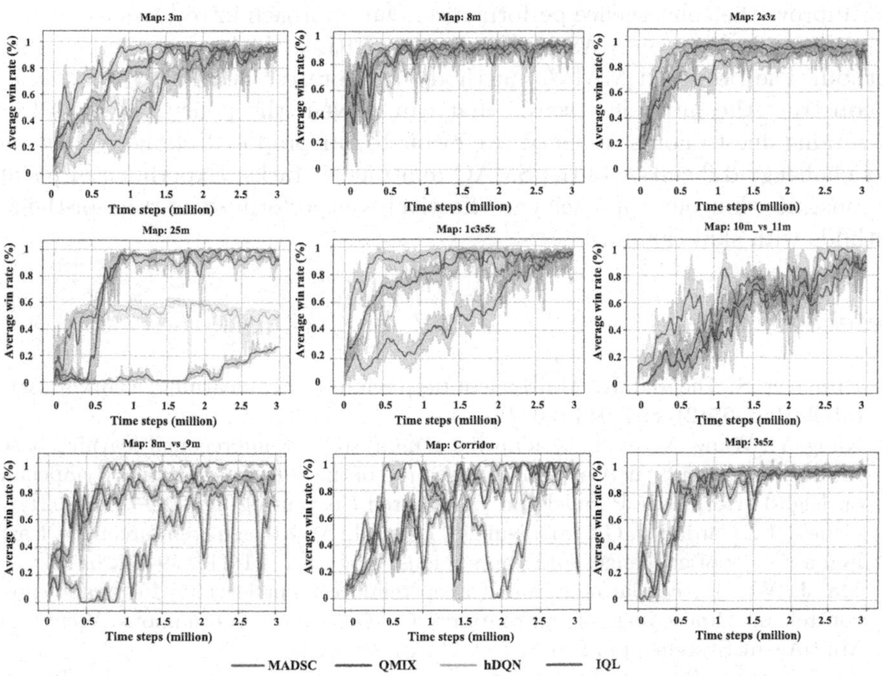

Fig. 4. Comparison of the win rates using various algorithms on SMAC tasks

hDQN can achieve an average 90% win rate after 0.5 million time steps. However, IQL does not perform well for map *2s3z*. For hard-level maps, e.g., *25* m, *10m_vs_11m*, *corridor*, our MADSC method can still perform well and converge to an average 90% win rate after 2.5 million time steps. The QMIX algorithm demonstrates a similar performance only on the *25m* task, but a deteriorated convergence performance on the *10m_vs_11m* and *corridor* tasks. However, due to the expanding size of thc action-state space and the unpredictable transition function of the state trajectories, the hDQN and IQL cannot converge within the three million training time steps. For normal-level maps, such as *1c3s5z* and *8m_vs_9m*, our proposed MADSC method outperforms all three algorithms. This is because our MADSC scheme allows the agents to diversify their choices of the state trajectories by using the mutual information-based intrinsic rewards, leading to efficient space exploration. In addition, our quest for high cumulative reward expectations aims to find optimal strategies based on accurate prediction performance. The exploration efficiency can prevent the agent's policy from falling into a local optimal and may eventually allow the policy network to converge to high accuracy with a greater probability.

6 Conclusions

This paper proposes a novel MARL method, MADSC, for cooperative multi-agents in macro-action level POMDPs to mitigate the nonstationary problem and improve the convergence performance. Our approach introduces the temporal abstraction layer on MAS by building on HRL with the deep skill chaining method. The proposed MADSC method incorporates the mutual information to construct the intrinsic rewards that can prevent the policy networks from converging due to non-stationary. We evaluate our proposed method using various challenge difficulties in the SMAC multi-agent tasks. Experimental results demonstrate that our approach can effectively outperform some state-of-the-art methods with skill chaining.

References

1. Gronauer, S., Diepold, K.: Multi-agent deep reinforcement learning: a survey. Artif. Intell. Rev. **55**(2), 895–943 (2022)
2. Kang, Y., Wang, X., et al.: Q-adaptive: a multi-agent reinforcement learning based routing on dragonfly network. In: Proceedings of the 30th International Symposium on High-Performance Parallel and Distributed Computing, pp. 189–200 (2021)
3. Canese, L., Cardarilli, G.C., Di Nunzio, L., et al.: Multi-agent reinforcement learning: a review of challenges and applications. Appl. Sci. **11**(11), 4948 (2021)
4. Ma, J., Wu, F.: Feudal multi-agent deep reinforcement learning for traffic signal control. In: Proceedings of International Conference on Autonomous Agents and MultiAgent Systems, pp. 816–824. AAMAS (2020)
5. Su, J., Adams, S.C., et al.: Value-decomposition multi-agent actor-critics. In: Proceedings of the AAAI Conference on Artificial Intelligence, pp. 11352–11360 (2021)

6. Sarafian, E., Tamar, A., Kraus, S.: Constrained policy improvement for efficient reinforcement learning. In: International Joint Conference on Artificial Intelligence, IJCAI, pp. 2863–2871 (2020)
7. Terry, J.K., Black, B., Grammel, N., et al.: Pettingzoo: gym for multi-agent reinforcement learning. In: Advances in Neural Information Processing Systems, NeurIPS, pp. 15032–15043 (2021)
8. Liu, Y., Hu, Y., Gao, Y., et al.: Value function transfer for deep multi-agent reinforcement learning based on N-step returns. In: Proceedings of the 28th International Joint Conference on Artificial Intelligence, IJCAI, pp. 457–463 (2019)
9. Phan, T., Belzner, L., et al.: Resilient multi-agent reinforcement learning with adversarial value decomposition. In: Proceedings of the AAAI Conference on Artificial Intelligence, pp. 11308–11316 (2021)
10. Danassis, P., Wiedemair, F., et al.: Improving multi-agent coordination by learning to estimate contention. In: Proceedings of the Thirtieth International Joint Conference on Artificial Intelligence, IJCAI, pp. 125–131 (2021)
11. Rashid, T., Samvelyan, M., et al.: QMIX: monotonic value function factorisation for deep multi-agent reinforcement learning. In: Proceedings of the International Conference on Machine Learning, ICML, vol. 80, pp. 4292–4301 (2018)
12. Bacon, P., Harb, J., Precup, D.: The option-critic architecture. In: Proceedings of the AAAI Conference on Artificial Intelligence, pp. 1726–1734 (2017)
13. Tang, H., Hao, J., Lv, T., et al.: Hierarchical deep multiagent reinforcement learning. CoRR abs/1809.09332 (2018)
14. Tessler, C., Givony, S., Zahavy, T., et al.: A deep hierarchical approach to lifelong learning in MineCraft. In: Proceedings of the AAAI Conference on Artificial Intelligence, vol. 31 (2017)
15. Andrychowicz, M., Crow, D., Ray, A., et al.: Hindsight experience replay. In: Advances in Neural Information Processing Systems, NeurIPS, pp. 5048–5058 (2017)
16. Bagaria, A., Konidaris, G.: Option discovery using deep skill chaining. In: International Conference on Learning Representations, ICLR (2020)
17. Sharma, A., Gu, S., Levine, S., et al.: Dynamics-aware unsupervised discovery of skills. In: International Conference on Learning Representations, ICLR (2019)
18. Sayin, M., Zhang, K., Leslie, D., et al.: Decentralized Q-learning in zero-sum markov games. In: Advances in Neural Information Processing Systems, NeurIPS, vol. 34, pp. 18320–18334 (2021)
19. Engstrom, L., Ilyas, A., Santurkar, S., et al.: Implementation matters in deep RL: a case study on PPO and TRPO. In: International Conference on Learning Representations, ICLR (2020)
20. Osband, I., Blundell, C., Pritzel, A., et al.: Deep exploration via bootstrapped DQN. In: Advances in Neural Information Processing Systems, NeurIPS, pp. 4026–4034 (2016)
21. Mnih, V., Badia, A.P., Mirza, M., Graves, A., et al.: Asynchronous methods for deep reinforcement learning. In: Proceedings of the International Conference on Machine Learning, ICML, vol. 48, pp. 1928–1937 (2016)
22. Kulkarni, T.D., Narasimhan, K., et al.: Hierarchical deep reinforcement learning: integrating temporal abstraction and intrinsic motivation. In: Advances in Neural Information Processing Systems, NeurIPS, pp. 3675–3683 (2016)
23. Mnih, V., Kavukcuoglu, K., Silver, D., et al.: Human-level control through deep reinforcement learning. Nature **518**(7540), 529–533 (2015)

24. Kim, J., Park, S., Kim, G.: Unsupervised skill discovery with bottleneck option learning. In: Proceedings of the International Conference on Machine Learning, ICML, vol. 139, pp. 5572–5582 (2021)
25. Lin, Y., Gou, Y., Liu, Z., et al.: COMPLETER: incomplete multi-view clustering via contrastive prediction. In: IEEE Conference on Computer Vision and Pattern Recognition, CVPR, pp. 11174–11183 (2021)
26. Hjelm, R.D., Fedorov, A., Lavoie-Marchildon, S., et al.: Learning deep representations by mutual information estimation and maximization. In: International Conference on Learning Representations, ICLR (2019)
27. Bachman, P., Hjelm, R.D., Buchwalter, W.: Learning representations by maximizing mutual information across views. In: Advances in Neural Information Processing Systems, NeurIPS, pp. 15509–15519 (2019)
28. Samvelyan, M., Rashid, T., de Witt, C.S., et al.: The StarCraft Multi-Agent Challenge. CoRR abs/1902.04043 (2019)

Natural Language Processing

Story Generation Based on Multi-granularity Constraints

Zhenpeng Guo, Jiaqiang Wan(✉), Hongan Tang, and Yunhua Lu

Chongqing University of Technology, Chongqing, China
guozhenpeng_xg@2020.cqut.edu.cn, {wanjiaqiang,yhlu}@cqut.edu.cn

Abstract. Generating a coherent story is a challenging task in natural language processing, which requires maintaining the coherence of plots and inter-sentence semantics throughout the generated text. While existing generation models can generate texts with good intra-sentence coherence, it is still difficult to plan a coherent plot and inter-sentence semantics throughout the text. In this paper, we propose a novel multi-granularity constraints text generation model, which constrains at the token level and sentence level, respectively. For the plot incoherence issue, the token-level constraint is added, which is a new plot guidance method to maintain the coherence of the plot while avoiding the introduction of extra exposure bias. For the problem of semantic incoherence, an auxiliary task of modelling the semantic relations between sentences is designed on the sentence-level constraint. Extensive experiments have shown that our model can generate more coherent stories than the baselines.

Keywords: Story generation · Multi-granularity constraints · Pre-training model

1 Introduction

Story generation, given the leading context, requires the model to generate a long, coherent natural language to describe a sensible sequence of events. Different from other text generation tasks, story generation is more challenging because the model not only needs to generate a story with a coherent plot but also should be concerned with the coherence between text units such as the sentences. Although existing generation models can generate texts with good intra-sentence coherence, it is still a struggle to maintain a coherent event sequence throughout the generated text, even using the powerful pretrained models [1,2] as illustrated in Fig. 1.

Prior works endeavored to tackle the incoherence issue in text generation have focused on the following aspects:

Model Structure Modifications: This method models the structure of human text discourse and then enhances the ability to model text by learning different levels of representations. Liet et al. [3] proposed a hierarchical RNN decoder for learning sentence-level representations. Shen et al. [4] further augmented

L. Fang et al. (Eds.): CICAI 2022, LNAI 13606, pp. 223–235, 2022.
https://doi.org/10.1007/978-3-031-20503-3_18

the hierarchical model using a method for learning multi-level latent variables. With this method, the model can learn more meaningful representations, and the ability to generate coherent text is improved.

Prior Knowledge Injection: Another line of works proposed injecting prior external knowledge into the model, making the generated content more relevant to the specific domain. For example, Guan et al. [5] and Xu et al. [6] used this method for commonsense story generation.

Controlled Generation: The final direction is to generate by controlling text attributes, such as keywords, plot, and character roles. Generally, the generation will be decomposed into multiple stages. The model first generates the attributes that need to be controlled and then expands it into the complete long text with fine detail. In terms of controlling the plot, Yao et al. [7] took the story's title as input, explicitly output keywords representing the story plot, and then went through these keywords to generate the whole story. Although this method has been widely used for story generation and achieved good performance in generating coherence texts, it is known to have the stage-level exposure bias [8], which can accumulate error through stages and impair the final generation quality.

Among the methods mentioned above, the controlled generation method has attracted much attention, and it has a good performance in generating coherent text, which is a research hotspot. Therefore this method is adopted in our model.

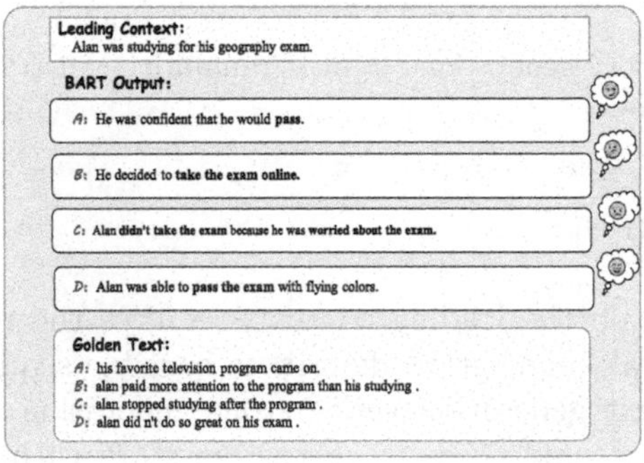

Fig. 1. Story examples written by the fine-tuned BART model. Although BART generates some relevant concepts (in **bold**), the generated still suffers from a severe incoherence problem.

In this paper, we propose MUGC, *a generation model based on **MU**lti-**G**ranularity **C**onstraints*, it is to tackle the incoherence issue in story generation. The whole model adopts an Encoder-Decoder generation framework and is constrained at the token level and sentence level, respectively. In terms of token level constraint, we adopt the controlled generation method to generate stories. In this constraint, to avoid introducing extra exposure bias and further improve the story's coherence, an implicit way is used to transmit the plot guidance information to the language model. In addition, most previous generative models were limited to token-level representation learning is insufficient to capture the semantic relationships between sentences [9]. For example, sentences C and D in Fig. 1 are semantically contradictory (e.g., "*Alan didn't take the exam*" but "*Alan was able to pass the exam with flying colours*"). Therefore, a task of modelling semantic relations between sentences is designed on sentence level constraint, which requires predicting the inter-sentence similarity.

The main contributions of this paper are summarized as follows:

- We propose a new multi-granularity constraints generation model named MUGC to tackle the incoherence issue in story generation, which constrains at the token level and sentence level, respectively.
- The experimental results on the ROCStories corpus demonstrate that MUGC can generate more coherent story and outperforms the baselines in both automatic metrics and human evaluation.

2 Related Works

2.1 Generation Framework

According to the characteristics of the model structure, the text generation models can be categorized into Auto-regressive Language Model, Hierarchical Encoder-Decoder, and Encoder-Decoder Framework.

The Auto-regressive Language Model is a standard left-to-right unidirectional language model that uses Decoder only for auto-regression decoding. However, this unidirectional language model does not capture context-relevant feature representations and is used primarily for sequential sequencing-complement problems, representing models such as GPT [10] and GPT-2 [2]. For the Hierarchical Encoder-Decoder architecture, such models use a hierarchical encoder to encode the source text at different levels and then decode it accordingly. When using RNN network modelling in the early stage, the hierarchical model can often improve performance by modelling local and global information with different granularities. After the proposed transformer [11] and pre-training language model, fully connected self-attention weakens this advantage. Relevant models include HMNet [12], Leader-Writer-network [13]. Finally, the Encoder-Decoder Framework uses an encoder to encode the source text and then uses a decoder to autoregressively decode the encoded text, which is widely applied in sequence to sequence tasks. The mainstream approach of the model using this framework is to combine it with the Transformer architecture. The representative models include BART [14], T5 [15], Pegasus [16], they are widely used in various generation tasks such as summarization and translation.

2.2 Controllable Story Generation

Controllable text generation refers to the generation of natural sentences with controllable attributes. In the story generation task, different story attributes can be controlled to generate, and the generation process is generally decomposed into multiple stages where the model first generates the controlled attributes and then expands it into a long text with a reasonable sequence of events. The controllable attributes can be the plot [7], a learnable skeleton [17], or an action sequence [18,19], which can also be generated by controlling the ending of the story [20], persona [21], and the topic sequence [22]. Although such methods improve the coherence of the story to a certain extent, the accumulation of stage exposure bias will eventually affect the quality of text generation. This multi-granularity constraints control generation method can avoid introducing extra exposure bias, which is an important research direction.

3 Methodology

3.1 Task Definition and Model Overview

Given the input, the model needs to generate a coherent story. To tackle this problem, typical generation models (e.g., BART) usually use a bidirectional encoder and a left-to-right decoder for autoregressive decoding to generate the text, whose training objective is to minimize the negative log-likelihood loss $\mathcal{L}_{LM}$ between the generated text and the human-written texts. The story generation task is defined as follows.

Definition 1. *(story generation): given the input $X = (x_1, x_2, \cdots, x_m)$ (e.g., a beginning or a prompt), the model generate a story with coherent plot $Y = (y_1, y_2, \ldots, y_n)$ (each x_i and y_i is a token), and the training objective is to minimize the negative log-likelihood loss $\mathcal{L}_{LM}$:*

$$\mathcal{L}_{LM} = -\sum_{t=1}^{n} \log P_{lm}\left(y_t \mid y_{<t}, X\right) \tag{1}$$

$$P_{lm}\left(y_t \mid y_{<t}, X\right) = \operatorname{softmax}\left(\boldsymbol{W}\mathbf{H}_t + \boldsymbol{b}\right) \tag{2}$$

$$\mathbf{H}_t = \operatorname{Decoder}\left(y < t, \{\mathbf{S}_i\}_{i=1}^{m}\right) \tag{3}$$

$$\{\mathbf{S}_i\}_{i=1}^{m} = \operatorname{Encoder}(X) \tag{4}$$

where H_t is the hidden state at the t-th position of the decoder, and S_i is the contextualized representation of x_i acquired from the encoder, W and b are the parameters to be trained.

However, it is difficult for the model to generate coherent texts due to the lack of plot-guiding information and the inability to capture the semantic relations between sentences. Therefore, to tackle this problem, we add constraints of different granularity, including token-level constraint and sentence-level constraint. Specifically, in terms of token-level constraint, we take the method of controlling the story plot to assist the language model in generating the story. This stage mainly includes two training tasks: keyword prediction and generation. The model first generates a keyword distribution that can represent the story

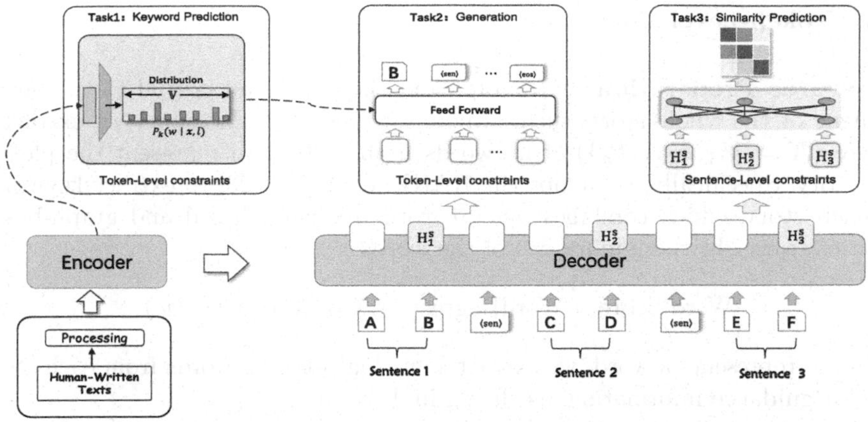

Fig. 2. An overview of **MUGC**, the constraint at the token level include two tasks: keyword prediction, which generates a keyword distribution that can represent the story plot (task1) and training to predict the next token under the guidance of the previous distribution (task2). Finally, the constraint at the sentence level is to predict inter-sentence semantic similarity (task3).

plot and then generates the whole story under the guide of this distribution. In the sentence-level constraint, we added a training task of similarity prediction between sentences, which can make the model capture the inter-sentence semantic relations.

Therefore, as mentioned above, our task is defined as follows:

Definition 2. *(multi-granularity constraints story generation): given the input* $X = (x_1, x_2, \cdots, x_m)$ *(e.g., a beginning or a prompt), the model first predicts a keyword distribution* $P_k(w \mid \boldsymbol{x}, p)$ *that represents the plot, and the loss function at this stage is* $\mathcal{L}_k$*. When decoding, the model generates the story text under the constraints of the distribution and predicts the similarity between the generated sentences, the loss functions are* $\mathcal{L}_{LM}$ *and* $\mathcal{L}_{Sen}$*, respectively. The final output of the model is a coherent story* $Y = (y_1, y_2, \ldots, y_n)$ *(each* x_i *and* y_i *is a token), and the training objective is to minimize* $\mathcal{L}_{total}$*:*

$$\mathcal{L}_{total} = \mathcal{L}_{LM} + \alpha \mathcal{L}_k + \lambda \mathcal{L}_{Sen} \tag{5}$$

where α and λ are adjustable scale factors. The overall framework of our model is shown in Fig. 2.

3.2 Token-Level Constraint

Keywords Preparation. In a story, some keywords are crucial to the development of the whole story, such as nouns, verbs, etc. Therefore, we extract a set K ($K = \{k_1, k_2 \dots k_n\}$) of keywords from a story to represent the plot of the story. Specifically, we adopt the RAKE algorithm [23] to extract keywords in each story, which combines several word frequency-based and graph-based metrics to weight the importance of the words:

$$\text{Word Score} = \text{WordDegree}(w)/\text{WordFrequency}(w) \tag{6}$$

where w represents a word, we extract some high-scoring words from each story as plot guidance information, as shown in Table 1.

Table 1. An example of extracting a keywords set from the story as plot guidance.

Title	Drained battery
Plot keywords (extracted)	Phone, friend, draining, battery, back, died
Story (human written)	Tom let his friend borrow his phone. The friend kept using it. He kept drain- ing the battery. Tom got it back way later. The phone died shortly after

Keywords Prediction. By obtaining the encoded information from Encoder to generate the plot guidance information, which is a keyword distribution $P_k(w \mid \boldsymbol{x}, p)$ over the whole vocabulary $\mathbb{V}$.

Specifically, we insert a special token p at the beginning of x and input them into the Encoder:

$$[\boldsymbol{S}_0, \boldsymbol{S}_1, \dots, \boldsymbol{S}_m] = \text{Encoder}\,(p, x_1, x_2, \dots, x_m) \tag{7}$$

where $\boldsymbol{S}_i (1 \leqslant i \leqslant m)$ is the hidden state at the position corresponding to x_i and S_0 represents the hidden state at the position of p. We then predict a keyword distribution $P_k(w \mid \boldsymbol{x}, p)$ over the whole vocabulary $\mathbb{V}$:

$$P_k(w \mid \boldsymbol{x}, p) = \text{softmax}\,(\mathbf{W}_k \boldsymbol{S}_c + \mathbf{b}_k) \tag{8}$$

where W_k and b_k are trainable parameters, and S_c summarizes the context embedding of the input information, we directly set $S_c = S_0$. The current stage training objective is to minimize the cross-entropy loss $\mathcal{L}_k$ as follows:

$$\mathcal{L}_k = -\sum_{i=1}^{|\mathbb{W}|} \hat{P}_k\,(w_i \mid p, \boldsymbol{x}) \log P_k\,(w_i \mid p, \boldsymbol{x}) \tag{9}$$

where w_i represents the i-th word on $\mathbb{V}$ and $\widehat{P}_k(w_i \mid p, x)$ is the ground truth distribution.

Generation. A left-to-right decoder is used to generate the whole story guided by the previously predicted keyword distribution $P_k(w \mid \boldsymbol{x}, p)$. This implicit guidance method can avoid introducing extra exposure bias and the training objective at this stage is to minimize the negative log-likelihood loss function $\mathcal{L}_{LM}$:

$$\mathcal{L}_{LM} = -\sum_{t=1}^{m} \log P\left(y_t \mid p, \boldsymbol{x}, y_{<t}\right) \tag{10}$$

where $P_k(w \mid \boldsymbol{x}, p)$ is obtained by combining the distribution $P_k(w \mid x, p)$ with the language model during the decoding process as follows:

$$\begin{aligned} P\left(y_t \mid l, \boldsymbol{x}, y_{<t}\right) = &P_{lm}\left(y_t \mid p, \boldsymbol{x}, y_{<t}\right) \cdot \left(1 - \boldsymbol{g}_t\right) \\ &+ P_k\left(y_t \mid p, \boldsymbol{x}\right) \cdot \boldsymbol{g}_t \end{aligned} \tag{11}$$

$$P_{lm}\left(y_t \mid p, \boldsymbol{x}, y_{<t}\right) = \operatorname{softmax}\left(\mathbf{W}_s \boldsymbol{H}_t + \mathbf{b}_s\right) \tag{12}$$

$$\boldsymbol{H}_t = \operatorname{Decoder}\left(y_{<t}, \{\boldsymbol{S}_i\}_{i=0}^{m}\right) \tag{13}$$

where both W_s and b_s are trainable parameters, P_{lm} is the probability distribution predicted by the language model over the vocabulary $\mathbb{V}$, and $\boldsymbol{g}_t \in \mathbb{R}^{|\mathbb{V}|}$is a gate vector denoting the weight of $P_k(w \mid x, p)$. $\boldsymbol{g}_t$ is computed as follows:

$$\boldsymbol{g}_t = \operatorname{sigmoid}\left(\mathbf{W}_g\left[\boldsymbol{r}_t; \boldsymbol{H}_t\right] + \boldsymbol{b}_g\right) \tag{14}$$

$$\boldsymbol{r}_t = \boldsymbol{W}_r P_k\left(y_t \mid p, \boldsymbol{x}\right) + \boldsymbol{b}_r \tag{15}$$

where W_g, b_g, W_r, and b_r are trainable parameters.

3.3 Sentence-Level Constraint

Semantically similar sentences have close representations in the vector space. Therefore, we can determine the semantic relations between sentences by learning the sentence representations. Assume that the target text Y contains K sentences from Y_1 to Y_K. We add a special sentence token $\langle \text{sen} \rangle$ at the end of each sentence and let $\mathbf{H}_k^S (1 \leqslant k \leqslant K)$ denote the hidden state of the decoder for the k-th sentence token, where the sentence token is used to aggregate the semantic information of each sentence during the decoding process. Specifically, the decoder of the MUGC is required to use $\mathbf{H}_i^{\mathrm{s}}$ and $\mathbf{H}_j^{\mathrm{s}}$ to predict the similarity score between any two sentences Y_i and Y_j. Since the learning of sentence representations has been well studied and has many powerful models such as

SentenceBERT [24], we use SentenceBERT similarity as the golden truth[1] in this task. Furthermore, to reduce the innate bias of SentenceBERT, the model does not need to fit the golden similarity exactly but allows the difference between the predicted scores and the golden similarity to be less than $\Delta \in [0, 1]$. The training objective in the sentence similarity prediction task is to minimize the loss function $\mathcal{L}_{\text{Sen}}$, which is calculated as follows:

$$\mathcal{L}_{Sen} = \frac{1}{K^2} \sum_{i=1}^{K} \sum_{j=1}^{K} \max\left(\left|p_{ij} - t_{ij}\right|, \Delta\right) \tag{16}$$

$$p_{ij} = \operatorname{sigmoid}\left(s_{ij} + s_{ji}\right)) \tag{17}$$

$$s_{ij} = \left(\mathbf{H}_i^{\mathrm{s}}\right)^{\mathrm{T}} \boldsymbol{W}^{\mathrm{s}} \mathbf{H}_j^{\mathrm{s}} \tag{18}$$

where W^{s} is the trainable parameter, S_{ij} is the intermediate variable that ensures p_{ij} is symmetric to i and j, t_{ij} is the golden similarity, and p_{ij} is the predicted similarity score.

In summary, based on token level constraint and increase the constraint on the sentence level. The token-level constraint enable the language model to generate stories with coherent plots based on the constraint guidance information. The sentence-level constraint enable the model to consider the semantic relations between the generated sentences when decoding to make the stories more coherent. Finally, the overall training objective of the model is to minimize the loss function $\mathcal{L}_{\text{total}}$.

4 Experimental Setup

4.1 Baselines

The experiments were conducted on the ROCStories corpus [25]. And MUGC is compared with the following baselines: **Seq2Seq:** It generates a text based on the input, which achieved by training the BART from scratch on the downstream dataset without pretraining. **Plan&Write:** It generates a sequence of keywords based on the input and then generates text based on the keywords [7]. **GPT-2 and BART:** They are obtained by fine-tuning the downstream dataset using only the language model as the objective.

Further, the ablation experiments are conducted on the test set to investigate the effect of different constraint levels. Finally, for a fair comparison, the sentence token is inserted into the training text of all baselines and set as the base version for all pretrained models.

[1] SentenceBERT calculates the similarity based on the cosine between the sentence embedding vectors, and we normalize the result to [0, 1].

4.2 Evaluation Metrics

Automatic Evaluation. Evaluate the model using the following automatic evaluation metrics: **(1) Perplexity (PPL):** The lower perplexity score means better overall fluency of the generated text. **(2) BLEU (B-n):** An evaluation of the n-gram overlap between the generated texts and the human-written texts [26], where we use $n = 1, 2$ to evaluate the models. **(3) Lexical Repetition (LR-n):** The metric calculates the percentage of text that repeats at least 4-gram of all generated texts, and we set $n = 2$. **(4) Distinct-4 (D-4):** We adopt distinct-4, a metric that measures generative diversity by the percentage of unique n-grams.

Human Evaluation. For human evaluation, MUGC is compared pair-wise with two strong baseline models (GPT-2 and BART). Furthermore, evaluate the models from the following aspects: **Coherence** to measure whether the plot of the story is coherent and logically correct, **Fluency** to determine whether the grammar within the sentence is correct, and **Relevance** to determine whether the content of the story is relevant to prompt. We randomly sampled 100 prompts from the test set and got a total of 300 stories from the two baseline models and ours. For each pair of stories, the three annotators need to give a preference (win, lose, tie) based on the coherence, fluency and relevance independently.

Table 2. Automatic evaluation results on TEST set. ↓ / ↑ means the lower/higher the better. The best performance is highlighted in bold. **w/o Tok** and **w/o Sen** means ablating the token-level and sentence-level, respectively.

Models	PPL↓	B-1↑	B-2↑	LR-2↓	D-4↑
Seq2Seq	18.14	0.302	0.130	0.28	0.663
Plan&Write	N/A	0.297	0.130	**0.201**	0.677
GPT-2	N/A	0.305	0.131	0.331	0.684
BART	9.83	0.307	0.133	0.307	0.699
MUGC	**9.52**	**0.343**	**0.168**	0.268	0.698
w/o Tok	9.68	0.3191	0.151	0.273	**0.703**
w/o Sen	9.59	0.331	0.162	0.264	0.691
Golden text	*N/A*	*N/A*	*N/A*	*0.058*	*0.891*

5 Results and Discussions

5.1 Automatic Evaluation and Human Evaluation

Table 2 shows the results of the automatic evaluation. Since the vocabulary used by Plan&Write and GPT-2 tokenize texts is different from BART, we do not provide the perplexity scores of these two models. As can be seen from the perplexity scores and BLEU scores, MUGC outperforms all baseline models, which indicates that the model has a better ability to model the texts in the test set and can generate more word overlaps with reference texts. It can also be seen that Plan&Write has less lexical repetition than the pre-trained models, which is consistent with the previous observation [6] because small models are better at learning short term statistics (e.g., n-gram) but not long term dependencies. And our approach does no harm to the generation diversity, as shown by the distinct scores. Furthermore, ablation experiments on the test set show that token-level constraints and sentence-level constraints are practical for generating coherent text, respectively. And both of them contribute to the ability to model the texts.

Table 3. Human evaluation results on TEST set. The scores indicate the percentages (%) of Win, Lose or Tie when comparing **MUGC** with a baseline.

Models	Coherence			Fluency			Logicality		
	Win (%)	Lose (%)	Tie (%)	Win (%)	Lose (%)	Tie (%)	Win (%)	Lose (%)	Tie (%)
MUGC vs. GPT-2	**43.5**	22.5	34	**34.5**	23	42.5	**42**	24.5	33.5
MUGC vs. BART	**52**	13	35	**38**	21	41	**46**	23	31

Table 3 presents the human evaluation results. It can be seen that MUGC achieves the best scores on all three metrics. Specifically, the results from the coherence and fluency evaluation can show that MUGC can generate more coherent stories than the baselines. Moreover, as can be seen from logicality evaluation, MUGC considers the semantic relationship between sentences and reduces the conflict of semantic logic.

5.2 Case Study

The generated cases are shown in Table 4, from which it can be seen that MUGC can generate more coherent stories than the baseline models. Specifically, although the baseline models can easily predict some words related to the input (e.g., *room*, *photo*, *wall*), the generated stories are unrelated to the leading context and do not have a coherent plot. For example, the text generated by GPT-2 in case 2 has a chaotic development of events and is irrelevant to the leading content. Also, it can be seen that in case 1, the text generated by BART suffers from conflicting semantic ("*going to decorate the house*" but "*going to have a firework display*"). By contrast, in these two generate cases, the texts generated by MUGC have a more coherent story plot, and the connections between sentences are more sensible. This result shows the effectiveness of multi-granularity constraints in generating coherent long text.

Table 4. Generated texts by different models. *Italic* words indicate the improper entities or events in terms of coherence in the context. And **bold** words denote the coherent event sequence.

Case 1	
Leading context:	The homeowners were going to decorate their house
Golden text:	They bought seven boxes of **christmas lights**. They **hung them from the roof and wrapped** them around the posts. **At night, the house looked beautiful!** they wanted to **decorate even more next year**
BART:	They were going to decorate the house. They were *going to have a firework display.* They were going to decorate it with the firework. The *firework would be a candle display*
GPT-2:	They decided to have a *small party.* They gathered all of their neighbors in the room. *They ate sandwiches for an hour.* It was*the most beautiful party* they ever had
MUGC:	**They were going to put up lights.** They **put them** in a big tree. **They had them all up.** They had a great time
Case 2	
Leading context:	The girl wanted to hang a photo on her wall
Golden text:	**She picked out the perfect frame for it**. She had her husband hang **the frame just like she wanted. She smiled every time she walked by that wall.** It was great to **keep memories** around the house
BART:	She drew a photo of herself *on a box.* She hung it *on a shelf to hang. She doesn't liked the picture* looked. *The girl liked it*
GPT-2:	*Dad made a mess* the wall. Dad brought the wall up. *She hung her backside mirror on the wall. Dad thanked her for taking care of her reflection*
MUGC:	**She went to the store to buy frames.** She **picked out the frames she wanted.** She **cut the frames** off. She **hung the photo on her wall**

6 Conclusion

This paper proposes a generation model based on multi-granularity constraints, named MUGC, to tackle the incoherence issue in story generation, which constrains at the token and sentence levels, respectively. Specifically, at the token level, the implicit plot guidance method can ensure plot coherence while avoiding the introduction of extra exposure bias; at the sentence level, we design an auxiliary task that enables the model to capture semantic relations between sentences when decoding. Experimental results on a story dataset demonstrate that MUGC can generate more coherent stories with related context than strong baselines.

Acknowledgments. This work was funded by Chongqing Technology Innovation and Application Development (Major Theme Special) Project, Development and Application of Cloud Service Platform for Intelligent Detection and Monitoring of Industrial Equipment, NO. cstc2019jscx-zdztzx0043.

References

1. Fan, A., Lewis, M., Dauphin, Y.: Hierarchical neural story generation. arXiv preprint arXiv:1805.04833 (2018)
2. Radford, A., Wu, J., Child, R., Luan, D., Amodei, D., Sutskever, I., et al.: Language models are unsupervised multitask learners. OpenAI Blog **1**(8), 9 (2019)
3. Li, J., Luong, M.T., Jurafsky, D.: A hierarchical neural autoencoder for paragraphs and documents. arXiv preprint arXiv:1506.01057 (2015)
4. Shen, D., et al.: Towards generating long and coherent text with multi-level latent variable models. arXiv preprint arXiv:1902.00154 (2019)
5. Guan, J., Huang, F., Zhao, Z., Zhu, X., Huang, M.: A knowledge-enhanced pretraining model for commonsense story generation. Trans. Assoc. Comput. Linguist. **8**, 93–108 (2020)
6. Xu, P., et al.: MEGATRON-CNTRL: controllable story generation with external knowledge using large-scale language models. arXiv preprint arXiv:2010.00840 (2020)
7. Yao, L., Peng, N., Weischedel, R., Knight, K., Zhao, D., Yan, R.: Plan-and-write: towards better automatic storytelling. In: Proceedings of the AAAI Conference on Artificial Intelligence, vol. 33, pp. 7378–7385 (2019)
8. Tan, B., Yang, Z., AI-Shedivat, M., Xing, E.P., Hu, Z.: Progressive generation of long text with pretrained language models. arXiv preprint arXiv:2006.15720 (2020)
9. Ribeiro, M.T., Wu, T., Guestrin, C., Singh, S.: Beyond accuracy: behavioral testing of NLP models with checklist. arXiv preprint arXiv:2005.04118 (2020)
10. Radford, A., Narasimhan, K., Salimans, T., Sutskever, I.: Improving language understanding by generative pre-training (2018)
11. Vaswani, A., et al.: Attention is all you need. In: Advances in Neural Information Processing Systems, vol. 30 (2017)
12. Zhu, C., Xu, R., Zeng, M., Huang, X.: A hierarchical network for abstractive meeting summarization with cross-domain pretraining. arXiv preprint arXiv:2004.02016 (2020)
13. Liu, C., Wang, P., Xu, J., Li, Z., Ye, J.: Automatic dialogue summary generation for customer service. In: Proceedings of the 25th ACM SIGKDD International Conference on Knowledge Discovery & Data Mining, pp. 1957–1965 (2019)
14. Lewis, M., et al.: BART: denoising sequence-to-sequence pre-training for natural language generation, translation, and comprehension. arXiv preprint arXiv:1910.13461 (2019)
15. Raffel, C., et al.: Exploring the limits of transfer learning with a unified text-to-text transformer. arXiv preprint arXiv:1910.10683 (2019)
16. Zhang, J., Zhao, Y., Saleh, M., Liu, P.: PEGASUS: pre-training with extracted gap-sentences for abstractive summarization. In: International Conference on Machine Learning, pp. 11328–11339. PMLR (2020)
17. Xu, J., Ren, X., Zhang, Y., Zeng, Q., Cai, X., Sun, X.: A skeleton-based model for promoting coherence among sentences in narrative story generation. arXiv preprint arXiv:1808.06945 (2018)

18. Fan, A., Lewis, M., Dauphin, Y.: Strategies for structuring story generation. arXiv preprint arXiv:1902.01109 (2019)
19. Goldfarb-Tarrant, S., Chakrabarty, T., Weischedel, R., Peng, N.: Content planning for neural story generation with aristotelian rescoring. arXiv preprint arXiv:2009.09870 (2020)
20. Peng, N., Ghazvininejad, M., May, J., Knight, K.: Towards controllable story generation. In: Proceedings of the First Workshop on Storytelling, pp. 43–49 (2018)
21. Chandu, K., Prabhumoye, S., Salakhutdinov, R., Black, A.W.: "My way of telling a story": persona based grounded story generation. In: Proceedings of the Second Workshop on Storytelling, pp. 11–21 (2019)
22. Huang, Q., Gan, Z., Celikyilmaz, A., Wu, D., Wang, J., He, X.: Hierarchically structured reinforcement learning for topically coherent visual story generation. In: Proceedings of the AAAI Conference on Artificial Intelligence, vol. 33, pp. 8465–8472 (2019)
23. Rose, S., Engel, D., Cramer, N., Cowley, W.: Automatic keyword extraction from individual documents. In: Text Mining: Applications and Theory, vol. 1, pp. 1–20 (2010)
24. Reimers, N., Gurevych, I.: Sentence-BERT: sentence embeddings using siamese BERT-networks. arXiv preprint arXiv:1908.10084 (2019)
25. Mostafazadeh, N., et al.: A corpus and cloze evaluation for deeper understanding of commonsense stories. In: Proceedings of the 2016 Conference of the North American Chapter of the Association for Computational Linguistics: Human Language Technologies, pp. 839–849 (2016)
26. Papineni, K., Roukos, S., Ward, T., Zhu, W.J.: BLEU: a method for automatic evaluation of machine translation. In: Proceedings of the 40th Annual Meeting of the Association for Computational Linguistics, pp. 311–318 (2002)

Chinese Word Sense Embedding with SememeWSD and Synonym Set

Yangxi Zhou, Junping Du(✉), Zhe Xue, Ang Li, and Zeli Guan

Beijing Key Laboratory of Intelligent Telecommunication Software and Multimedia, School of Computer Science, Beijing University of Posts and Telecommunications, Beijing 100876, China

junpingdu@126.com, {zhouyx,xuezhe,david.lee,guanzeli}@bupt.edu.cn

Abstract. Word embedding is a fundamental natural language processing task which can learn feature of words. However, most word embedding methods assign only one vector to a word, even if polysemous words have multi-senses. To address this limitation, we propose SememeWSD Synonym (SWSDS) model to assign a different vector to every sense of polysemous words with the help of word sense disambiguation (WSD) and synonym set in OpenHowNet. We use the SememeWSD model, an unsupervised word sense disambiguation model based on OpenHowNet, to do word sense disambiguation and annotate the polysemous word with sense id. Then, we obtain top n synonyms of the word sense from OpenHowNet and calculate the average vector of synonyms as the vector of the word sense. In experiments, We evaluate the SWSDS model on semantic similarity calculation with Gensim's wmdistance method. It achieves improvement of accuracy. We also examine the SememeWSD model on different BERT models to find the more effective model.

Keywords: Word sense disambiguation · Word sense embedding · Synonym set

1 Introduction

The purpose of word embedding is to learn the vector representation of a word in its context [1]. The learned word vectors can be used in subsequent natural language processing tasks, such as semantic similarity calculating [2–4] and sentiment analysis [5–7]. The quality of word vectors learned by word embedding method can directly influence the performance of these tasks. Therefore, improving the performance of word embedding is a critical issue for natural language understanding.

Polysemous words are common in natural languages. However, most traditional word embedding methods assume that each word corresponds to only one word vector, even if it is a polysemous word. It means that they are single-prototype word vectors. For example, word2vec (Mikolov et al., 2013) [8] is the most popular word embedding method that strikes a good balance between effectiveness and efficiency and assigns only one vector to each word. "Apple" is a

L. Fang et al. (Eds.): CICAI 2022, LNAI 13606, pp. 236–247, 2022.
https://doi.org/10.1007/978-3-031-20503-3_19

typical polysemous word that has many senses. As suggested in Fig. 1, on the one hand, "Apple" means "Computer" in the context "I am using an apple computer". On the other hand, it means "Fruit" in the context "Apple is rich in nutrients". As a result, single prototype word vectors cannot convey the different senses of polysemous words in different contexts.

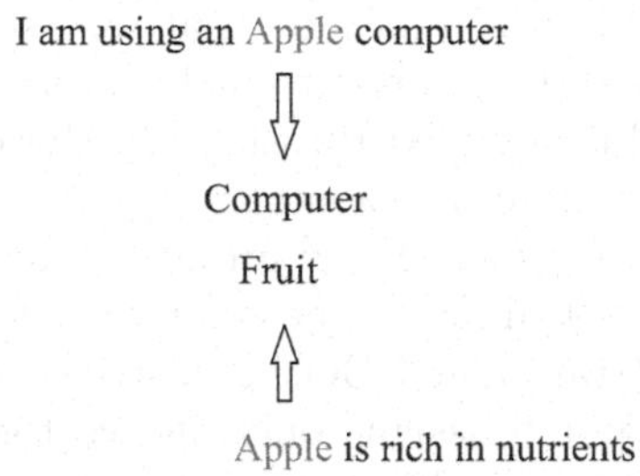

Fig. 1. Different senses of "Apple" in different contexts.

To resolve issues mentioned above, Huang et al. (2012) [9] present a multi-prototype model based on neural network. Furthermore, Chen et al. (2014) [10] combine word sense representation with word sense disambiguation and they make use of the synonyms set in WordNet [11]. Word sense disambiguation aims to determine the specific sense of polysemous words in the context. In reality, OpenHowNet is the most famous Chinese/English sememe knowledge base which also annotate each word sense with the synonym set [12]. From previous studies, we learn that word sense disambiguation is beneficial to word sense embedding (WSE) and synonym set can convey the sense of the polysemous word to a certain extent.

In this paper, we intend to further incorporate word sense disambiguation and synonym set in OpenHowNet into word sense embedding for Chinese. To prove its feasibility, we present a SememeWSD-Synonym (SWSDS) model, which disambiguates polysemous words and represents it by synonym set. In experiments, we further study the SememeWSD model [13] on WSD dataset and we also evaluate our model on semantic similarity. The experiment results show that our model outperforms the baseline method. This means that our model can improve word sense embedding for subsequent tasks with the help of word sense disambiguation and synonym set.

The key contributions of this paper can be concluded as follows:

(1) We built a polysemous words' dictionary based on OpenHowNet to identify polysemous words in a given sentence.
(2) We conducted extensive experiments on SememeWSD and found a more effective BERT model for word sense disambiguation dataset.
(3) To the best of our knowledge, this is the first work to utilize synonym set in OpenHowNet to improve word sense embedding.

2 Related Work

2.1 Word Embedding

With the rapid development of natural language processing, word embedding has become the most important method to obtain the effective representations of text words [14–16].

The one-hot representation is a method that represents each text word with a binary vector. As the construction is extremely simple, the one-hot representation has been widely used in some NLP tasks [17]. However, this representation method is likely to suffer from dimension curse [18].

To address this issue, Bengio et al. [18] presented the concept of word embedding. Afterwards, Mikolov et al. [8] proposed word2vec model based on neural network, including Skip-Gram and CBOW. It strikes a good balance between effectiveness and efficiency and assigns only one vector to each word. But the sense of a text word changes with the context. To solve the problem, Pennington et al. [19] present GloVe to combine the merits of the matrix factorization and the prediction-based methods. Embeddings from Language Model (EMLO) is the implementation of the deep contextualized word representations proposed by Matthew et al. [20]. Huang et al. [9] utilize multi-prototype vector models to learn word embedding and arrange a distinct vector for each word sense.

2.2 Word Sense Disambiguation and Word Sense Embedding

With the popularity of the natural language processing, word sense disambiguation has gradually emerged into our vision and becomes increasingly important to many tasks [21–23]. It aims to identify word sense in a context which contains two approaches. Supervised methods utilize classifiers for word sense disambiguation [24]. It requires large-scale human annotation data.

Instead, knowledge-based methods base on large external knowledge bases to find the most likely sense of a word. Agirre et al. [25] present a WSD algorithm based on random walks over large Lexical Knowledge Bases. Ustalov et al. [26] propose a model that first obtains sense embeddings from the word embeddings of the corresponding senses' synonyms in WordNet and then selects the sense that has the closest embedding similarity with the context. Chen et al. [10] take synsets in WordNet into consideration and do word sense disambiguation and word sense representation learning concurrently. Hou et al. propose an unsupervised HowNet-based word sense disambiguation model with the help of BERT model which is named SememeWSD. They also build an OpenHowNet-based WSD datasets.

In a word, many researchers have made progress in word embedding and word sense disambiguation recently [27–29]. In this paper, we utilize the SememeWSD to do word sense disambiguation and do word sense embedding with synonym set in OpenHowNet.

3 Methodology

We now describe the SWSDS model, our proposed an approach for word sense embedding. We first give a brief introduction to synonyms, sememes and senses in OpenHowNet and SememeWSD. Then we present the architecture of our model and explain how the model works.

3.1 OpenHowNet and SememeWSD

In recent years, HowNet has been successfully applied to various natural language processing tasks. The meaning of all words in HowNet can be expressed by a set of sememes. Sememe is defined as the smallest semantic unit of human language. Openhownet is derived from HowNet.OpenHowNet originates from HowNet, which is the most famous sememe knowledge base. It defines about 2,000 sememes and uses them to annotate senses of more than 100,000 Chinese words and phrases [12]. The senses of all the words in HowNet can be represented by a set of sememes. Sememe is defined as the minimum semantic unit of human language.

Sememe annotations in OpenHowNet for each sense looks like a tree, as illustrated in Fig. 2. The word "Apple" has two senses including "Computer" and "Fruit". For the first sense "Computer", it is annotated with 2 sememes while the second is annotated with only 1 sememe.

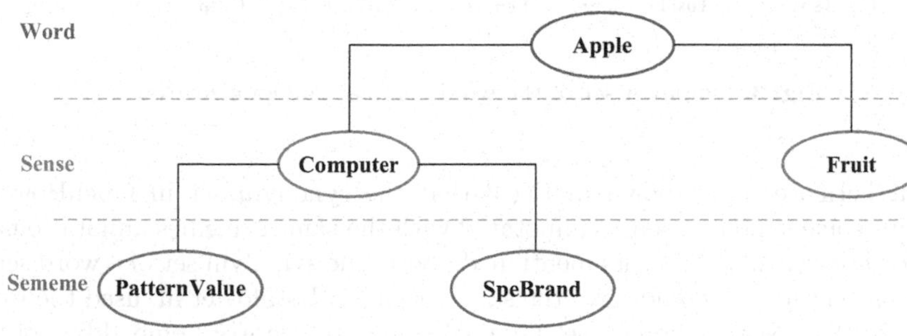

Fig. 2. Sememe annotations of the word "Apple" in OpenHowNet.

According to the definition of sememe and the principle of OpenHowNet, sememes of a sense can convey its sense, and two senses with the same sememes are supposed to have the same meaning.

SememeWSD is an unsupervised word sense disambiguation method based on OpenHowNet and large-scale pretraining model Bert proposed by Hou et al. SememeWSD finds the synonym set of the word meaning in OpenHowNet for each word meaning of the target polysemous word and selects a series of synonyms as substitutes, and these substitutes contain the original set of synonyms with the target semantics.

The Bert model can calculate the probability of a word's vacancy in a sentence, that is, the MLM (masked language model) prediction score. Then the average MLM prediction score of all substitute words of a word meaning can reflect the probability of the target word taking this word meaning in the context. Therefore, the principle of SememeWSD is to calculate the MLM prediction score of each substitute word and then calculate the average score. Finally, SememeWSD selects the word meaning with the highest average score, that is, the highest probability in a given context, as the word meaning of the target polysemous word. For example, for "the apple is very sweet", the Bert model can give the probability of any word in the "(mask) is very sweet", while SememeWSD is to calculate the MLM prediction score for a series of substitute words (the substitute words for "Apple = fruit" are "orange", "mango", "watermelon", etc., "Apple = computer" are "Toshiba", "Huawei", "Samsung", etc. Then take the average score to get a higher score of "Apple = fruit", so as to judge the meaning of "Apple = fruit" in "the apple is very sweet".

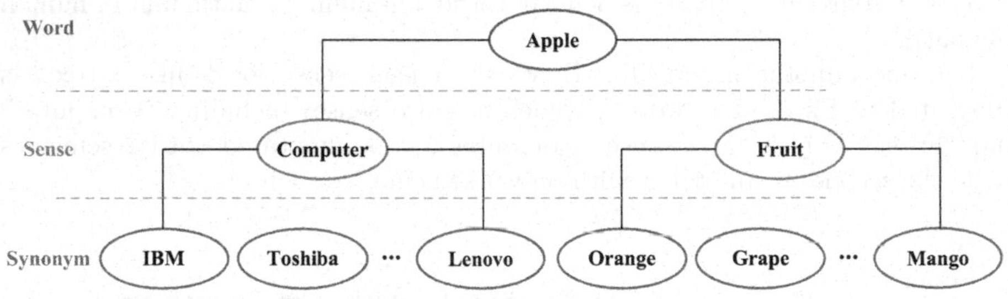

Fig. 3. Synonym set of the word "Apple" in OpenHowNet.

Furthermore, it is shown in Fig. 3 that the synonym set in OpenHowNet contains some synonym senses annotated with the same sememes annotations as the word sense. As the factors mentioned above, the synonym set of a word sense can represent its sense and have the same meaning. Ustalov et al. used the word vector of WordNet's synonym set (synset) to obtain the word embedding of the target word. In this paper, we utilize the word vector of the top n synonyms from the pre-trained Word2Vec model and calculate the average vector of synonyms as the word vector of the word sense.

3.2 SWSDS Model

The SWSDS model can be described as a 3-stage process: preprocess, word sense disambiguation and word sense embedding.

Preprocess. The original text dataset needs to be preprocessed before the word sense disambiguation and word sense embedding. First, we use the Jieba word segmentation tool to do word segmentation and part of speech tagging; After that, we identify polysemous words.

In order to automatically identify polysemous words in the text, we also established a polysemous word dictionary based on OpenHowNet as shown in Table 1. For words with multiple meanings in openhownet, we added the polysemous word dictionary and marked the part of speech, and scanned text after word segmentation and marked the part of speech one by one, If the keyword of the query polysemous word dictionary has this word and the part of speech is the same, it indicates that this word is a polysemous word, and word sense disambiguation is required.

Table 1. The polysemous words' dictionary.

Full	Verb
Insist	Verb
Art	Noun
...	...
Question	Noun

After recognizing the polysemous words, we also need to convert the original text data into a data format that can be processed by SWSDS model, as shown in Table 2. It mainly includes fields such as "context" (marking polysemous words as (target)) for MLM prediction by Bert in sememewsd model), "part of speech" (corresponding part of speech), "target word" (target word to be disambiguated), "target position" (position of target word in the context), "target word pos" (part of speech of target word).

Table 2. Require data format of "Apple is rich in nutrients".

Context	'(targe)', 'is', 'rich', 'in', 'nutrients'
Part of speech	'n', 'v', 'a', 'p', 'n'
Target word	Apple
Target position	0
Target word pos	n

Word Sense Disambiguation. Given large amounts of text data, we first transform the text preprocessing into the data form required by the SememeWSD model; Then, we use SememeWSD for word sense disambiguation, and mark polysemous words with the ID of the word meaning in OpenHowNet in the context; Finally, we get an annotated sentence. Example of word sense disambiguation: the system automatically recognizes the polysemous word "apple" in the context "the apple is very sweet", and converts it into the required data format through preprocessing: "The, (target), is, very, sweet"; Then input SememeWSD model for word sense disambiguation. The disambiguation result shows that the polysemous word "apple" has the meaning of "fruit" and is annotated with its ID "244397" in OpenHowNet. That is to say, the context "the apple is very sweet" changes to "the apple = 244397 is very sweet".

Word Sense Embedding. Given annotated words in the context and pre-trained word2vec model, we propose a simple algorithm to obtain the word sense vector of each annotated word: First, we acquire top n synonyms set S from OpenHowNet that have the same sememes annotation as the annotated word w'. Then, we obtain vectors $(V_i, i = 1...n)$ of these n synonyms with the pre-trained word2vec model. Last, we calculate the average vector $V_{average}$ of these n vectors as the sense vector $V_{w'}$ of the word annotated with sense w' and add it to the pretrained word2vec model. Formally, the sense vector $V_{w'}$ is computed as follows:

$$V_{w'} = V_{average} = \frac{1}{n}\sum_{i=1}^{n} V_i$$

After calculation, we obtain the vector of the word sense "Apple = 244397" and add it to pretrained vectors. Finally, we finish the word sense embedding and have vectors model including "Apple = 244397".

4 Experiment

In this section, we first experiment the SememeWSD model on the OpenHowNet-based WSD dataset, showing that there are more effective BERT model for SememeWSD. Then we evaluate our SWSDS model on LCQMC (A Large scale Chinese Question Matching Corpus) [30] dataset with semantic similarity calculation.

4.1 SWSDS Experiment

We can see that sememewsd internally uses the Bert base Chinese model, and there are many Chinese oriented BERT pre training models. So we thought of using the openhownet based WSD dataset to evaluate different BERT models [31]: TinyBERT, AlBERT-tiny [32], AlBERT-base [32], DistilBERT [33] and Chinese-BERT-wwm [34]. We use the HowNet-based WSD dataset to evaluate different BERT models. It is proposed by Hou et al. in their paper which is built on the Chinese Word Sense Annotated Corpus used in SemEval-2007 task 5 [35], including 2, 969 items for 36 target polysemous words (17 nouns and 19 verbs).

Table 3 shows that the Chinese-BERT-wwm model is superior to all other models in terms of Micro-F1, Macro-F1 scores of nouns and Macro-F1 scores of verbs. At the same time, TinyBERT does not seem to be suitable for the word

Table 3. WSD F1 scores of Chinese BERT model.

Model	Nouns		Verbs	
	Micro-F1	Macro-F1	Micro-F1	Macro-F1
Random	38.5	36.1	23.2	22.8
Dense	52.3	39.0	35.1	33.0
BERT-base	53.7	41.7	52.4	48.0
TinyBERT	36.0	24.7	20.4	16.3
AlBERT-tiny	47.9	36.1	38.9	33.0
AlBERT-base	51.9	39.3	42.3	37.4
DistilBERT	52.0	39.1	41.6	35.2
Chinese-BERT-wwm	**54.4**	**42.3**	**53.6**	**49.2**

sense disambiguation algorithm of SememeWSD, and has obtained the worst score, even lower than the random method.

Table 4 illustrates that the accuracy of the Chinese-BERT-wwm model reached the highest 53.9%. Tinybert model has the lowest accuracy of 26.4%. Since we are conducting research, we used Chinese-BERT-wwm with highest accuracy rather than Bert-Base model in SWSDS model.

Table 4. WSD results and speed of Chinese BERT model.

Model	All		
	Micro-F1	Macro-F1	Accuracy
Random	29.1	29.1	29.1%
Dense	41.7	35.8	41.7%
BERT-base	52.9	45.0	53.0%
TinyBERT	26.4	20.3	26.4%
AlBERT-tiny	42.4	34.5	42.4%
AlBERT-base	46.0	38.3	46%
DistilBERT	45.6	37.1	45.6%
Chinese-BERT-wwm	**53.8**	**46.0**	**53.9%**

4.2 Effectiveness Evaluation of SWSDS Model

In this section, We conduct text similarity experiments on the SWSDS model to observe the improvement of the algorithm on the quality of the word sense vector.

Experimental Setting. We choose LCQMC dataset to evaluate our model. LCQMC is a semantic matching dataset, whose goal is to determine whether the semantics of two questions are the same. Therefore, a word vector with higher accuracy of text similarity calculation means higher word embedding quality.

Word Mover's Distance (WMD) is based on recent results in word embeddings that learn semantically meaningful representations for words from local co-occurrences in sentences. WMD leverages the results of advanced embedding techniques like word2vec and Glove, which generates word embeddings of unprecedented quality and scales naturally to very large data sets. These embedding techniques demonstrate that semantic relationships are often preserved in vector operations on word vectors. For example, vector (Berlin) – vector (Germany) + vector (France) is close to vector (Paris).

Word Mover's Distance (WMD), suggests that distances and between embedded word vectors are to some degree semantically meaningful. It utilizes this property of word vector embeddings and treats text documents as a weighted point cloud of embedded words. The distance between two text documents A and B is calculated by the minimum cumulative distance that words from the text document A needs to travel to match exactly the point cloud of text document B.

We use wmdistance method in Gensim [36] library to calculate semantic similarity. The wmdistance method is implementation of the Word Mover's Distance (WMD) algorithm [37] which is a novel distance function between text documents. The WMD distance measures the semantic dissimilarity between two text documents.

Baseline Method. We compare our model with a high-quality word2vec model pre-trained by Tencent AILab [38]. We use the pretrained word vector mentioned above to calculate the word shift distance. The word vector provides two versions of 100 dimensional vector and 200 dimensional vector for more than 12 million Chinese words and phrases. These words and phrases are pretrained on large-scale high-quality data corpus.

The specific text similarity calculation rules are as follows: for two sentences with the calculated word mover's distance less than 0.5, we consider them to be similar and mark them as "1"; otherwise, we consider them to be dissimilar and mark them as "0"; Based on this method, the similar tags predicted by each pair of text in the LCQMC dataset is calculated, compared with the tags in the test set, and the accuracy rate is calculated. Table 5 shows that the accuracy rates of the 100 dimensional word vector and the 200 dimensional word vector of the baseline method are 67.9% and 70.4% respectively.

Then, based on the pre-training word vectors used by the Baseline method, we add the SWSDS model to obtain the word sense vectors. The accuracy of the 100-dimensional word vectors and the 200-dimensional word vectors is 71.9% and 74.0%, respectively. Therefore, the semantic word vectors obtained by our model are 4% and 3.6% higher than the pre-training 100-dimensional word vectors and 200-dimensional word vectors used by the Baseline method, respectively.

Table 5. Semantic similarity calculation results.

Method	Accuracy
WMD-100	67.9%
WMD-100+SWSDS	71.9%
WMD-200	70.4%
WMD-200+SWSDS	74.0%

5 Conclusion

In this paper, we present a word sense embedding method based on word sense disambiguation and synonym set. In addition, we build a polysemous words dictionary to identify polysemous words in sentences.

Experimental results show that our model improves the performance of semantic similarity compared to the baseline method. This means that the word sense embedding based on word sense disambiguation and synonym set can more accurately represent the semantics of polysemous words.

Acknowledgment. This work was supported by the National Natural Science Foundation of China (62192784, 62172056), and CAAI-Huawei MindSpore Open Fund under grand CAAIXSJLJJ-2021-057A.

References

1. Jiao, Q., Zhang, S.: A brief survey of word embedding and its recent development. In: 2021 IEEE 5th Advanced Information Technology, Electronic and Automation Control Conference (IAEAC) (2021)
2. Li, A., et al.: Scientific and technological information oriented semantics-adversarial and media-adversarial cross-media retrieval (2022)
3. Kou, F., Du, J., He, Y., Ye, L.: Social network search based on semantic analysis and learning. CAAI Trans. Intell. Technol. **1**(4), 293–302 (2016)
4. Xu, M., Du, J., Xue, Z., Kou, F., Xu, X.: A semi-supervised semantic-enhanced framework for scientific literature retrieval. Neurocomputing **461**(2), 450–461 (2021)
5. Balikas, G., Moura, S., Amini, M.R.: Multitask learning for fine-grained Twitter sentiment analysis. In: International ACM SIGIR Conference on Research & Development in Information Retrieval, pp. 1005–1008 (2017)
6. Ma, Z., Du, J., Zhou, Y.: Sentiment analysis based on evaluation of tourist attractions. In: Jia, Y., Du, J., Li, H., Zhang, W. (eds.) Proceedings of the 2015 Chinese Intelligent Systems Conference. LNEE, pp. 375–382. Springer, Heidelberg (2016). https://doi.org/10.1007/978-3-662-48386-2_39
7. Ye, H., Du, J.: Opinion leader mining of social network combined with hierarchical sentiment analysis. In: Deng, Z. (ed.) CIAC 2017. LNEE, vol. 458, pp. 639–646. Springer, Singapore (2018). https://doi.org/10.1007/978-981-10-6445-6_70
8. Mikolov, T., Chen, K., Corrado, G., Dean, J.: Efficient estimation of word representations in vector space. arXiv preprint arXiv:1301.3781 (2013)

9. Huang, E.H., Socher, R., Manning, C.D., Ng, A.Y.: Improving word representations via global context and multiple word prototypes. In: Proceedings of the 50th Annual Meeting of the Association for Computational Linguistics (Volume 1: Long Papers), pp. 873–882 (2012)
10. Chen, X., Liu, Z., Sun, M.: A unified model for word sense representation and disambiguation. In: Proceedings of the 2014 Conference on Empirical Methods in Natural Language Processing (EMNLP), pp. 1025–1035 (2014)
11. Miller, G.A., Beckwith, R., Fellbaum, C., Gross, D., Miller, K.J.: Introduction to WordNet: an on-line lexical database. Int. J. Lexicogr. **3**(4), 235–244 (1990)
12. Qi, F., Yang, C., Liu, Z., Dong, Q., Sun, M., Dong, Z.: OpenHowNet: an open sememe-based lexical knowledge base. arXiv preprint arXiv:1901.09957 (2019)
13. Hou, B., Qi, F., Zang, Y., Zhang, X., Liu, Z., Sun, M.: Try to substitute: an unsupervised Chinese word sense disambiguation method based on HowNet. In: Proceedings of the 28th International Conference on Computational Linguistics, pp. 1752–1757 (2020)
14. Xue, Z., Du, J., Du, D., Lyu, S.: Deep low-rank subspace ensemble for multi-view clustering. Inf. Sci. **482**, 210–227 (2019)
15. Xue, Z., Du, J., Zheng, C., Song, J., Ren, W., Liang, M.: Clustering-induced adaptive structure enhancing network for incomplete multi-view data. In: IJCAI, pp. 3235–3241 (2021)
16. Xu, L., Du, J., Li, Q.: Image fusion based on nonsubsampled contourlet transform and saliency-motivated pulse coupled neural networks. Math. Probl. Eng. **2013** (2013)
17. Sebastiani, F.: Machine learning in automated text categorization. ACM Comput. Surv. (CSUR) **34**(1), 1–47 (2002)
18. Bengio, Y., Ducharme, R., Vincent, P.: A neural probabilistic language model. In: Advances in Neural Information Processing Systems, vol. 13 (2000)
19. Pennington, J., Socher, R., Manning, C.D.: GloVe: global vectors for word representation. In: Proceedings of the 2014 Conference on Empirical Methods in Natural Language Processing (EMNLP), pp. 1532–1543 (2014)
20. Peters, M., Neumann, M., Iyyer, M., Gardner, M., Zettlemoyer, L.: Deep contextualized word representations (2018)
21. Iacobacci, I., Pilehvar, M.T., Navigli, R.: Embeddings for word sense disambiguation: an evaluation study. In: Proceedings of the 54th Annual Meeting of the Association for Computational Linguistics (Volume 1: Long Papers), pp. 897–907 (2016)
22. Raganato, A., Bovi, C.D., Navigli, R.: Neural sequence learning models for word sense disambiguation. In: Proceedings of the 2017 Conference on Empirical Methods in Natural Language Processing, pp. 1156–1167 (2017)
23. Lin, P., Jia, Y., Du, J., Yu, F.: Average consensus for networks of continuous-time agents with delayed information and jointly-connected topologies. In: 2009 American Control Conference, pp. 3884–3889. IEEE (2009)
24. Lee, Y.K., Ng, H.T., Chia, T.K.: Supervised word sense disambiguation with support vector machines and multiple knowledge sources. In: Proceedings of SENSEVAL-3, the Third International Workshop on the Evaluation of Systems for the Semantic Analysis of Text, pp. 137–140 (2004)
25. Agirre, E., de Lacalle, O.L., Soroa, A.: Random walks for knowledge-based word sense disambiguation. Comput. Linguist. **40**(1), 57–84 (2014)
26. Ustalov, D., Teslenko, D., Panchenko, A., Chernoskutov, M., Biemann, C., Ponzetto, S.P.: An unsupervised word sense disambiguation system for under-resourced languages. arXiv preprint arXiv:1804.10686 (2018)

27. Niu, Y., Xie, R., Liu, Z., Sun, M.: Improved word representation learning with sememes. In: Proceedings of the 55th Annual Meeting of the Association for Computational Linguistics (Volume 1: Long Papers), pp. 2049–2058 (2017)
28. Barba, E., Procopio, L., Navigli, R.: ConSeC: word sense disambiguation as continuous sense comprehension. In: Proceedings of the 2021 Conference on Empirical Methods in Natural Language Processing, pp. 1492–1503 (2021)
29. Bevilacqua, M., Pasini, T., Raganato, A., Navigli, R., et al.: Recent trends in word sense disambiguation: a survey. In: Proceedings of the Thirtieth International Joint Conference on Artificial Intelligence, IJCAI-2021. International Joint Conference on Artificial Intelligence, Inc. (2021)
30. Liu, X., et al.: LCQMC: a large-scale Chinese question matching corpus. In: Proceedings of the 27th International Conference on Computational Linguistics, pp. 1952–1962 (2018)
31. Devlin, J., Chang, M.W., Lee, K., Toutanova, K.: BERT: pre-training of deep bidirectional transformers for language understanding. arXiv preprint arXiv:1810.04805 (2018)
32. Lan, Z., Chen, M., Goodman, S., Gimpel, K., Sharma, P., Soricut, R.: ALBERT: a lite BERT for self-supervised learning of language representations. arXiv preprint arXiv:1909.11942 (2019)
33. Sanh, V., Debut, L., Chaumond, J., Wolf, T.: DistilBERT, a distilled version of BERT: smaller, faster, cheaper and lighter. arXiv preprint arXiv:1910.01108 (2019)
34. Cui, Y., Che, W., Liu, T., Qin, B., Yang, Z.: Pre-training with whole word masking for Chinese BERT. IEEE/ACM Trans. Audio Speech Lang. Process. **29**, 3504–3514 (2021)
35. Jin, P., Wu, Y., Yu, S.: SemEval-2007 task 05: multilingual Chinese-English lexical sample. In: Proceedings of the Fourth International Workshop on Semantic Evaluations (SemEval-2007), pp. 19–23 (2007)
36. Rehurek, R., Sojka, P.: Software framework for topic modelling with large corpora. In: Proceedings of the LREC 2010 Workshop on New Challenges for NLP Frameworks. Citeseer (2010)
37. Kusner, M., Sun, Y., Kolkin, N., Weinberger, K.: From word embeddings to document distances. In: International Conference on Machine Learning, pp. 957–966. PMLR (2015)
38. Song, Y., Shi, S., Li, J., Zhang, H.: Directional skip-gram: explicitly distinguishing left and right context for word embeddings. In: Proceedings of the 2018 Conference of the North American Chapter of the Association for Computational Linguistics: Human Language Technologies, Volume 2 (Short Papers), pp. 175–180 (2018)

Nested Named Entity Recognition from Medical Texts: An Adaptive Shared Network Architecture with Attentive CRF

Junzhe Jiang[1,2], Mingyue Cheng[1,2], Qi Liu[1,2(✉)], Zhi Li[3], and Enhong Chen[1,2]

[1] Anhui Province Key Laboratory of Big Data Analysis and Application, University of Science and Technology of China, Hefei, China
{jzjiang,mycheng}@mail.ustc.edu.cn, {qiliuql,cheneh}@ustc.edu.cn
[2] State Key Laboratory of Cognitive Intelligence, Hefei, China
[3] Shenzhen International Graduate School, Tsinghua University, Shenzhen, China
zhilizl@sz.tsinghua.edu.cn

Abstract. Recognizing useful named entities plays a vital role in medical information processing, which helps drive the development of medical area research. Deep learning methods have achieved good results in medical named entity recognition (NER). However, we find that existing methods face great challenges when dealing with the nested named entities. In this work, we propose a novel method, referred to as ASAC, to solve the dilemma caused by the nested phenomenon, in which the core idea is to model the dependency between different categories of entity recognition. The proposed method contains two key modules: the adaptive shared (AS) part and the attentive conditional random field (ACRF) module. The former part automatically assigns adaptive weights across each task to achieve optimal recognition accuracy in the multi-layer network. The latter module employs the attention operation to model the dependency between different entities. In this way, our model could learn better entity representations by capturing the implicit distinctions and relationships between different categories of entities. Extensive experiments on public datasets verify the effectiveness of our method. Besides, we also perform ablation analyses to deeply understand our methods.

Keywords: Medical named entity recognition · Adaptive shared mechanism · Attentive conditional random fields · Information processing

1 Introduction

Natural language documents in the medical field, such as medical textbooks, medical encyclopedias, clinical cases, and test reports, contain much medical expertise and terminology. The key idea of understanding medical data is to extract critical knowledge from the medical text accurately [21]. Therefore, accurate and rapid extraction of medical entities and transformation of these unstructured data into structured domain knowledge graphs are crucial for obtaining and

L. Fang et al. (Eds.): CICAI 2022, LNAI 13606, pp. 248–259, 2022.
https://doi.org/10.1007/978-3-031-20503-3_20

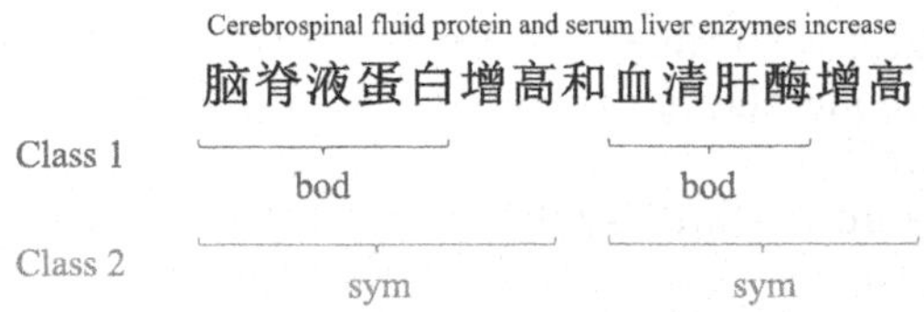

Fig. 1. An example of nested NER in Chinese medical texts, while 'bod' and 'sym' represent *body* and *symptom* entities, respectively.

exploiting medical information. Named entity recognition (NER) based on deep learning applies machines to read medical texts, which significantly improve the efficiency and quality of medical research and serve downstream subtasks [16].

However, there are often cases of nested named entities in texts in the medical field. Figure 1 displays an example of nested NER in Chinese medical texts, where the *symptom* named entity contains the *body* named entity. The foremost sequence labeling method [1,9] is only valid for non-nested entities (or flat entities). Previous studies have given some solutions, which regard NER either as question answering [25], span classification [12], dependency parsing tasks [32], or discrete joint model [11]. However, despite the success of span prediction based systems, these methods suffer from some different weaknesses [26]. First, their method suffers from the boundary inconsistency problem due to the separate decoding procedures. Then, due to numerous low-quality candidate spans, these methods require high computational costs. Next, it is hard to identify long entities frequently found in medical texts because the length of the span enumerated during training is not infinite.

Different from the above researches, layered methods solve this task through multi-level sequential labeling [6,27], which can capture the dependencies between adjacent word-level labels and maximize the probability of predicted labels over the whole sentence [31]. However, these methods always simply divide entities into several levels, where the term level indicates the depth of entity nesting, and sequential labeling is performed repeatedly. Therefore, the differences between entity categories have not been noticed, since just identifying the results of each layer independently. Moreover, there are often implicit relationships between the recognition results of different entity categories, such as *symptom* and *body*, but it is easy to be ignored or only pass results from lower layers to upper layers without the reverse way.

In this paper, we propose a **A**daptive **S**hared network architecture with **A**ttentive **C**onditional random fields namely **ASAC** based on the end-to-end sequence labeling, which can handle the nested NER for medical texts. Our model not only takes advantage of sequence labeling, but also equally considers the implicit distinctions and relationships between entities in different layers. More specifically, we introduce the adaptive shared mechanism for pre-trained language model at encoding, to adaptively learn the difference between different entities recognition tasks [22]. This method has a better fitting ability than the hard-shared method [3,29], which shares different modules according to a fixed strategy. Besides, we utilize attentive conditional random fields to explore the relationships among the multi-layer learning results at decoding based on attention mechanism [20].

Our main contributions are as follows:

- We introduce a novel ASAC model based on deep learning methods, which can efficiently extract named entities from medical texts. We leverage effective mechanisms to enable the model to adaptively discover distinctions and connections between multiple entity recognition tasks in parallel, in order to enhance the performance of the NER task.
- We adopt the Adaptive Shared (AS) mechanism [29] to adaptively select the output of each layer of the pre-trained model for encoding the input texts, thus obtaining the different characteristics of different entity categories. Through this mechanism, we can learn the contextual features from information contained in different layers of the pre-trained language model, for the downstream corresponding tasks.
- We exploit the Attentive Conditional Random Fields (ACRF) model in the decoding stage to get the labels. It can use the *Viterbi* decoding outputs of the other entity recognition tasks as a query. Then, input the query through the attention mechanism as the residual to the origin CRF for deviation correction. In this way, the recognition results of other levels can be integrated to improve the recognition effect of nested named entities.
- We evaluate our proposed model on a public dataset of Chinese medical NER, namely the Chinese Medical Entity Extraction dataset [8] (CMeEE). Extensive experiment results show the effectiveness of the proposed model.

2 Related Work

In this section, we review the related work on the current approaches for NER, including the pre-trained language model, nested NER algorithm, and the Chinese NER in the medical domain.

2.1 Chinese Medical NER

With the increasing demand for NER in the medical field, there have emerged some related medical competitions and datasets, such as CHIP[4] and CCKS[5]. However, due to privacy or property rights protection issues, the medical NER corpora, especially the Chinese medical NER corpora, is particularly scarce [21]. There is no other officially authorized way for these datasets. So most of the research on Chinese medical NER is based on China Conference on Knowledge Graph and Semantic Computing (CCKS) datasets.

When it comes to solutions, Chinese NER methods based on deep learning have gradually become dominant and have achieved continuous performance

[4] http://www.cips-chip.org.cn/.

[5] http://sigkg.cn/ccks2022/.

improvements [13]. Cai et al. [2] added attention layers between the character representation and the context encoder. Zhu et al. [33] used a convolutional attention layer between the feature representation and the encoder.

2.2 Nested NER

Previously, some researchers have also discussed the nested NER based on deep learning. There are various solutions, mainly including the following: (a) transform the decoding process into multi-classification decoding [19,28]. (b) span-based methods which treat NER as a classification task on a span with the innate ability to recognize nested named entities [7,12]. (c) use other modeling methods instead of sequence labeling and span-based methods, such as machine reading comprehension task [18], constructing a hyper-graph [30], etc.

2.3 Pre-trained Model

In recent years, the emergence of the pre-trained model has brought NLP into a new era. Many researches [21] have shown that the pre-trained model trained on a very large corpus can learn a lot of language text representation suitable for different domains. This approach can often help different downstream NLP tasks, including NER. Among various well-known pre-trained language models, the BERT model [5] with excellent results is widely used. BERT is a bi-directional encoder based on transformer. It can mine the representation information contained in the context through the self-supervised learning task. Some researchers have carried out more research on the basis of BERT and put forward many improved models, such as RoBERTa [23], AlBERT [15], and so on.

3 Methodology

Our ASAC model can be divided into two modules. The former part is an adaptive shared pre-trained language model for encoding the input texts to capture the distinctions between pre-defined different entity categories. The latter part is the attentive conditional random field for decoding to acquire the relationships of recognition results between the parallel tasks. The input texts pass through the former module and obtain the encoding features matched with different entity categories according to the pre-defined entity categories classes. Afterwards the attentive conditional random fields make the model learns the residual value according to the label results of other classes, and uses the attention mechanism to correct the output of the original conditional random fields. Figure 2 illustrates an overview of the model structure.

3.1 Adaptive Shared Pre-trained Model

The data representation method follows the paper of Devlin et al. [5]. We use special '[CLS]' and '[SEP]' tokens at the beginning and end of sentences respectively, and add '[PAD]' tokens at the end of sentences to make their lengths equal to the maximum sequence length. All models are character-level based.

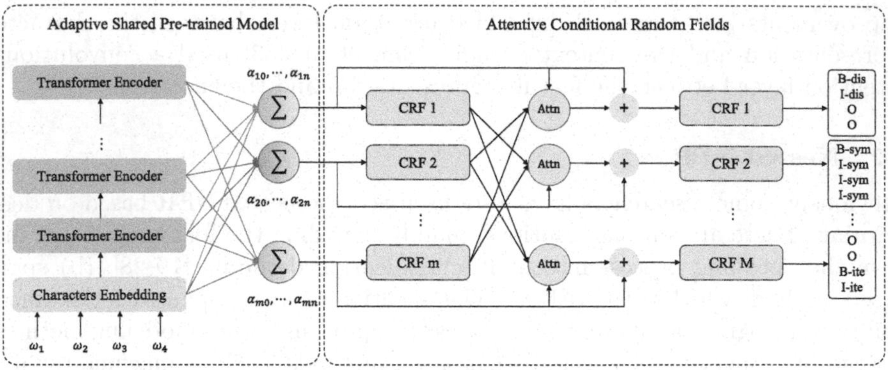

Fig. 2. The architecture of our model. The dotted lines mean these components are shared across levels. α_{ij} denotes the adaptive weight, *Attn* represents the attention mechanism.

The adaptive shared mechanism [29] will assign a learnable weight α_{ij} to each transformer encoder layer of the BERT pre-trained model, and update the value of the weight during backward propagation. i indicates that the number of entity categories classes, and the number of transformer encoder layers of the pre-trained model defines the max value of j. They can be calculated as:

$$h_{j+1} = \text{TransformerEncoder}(h_j), \tag{1}$$

$$E_i = \sum_{j=0}^{n} \alpha_{ij} h_j, \tag{2}$$

where n denotes the total number of transformer encoder layers of the pre-trained language model, E_i denotes the encoder results of ith entity categories classes. In addition, in order to ensure that the weight can better reflect the actual action of each layer after each batch of learning, and prevent the weight from disappearing or exploding, the $softmax$ function is used to output the corrected value of the weight after calculation, as follows:

$$\alpha_{ij}^* = \text{softmax}(\alpha_{ij}) = \frac{\exp^{\alpha_{ij}}}{\sum_{j=0}^{n} \exp^{\alpha_{ij}}}. \tag{3}$$

3.2 Attentive Conditional Random Fields

Conditional random fields (CRF) were proposed by Lafferty et al. [14] in 2001. It has mixed the characteristics of maximum entropy model and hidden Markov model. CRF has been widely used in state-of-the-art NER models to help make better decisions, which considers strong label dependencies by adding transition scores between neighboring labels. *Viterbi* algorithm is applied to search for the label sequence with the highest probability during the decoding process.

For $y = \{y_1, ..., y_N\}$ being a sequence of predictions with length N. Its score is defined as follows.

$$s(x, y) = \sum_{i=0}^{N-1} T_{y_i, y_{i+1}} + \sum_{i=1}^{N} H_{i, y_i}, \tag{4}$$

where $T_{y_i, y_{i+1}}$ represents the transmission score from y_i to y_{i+1}, H_{i, y_i} is the score of the jth tag of the ith word from the adaptive shared encoder. CRF model defines a family of conditional probability $p(y|x)$ over all possible tag sequences y:

$$p(y|x) = \frac{\exp^{s(x,y)}}{\sum_{\tilde{y} \in y} \exp^{s(x,\tilde{y})}}, \tag{5}$$

during training phase, we consider the maximum log probability of the correct predictions. While decoding, we search the tag sequences with maximum score:

$$y^* = \arg\max_{\tilde{y} \in y} \text{score}(x, \tilde{y}). \tag{6}$$

Supposed that there are m entity categories classes that we defined in advance. It means that we have m parallel CRFs. For each CRF, the inference results of other parallel CRFs are used as the query of the attention mechanism. Here, set C denotes the *Viterbi* decoding results of all CRFs except the current one, and d_l indicates the max sequence length of each input text. We calculate the attention value of ith CRF as:

$$Q_i = \sum_{y_i^* \in C} W_f [y^*_{i[0,N]}; c_{(N,d_l]}] + b_f, \tag{7}$$

where $W_f \in \mathbb{R}^{d_t \times d_l}$, $b_f \in \mathbb{R}^{d_t \times d_l}$. c denotes the padding constant, and is set to 0 in our paper. d_t denotes the number of tags plus the padding zero. Then we exploit attention mechanism to explicitly learn the dependencies between the origin score and the parallel results and capture the inner structure information of sentence.

$$R_i = \text{Attention}(Q_i^T, K_i, V_i) = \text{softmax}\left(\frac{Q_i^T K_i}{\sqrt{d_t}}\right) V_i \tag{8}$$

where $K_i \in \mathbb{R}^{d_t \times d_l}$, $V_i \in \mathbb{R}^{d_t \times d_l}$ are keys matrix and value matrix, respectively. In our setting, $K_i = V_i = H_i$, denotes that the input of ith CRF. Then we plus the residual R_i and the origin H_i to i^{th} CRF again, and get the final prediction label list.

$$\begin{aligned} y^* &= \arg\max_{\tilde{y} \in y} \text{s}(x, \tilde{y}) \\ &= \arg\max_{\tilde{y} \in y} \textstyle\sum_{i=0}^{N-1} T_{y_i, y_{i+1}} + \sum_{i=1}^{N} (R + H)_{i, y_i}. \end{aligned} \tag{9}$$

4 Experiments

4.1 Dataset

We conduct experiments on the Chinese medical NER, namely the Chinese Medical Entity Extraction dataset (CMeEE) [8], which is a subtask of Chinese Biomedical Language Understanding Evaluation[6] (CBLUE). While absorbing the previous CHIP/CCKS/CCL and other academic evaluation tasks, the dataset also appropriately increases the industry data. Hence, this dataset can well cover the task of Chinese NER in the medical field.

CMeEE dataset contains 20,000 texts and we randomly split these data into training, development, and test sets by 14:3:3, respectively. The goal of this task is to detect and extract the named entities from Chinese medical texts, and divide them into one of the nine pre-defined categories. The statistical results of the nine types of named entities are shown in Table 1. Dataset provider points that all nested named entities are allowed in the *sym* entity category, and other eight types of entities are allowed inside the entity. So we divide these nine entity categories into two classes: one contains the *sym* entity category and the other one contains eight other entity categories.

Table 1. Category and number of entities in the experimental dataset.

Categories	dis	sym	pro	equ	dru	ite	bod	dep	mic	Total
Number	20,778	16,399	8,389	1,126	5,370	3,504	23,580	458	2,492	**82,096**

4.2 Experimental Setup

Precision (P), Recall (R) and F-score ($F1$) are used to evaluate the predicted entities. An entity is confirmed correct when the predicted category and predicted span both are completely correct. Our model is implemented with PyTorch and we run experiments on NVIDIA Tesla V100 with 32 GB memory.

We use the AdamW optimizer for parameter optimization [24]. Most of the model hyper-parameters are listed in Table 2. The pre-trained model used in this paper comes from [4], namely *BERT-wwm-ext, Chinese*, with 12-layer, 768-hidden, 12-heads and 110M parameters.

4.3 Results and Comparisons

Table 3 shows the overall comparisons of different models for the end-to-end Chinese medical NER. We conduct several neural models as baselines. First is the BiLSTM-CRF, which adopts BiLSTM [9] for encoding input texts and a single CRF for decoding entity labels. And BERT-MLP [4], which employs the

[6] https://tianchi.aliyun.com/cblue.

Table 2. Architecture hyper-parameters.

Hyper-parameters	Values
Batch size	16
Embedding size	768
Max sequence length	128
Dropout rate	0.1
LSTM hidden size	500
BERT learning rate	4e−5
ACRF learning rate	2e−4
AdamW weight decay	1e−5

linear layer for label classification above the BERT pre-trained language model. Another one is BERT-CRF [17], which uses CRF for decoding on the basis of the BERT encoding features.

Table 3. Overall comparisons of different models on CMeEE dataset, while bold ones indicate the best F_1 of each categories.

Categories	BiLSTM-CRF			BERT-MLP			BERT-CRF			ASAC		
	P	R	F_1	P	R	F_1	P	R	F_1	P	R	F_1
dis	56.8	55.6	56.2	63.7	64.3	64.0	63.9	64.1	64.0	64.0	65.9	**64.9**
sym	38.2	37.0	37.6	47.5	51.4	49.4	49.6	51.5	50.5	52.0	50.5	**51.5**
pro	40.9	40.1	40.5	53.7	51.8	52.7	54.2	56.2	55.2	55.3	56.6	**55.9**
equ	50.5	24.3	32.9	65.4	64.0	**64.7**	61.2	67.7	64.3	60.7	68.8	64.5
dru	58.8	55.2	56.9	63.8	71.4	67.4	67.3	72.3	**69.7**	66.7	71.2	68.9
ite	32.6	14.1	19.7	32.1	32.0	32.0	38.6	29.0	33.1	38.8	37.6	**38.2**
bod	54.4	42.5	47.7	58.9	51.8	55.1	61.4	52.6	56.6	59.7	62.9	**61.2**
dep	60.0	30.9	40.8	58.1	52.9	**55.4**	57.4	45.9	51.0	58.3	51.5	54.7
mic	68.4	56.4	61.8	67.9	68.1	68.0	66.8	67.4	67.1	67.3	69.2	**68.2**
Overall	50.2	44.2	47.0	56.8	56.0	56.4	58.6	56.4	57.5	58.8	60.3	**59.5**

On the CMeEE dataset, our method improves the F_1 scores by 12.5, 3.1, and 2.0 respectively, compared with BiLSTM-CRF, BERT-MLP, and BERT-CRF. Furthermore, our approach achieves better results in 6 of the 9 entity categories, surpassing other baseline models.

4.4 Ablation Studies

In this paper, we introduce the interactions of adaptive shared mechanism and attentive conditional random fields to respectively help better predict nested named entities in the dataset. We implement an ablation study to verify the effectiveness of the interactions. We conduct four experiments: (a) our model with adaptive shared mechanism and attentive conditional random fields. (b) without adaptive shared mechanism: we skip Eq. (2)–(3) and only used the final output layer of the BERT-based pre-trained model for encoding. (c) without attentive conditional random fields: we skip Eq. (7)–(9) and use separate CRFs to predict the labels of different category classes, and combine the results. (d) without adaptive shared mechanism and attentive conditional random fields: a combination of (b) and (c), different from BERT-CRF, there are two independent CRFs for decoding entities from two category classes. Table 4 shows the experimental results. Experimental results show that adaptive shared mechanism and attentive conditional random fields both make a positive effect on prediction, while attentive conditional random fields are a little more effective. Compared to abandoning the above two methods, our model received a 1.12 F1 score boost.

Table 4. Ablation study on CMeEE dataset.

	P	R	F_1
Ours	**58.78**	**60.30**	**59.53**
Ours (w/o AS)	58.38	59.74	59.05
Ours (w/o ACRF)	58.14	59.73	58.89
Ours (w/o AS, ACRF)	58.59	58.24	58.41

Analysis of the Adaptive Shared Mechanism. Some researches [10] show that different context features are stored in different layers of the BERT-based pre-trained language model. Therefore, we reasonably infer that different features of hidden layers can variously affect the recognition of different entity categories. As mentioned above, we divide entity categories into two classes. Under the adaptive shared mechanism, there are two groups of weights: $\{\alpha_{1,0}, ..., \alpha_{1,11}\}$ for class 1, $\{\alpha_{2,0}, ..., \alpha_{2,11}\}$ for class 2 to extract the different layers of the pre-trained model. Figure 3 shows the weights of two classes after training.

According to the data shown in the figure, we notice that class 1 prefers the upper layers outputs of the BERT-based model while middle layers outputs have a greater impact on class 2. Therefore, the adaptive shared mechanism enables the model to learn the context features that match the nested entity category better, which is conducive to subsequent decoding.

Analysis of the Attentive Conditional Random Fields. In the decoding module, we count the changed labels between the outputs of the initial CRF and the outputs plus the residual of the CRF after the attention mechanism. Table 5 evaluates the performance of attentive conditional random fields. In a total of 155,658 tokens of 3,000 test cases, 4,816 and 1,392 prediction labels for the two classes have changed, where the positive changes are exceeded 60%.

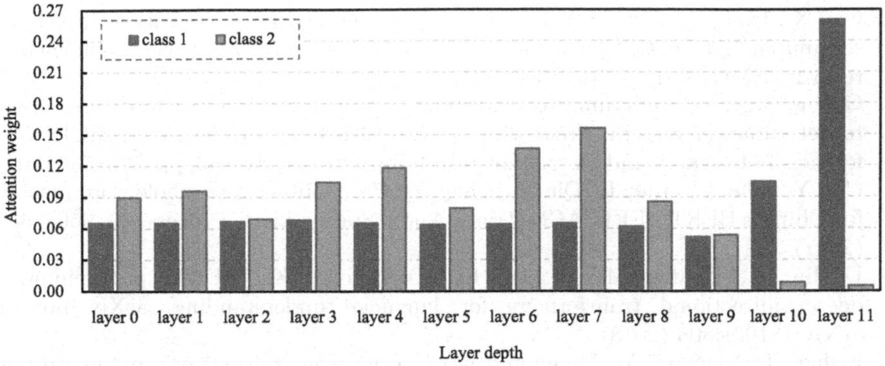

Fig. 3. The adaptive shared weights of different layers and entity category classes of the pre-trained model.

Table 5. Statistics of labels positive changed after attentive conditional random fields.

	Class 1	Class 2
Changed	4,816	1,392
Positive	3,170	857
Positive ratio	65.8%	61.5%

5 Conclusion

In this paper, we proposed a novel end-to-end model ASAC for Chinese nested named entity recognition in the medical domain, which can adaptively learn the distinctions and connections between different entities recognition tasks based on sequence labeling. First, we introduce the adaptive shared mechanism to get features of different layers of the BERT-based model to different nested entity category classes. Moreover, by constructing attentive conditional random fields, our model can exploit the encoding features to decode corresponding labels by blending other predictions via the attention mechanism. Extensive experiments conducted on publicly available datasets show the effectiveness of the proposed method. In the future, we will apply this novel model to more different types of datasets to verify validity.

Acknowledgment. This research was partially supported by grants from the National Natural Science Foundation of China (Grant No. 61922073) and the Joint Fund for Medical Artificial Intelligence (Grant No. MAI2022C007).

References

1. Akbik, A., Blythe, D., Vollgraf, R.: Contextual string embeddings for sequence labeling. In: Proceedings of the 27th International Conference on Computational Linguistics, pp. 1638–1649 (2018)

2. Cai, X., Dong, S., Hu, J.: A deep learning model incorporating part of speech and self-matching attention for named entity recognition of Chinese electronic medical records. BMC Med. Inform. Decis. Making **19**(2), 101–109 (2019)
3. Cheng, M., et al.: Learning recommender systems with implicit feedback via soft target enhancement. In: Proceedings of the 44th International ACM SIGIR Conference on Research and Development in Information Retrieval, pp. 575–584 (2021)
4. Cui, Y., Che, W., Liu, T., Qin, B., Yang, Z.: Pre-training with whole word masking for Chinese BERT. IEEE/ACM Trans. Audio Speech Lang. Process. **29**, 3504–3514 (2021)
5. Devlin, J., Chang, M.W., Lee, K., Toutanova, K.: BERT: pre-training of deep bidirectional transformers for language understanding. arXiv preprint arXiv:1810.04805 (2018)
6. Fisher, J., Vlachos, A.: Merge and label: a novel neural network architecture for nested NER. arXiv preprint arXiv:1907.00464 (2019)
7. Fu, J., Huang, X.J., Liu, P.: SpanNER: named entity re-/recognition as span prediction. In: Proceedings of the 59th Annual Meeting of the Association for Computational Linguistics and the 11th International Joint Conference on Natural Language Processing (Volume 1: Long Papers), pp. 7183–7195 (2021)
8. Hongying, Z., Wenxin, L., Kunli, Z., Yajuan, Y., Baobao, C., Zhifang, S.: Building a pediatric medical corpus: word segmentation and named entity annotation. In: Liu, M., Kit, C., Su, Q. (eds.) CLSW 2020. LNCS (LNAI), vol. 12278, pp. 652–664. Springer, Cham (2021). https://doi.org/10.1007/978-3-030-81197-6_55
9. Huang, Z., Xu, W., Yu, K.: Bidirectional LSTM-CRF models for sequence tagging. arXiv preprint arXiv:1508.01991 (2015)
10. Jawahar, G., Sagot, B., Seddah, D.: What does BERT learn about the structure of language? In: ACL 2019–57th Annual Meeting of the Association for Computational Linguistics (2019)
11. Ji, Z., Xia, T., Han, M., Xiao, J.: A neural transition-based joint model for disease named entity recognition and normalization. In: Proceedings of the 59th Annual Meeting of the Association for Computational Linguistics and the 11th International Joint Conference on Natural Language Processing (Volume 1: Long Papers), pp. 2819–2827 (2021)
12. Jiang, Z., Xu, W., Araki, J., Neubig, G.: Generalizing natural language analysis through span-relation representations. In: Proceedings of the 58th Annual Meeting of the Association for Computational Linguistics. Association for Computational Linguistics, July 2020
13. Kou, F., et al.: A semantic modeling method for social network short text based on spatial and temporal characteristics. J. Comput. Sci. **28**, 281–293 (2018)
14. Lafferty, J., McCallum, A., Pereira, F.C.: Conditional random fields: probabilistic models for segmenting and labeling sequence data (2001)
15. Lan, Z., Chen, M., Goodman, S., Gimpel, K., Sharma, P., Soricut, R.: ALBERT: a lite BERT for self-supervised learning of language representations. arXiv preprint arXiv:1909.11942 (2019)
16. Li, J., Sun, A., Han, J., Li, C.: A survey on deep learning for named entity recognition. IEEE Trans. Knowl. Data Eng. **34**(1), 50–70 (2020)
17. Li, X., Zhang, H., Zhou, X.H.: Chinese clinical named entity recognition with variant neural structures based on BERT methods. J. Biomed. Inform. **107**, 103422 (2020)
18. Li, X., Feng, J., Meng, Y., Han, Q., Wu, F., Li, J.: A unified MRC framework for named entity recognition. In: Proceedings of the 58th Annual Meeting of the Association for Computational Linguistics, pp. 5849–5859 (2020)

19. Li, Z., Wu, B., Liu, Q., Wu, L., Zhao, H., Mei, T.: Learning the compositional visual coherence for complementary recommendations. arXiv preprint arXiv:2006.04380 (2020)
20. Liang, Z., Du, J., Li, C.: Abstractive social media text summarization using selective reinforced seq2seq attention model. Neurocomputing **410**, 432–440 (2020)
21. Liu, P., Guo, Y., Wang, F., Li, G.: Chinese named entity recognition: the state of the art. Neurocomputing **473**, 37–53 (2022)
22. Liu, Q., et al.: Exploiting cognitive structure for adaptive learning. In: Proceedings of the 25th ACM SIGKDD International Conference on Knowledge Discovery & Data Mining, pp. 627–635 (2019)
23. Liu, Y., et al.: RoBERTa: a robustly optimized BERT pretraining approach. arXiv preprint arXiv:1907.11692 (2019)
24. Loshchilov, I., Hutter, F.: Decoupled weight decay regularization. arXiv preprint arXiv:1711.05101 (2017)
25. Mengge, X., Yu, B., Zhang, Z., Liu, T., Zhang, Y., Wang, B.: Coarse-to-fine pre-training for named entity recognition. In: Proceedings of the 2020 Conference on Empirical Methods in Natural Language Processing (EMNLP), pp. 6345–6354 (2020)
26. Shen, Y., Ma, X., Tan, Z., Zhang, S., Wang, W., Lu, W.: Locate and label: a two-stage identifier for nested named entity recognition. In: Proceedings of the 59th Annual Meeting of the Association for Computational Linguistics and the 11th International Joint Conference on Natural Language Processing (Volume 1: Long Papers), pp. 2782–2794 (2021)
27. Shibuya, T., Hovy, E.: Nested named entity recognition via second-best sequence learning and decoding. Trans. Assoc. Comput. Linguist. **8**, 605–620 (2020)
28. Straková, J., Straka, M., Hajic, J.: Neural architectures for nested NER through linearization. In: Proceedings of the 57th Annual Meeting of the Association for Computational Linguistics, pp. 5326–5331 (2019)
29. Sun, X., Panda, R., Feris, R., Saenko, K.: AdaShare: learning what to share for efficient deep multi-task learning. In: Advances in Neural Information Processing Systems, vol. 33, pp. 8728–8740 (2020)
30. Wang, B., Lu, W., Wang, Y., Jin, H.: A neural transition-based model for nested mention recognition. In: Proceedings of the 2018 Conference on Empirical Methods in Natural Language Processing, pp. 1011–1017 (2018)
31. Wang, Y., Shindo, H., Matsumoto, Y., Watanabe, T.: Nested named entity recognition via explicitly excluding the influence of the best path. J. Nat. Lang. Process. **29**(1), 23–52 (2022)
32. Yu, J., Bohnet, B., Poesio, M.: Named entity recognition as dependency parsing. In: Proceedings of the 58th Annual Meeting of the Association for Computational Linguistics. Association for Computational Linguistics, July 2020
33. Zhu, Y., Wang, G.: CAN-NER: convolutional attention network for Chinese named entity recognition. In: Proceedings of the 2019 Conference of the North American Chapter of the Association for Computational Linguistics: Human Language Technologies, Volume 1 (Long and Short Papers), pp. 3384–3393 (2019)

CycleResume: A Cycle Learning Framework with Hybrid Attention for Fine-Grained Talent-Job Fit

Zichen Zhang[1,2], Yong Luo[1,2(✉)], Yonggang Wen[3(✉)], and Xinwen Zhang[4]

[1] National Engineering Research Center for Multimedia Software, School of Computer Science, Institute of Artificial Intelligence and Hubei Key Laboratory of Multimedia and Network Communication Engineering, Wuhan University, Wuhan, China
{zhangzichen,luoyong}@whu.edu.cn
[2] Hubei Luojia Laboratory, Wuhan, China
[3] School of Computer Science and Engineering, Nanyang Technological University, Singapore, Singapore
ygwen@ntu.edu.sg
[4] Hiretual, Mountain View, CA, USA
xinwenzhang@hiretual.com

Abstract. Online recruitment platforms are becoming powerful tools for job hunters to look for jobs and for companies to search for outstanding candidates. To liberate the workforce for resume assessment, there is an increasing attention on developing learning-based systems for automatic talent-job fit. Existing approaches suffer from several drawbacks, such as simply ignoring the interactions between talent's skills and experiences and only utilizing the final output representations of resume and job-post for matching. To address these issues, we propose a new deep cycle learning-based framework for talent-job fit. In particular, we first propose a noise-incorporated co-attention mechanism to measure resume consistency in a fine-grained manner. Then a multi-scale CNN is utilized for job representation learning, and the matching score between resume and job is measured via multi-scale attention. To evaluate the effectiveness of the proposed approach, we conduct extensive experiments on a self-collected resume-job dataset. The results show that our method achieves state-of-the-art performance, and we demonstrate that the proposed modules can be utilized as off-the-shelf blocks for similar application tasks.

Keywords: Talent-job fit · Cycle learning · Hybrid attention

1 Introduction

Nowadays, online recruitment platforms attract both job hunters and companies' attention owing to their convenience. As reported by LinkedIn[1], 20 million jobs

[1] https://www.businessofapps.com/data/linkedin-statistics/.

L. Fang et al. (Eds.): CICAI 2022, LNAI 13606, pp. 260–271, 2022.
https://doi.org/10.1007/978-3-031-20503-3_21

have invited over 122 million out of 660 million users to an interview via the platform, and 35.5 million of whom successfully got a job only for the year of 2019. With the growing number of job posts and user, to build a machine learning system for automated talent-job fit becomes increasingly important. However, it is a non-trivial task due to the huge amount of online recruitment data and relatively sparse job-user interactions.

Existing methods typically regard the task of talent-job fit as a text matching problem [6,21], and focus on improving the representations of job descriptions, and user resumes for their matching. Zhu et al. [22] design a 2-way Convolutional Neural Network (CNN) to encode resume and job, respectively. In [8], RNN and BiLSTM are introduced to capture textual sequential information. [16] highlights the varying key points of jobs and resumes—experienced human resources (HRs) are able to change their focus with various areas—and thus, jointly model resumes and job descriptions with latent preference representations [18,19]. For further interactions, [2] utilize a hierarchical attention network [17], and [4] study the effectiveness of adversarial learning. However, their performance is still unsatisfactory without considering the following challenges:

Quality of Resume. A high-quality resume usually reflects the excellence of candidates. If a resume is in low quality, it should be less considered for matching. A crucial aspect of measuring quality is whether the resume is consistent with itself. For instance, a bad candidate may add "*good at python programming*" up to the resume, even if he/she is not. We thus have reasons to give more credits to consistent resume, such that if "*good at python programming*" is stated in a skill section, there should be relative proof like "*graduate as CS student and use python to complete final year project*" or "*work as a python developer for three years*" appearing in the following section of resumes.

Matching Efficiency. Although the joint modeling of jobs and resumes improve the representations by discriminating different sentence importance from both sides, it greatly increases the computational complexity. Take LinkedIn as an example, which has over 660 million users and 20 million jobs, the above methods have to learn embeddings for each user-job pair in real-time to compute the interactive attentions, leading to a $6.6 * 2 * 10^{8+7}$ search space. It is thus impractical to provide timely talent-job fit service in online deployment [9,11].

In this paper, we propose a novel model, CycleResume for talent-job fit, that aims to improve the job and resume representation learning by considering the resume quality and computational efficiency. To address the first challenge, we learn consistency-aware resume representations by examining the resume consistency with the help of structures' specific meaning. As shown in Fig. 1, we extract various skills and working experiences from corresponding sections and design a noise-incorporated co-attention scheme to enhance the embeddings through mutual verification. In this way, features with no support information are represented by adding a noise vector to reduce attention weights. Meanwhile, the original information is preserved by utilizing the residual connection to get a fine-grained feature matrix for representing resumes.

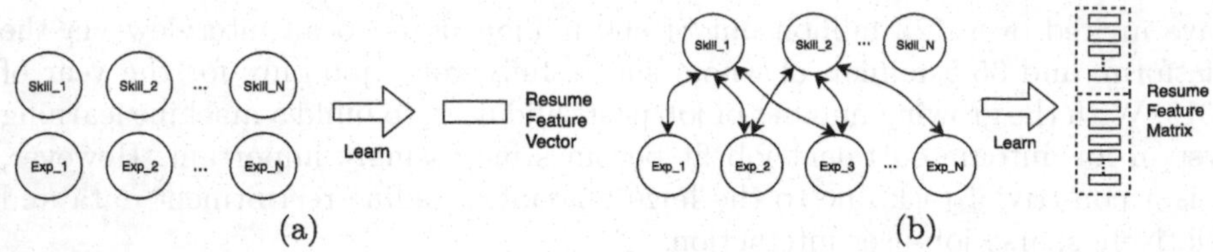

Fig. 1. Illustration of resume consistency measurement: (a) previous works that can not capture the consistency information, and (b) our proposed scheme. A resume can contain many skills as well as multiple work experiences, and a positive candidate resume should have skills that can be approved by work experience or vice versa.

To address the second issue of matching efficiency, instead of joint modeling jobs and resumes via mutual attention, we decouple the job and resume representation learning. Thus, their embeddings can be computed offline and reused when matching, significantly reducing the search space. To maintain the high-quality of representations, we (1) utilize a multi-layer 1D-CNN and stores outputs from each layer as the multi-scale job embeddings, and (2) incorporate interactions in the final matching layer via an attended resume-job fit module, which calculates a co-attention map and applies it to integrate all levels as final resume-job representation for prediction.

The importance and originality of this study lie in the following aspects: 1) We propose that resume consistency measurement should be incorporated into talent-job fit task and design a noise-incorporated co-attention module for learning this feature; 2) We develop a multi-level job representation sub-network based on CNN, and this can be utilized to generate multi-level attention maps to improve the talent-job representation; 3) We collect a dataset with authoritative annotations from expert human resource managers (HRs). We conduct a series of experiments, and the results demonstrate effectiveness of our approach as compared with six state-of-the-art (SOTA) models. Specifically, our model improves the previous SOTA from 73.9% recall to 79.0%. An ablation study is also provided to verify the effectiveness of each key module.

2 Framework Design

In this section, we will formulate the talent-job fit task and describe the proposed architecture.

2.1 Problem Formulation

We formulate the talent-job fit task into a regression problem. Specifically, given a resume-job pair, denoted by $\{R, J\}$, where R is a resume and J is a job post, as input, our target is to learn a model to output their similarity score, $\hat{y}$. After discussing with many HRs, in this work, we make the following assumptions.

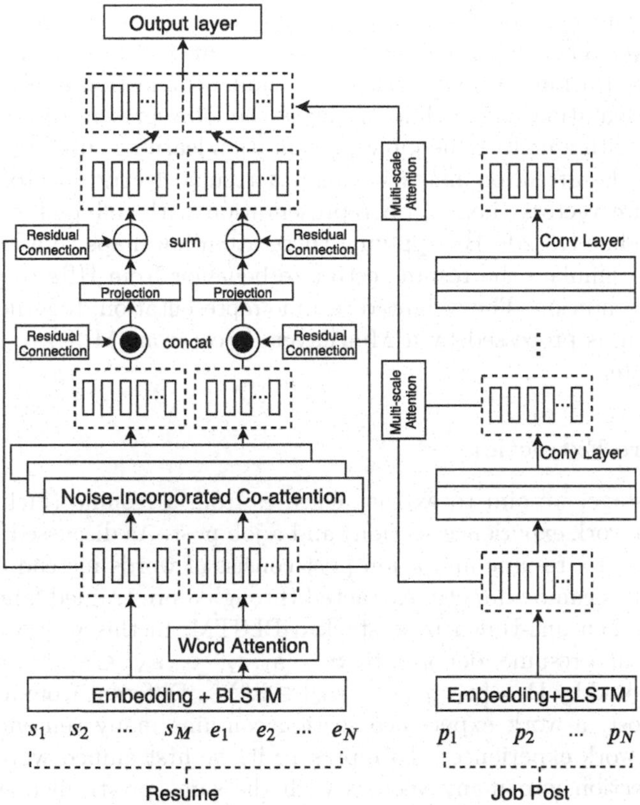

Fig. 2. The proposed CycleResume framework. It contains three phases: feature extraction, resume and job representation learning, and fit prediction.

Resumes of talents usually include two important sections: skill and work experience. The skill section may contains many words (such as "*python*" and "*machine learning*"), which can be denoted as $\{s_1, s_2, \cdots, s_M\}$, while the work experience section includes detailed description (such as "*I worked as a machine learning research engineer in LinkedIn*"), which can be denoted by $\{e_1, e_2, \cdots, e_N\}$ where e_n is one work experience. Hence, a resume R can be represented by $R = \{\{s_1, s_2, \cdots, s_M\}, \{e_1, e_2, \cdots, e_N\}\}$. Similarly, a job post P is represented by $P = \{p_1, p_2, \cdots, p_K\}$ where p_k is a word in job description. We aim to map a resume-job pair $\{R, J\}$ to a similarity score $\hat{y}$ ranging from 0 to 1 (A larger value means a better match).

2.2 Architecture Overview

As shown in Fig. 2, our framework processes a resume-job pair in three stages: feature extraction, resume-job representation learning, and resume-job matching. Specifically, we first utilize the pre-trained language models such as [7] to get the initial feature vectors of a resume-job pair and then fine-tuned them through a BLSTM layer. Then, we input the obtained resume features into a cycle resume representation learning sub-network with residual connection to get fine-grained resume representation. Meanwhile, we learn the job representation under varied granularities with a multi-scale job representation learning sub-network. Finally, we calculate the attention maps between a resume feature vector and multiple job feature vectors (i.e., a job representation with different scales) via an attention module [15,20]. By applying these attention maps to a resume representation, we simulate the resume estimate behavior from HRs to get the final resume representation. The weighted resume representation, together with a job representation, is processed by a Multilayer Perceptron (MLP) to get the final similarity score.

2.3 Feature Extraction

In the first stage, we aim to extract features from a resume (including a skill section and a work experience section) and a job post. As discussed before, both a skill section of a resume and a job post consist of words instead of sentences, so their features can be directly extracted through a pre-trained language model from [7] and then fine-tuned by a stacked BLSTM. In this way, we obtain the skill features in a resume, denoted by $\mathbf{S} = [\mathbf{s}_1, \mathbf{s}_2, \cdots, \mathbf{s}_N] \in \mathbb{R}^{d \times N}$, and job post features, denoted by $\mathbf{P} = [\mathbf{p}_1, \mathbf{p}_2, \cdots, \mathbf{p}_K] \in \mathbb{R}^{d \times K}$. Different from a skill section and a job post, a work experience section contains many sentences, presenting multiple work experiences. To represent it, we first embed words in a work experience section into many vectors with the same pre-trained model. Then we utilize a simple word attention mechanism from [5] to fuse vectors from the same experience into one vector (representing a work experience), denoted by $\mathbf{E} = [\mathbf{e}_1, \mathbf{e}_2, \cdots, \mathbf{e}_M] \in \mathbb{R}^{d \times M}$.

2.4 Fine-Grained Resume Representation Learning

We learn a resume representation with a cycle learning sub-network, including two modules: the multi-head noise incorporated co-attention for learning a consistency feature and the residual connection for a more fine-grained resume representation.

Vanilla Co-attention. Vanilla co-attention [12,14] aims to measure the weighting map between two related features. For example, given skill features $\boldsymbol{S}$ and work experience features $\boldsymbol{E}$, the affinity matrix between them is calculated as follows,

$$\mathbf{A} = \mathbf{S}^{\mathsf{T}}\mathbf{W}\mathbf{E}, \tag{1}$$

where $\boldsymbol{W}$ is a learnable weight parameter. However, it is one-side solution to measure the resume consistency just in a single direction, which means that we need to derive attention maps both on skills conditioned by experiences and experiences conditioned on skills. Hence, we generate positive weights by using softmax function in column-wise and row-wise as $\mathbf{A^S} = f(\mathbf{A}^\mathsf{T})$ and $\mathbf{A^E} = f(\mathbf{A})$, where f is the softmax function.

Noise-Incorporated Co-attention. As mentioned before, the authenticity of resumes is crucial to matching results of talent-job task, but the basic co-attention mechanism cannot handle the poor-authenticity resumes like skills or experiences without supporting materials. This causes that attention maps mislead the network training. To tackle this problem, we generate $T \in [1, ...N]$ noise vectors, which are able to store those inconsistent skills and experiences. These T noise vectors combined with skill and experience features are used to calculate attention maps (Here we set $T = 1$). The new skill and experience features are denoted by $\mathbf{S}'$ and $\mathbf{E}'$, respectively. In this way, we can get noise-incorporated attention maps $\mathbf{A^{S'}}$ and $\mathbf{A^{E'}}$.

Multi-head Co-attention Extension. Inspired by works [13], we propose multi-head co-attention to improve the robustness of the framework. Multi-head co-attention firstly projects $\boldsymbol{S}'$ and $\boldsymbol{E}'$ into several low-dimension features through linear layers to learn abundant information. For example, we use q to denote the number of low-dimension features and their dimensions are $d^q(d/q)$. Thus, each affinity matrix $\boldsymbol{A}_i$ is calculated as follows,

$$\mathbf{A}_i = (\mathbf{W}_i^{\mathbf{S}'}\mathbf{S}')^\mathsf{T}(\mathbf{W}_i^{\mathbf{E}'}\mathbf{E}'), \tag{2}$$

where $\mathbf{W}_i^{\mathbf{S}'}$ and $\mathbf{W}_i^{\mathbf{E}'}$ are learnable weight parameters and $i = 1, \ldots, q$. Then, similar to previous processing, each attention map is generated by,

$$\mathbf{A}_i^{\mathbf{S}} = f(\frac{\mathbf{A}_i^\mathsf{T}}{\sqrt{d^q}}),\ \mathbf{A}_i^{\mathbf{E}} = f(\frac{\mathbf{A}_i}{\sqrt{d^q}}), \tag{3}$$

where f is the softmax function. Next, the average attention map is obtained as $\mathbf{A^S} = \frac{1}{q}\sum_{i=1}^{q}\mathbf{A}_i^{\mathbf{S}}$ and $\mathbf{A^E} = \frac{1}{q}\sum_{i=1}^{q}\mathbf{A}_i^{\mathbf{E}}$.

Since we already get the multi-head noise incorporated attention maps, we perform the dot-product attention operation to get consistency features $\boldsymbol{S}^c$ and $\boldsymbol{E}^c$. As we employ noise vectors to store the inconsistent skills or experiences, we ignore them in this final stage. Thus the consistency feature calculation can be formed by,

$$\mathbf{S}^c = \mathbf{S}'\mathbf{A^S}[1:M,:]^\mathsf{T},\ \mathbf{E}^c = \mathbf{E}'\mathbf{A^E}[1:N,:]^\mathsf{T}. \tag{4}$$

Resume Representation Learning. As illustrated in the left part of Fig. 2, we employ a residual connection network to learn fine-grained resume feature matrix after getting consistency features, $\mathbf{S}^c$ and $\mathbf{E}^c$. These features are concatenated to original skill features $\mathbf{S}$ and experience features $\mathbf{E}$. The combined features will be projected back to the original dimensional space by a fully connected layer with *tanh* activation function and then added to original features, as follows,

$$\mathbf{S}^{res} = \mathcal{F}(\mathbf{S}^c) + \mathbf{S},\ \mathbf{E}^{res} = \mathcal{F}(\mathbf{E}^c) + \mathbf{E}. \tag{5}$$

Finally, we concatenate $\mathbf{S}^{res}$ and $\mathbf{E}^{res}$ into one feature matrix $\mathbf{R}$ in column-wise manner to represent more fine-grained resume representation as follows,

$$\mathbf{R} = \mathbf{S}^{res} \odot \mathbf{E}^{res}. \tag{6}$$

2.5 Multi-scale Job-Post Representation Learning

In this section, we aim to learn multiple job features from N scales to capture information in a job description with different granularities. After the first feature extraction stage, we obtain the initial job post features $\mathbf{P} \in \mathbb{R}^{d \times K}$ where d is the dimensions of the features, and K is the number of words in a job description. To understand different level information (e.g., words, phrases, and sentences) in a job-post, we employ multi-level CNN for learning a job representation.

A convolutional layer contains multiple convolutional filters, which work as sliding windows processing across features $\mathbf{P}$. If the filter size is n, we can get high-level n-gram representations. After processed by the first convolutional layer with F filters, we obtain a new job features $\boldsymbol{J}_1 \in \mathbb{R}^{X \times F}$, where X depends on the padding size and stride size. We then stack multiple convolutional layers to extract job descriptions from different levels as follows,

$$\mathbf{J}_i = f_i(\mathbf{J}_{i-1}), i = 1, \ldots, N, \tag{7}$$

where $\mathbf{J}_i$ is job representation outputted from the i-th convolutional layer f_i.

2.6 Attention Based Resume-Job Fit Prediction

This sub-network aims to measure the similarity score between a resume-job pair, and it consists of two modules, the attention based resume review and the final similarity prediction. As we already get multi-scale job representations from multiple convolutional layers, the attention maps between a resume representation $\mathbf{R}$ and a job representation $\mathbf{J}_i$ can be calculated as follows,

$$\mathbf{P}_i = \mathbf{R}\mathbf{J}_i^\top, P \in \mathbb{R}^{(N+M) \times K}, \ \mathbf{P}_i^R = f(\mathbf{P}_i), \tag{8}$$

where f is the softmax function. These attention maps $\mathbf{P}_i$ are applied back to resumes to obtain the multi-level attended resume features but not apply to job features. Then *max* pooling is utilized to transform the attended resume features to discriminative features $\{\mathbf{r}_1, \mathbf{r}_2, \ldots, \mathbf{r}_N\}$. We also take the output from the last layer of job sub-network and *max* pool it to a vector as the job representations $\mathbf{j}_N$. We concatenate all of these vectors as the final resume-job pair representation $\mathbf{p} = [\mathbf{j}_N; \mathbf{r}_1; \mathbf{r}_2; \ldots; \mathbf{r}_N]$ and feed the representation into a two-layer MLP for similarity scores prediction, i.e.,

$$\hat{y} = f_{MLP}(\mathbf{p}). \tag{9}$$

2.7 Training Methodology

Considering that the groundtruth in our dataset is either 1 (that means positive fit) or 0 (that means a negative fit), We train our framework via the cross-entropy loss function,

$$\mathcal{L} = -\frac{1}{N}\sum_{i=1}^{N} y_i log(\hat{y}_i) + (1 - y_i)log(1 - \hat{y}_i), \tag{10}$$

where N is the batch size, y is the groundtruth and $\hat{y}$ is the prediction score.

3 Performance Evaluation

In this section, we first provide detailed descriptions of our self-collected resume assessment dataset. Then we introduce four metrics used to evaluate the performance of different talent-job fit frameworks. Finally, experiment settings and experiments results are presented to justify the proposed method.

3.1 Dataset Description

We build a large-scale dataset using the resume and job post information provided by a private recruitment company. We first remove sensitive personal information in resumes to protect users' privacy and only keep the information such as skills and work experiences. Then we filter job posts and preserve necessary requirements information. Finally, We get 32,082 resumes and 811 job-posts in total. Label information is also provided by the recruitment company. For each job post, they first collect related resumes by applying their search system and then invite HRs to label these resumes as "positive-fit" or "negative-fit" manually. In this way, we finally obtain 38,724 resume-job pairs including 23,530 positive pairs and 15,194 negative pairs.

3.2 Evaluation Metric

We utilize **precision, recall, F1** and **accuracy** as the metric of the talent-job fit task like work [8,22]. Since we apply *Sigmoid* function as the final activation function, the final output is a continuous value ranging form 0 to 1. A threshold is set empirically to filter the value (i.e., the value larger than 0.5 will be positive fit and less than 0.5 is negative fit). **AUC**, the area under the curve **ROC** (receiver operating characteristic, provides an aggregate measure of performance under all possible thresholds and is also utilized in our experiments.

3.3 Experiment Setting

Baselines. We utilize several non-deep-learning based algorithms such as AdaBoost, gradient boosting decision tree (GBDT), random forest (RF) and logistic regression (LR) as baselines like work [8,22]. For fair comparisons, we fed all embedded information used in our framework such as skills, work experiences and job-post into these shallow learning models for binary-classification. We further compare two competitive deep learning based approaches:

Table 1. Ablation study. Our framework achieves the best performance under four noise vectors, two attentions and three CNN layers. * means the best setting.

Model variants	Setting	Accuracy	F1
Incorporated noise vectors	2	0.761	0.804
	3	0.753	0.787
	4*	0.772	0.807
	5	0.761	0.800
Multi-head attentions	1	0.769	0.805
	2*	0.772	0.807
	4	0.762	0.797
	8	0.763	0.800
CNN layers	2	0.770	0.802
	3*	0.772	0.807
	4	0.762	0.800

- **PJFNN** [22]. PJFNN is a CNN based network [3,10] for talent-job fit task.
- **APJFNN** [8]. APJFNN adopts a hierarchical RNN [1,17] as the basic backbone network to learn how to match resume and job pairs end-to-end.

However, neither of above deep learning based methods consider consistency feature and provide fine-grained resume-job representations for matching.

3.4 Evaluation Result

We present the experiment results to evaluate our framework. We show the ablation study and hyper-parameter influences to verify the robustness of the proposed framework. We finally compare our method with other baselines to justify is effectiveness.

Ablation Study. Since our method contains several modules, we first perform the ablation study to check their contributions. Table 1 presents the performance of the proposed framework under different settings. First, incorporating noise vectors to co-attention can measure the inconsistency features. When setting the number to 4, we can get the best performance. We attribute this to the distribution of the number of skills and work experiences in the dataset. In our dataset, four-vectors can memory inconsistency skills and work experiences. Second, the number of noise-incorporated attention should be adjusted carefully. The value is large or less than two accounts for overfitting and underfitting that damage performance. Third, we use various job representation from multiple CNN layers to navigate resume features. The job sub-network with three layers perform best. The network with fewer layers can not output enough job features to match resume features comprehensively but with more layers will adopt too many parameters to networks which will account for overfitting.

Table 2. Overall performance of all methods. Our proposed CycleResume performs better than both shallow learning and deep learning based methods. It is clear that deep learning based methods are better than non-deep learning methods owning to their end-to-end representation learning. CycleResume's success proves the effectiveness of the fine-grained resume and job representation.

Method	Accuracy	Precision	Recall	F1
AdaBoost	0.677	0.732	0.658	0.693
LR	0.705	0.749	0.707	0.727
RF	0.712	0.747	0.705	0.725
GBDT	0.734	0.791	0.730	0.760
PFJFNN	0.744	0.825	0.712	0.769
APJFNN	0.756	0.826	0.739	0.780
Ours (BERT)	0.770	**0.842**	0.764	0.801
Ours	**0.772**	0.828	**0.790**	**0.807**

Compare to Existing Methods. Table 2 shows the best performance of the proposed framework on the test set along with other baselines. Among all shallow learning-based methods, GBDT performs best, which means the tree-based method can capture more discriminative feature. Deep learning-based methods outperform than those shallow learning-based methods which demonstrates the power of deep resume and job representation. APJFNN adopts the ability-aware module presents noticeable results. Compare to those deep learning methods, our CycleResume further provides the consistency feature as well as fine-grained job-resume matching. The best performance under all evaluation metrics verifies our design. Further analysis shows that the recall our model achieves is higher than the others a lot, which means that the proposed framework can mine positive fit pairs deeply and predict them more correctly. We also try the latest BERT model for feature extraction, and the results show that its performance decreases a little. Many reasons can account for this result. For instance, the BERT model first tokenizes the input words into smaller units, and this operation may not be suitable for the resume-job fit task as many words in this area have unique purposes. We leave this exploration for future work.

4 Conclusion and Future Work

Nowadays, the resume assessment platforms supported by deep learning have been widely deployed to liberate HRs from tedious work. However, previous works pay less attention to resume consistency and only assess resumes according to job features in a coarse manner. In this paper, we develop a new neural architecture, termed CycleResume, for talent-job fit. It models the resume consistency as well as skill and work experience features with a cycle learning sub-network in a fine-grained manner. Meanwhile, the framework extracts multi-level job features with a CNN based sub-network. These job features are applied to the resume feature matrix to generate attended joint resume-job representation. Finally, we utilize the obtained representation for resume-job matching. We conduct extensive experiments on a self-collected real-world dataset, and our framework outperforms all SOTA baselines in terms of all metrics. In the future, we intend to use the developed modules as basic blocks to build an AutoML based framework for ads understanding task.

Acknowledgements.. This work is supported by the Special Fund of Hubei Luojia Laboratory under Grant 220100014.

References

1. Adhikari, A., Ram, A., Tang, R., Lin, J.: Rethinking complex neural network architectures for document classification. In: Proceedings of the 2019 Conference on ACL, pp. 4046–4051 (2019)
2. Bian, S., Zhao, W.X., Song, Y., Zhang, T., Wen, J.R.: Domain adaptation for person-job fit with transferable deep global match network. In: Proceedings of the 2019 Conference on Empirical Methods in Natural Language Processing and the 9th International Joint Conference on Natural Language Processing (EMNLP-IJCNLP), pp. 4812–4822 (2019)
3. Kim, Y.: Convolutional neural networks for sentence classification. arXiv preprint arXiv:1408.5882 (2014)
4. Luo, Y., Zhang, H., Wen, Y., Zhang, X.: ResumeGAN: an optimized deep representation learning framework for talent-job fit via adversarial learning. In: Proceedings of the 28th ACM International Conference on Information and Knowledge Management, pp. 1101–1110 (2019)
5. Luong, T., Pham, H., Manning, C.D.: Effective approaches to attention-based neural machine translation. In: Proceedings of the 2015 Conference on EMNLP, pp. 1412–1421 (2015)
6. Malinowski, J., Keim, T., Wendt, O., Weitzel, T.: Matching people and jobs: a bilateral recommendation approach. In: Proceedings of the 39th Annual Hawaii International Conference on System Sciences (HICSS 2006), vol. 6, p. 137c. IEEE (2006)
7. Peters, M.E., et al.: Deep contextualized word representations. arXiv preprint arXiv:1802.05365 (2018)
8. Qin, C., et al.: Enhancing person-job fit for talent recruitment: an ability-aware neural network approach. In: The 41st International ACM SIGIR Conference on Research & Development in Information Retrieval, pp. 25–34. ACM (2018)
9. Ramanath, R., et al.: Towards deep and representation learning for talent search at Linkedin. In: Proceedings of the 27th ACM CIKM, pp. 2253–2261. ACM (2018)
10. Rao, J., Yang, W., Zhang, Y., Ture, F., Lin, J.: Multi-perspective relevance matching with hierarchical convnets for social media search. In: Proceedings of the AAAI Conference on Artificial Intelligence, vol. 33, pp. 232–240 (2019)
11. Shen, D., Zhu, H., Zhu, C., Xu, T., Ma, C., Xiong, H.: A joint learning approach to intelligent job interview assessment. In: IJCAI, pp. 3542–3548 (2018)
12. Tay, Y., Luu, A.T., Hui, S.C.: Hermitian co-attention networks for text matching in asymmetrical domains. In: IJCAI, 4425–4431, July 2018
13. Vaswani, A., et al.: Attention is all you need. In: Advances in Neural Information Processing Systems, pp. 5998–6008 (2017)

14. Xiong, C., Zhong, V., Socher, R.: Dynamic coattention networks for question answering. In: ICLR, pp. 1–14 (2016). http://arxiv.org/abs/1611.01604
15. Xu, Y., Zhang, Q., Zhang, J., Tao, D.: ViTAE: vision transformer advanced by exploring intrinsic inductive bias. In: Advances in Neural Information Processing Systems, vol. 34 (2021)
16. Yan, R., Le, R., Song, Y., Zhang, T., Zhang, X., Zhao, D.: Interview choice reveals your preference on the market: to improve job-resume matching through profiling memories. In: Proceedings of the 25th ACM SIGKDD International Conference on Knowledge Discovery & Data Mining, pp. 914–922 (2019)
17. Yang, Z., Yang, D., Dyer, C., He, X., Smola, A., Hovy, E.: Hierarchical attention networks for document classification. In: Proceedings of the 2016 Conference on ACL, pp. 1480–1489 (2016)
18. Zhan, Y., Yu, J., Yu, T., Tao, D.: On exploring undetermined relationships for visual relationship detection. In: Proceedings of the IEEE/CVF Conference on Computer Vision and Pattern Recognition, pp. 5128–5137 (2019)
19. Zhan, Y., Yu, J., Yu, T., Tao, D.: Multi-task compositional network for visual relationship detection. Int. J. Comput. Vis. **128**(8), 2146–2165 (2020). https://doi.org/10.1007/s11263-020-01353-8
20. Zhang, Q., Xu, Y., Zhang, J., Tao, D.: VSA: learning varied-size window attention in vision transformers. arXiv preprint arXiv:2204.08446 (2022)
21. Zhang, Y., Yang, C., Niu, Z.: A research of job recommendation system based on collaborative filtering. In: 2014 Seventh International Symposium on Computational Intelligence and Design, vol. 1, pp. 533–538. IEEE (2014)
22. Zhu, C., et al.: Person-job fit: adapting the right talent for the right job with joint representation learning. ACM Trans. Manag. Inf. Syst. (TMIS) **9**(3), 12 (2018)

Detecting Alzheimer's Disease Based on Acoustic Features Extracted from Pre-trained Models

Kangdi Mei[1], Zhiqiang Guo[1], Zhaoci Liu[1], Lijuan Liu[1,2], Xin Li[1,2], and Zhenhua Ling[1,3(✉)]

[1] National Engineering Research Center of Speech and Language Information Processing, University of Science and Technology of China, Hefei, People's Republic of China
zhling@ustc.edu.cn

[2] iFlytek Research, Hefei, People's Republic of China

[3] Interdisciplinary Research Center for Linguistic Science, University of Science and Technology of China, Hefei, People's Republic of China

Abstract. In this paper, we study the performance of Alzheimer's disease (AD) detection using two different high-level acoustic features extracted from pre-trained models, i.e., bottleneck features obtained by supervised learning and wav2vec 2.0 representations obtained by self-supervised learning. We exploit early fusion at the frame level and late fusion at the score level to combine these two features. Moreover, the silence-related information extracted based on voice activity detection (VAD) is integrated to further optimize the detection results. Experiments on the INTERSPEECH 2020 ADRess Challenge dataset show that the bottleneck features and wav2vec 2.0 representations perform better in the detection of AD class and non-AD class respectively, while the late fusion provides a higher accuracy than both of them, which suggests that there exists complementary information between these two features. The integration of the silence-related information improves the fusion system even further. Our highest accuracy on AD detection is 79.2%, which achieves the state-of-the-art performance of detecting AD using only audio data on this dataset.

Keywords: Alzheimer's disease · Speech analysis · Pre-trained acoustic representation · Feature fusion

1 Introduction

Alzheimer's disease (AD) is the most common cause of dementia and also the most prevalent neurodegenerative disease. It can gradually cause irreversible

This work was partially funded by the National Nature Science Foundation of China (Grant No. 62106246) and China Postdoctoral Science Foundation (Grant No. 2021M693101).

L. Fang et al. (Eds.): CICAI 2022, LNAI 13606, pp. 272–283, 2022.
https://doi.org/10.1007/978-3-031-20503-3_22

damage to the brain, slowly destroy memory and thinking abilities, and eventually leave the patient incapable of performing the simplest tasks [1]. As global aging increases and the social problems caused by AD become more and more serious, the early detection of AD is becoming increasingly significant [2]. Nowadays several biochemical methods have been developed for early assessment, but they are usually costly, demanding and time-consuming. At the early stages of the disease, AD patients suffer from language impairment and cognitive decline. They have difficulty in word finding and word retrieval, which makes their performances on semi-structured speech tasks, such as picture description, different from healthy people. Therefore, using speech analysis and natural language processing (NLP) techniques shows the potential to detect AD in a quick, non-invasive and cost-effective way.

The ADRess Challenge [3] at INTERSPEECH 2020 defined a shared task for the systematic comparison of different approaches to the detection of AD based on spontaneous speech. It provided a benchmark dataset of spontaneous speech that was acoustically preprocessed and balanced in both age and gender. The dataset contained speech recordings and corresponding manual transcriptions from 156 participants' description of the picture "Cookie Theft" which was widely used in Boston Diagnostic Aphasia Examination [4]. Many studies on AD detection from speech and text transcriptions have been conducted on this ADRess Challenge dataset. Pappagari et al. [5] used an x-vector model to characterize speech signals and exploited a pre-trained BERT model to handle manual transcriptions, achieving 75.0% accuracy on the test dataset. Martinc et al. [6] combined the predictive ability of acoustic features, term frequency-inverse document frequency (TF-IDF) features, word embeddings and linguistic readability-related information. This method achieved a classification accuracy of 77.1% on the test dataset. Rohanian et al. [7] reported 79.2% detection accuracy by employing multi-model fusion with gating using audio, lexical and disfluency features. Balagopalan et al. [8] designed specific features based on domain knowledge and fine-tuned a BERT-based sequence classification model, increasing the accuracy of the test dataset to 83.3%.

One deficiency of above methods is that they all relied on the texts of manual for classification, which makes these methods incapable of fully automated AD detection. One solution is to replace manual transcription with transcripts generated by automatic speech recognition (ASR). However, because of a high ASR error rate and other uncertain errors, the recognized transcriptions usually degrade the performance of AD detection [9]. Another solution is to build classifiers using only acoustic characteristics of speech. The acoustic features used in the ADRess Challenge are mostly low-level features that are artificially designed or calculated, which often incorporate too much irrelevant information and contain limited semantic information, resulting in a lower accuracy of AD detection than using manual transcriptions. In previous studies [5,7], the detection accuracy dropped from 75.0% and 79.2% to 66.7% and 66.6% when using only speech data. Not only that, in the baseline paper given by the ADRess Challenge [3], the accuracy of AD classification using linguistic features extracted based on manual transcription could reach 75.0%, while that was only 62.5% when using low-level acoustic features. The work of Martinc et al. [6] also shows that it is difficult to achieve high accuracy on AD detection tasks with low-level acoustic features. Similarly, although Balagopalan et al. [8] achieved a high accuracy of 83.3% on the test dataset, the features they used were mainly extracted based on the semantic and lexical-grammatical information from the manual transcription. Therefore, in the classification task of the ADRess Challenge, it still faces technical challenges to obtain high detection accuracy when using only speech data.

In this paper, we study the performance on AD detection of two different high-level acoustic features extracted from pre-trained models, i.e., bottleneck features obtained by supervised learning and wav2vec 2.0 representations obtained by self-supervised learning. Compared with low-level acoustic features, bottleneck features and wav2vec 2.0 representations can provide intermediate representations between raw acoustic features and text transcriptions, and contain more semantic information. A fully automatic AD detection method based on bottleneck features and wav2vec 2.0 representations is designed to detect AD from speech without relying on human transcription. We first build detection models with bottleneck features and wav2vec 2.0 representations as input to detect AD. Then early fusion at the frame level and late fusion at the score level are exploited to combine these two features. Moreover, the silence-related information extracted based on voice activity detection (VAD) is integrated to further optimize the detection results. Experimental results on the ADRess Challenge dataset demonstrate the effectiveness of our approaches.

2 Methods

2.1 Extracting Bottleneck Features and Wav2vec 2.0 Representations from Pre-trained Models

Bottleneck Features. Bottleneck features were initially proposed by utilizing a multi-layer perceptron (MLP) model for a speech recognition tasks [10,11]. The model contained a 5-layer network with 3 hidden layers, one of which had only 39 units while other layers had 2048 units. This small hidden layer, called the bottleneck layer, created a constraint in the network such that the information obtained from training was forced into a low-dimensional representation, which was extracted as bottleneck features. In order to further improve the performance of bottleneck features, there have been studies using pre-training models as the feature extractor to replace traditional MLPs [11]. In this paper, the pre-trained model we used to extract bottleneck features is a speaker-independent long short-term memory (LSTM) model for ASR, which was trained using about 3000 h hours of labeled English recordings on an internal dataset of iFLYTEK company [12]. The dimension at each frame of bottleneck features we have obtained is 512 and the frame shift is 40 ms.

Wav2vec 2.0 Representations. Wav2vec 2.0 representations are extracted from the wav2vec 2.0 model proposed by Facebook [13]. The initial wav2vec model [14] utilizes a group of CNNs to obtain acoustic implicit representations from raw speech samples, and uses these representations to train on a noise contrastive binary classification task defined by self-supervised learning. On the basis of the original version, the wav2vec 2.0 model employs the Transformer architecture to obtain contextualized representations, and adds a quantization module for the contrastive task. The pre-trained model we make use of is the *wav2vec2-large-xlsr-53-english* model released by huggingface[1]. The dimension at each frame of wav2vec 2.0 representations is 1024 and the frame shift is 20 ms.

2.2 AD Detection Model

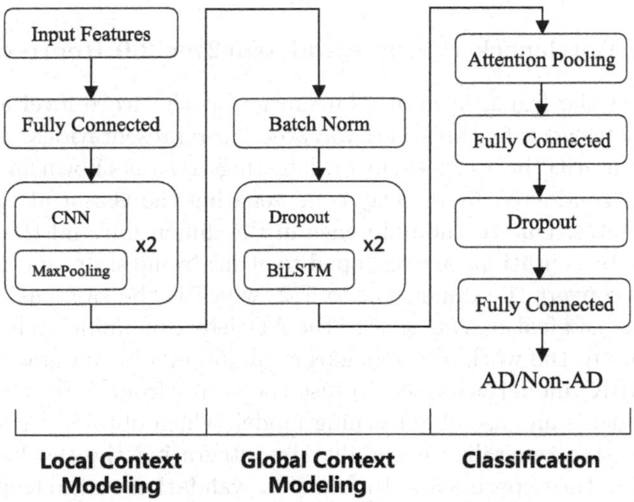

Fig. 1. The structure of the AD detection model for bottleneck features [12].

Bottleneck Features. To fully utilize the bottleneck features, the AD detection model designed by Liu et al. [12] is adopted in our study. The model structure is shown as Fig. 1, which contains three blocks: local context modeling, global context modeling, attention pooling and classification. In the block of local context modeling, two convolutional neural network (CNN) layers are exploited to extract local context information contained by the input features. The global context modeling block mainly employs LSTM units to obtain the global context information among the bottleneck features at different frames, and each LSTM layer is bidirectional to integrate both historical and future sequence information. Then, an attention pooling layer is used [15,16] before the fully connected layers to compute the importance of each frame and adaptively extract the key information for classification. On the basis of the original model, we replace the

[1] https://huggingface.co/models.

kernels of the two CNN layers with 64 and 128, respectively. The model is trained to minimize the cross-entropy loss function using the Adam optimizer.

Wav2vec 2.0 Representations. For the specific AD detection task, the wav2vec 2.0 pre-trained model is fine-tuned by freezing the *feature_extractor* layer [17], adding an average pooling layer after the Transformers, and then connecting a linear classification block. The classification block contains two layers: one is a linear layer consisting of 1024 units and Tanh activation function, and the other is a binary fully connected layer for AD classification. The loss function is also cross-entropy and the optimizer is *Transformer.Trainer* which is provided by Facebook in pytorch.

2.3 Fusing Bottleneck Features and Wav2vec 2.0 Representations

Early fusion at the frame level and late fusion at the score level are exploited to combine bottleneck features and wav2vec 2.0 representations, and to study the complementarity between them in detecting AD (as shown in Fig. 2). The early fusion is conducted by directly concatenating the vector at each frame of these two features. Due to the difference in the dimension and the frame shift, wav2vec 2.0 representations are reshaped to 40 ms frame shift, and a linear layer is inserted to convert the dimension to 512, which is the same as BN features. Then the combined features are sent to the AD detection model in Fig. 1. For late fusion, we refer to the work of Pappagari et al. [5], in which a gradient boosting regression (GBR) model was used to fuse the scores from different approaches. The GBR model is an ensemble learning model, which obtains a comprehensive model with better generalization ability by integrating the results of multiple weak models. In the experiments, 10-fold cross-validation on the training dataset was conducted to obtain the GBR model training data, which was composed of the scores from the evaluation subsets of all folds in the 10-fold cross validation.

2.4 Utilizing Silence-Related Information

Studies have shown that the silence-related information is helpful for AD detection [18]. To complement the bottleneck features and wav2vec 2.0 representations, the silence information extracted from speech recordings is integrated with them. Before calculating silence features, the energy-based VAD algorithm is employed to distinguish whether a speech frame is silent or not. The silence features contain the proportion of silent segments in the entire audio, the number of silent segments, and the minimum silent segments length. When extracting the silence information, only the silence segments longer than 1.0 s are considered (if there are no segments longer than 1.0 s, then the longest silent segment is taken), and the silences at the start and end of the recordings are removed if they exist. Finally, the 3-dimension silence features are integrated into the basic AD detection model by connecting them to the output of the last pooling layer before fully connected layers.

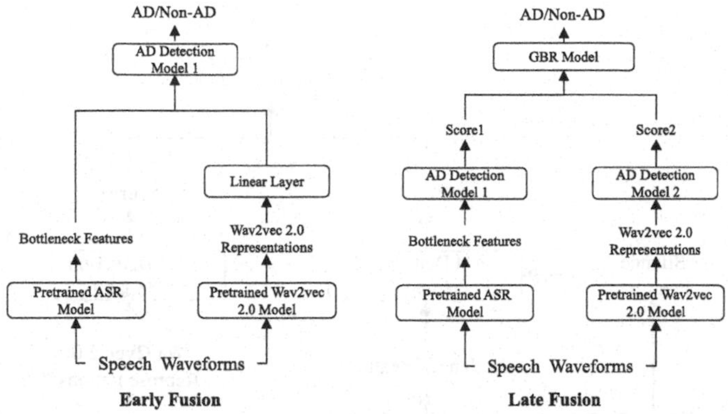

Fig. 2. The stucture of early fusion on the frame level (left) and late fusion on the score level (right). The AD detection model 1 exploited for bottleneck features and the AD detection model 2 employed for wav2vec 2.0 representations are both introduced in Sect. 2.2.

Furthermore, in addition to integrating the silence-related information with bottleneck features and wav2vec 2.0 representations, the feature fusion and silence features are combined to improve AD detection performance. We tried different positions where silence features were integrated in the experiment. For example, Fig. 3 shows the model which combines the silence features to bottleneck features, and then performs late fusion with the wav2vec 2.0 representations.

3 Experiments

3.1 Dataset

The dataset used in this work is the ADRess Challenge dataset [3], which contains speech recordings and manual transcriptions from speakers who were asked to describe the Cookie Theft picture as much as possible. Only the speech samples are adopted in the dataset for AD detection. The dataset includes two groups of speakers: those diagnosed with AD (ad group) and the age- and sex-matched control speakers (cc group), and is divided into a training set and a test set. Each group in the training set is composed of 24 male and 30 female participants, while that in the test set consists of 11 male and 13 female participants. The average duration of speech recordings in the training and test sets is 72.10 s and 82.51 s, respectively.

Due to the limited amount of training data, the ADRess Challenge adopts a classification principle based on the majority voting results of segments. Specifically, the approach is to divide a speech sample into several segments firstly, then count the prediction results of each segment, and finally select the majority label as the classification result. In the experiments, each speech sample was

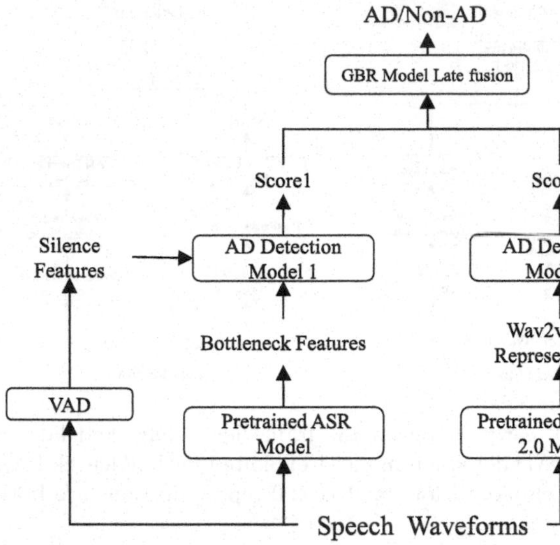

Fig. 3. An example model structure of combining feature fusion and silence features. The AD detection model 1 exploited for bottleneck features and the AD detection model 2 employed for wav2vec 2.0 representations are both introduced in Sect. 2.2.

divided into 10 s long segments. Moreover, the dataset was pre-processed before the segmentation for better detection performance. We discarded almost pure mute samples containing no speech, amplified the audios with average energy below 30 db, and removed the parts where interviewer speak. Finally, the training set is composed of 42 AD samples and 42 non-AD samples, and comprises 400 segments; the test set is composed of 24 AD samples and 24 non-AD samples, and comprises 223 segments.

3.2 Training Setup

Using Bottleneck Features and Wav2vec 2.0 Representations. For bottleneck features, the batch size of the proposed model in Sect. 2.2 was set to 16, the epoch number was 100, and the learning rate was 1×10^{-5} initially and decayed to 5×10^{-6} with epochs. For wav2vec 2.0 representations, we set the batch size to 2, the epoch number to 4, and the learning rate to 1×10^{-5}.

Feature Fusion. In early fusion of two features, we employed the approach introduced in Sect. 2.3 and adopted the same detection model as bottleneck features. The epoch number was changed to 55 and the learning rate was fixed at 1×10^{-5}. The late fusion was conducted by using GBR model at the score level. The learning rate of GBR model was set to 0.01, the estimator number was 200, and the minimum sample number of leaf nodes was 10.

Silenced-Related Information Integration. For bottleneck features and wav2vec 2.0 representations, the silence features were concatenated to the output of the attention pooling layer and the average pooling layer, respectively. The batch size for bottleneck features was set to 16, the epoch number was 120, and the learning rate was 1×10^{-5}; The batch size for wav2vec 2.0 representations was set to 2, the epoch number was 4, and the learning rate was 1×10^{-5}.

Moreover, these two features, feature fusion and silence information were combined for further study. In our experiments, we found that early fusion hardly improved the detection performance, thus only the silence feature integration on the late fusion models was performed. There approaches of the combination were used (see in Table 1) and the learning rates of GBR model were 0.006, 0.01 and 0.01, respectively.

3.3 Evaluation

The ADRess training set was used to performed the 10-fold cross validation, which was done speaker-independently. Then the hyperparameters in model training were selected based on the performance of cross-validation results. The evaluation results were obtained by testing models trained with the complete train set on the test set. For evaluation metrics, we present our results in terms of accuracy, precision, recall, and F1 score, which was the same as proposed in [3].

4 Results and Discussion

In this section, we reported our classification results on the 10-fold cross validation and the test dataset (both see in Table 1). In the table, the baseline result on the test set was proposed by the ADRess Challenge [3] and it was obtained based on conventional acoustic features. For simplicity, in cross validation results only the detection accuracy was reported. It is worth nothing that all the results were computed based on classifying the samples instead of the segments.

4.1 Bottleneck Features and Wav2vec 2.0 Representations

Compared with the conventional acoustic features used in baseline, both of the bottleneck features and wav2vec 2.0 representations provide a higher detection accuracy, which suggests that these pre-trained features contain more information useful for AD detection. The accuracy of bottleneck feature and wav2vec 2.0 representations is 71.4%, 73.8% respectively on cross-validation, and 75.0%, 72.9% respectively on the test set. Moreover, on the test set, the recall is much higher on cc class than on ad class for wav2vec 2.0 representations, while the performance is opposite for bottleneck features. From the comprehensive evaluation of F1 score, the bottleneck features perform better in the detection of the cc class, while the wav2vec 2.0 representations have more advantages in the detection of the ad class.

Table 1. The results of AD detection using different methods. "BN" and "W2V" Characterize bottleneck features and wav2vec 2.0 representations respectively, while "EF" and "LF" stand for early fusion and late fusion respectively. "+Sil" means silence feature integration.

	Cross validation	Test dataset				
Methods	Accuracy	Class	Precision	Recall	F1-score	Accuracy
Baseline [3]	\	cc	0.67	0.50	0.57	0.625
		ad	0.60	0.75	0.67	
BN	0.714	cc	0.773	0.708	0.739	**0.750**
		ad	0.731	0.792	0.760	
W2V	**0.738**	cc	0.690	0.833	0.755	0.729
		ad	0.789	0.625	0.698	
EF-BN & W2V	0.750	cc	0.739	0.708	0.723	0.729
		ad	0.720	0.750	0.735	
LF-BN & W2V	**0.774**	cc	0.783	0.750	0.766	**0.771**
		ad	0.760	0.792	0.776	
BN+Sil	0.762	cc	0.818	0.750	0.783	**0.792**
		ad	0.769	0.833	0.800	
W2V+Sil	0.738	cc	0.750	0.750	0.750	0.750
		ad	0.750	0.750	0.750	
LF-(BN+Sil) & W2V	**0.821**	cc	0.818	0.750	0.783	**0.792**
		ad	0.769	0.833	0.800	
LF-BN &(W2V+Sil)	0.774	cc	0.818	0.750	0.783	**0.792**
		ad	0.769	0.833	0.800	
LF-(BN+Sil) & (W2V+Sil)	0.786	cc	0.818	0.750	0.783	**0.792**
		ad	0.769	0.833	0.800	

4.2 Feature Fusion

On the basis of the performance of bottleneck features and wav2vec2.0 features in Table 1, it is proven that the early fusion hardly improved the results, while the late fusion has a significant improvement. The late fusion improves the value of accuracy on cross-validation by 3.6% for wav2vec 2.0 representations and on the test set by 2.7% for bottleneck features, suggesting that these two features can be effectively complemented by the GBR model employed in the late fusion. In the detection task, some samples present opposite results in two feature models, while the GBR model can learn from these independent errors, thereby improving the detection accuracy. In addition, the recall and F1 score of both the early fusion and the late fusion are relatively close on the ad class and cc class, which further indicates that the feature fusion does work.

4.3 Silence-Related Information Integration

After integrating silence-related information, on cross-validation and the test set results, the accuracy of bottleneck features increases from 71.4% and 75.0% to 76.2% and 79.2%, and the accuracy of wav2vec 2.0 representations also increases from 73.8% and 72.9% to 73.8% and 75.0% (see in Table 1). In the aspects of precision, recall and F1-score, the performance on the ad class and the cc class is closer than the original features, which also indicates that there is a complementary effect between silence features and the two acoustic features. Therefore, a comprehensive study of the role of all the features was conducted by utilizing the approach of late fusion which has a great performance in Sect. 4.2. The results on the late fusion of "LF-(BN+Sil) & W2V", "LF-BN & (W2V+Sil)", "LF-(BN+Sil) & (W2V+Sil)", all present the same accuracy of 79.2%, which is the best detection performance in experiments. But on cross-validation, "LF-(BN+Sil) & W2V" achieves the highest accuracy of 82.1%. Therefore, according to the results of cross-validation and the test set, the approach of connecting the silence information to bottleneck features, and then performing late fusion with the wav2vec 2.0 representations, is more conducive to the AD detection.

4.4 Comparison with Related Work

In order to further evaluate the performance of our methods, we next compare our best result on the test set in this paper with the results of the related work in the ADRess Challenge.

Table 2. The performance comparison between our method and other methods using different modal inputs on the ADReSS test set.

Methods		Accuracy
Speech	Our method	0.792
	(Pappagari et al., 2020) [5]	0.667
	(Rohanian et al., 2020) [7]	0.667
	(Edwards et al., 2020) [19]	0.604
	(Koo et al., 2020) [20]	0.729
Text	(Edwards et al., 2020) [19]	0.771
	(Pompili et al., 2020) [21]	0.729
	(Searle et al., 2020) [22]	0.813
Speech + Text	(Pappagari et al., 2020) [5]	0.750
	(Martinc et al., 2020) [6]	0.771
	(Rohanian et al., 2020) [7]	0.792
	(Balagopalan et al., 2020) [8]	0.833

As shown in Table 2, in the ADRess Challenge, the relatively great detection results obtained by using both speech and text are 75.0%, 77.1%, 79.2%, and

83.3% [5–8]. The highest accuracy of the methods implemented in this paper is 79.2%, which is close to them and even exceeds some of them. Among the existing approaches in the ADRess Challenge, the accuracy of AD detection using only text is 77.1%, 72.9% and 81.3% [19,21,22], while using only speech the accuracy is merely 66.7%, 66.7%, 60.4% and 72.9% [5,7,19,20]. Then it can be seen that using manual transcriptions is more advantageous than using speech in AD detection. In addition, according to the statistical results in Table 2, when only the speech data is utilized, all the detection accuracy in the ADRess Challenge does not exceed 75.0%, while our accuracy achieves 79.2% through the combination of pre-trained features and silence-related information. This further demonstrates the effectiveness of our proposed method.

5 Conclusions

In this paper, we have studied the performance on AD detection of bottleneck features and wav2vec 2.0 representations, both of which were obtained from pre-trained models. Through the experiments on the ADRess Challenge dataset, we find that these two features contain more abundant speech information for AD classification than low-level acoustic features. At the same time, there is complementary information between bottleneck features and wav2vec 2.0 representations, which can be combined with each other through the late fusion utilizing a GBR model at the score level. After integrating the silence-related information on the basis of feature fusion, our method has achieved an accuracy of 79.2%, which is the state-of-the-art performance of detecting AD using only audio data on this dataset. In the near future, we will employ other pre-trained acoustic features for AD detection and continue to improve the detection accuracy by fusing these features more effectively and designing better AD detection models.

References

1. Szatloczki, G., Hoffmann, I., Vincze, V., Kalman, J., Pakaski, M.: Speaking in Alzheimer's disease, is that an early sign? Importance of changes in language abilities in Alzheimer's disease. Front. Aging Neurosci. **7**, 195 (2015)
2. Rajan, K.B., Weuve, J., Barnes, L.L., Wilson, R.S., Evans, D.A.: Prevalence and incidence of clinically diagnosed Alzheimer's disease dementia from 1994 to 2012 in a population study. Alzheimer's Dementia **15**(1), 1–7 (2019)
3. Luz, S., Haider, F., de la Fuente, S., Fromm, D., MacWhinney, B.: Alzheimer's dementia recognition through spontaneous speech: the ADReSS challenge. In: INTERSPEECH, pp. 2172–2176 (2020)
4. Goodglass, H., Kaplan, E.: Boston diagnostic aphasia examination booklet. Lea & Febiger (1983)
5. Pappagari, R., Cho, J., Moro-Velazquez, L., Dehak, N.: Using state of the art speaker recognition and natural language processing technologies to detect Alzheimer's disease and assess its severity. In: INTERSPEECH, pp. 2177–2181 (2020)

6. Martinc, M., Pollak, S.: Tackling the ADReSS challenge: a multimodal approach to the automated recognition of Alzheimer's dementia. In: INTERSPEECH, pp. 2157–2161 (2020)
7. Rohanian, M., Hough, J., Purver, M.: Multi-modal fusion with gating using audio, lexical and disfluency features for Alzheimer's dementia recognition from spontaneous speech. In: INTERSPEECH, pp. 2187–2191 (2020)
8. Balagopalan, A., Eyre, B., Rudzicz, F., Novikova, J.: To BERT or not to BERT: comparing speech and language-based approaches for Alzheimer's disease detection. In: INTERSPEECH, pp. 2167–2171 (2020)
9. Pan, Y., Mirheidari, B., Reuber, M., Venneri, A., Blackburn, D., Christensen, H.: Automatic hierarchical attention neural network for detecting AD. In: Proceedings of Interspeech 2019, pp. 4105–4109. International Speech Communication Association (ISCA) (2019)
10. Grézl, F., Karafiát, M., Kontár, S., Cernocky, J.: Probabilistic and bottle-neck features for LVCSR of meetings. In: 2007 IEEE International Conference on Acoustics, Speech and Signal Processing-ICASSP 2007, vol. 4, pp. IV-757. IEEE (2007)
11. Yu, D., Seltzer, M.L.: Improved bottleneck features using pretrained deep neural networks. In: Twelfth Annual Conference of the International Speech Communication Association (2011)
12. Liu, Z., Guo, Z., Ling, Z., Li, Y.: Detecting Alzheimer's disease from speech using neural networks with bottleneck features and data augmentation. In: ICASSP 2021–2021 IEEE International Conference on Acoustics, Speech and Signal Processing (ICASSP), pp. 7323–7327. IEEE (2021)
13. Baevski, A., Zhou, Y., Mohamed, A., Auli, M.: wav2vec 2.0: A framework for self-supervised learning of speech representations. Adv. Neural Inf. Process. Syst. **33**, 12449–12460 (2020)
14. Schneider, S., Baevski, A., Collobert, R., Auli, M.: wav2vec: unsupervised pre-training for speech recognition. Proc. Interspeech **2019**, 3465–3469 (2019)
15. Raffel, C., Ellis, D.P.: Feed-forward networks with attention can solve some long-term memory problems. arXiv preprint arXiv:1512.08756 (2015)
16. Zhou, P., et al.: Attention-based bidirectional long short-term memory networks for relation classification. In: Proceedings of the 54th Annual Meeting of the Association for Computational Linguistics (volume 2: Short papers), pp. 207–212 (2016)
17. Conneau, A., Baevski, A., Collobert, R., Mohamed, A., Auli, M.: Unsupervised cross-lingual representation learning for speech recognition. arXiv preprint arXiv:2006.13979 (2020)
18. Al-Hameed, S., Benaissa, M., Christensen, H.: Simple and robust audio-based detection of biomarkers for Alzheimer's disease. In: 7th Workshop on Speech and Language Processing for Assistive Technologies (SLPAT), pp. 32–36 (2016)
19. Edwards, E., Dognin, C., Bollepalli, B., Singh, M.K., Analytics, V.: Multiscale system for Alzheimer's dementia recognition through spontaneous speech. In: INTERSPEECH, pp. 2197–2201 (2020)
20. Koo, J., Lee, J.H., Pyo, J., Jo, Y., Lee, K.: Exploiting multi-modal features from pre-trained networks for Alzheimer's dementia recognition. In: INTERSPEECH, pp. 2217–2221 (2020)
21. Pompili, A., Rolland, T., Abad, A.: The INESC-ID multi-modal system for the ADReSS 2020 challenge. In: INTERSPEECH, pp. 2202–2206 (2020)
22. Searle, T., Ibrahim, Z., Dobson, R.: Comparing natural language processing techniques for Alzheimer's dementia prediction in spontaneous speech. In: INTERSPEECH, pp. 2192–2196 (2020)

Similar Case Based Prison Term Prediction

Siying Zhou[1], Yifei Liu[2], Yiquan Wu[2(✉)], Kun Kuang[2], Chunyan Zheng[1(✉)], and Fei Wu[2,3,4]

[1] Guanghua Law School, Zhejiang University, Hangzhou 310008, China
{zhousiying,boxzheng}@zju.edu.cn
[2] Institute of Artificial Intelligence, Zhejiang University, Hangzhou 310007, China
{liuyifei,wuyiquan,kunkuang}@zju.edu.cn, wufei@cs.zju.edu.cn
[3] Shanghai Institute for Advanced Study of Zhejiang University, Shanghai 201203, China
[4] Shanghai AI Laboratory, Shanghai 200232, China

Abstract. Legal judgment (e.g., charge, law article and prison term) prediction is an important task for Legal AI, aiming to assist the judges to get the legal judgment and improve the efficiency of judges. Recently, many existing works have been proposed to promote the performance of legal judgment prediction. While most of them pay attention to confusing charges identification and article recommendation, neglecting the prison term prediction could limit the overall performance (e.g., the accuracy of prison term is less than 50%). In this paper, we focus on the task of prison term prediction. According to the principle of **"treating like cases alike"** stressed by the Supreme People's Court of China, we propose a similar case based prison term prediction (SPTP) method, consisting of a prison term prediction module and a prison term rectification module. The prison term prediction module uses the basic prison term prediction method to get an initial prison term, and then the prison term rectification module rectifies the initial prison term to acquire the final prison term prediction. Specifically, the rectification module includes a similar case retrieval part and a sentencing rectification part. Extensive experiments show the effectiveness of our method under the quantitative evaluation metrics.

Keywords: Deep learning · Natural language processing · Prison term prediction

1 Introduction

Legal Judgment Prediction(LJP) is a typical application in the field of legal artificial intelligence, which has largely accelerated the process of judicial intelligence. As shown in Fig. 1, given the fact description, LJP aims to predict the law article, the charge and the prison term respectively. In recent years, many approaches [8,10,19,20,22,24] and datasets [3,7,23] have been proposed to promote the development of LJP. However, most of them pay attention to

L. Fang et al. (Eds.): CICAI 2022, LNAI 13606, pp. 284–297, 2022.
https://doi.org/10.1007/978-3-031-20503-3_23

Fact Description	**The trial found that defendant A illegally possessed one gunpowder gun knowing that the state expressly prohibited individuals from illegally possessing firearms. On April 19, 2016, A was seized and returned to the case at his home. The gun involved was identified as a homemade firearm powered by gunpowder capable of firing projectiles and capable of normal firing.**
Article	Article 128
Charge	Crime of illegal possession and concealment of firearms and ammunition
Prison Term	12 months

Fig. 1. An example case of the LJP task. Given the fact description, LJP aims to predict the law article, charge, and prison term.

the confusing charges identification and article recommendation, ignoring the importance of the prison term prediction, which is the most challenging task in LJP (e.g., the accuracy of prison term is less than 50%).

In this paper, we explore the task of prison term prediction by introducing the similar case into it. In 2020, the Supreme People's Court of China stressed the principle of **"treating like cases alike"** in judicial practice, pointing out the importance of similar cases in prison term prediction. Recently, some research [1,15,16] and datasets [11,12] have been proposed in the field of similar case retrieval. However, most of them formulate the task as a text matching task and none of them utilize these retrieved legal cases in prison term prediction.

To achieve the above principle, we propose a novel similar case based prison term prediction (SPTP) method. SPTP consists of a prison term prediction module and a prison term rectification module. The prison term prediction module uses the basic prison term prediction method to get an initial prison term, and then the prison term rectification module rectifies the initial prison term to acquire the final prison term prediction. Specifically, the rectification module includes a similar case retrieval part and a sentencing rectification part. SPTP is an intuitive solution to the problem that existing work neglects the usefulness of similar cases for pending cases. Since the judgments of cases are time-related and region-dependent [21], we introduce the background information of criminal cases into our model.

To sum up, our contributions are as follows:

1. We investigating the problem of prison term prediction by introducing the similar case, which is in line with the spirit of the guidance[1] issued by the Supreme People's Court of China.
2. We propose a novel SPTP method to jointly optimize a prison term prediction module and a prison term rectification module. The rectification module uses similar cases to automatically rectify the initial prison term predicted by the prediction module.

[1] https://www.court.gov.cn/zixun-xiangqing-244021.html

3. We conduct extensive experiments on two legal datasets to verify the effectiveness of our method.
4. To motivate other researchers to investigate this novel but important problem, we make the code and datasets publicly available[2].

2 Related Work

2.1 Legal Judgment Prediction

Legal Judgment Prediction is a fundamental task of legal intelligence, especially in civil law systems. Its subtasks in the context of criminal law generally contain law article prediction, charge prediction and prison term prediction [27]. Charge prediction often appears as the focus of research and is generally treated as a text classification problem [10,24,26]. In addition, the law article prediction task [10] and the confusing article task [24] are also the common challenges.

Specifically, Luo et al. [10] proposed an attention-based neural network model that exploits the semantics of the law articles for charge prediction with the weighted relevant law articles used as a legal basis to support charge prediction. TopJudge [26] is a topological multi-task learning model that captures dependencies between subtasks in legal judgment prediction. LADAN [24] is an attention-based model that solves the confusing charge prediction problem by mining the similarity between fact descriptions and law articles and uses a graph distillation operator to learn the discriminative features between confusing charges.

2.2 Prison Term Prediction

The purpose of prison term prediction, as one of the subtasks of LJP, is to predict the appropriate prison term based on the facts of the cases. Many LJP methods [6,24,26] contain prison term prediction tasks. To the best of our knowledge, prison term prediction is often treated as one of the subtasks, and rarely as a separate research focus. CPTP [4] is the sole work within our research that focuses on prison term prediction. Given the charge, CPTP uses a deep gating network to progressively filter and aggregates information specified by the charge, which helps the prison term prediction.

The prison term prediction task is often formulated as a classification problem with the prison terms in the form of non-overlapping intervals divided by years [24,26]. The above setting lacks some of its practicality since the minimum unit in judicial practice is a month rather than a year.

The proposed SPTP method in this paper focuses on the prison term prediction task, and the output is the prison term in months. This is an intuitive solution to the problem that existing work neglects the usefulness of similar cases for pending cases.

[2] https://github.com/6666ev/SPTP.git

2.3 Similar Case Retrieval

Similar case retrieval is another important application of legal intelligence in common law systems. Shao [14] argued that although similar case retrieval could be regarded as a text matching task in a specific direction, it was still far from the traditional text matching task due to its lengthy and complex text, complicated relevance criteria, and high judgment cost. According to the previous work [2,11, 25], we select similar cases on the basis of fact description. Besides, the method proposed in this paper uses a more practical perspective to guide similar case retrieval, i.e., the difference between the prison terms of the pending case and the similar case is used as a supervisory signal.

3 Problem Definition

In this section, we define the problem of prison term prediction.

First, we define fact description, prison term, and similar case. Fact description consists of several descriptive sentences that summarizes the facts in a case. Here, we denote the **fact description** as $f = \left\{w_t^f\right\}_{t=1}^{l_f}$, where l_f is the number of words in the fact description and we denote the **prison term** as p, which is a positive number in months.

Then, We name the case to be predicted as the pending case to distinguish from similar cases. We denote the **pending case** as pc, and denote the **similar cases** of the pending case as $sc = \{sc_k\}_{k=1}^{n_{sc}}$, where n_{sc} is the number of similar cases. Also, We introduce the concept of background information of criminal cases, which includes charge, region and year. We denote the the **charge** of the pending case as c, the **year** in which the pending case occurred as y, the **region** in which the pending case occurred as r. Also, we define the concatenation of the fact description and background information as the **synthesized fact**, denoted as $sf = \{f, c, y, r\}$.

Thus, the problem of prison term prediction can be formalized as: given the synthesized fact of the pending case $sf = \{f, c, y, r\}$, the model should output the predicted prison term p. In our method, the model uses an extra rectification module to rectify the prison term.

4 Method

To rectify the value of the prison term, we introduce the similar case retrieval into prison term prediction. Specifically, we design the SPTP model as shown in Fig. 2, which consists of a prison term prediction module and a prison term rectification module. The left part is the term prediction module, where the prison term is obtained from the encoder-predictor manner and used as the final result, which is also our basic setting. The right part of the figure is the innovative part of our work. The following is a detailed description of each module in SPTP.

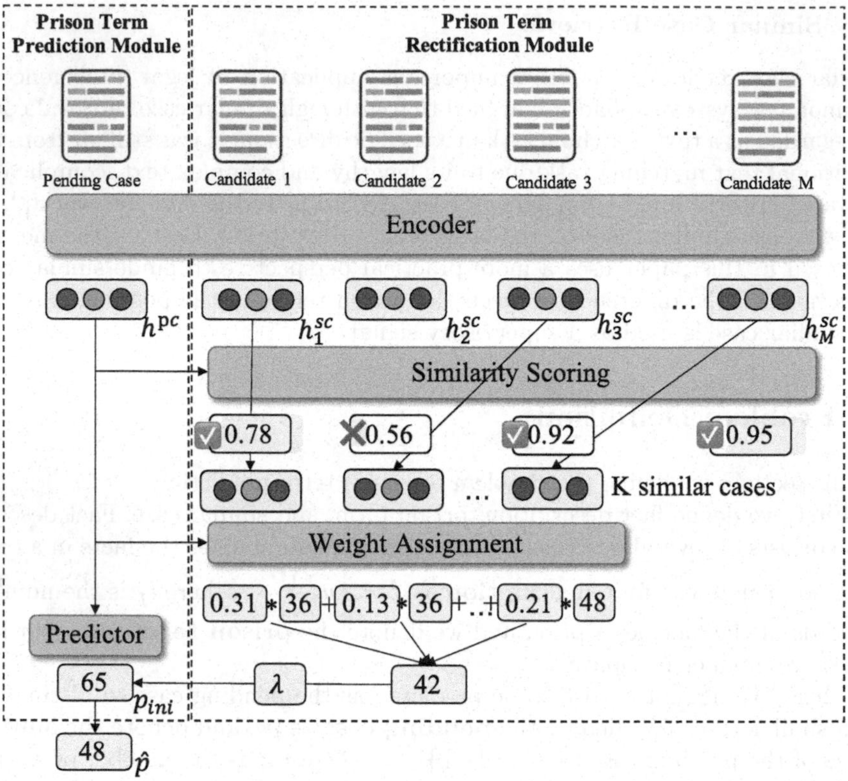

Fig. 2. The architecture of SPTP, which consists of a prison term prediction module and a prison term rectification module.

4.1 Prison Term Prediction Module

The prison term prediction module is exactly a basic model which consists of an Encoder and a Predictor to predict the initial prison term.

Encoder. The input of the encoder is the concatenation of the fact description, charge, year, and region of the pending case *pc*, denoted as $sf^{pc} = \{f, c, y, r\}$. The synthesized fact description is fed into the encoder to get the representation

$$h^{pc} = \text{Encoder}(sf^{pc}). \tag{1}$$

The encoder can be CNN [9], LSTM [17], Transformer [18], and other structures.

Predictor. The predictor takes the representation h encoded by the encoder as the input and goes through a fully-connected layer to get the initially predicted prison term

$$p_{ini} = \text{FC}(h^{pc}), \tag{2}$$

where FC is the fully-connected layer.

4.2 Prison Term Rectification Module

The prison term rectification module aims to rectify the initial prison term generated by the prediction module, and it consists of two part: similar case retrieval and sentencing rectification.

Similar Case Retrieval Part. The similar case retrieval is conducted in the following three steps:

(1) Candidate Case Selection. In this step, we perform the initial screening of similar cases using BM25. After the initial screening, each case has M candidates. We describe this step in detail in Sect. 5.1.

(2) Best Similar Case Selection.

After M candidate cases have been obtained, we input these case into the same encoder to obtain the representation of the similar cases $h^{sc} = \{h_i^{sc}\}_{i=1}^{M}$.

The similarity scores between the pending case and M similar cases are calculated separately by a similarity scoring function. The choice of similarity scoring function is discussed in the Appendix B. Here, we adopt the cosine as the scoring function. The similar cases are sorted according to the similarity scores in descending order, and the K most relevant similar cases are selected. (3) Similar Case Weighting. The weight w_i of ith similar case is obtained by inputting the representation of pending case h^{pc} and these K most relevant similar cases h^{sc} into the multi-head attention block [18], where h^{pc} is taken as query, h^{sc} is taken as key and value.

Sentencing Rectification Part. In this part, we use the initially predicted prison term p_{ini} and obtained weight w_i to get the rectified prison term $\hat{p}$.

First, the difference between the prison terms of the pending case and the K similar cases is summed weighted by their corresponding weights to obtain the rectification value v_{sum} for the similar cases:

$$v_{sum} = \sum_{i=1}^{K} w_i * (p_i - p_{ini}). \tag{3}$$

where p_{ini} is the initially predicted prison term for this case, p_i is the prison term for a certain similar case, and w_i is its corresponding weight.

Second, the prison term is rectified according to the rectification value v_{sum} to obtain the final prison term $\hat{p}$:

$$\hat{p} = p_{ini} + \lambda * v_{sum}, \tag{4}$$

where λ is the hyperparameter to control the rectification strength.

The pseudocode of the whole procedure is shown in Algorithm 1.

Algorithm 1: The algorithm of SPTP

Input : dataset D, the number of candidates M, the number of best similar cases K, hyperparameter λ, pending case pc, each case contains 4 features, including fact description f, charge c, year y, region r and prison term p, where the 4 features make up the synthesized fact sf .

Output: predicted prison term $\hat{p}$.

set^{sc} = DataBaseFilter $(D, f^{pc}, c^{pc}, y^{pc}, r^{pc})$;
$candidates$ = BM25(set^{sc}, f^{pc}, M) ;
h^{pc} = Encoder(sf^{pc}) ;
h^{sc} = [Encoder(sf^{sc}) **for** sc **in** $candidates$] ;
p = [p^{sc} **for** sc **in** $candidates$] ;
for $i = 1$ **to** M **do**
    w_i = GetAttentionWeight(h^{pc}, h_i^{sc}) ;
end
$index$ = TopK(w, K) ;
w = GatherByIndex(w, $index$) ;
p = GatherByIndex(p, $index$) ;
p_{ini} = Predictor(h^{pc}) ;
$\hat{p} = p_{ini} + \lambda * \sum_{i=1}^{K} w_i * (p_i - p_{ini})$;
return $\hat{p}$

4.3 Training and Inference

Training. The loss of the prison term prediction module is:

$$\mathcal{L}_p = \text{square}\left(\log\left(\hat{p}+1\right) - \log(p+1)\right), \tag{5}$$

where p is the ground truth, and $\hat{p}$ is the value of prison term prediction.

The ground-truth *similar_score* is obtained by:

$$similar\_score = 2 * exp(-\frac{x^2}{500}) - 1, \tag{6}$$

where x is the difference between the prison terms of the pending case and the similar case.

The loss of similar case ranking process is:

$$\mathcal{L}_s = \Sigma_i^M |\hat{s}_i - similar\_score(|p^{pc} - p_i^{sc}|)|, \tag{7}$$

where $\hat{s}_i$ is the predicted similar score between pending case and ith similar case, and the ground-truth similar score is calculated by Formula (6), p^{pc} and p_i^{sc} denote ground-truth prison terms of pending case and ith similar case respectively.

Our final loss function is:

$$\mathcal{L} = \alpha * \mathcal{L}_p + \beta * \mathcal{L}_s, \quad (8)$$

where α and β are the weight factors for each loss.

Inference. During inference, we input the synthesized fact description of a certain case to get the initial prison term p_{ini} and similar cases sc. By combining the sc, we can obtain the rectified prison term $\hat{p}$.

5 Experiments

5.1 Dataset Construction

Since there is no publicly available dataset that meets our task, we process the dataset from the competition CAIL2018[3] and LAIC2021[4] in the following steps:

(1) Data Cleaning. The main function of this section is to sift irregular characters from the text of the fact description using rules such as regular expressions and stop words list.

(2) Background Information Abstraction. We extract the year and region information by regular expressions from the fact description text. Thus, each record of the dataset contains five field values: fact description, charge, year, region, and prison term.

(3) Data Filtering. Considering the scope of similar case search introduced in Article 4 of the Guidance of the Supreme People's Court on Uniform Application of Law to Strengthen Case Searching issued by the Supreme People's Court of China, we regard two cases as similar cases based on the initial criteria that the charges, years and regions of them are the same. In order to ensure that high-quality similar cases can be extracted in the subsequent steps, we only keep the cases with the number of similar cases greater than or equal to 50. After the above steps, the records in the LAIC dataset are 21,354, and the records in the CAIL-small dataset are 21,178.

(4) Candidate Case Selection. We divide the data into different sets according to the criteria of having the same charge, year and region. For each record in each set, the fact description is matched using BM25 [13] to obtain the most relevant M similar cases and store them as an ordered sequence of serial numbers as a field, called as similar case indexes(sc_idx).

The detailed statistics of the datasets are shown in Table 1.

[3] The dataset can be downloaded from https://github.com/china-ai-law-challenge/CAIL2018.

[4] The dataset can be downloaded from http://data.court.gov.cn/pages/laic2021.html.

Table 1. The statistics of datasets.

Dataset	LAIC	CAIL
#Training Set Cases	17083	16183
#Validation Set Cases	2135	1761
#Testing Set Cases	2136	3234
Avg. # tokens in fact	213.96	258.13
#Charge	16	70
#Year	8	9
#Region	27	27

5.2 Evaluation Metrics

The four evaluation metrics that appear in this paper are as follows.

- Acc25(accuracy) is proposed in the LAIC 2021 competition where the predicted value will be considered as correct if it is within the upper and lower 25% range of the correct value. The calculation formula is shown in Formula (9):

$$\text{Acc25} = \frac{|\hat{y}_i - y_i|}{y_i} \leq 0.25. \tag{9}$$

- LogDis(log distance) takes the valuc range $[0, 2.477]$ in this experiment. The smaller the value, the better the effect. The calculation formula is shown in Formula (10):

$$\delta_j = |\log(y_j + 1) - \log(\hat{y}_j + 1)|. \tag{10}$$

- Correlation(Spearman's rank correlation coefficient) takes the value range $[-1, 1]$. The calculation formula is shown in Formula (11):

$$\rho = \frac{\sum_i (x_i - \bar{x})(y_i - \bar{y})}{\sqrt{\sum_i (x_i - \bar{x})^2 \sum_i (y_i - \bar{y})^2}}. \tag{11}$$

- P@k(precision@k) indicates the probability of correctly predicting a positive sample among the first k results. The value range is $[0, 1]$ and the formula is shown in Formula (12):

$$\text{Precision@ } k = \frac{TP@k}{TP@k + FP@k}. \tag{12}$$

5.3 Baselines

To evaluate the performance of our model on prison term prediction, we adopt the following models as baselines:

- **CNN** [9] extracts text local features through several convolutional operations with different convolutional kernel sizes and pooling operations for the text classification task.

- **LSTM** [17] solves the problem of RNN short-term memory by adding gating on the basis of the RNN model.
- **Transformer** [18] is a model using a self-attention mechanism with parallelizable training which outperforms the traditional RNN in terms of both inference time and semantics extraction capability.
- **BERT** [5] is a masked language model which is pre-trained on a large corpus and could be fine-tuned on downstream NLP tasks.
- **TopJudge** [26] is a topological multitask learning model that captures dependencies between subtasks in LJP.
- **LADAN** [24] uses a graph distillation operator to extract discriminative features for distinguishing confusing charges and law articles.
- **CPTP** [4] is a charge-based, prison term prediction model with a deep gating network that progressively filters and aggregates information specified by the charge.

5.4 Experimental Details

Word2Vec Pre-training. In our experiments, we use Stanford NLP Gensim to perform unsupervised word vector pre-training on the dataset CAIL-large, which contains about 1.5 million legal documents.

Hyperparameter Settings. This paragraph describes our choice of the hyperparameters used in the experiment. We set the hyperparameters λ, α and β as 0.3, 1 and 100 respectively. The number of candidates is M. The value of M is positively correlated with the scope of the case search and is proportional to the time complexity in the best similar case selection step. In this experiment, we set M as 10. For the choice of K, the smaller value of K, the fewer cases could be used to support the pending case, while the higher quality similar cases could be utilized. On balance, we set K as 5.

5.5 Experimental Results

Table 2. Relevance of the prison term rectification and similar case retrieval.

Dataset	Model	LogDis↓	Acc25	Correlation	P@5
LAIC	SPTP+CNN [9]	2.139	0.3600	0.2293	0.5868
	SPTP+LSTM [17]	2.366	0.2495	0.1524	0.5610
	SPTP+Transformer [18]	**1.988**	**0.4480**	**0.5076**	**0.7127**
CAIL	SPTP+CNN [9]	1.563	0.3772	0.1422	0.5464
	SPTP+LSTM [17]	1.642	0.3327	0.0805	0.5259
	SPTP+Transformer [18]	**1.478**	**0.4335**	**0.3945**	**0.6536**

Table 3. Results of prison term prediction.

Model	LAIC		CAIL	
	LogDis↓	Acc25	LogDis↓	Acc25
LSTM [17]	2.236	0.3596	1.591	0.3791
CNN [9]	2.186	0.3511	1.568	0.3701
Transformer [18]	2.091	0.4134	1.636	0.3565
BERT [5]	2.097	0.4178	1.677	0.3532
TopJudge [26]	2.246	0.3324	1.618	0.3670
LADAN [24]	2.849	0.4180	2.238	0.3870
CPTP [4]	2.085	0.4143	1.555	0.3958
SPTP+LSTM [17]	2.366	0.2495	1.642	0.3327
SPTP+CNN [9]	2.139	0.3600	1.563	0.3772
SPTP+Transformer [18]	**1.988**	**0.4480**	**1.478**	**0.4335**
Avg-of-K-best	**1.448**	**0.7112**	**1.036**	**0.7359**

Table 2 shows the results after rectifying the prison term using similar cases, and we have the following observations:

1. Compared with SPTP+CNN and SPTP+LSTM, SPTP+Transformer achieves prison term prediction results have higher accuracy, with better retrieval performance. This indicates that the model with better ranking results achieves superior performance in prison term prediction.
2. It also proves that good similar case selection can indeed help prison term prediction.

We compare our methods with some baselines and the results are shown in Table 3. We find that:

1. Compared to other baselines, SPTP+Transformer achieves the best performance with the LogDis and Acc25 values on both LAIC and CAIL-small datasets, which demonstrates the effectiveness of our method.
2. Compared to CNN and Transformer, SPTP+CNN and SPTP+Transformer achieves better performance on prison term prediction, which proves the positive effectiveness of the proposed prison term rectification module.
3. Compared to LSTM, the SPTP+LSTM achieves a worse performance, which is probably because the textual representation of judicial long texts is difficult and the LSTM is relatively simple in structure and cannot model the vector representation of judicial texts well, leading to a poor performance in similar case retrieval. Thus, the effect of prison term rectification is affected as well, which makes the accuracy of prison term decrease.
4. In Avg-of-K-best, we use average prison term of the real best K similar cases, which shows the high upper bound of our method (e.g. Acc25 is over 70%). It proves that similar cases have huge potential to improve prison term prediction.

6 Ethical Discussion

While AI is gaining adoption in legal justice, any subtle statistical miscalculation may trigger serious consequences.

We claim that the proposed algorithm is designed for assisting the trial judges in decision making. This work is an algorithmic investigation, but such an algorithm should never 'replace' human judges. The judgment should be the final safeguard to protect social justice and individual fairness.

7 Conclusion and Future Work

In this paper, we investigating the problem of prison term prediction by introducing the similar case into it. We propose a novel SPTP method which consists of a basic prison term prediction module and an extra prison term rectification module. In the rectification module, we first make similar case retrieval and then rectify the initial prison term generated by the prediction module. The experimental results show the effectiveness of our method.

Based on the SPTP method, in the future we can explore following directions: (1) Investigate more effective encoders for encoding long legal documents. (2) Explore more sophisticated retrieval methods. (3) Automatically generate similar case reports.

Acknowledgements. This work was supported in part by Program of Zhejiang Province Science and Technology (2022C01044), National Key Research and Development Program of China (2021YFC3340300), Key R & D Projects of the Ministry of Science and Technology (2020YFC0832500), and the Fundamental Research Funds for the Central Universities (226–2022–00142, 226–2022–00051).

References

1. Althammer, S., Askari, A., Verberne, S., Hanbury, A.: Dossier@ coliee 2021: leveraging dense retrieval and summarization-based re-ranking for case law retrieval. arXiv preprint arXiv:2108.03937 (2021)
2. Bhattacharya, P., Ghosh, K., Pal, A., Ghosh, S.: Hier-SPCNet: a legal statute hierarchy-based heterogeneous network for computing legal case document similarity, pp. 1657–1660. Association for Computing Machinery, New York, NY, USA (2020). https://doi.org/10.1145/3397271.3401191
3. Chalkidis, I., Fergadiotis, M., Malakasiotis, P., Androutsopoulos, I.: Large-scale multi-label text classification on EU legislation. arXiv preprint arXiv:1906.02192 (2019)
4. Chen, H., Cai, D., Dai, W., Dai, Z., Ding, Y.: Charge-based prison term prediction with deep gating network. In: Proceedings of the 2019 Conference on Empirical Methods in Natural Language Processing and the 9th International Joint Conference on Natural Language Processing (EMNLP-IJCNLP), pp. 6362–6367. Association for Computational Linguistics, Hong Kong, China (2019). https://doi.org/10.18653/v1/D19-1667. http://aclanthology.org/D19-1667

5. Devlin, J., Chang, M.W., Lee, K., Toutanova, K.: BERT: pre-training of deep bidirectional transformers for language understanding. arXiv preprint arXiv:1810.04805 (2018)
6. Dong, Q., Niu, S.: Legal judgment prediction via relational learning, pp. 983–992 Association for Computing Machinery, New York, NY, USA (2021). https://doi.org/10.1145/3404835.3462931
7. Duan, X., et al.: CJRC: a reliable human-annotated benchmark dataset for Chinese judicial reading comprehension. In: Sun, M., Huang, X., Ji, H., Liu, Z., Liu, Y. (eds.) CCL 2019. LNCS (LNAI), vol. 11856, pp. 439–451. Springer, Cham (2019). https://doi.org/10.1007/978-3-030-32381-3_36
8. Gan, L., Kuang, K., Yang, Y., Wu, F.: Judgment prediction via injecting legal knowledge into neural networks. In: Proceedings of the AAAI Conference on Artificial Intelligence, vol. 35, no.14, pp. 12866–12874 (2021). http://ojs.aaai.org/index.php/AAAI/article/view/17522
9. Kim, Y.: Convolutional neural networks for sentence classification. In: Proceedings of the 2014 Conference on Empirical Methods in Natural Language Processing (EMNLP), pp. 1746–1751. Association for Computational Linguistics, Doha, Qatar (2014). https://doi.org/10.3115/v1/D14-1181. http://aclanthology.org/D14-1181
10. Luo, B., Feng, Y., Xu, J., Zhang, X., Zhao, D.: Learning to predict charges for criminal cases with legal basis. In: Proceedings of the 2017 Conference on Empirical Methods in Natural Language Processing, pp. 2727–2736. Association for Computational Linguistics, Copenhagen, Denmark (2017).https://doi.org/10.18653/v1/D17-1289. http://aclanthology.org/D17-1289
11. Ma, Y., Shao, Y., Wu, Y., Liu, Y., Zhang, R., Zhang, M., Ma, S.: LeCaRD: a legal case retrieval dataset for Chinese law system, pp. 2342–2348 Association for Computing Machinery, New York, NY, USA (2021). https://doi.org/10.1145/3404835.3463250
12. Rabelo, J., Kim, M.-Y., Goebel, R., Yoshioka, M., Kano, Y., Satoh, K.: COLIEE 2020: methods for legal document retrieval and entailment. In: Okazaki, N., Yada, K., Satoh, K., Mineshima, K. (eds.) JSAI-isAI 2020. LNCS (LNAI), vol. 12758, pp. 196–210. Springer, Cham (2021). https://doi.org/10.1007/978-3-030-79942-7_13
13. Robertson, S., Zaragoza, H.: The probabilistic relevance framework: BM25 and Beyond. Now Publishers Inc. (2009)
14. Shao, Y.: Towards legal case retrieval, p. 2485. Association for Computing Machinery, New York, NY, USA (2020). https://doi.org/10.1145/3397271.3401457
15. Shao, Y., et al.: BERT-PLI: modeling paragraph-level interactions for legal case retrieval. In: IJCAI, pp. 3501–3507 (2020)
16. Shao, Y., Wu, Y., Liu, Y., Mao, J., Zhang, M., Ma, S.: Investigating user behavior in legal case retrieval. In: Proceedings of the 44th International ACM SIGIR Conference on Research and Development in Information Retrieval, pp. 962–972 (2021)
17. Sutskever, I., Vinyals, O., Le, Q.V.: Sequence to sequence learning with neural networks. In: Proceedings of the 27th International Conference on Neural Information Processing Systems - Vol. 2, pp. 3104–3112. NIPS2014, MIT Press (2014)
18. Vaswani, A., et al.: Attention is all you need. In: Advances in Neural Information Processing Systems 30 (2017)
19. Wang, P., Fan, Y., Niu, S., Yang, Z., Zhang, Y., Guo, J.: Hierarchical matching network for crime classification. In: Proceedings of the 42nd International ACM SIGIR Conference on Research and Development in Information Retrieval, pp. 325–334 (2019)

20. Wang, P., Yang, Z., Niu, S., Zhang, Y., Zhang, L., Niu, S.: Modeling dynamic pairwise attention for crime classification over legal articles. In: The 41st International ACM SIGIR Conference on Research Development in Information Retrieval, pp. 485–494 (2018)
21. Wang, Y., et al.: Equality before the law: legal judgment consistency analysis for fairness. arXiv preprint arXiv:2103.13868 (2021)
22. Wu, Y., et al.: De-biased court's view generation with causality. In: Proceedings of the 2020 Conference on Empirical Methods in Natural Language Processing (EMNLP), pp. 763–780 (2020)
23. Xiao, C., et al.: CAIL 2018: a large-scale legal dataset for judgment prediction. arXiv preprint arXiv:1807.02478 (2018)
24. Xu, N., Wang, P., Chen, L., Pan, L., Wang, X., Zhao, J.: Distinguish confusing law articles for legal judgment prediction. In: Proceedings of the 58th Annual Meeting of the Association for Computational Linguistics, pp. 3086–3095. Association for Computational Linguistics (2020). https://doi.org/10.18653/v1/2020.acl-main.280. http://aclanthology.org/2020.acl-main.280
25. Yao, F., et al.: Leven: a large-scale Chinese legal event detection dataset. arXiv preprint arXiv:2203.08556 (2022)
26. Zhong, H., Guo, Z., Tu, C., Xiao, C., Liu, Z., Sun, M.: Legal judgment prediction via topological learning. In: Proceedings of the 2018 Conference on Empirical Methods in Natural Language Processing, pp. 3540–3549. Association for Computational Linguistics, Brussels, Belgium (2018). https://doi.org/10.18653/v1/D18-1390. http://aclanthology.org/D18-1390
27. Zhong, H., Xiao, C., Tu, C., Zhang, T., Liu, Z., Sun, M.: How does NLP benefit legal system: a summary of legal artificial intelligence. In: Proceedings of the 58th Annual Meeting of the Association for Computational Linguistics, pp. 5218–5230. Association for Computational Linguistics (2020)

Interactive Fusion Network with Recurrent Attention for Multimodal Aspect-based Sentiment Analysis

Jun Wang[1,3], Qianlong Wang[1,3], Zhiyuan Wen[1,3], Xingwei Liang[2,3], and Ruifeng Xu[1,3](✉)

[1] Harbin Institute of Technology (Shenzhen), Shenzhen, China
[2] Konka Research Institute, Shenzhen, China
[3] Joint Lab of HIT-KONKA, Shenzhen, China
21BF51014@stu.hit.edu.cn, xuruifeng@hit.edu.cn

Abstract. The goal of multimodal aspect-based sentiment analysis is to comprehensively utilize data from different modalities (*e.g.*,, text and image) to identify aspect-specific sentiment polarity. Existing works have proposed many methods for fusing text and image information and achieved satisfactory results. However, they fail to filter noise in the image information and ignore the progressive learning process of sentiment features. To solve these problems, we propose an interactive fusion network with recurrent attention. Specifically, we first use two encoders to encode text and image data, respectively. Then we use the attention mechanism to obtain the semantic information of the image at the token level. Next, we employ GRU to filter out the noise in the image and fuse information from different modalities. Finally, we design a decoder with recurrent attention to progressively learn aspect-specific sentiment features for classification. The results on two Twitter datasets show that our method outperforms all baselines.

Keywords: Multimodal aspect-based sentiment analysis · Attention mechanism · Progressively learning

1 Introduction

Aspect-based sentiment analysis (ABSA) is a fine-grained sentiment analysis task whose purpose is to identify the sentiment polarity corresponding to a particular aspect term. For example, in the text "*the dishes taste good, but the queue time is too long*", the sentiment polarities of "*dishes*" and "*queue time*" are positive and negative, respectively.

With the rapid development of mobile networks, more people use multimodal content (the most common form is image-text pair) on social platforms instead of just text. Analyzing the sentiment contained in multimodal data has become an important research direction. Multimodal aspect-based sentiment analysis (MABSA) comprehensively considers data from multiple modalities to judge

L. Fang et al. (Eds.): CICAI 2022, LNAI 13606, pp. 298–309, 2022.
https://doi.org/10.1007/978-3-031-20503-3_24

the sentiment polarity of a specific aspect. Taking Fig. 1 as an example, if we only consider the text, we cannot clearly judge the sentiment polarity of the aspect term "*Cuba*". However, if we take into account the beautiful scenery in the image, it is easy to conclude that the sentiment polarity of "*Cuba*" is positive.

Fig. 1. An example of MABSA. The aspect is marked in orange. (Color figure online)

To address the challenges on ABSA posed by multimodal data, many researchers have proposed solutions from different perspectives. For example, Xu et al. [18] proposed a multi-interactive memory network to capture the interaction between text and image. Besides, Zhang et al. [21] designed a discriminant matrix to fuse the information in images and texts. Zhou et al. [22] designed an adversarial training strategy to align the semantic space of image and text representations, resulting in better multimodal feature fusion results. Although the above methods have achieved satisfactory results, they still have the following shortcomings: (1) they directly interact text with image information without considering image noise. (2) they model MABSA as a classification task without considering the recurrent progressive learning of sentiment features.

To solve these two shortcomings, we propose an interactive fusion network with recurrent attention (IFNRA) for MABSA. Our model consists of three parts: encoder, interactive fusion module, and decoder with recurrent attention. To be specific, we first extract text and image features using BERT [3] and Bottom-Up Attention Model [1] (BUA), respectively. Then we use the attention mechanism to obtain the image semantic information of each token. The acquired image semantic information inevitably contains some noise. To filter out image noise and obtain multimodal representations, we use GRU [2] to fuse beneficial image information and textual information. Finally, we design a decoder with recurrent attention mechanism. Each cycle extracts required information from the multimodal representations to continuously update and improve the sentiment feature of the specific aspect. The classifier infers the sentiment polarity based on the output of the last cycle. We conduct experiments on the Twitter datasets. The results show that our method[1] can achieve better performance than baselines.

2 Related Work

Aspect-based Sentiment Analysis. Early research on ABSA mainly used traditional machine learning methods. Constructing features based on text and external knowledge such as lexical resources was the focus of works [7,9,10,16]. The selection and construction of features largely determined the performance of the model. In recent years, neural networks have achieved rapid development,

[1] The source code is publicly released at https://github.com/0wj0/IFNRA.

exhibiting powerful feature learning capabilities. Dong et al. [4] introduced recurrent neural networks to the ABSA and obtained sentiment polarities of aspect terms according to contextual and syntactic relations. Tang et al. [14] incorporated aspect information into LSTMs to establish connections between aspect words and context words without relying on external knowledge. Furthermore, Wang et al. [17] used the attention mechanism to obtain information that is highly correlated with the sentiment polarity of aspect words at the word and clause levels. With the development and widespread use of pre-trained language models, Phan et al. [12] used BERT for contextual embedding and combined part-of-speech information and syntactic information for ABSA.

Multimodal Aspect-based Sentiment Analysis. Different from traditional ABSA, MABSA comprehensively considers the impact of multi-modalities data on the sentiment polarity. The characteristics of social media data (such as the inconsistency between pictures and texts) pose certain challenges on MABSA. To solve the above problems, Xu et al. [18] first proposed the MABSA. They used the attention mechanism to extract information related to specific aspects in each modality, and designed a multi-hop memory network to make the information of each modality fully interact. Yu et al. [19] used additional BERT layers to obtain features of image regions that are closely related to aspects. In addition, Zhang et al. [21] designed a discriminant matrix to fuse the complementary information between different modalities, so that the representation can contain richer semantic information. Zhou et al. [22] designed an adversarial training strategy to align the semantic space of image and text representations, and then made any two of text, image and aspect interact with each other through the multimodal interaction layer.

Unlike previous works, we propose an interactive fusion network with recurrent attention. Our model filters noise in image through GRU and uses recurrent attention to progressively learn different sentiment features of specific aspects.

3 Model Architecture

3.1 Task Definition

Given a text $T = \{w_1, w_2, ..., w_n\}$, a related image I, and a specific aspect $A = \{a_1, a_2, ..., a_m\}$, the goal of the task is to predict the sentiment polarity $c \in C$ of the aspect. Here, A is a subsequence of T, n and m denote the number of tokens in T and A, respectively. $C = \{Negative, Neutral, Positive\}$.

3.2 Overview

As shown in Fig. 2, our proposed interactive fusion network consists of an encoder, an interactive fusion module, and a decoder with recurrent attention. We use BERT and BUA as encoders, which encode text and images into embedding representations, respectively. After obtaining representations of text and

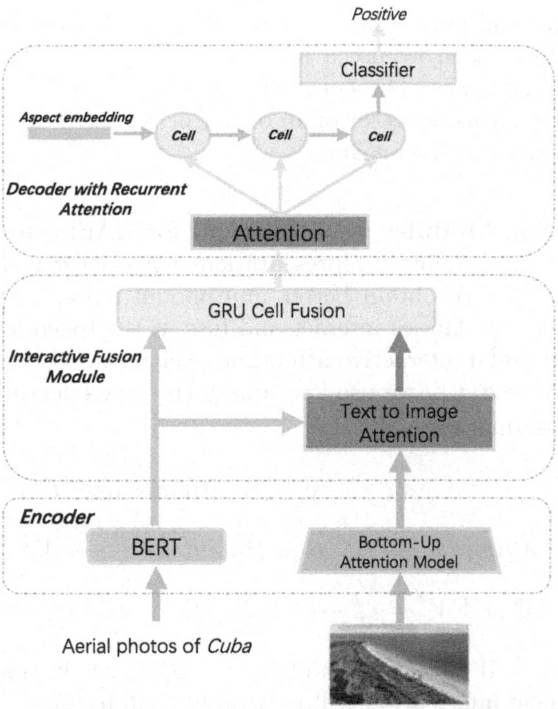

Fig. 2. The overview of our interactive fusion network.

images, we use an interactive fusion module to fuse features from different modalities to obtain multimodal representations. Finally, we complete the classification task using a decoder with the recurrent attention mechanism, which recurrently learns aspect-specific sentiment features and uses feature from the last step to judge the sentiment polarity.

3.3 Model Architecture

Encoder. For each token w_i in the text T, we use BERT to obtain its embedding $e_i^w \in \mathbb{R}^d$ containing contextual semantic information:

$$E^w = BERT(T) \tag{1}$$

The encoded text is $E^w = \{e_1^w, e_2^w, ..., e_n^w\}$, and the encoded aspect is $E^a = \{e_1^a, e_2^a, ..., e_m^a\}$. Here, E^a is a subsequence of E^w. d denotes the dimension of the hidden state of BERT.

The convolutional neural network evenly divides the image into several regions of equal size, and outputs the features of each region. However, this method destroys the semantic information expressed by the entire object or region in the original image. To solve the above problems, we choose the pre-trained BUA model as the image encoder. For the given image I, the BUA model can detect objects and other salient regions in the image, and output the corresponding features. Then we use a linear layer to project the feature vectors into a d-dimensional space. The encoded image is $E^I = \{e_1^I, e_2^I, ..., e_k^I\}$. Here, $e_i^I \in \mathbb{R}^d$ denotes the feature vector of an object or region. k denotes the number of objects and regions in the image.

Interactive Fusion Module. It is challenging for MABSA to use the features of multiple modalities to obtain representations with richer and more complete semantic information. To obtain better multimodal representation, we let the features of different modalities interact and fuse at the token level. Specifically, we design a multimodal interactive attention mechanism. For each token vector e_i^w in the encoded text E^w, we use it to query the image features E^I to obtain relevant image information v_i^{img}.

$$v_i^{img} = Attention(e_i^w, E^I) \tag{2}$$

$$Attention(e_i^w, E^I) = softmax(\frac{Q_i^w (K^I)^T}{\sqrt{d}})V^I \tag{3}$$

$$Q_i^w, K^I, V^I = e_i^w \cdot W^{Qw}, E^I \cdot W^{KI}, E^I \cdot W^{VI} \tag{4}$$

Here, $W^{Qw} \in \mathbb{R}^{d \times d}$, $W^{KI} \in \mathbb{R}^{d \times d}$ and $W^{VI} \in \mathbb{R}^{d \times d}$ are learnable parameters. The obtained image information will inevitably contain noise. We use GRU to filter the noise and fuse multi-modal information. For token w_i, we take its token vector e_i^w and its image information v_i^{img} as the initial hidden state and input of the GRU cell, respectively. Its multimodal representation e_i^{wI} is the updated hidden state. The formula is as follows:

$$e_i^{wI} = GRUCell(v_i^{img}, e_i^w) \tag{5}$$

All token-level representations constitute the fusion result of multi-modal information, i.e., $E^{wI} = \{e_1^{wI}, e_2^{wI}, ..., e_n^{wI}\}$.

Decoder. To extract aspect-specific information from multimodal fusion representations, we design a decoder with recurrent attention, which considers the recurrent learning process of different attention features. Specifically, we take the average of all word vectors in the encoded aspect E^a as the initial aspect representation h_0. At the t-th time step, the multimodal fusion representation E^{wI} is queried with the aspect representation h_{t-1} produced at the previous time step to obtain the multimodal information v_t^{mul}. The aspect representation is continuously updated according to the acquired multimodal information through the GRU[2].

[2] We create an instance of GRU cell for each time step.

$$v_t^{mul} = Attention(h_{t-1}, E^{wI}), t \geq 1 \tag{6}$$

$$Attention(h_{t-1}, E^{wI}) = softmax(\frac{Q_{t-1}^h (K^{wI})^T}{\sqrt{d}}) V^{wI}, t \geq 1 \tag{7}$$

$$Q_{t-1}^h = h_{t-1} \cdot W^{Qh}, t \geq 1 \tag{8}$$

$$K^{wI}, V^{wI} = E^{wI} \cdot W^{KwI}, E^{wI} \cdot W^{VwI} \tag{9}$$

$$h_t = GRUCell_t(v_t^{mul}, h_{t-1}), t \geq 1 \tag{10}$$

Here, $h_0 = \frac{1}{m}\sum_i e_i^a$, $W^{Qh} \in \mathbb{R}^{d \times d}$, $W^{KwI} \in \mathbb{R}^{d \times d}$ and $W^{VwI} \in \mathbb{R}^{d \times d}$ are learnable parameters. After updating and refining for N time steps, the multimodal aspect representation produced at the last time step is fed into a softmax layer to predict the probability distribution of its sentiment polarity.

$$p = softmax(W \cdot h_N + b) \tag{11}$$

Here, W and b are learnable parameters.

3.4 Loss Function and Optimizer

We take the cross-entropy between predicted results and targets as the loss function, which is calculated as follows:

$$loss = -\frac{1}{M} \sum_{j=1}^{M} \sum_{c \in C} q_j(c) log(p_j(c)) \tag{12}$$

Here, M denotes the total number of samples, $q_j(c)$ and $p_j(c)$ denote the true and predicted probability that sample j belongs to class c, respectively. We use the Adam [8] algorithm to minimize loss.

4 Experiments

4.1 Datasets and Settings

Datasets. We use the Twitter-15 and Twitter-17 datasets [19] to evaluate the effectiveness of the proposed method. The Twitter-15 and Twitter-17 are multimodal datasets where each sample contains a twitter text, a text-related image, a specific aspect, and its corresponding sentiment polarity label. There are three sentiment polarity labels: *positive*, *neutral*, and *negative*. The basic statistics of the Twitter-15 and Twitter-17 datasets are shown in Table 1.

Table 1. The basic statistics of datasets.

	Twitter-15			Twitter-17		
	Train	Dev	Test	Train	Dev	Test
Negative	368	149	113	416	144	168
Neutral	1883	670	607	1638	517	573
Positive	928	303	317	1508	515	493
Total	3179	1122	1037	3562	1176	1234
Avg. length	16.72	16.74	17.05	16.21	16.37	16.38

Experimental Settings. We select the pre-trained BERT-base[3] model as the text encoder. Its number of layers is 12, the size of the hidden layer state is 768, and the number of attention headers is 12. For the Bottom-Up Attention model, we use the pre-trained dynamic 10–100 model[4], setting the minimum and maximum number of features to 3 and 36, respectively. The BUA model is only used to extract image features and does not participate in training. The maximum value of time steps in the decoder is set to 3, *i.e.*, $N = 3$. Besides, the batch size is set to 32 and the learning rate is 2e-5. For each experiment, we run it 5 times with different random seeds and take the average as the final result. All experiments are performed on an NVIDIA GeForce RTX 3090 GPU. Following Yu et al. [19], we use accuracy and Macro-F1 as evaluation metrics.

4.2 Baselines

To verify the effectiveness of our model, we compare our model with several baseline methods:

(1) **RES-Target** uses ResNet [6] to encode images and BERT to encode aspect words, and then concatenates the encoding results of images and aspect words as features for sentiment polarity classification.
(2) **MGAN** [5] combines coarse- and fine-grained attention mechanisms to capture the interaction between aspect and context.
(3) **BERT** [3] uses a multi-layer transformer encoder [15] to obtain dynamic word vector representations.
(4) **BERT+BL** adds an additional BERT layer on top of the BERT-base model.
(5) **MIMN** [18] uses the attention mechanism to extract information in each modality, and designs a multi-hop memory network to make the information fully interact.
(6) **Res-MGAN** concatenates the image encoding results from ResNet and the text encoding results from MGAN.
(7) **Res-MGAN-TFN** uses tensor fusion network (TFN) [20] to fuse the image encoding results from ResNet and the text encoding results from MGAN.
(8) **Res-BERT+BL** concatenates the image encoding results from ResNet and the text encoding results from BERT+BL.

[3] https://huggingface.co/google/bert_uncased_L-12_H-768_A-12.

[4] https://github.com/MILVLG/bottom-up-attention.pytorch.

Table 2. Results of different models on the Twitter datasets. The best results are in **bold** and the second best results are *underlined.*

Modality	Model	Twitter-15		Twitter-17	
		Acc	Mac-F1	Acc	Mac-F1
Visual	RES-Target [6]	59.88	46.48	58.59	53.98
Text	MGAN [5]	71.17	64.21	64.75	64.16
	BERT [3]	74.15	68.86	68.15	65.23
	BERT+BL	74.25	<u>70.04</u>	68.88	66.12
Text+Visual	MIMN [18]	73.53	66.49	67.22	63.85
	Res-MGAN [5]	71.65	63.88	66.37	63.04
	Res-MGAN-TFN [20]	70.30	64.14	64.10	59.13
	Res-BERT+BL	<u>75.02</u>	69.21	<u>69.20</u>	<u>66.48</u>
	IFNRA(ours)	**76.03**	**71.79**	**70.78**	**69.48**

4.3 Main Results

Table 2 presents performance comparison among different models. As can be seen, our model outperforms all baselines, which shows the effectiveness of the proposed model. Besides, we can find that the multimodal model achieves better accuracy than the single-modal model, which indicates the complementarity between the different modal data. Furthermore, the text-only models perform better than the image-only model, suggesting that MABSA has a greater dependence on text.

Table 3. Results of ablation experiments. The best results are in **bold** and the second best results are *underlined.*

Model	Twitter-15		Twitter-17	
	Acc	Mac-F1	Acc	Mac-F1
IFNRA	**76.03**	**71.79**	**70.78**	**69.48**
IFNRA *w/ txt2img_attn*	<u>75.60</u>	70.63	69.9	68.41
IFNRA *w/ fuse*	74.95	<u>70.65</u>	69.81	68.46
IFNRA *w/ classify*	74.97	70.61	<u>70.10</u>	<u>68.57</u>

4.4 Ablation Study

To verify the effectiveness of different modules, we conduct ablation experiments by constructing several variants: (1) **IFNRA** *w/ txt2img_attn* does not use GRU to fuse text and image information. It directly inputs the result $R = \{v_1^{img}, v_2^{img}, ..., v_n^{img}\}$ of Text-to-Image Attention to the decoder. (2) **IFNRA** *w/ fuse* directly adds the word vector e_i^w and the image information v_i^{img} to obtain the multimodal representation e_i^{wI} instead of going through the GRU.

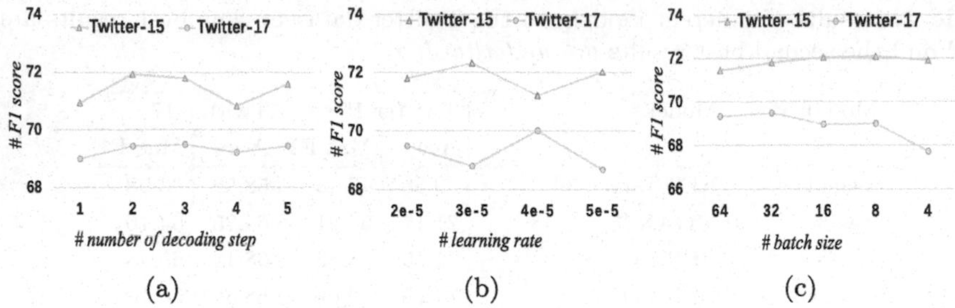

Fig. 3. Hyper-parameters sensitivity analysis.

(3) **IFNRA** *w/ classify* directly uses the multimodal information v_1^{img} to judge the sentiment polarity without using recurrent attention.

The results of ablation experiment are shown in Table 3. We find that the performance of the complete model is better than that of all variants, which shows that each module in IFNRA plays an important role. Specifically, comparing IFNRA *w/ txt2img_attn* with IFNRA can reflect the importance of interactive fusion of multimodal information. In addition, comparing IFNRA *w/ fuse* with IFNRA highlights the role of GRU in filtering noise and fusing multimodal data. Moreover, the effectiveness of the recurrent attention learning in decoder can be shown by comparing IFNRA *w/ classify* with IFNRA.

4.5 Discussion

Effect of Different Decoding Steps on Performance. To explore the effect of the number of decoding steps N on performance, we change N from 1 to 5. The experimental results are shown in Fig. 3a. We can find that the performance of the model gradually improves as the increase of the number of steps, which shows that increasing the number of steps in the recurrent decoding can obtain more effective attention features. Furthermore, when the number of steps continues to increase, the model performance peaks and then declines, indicating that increasing the number of steps too much can lead to too many parameters and over-fitting.

Effect of Different Learning Rates on Performance. To explore the effect of the learning rate on performance, we change the learning rate from 2e-5 to 5e-5. The experimental results are shown in Fig. 3b. As the learning rate changes, the model performance fluctuates only slightly. It means that a small change in the learning rate will not seriously affect the performance.

Effect of Different Batch Sizes on Performance. To explore the effect of the batch size on performance, we change the batch size from 4 to 64. The experimental results are shown in Fig. 3c. It can be found that when the batch

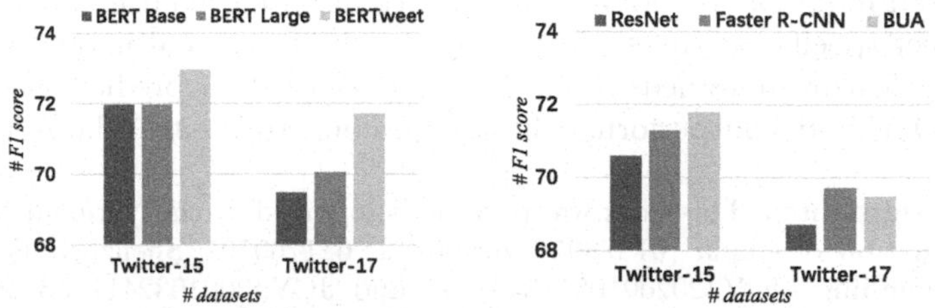

(a) Pre-trained language models. (b) Pre-trained vision models.

Fig. 4. The effect of pre-trained models on F1.

size is moderate (i.e. 32, 16 and 8), the overall performance is better. It indicates that a larger or smaller batch size could affect the generalizability of the model and the convergence of the parameters.

Effect of Different Pretrained Models on Performance. To explore the impact of different pre-trained models on performance, we replace different text and image encoders and conduct experiments. The results are shown in Fig. 4a and Fig. 4b. We can find that compared with BERT-base, although BERT Large[5] has a huge amount of parameters, the improvement brought by it is limited. We also note that BERTweet[6] [11] improves performance significantly, illustrating the importance of incorporating domain expertise into pre-trained language models. Furthermore, using Faster R-CNN[7] [13] as the image encoder can achieve better performance compared to using ResNet[8]. This shows that evenly dividing the image and extracting features will cause more loss of semantic information, and processing the objects or salient regions in the image as a whole can achieve better results. While BUA detects an object, it also predicts related properties (*e.g.*, color), which makes the image features contain more information. Therefore, BUA can achieve better overall performance than Faster R-CNN.

5 Conclusion

In this paper, we propose an interactive fusion network with recurrent attention to solve the MABSA. Specifically, we use BERT and BUA model to obtain features for the text and image, respectively. Furthermore, we design an interactive fusion module to obtain the semantic information of the image at the token level. We use GRU to filter out noise in image information and fuse data from different

[5] https://huggingface.co/bert-large-uncased.
[6] https://huggingface.co/vinai/bertweet-base.
[7] https://github.com/facebookresearch/detectron2/blob/main/configs/COCO-Detection/faster_rcnn_R_101_FPN_3x.yaml.
[8] https://download.pytorch.org/models/resnet50-0676ba61.pth.

modalities. Finally, we design a recurrent attention mechanism to progressively learn aspect-specific sentiment features from the fused multimodal representations. The feature obtained in the last step is used to predict the sentiment polarity. Our model outperforms all baselines on two Twitter datasets.

Acknowledgments.. This work was partially supported by the National Natural Science Foundation of China (61876053, 62006062, 62176076), Shenzhen Foundational Research Funding (JCYJ20200109113441941 and JCYJ2021032411 5614039), Joint Lab of HIT and KONKA.

References

1. Anderson, P., et al.: Bottom-up and top-down attention for image captioning and visual question answering. In: Proceedings of CVPR, pp. 6077–6086 (2018). https://openaccess.thecvf.com/content_cvpr_2018/papers/Anderson_Bottom-Up_and_Top-Down_CVPR_2018_paper.pdf
2. Cho, K., van Merriënboer, D., Bougares, F., Schwenk, H., Bengio, Y.: Learning phrase representations using RNN encoder-decoder for statistical machine translation. In: Proceedings of EMNLP, pp. 1724–1734 (2014). https://aclanthology.org/D14-1179.pdf
3. Devlin, J., Chang, M.W., Lee, K., Toutanova, K.: BERT: pre-training of deep bidirectional transformers for language understanding. In: Proceedings of NAACL, pp. 4171–4186 (2019). https://aclanthology.org/N19-1423/
4. Dong, L., Wei, F., Tan, C., Tang, D., Zhou, M., Xu, K.: Adaptive recursive neural network for target-dependent twitter sentiment classification. In: Proceedings of ACL, pp. 49–54 (2014). https://aclanthology.org/P14-2009.pdf
5. Fan, F., Feng, Y., Zhao, D.: Multi-grained attention network for aspect-level sentiment classification. In: Proceedings of EMNLP, pp. 3433–3442 (2018). https://aclanthology.org/D18-1380/?ref=githubhelp.com
6. He, K., Zhang, X., Ren, S., Sun, J.: Deep residual learning for image recognition. In: Proceedings of CVPR, pp. 770–778 (2016). https://openaccess.thecvf.com/content_cvpr_2016/papers/He_Deep_Residual_Learning_CVPR_2016_paper.pdf
7. Jiang, L., Yu, M., Zhou, M., Liu, X., Zhao, T.: Target-dependent twitter sentiment classification. In: Proceedings of ACL, pp. 151–160 (2011). https://aclanthology.org/P11-1016.pdf
8. Kingma, D.P., Ba, J.: Adam: a method for stochastic optimization. In: Proceedings of ICLR (Poster) (2015). https://openreview.net/forum?id=8gmWwjFyLj
9. Kiritchenko, S., Zhu, X., Cherry, C., Mohammad, S.: NRC-Canada-2014: detecting aspects and sentiment in customer reviews. In: Proceedings of SemEval, pp. 437–442 (2014). https://aclanthology.org/S14-2076.pdf
10. Mohammad, S., Kiritchenko, S., Zhu, X.: NRC-Canada: building the state-of-the-art in sentiment analysis of tweets. In: Proceedings of SemEval, pp. 321–327 (2013). https://aclanthology.org/S13-2053.pdf
11. Nguyen, D.Q., Vu, T., Nguyen, A.T.: BERTweet: a pre-trained language model for English Tweets. In: Proceedings of EMNLP, pp. 9–14 (2020). https://aclanthology.org/2020.emnlp-demos.2.pdf
12. Phan, M.H., Ogunbona, P.O.: Modelling context and syntactical features for aspect-based sentiment analysis. In: Proceedings of ACL, pp. 3211–3220 (2020). https://aclanthology.org/2020.acl-main.293/?ref=githubhelp.com

13. Ren, S., He, K., Girshick, R., Sun, J.: Faster R-CNN: Towards real-time object detection with region proposal networks. In: NIPS 28 (2015)
14. Tang, D., Qin, B., Feng, X., Liu, T.: Effective LSTMs for target-dependent sentiment classification. In: Proceedings of COLING, pp. 3298–3307 (2016). https://aclanthology.org/C16-1311/?ref=githubhelp.com
15. Vaswani, A., et al.: Attention is all you need. In: NIPS 30 (2017). https://proceedings.neurips.cc/paper/2017/file/3f5ee243547dee91fbd053c1c4a845aa-Paper.pdf
16. Wagner, J., et al.: DCU: aspect-based polarity classification for Semeval task 4. In: Proceedings of COLING, pp. 223–229 (2014). https://aclanthology.org/S14-2.pdf#page=243
17. Wang, J., et al.: Aspect sentiment classification with both word-level and clause-level attention networks. In: Proceedings of IJCAI, vol. 2018, pp. 4439–4445 (2018). www.ijcai.org/proceedings/2018/0617.pdf
18. Xu, N., Mao, W., Chen, G.: Multi-interactive memory network for aspect based multimodal sentiment analysis. In: Proceedings of AAAI, vol. 33, pp. 371–378 (2019). https://ojs.aaai.org/index.php/AAAI/article/view/3807/3685
19. Yu, J., Jiang, J.: Adapting BERT for target-oriented multimodal sentiment classification. In: Proceedings of IJCAI (2015). www.ijcai.org/Proceedings/2019/0751.pdf
20. Zadeh, A., Chen, M., Poria, S., Cambria, E., Morency, L.P.: Tensor fusion network for multimodal sentiment analysis. In: Proceedings of EMNLP, pp. 1103–1114 (2017)
21. Zhang, Z., Wang, Z., Li, X., Liu, N., Guo, B., Yu, Z.: ModalNet: an aspect-level sentiment classification model by exploring multimodal data with fusion discriminant attentional network. World Wide Web **24**(6), 1957–1974 (2021). https://link.springer.com/article/10.1007/s11280-021-00955-7
22. Zhou, J., Zhao, J., Huang, J.X., Hu, Q.V., He, L.: MASAD: a large-scale dataset for multimodal aspect-based sentiment analysis. Neurocomputing **455**, 47–58 (2021). www.sciencedirect.com/science/article/pii/S0925231221007931

Developing Relationships: A Heterogeneous Graph Network with Learnable Edge Representation for Emotion Identification in Conversations

Zhenyu Li[1,3], Geng Tu[1,3], Xingwei Liang[2,3(✉)], and Ruifeng Xu[1,3(✉)]

[1] School of Computer Science and Technology, Harbin Institute of Technology (Shenzhen), Shenzhen, China
190110709@stu.hit.edu.cn

[2] Konka Research Institute, Shenzhen, China
liangxingwei@konka.com

[3] Joint Lab of HIT-KONKA, Shenzhen, China
xuruifeng@hit.edu.cn

Abstract. Emotion recognition in conversations (ERC) aims to predict the emotion of utterances. Modeling context dependencies is the critical challenge of the task. Existing efforts in ERC are mainly based on the sequence and graph models. The graph models can better capture structured information than the sequence models. Unfortunately, there are few suitable aggregation strategies for ERC models based on high-dimensional edge features. Moreover, the adjustment of edge representation in graph-based models has been ignored for a long time. Based on this, we propose a learnable edge message-passing model based on a heterogeneous dialog graph. The model first calculates the attention weights between utterance nodes and between nodes and edges separately and then learns contextual utterance representations through these learnable edge representations. Additionally, we conducted our experiment on four public datasets and achieved advanced results.

Keywords: Conversational emotion identification · Graph transformer · Learnable edge

1 Introduction

Conversation is the most common and effective way to convey information and emotion between people. With the rapid development of the Internet, social media platforms such as YouTube and Twitter have generated massive comment and conversation data, attracting more and more researchers to participate in data mining and emotion recognition [5,9]. Emotion recognition in conversations (ERC) can not only be helpful in chatbots to achieve emotional responses driven by user emotions [1,16] but also contributes to recommendation systems [17].

L. Fang et al. (Eds.): CICAI 2022, LNAI 13606, pp. 310–322, 2022.
https://doi.org/10.1007/978-3-031-20503-3_25

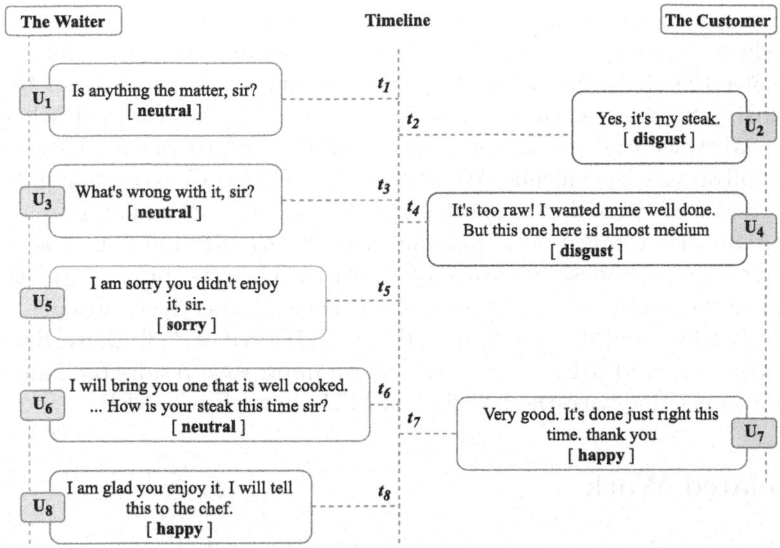

Fig. 1. An abridged dialogue between the waiter and the customer in a restaurant from DailyDialog [11]. There are two speakers: the waiter and the customer. Each utterance at time t_i is denoted by U_i. Emotion labels are marked in the form of *"[happy]."*

Compared with vanilla sentiment analysis, modeling the context in conversations is the key challenge in ERC. Fig. 1 illustrates an example of emotion recognition in a dialogue. When the emotion analysis is performed out of context, the short utterance *"Yes, it's my steak"* often obtains neutral results. Only by combining U_2 and U_4 can we know the emotional tendency of this utterance.

Existing work in ERC can be divided into two categories by context modeling: sequence-based and graph-based models. *Sequence models* [6,13,14] often use LSTMs or GRUs to extract dialog-level and speaker-level information. *Graph models* [5,7,19] pay attention to the correlation between utterances, and can model the complex contextual structure between dialog-level and speaker-level by making various associations between utterances.

Utterances information aggregation algorithm is as important as a sufficiently reasonable dialog graph modeling. The graph models above often use attention weights as edge features, and use methods such as Graph Attention Network (GAT) [22] or modified versions to aggregate information between utterances nodes weighted. However, models like SKAIG [9] use more complex common sense edge features, which is difficult for simple GAT to use complex edge features to generate effective attention weights and aggregate context information. Although the Unified Graph Transformer (UGT) [20] can allow edge features to participate in attention calculation, it still cannot allow edge features to be dynamically learned, which makes context modeling solidified. CensNet [8], NENN [24], and EGAT [23] propose to regard edges as nodes and update edges with neighbor edges and neighbor nodes. These methods need to construct edge-to-edge subgraphs based on the original graph, which mixes the features of different edges since neighboring edges cannot be guaranteed to be similar.

To solve the above problems, we propose a new graph transformer based on the Transformer [21], Learnable Edge Message Passing Network (LEMPN), and build an abstract dialogue graph framework adapted to the model for subsequent applications. Specifically, We first use RoBERTa [12] to encode utterances in conversations. Second, we define a simple and abstract dialog graph model, which treats all utterances as nodes, and regards any edge relationships between sentences as high-dimensional learnable features. Then, we propose an attention mechanism for sentence nodes and adjacent edges, respectively. Finally, we test our model on four high-frequency used datasets: IEMOCAP [3], DailyDialog [11], EmoryNLP [25], and MELD [15]. We achieve impressive results on DailyDialog and are competitive with the baseline on MELD and EmoryNLP.

2 Related Work

Emotion recognition in conversation is still a research hotspot, arousing broad interest among NLP researchers. There are many works in context modeling and can be divided into sequence-based and graph-based models.

2.1 Sequence-based Models

Sequence models [6,13,14] often use LSTM or GRU to extract sequential context information. ICON [6] extracts the speaker-level context representations of each speaker from the dialog, then use these representations to build dialog-level context, and stores the dialog-level features into the multi-hop memory. DialogueRNN [13] uses GRU to model the dialog-level context first and then uses attention to model the speaker-level context of the listener and speaker. COSMIC [4] makes improvements based on DialogueRNN and introduces external common sense knowledge to improve model performance. From the message passing view, sequence methods are limited to updating utterances stated only with the adjacent utterances. It is difficult for them to extract complex associations between utterances, so many graph-based models have been proposed.

2.2 Graph-based Models

Graph models [5,7,19] pay attention to the correlation between utterances and can model the complex contextual relationship between dialog-level and speaker-level at a deeper level by making various associations between utterances and utterances. For example, the DialogueGCN [5] builds a directed graph based on dialogue, uses edges representing relationships to connect related utterances, and allows the edges to carry the attention scores between utterances and weighted aggregate. RGAT [7] adds position encoding on the basis of DialogueGCN. For DAG-ERC [19], a directed acyclic graph is established in terms of timing and information transmission. SKAIG [9] establishes utterances associations based on psychological common sense implicit in dialogues, and uses a Unified Graph Transformer (UGT) [20] to aggregate information. From the message passing view, these models build a directed graph of conversations, use edges to represent the direction of message delivery, additional relation or type information, and aggregates based on computable attention edge weights using methods like GAT [22] or modified version. The edge features of SKAIG [9] have semantics and are more informative than the simple edge weights of other models. Its edge features are common sense from COMET [2], which implies psychological state change momentum.

Simple weighted aggregation cannot fully use edge features, so we propose a Learnable Edge Message Passing Network (LEMPN) based on a graph transformer, which innovatively calculates the attention weights from nodes to edges and makes edge features adaptive through learnable edge representations. As far as we know, our work is the first application of an edge learnable graph transformer in emotion identification in conversations.

3 Methodology

3.1 Problem Definition

A conversation is defined as a sequence of utterances $\{u_1, u_2, ..., u_N\}$, where N is the number of utterances in conversation. Each utterance u_i contains L_i words as $\{w_1, w_2, ..., w_{L_i}\}$. Additionally, a conversation have P people involved, and the utterance u_i expressed by its corresponding speaker $S_j \in \{S_1, S_2, ..., S_P\}$. The objective of emotion identification in conversation is to classify all utterances in a conversation to their correct emotion label $\{E_1, E_2, ..., E_M\}$, where M is the number of the labels.

3.2 Utterance Independent Encoder

RoBERTa [12] is widely used in ERC for extracting utterances features. We feed utterance $u_i = \{w_{i,1}, w_{i,2}, ..., w_{i,L_i}\}$ into RoBERTa, and obtain the hidden states of the last layer.

$$F_i = RoBERTa(w_{i,1}, ..., w_{i,L_i}) \tag{1}$$

where $F_i \in \mathbb{R}^{L_i \times d_w}$. d_w is the dimension of the hidden states of each word.

High-dimensional features will introduce more parameters, which is not conducive to feature expression. Therefore we introduce a max-pooling and linear projection operation to reduce the dimension following Li et al. [9]:

$$f_i = Linear(MaxPooling(F_i)) \tag{2}$$

where $f_i \in \mathbb{R}^{d_h}$ is the representation of the utterance, and d_h is the dimension of the representation. After all utterances are encoded, we obtain the context-independent conversation representation of conversation $C \in \mathbb{R}^{N \times d_h}$.

3.3 Dialogue Graph Modeling

In this section, we define a dialog abstract directed graph framework with edge features for introducing our message passing model below. Generally, a dialogue graph can be denoted as $\mathcal{G} = (\mathcal{V}, \mathcal{E})$. $v_i \in \mathcal{V}$ is an utterance node. $e_{ij} \in \mathcal{E}$ is an directed edge from node v_i to v_j.

Vertices. The whole conversation has been encoded as C and we set the initial node feature of v_i as $h_i = f_i \in C$. The node will aggregate the contextual information from its adjacent nodes and edges, then update its features.

Edges. The meaning of a directed edge e_{ij} is to express the temporal, semantics, speaker, and other contextual relations between node v_i and v_j, and to indicate the direction of information transmission. The rules of edge selection between nodes depend on the specific implementation. For acyclic graphs or cyclic graphs, fully connected graphs, or graphs with limited window size, message passing can always be used to aggregate node information.

Edges can express different types of contextual relationships through their features through relative position encoding, relation embedding, etc. But recent works often ignore edges themselves can carry meaningful information at higher dimensions. Therefore, we propose the following model to fill in the field gap.

3.4 Learnable Edge Message Passing Network

Learnable Edge Message Passing Network, namely LEMPN, uses separate attention to node and edge to make better use of edge features and enable self-learning of edge features based on the former abstract graph.

Graph Transformer. Transformer [21] has proven to be very effective in NLP, and we extend its multi-head attention mechanism to node-to-node and node-to-edge in pair drawing. Specially, in layer l, given node features of an conversation $H = \{h_1, h_2, ...h_N\}$, we construct multi-head attention function from v_i to v_j as following:

$$f_{q,t}(h_i) = W_{q,t}h_i + b_{q,t} \tag{3}$$

$$f_{k,t}(h_j) = W_{k,t}h_j + b_{k,t} \tag{4}$$

$$\alpha_{ij,t} = Softmax\left(\frac{\langle f_{q,t}(h_i),\ f_{k,t}(h_j)\rangle}{\sum_{n\in\mathcal{N}(i)}\langle f_{q,t}(h_i),\ f_{k,t}(h_n)\rangle}\right) \tag{5}$$

where $f_{q,t}(h_i) \in \mathbb{R}^{d_h}$ and $f_{k,t}(h_j) \in \mathbb{R}^{d_h}$, and d_h is the dimension of node feature. $W_{q,t}, W_{k,t}, b_{q,t}, b_{k,t}$ is trainable weights and bias, and $\langle a, b\rangle = \frac{a\cdot b^\top}{\sqrt{d_t}}$, where d_t is the dimension of each head. For the t-th head attention, we use $f_{q,t}(h_i)$ and $f_{k,t}(h_j)$ to transform v_i feature h_i and v_j feature h_j into query vector and key vector. In this way, the information weight of each adjoining node during message transmission is determined. Compared with directly processing edge features with inter-node attention weight representation or letting edge features directly participate in attention calculation, we consider that high-dimensional edge features may have semantic meanings and should be treated separately. Similar to the above, we compute the node-to-edge attention score in the same way as follows:

$$g_{k,t}(e_{ji}) = M_{k,t}e_{ji} + p_{k,t} \tag{6}$$

where $g_{k,t}(e_{ji}) \in \mathbb{R}^{d_e}$, and d_e is the dimension of edge features. $M_{k,t}, p_{k,t}$ are also trainable weight and bias. e_{ji} is a directed edge from v_j to v_i. For the t-th head attention, the key vector of edge e_{ij} is $g_{k,t}(e_{ji})$. Then, we calculate the attention weight of the v_i to all the e_{ji} pointing to it as $\beta_{ij,t}$ by using the edge feature, which supplements the information weight of each adjacent edge when the message is passed.

$$\beta_{ij,t} = Softmax\left(\frac{\langle f_{q,t}(h_i),\ g_{k,t}(e_{ji})\rangle}{\sum_{n\in\mathcal{N}(i)}\langle f_{q,t}(h_i),\ g_{k,t}(e_{ni})\rangle}\right) \tag{7}$$

Message Passing. The so-called *message passing* means that after calculating the attention weight, the information of adjacent nodes and adjacent edges is aggregated and the node feature is updated. We first compute the value matrix for nodes and edges. Then we aggregate the contextual information of adjacent nodes and edges separately, and simply add them together to construct message msg_i to v_i as follows:

$$f_{v,t}(h_j) = W_{v,t}h_j + b_{v,t} \tag{8}$$

$$g_{v,t}(e_{ji}) = M_{v,t}e_{ji} + p_{v,t} \tag{9}$$

$$msg_i = \sum_{n\in\mathcal{N}(i)} \mathrm{Concat_t}\left(\alpha_{in,t}f_{v,t}(h_n)\right) + \sum_{n\in\mathcal{N}(i)} \mathrm{Concat_t}\left(\beta_{in,t}g_{v,t}(e_{ni})\right) \tag{10}$$

where $f_{v,t}(h_j) \in \mathbb{R}^{d_h}$ and $g_{v,t}(e_{ji}) \in \mathbb{R}^{d_e}$. $\mathrm{Concat_t}(\cdot)$ concats all heads. When $d_h \neq d_e$, a projection for edge features is additionally needed.

Node Update. In this part, we use a gated residual connection between layers inspired by UGT [20] to prevent our model from over-smoothing. In layer l, we calculate the v_i new feature as follows:

$$\gamma_i = Sigmoid(W_1[msg_i, o_i, msg_i - o_i]) \tag{11}$$

$$\hat{h}_i = (1-\gamma_i)msg_i + \gamma_i o_i \tag{12}$$

where o_i is an trainable linear projection of h_i.

Edge Update. From the perspective of an edge, the context of the edge e_{ji} comes from v_i and v_j. So the edge features need to be adaptive according to the changes of the nodes at both ends.

$$f_{qk}(h_i, h_j) = \frac{f_{q,t}(h_i) \circ f_{k,t}(h_j)}{\sqrt{d_t}} \tag{13}$$

$$g_{qk}(h_i, e_{ji}) = \frac{f_{q,t}(h_i) \circ g_{k,t}(e_{ji})}{\sqrt{d_t}} \tag{14}$$

$$\hat{e}_{ji} = e_{ji} + W_e[g_{v,t}(e_{ji})f_{qk}(h_i, h_j);\ g_{v,t}(e_{ji})g_{qk}(h_i, e_{ji})] + b_e \tag{15}$$

where $f_{qk}(h_i, h_j) \in \mathbb{R}^{d_h}$, and $g_{qk}(h_i, e_{ji}) \in \mathbb{R}^{d_e}$. $a \circ b$ is Hadamard product, namely element-wise multiply. W_e, b_e are trainable. This formula weights the point-to-point and point-to-edge aspects of edge features separately through the query and key results of adjacent points and edges, and finally aggregates them linearly.

Layer Passing. We learn point-wise feed forward network (FFN) from Transformer [21] to update node features and edge features respectively to enhance the nonlinear ability of the model. The node feature h_i and e_{ji} on layer l will be updated to h_i^+ and e_{ji}^+ as follows:

$$h_i^+ = LayerNorm(\hat{h}_i + Linear_1(ReLU(Linear_2(\hat{h}_i)))) \tag{16}$$

$$e_{ji}^+ = LayerNorm(\hat{e}_{ji} + Linear_3(ReLU(Linear_4(\hat{e}_{ji})))) \tag{17}$$

3.5 Emotion Classifier

After sufficient feature extraction by the upstream model, we use the following linear units for classification.

$$Z = Softmax(W_z H + b_z) \tag{18}$$

where H is the conversation feature matrix of the graph last layer, and $W_z \in \mathbb{R}^{M \times d_h}$, $b_z \in \mathbb{R}^M$ are trainable. The cross-entropy loss is utilized to train the model base on conversation as follows:

$$loss = -\frac{1}{N}\sum_{i=1}^{N}\sum_{e=1}^{M} z_i log(Z_i) \tag{19}$$

where z_i is the one-hot vector of utterance true emotion label.

4 Experimental Settings

4.1 Datasets

We evaluate our LEMPN on four widely used conversational emotion identification datasets. The statistics of them are shown in Table 1.

IEMOCP. [3] is a dataset of two-person conversations, which was recorded from ten actors in dyadic sessions. Each utterance in it is annotated with an emotion label as *neutral, happy, sad, angry, excited, frustrated.* The dataset is not divided into training and validation sets, so we divide it the same as Li et al. [9].

DailyDialog. [11] is a two-way dialogue in various scenarios, including 7 emotions: *neutral, happiness, sadness, anger, surprise, disgust, fear.* 83% of the data is marked as neutral.

EmoryNLP. [25] is taken from the TV series *Friends* and contains multiple conversations. Each sentence is marked with one of the *neutral, mad, sad, scared, powerful, peaceful, joyful* emotional labels.

MELD. [15] is also collected from *Friends*, and its emotional label is consistent with DailyDialog.

Table 1. Statistics of IEMOCAP, DailyDialog, MELD, EmoryNLP

Dataset	#Conversations			#Utterances			#Metrics
	Train	Dev	Test	Train	Dev	Test	
IEMOCAP	120		31	5810		1623	Weighted-F1
DailyDialog	11118	1000	1000	87170	8069	7740	Micro-F1 & Macro F1
EmoryNLP	659	89	79	7551	954	984	Weighted-F1
MELD	1039	114	280	9989	1109	2610	Weighted-F1

4.2 Compared Methods

We compared our model with the following methods and baselines.

Sequence-based Models. ICON [13], DialogueRNN [13], Dialogue-RNN with RoBERTa which is also implemented by Li et al. [4], HiTrans [10], DialogXL [18] and COSMIC [4].

Graph-based Models. DialogueGCN [5], RGAT [7], DAG-ERC [19], and the SKAIG [9] with UGT [20] using the same train settings to control variables.

Baselines. We also compared our work with RoBERTa [12] followed by a simple classifier and RoBERTa-Transformer, which replaces the graph transformer with transformer. They are implemented by Li et al. [4].

5 Results and Discussions

5.1 Overall Results

Overall experiment results are illustrated in Table 2. We can notice that on DailyDialog, our model achieves out-performed results, higher than the second-placed SKAIG with UGT +1.62 (2.7%) and +0.39 (0.7%) on Micro-F1 and Macro-F1, respectively. As Table 1 shows, DailyDialog has 8 to 15 times as many sentences as other datasets. Our model separately computes attention for nodes and edges while additionally learning edge features with a linear layer, which increases the parameters of the model. In this way, our model can learn better from the dataset.

On MELD, our model still performs well. Weighted F1 is +0.15 (0.2%) higher than the second-placed COSMIC model. MELD has the same sentiment label as DailyDialog, but with fewer data. The best graph-based model (SKAIG + UGT) has lower results than the best sequence-based model (COSMIC), according to Table 2. This is because MELD comes from episodes, frequent multi-party conversations, and discontinuous sampling problems, making the graph model structure on it broken and performing poorly. Our model fixes discontinuities to some extent by adaptable edge weights. But still, small datasets make our model less effective. EmoryNLP also comes from the same series as MELD, but with different tags and fewer data. Our model's Weighted F1 is slightly higher than SKAIG with UGT but lower than DAG-ERC by about -0.1 (0.3%) because its overall data size is smaller than MELD.

Table 2. Results of our method and other baselines

Methods	IEMOCAP	DailyDialog		EmoryNLP	MELD
	Weighted-F1	Micro-F1	Macro-F1	Weighted-F1	Weighted-F1
RoBERTa	55.67	55.16	48.20	37.00	62.75
+ Transformer	63.78	58.28	47.00	37.50	64.59
ICON	63.50	–	–	–	–
DialogueRNN	62.57	55.95	41.80	31.70	57.03
+ RoBERTa	64.76	57.32	49.65	37.44	63.61
HiTrans	64.50	–	–	36.75	61.94
DialogXL	65.94	54.93	–	34.73	62.41
COSMIC	65.28	58.48	51.05	38.11	65.21
DialogueGCN	64.18	–	–	–	58.10
RGAT	65.22	54.31	–	34.42	60.91
DAG-ERC	**68.03**	59.33	–	**39.02**	63.65
SKAIG + UGT	66.96	59.75	51.95	38.88	65.18
SKAIG + LEMPN	66.14	**61.37**	**52.34**	38.91	**65.36**

Our model doesn't perform well on IEMOCAP, with -0.82 (1.2%) lower than the baseline model SKAIG with UGT and -1.89 (2.8%) lower than the best method DAG-ERC. Considering that the amount of data in IEMOCAP is the least due to a large number of parameters, our model has less generalization ability on small datasets. Moreover, the dialogue length is much larger than other dialogues, which makes IEMOCAP rich in contextual information. Our learnable edge features can supplement contextual information in small dialogues with less contextual information but may do the opposite in long dialogues. We conducted a detailed analysis in 5.3 and confirmed our conjecture.

5.2 Ablation Studies

In Table 3, we investigated the effectiveness of node-to-node with node-to-edge attention and learnable edge features.

First, we test the case "Edge Feature through FFN," which is equivalent to updating an edge feature using an FFN without node features involved. Interestingly, the trends on the two datasets are consistent, with a decrease of 2.53 (4.12%) and 1.4 (2.1%), respectively. This shows that FFN needs a good combination of edge feature learning and representation to obtain good results. From another point of view, enhancing the nonlinearity of edge features with FFN may not be the best choice.

Table 3. Ablation studies on best-performed dataset. Metrics for DailyDialog is Micro-F1, and Weighted-F1 For MELD.

	Node + Edge attention			Node Attention	UGT
	Learnable edge feature	Edge feature through FFN	Unlearnable Edge Feature	Unlearnable edge feature	
DailyDialog	**61.37**	58.84	59.98	59.67	59.75
MELD	**65.36**	63.96	64.96	64.63	65.18

Second, we test the case "Unlearnable Edge Feature," where the edge features are fixed and completely unlearnable. Comparing the results, we can find that the unlearnable edge feature reduces the evaluation indicators of the two datasets by 1.39 (2.3%) and 0.4 (0.6%), respectively. Both DailyDialog and MELD are based on short daily dialogues, and we believe that learnable edge features can fully capture the underlying contextual information in such cases.

Third, we do tests when only using node-to-node attention without node-to-edge attention. UGT calculates the attention weights as $Attention(h_i, h_i + e_{ji})$ and this attention mechanism is between node-edge separate attention and node attention only. Comparing the results, we find that the absence of node-to-edge attention leads to a 0.31 (0.5%) and 0.33 (0.5%) drop, respectively, on both datasets.

5.3 Error Analysis

In this part, we analyze from the bottom up which part reduces the model effect. According to Table 4, we can find out that our model results are 0.38 (0.6%) higher than the SKAIG with UGT baseline with a separate attention mechanism only. However, the addition of FFN caused a substantial drop in our model results by about 3.19 (4.7%), which is the highest reduction of all the four datasets. In addition, even though we introduce learnable edge features after removing FFN, the model performance still drops by about 0.83 (1.2%).

Table 4. Error analysis on the best and worst-performing datasets. Metrics are Micro-F1 for DailyDialog and Weighted-F1 for IEMOCAP. Data with * are trained and predicted without FFN.

	UGT	Node attention + Edge attention		
		Unlearnable edge feature	Edge feature through FFN	Learnable edge feature
DailyDialog	59.75	59.98	58.84	**61.37**
IEMOCAP	66.96	**67.34**	64.15	66.54*

Combining all the experimental results, we find that FFN always causes the results to drop no matter in the largest dataset DailyDialog or the smallest dataset, IEMOCAP. We speculate that it may not be appropriate to eliminate edge feature nonlinearity by FFN. Also, the average conversation length of 48 is much longer than the 12 of other datasets. We believe that the context information span of long dialogue is larger than that of short dialogue, and our edge feature update is limited to adjacent nodes, which is more suitable for capturing short-distance context information. Therefore, it can be necessary to adapt the edge learning approach to the dialogue form.

5.4 Conclusion

In this paper, we propose a Learnable Edge Message Passing Model based on a heterogeneous dialog graph, which calculates the node-to-node and node-to-edge attentions separately and updates edge features. The results show our attention mechanisms are generally effective. The learnable edge feature performs differently on different datasets and may need to be improved according to the types of datasets. Although there are few related works in the field, we will strive to find better edge learning methods and use external data to build better graph models in the future.

Acknowledgments. This work was partially supported by the National Nat- ural Science Foundation of China (61876053, 62006062, 62176076), Shenzhen Foundational Research Funding (JCYJ20200109113441941 and JCYJ2021032411 5614039), Joint Lab of HIT and KONKA.

References

1. Adikari, A., De Silva, D., Alahakoon, D., Yu, X.: A cognitive model for emotion awareness in industrial chatbots. In: Proceedings of INDIN, vol. 1, pp. 183–186 (2019)
2. Bosselut, A., Rashkin, H., Sap, M., Malaviya, C., Celikyilmaz, A., Choi, Y.: COMET: commonsense transformers for automatic knowledge graph construction. In: Proceedings of ACL, pp. 4762–4779 (2019)
3. Busso, C., et al.: IEMOCAP: interactive emotional dyadic motion capture database. Lang. Resour. Eval. **42**(4), 335–359 (2008)
4. Ghosal, D., Majumder, N., Gelbukh, A., Mihalcea, R., Poria, S.: Cosmic: commonsense knowledge for emotion identification in conversations. In: Findings of ACL: EMNLP 2020, pp. 2470–2481 (2020)
5. Ghosal, D., Majumder, N., Poria, S., Chhaya, N., Gelbukh, A.: DialogueGCN: a graph convolutional neural network for emotion recognition in conversation. In: Proceedings of EMNLP-IJCNLP, pp. 154–164 (2019)
6. Hazarika, D., Poria, S., Mihalcea, R., Cambria, E., Zimmermann, R.: Icon: interactive conversational memory network for multimodal emotion detection. In: Proceedings of EMNLP, pp. 2594–2604 (2018)
7. Ishiwatari, T., Yasuda, Y., Miyazaki, T., Goto, J.: Relation-aware graph attention networks with relational position encodings for emotion recognition in conversations. In: Proceedings of EMNLP, pp. 7360–7370 (2020)
8. Jiang, X., Ji, P., Li, S.: CensNet: convolution with edge-node switching in graph neural networks. In: Proceedings of IJCAI, pp. 2656–2662 (2019)
9. Li, J., Lin, Z., Fu, P., Wang, W.: Past, present, and future: conversational emotion recognition through structural modeling of psychological knowledge. In: Findings of ACL: EMNLP 2021, pp. 1204–1214 (2021)
10. Li, J., Ji, D., Li, F., Zhang, M., Liu, Y.: HiTrans: a transformer-based context-and speaker-sensitive model for emotion detection in conversations. In: Proceedings of COLING, pp. 4190–4200 (2020)
11. Li, Y., Su, H., Shen, X., Li, W., Cao, Z., Niu, S.: DailyDialog: a manually labelled multi-turn dialogue dataset. In: Proceedings of IJCNLP, pp. 986–995 (2017)
12. Liu, Y., et al.: Roberta: a robustly optimized BERT pretraining approach. arXiv preprint arXiv:1907.11692 (2019)
13. Majumder, N., Poria, S., Hazarika, D., Mihalcea, R., Gelbukh, A., Cambria, E.: DialogueRNN: an attentive RNN for emotion detection in conversations. In: Proceedings of AAAI, vol. 33, pp. 6818–6825 (2019)
14. Poria, S., Cambria, E., Hazarika, D., Majumder, N., Zadeh, A., Morency, L.P.: Context-dependent sentiment analysis in user-generated videos. In: Proceedings of ACL, pp. 873–883 (2017)
15. Poria, S., Hazarika, D., Majumder, N., Naik, G., Cambria, E., Mihalcea, R.: Meld: a multimodal multi-party dataset for emotion recognition in conversations. In: Proceedings of ACL, pp. 527–536 (2019)
16. Poria, S., Majumder, N., Mihalcea, R., Hovy, E.: Emotion recognition in conversation: research challenges, datasets, and recent advances. IEEE Access **7**, 100943–100953 (2019)
17. Shen, T., et al.: PEIA: personality and emotion integrated attentive model for music recommendation on social media platforms. In: Proceedings of AAAI, vol. 34, pp. 206–213 (2020)

18. Shen, W., Chen, J., Quan, X., Xie, Z.: DialogXL: All-in-one XLNet for multi-party conversation emotion recognition. In: Proceedings of AAAI, vol. 35, pp. 13789–13797 (2021)
19. Shen, W., Wu, S., Yang, Y., Quan, X.: Directed acyclic graph network for conversational emotion recognition. In: Proceedings of ACL-IJCNLP, pp. 1551–1560 (2021)
20. Shi, Y., Huang, Z., Wang, W., Zhong, H., Feng, S., Sun, Y.: Masked label prediction: unified massage passing model for semi-supervised classification. In: Proceedings of IJCAI (2021)
21. Vaswani, A., et al.: Attention is all you need. In: NIPS 30 (2017)
22. Velickovic, P., Cucurull, G., Casanova, A., Romero, A., Lio, P., Bengio, Y.: Graph attention networks. Stat **1050**, 20 (2017)
23. Wang, Z., Chen, J., Chen, H.: EGAT: edge-featured graph attention network. In: Proceedings of ICANN, pp. 253–264 (2021)
24. Yang, Y., Li, D.: NENN: Incorporate node and edge features in graph neural networks. In: Proceedings of ACML, pp. 593–608 (2020)
25. Zahiri, S.M., Choi, J.D.: Emotion detection on tv show transcripts with sequence-based convolutional neural networks. In: Workshops of AAAI (2018)

Optimization

Adaptive Differential Evolution Algorithm with Multiple Gaussian Learning Models

Genghui Li, Qingyan Li, and Zhenkun Wang(✉)

Southern University of Science and Technology, Shenzhen 518055, China
genghuili2-c@my.cityu.edu.hk, 12032674@mail.sustech.edu.cn,
wangzhenkun90@gmail.com

Abstract. The search efficiency of the differential evolution (DE) algorithm significantly depends on its mutation strategies and control parameters. The multiple Gaussian learning model-based parameter adaptation mechanism is useful in choosing suitable parameters. Naturally, it is worth studying whether this mechanism is also beneficial for the automatic selection of the appropriate mutation strategies. To this end, this paper proposes an adaptive Differential Evolution algorithm with Multiple Gaussian Learning Models, known as MGLMDE, which includes two adaptation mechanisms, i.e., the multiple Gaussian learning model-based parameter adaptation mechanism (MGLMP) and multiple Gaussian learning model-based strategy adaptation mechanism (MGLMS). MGLMP and MGLMS determine the future mutation strategies and control parameters, respectively, using multiple Gaussian models to learn from successful historical memories. The linear population size reduction (LPSR) mechanism is used to control the population size. The proposed algorithm is evaluated via comparisons with some powerful DE methods on lots of test problems. The experimental results demonstrate that the proposed method is better than or at least highly competitive with the state-of-the-art DE algorithms.

Keywords: Differential evolution · Multiple Gaussian learning models · Numerical optimization · Parameter adaptation · Strategy adaptation

1 Introduction

Evolutionary algorithm (EA) is one of effective heuristics for dealing with optimization problems. This paper focuses on differential evolution (DE) [23] since it is a highly effective and competitive EA for solving multifarious complex optimization problems. However, the performance of DE significantly relies upon the offspring generation strategies (i.e., mutation and crossover strategies) and the

This work was supported by the National Natural Science Foundation of China under Grant 62206120 and Grant 62106096.

L. Fang et al. (Eds.): CICAI 2022, LNAI 13606, pp. 325–336, 2022.
https://doi.org/10.1007/978-3-031-20503-3_26

associated control parameters (i.e., population size NP, scaling factor F, and crossover rate CR) [7,15]. Selection of suitable offspring generation strategies and control parameters is vital to the performance of DE, but it is a problem-dependent task [22]. Therefore, to find a satisfactory solution for a problem at hand, it is usually necessary to use the time-consuming and tedious trial-and-error method to determine the appropriate strategies and their parameter values. To address this troublesome issue, many effective techniques have been proposed. These approaches can be roughly divided into four categories (i.e., adaptation or self-adaptation technique [4,22,32], composition technique [28,30], ensemble technique [18,19], and multi-population technique [2,6,29]). Specifically, Tanabe and Fukunaga proposed a multiple Gaussian learning model-based parameter adaptation mechanism (MGLMP) for tuning the scaling factor F and crossover rate CR in SHADE [24,25]. They further used linear population size reduction (LPSR) to control the population size NP in L-SHADE [26]. SHADE and L-SHADE are highly successful in CEC2013 competition [13,17] and CEC2014 competition [14,16], respectively. However, both SHADE [25] and L-SHADE [26] only use one offspring generation strategy, which may restrict their performance to a certain extent. Therefore, some enhanced versions of SHADE and L-SHADE with multiple strategies have been proposed, such as SHADE4 [5], L-SHADE44 [21], LSHADE-SPACMA [20], and LSHAD-CLM [33]. In addition, [11] shows that some well-developed adaptation mechanisms could be used for both strategy selection and parameter adjustment, which inspires us to design a multiple Gaussian learning model-based strategy adaptation mechanism (MGLMS) for DE. Based on these considerations and with the goal of improving the performance of DE by online determination of the mutation strategy and control parameters, an adaptive differential evolution algorithm based on multiple Gaussian learning models is proposed (called MGLMDE). In MGLMDE, the mutation strategies and associated control parameters are generated by the multiple Gaussian learning models, the parameters of which are adaptively adjusted by learning from the historical memories. Moreover, the population size is also regulated by the linear population size reduction mechanism. To demonstrate the performance of the proposed method, MGLMDE is compared with some outstanding DE variants on lots of test problems.

2 Basic DE Algorithm

DE conducts a series of operations on a population $\mathbf{P}^G = \{X_1^G, \ldots, X_{NP}^G\}$ that consists of NP individuals. Each individual represents a possible solution to the optimization problem, and it is denoted by a vector containing D variables. In detail, $X_i^G = (x_{i,1}^G, \ldots, x_{i,D}^G)$ is the ith individual (or solution) of the population in the Gth generation. G denotes the generation count, NP is the population size, and it is generally fixed in the entire evolution process. Without loss of generality, in this paper, we use DE to address the minimum optimization problems with boundary constraints, and it can be described as follows,

$$\begin{array}{ll} \text{minimize} & f(X) \\ \text{subject to} & X \in \boldsymbol{\Omega}\text{'} \end{array} \tag{1}$$

where $f(\cdot)$ represents the optimization problem. $\boldsymbol{\Omega} = [x_1^l, \ldots, x_D^l] \times [x_1^u, \ldots, x_D^u]$ is the search space, x_j^l and x_j^u represent the lower and upper boundary of the jth variable, respectively, and D is the number of variables.

The detailed steps of DE are described as follows.

Step 1: Initialization: Set $G = 1$, and determine the population size NP, scaling factor F, crossover rate CR, and the maximum number of generations $G_{\max}$. Generate an initial random population as follows,

$$x_{i,j}^G = x_j^l + rand_{i,j}(0,1) \cdot (x_j^u - x_j^l), \tag{2}$$

where $i = 1, \ldots, NP$ and $j = 1, \ldots, D$; and $rand_{i,j}(0,1)$ is a random number in the range of $[0,1]$.

Step 2: For each individual X_i^G $(i = 1, \ldots, NP)$, conduct mutation, crossover, and selection operations to evolve the population to the next generation.

Step 2.1 Mutation: For each X_i^G in the population, the mutant vector (also known as the donor vector) V_i^G is generated as

$$V_i^G = X_{r1}^G + F \cdot (X_{r2}^G - X_{r3}^G), \tag{3}$$

where X_{r1}^G, X_{r2}^G, and X_{r3}^G are all randomly selected from the population, different from each other, and distinct to X_i^G. Additionally, $F \in [0,1]$ is the scaling factor, and it is used to control the amplification of the difference vector. Many other types of mutation operators also exist, as referred in [18,22,28].

Step 2.2 Crossover: After mutation, the trial vector U_i^G is generated by combining the components of target vector X_i^G and mutant vector V_i^G based on binary crossover operator, and it is expressed as follows,

$$u_{i,j}^G = \begin{cases} v_{i,j}^G & \text{if} \quad rand_{i,j}(0,1) \le CR \quad \text{or} \quad j == j_{rand} \\ x_{i,j}^G & \text{otherwise} \end{cases}, \tag{4}$$

where $rand_{i,j}(0,1)$ is a uniform random number in the range of $[0,1]$, j_{rand} is an integer randomly chosen from $[1, D]$, and $CR \in [0,1]$ is the crossover rate.

Step 2.2 Selection : To realize the natural rule of survival of the fittest, the target vector X_i^G competes with its trial vector U_i^G. The selection operation is expressed as follows,

$$X_i^{G+1} = \begin{cases} U_i^G & \text{if} \quad f(U_i^G) \le f(X_i^G) \\ X_{i,j}^G & \text{otherwise} \end{cases}. \tag{5}$$

Step 3: $G = G + 1$. Return to **Step 2** if the stop condition is not met.

3 Proposed Method: MGLMDE

JADE [32] proposed a simple but highly efficient adaptation mechanism for automatic adjustment of the control parameters (F and CR) of DE, and it was further improved by SHADE [25] and L-SHADE [26]. However, JADE, SHADE, and L-SHADE use a single mutation strategy during the entire evolution process, which might limit their performance. Therefore, combined with a highly efficient strategy selection mechanism offers a promising approach to further performance improvement. In addition, the prominent performance of SHADE and L-SHADE primarily relies on their multiple Gaussian learning model-based parameter adaptation mechanism (MGLMP), which motivates us to establish a multiple Gaussian learning model-based strategy adaptation mechanism (MGLMS).

3.1 Multiple Gaussian Learning Model-based Strategy Adaptation Mechanism (MGLMS)

Suppose that there are K mutation strategies in the strategy pool. For the ith individual X_i^G, its mutation strategy is obtained as [11],

$$S_i = \min(\lfloor \eta_i \cdot K + 1 \rfloor, K), \tag{6}$$

$$\eta_i = \text{randn}(M_{\eta,ri}, 0.1), \tag{7}$$

where $\lfloor x \rfloor$ is the maximum integer that is less than x, and $\eta \in [0,1]$ is a strategy control parameter generated by a Gaussian model with mean value $M_{\eta,ri}$ and standard deviation 0.1. In the case of $\eta_i > 1$, it is truncated to 1, and in the case of $\eta_i < 0$, it is truncated to 0. The expectation value $M_{\eta,i}$ of Gaussian distribution is randomly selected from a historical strategy memory with H entries, as shown in Fig. 1. In the beginning, $M_{\eta,i}$, $(i = 1, \ldots, H)$ is set to 0.5.

Index	1	2	......	H-1	H
M_η	$M_{\eta,1}$	$M_{\eta,2}$	......	$M_{\eta,H-1}$	$M_{\eta,H}$

Fig. 1. Historical strategy memory.

In each generation, each individual X_i^G selects a mutation strategy via a strategy control parameter η_i from the candidate strategy pool. If the offspring of X_i^G is better than X_i^G, then the corresponding strategy control parameter η_i is saved in a set S_η. At the end of each generation, the historical strategy memory is updated as follows.

$$M_{\eta,k}^{G+1} = \begin{cases} \text{mean}_{WL}(S_\eta) & \text{if} \quad S_\eta \neq \varnothing \\ M_{\eta,k}^{G} & \text{otherwise} \end{cases}, \tag{8}$$

$$\text{mean}_{WL}(S_\eta) = \frac{\sum_{s=1}^{|S_\eta|} w_s \cdot S_{\eta,s}^2}{\sum_{s=1}^{|S_\eta|} w_s \cdot S_{\eta,s}}, \tag{9}$$

$$w_s = \frac{|f(U_s^G) - f(X_s^G)|}{\sum_{s=1}^{|S_\eta|} |f(U_s^G) - f(X_s^G)|}, \tag{10}$$

where $S_{\eta,s}$ denotes the s-th element in set S_η.

As shown in (8), the index k ($k \in [1, H]$) is used to confirm the updated position in the memory. In the initialization, k is set to 1, and k is incremented by 1 whenever an entry is updated. If $k > H$, k is set to 1. Note that when all individuals fail to produce an offspring that is better than itself, i.e., $S_\eta = \varnothing$, the memory is not updated, and k remains unchanged [25]. Moreover, the use of weight aims to make the memory preferentially approach parameter values that can deliver a large improvement.

Moreover, to implement MGLMS, we must establish a candidate strategy pool. As mentioned previously, many mutation strategies are available, and different strategies have distinct characteristics. However, no theoretical researches exist on how to determine the optimal size of the strategy pool and how to select the strategies to form the strategy pool [11], but a basic principle is that each strategy in the strategy pool should show different search behaviors. In this paper, we choose three strategies as an example to form the candidate strategy pool, as shown in (11)-(13).

1. DE/rand-pbest/1 (short for M1):

$$V_i^G = \frac{R_{r1} \cdot X_{r1}^G}{R_{r1} + R_{pbest}} + \frac{R_{pbest} \cdot X_{pbest}^G}{R_{r1} + R_{pbest}} + F_i \cdot (X_{r2}^G - X_{r3}^G). \tag{11}$$

2. DE/rand-to-pbest/1 without the archive (short for M2):

$$V_i^G = X_{r1}^G + F_i \cdot (X_{pbest}^G - X_{r1}^G) + F_i \cdot (X_{r2}^G - X_{r3}^G). \tag{12}$$

3. DE/current-to-pbest/1 with the archive (short for M3):

$$V_i^G = X_i^G + F_i \cdot (X_{pbest}^G - X_i^G) + F_i \cdot (X_{r2}^G - X_{\tilde{r}3}^G). \tag{13}$$

where X_{r1}^G, X_{r2}^G, and X_{r3}^G are randomly selected from the population, X_{pbest}^G is randomly selected from the top $\lceil p \cdot NP \rceil$ $(0 < p < 1)$ individuals and NP denotes the population size, $\lceil x \rceil$ is the minimum integer that is larger than x, $X_{\tilde{r}3}^G$ is randomly selected from the union of the population $\mathbf{P}$ and an external archive $\mathbf{A}$ that stores the eliminative target vectors, R_i denotes the ranking of the ith individual in the population based on its fitness value, and it is calculated as follows [10,34],

$$R_i = NP - I_i + 1, \tag{14}$$

where I_i represents the index of the ith individual after sorting from best to worst. For example, the index of the current best individual is 1, and thus its ranking is NP. In contrast, the ranking of the current worst individual is 1. Besides, p is set as

$$p = 0.2 - (0.2 - 0.05) \cdot \frac{FES}{FES_{\max}}, \tag{15}$$

where FES and $FES_{\max}$ denote the current number of function evaluations and the maximum number of function evaluations, respectively.

The reasons for selecting these three strategies are twofold. First, M1 is a variant of DE/rand/1 [22,23,28] and DE/lbest/1 [31], which can obtain a trade-off between exploration and exploitation. Moreover, as claimed in [34], exploiting the collective information of different individuals can contribute to the enhancement of the search efficiency. Second, M2 and M3 show good performance independently [12,32], and they are used in many improved DE methods [8,11,35].

3.2 Multiple Gaussian Learning Model-based Parameter Adaptation Mechanism (MGLMP)

The target individual X_i^G $(i = 1, \ldots, NP)$ obtains its scaling factor F and crossover rate CR at the Gth generation in the following manner,

$$F_i^G = \text{randc}_i(M_{F,ri}, 0.1), \tag{16}$$

$$CR_i^G = \begin{cases} 0 & \text{if } M_{CR,ri} = 0 \\ \text{randn}_i(M_{CR,ri}, 0.1) & \text{otherwise} \end{cases}, \tag{17}$$

where $\text{randc}_i(M_{F,ri}, 0.1)$ is a random number generated by the Cauchy model with position parameter $M_{F,ri}$ and standard deviation 0.1. $\text{randn}_i(M_{CR,ri}, 0.1)$ is a random number produced by the Gaussian model with expectation $M_{CR,ri}$ and standard deviation 0.1. When $F_i^G > 1$, it is truncated to 1. When $F_i^G < 0$, it is regenerated using (16). Similarly, when $CR_i^G > 1$ or $CR_i^G < 0$, it is truncated to 1 or 0, respectively. $M_{F,ri}$ and $M_{CR,ri}$ are randomly chosen from the historical parameter memory with H elements as shown in Fig. 2. In the beginning, the values of the memory are all set to 0.5 (i.e., $M_{F,k} = M_{CR,k} = 0.5$, $1 \le k \le H$).

Index	1	2	......	H-1	H
M_F	$M_{F,1}$	$M_{F,2}$	......	$M_{F,H-1}$	$M_{F,H}$
M_{CR}	$M_{CR,1}$	$M_{CR,2}$	......	$M_{CR,H-1}$	$M_{CR,H}$

Fig. 2. Historical parameter memory.

In each generation, if target individual X_i^G uses F_i^G and CR_i^G to generate a better offspring U_i^G (i.e., $f(U_i^G) < f(X_i^G)$), the corresponding F_i^G and CR_i^G are stored in the sets S_F and S_{CR}, respectively. At the end of each generation, the historical parameter memory is updated as follows,

$$M_{F,k}^{G+1} = \begin{cases} \text{mean}_{WL}(S_F) \text{ if } & S_F \neq \varnothing \\ M_{F,k}^G & \text{otherwise} \end{cases}, \tag{18}$$

$$M_{CR,k}^{G+1} = \begin{cases} \text{mean}_{WL}(S_{CR}) \text{ if } & S_{CR} \neq \varnothing \cap M_{CR,k}^G \neq 0 \\ 0 \quad \text{if } & S_{CR} \neq \varnothing \cap M_{CR,k}^G = 0 \\ M_{CR,k}^G \quad \text{if } & S_{CR} = \varnothing \end{cases}, \tag{19}$$

where $\text{mean}_{WL}(\cdot)$ denotes the weighted Lehmer mean, and it is calculated as in (9). Index k is set in the same way as that in MGLMS.

3.3 Linear Population Size Reduction (LPSR)

Population size NP works as an algorithmic parameter, which is also vital to the performance of DE. LPSR has been shown to be highly effective in enhancing the performance of DE [26,27]. To leverage its advantages, we retain LPSR in our proposed method. More specifically, the population size NP is dynamically regulated in the following manner [26],

$$NP^{G+1} = \text{round}(NP_{\text{init}} - \frac{FES}{FES_{\max}}(NP_{\text{init}} - NP_{\min})), \tag{20}$$

where round(x) is a rounding function, NP_{init} denotes the initial (or maximum) population size, and $NP_{\min}$ is the smallest population size to which the mutation strategies can be applied. In our proposed algorithm, $NP_{\min}$ is set to 4. In the beginning, NP is set to NP_{init}, and at the end of each generation, the $NP^G - NP^{G+1}$ worst-ranking individuals based on fitness value are deleted from the population. Moreover, in this paper, the maximum size of the archive $\mathbf{A}$ is set to the initial population size NP_{init}. If the archive size exceeds NP_{init}, the redundant individuals are randomly deleted from the archive $\mathbf{A}$.

3.4 MGLMDE

By combining MGLMS, MGLMP, and LPSR with DE, the entire proposed MGLMDE is described in **Algorithm 1**.

4 Experiments

4.1 Test Problems and Experiment Setup

The proposed algorithm MGLMDE is tested on a set of 30 benchmark optimization problems with $D = 30$, $D = 50$, and $D = 100$ from the special session and competition on the real-parameters of IEEE CEC 2014. The benchmark problems show various characteristics, and thier details can be found in [16]. The performance of MGLMDE is compared with that of some state-of-the-art DE variants, i.e., L-SHADE [26], MPEDE [29], ZEPDE [9], ETI-JADE [8], JADE-sort [35], sTDE-dR [2], and dDSF-EA [3]. For a fair comparison, the parameter settings of the compared algorithms are the same as those used in the original papers. The maximum number of functions evaluations ($FES_{\max}$) is used as the termination criteria, which is set to $10000D$. To obtain stable results, each algorithm conducts 30 independent runs on each function. The function error $f(X_{best}) - f(X^*)$ is recorded for each independent run of each algorithm on each function, where X_{best} denotes the best solution obtained by the algorithm, and X^* is the actual best solution of the test function. Moreover, to effectively analyze the experimental results obtained by all algorithms, two nonparametric statistical tests (i.e., the Wilcoxon rank sum test and the Friedman test) with a significance level of 0.05 are applied in the experiments. More specifically, the

Algorithm 1: MGLMDE algorithm

Input: Objective function f, search space $\mathbf{\Omega}$, parameter NP_{init} and H

Set $G = 1$, $NP^G = NP_{\text{init}}$, $\mathbf{A} = \varnothing$, $k = 1$, and $K = 3$
Randomly Generate an initial population in the search space as (2)
Set $M_{\eta,h} = M_{F,h} = M_{CR,h} = 0.5$, $1 \le h \le H$
while the termination criterion is not met
  $S_\eta \ne \varnothing$, $S_F \ne \varnothing$, $S_{CR} \ne \varnothing$
  for $i = 1$ **to** NP^G
    $ri = \text{randint}(1, H)$
    $\eta_i = \text{randn}_i(M_{\eta,ri}, 0.1)$
    $S_i^G = \min(\text{floor}(\eta_i \cdot K) + 1, K)$
    $ri = \text{randint}(1, H)$
    Generate F_i^G and CR_i^G as (16) and (17), respectively
    if $S_i = 1$
      Generate mutation vector V_i^G as (11)
    elseif $S_i = 2$
      Generate mutation vector V_i^G as (12)
    elseif $S_i = 3$
      Generate mutation vector V_i^G as (13)
    end if
    Generate trial vector U_i^G as (4)
  end for
  for $i = 1$ **to** NP^G
    if $f(U_i^G) \le f(X_i^G)$
      $X_i^{G+1} = U_i^G$
    else
      $X_i^{G+1} = X_i^G$
    end if
    if $f(U_i^G) < f(X_i^G)$
      $X_i^G \to \mathbf{A}$, $\eta_i \to S_\eta$, $F_i^G \to S_F$, $CR_i^G \to S_{CR}$
    end if
  end for
  If Archive size is larger than NP_{init}, randomly delete redundant individuals
  Update memories M_η, M_F, and M_{CR} as (8), (18), and (19), respectively
  Calculate NP^{G+1} as (20)
  Sort all individuals in population based on fitness
  Delete the $NP^G - NP^{G+1}$ worst individuals
  $G = G + 1$
end while

Output: The best solution in the population

Wilcoxon rank sum test is conducted on the experimental results of 30 independent runs between the compared algorithms and MGLMDE for each function, and the test result is recorded as +/=/− which means that MGLMDE is better than, similar to, and worse than the compared algorithm on the corresponding function. The Friedman test [1] is conducted on 30 average function error results (30 problems) for all compared algorithms, which is used to show the comprehensive optimization performance of the compared algorithms on all test functions.

4.2 Experimental Results

The results of the Wilcoxon rank sum test and the Friedman test are shown in Tables 1 and 2, respectively. It can be observed from Table 1 that MGLMDE performs significantly better than the compared algorithms on most CEC2014 functions. More specifically, for the 30*D* functions, MGLMDE is significantly better than LSHADE, MPEDE, ZEPDE, ETI-JADE, JADE-sort, *s*TDE-*d*R, and *d*DSF-EA on 18, 22, 23, 18, 21, 15, and 12 functions, respectively. Moreover, for 50*D* functions, MGLMDE is significantly superior to LSHADE, MPEDE, ZEPDE, ETI-JADE, JADE-sort, *s*TDE-*d*R, and *d*DSF-EA on 15, 21, 21, 17, 22, 14, and 16 cases, respectively. With respect to 100*D* functions, MGLMDE significantly outperforms LSHADE, MPEDE, ZEPDE, ETI-JADE, JADE-sort, *s*TDE-*d*R, and *d*DSF-EA on 14, 23, 24, 21, 24, 15, and 11 out of 30 functions.

Furthermore, Table 2 indicates that based on the *mean function error value*, the overall performance of MGLMDE is the best among all of the compared algorithms according to the ranking of the Friedman test. The second best algorithm is *d*DSF-EA. In summary, we can conclude that for 30*D* CEC2014 problems, the average performance of MGLMDE is significantly better than those of MPEDE, ZEPDE, ETI-JADE, JADE-sort, and *s*TDE-*d*R, and it is similar to those of L-SHADE and *d*DSF-EA. For 50*D* and 100*D* CEC2014 problems, the average performance of MGLMDE is significantly superior to those of MPEDE, ZEPDE,

Table 1. Comparison of MGLMDE and seven state-of-the-art algorithms based on the Wilcoxon ranksum test

MGLMDE	30*D*			50*D*			100*D*		
vs	+	=	−	+	=	−	+	=	−
L-SHADE [26]	18	8	4	15	7	8	14	8	8
MPEDE [29]	22	5	3	21	5	4	23	5	2
ZEPDE [9]	23	5	2	21	3	6	24	3	3
ETI-JADE [8]	18	6	6	17	3	10	21	1	8
JADE-sort [35]	21	7	2	22	4	4	24	4	3
*s*TDE-*d*R [2]	15	12	3	14	12	4	15	7	8
*d*DSF-EA [3]	12	3	5	16	10	4	11	10	9

Table 2. Rankings of all compared algorithms based on the Friedman test on the mean function error value

Ranking	30D	50D	100D
L-SHADE [26]	3.67	3.75	3.93
MPEDE [29]	5.62	5.65	5.37
ZEPDE [9]	5.48	6.00	5.68
ETI-JADE [8]	5.17	4.48	4.68
JADE-sort [35]	5.38	5.78	5.78
sTDE-dR [2]	3.85	4.32	3.88
dDSF-EA [3]	3.43	3.45	3.61
MGLMDE	**2.77**	**3.03**	**3.05**

and JADE-sort, and is highly comparable to those of L-SHADE, ETI-JADE, sTDE-dR, and dDSF-EA.

5 Conclusion

Inspired by the multiple Gaussian model-based parameter adaptation mechanism (MGLMP), we proposed a new adaptive operator selection method, MGLMS, for automatic choice of suitable mutation strategies in an online manner for DE. This approach adopts a strategy control parameter that is generated by Gaussian models to determine the strategy used by the individuals. Moreover, the successful strategy control parameter values are used to update the expectations of the Gaussian models. In this manner, the adaptation process is established. By combining MGLMS with MGLMP and LPSR, we developed a new DE variant known as MGLMDE. The experimental results on a lot of test problems show that MGLMDE is significantly better than or at least highly competitive with some outstanding DE variants. In future, we will plan to examine whether MGLMS and MGLMP can be effectively used on constraint optimization problems and multi-objective optimization problems.

References

1. Alcalá-Fdez, J., et al.: Keel: a software tool to assess evolutionary algorithms for data mining problems. Soft. Comput. **13**(3), 307–318 (2009)
2. Ali, M.Z., Awad, N.H., Suganthan, P.N., Reynolds, R.G.: An adaptive multipopulation differential evolution with dynamic population reduction. IEEE Trans. Cybern. **47**(9), 2768–2779 (2017)
3. Awad, N.H., Ali, M.Z., Suganthan, P.N., Jaser, E.: A decremental stochastic fractal differential evolution for global numerical optimization. Inf. Sci. **372**, 470–491 (2016)

4. Brest, J., Greiner, S., Boskovic, B., Mernik, M., Zumer, V.: Self-adapting control parameters in differential evolution: a comparative study on numerical benchmark problems. IEEE Trans. Evol. Comput. **10**(6), 646–657 (2006)
5. Bujok, P.: Success-history based differential evolution with adaptation by competing strategies. In: 2016, Manuscript Submitted for Publication in Swarm and Evolutionary Computation (2016)
6. Cui, L., Li, G., Lin, Q., Chen, J., Lu, N.: Adaptive differential evolution algorithm with novel mutation strategies in multiple sub-populations. Comput. Oper. Res. **67**, 155–173 (2016)
7. Cui, L., et al.: Adaptive multiple-elites-guided composite differential evolution algorithm with a shift mechanism. Inf. Sci. **422**, 122–143 (2018)
8. Du, W., Leung, S.Y.S., Tang, Y., Vasilakos, A.V.: Differential evolution with event-triggered impulsive control. IEEE Trans. Cyber. **47**(1), 244–257 (2017)
9. Fan, Q., Yan, X.: Self-adaptive differential evolution algorithm with zoning evolution of control parameters and adaptive mutation strategies. IEEE Trans. Cyber. **46**(1), 219–232 (2016)
10. Gong, W., Cai, Z.: Differential evolution with ranking-based mutation operators. IEEE Trans. Cyber. **43**(6), 2066–2081 (2013)
11. Gong, W., Cai, Z., Ling, C.X., Li, H.: Enhanced differential evolution with adaptive strategies for numerical optimization. IEEE Trans. Syst. Man Cyber. Part B (Cybern.) **41**(2), 397–413 (2011)
12. Gong, W., Cai, Z., Wang, Y.: Repairing the crossover rate in adaptive differential evolution. Appl. Soft Comput. **15**, 149–168 (2014)
13. I. Loshchilov, T.S., Liao, T.: Ranking results of CEC13 special session and competition on real-parameter single objective optimization (2013)
14. J. J. Liang, B.Y.Q., Suganthan, P.N.: Ranking results of CEC14 special session and competition on real-parameter single objective optimization (2014)
15. Li, G., et al.: A novel hybrid differential evolution algorithm with modified CODE and JADE. Appl. Soft Comput. **47**, 577–599 (2016)
16. Liang, J., Qu, B., Suganthan, P.: Problem definitions and evaluation criteria for the CEC 2014 special session and competition on single objective real-parameter numerical optimization. Zhengzhou University, Zhengzhou China and Technical Report, Nanyang Technological University, Singapore, Computational Intelligence Laboratory (2013)
17. Liang, J., Qu, B., Suganthan, P., Hernández-Díaz, A.G.: Problem definitions and evaluation criteria for the CEC 2013 special session on real-parameter optimization. Zhengzhou University, Zhengzhou, China and Nanyang Technological Uni. Singapore, Technical Report. Comput. Intell. Lab. **201212**(34), 281–295 (2013)
18. Mallipeddi, R., Suganthan, P.N., Pan, Q.K., Tasgetiren, M.F.: Differential evolution algorithm with ensemble of parameters and mutation strategies. Appl. Soft Comput. **11**(2), 1679–1696 (2011)
19. Mallipeddi, R., Suganthan, P.N.: Ensemble differential evolution algorithm for cec2011 problems. In: 2011 IEEE Congress of Evolutionary Computation (CEC), pp. 1557–1564. IEEE (2011)
20. Mohamed, A.W., Hadi, A.A., Fattouh, A.M., Jambi, K.M.: LSHADE with semi-parameter adaptation hybrid with CMA-ES for solving CEC 2017 benchmark problems. In: 2017 IEEE Congress on Evolutionary Computation (CEC), pp. 145–152. IEEE (2017)
21. Poláková, R., Tvrdík, J., Bujok, P.: Evaluating the performance of l-shade with competing strategies on CEC 2014 single parameter-operator test suite. In: 2016 IEEE Congress on Evolutionary Computation (CEC), pp. 1181–1187. IEEE (2016)

22. Qin, A.K., Huang, V.L., Suganthan, P.N.: Differential evolution algorithm with strategy adaptation for global numerical optimization. IEEE Trans. Evol. Comput. **13**(2), 398–417 (2008)
23. Storn, R., Price, K.: Differential evolution-a simple and efficient heuristic for global optimization over continuous spaces. J. Global Optim. **11**(4), 341–359 (1997)
24. Tanabe, R., Fukunaga, A.: Evaluating the performance of shade on CEC 2013 benchmark problems. In: 2013 IEEE Congress on Evolutionary Computation, pp. 1952–1959 (2013)
25. Tanabe, R., Fukunaga, A.: Success-history based parameter adaptation for differential evolution. In: 2013 IEEE Congress on Evolutionary Computation, pp. 71–78 (2013)
26. Tanabe, R., Fukunaga, A.S.: Improving the search performance of shade using linear population size reduction. In: 2014 IEEE Congress on Evolutionary Computation (CEC) (2014)
27. Tian, M., Gao, X., Dai, C.: Differential evolution with improved individual-based parameter setting and selection strategy. Appl. Soft Comput. **56**, 286–297 (2017)
28. Wang, Y., Cai, Z., Zhang, Q.: Differential evolution with composite trial vector generation strategies and control parameters. IEEE Trans. Evol. Comput. **15**(1), 55–66 (2011)
29. Wu, G., Mallipeddi, R., Suganthan, P., Wang, R., Chen, H.: Differential evolution with multi-population based ensemble of mutation strategies. Inf. Sci. **329**, 329–345 (2016)
30. Yang, X., Liu, G.: Self-adaptive Clustering-Based Differential Evolution with New Composite Trial Vector Generation Strategies. In: Gaol, F., Nguyen, Q. (eds.) Proceedings of the 2011 2nd International Congress on Computer Applications and Computational Science. Advances in Intelligent and Soft Computing, vol. 144. Springer, Heidelberg (2012). https://doi.org/10.1007/978-3-642-28314-7_35
31. Yu, W., et al.: Differential evolution with two-level parameter adaptation. IEEE Trans. Cyber. **44**(7), 1080–1099 (2014)
32. Zhang, J., Sanderson, A.C.: JADE: adaptive differential evolution with optional external archive. IEEE Trans. Evol. Comput. **13**(5), 945–958 (2009)
33. Zhao, F., Zhao, L., Wang, L., Song, H.: A collaborative LSHADE algorithm with comprehensive learning mechanism. Appl. Soft Comput. **96**, 106609 (2020)
34. Zheng, L.M., Zhang, S.X., Tang, K.S., Zheng, S.Y.: Differential evolution powered by collective information. Inf. Sci. **399**, 13–29 (2017)
35. Zhou, Y., Yi, W., Gao, L., Li, X.: Adaptive differential evolution with sorting crossover rate for continuous optimization problems. IEEE Trans. Cyber. **47**(9), 2742–2753 (2017)

Online Taxi Dispatching Algorithm Based on Quantum Annealing

Chao Wang, Tongyu Ji(✉), and Suming Wang

Shanghai University, Shanghai, China
wangchao@shu.edu.cn, 624460914@qq.com

Abstract. Quantum annealing techniques are mainly used to solve optimization and sampling problems, which have a greater potential to obtain globally optimal solutions for specific combinatorial optimization problems due to their unique quantum tunneling properties. Firstly, this paper transforms the taxi repositioning problem into a combinatorial optimization problem that can be represented by a QUBO model with exponentially growing path solutions. The optimal solution aims at repositioning the online taxis to optimize the passenger waiting time and the average daily revenue of taxi drivers. Secondly, this paper proposes a QUBO formulation to solve a specific taxi repositioning problem, which optimizes the traffic distribution of taxis based on the existing spatial transfer probability to obtain the unique target region selection result. Finally, the simulation results on the real trajectory data of taxis in Chengdu in November 2016 show that the solution is nearly optimal, and the quantum annealing algorithm optimizes about 7% of the indicator than the simulated annealing algorithm by making each vehicle select the unique region to accurately reduce the passenger waiting time while improving the driver's revenue.

Keywords: Quantum annealing · Taxi repositioning · QUBO model

1 Introduction

With the growing demand for travel and the booming development of mobile Internet technology, online taxi have become a popular way in people's daily life today, such as Didi and Uber, through which users send real-time demand on their cell phones, and then the platform dynamically matches passengers with available drivers through intelligent algorithms [1]. Compared with the traditional way of finding passengers with the experience of drivers, big data and intelligent algorithm-guided online taxi repositioning have led to a significant reduction in passenger waiting time and have improved the revenue of drivers and the operational efficiency of urban transportation to a certain extent.

Taxi repositioning is the core module of the online car-hailing system [2], which needs to manage all taxis in the city, assign target areas and routes to idle taxis to repositioning them to potential passenger locations, while coordinating the traffic of all parties to prevent them from competing with each other for passengers, maximizing

L. Fang et al. (Eds.): CICAI 2022, LNAI 13606, pp. 337–347, 2022.
https://doi.org/10.1007/978-3-031-20503-3_27

the use of online taxi resources, balancing the taxi supply and passenger demand, and minimizing passenger Waiting time and passenger travel satisfaction. But the reality is that the imbalance between supply and demand still exists, and during certain peak hours, passenger demand surges, often requiring a half-hour advance queue to be matched with a vehicle at the target time, while at the same time there is still the problem of drivers cruising for too long to receive orders, resulting in low daily revenue. This kind of supply and demand imbalance happens every day in big cities, which affects both the user experience and reduces the driver and platform revenue. Therefore, it is important to establish an effective online taxi planning algorithm for online car platforms and even for the transportation of the whole city.

Taxi repositioning problems can use traversal search to reach a globally optimal solution in simple scenarios, but as the problem scales up, the search space quickly exceeds the maximum arithmetic power level of current computers. Many intelligent algorithms have been used to solve such problems, such as metaheuristics [3] and reinforcement learning [4], but there is always the problem of being prone to fall into local optimal solutions. In contrast, the quantum annealing algorithm [5] has a greater probability of jumping out of local suboptimality to reach the global optimal solution because of its unique quantum tunneling effect. The quantum annealer solves the optimization problem by establishing the QUBO model, which minimizes the quadratic polynomial in binary variables to obtain the lowest energy state of the system based on the principle of adiabatic annealing. Compared with traditional machine learning algorithms, quantum annealing can overcome the characteristics of traditional machine learning such as poor robustness and sensitivity to initial points. A large number of applications have been realized using the advantages of quantum annealing, such as traffic optimization, quantum chemistry [6], and code-breaking [7]. Neukart et al. successfully used quantum annealing to balance the taxi flow in Beijing in 2017 [8], and then the Volkswagen Group used quantum annealing to optimize realistic traffic flow for the first time during the Global Network Summit in Lisbon, Portugal, in 2019 to guide the 9 buses' routing decisions during the Congress [9]. In this paper, we propose a formulation using D-Wave quantum annealer to solve the online taxi repositioning by finding the target region of online taxi, and the objective of the formulation is to minimize the vehicle congestion level and the total vehicle idling probability to optimize the global reposition problem.

2 Modeling

2.1 D-Wave Quantum Annealing

The D-Wave quantum annealer solves the problem by building a binary quadratic model that models the problem as an Ising model or a quadratic unconstrained binary optimization (QUBO) model [10]. The BQM takes the form of an Ising model or a QUBO that consists of an objective function and a constraint function. The QUBO maps the problem into a form that takes values of the Boolean variables such that their values minimize the objective function, the target optimized function is:

$$Obj(x, Q) = x^T \cdot Q \cdot x \tag{1}$$

where x is a binary vector of length N and Q is an $N \times N$ of upper triangular matrices, and the detailed relationship between Q and x is given by the following equation:

$$f(x_1, \cdots, x_n) = \sum_{i=1}^{n} Q_{i,i}x_i + \sum_{1 \le i < j \le n} Q_{i,j}x_i x_j \tag{2}$$

where $Q_{i,i}$ are the linear coefficients, and $Q_{i,j}$ is the quadratic term coefficient, the values of the binary variables are searched in the obtained QUBO formula and combined to find the energy value of the QUBO formula.

2.2 Transformation of the Online Taxi Repositioning Problem into a QUBO Model

The goal of the online taxi repositioning problem is to minimize customer waiting time and improve the efficiency of taxi operations by giving the respective required driving goals to find potential customers in a set of vehicles [11]. To prevent congestion and reduce the phenomenon of fight for order, We assume that the time to cross a load is proportional to the number of cars currently on that load and that the loss of repositioning for drivers who fail in the competition for the order is proportional to the square of the sum of all drivers who fail to compete in the current region, with the loss of competition in a single region determined by a quadratic function of the number of currently empty drivers. To solve the repositioning problem, we formulate about 10 to 17 feasible target regions for each possible location where a taxi is located based on the open source dataset of Didi [12] and require that each taxi must choose one of these regions for repositioning and must satisfy two constraints of minimum total idle rate and minimum order-squatting loss.

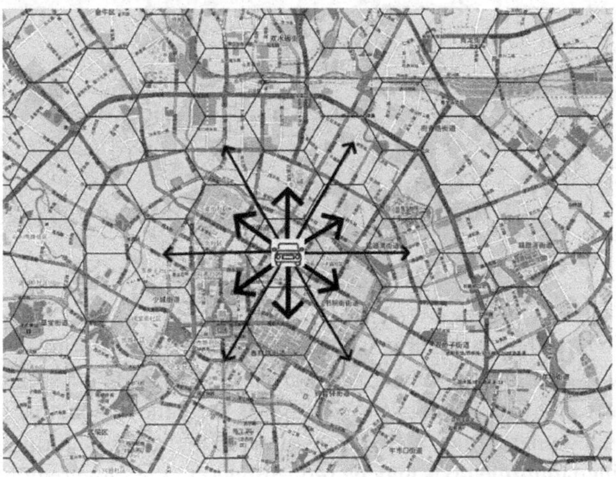

Fig. 1. Candidate repositioning area of a taxi.

Figure 1 shows a roadmap of the possible movements of a taxi with twelve target areas. Each taxi to be repositioned in the whole system has ten to twenty target areas to choose from, but only one area can be selected to go at a time. Taxis are represented by variable i = {1, 2, ..., n}, and repositioning area are represented by variable j = {1, 2, ..., k}.

To achieve the unique area selection each time, we define a binary variable q_{ij} denotes that taxi i reposition target is region j. Since each time a unique region is selected, it is required that only one of the reposition target variables of all vehicles is true, for which we implement the following constraint to ensure that vehicle i selects a unique region:

$$\begin{aligned}\left(\sum_{j=1}^{k} q_{ij} - 1\right)^2 &= 2q_{i2} \cdot q_{i1} + 2q_{i3} \cdot (q_{i1} + q_{i2}) + \\ &\cdots + 2q_{ik} \cdot \left(q_{i1} + \cdots + q_{i(k-1)}\right) + {q_{i1}}^2 + \\ &\cdots + {q_{ik}}^2 - 2(q_{i1} + \cdots + q_{ik}) + 1 = 0\end{aligned} \tag{3}$$

Simplified using the binary rule $x^2 = x$, we can get that:

$$2q_{i2} \cdot q_{i1} + 2q_{i3} \cdot (q_{i1} + q_{i2}) + \cdots + 2q_{ik} \cdot \left(q_{i1} + \cdots + q_{i(k-1)}\right) - (q_{i1} + \cdots + q_{ik}) + 1 = 0 \tag{4}$$

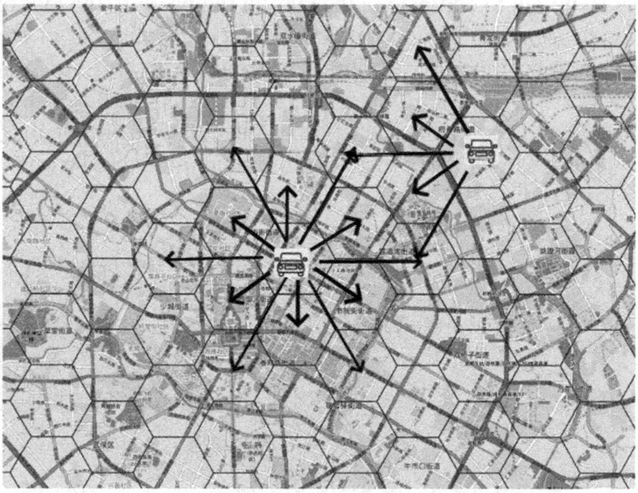

Fig. 2. Case of repositioning area overlap.

Figure 2 shows a schematic diagram of the possible conflicts during the repositioning, in which two vehicles have potential confliction in a total of two regions during the repositioning process(we define if two taxis target is the same region, that's a confliction), while in the actual scenario there may be potential confliction for multiple vehicles in the same region. Since multiple taxis may be repositioned to the same region, the following constraint function is given to balance the taxi confliction and avoid congestion, limiting the number of drivers with region j as the target:

$$Cost(j) = \left(\sum_{i=1}^{n} q_{ij}\right)^2 \tag{5}$$

Expanding it we can come to:

$$\begin{aligned} Cost(j) &= \left(q_{1j} + q_{2j} + \cdots + q_{nj}\right)^2 \\ &= 2q_{1j} \cdot q_{2j} + 2\left(q_{1j} + q_{2j}\right) \cdot q_{3j} + \cdots + 2\left[q_{1j} + \cdots + q_{(n-1)j}\right] + q_{1j} + \cdots + q_{nj} \end{aligned} \tag{6}$$

Also simplified using the binary rule $x^2 = x$. This constraint limits the number of drivers to be repositioned to the same area to a relative number, which ensures that there will be no congestion in popular areas and prevents vehicle congestion to a certain extent.

In the classical taxi scenario, the target driving region of the driver will refer to the past experience and tend to move to the area which more popular in the past, while in the online taxi repositioning scenario, the idle rate is provided as data to the algorithm to guide the driver's next decision. The higher idle rate means the lower number of orders in the area, higher probability driver idle when passing by. Based on the data set of Chengdu in November 2016 provided by Didi, we can come to the cost function:

$$Cost(i) = \sum_{j=1}^{k} p_{ij} \cdot q_{ij} \tag{7}$$

where p_{ij} is the idle probability from the region of vehicle i to the target region j, and q_{ij} is the binary variable from taxi i to target region j. Value is 1 means vehicle i will be repositioned to region j, and value is 0 means it will not.

The final cost function will be defined as:

$$Obj = \lambda_1 Cost_1 + \lambda_2 Cost_2 + \lambda_3 Cost_3 \tag{8}$$

Bringing in the components, the optimization problem model is obtained as follows:

$$Obj = \lambda_1 \sum_{i=1}^{n} \left(\sum_{j=1}^{k} q_{ij} - 1 \right)^2 + \lambda_2 \sum_{j=1}^{k} \left(\sum_{i=1}^{n} q_{ij} \right)^2 + \lambda_3 \sum_{i=1}^{n} \sum_{j=1}^{k} p_{ij} \cdot q_{ij} \tag{9}$$

where λ becomes the Lagrangian parameter, and the value of λ determines whether the final result obtained is a valid value. By our calculation, when λ_1 is 4, λ_2 and λ_3 is 1 the result is obtained as valid. The QUBO model consisting of cost and constraint matrices in the next subsection will be solved by the D-Wave real quantum annealer.

3 Experiments and Results

This experiment is based on the real dataset of Chengdu in November 2016, which is open-sourced by Didi, and this dataset contains roughly 100G real online taxi trajectory data, as well as the idle probability of taxi grid id in Chengdu city, the data of online taxi orders in November, etc. The trajectory data contains order number, driver id, driver coordinates, current time, etc. In this paper, we assume that the last coordinates of each order is the drop-off point, and the driver starts dispatching at this point, and selects all the drivers whose orders end in a certain time slot starts the global repositioning.

The result of annealing is the binary quantum bit combination form of the solution that minimizes the cost function, and each quantum bit combination will represent the route selection of each taxi. When the quantum bit takes the value of 1 represents the taxi target is the region represented by this quantum bit. In this paper, We start quantum annealing repositioning for all vacant online vehicles in Chengdu city at 10:00 on November 1, 2016, and the experimental steps are as follows.

1. Data pre-processing, cleaning the original data and get the data of the taxi to be repositioned at the corresponding time.
2. We get current area of all the vehicles to be repositioned, then get the idle probability of current area to other nearby areas.
3. Use the data to build a QUBO model consisting of a cost function and a constraint function
4. Creation of BQM for quantum annealer execution from QUBO model
5. Use Leap platform to access D-Wave quantum annealer to solve BQM and get binary quantum bit combinations
6. The final reposition area of each taxi is obtained by mapping binary quantum bits to map.

3.1 Algorithm Comparison

The data of the drop-off point at 10:05 on November 1, 2016 is selected as the initial dispatch point, and the drop-off point data obtained based on this sample is shown in Fig. 3. The figure shows that the downtown area at the bottom right is more dense than the drop-off point at the edge area, which shows that it is particularly important to solve the competition of order taking between taxis, and it is a great challenge to the robustness of the algorithm.

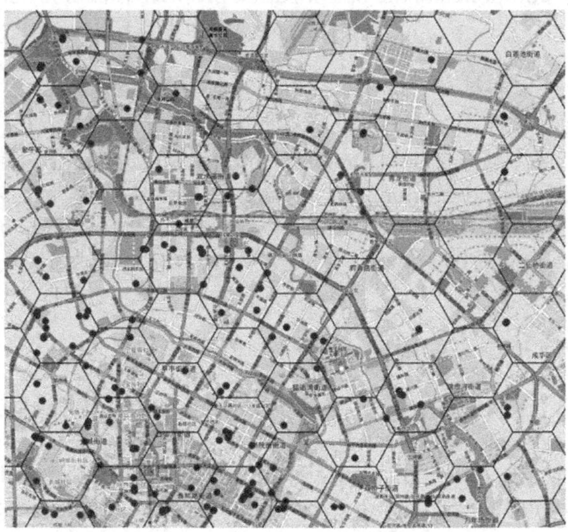

Fig. 3. Distribution map of drop off point.

The data were transformed into BQM executed on a D-Wave quantum annealer with 10 samples and a maximum number of iterations is 10, and the following results were obtained, which in comparison with the results obtained by simulated annealing show that the results generated by D-Wave quantum annealing always provide lower cost values in each data set (Tables 1 and 2).

Table 1. Table of the energy of quantum annealing.

	1-04417fb63aed6723	...	99-fa7248c2a1040346	Energy	Num_oc.
6	0	...	0	−538.012769	1
8	0	...	0	−536.212814	1
9	0	...	0	−536.205022	1
0	1	...	0	−534.17318	1
5	0	...	0	−534.165134	1
7	1	...	0	−534.098891	1
2	0	...	0	−534.035086	1
3	0	...	0	−532.23376	1
4	1	...	0	−532.049337	1
1	0	...	0	−531.964491	1

Table 2. Table of the energy of simulated annealing.

	1-04417fb63aed6723	...	99-fa7248c2a1040346	Energy	Num_oc.
3	0	...	0	−518.916927	1
9	0	...	0	−517.140948	1
1	0	...	0	−517.055956	1
7	0	...	0	−515.132892	1
4	0	...	0	−514.905992	1
0	0	...	0	−513.400791	1
8	0	...	0	−513.215483	1
6	0	...	0	−513.142004	1
2	0	...	0	−512.691017	1
5	0	...	0	−508.956634	1

And then five time periods of data (10:05 a.m. from November 1 to 5, 2016, respectively) were selected for repositioning operations, and the results obtained from D-wave quantum annealing were compared with those obtained from simulated annealing, and the average energy was obtained as (Table 3):

Table 3. Comparison of five groups of experimental results.

	D-Wave quantum annealing	Simulated annealing
Group 1	−534.315	−514.456
Group 2	−329.531	−311.929
Group 3	−467.285	−455.930
Group 4	−788.163	−750.498
Group 5	−272.105	−239.580

The cost values vary widely with the QUBO equation due to the different number of vehicles ending orders in each time period, but the table shows that using the quantum annealing method to solve the taxi repositioning problem in each iteration always results in an optimal solution, and the energy is on average about four percent lower than that of simulated annealing, proving that quantum annealing has a higher probability of jumping out of the local suboptimal solution.

3.2 Comparison of Arithmetic Cases

In this subsection, this paper creates a taxi simulation revenue environment using a real order dataset to test the algorithm effectiveness. This order dataset includes order start time, start location, end time, end location, and order estimated revenue coefficients, and a greedy algorithm [13] is used to perform dynamic order matching, which can simulate the sum of all drivers' revenue in a certain period of time, and the real revenue environment simulation steps are as follows.

1. The current vacant vehicles are repositioned to get the target location of the vehicles. Assuming that the time taken by urban taxis from the center of one hexagonal grid to the center of the adjacent hexagonal grid is 2 min (the speed limit of urban vehicles is 40 km/h and the spacing between hexagonal grids is 1.4 km), the algorithm is called once every two minutes to recalculate the target area for reposition.
2. At the beginning of the order, it matchs to the closest driver (randomly assigned if there is competition), and the assigned driver must continue to participate in repositioning only after this order is completed (the end of the order time), and the remaining drivers in the scene who not get the order and the drivers who become empty at this moment continue to be repositioned.
3. After a period of time all driver earnings are added up and used to evaluate the results.

Firstly, the vehicles from 10:00 to 11:00 on November 1, 2016 were chosen to start the real scenario simulation using two algorithms respectively, and the gain curve with time is shown in Fig. 4.

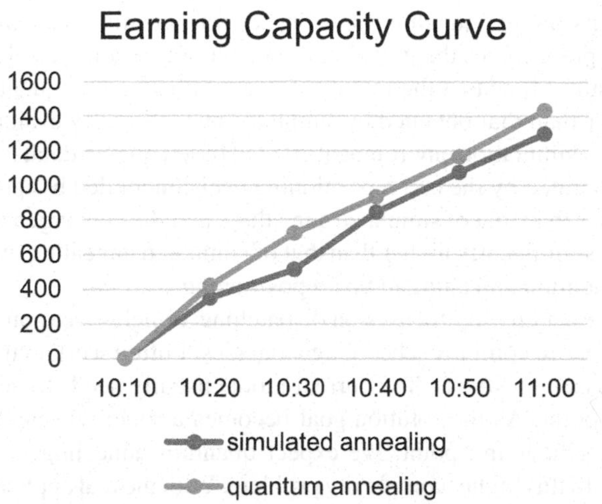

Fig. 4. Profit curve comparison.

The Fig. 4 Shows that the D-Wave quantum annealing algorithm repositioning reward is greater than the simulated annealing algorithm repositioning reward from 10:00 to 11:00, and the final reward is about 9.4% higher than simulated annealing, which is related to the fact that the quantum annealing algorithm can reach the low energy state more easily.

Thereafter, five timeslots were selected to simulate real scenarios for both algorithms, and we can come to the following test results (Table 4):

Table 4. Table of earning capacity curve.

	Oct.1st	Oct.2nd	Oct.3rd	Oct.4th	Oct.5th
Quantum annealing	1433.5	1327.04	1378.34	1554.58	1012.19
Simulated annealing	1298.16	1284.25	1233.87	1482.35	964.82

The table shows the total driver order reward factor obtained from real simulations conducted from 10:00 a.m. to 11:00 a.m. on November 1–5, which is proportional to the order revenue. Quantum annealing always gives better repositioning results compared to simulated annealing, and drivers who are repositioned by quantum annealing gain about 7% more than simulated annealing.

4 Conclusion

This paper proposes a quantum annealing algorithm for taxi repositioning, transforming the taxi repositioning model into the QUBO equation, and solving this combinatorial optimization problem using the Leap platform connected to a D-Wave real quantum computer. With its unique quantum tunneling effect, quantum annealing has a higher probability to explore to get the global optimal solution in a huge solution space. The experimental results show that the minimum energy obtained by quantum annealing is about 3.5% lower than that obtained by simulated annealing. By evaluating the experimental results in a simulation environment created based on real data, it is found that the total revenue generated by the taxi repositioning decision guided by quantum annealing is about 7% higher than that of simulated annealing, and the convergence speed of quantum annealing is significantly higher than that of simulated annealing, which verifies the availability of quantum annealing in taxi repositioning.

The current research only focuses on dispatching vehicles from one area to another but cannot make more complex scheduling strategies. Future work may consider giving detailed routes based on streets. The current model usually needs to define each region as a scheduling point. As the solution goal becomes a route, we can try to define the route node as a scheduling point. We expect quantum annealing to provide a more efficient solution to this higher complexity problem. With the real application of quantum annealing to bus route selection during the Lisbon Network Summit in Portugal in 2019, its potential is gradually emerging.

Appendix

See Table 5.

Table 5. Meanings of notations.

Notation	Size	Meaning
x	Constant	Binary coding vector of length N
i	Constant	The number of taxi
j	Constant	The number of region
Q	$i \times j$	Coefficient matrix
q_{ij}	Constant	The Boolean variable of if taxi i selected region j
p_{ij}	Constant	The idle probability from region i (the region where taxi i on) to the region j
λ	Constant	The lagrange coefficient

References

1. Truong, M., Purdy, D., Mawas, R.: U.S. Patent Application No. 14/793,593 (2017)
2. Seow, K.T., Dang, N.H., Lee, D.H.: A collaborative multiagent taxi-dispatch system. IEEE Trans. Autom. Sci. Eng. **7**(3), 607–616 (2009)
3. Jung, J., Jayakrishnan, R., Park, J.Y.: Dynamic shared-taxi dispatch algorithm with hybrid-simulated annealing. Comput.-Aided Civ. Infrastruct. Eng. **31**(4), 275–291 (2016)

4. Lin, K., Zhao, R., Xu, Z., Zhou, J.: Efficient large-scale fleet management via multi-agent deep reinforcement learning. In: Proceedings of the 24th ACM SIGKDD International Conference on Knowledge Discovery & Data Mining, pp. 1774–1783, July 2018
5. Finnila, A.B., Gomez, M.A., Sebenik, C., Stenson, C., Doll, J.D.: Quantum annealing: a new method for minimizing multidimensional functions. Chem. Phys. Lett. **219**(5–6), 343–348 (1994)
6. Perdomo-Ortiz, A., Dickson, N., Drew-Brook, M., Rose, G., Aspuru-Guzik, A.: Finding low-energy conformations of lattice protein models by quantum annealing. Sci. Rep. **2**(1), 1–7 (2012)
7. Peng, W., et al.: Factoring larger integers with fewer qubits via quantum annealing with optimized parameters. Sci. China Phys. Mech. Astron. **62**(6), 1–8 (2019). https://doi.org/10.1007/s11433-018-9307-1
8. Neukart, F., Compostella, G., Seidel, C., Von Dollen, D., Yarkoni, S., Parney, B.: Traffic flow optimization using a quantum annealer. Frontiers in ICT **4**, 29 (2017)
9. Yarkoni, S., et al.: Quantum shuttle: traffic navigation with quantum computing. In: Proceedings of the 1st ACM SIGSOFT International Workshop on Architectures and Paradigms for Engineering Quantum Software, pp. 22–30, November 2020
10. Lewis, M., Glover, F.: Quadratic unconstrained binary optimization problem preprocessing: theory and empirical analysis. Networks **70**(2), 79–97 (2017)
11. Duan, Y., Wang, N., Wu, J.: Optimizing order dispatch for ride-sharing systems. In: 2019 28th International Conference on Computer Communication and Networks (ICCCN), pp. 1–9. IEEE, July 2019
12. Data source: Didi chuxing gaia open dataset initiative. https://gaia.didichuxing.com
13. Duan, Y., Wu, J., Zheng, H.: A greedy approach for carpool scheduling optimisation in smart cities. Int. J. Parallel Emergent Distrib. Syst. **35**(5), 535–549 (2020)

A Snapshot Gradient Tracking for Distributed Optimization over Digraphs

Keqin Che and Shaofu Yang(✉)

School of Computer Science and Engineering, Southeast University, Nanjing, China
{213180322,sfyang}@seu.edu.cn

Abstract. This paper addresses distributed optimization problems over digraphs in which multiple agents cooperatively minimize the finite sum of their local objective functions via local communication. To improve the computation efficiency, we propose a novel algorithm named SSGT-PP by combining Snapshot Gradient Tracking technique with Push-Pull method. In SSGT-PP, agents compute the full-gradient intermittently under the control of a random variable, so that the frequency of gradient computation is reduced. As a result, the proposed algorithm can save computing resources especially for large-scale distributed optimization problems. We theoretically show that SSGT-PP can achieve linear convergence rate on strongly convex functions. Finally, we substantiate the effectiveness of SSGT-PP by numerical experiments.

Keywords: Distributed optimization · Directed networks · Gradient tracking · Linear convergence

1 Introduction

In this paper, we consider the n-nodes (agents) distributed optimization problem:

$$\min_{\mathbf{x}\in\mathbb{R}^d} f(\mathbf{x}) \triangleq \frac{1}{n}\sum_{i=1}^{n} f_i(\mathbf{x}), \tag{1}$$

where $\mathbf{x}$ is the global decision vector and f_i is the local objective function only known by node i. All nodes exchange information with others via communication networks to seek for the optimal solution of problem (1). Distributed optimization problems arise in various engineering applications such as smart grids [24], machine learning [1], and intelligence system [16].

To solve problem (1), plenty of gradient-based distributed optimization algorithms have been proposed since the pioneer algorithm DGD [9], which sublinearly converges to the optimal solution by using a diminishing stepsize. To

This work was supported in part by the National Natural Science Foundation of China under Grant 62176056, and is supported in part by Young Elite Scientists Sponsorship Program by CAST, 2021QNRC001.

L. Fang et al. (Eds.): CICAI 2022, LNAI 13606, pp. 348–360, 2022.
https://doi.org/10.1007/978-3-031-20503-3_28

improve the convergence rate, distributed algorithms with constant stepsize are developed based on primal-dual formulation [5,18], which can attain linear convergence rate. Another approach for developing algorithms with constant stepsize is to use gradient tracking technique, such as [4,10,12,13,19,22]. In these algorithms, an auxiliary variable is introduced to track the average gradient of all nodes at each iteration and linear convergence rate is also guaranteed. Due to the limited computation ability of each node and the requirement to save the implementation time, many efforts have been devoted to improve the efficiency of local computation. One common used method is stochastic gradient method [10,22], especially in the scenario of big data. More recently, Ref. [19] proposes a novel technique for reducing gradient computation complexity, named snapshot gradient tracking, in which intermittent gradient computation is required.

All aforementioned algorithms are developed over undirected or directed but balanced graphs. However, in practice, general directed (may not be balanced) graphs are prevalent. Thus, distributed algorithms over digraphs are imperative. A main challenge in the scenario of general digraphs is that the "unbalance" of communication graph will lead to a drift of optimal solution. To overcome this challenge, push-sum protocol [3] is introduced into distributed subgradient descent algorithm [8], and is further combined with gradient tracking method to develop distributed algorithms with linear convergence rate over general digraphs [7,20,23]. It is worthy noting that push-sum protocol requires an additional computation and communication. To avoid the additional subsystem, recently, Ref. [11,21] propose push-pull protocol based distributed optimization algorithms by utilizing a row-stochastic matrix for consensus and a column stochastic matrix for gradient tracking. These algorithms are also shown to be linear convergent. Further extensions of push-pull based algorithms have been reported in the scenario of time-varying networks [17] and asynchronous settings [25]. Despite of the progress, complexity analysis of local computation over digraphs remains rare. Some stochastic methods combined with the push-sum protocol are proposed for reducing gradient computation complexity, such as [6,14,15]. These stochastic methods are more suitable for the scenario that a mass of data distributed in the communication network leads to unbearable full-gradient computation for each node. However, when full-gradient is allowed, stochastic methods have some disadvantages compared to non-stochastic ones, such as more communication cost for reaching the same convergence precision, and extra resources brought by variance reduced techniques. Consequently, it is meaningful to develop an algorithm efficient in gradient computation complexity based on full-gradient. As far as we know, there is little literature achieving this objective over digraphs.

Motivated by above discussions, we develop a novel distributed optimization algorithm SSGT-PP by combining push-pull protocol with snapshot gradient tracking, which is suitable for general digraphs with reduced complexity of local gradient computation. We theoretically prove its linear convergence rate on smooth strongly convex functions and explicitly characterize the gradient computation complexity $\mathcal{O}\left(\frac{\kappa^2}{\theta}\log\frac{1}{\epsilon}\right)$, where κ represents the condition number of functions, and θ represents the condition number of digraphs. It shows that

SSGT-PP is efficient in gradient computation complexity. Finally, we experimentally demonstrate the efficiency of SSGT-PP.

Notation. $\mathbf{1}$ represents the all-ones column vector, $\mathbf{I}$ represents the identity matrix and $\mathbf{O}$ represents the all zero matrix. The norms $\|\cdot\|$ appear in this paper all denote l_2-norm.

2 Preliminaries

In this section, we introduce the setting of communication in our algorithm and requirements on objective functions.

We consider the scenario of directed communication network, which is described as a digraph $\mathcal{G} = (\mathcal{V}, \mathcal{E})$, where $\mathcal{V} = \{1, 2, \ldots, n\}$ is the set of nodes and $\mathcal{E} \subset \mathcal{V} \times \mathcal{V}$ is the set of edges associating the ordered nodes. The adjacent weighted matrix $\mathbf{W} = [w_{ij}]_{n\times n}$ associated with $\mathcal{G}$ is defined as $w_{ij} > 0$ iff $(j, i) \in \mathcal{E}$, and $w_{ij} = 0$ otherwise. The root node in graph $\mathcal{G}_{\mathbf{W}}$ is defined as the one which can reach all the other nodes through a directed path. The set of all root nodes in $\mathcal{G}_{\mathbf{W}}$ is denoted as $\mathcal{R}_{\mathbf{W}}$. In our algorithm, all nodes communicate with others through two communication networks $\mathcal{G}_{\mathbf{R}}$ and $\mathcal{G}_{\mathbf{C}}$, whose adjacent weighted matrices are $\mathbf{R}$ and $\mathbf{C}$, respectively. The two communication networks satisfy the following assumption.

Assumption 1 (Communication). *The digraphs $\mathcal{G}_{\mathbf{R}}$ and $\mathcal{G}_{\mathbf{C}}$ satisfy*

(a) Both $\mathcal{G}_{\mathbf{R}}$ and $\mathcal{G}_{\mathbf{C}^{\mathrm{T}}}$ contain at least one spanning tree and $\mathcal{R}_{\mathbf{R}} \cap \mathcal{R}_{\mathbf{C}^{\mathrm{T}}} = \emptyset$.
(b) $\mathbf{R}$ is row-stochastic and $\mathbf{C}$ is column-stochastic, i.e., $\mathbf{R1} = \mathbf{1}$ and $\mathbf{1}^{\mathrm{T}}\mathbf{C} = \mathbf{1}^{\mathrm{T}}$.

The following assumptions for the objective functions in problem (1) are required throughout this paper.

Assumption 2 (Smoothness). *Each $f_i(\mathbf{x})$ is L_i-smooth, i.e., $\|\nabla f_i(\mathbf{y}) - \nabla f_i(\mathbf{x})\| \leq L_i \|\mathbf{y} - \mathbf{x}\|$, $\forall \mathbf{x}, \mathbf{y} \in \mathbb{R}^d$.*

Assumption 3 (Strong convexity). *Each $f_i(\mathbf{x})$ is μ-strongly convex, i.e.,*

$$f_i(\mathbf{y}) \geq f_i(\mathbf{x}) + \langle \nabla f_i(\mathbf{x}), \mathbf{y} - \mathbf{x} \rangle + \frac{\mu}{2} \|\mathbf{y} - \mathbf{x}\|^2, \quad \forall \mathbf{x}, \mathbf{y} \in \mathbb{R}^d.$$

Assumption 1(a) unifies different types of distributed architecture [11]. Under Assumption 2, we can obtain that f is $\bar{L}$-smooth, where $\bar{L} = \frac{1}{n}\sum_{i=1}^{n} L_i$. Define $\hat{L} = \max_i L_i$. The following result is instrumental for analysis of our algorithm.

Lemma 1 ***(see [2] and [11]).*** *Under Assumption 1, it holds that*

(a) $\mathbf{R}$ has a unique non-negative left eigenvector $\mathbf{u}$ satisfying $\mathbf{u}^{\mathrm{T}}\mathbf{R} = \mathbf{u}^{\mathrm{T}}$ and $\mathbf{u}^{\mathrm{T}}\mathbf{1} = n$. $\mathbf{C}$ has a unique non-negative right eigenvector $\mathbf{v}$ satisfying $\mathbf{Cv} = \mathbf{v}$ and $\mathbf{1}^{\mathrm{T}}\mathbf{v} = n$.
(b) $\sigma_{\mathbf{R}} \triangleq \|\mathbf{R} - \frac{1}{n}\mathbf{1}\mathbf{u}^{\mathrm{T}}\| < 1$ and $\sigma_{\mathbf{C}} \triangleq \|\mathbf{C} - \frac{1}{n}\mathbf{v}\mathbf{1}^{\mathrm{T}}\| < 1$.

Algorithm 1. SSGT-PP

Require: initial position: $\mathbf{x}_i^0 = \mathbf{q}_i^0$, $\mathbf{y}_i^0 = \nabla f_i(\mathbf{x}_i^0)$ $\forall i$; gossip matrices: $\mathbf{R} = [r_{ij}]_{n\times n}$ and $\mathbf{C} = [c_{ij}]_{n\times n}$; probability: $p \in (0,1)$; stepsize: $\alpha > 0$.
Ensure: $\mathbf{x}_i^K$ or $\mathbf{q}_i^K$, $\forall i$.
for $k = 0, 1, 2 \ldots K$ **do**
    Sample $\xi^k \sim \text{Bernoulli}(p)$.
    if $\xi^k == 0$ **then**
        $\mathbf{y}_i^{k+1} = \sum_{j=1}^{n} c_{ij}\mathbf{y}_j^k$
        $\mathbf{q}_i^{k+1} = \mathbf{q}_i^k$
    else
        $\mathbf{y}_i^{k+1} = \sum_{j=1}^{n} c_{ij}\mathbf{y}_j^k + \nabla f_i(\mathbf{x}_i^k) - \nabla f_i(\mathbf{q}_i^k)$
        $\mathbf{q}_i^{k+1} = \mathbf{x}_i^k$
    end if
    $\mathbf{x}_i^{k+1} = \sum_{j=1}^{n} r_{ij}(\mathbf{x}_j^k - \alpha\mathbf{y}_j^k)$
end for

3 Our Algorithm and Main Result

In this section, we formulate our algorithm SSGT-PP and then present its convergence result.

Recalling the Push-Pull algorithm [11], each node i holds a local copy $\mathbf{x}_i \in \mathbb{R}^d$ of $\mathbf{x}$ and a gradient tracker $\mathbf{y}_i \in \mathbb{R}^d$, and then *pulls* $\mathbf{x}_j$ from neighbors for consensus and *pushes* $\mathbf{y}_i$ to neighbors for gradient tracking. The "pulls" and "pushes" are executed over two independent communication graphs $\mathcal{G}_\mathbf{R}$ and $\mathcal{G}_\mathbf{C}$. Specifically, at kth iteration, the update of $\mathbf{x}_i$ and $\mathbf{y}_i$ is given by

$$\mathbf{x}_i^{k+1} = \sum_{j=1}^{n} r_{ij}(\mathbf{x}_j^k - \alpha\mathbf{y}_j^k), \tag{2a}$$

$$\mathbf{y}_i^{k+1} = \sum_{j=1}^{n} c_{ij}\mathbf{y}_j^k + \nabla f_i(\mathbf{x}_i^{k+1}) - \nabla f_i(\mathbf{x}_i^k), \tag{2b}$$

where $\mathbf{R} = [r_{ij}]_{n\times n}$ and $\mathbf{C} = [c_{ij}]_{n\times n}$ are adjacent matrices of $\mathcal{G}_\mathbf{R}$ and $\mathcal{G}_\mathbf{C}$ respectively. It can be observed that the gradient is computed at each iteration for tracking in (2). Aiming to reduce the gradient computation cost, we introduce *snapshot gradient tracking* [19] to Push-Pull algorithm (2) and propose SSGT-PP, whose pseudo code is shown in Algorithm 1. In SSGT-PP, each node additionally holds a snapshot point $\mathbf{q}_i \in \mathbb{R}^d$ to record the history of the $\mathbf{x}_i$ and shares a stochastic parameter $\xi \sim \text{Bernoulli}(p)$ to decide the execution time of gradient computation.

Initially, $\mathbf{x}_i^0 = \mathbf{q}_i^0$, $\mathbf{y}_i^0 = \nabla f_i(\mathbf{x}_i^0)$. If $\xi^k = 0$, nodes just sum up $\mathbf{y}_j^k$ "pushed" from their neighbors without computing gradient, and keep $\mathbf{q}_i^k$ unchanged. Otherwise, nodes proceed gradient computation and update $\mathbf{q}_i^k$.

As will be shown in Lemma 1, all nodes here actually track the historical average of global gradient value rather than the latest one. It is worthy noting that $\mathbf{q}_i$ is not necessary when we implement SSGT-PP. In fact, compared with (2), SSGT-PP only requires few extra storage for ξ^k but can reduce about $1-p$ gradient computations in expectation. Regarding SSGT-PP, we have the following convergence result.

Theorem 1. *Supposed that Assumptions 1–3 hold, if the positive step α satisfies*

$$\alpha \leq \min\left\{\frac{2n}{(\mu+\bar{L})\mathbf{u}^{\mathrm{T}}\mathbf{v}}, \frac{(1-\sigma_{\mathbf{R}})n\min\{1,\mu\}}{64\,\|\mathbf{u}\|\,\|\mathbf{v}\|\max\{1,\hat{L}^2\}}\cdot\min\left\{p, \frac{(1-\sigma_{\mathbf{C}})\mathbf{u}^{\mathrm{T}}\mathbf{v}}{4\sqrt{n}\,\|\mathbf{u}\|}\right\}\right\},$$

then SSGT-PP converges in expectation to an optimal solution $\mathbf{x}^*$ at the linear rate $\mathcal{O}(\rho^k)$ with $\rho = 1-\min\left\{\frac{\mathbf{u}^{\mathrm{T}}\mathbf{v}}{4n}\alpha\mu, \frac{11(1-\sigma_{\mathbf{R}})}{16}, \frac{9(1-\sigma_{\mathbf{C}})}{16}, \frac{17p}{32}\right\} < 1$.

Complexity of Communication and Computation. Define the condition number of functions as $\kappa = \frac{\hat{L}}{\mu}$ and the condition number of graph as $\theta^2 = (1-\sigma_{\mathbf{R}})(1-\sigma_{\mathbf{C}})$ with specially $p = \frac{1-\sigma_{\mathbf{C}}}{2}$. When the local objective functions satisfy $\mu < 1$ and $\hat{L} > 1$, then $\alpha \leq \frac{c\theta^2\mu}{\hat{L}^2}$, where c is a constant. According to Theorem 1, the convergence rate is determined by $\rho = 1-\frac{\mathbf{u}^{\mathrm{T}}\mathbf{v}}{4n}\alpha\mu \geq 1-\tilde{c}\frac{\theta^2}{\kappa^2}$, where $\tilde{c}$ is a constant. In this case, SSGT-PP achieves an ϵ-optimal solution in $\mathcal{O}\left(\frac{\kappa^2}{\theta^2}\log\frac{1}{\epsilon}\right)$ communications, and $\mathcal{O}\left(\frac{\kappa^2}{1-\sigma_{\mathbf{R}}}\log\frac{1}{\epsilon}\right)$ gradient computations. Note that $(1-\sigma_{\mathbf{R}})$ can be regarded as θ, the special first power dependence on the graph property. It is interesting to find that the gradient computation complexity exhibits dependence on the properties of $\mathbf{R}$ rather than that of $\mathbf{C}$.

Stepsize. For the step size α, Theorem 1 still holds if we take α to be an upper bound of the α_is. It is worthy noting that the upper bound on stepsize is still conservative, which will be shown in numerical experiments.

4 Convergence Analysis

In this section, we prove the convergence results for SSGT-PP. Before proceeding, we write SSGT-PP into an aggregated form for simplicity. Define

$$\mathbf{X}^k \triangleq [\mathbf{x}_1^k, \mathbf{x}_2^k, \ldots, \mathbf{x}_n^k]^{\mathrm{T}} \in \mathbb{R}^{n\times d}, \qquad \mathbf{Y}^k \triangleq [\mathbf{y}_1^k, \mathbf{y}_2^k, \ldots, \mathbf{y}_n^k]^{\mathrm{T}} \in \mathbb{R}^{n\times d},$$

$$\mathbf{Q}^k \triangleq [\mathbf{q}_1^k, \mathbf{q}_2^k, \ldots, \mathbf{q}_n^k]^{\mathrm{T}} \in \mathbb{R}^{n\times d}, \qquad F(\mathbf{X}^k) \triangleq \sum_{i=1}^n f_i(\mathbf{x}_i^k),$$

$$\nabla F(\mathbf{X}^k) \triangleq \left[\nabla f_1(\mathbf{x}_1^k), \nabla f_2(\mathbf{x}_2^k), \ldots, \nabla f_n(\mathbf{x}_n^k)\right]^{\mathrm{T}} \in \mathbb{R}^{n\times d}.$$

Then, SSGT-PP can be written as follows:

$$\mathbf{X}^{k+1} = \mathbf{R}(\mathbf{X}^k - \alpha\mathbf{Y}^k), \tag{3a}$$

$$\mathbf{Q}^{k+1} = \xi^k\mathbf{X}^k + (1-\xi^k)\mathbf{Q}^k, \tag{3b}$$

$$\mathbf{Y}^{k+1} = \mathbf{C}\mathbf{Y}^k + \xi^k\left(\nabla F(\mathbf{X}^k) - \nabla F(\mathbf{Q}^k)\right). \tag{3c}$$

Based on the properties of $\mathbf{R}$ and $\mathbf{C}$ (see (a) and (b) in Lemma 1), we further define $\overline{\mathbf{x}}^k \triangleq \frac{1}{n}(\mathbf{X}^k)^{\mathrm{T}}\mathbf{u}$, $\overline{\mathbf{q}}^k \triangleq \frac{1}{n}(\mathbf{Q}^k)^{\mathrm{T}}\mathbf{u}$, $\overline{\mathbf{y}}^k \triangleq \frac{1}{n}(\mathbf{Y}^k)^{\mathrm{T}}\mathbf{1}$, and $\overline{\mathbf{X}}^k \triangleq \mathbf{1}(\overline{\mathbf{x}}^k)^{\mathrm{T}}$, $\overline{\mathbf{Q}}^k \triangleq \mathbf{1}(\overline{\mathbf{q}}^k)^{\mathrm{T}}$, $\overline{\mathbf{Y}}^k \triangleq \mathbf{v}(\overline{\mathbf{y}}^k)^{\mathrm{T}}$.

Then, by further using Lemma 1, it can be checked that $\left(\mathbf{R}-\frac{1}{n}\mathbf{1}\mathbf{u}^{\mathrm{T}}\right)\overline{\mathbf{X}}^k = \mathbf{O}$ and $\left(\mathbf{C}-\frac{1}{n}\mathbf{v}\mathbf{1}^{\mathrm{T}}\right)\overline{\mathbf{Y}}^k = \mathbf{O}$, which imply

$$\left\|\left(\mathbf{R}-\frac{1}{n}\mathbf{1}\mathbf{u}^{\mathrm{T}}\right)\mathbf{X}^k\right\| \le \sigma_{\mathbf{R}}\left\|\mathbf{X}^k-\overline{\mathbf{X}}^k\right\|, \tag{4}$$

$$\left\|\left(\mathbf{C}-\frac{1}{n}\mathbf{v}\mathbf{1}^{\mathrm{T}}\right)\mathbf{Y}^k\right\| \le \sigma_{\mathbf{C}}\left\|\mathbf{Y}^k-\overline{\mathbf{Y}}^k\right\|. \tag{5}$$

The two inequalities (4) and (5) reflect the information fusion rate, which are affected by the structures of $\mathcal{G}_{\mathbf{R}}$ and $\mathcal{G}_{\mathbf{C}}$. In addition, according to (3), we also have the following result. It shows that $\overline{\mathbf{y}}^k$ tracks the average of the historical gradient depending on the snapshot point $\mathbf{q}^k$.

Lemma 1 (see [19]). $\forall k \ge 0$, $\overline{\mathbf{y}}^{k+1} = \frac{1}{n}\left(\nabla F(\mathbf{Q}^{k+1})\right)^{\mathrm{T}}\mathbf{1}$.

The next lemma is important for proving the convergence results of our algorithm. Denote the optimal solution as $\mathbf{x}^* \in \mathbb{R}^d$ and $\mathbf{X}^* \triangleq \mathbf{x}^*\mathbf{1}^{\mathrm{T}}$.

Lemma 2 (see [12]). *Under Assumptions 2 and 3, if* $\eta \in (0, \frac{2}{\mu+L}]$, $\lambda = 1-\eta\mu < 1$, *and* $\mathbf{x}_+ = \mathbf{x} - \eta\nabla f(\mathbf{x}) + \varepsilon$, *then* $\|\mathbf{x}_+ - \mathbf{x}^*\| \le \lambda\|\mathbf{x}-\mathbf{x}^*\| + \|\varepsilon\|$.

Now we are ready to prove Theorem 1. We begin with a brief description of our proof framework. We will first bound several errors in terms of the linear combination of their previous values. Then, we will construct a linear inequality to drive the linear convergence rate in expectation.

We first bound the average gradient $\mathbb{E}\left\|\overline{\mathbf{y}}^k\right\|$. According to Lemma 1, we have

$$\begin{aligned}
\mathbb{E}\left\|\overline{\mathbf{y}}^k\right\| &= \mathbb{E}\left\|\frac{1}{n}\mathbf{1}^{\mathrm{T}}\nabla F(\mathbf{Q}^k)\right\| \\
&= \mathbb{E}\left\|\frac{1}{n}\mathbf{1}^{\mathrm{T}}\left(\nabla F(\mathbf{Q}^k) - \nabla F(\overline{\mathbf{X}}^k) + \nabla F(\overline{\mathbf{X}}^k) - \nabla F(\mathbf{X}^*)\right)\right\| \\
&\le \frac{\hat{L}}{\sqrt{n}}\mathbb{E}\left\|\mathbf{Q}^k - \overline{\mathbf{Q}}^k\right\| + \hat{L}\mathbb{E}\left\|\overline{\mathbf{x}}^k - \overline{\mathbf{q}}^k\right\| + \hat{L}\mathbb{E}\left\|\overline{\mathbf{x}}^k - \mathbf{x}^*\right\|.
\end{aligned} \tag{6}$$

Next we bound the consensus error $\mathbb{E}\left\|\mathbf{X}^{k+1} - \overline{\mathbf{X}}^{k+1}\right\|$, the delayed consensus error: $\mathbb{E}\left\|\mathbf{Q}^{k+1} - \overline{\mathbf{Q}}^{k+1}\right\|$, the gradient tracking error $\mathbb{E}\left\|\mathbf{Y}^{k+1} - \overline{\mathbf{Y}}^{k+1}\right\|$, the delay gap $\mathbb{E}\left\|\overline{\mathbf{x}}^{k+1} - \overline{\mathbf{q}}^{k+1}\right\|$ and the optimality gap $\mathbb{E}\left\|\overline{\mathbf{x}}^{k+1} - \mathbf{x}^*\right\|$.

Bounding $\left\|\mathbf{X}^{k+1}-\overline{\mathbf{X}}^{k+1}\right\|$: It follows from (3a) that

$$\begin{aligned}
&\mathbb{E}\left\|\mathbf{X}^{k+1}-\overline{\mathbf{X}}^{k+1}\right\| \\
&=\mathbb{E}\left\|\left(\mathbf{R}-\frac{1}{n}\mathbf{1}\mathbf{u}^{\mathrm{T}}\right)\mathbf{X}^{k}+\overline{\mathbf{X}}^{k}-\overline{\mathbf{X}}^{k}-\alpha\left(\mathbf{R}-\frac{1}{n}\mathbf{1}\mathbf{u}^{\mathrm{T}}\right)\mathbf{Y}^{k}\right\| \\
&\overset{(a)}{\leq}\sigma_{\mathbf{R}}\mathbb{E}\left\|\mathbf{X}^{k}-\overline{\mathbf{X}}^{k}\right\|+\alpha\sigma_{\mathbf{R}}\mathbb{E}\left\|\mathbf{Y}^{k}-\overline{\mathbf{Y}}^{k}+\mathbf{v}(\overline{\mathbf{y}}^{k})^{\mathrm{T}}\right\| \\
&\leq\sigma_{\mathbf{R}}\mathbb{E}\left\|\mathbf{X}^{k}-\overline{\mathbf{X}}^{k}\right\|+\alpha\sigma_{\mathbf{R}}\left(\mathbb{E}\left\|\mathbf{Y}^{k}-\overline{\mathbf{Y}}^{k}\right\|+\|\mathbf{v}\|\,\mathbb{E}\left\|\overline{\mathbf{y}}^{k}\right\|\right), \\
&\leq\sigma_{\mathbf{R}}\mathbb{E}\left\|\mathbf{X}^{k}-\overline{\mathbf{X}}^{k}\right\|+\frac{\hat{L}\alpha\sigma_{\mathbf{R}}\|\mathbf{v}\|}{\sqrt{n}}\mathbb{E}\left\|\mathbf{Q}^{k}-\overline{\mathbf{Q}}^{k}\right\|+\alpha\sigma_{\mathbf{R}}\mathbb{E}\left\|\mathbf{Y}^{k}-\overline{\mathbf{Y}}^{k}\right\| \\
&\quad+\hat{L}\alpha\sigma_{\mathbf{R}}\|\mathbf{v}\|\,\mathbb{E}\left\|\overline{\mathbf{x}}^{k}-\overline{\mathbf{q}}^{k}\right\|+\hat{L}\alpha\sigma_{\mathbf{R}}\|\mathbf{v}\|\,\mathbb{E}\left\|\overline{\mathbf{x}}^{k+1}-\mathbf{x}^{*}\right\|. \qquad (7)
\end{aligned}$$

where $\overset{(a)}{\leq}$ utilizes inequality (4).

Bounding $\mathbb{E}\left\|\mathbf{Q}^{k+1}-\overline{\mathbf{Q}}^{k+1}\right\|$: It follows from (3b) that

$$\begin{aligned}
\mathbb{E}\left\|\mathbf{Q}^{k+1}-\overline{\mathbf{Q}}^{k+1}\right\|&=\mathbb{E}\left\|\xi^{k}\left(\mathbf{X}^{k}-\overline{\mathbf{X}}^{k}\right)+(1-\xi^{k})\left(\mathbf{Q}^{k}-\overline{\mathbf{Q}}^{k}\right)\right\| \\
&\overset{(a)}{=}p\mathbb{E}\left\|\mathbf{X}^{k}-\overline{\mathbf{X}}^{k}\right\|+(1-p)\mathbb{E}\left\|\mathbf{Q}^{k}-\overline{\mathbf{Q}}^{k}\right\|, \qquad (8)
\end{aligned}$$

where $\overset{(a)}{=}$ utilizes $\mathbb{E}\xi^{k}=p$.

Bounding $\mathbb{E}\left\|\mathbf{Y}^{k+1}-\overline{\mathbf{Y}}^{k+1}\right\|$: It follows from (3c) that

$$\begin{aligned}
&\mathbb{E}\left\|\mathbf{Y}^{k+1}-\overline{\mathbf{Y}}^{k+1}\right\| \\
&=\mathbb{E}\left\|\left(\mathbf{C}-\frac{1}{n}\mathbf{v}\mathbf{1}^{\mathrm{T}}\right)\mathbf{Y}^{k}+\overline{\mathbf{Y}}^{k}-\overline{\mathbf{Y}}^{k}+\xi^{k}\left(\mathbf{I}-\frac{1}{n}\mathbf{v}\mathbf{1}^{\mathrm{T}}\right)\left(\nabla F(\mathbf{X}^{k})-\nabla F(\mathbf{Q}^{k})\right)\right\| \\
&\overset{(a)}{\leq}\sigma_{\mathbf{C}}\mathbb{E}\left\|\mathbf{Y}^{k}-\overline{\mathbf{Y}}^{k}\right\|+2p\mathbb{E}\left\|\nabla F(\mathbf{X}^{k})-\nabla F(\mathbf{Q}^{k})\right\|, \qquad (9)
\end{aligned}$$

where $\overset{(a)}{\leq}$ follows from (5) and $\left\|\mathbf{I}-\frac{1}{n}\mathbf{v}\mathbf{1}^{\mathrm{T}}\right\|\leq 2$. Regarding $\mathbb{E}\left\|\nabla F(\mathbf{X}^{k})-\nabla F(\mathbf{Q}^{k})\right\|$, it follows from Assumption 2 that $\mathbb{E}\left\|\nabla F(\mathbf{X}^{k})-\nabla F(\mathbf{Q}^{k})\right\|\leq\hat{L}\mathbb{E}\left\|\mathbf{X}^{k}-\mathbf{Q}^{k}\right\|$, where $\mathbb{E}\left\|\mathbf{X}^{k}-\mathbf{Q}^{k}\right\|$ is further estimated as below:

$$\begin{aligned}
\mathbb{E}\left\|\mathbf{X}^{k}-\mathbf{Q}^{k}\right\|&=\mathbb{E}\left\|\mathbf{X}^{k}-\overline{\mathbf{X}}^{k}+\overline{\mathbf{X}}^{k}-\overline{\mathbf{Q}}^{k}-\left(\mathbf{Q}^{k}-\overline{\mathbf{Q}}^{k}\right)\right\| \\
&\leq\mathbb{E}\left\|\mathbf{X}^{k}-\overline{\mathbf{X}}^{k}\right\|+\sqrt{n}\mathbb{E}\left\|\overline{\mathbf{x}}^{k}-\overline{\mathbf{q}}^{k}\right\|+\mathbb{E}\left\|\mathbf{Q}^{k}-\overline{\mathbf{Q}}^{k}\right\| \qquad (10)
\end{aligned}$$

Bounding $\mathbb{E}\left\|\overline{\mathbf{x}}^{k+1}-\overline{\mathbf{q}}^{k+1}\right\|$: Recalling the evolution of $\overline{\mathbf{x}}^{k+1}$ and $\overline{\mathbf{q}}^{k+1}$, we have

$$\begin{aligned}\mathbb{E}\left\|\overline{\mathbf{x}}^{k+1}-\overline{\mathbf{q}}^{k+1}\right\| &= \mathbb{E}\left\|(1-\xi^k)(\overline{\mathbf{x}}^k-\overline{\mathbf{q}}^k)^{\mathrm{T}}-\frac{\alpha}{n}\mathbf{u}^{\mathrm{T}}\left(\mathbf{Y}^k-\overline{\mathbf{Y}}^k+\mathbf{v}(\overline{\mathbf{y}}^k)^{\mathrm{T}}\right)\right\| \\ &\leq (1-p)\mathbb{E}\left\|\overline{\mathbf{x}}^k-\overline{\mathbf{q}}^k\right\|+\alpha\frac{\|\mathbf{u}\|}{n}\mathbb{E}\left\|\mathbf{Y}^k-\overline{\mathbf{Y}}^k\right\|+\hat{\alpha}\mathbb{E}\left\|\overline{\mathbf{y}}^k\right\| \\ &\leq \left((1-p)+\hat{\alpha}\hat{L}\right)\mathbb{E}\left\|\overline{\mathbf{x}}^k-\overline{\mathbf{q}}^k\right\|+\alpha\frac{\|\mathbf{u}\|}{n}\mathbb{E}\left\|\mathbf{Y}^k-\overline{\mathbf{Y}}^k\right\| \\ &\quad+\hat{\alpha}\frac{\hat{L}}{\sqrt{n}}\mathbb{E}\left\|\mathbf{Q}^k-\overline{\mathbf{Q}}^k\right\|+\hat{\alpha}\hat{L}\mathbb{E}\left\|\overline{\mathbf{x}}^{k+1}-\mathbf{x}^*\right\|, \end{aligned} \tag{11}$$

Bounding $\mathbb{E}\left\|\overline{\mathbf{x}}^{k+1}-\mathbf{x}^*\right\|$: Recalling the evolution of $\overline{\mathbf{x}}$, we have

$$\begin{aligned}\overline{\mathbf{x}}^{k+1} &= \overline{\mathbf{x}}^k-\frac{\alpha}{n}(\mathbf{Y}^k)^{\mathrm{T}}\mathbf{u} \\ &= \overline{\mathbf{x}}^k-\frac{\alpha}{n}\left(\mathbf{Y}^k-\mathbf{v}(\overline{\mathbf{y}}^k)^{\mathrm{T}}+\mathbf{v}(\overline{\mathbf{y}}^k)^{\mathrm{T}}\right)^{\mathrm{T}}\mathbf{u} \\ &= \overline{\mathbf{x}}^k-\hat{\alpha}\overline{\mathbf{y}}^k-\frac{\alpha}{n}\left(\mathbf{Y}^k-\overline{\mathbf{Y}}^k\right)^{\mathrm{T}}\mathbf{u} \\ &= \overline{\mathbf{x}}^k-\hat{\alpha}\nabla f(\overline{\mathbf{x}}^k)+\Xi^k \end{aligned} \tag{12}$$

where $\hat{\alpha} \triangleq \frac{\alpha}{n}\mathbf{u}^{\mathrm{T}}\mathbf{v}>0$ and $\Xi^k \triangleq \hat{\alpha}(\nabla f(\overline{\mathbf{x}}^k)-\overline{\mathbf{y}}^k)-\frac{\alpha}{n}(\mathbf{Y}^k-\overline{\mathbf{Y}}^k)^{\mathrm{T}}\mathbf{u}$. According to Lemma 2, we have

$$\mathbb{E}\left\|\overline{\mathbf{x}}^{k+1}-\mathbf{x}^*\right\| \leq \lambda\mathbb{E}\left\|\overline{\mathbf{x}}^k-\mathbf{x}^*\right\|+\mathbb{E}\left\|\Xi^k\right\|, \tag{13}$$

where $\lambda=1-\hat{\alpha}\mu<1$ when $\hat{\alpha}<\frac{2}{\mu+\bar{L}}$. The term $\left\|\Xi^k\right\|$ is bounded as below:

$$\begin{aligned}\left\|\Xi^k\right\| &= \left\|\hat{\alpha}\left(\nabla f(\overline{\mathbf{x}}^k)-\overline{\mathbf{y}}^k\right)-\frac{\alpha}{n}\left(\mathbf{Y}^k-\overline{\mathbf{Y}}^k\right)^{\mathrm{T}}\mathbf{u}+\hat{\alpha}\left(\frac{(\nabla F(\overline{\mathbf{Q}}^k))^{\mathrm{T}}\mathbf{1}}{n}-\frac{(\nabla F(\mathbf{Q}^k))^{\mathrm{T}}\mathbf{1}}{n}\right)\right\| \\ &\overset{(a)}{\leq} \hat{\alpha}\bar{L}\left\|\overline{\mathbf{x}}^k-\overline{\mathbf{q}}^k\right\|+\alpha\frac{\|\mathbf{u}\|}{n}\left\|\mathbf{Y}^k-\overline{\mathbf{Y}}^k\right\|+\frac{\hat{\alpha}\hat{L}}{\sqrt{n}}\left\|\mathbf{Q}^k-\overline{\mathbf{Q}}^k\right\|, \end{aligned} \tag{14}$$

where $\overset{(a)}{\leq}$ utilizes Lemma 1.

Define

$$\begin{aligned}\Phi^{k+1} \triangleq\ & c_1\mathbb{E}\left\|\overline{\mathbf{x}}^{k+1}-\mathbf{x}^*\right\|+c_2\mathbb{E}\left\|\mathbf{X}^{k+1}-\overline{\mathbf{X}}^{k+1}\right\|+c_3\mathbb{E}\left\|\mathbf{Y}^{k+1}-\overline{\mathbf{Y}}^{k+1}\right\| \\ &+c_4\mathbb{E}\left\|\mathbf{Q}^{k+1}-\overline{\mathbf{Q}}^{k+1}\right\|+c_5\mathbb{E}\left\|\overline{\mathbf{x}}^{k+1}-\overline{\mathbf{q}}^{k+1}\right\| \end{aligned} \tag{15}$$

where $c_1=\frac{8\hat{L}^2\sqrt{n}\|\mathbf{u}\|\|\mathbf{v}\|}{\mu\mathbf{u}^{\mathrm{T}}\mathbf{v}}$, $c_2=4pL$, $c_3=\frac{1-\sigma_{\mathbf{R}}}{8}$, $c_4=\hat{L}(1-\sigma_{\mathbf{R}})$, $c_5=2\hat{L}\sqrt{n}(1-\sigma_{\mathbf{R}})$. Then we have the following result.

Lemma 3. *Under the conditions in Theorem 1, it holds that* $\Phi^{k+1}\leq\rho\Phi^k$.

Proof. Plugging (13), (7), (9), (8) and (11) into (15), we have

$$\Phi^{k+1} \leq C_1\mathbb{E}\left\|\overline{\mathbf{x}}^k - \mathbf{x}^*\right\| + C_2\mathbb{E}\left\|\mathbf{X}^k - \overline{\mathbf{X}}^k\right\| + C_3\mathbb{E}\left\|\mathbf{Y}^k - \overline{\mathbf{Y}}^k\right\| + C_4\mathbb{E}\left\|\mathbf{Q}^k - \overline{\mathbf{Q}}^k\right\| + C_5\mathbb{E}\left\|\overline{\mathbf{x}}^k - \overline{\mathbf{q}}^k\right\| \tag{16}$$

where

$$\begin{aligned}
C_1 &\triangleq c_1\lambda + c_2\hat{L}\alpha\sigma_{\mathbf{R}}\left\|\mathbf{v}\right\| + c_5\hat{\alpha}\hat{L},\\
C_2 &\triangleq c_2\sigma_{\mathbf{R}} + c_3 2p\hat{L} + c_4 p,\\
C_3 &\triangleq c_1\alpha\frac{\left\|\mathbf{u}\right\|}{n} + c_2\alpha\sigma_{\mathbf{R}} + c_3\sigma_{\mathbf{C}} + c_5\alpha\frac{\left\|\mathbf{u}\right\|}{n},\\
C_4 &\triangleq c_4(1-p) + (c_1+c_5)\frac{\hat{\alpha}\hat{L}}{\sqrt{n}} + c_2\frac{\hat{L}\alpha\sigma_{\mathbf{R}}\left\|\mathbf{v}\right\|}{\sqrt{n}} + c_3 2p\hat{L},\\
C_5 &\triangleq c_5(1-p) + (c_1+c_5)\hat{\alpha}\hat{L} + c_2\hat{L}\alpha\sigma_{\mathbf{R}}\left\|\mathbf{v}\right\| + c_3 2p\hat{L}\sqrt{n}.
\end{aligned}$$

Next, we will show that $C_i \leq \rho c_i, i = 1, \ldots, 5$.

For C_1: since $\hat{\alpha} \leq \frac{2}{\mu+L}$, which implies $\alpha \leq \frac{2n}{(\mu+L)\mathbf{u}^{\mathrm{T}}\mathbf{v}}$,

$$C_1 \overset{(a)}{\leq} c_1\left(1 - \hat{\alpha}\mu + \frac{4}{8}\hat{\alpha}\mu + \frac{2}{8}\hat{\alpha}\mu\right) = c_1\left(1 - \frac{\hat{\alpha}\mu}{4}\right) \leq \rho c_1. \tag{17}$$

where $\overset{(a)}{\leq}$ utilizes $\left\|\mathbf{u}\right\| = \sqrt{\mathbf{u}^{\mathrm{T}}\mathbf{u}} \geq \frac{\mathbf{u}^{\mathrm{T}}\mathbf{1}}{\sqrt{n}} = \sqrt{n}$.

For C_2:

$$C_2 = c_2\left(\sigma_{\mathbf{R}} + \frac{1-\sigma_{\mathbf{R}}}{16} + \frac{1-\sigma_{\mathbf{R}}}{4}\right) = c_2\left(1 - \frac{11(1-\sigma_{\mathbf{R}})}{16}\right) \leq \rho c_2. \tag{18}$$

For C_3**:** since $\alpha \leq \frac{(1-\sigma_{\mathbf{R}})(1-\sigma_{\mathbf{C}})\sqrt{n}\mathbf{u}^{\mathrm{T}}\mathbf{v}\min\{1,\mu\}}{256\left\|\mathbf{u}\right\|^2\cdot\left\|\mathbf{v}\right\|\max\{1,\hat{L}^2\}}$, then

$$\begin{aligned}
C_3 &\overset{(a)}{\leq} c_3\left(\frac{64\hat{L}^2(1-\sigma_{\mathbf{C}})}{256\mu}\cdot\frac{\min\{1,\mu\}}{\max\{1,\hat{L}^2\}} + \frac{32\hat{L}(1-\sigma_{\mathbf{C}})}{256\max\{1,\hat{L}^2\}} + \sigma_{\mathbf{C}} + \frac{16\hat{L}(1-\sigma_{\mathbf{C}})}{256\max\{1,\hat{L}^2\}}\right)\\
&\leq c_3\left(1 - \frac{9(1-\sigma_{\mathbf{C}})}{16}\right) \leq \rho c_3.
\end{aligned} \tag{19}$$

where $\overset{(a)}{\leq}$ follows from $\mathbf{u}^{\mathrm{T}}\mathbf{v} \leq \left\|\mathbf{u}\right\|\left\|\mathbf{v}\right\|$, $\left\|\mathbf{u}\right\| \geq \sqrt{n}$ and $\left\|\mathbf{v}\right\| \geq \sqrt{n}$.

For C_4: since $\alpha \leq \frac{(1-\sigma_{\mathbf{R}})pn\min\{1,\mu\}}{64\|\mathbf{u}\|\cdot\|\mathbf{v}\|\max\{1,\hat{L}^2\}}$, then we have

$$\begin{aligned} C_4 &= c_4\left(\frac{8\hat{L}^2\|\mathbf{u}\|\|\mathbf{v}\|}{\mu n(1-\sigma_{\mathbf{R}})}\alpha + \frac{4p\hat{L}\sigma_{\mathbf{R}}\|\mathbf{v}\|}{\sqrt{n}(1-\sigma_{\mathbf{R}})}\alpha + \frac{p}{4} + 1 - p + \frac{2\hat{L}\mathbf{u}^{\mathrm{T}}\mathbf{v}}{n}\alpha\right) \\ &\leq c_4\left(\frac{8p\hat{L}^2}{64\mu}\cdot\frac{\min\{1,\mu\}}{\max\{1,\hat{L}^2\}} + \frac{4p\hat{L}\sqrt{n}}{64\|\mathbf{u}\|\max\{1,\hat{L}^2\}} + \frac{p}{4} + 1 - p + \frac{2\hat{L}p}{64\max\{1,\hat{L}^2\}}\right) \\ &\leq c_4\left(1 - \frac{17p}{32}\right) \leq \rho c_4. \end{aligned} \tag{20}$$

For C_5: since $\alpha \leq \frac{(1-\sigma_{\mathbf{R}})pn\min\{1,\mu\}}{64\|\mathbf{u}\|\cdot\|\mathbf{v}\|\max\{1,\hat{L}^2\}}$, then we have

$$\begin{aligned} C_5 &= c_5\left(\frac{4\hat{L}^2\|\mathbf{u}\|\|\mathbf{v}\|}{\mu n(1-\sigma_{\mathbf{R}})}\alpha + \frac{2p\hat{L}\sigma_{\mathbf{R}}\|\mathbf{v}\|}{\sqrt{n}(1-\sigma_{\mathbf{R}})}\alpha + \frac{p}{8} + 1 - p + \frac{\mathbf{u}^{\mathrm{T}}\mathbf{v}\hat{L}}{n}\alpha\right) \\ &\leq c_5\left(\frac{4p\hat{L}^2}{64\mu}\cdot\frac{\min\{1,\mu\}}{\max\{1,\hat{L}^2\}} + \frac{2p\hat{L}\sqrt{n}}{64\|\mathbf{u}\|\max\{1,\hat{L}^2\}} + \frac{p}{8} + 1 - p + \frac{p\hat{L}}{64\max\{1,\hat{L}^2\}}\right) \\ &\leq c_5\left(1 - \frac{57p}{64}\right) \leq \rho c_5. \end{aligned} \tag{21}$$

Plugging (17)-(21) into (16), we have $\Phi^{k+1} \leq \rho\Phi^k$, which completes the proof.

According to Lemma 2 and Lemma 3, we have

$$\mathbb{E}\left\|\mathbf{X}^k - \mathbf{X}^*\right\| \leq \left\|\mathbf{X}^k - \overline{\mathbf{X}}^k\right\| + \mathbb{E}\left\|\overline{\mathbf{X}}^k - \mathbf{X}^*\right\| \leq \rho^k\left(\frac{2\hat{L} + \mu p}{8p\hat{L}^2}\Phi^0\right),$$

which implies that $\mathbb{E}\left\|\mathbf{X}^k - \mathbf{X}^*\right\|$ converges linearly, and then the proof of Theorem 1 has completed.

5 Numerical Experiments

In this section, we randomly select two classes images of MINIST dataset for binary logistic regression classification. The logistic regression problem is described as

$$f(\mathbf{x}) = \min_{\mathbf{x}\in\mathbb{R}^d}\frac{1}{n}\sum_{i=1}^{n}\frac{\mu}{2}\|\mathbf{x}\|^2 + \frac{1}{s_i}\sum_{j=1}^{s_i}\log\left(1 + e^{-l_{i,j}\mathbf{a}_{i,j}^{\mathrm{T}}\mathbf{x}}\right),$$

where $\mathbf{x}$ is unknown parameter. s_i represents the number of images available to each node. $(\mathbf{a}_{i,j}, l_{i,j}) \in \mathbb{R}^d \times \{-1, 1\}$ where $\mathbf{a}_{i,j} \in \mathbb{R}^d$ is the feature vector and $l_{i,j}$ is the corresponding label. We use $n = 16$, $s_i = 750$ and $\mu = 0.01$. We conduct the experiments on an exponential digraph which is the same as that described in [14]. We compare SSGT-PP with Push-Pull, Push-DIGing [7], S-ADDOPT [15] and Push-SAGA [14] against the number of iterations and the

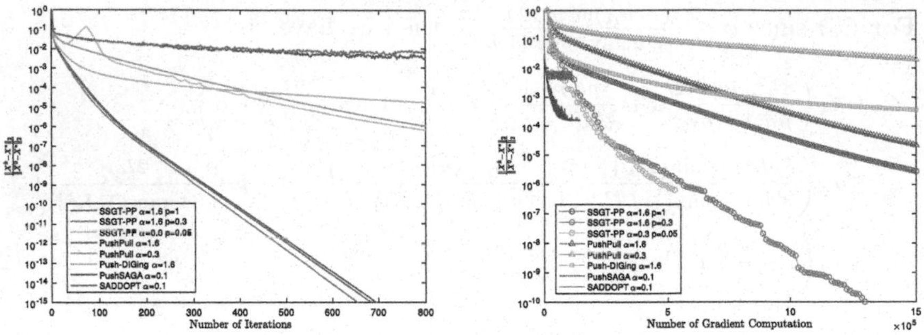

Fig. 1. Performance comparison against number of iterations(see the left figure (a)) and number of gradient computation(see the right figure (b))

number of gradient computation in Fig. 1. The residual is measured by $\frac{\|\mathbf{X}^k - \mathbf{X}^*\|}{\|\mathbf{X}^0 - \mathbf{X}^*\|}$. The stepsize α and p are hand-tuned for all algorithms to give faster convergence and more accurate solution.

We observed that, in Fig. 1(a), SSGT-PP performs the same or even better convergence effect than Push-Pull and Push-DIGing when the same stepsize is chosen. Figure 1(b) shows that SSGT-PP can converge fast with fewer gradient computation times compared to Push-Pull and Push-DIGing. We also surprisingly observe that in Fig. 1(b), SSGT-PP with small p can compute as few gradients as S-ADDOPT and Push-SAGA to reach the same precision. It verifies the low gradient computation complexity of SSGT-PP.

6 Conclusion

In this paper, we propose a gradient-efficient algorithm over digraphs, namely SSGT-PP. By combining the snapshot gradient tracking technique with push-pull protocol, SSGT-PP allows agents to intermittently compute full-gradient, under the control of ξ^k. SSGT-PP achieves linear convergence rate on smooth strongly convex functions. We also analyse its gradient computation complexity $\mathcal{O}\left(\frac{\kappa^2}{(1-\sigma_{\mathbf{R}})} \log \frac{1}{\epsilon}\right)$, and the communication complexity $\mathcal{O}\left(\frac{\kappa^2}{(1-\sigma_{\mathbf{R}})(1-\sigma_{\mathbf{C}})} \log \frac{1}{\epsilon}\right)$. By numerical experiments, we demonstrate the effectiveness of SSGT-PP compared to the related algorithms for the logistic regression problem. Future investigations may aim at extending SSGT-PP to the asynchronous setting.

References

1. Assran, B.M., Aytekin, A., Feyzmahdavian, H.R., Johansson, M., Rabbat, M.G.: Advances in asynchronous parallel and distributed optimization. Proc. IEEE **108**(11), 2013–2031 (2020)
2. Horn, R.A., Johnson, C.R.: Matrix Analysis (1990)

3. Kempe, D., Dobra, A., Gehrke, J.: Gossip-based computation of aggregate information. In: Annual Symposium on Foundations of Computer Science, pp. 482–491 (2003)
4. Li, H., Lin, Z.: Accelerated gradient tracking over time-varying graphs for decentralized optimization (2021). http://arxiv.org/abs/2104.02596
5. Li, Z., Shi, W., Yan, M.: A decentralized proximal-gradient method with network independent step-sizes and separated convergence rates. IEEE Trans. Signal Process. **67**, 4494–4506 (2019)
6. Nedic, A., Olshevsky, A.: Stochastic gradient-push for strongly convex functions on time-varying directed graphs. IEEE Trans. Autom. Control **61**, 3936–3947 (2016)
7. Nedić, A., Olshevsky, A., Shi, W.: Achieving geometric convergence for distributed optimization over time-varying graphs. SIAM J. Optim. **27**, 2597–2633 (2017)
8. Nedić, A., Olshevsky, A.: Distributed optimization over time-varying directed graphs. IEEE Trans. Autom. Control **60**, 601–615 (2015)
9. Nedić, A., Ozdaglar, A.: Distributed subgradient methods for multi-agent optimization. IEEE Trans. Autom. Control **54**, 48–61 (2009)
10. Pu, S., Nedić, A.: Distributed stochastic gradient tracking methods. Math. Program. **187**, 409–457 (2021)
11. Pu, S., Shi, W., Xu, J., Nedic, A.: Push-pull gradient methods for distributed optimization in networks. IEEE Trans. Autom. Control **66**, 1–16 (2021)
12. Qu, G., Li, N.: Harnessing smoothness to accelerate distributed optimization. IEEE Trans. Control Network Syst. **5**(3), 1245–1260 (2018)
13. Qu, G., Li, N.: Accelerated distributed nesterov gradient descent. IEEE Trans. Autom. Control **65**, 2566–2581 (2020)
14. Qureshi, M.I., Xin, R.: Push-SAGA: a decentralized stochastic algorithm with variance reduction over directed graphs (2020). http://arxiv.org/abs/2008.06082
15. Qureshi, M.I., Xin, R., Kar, S., Khan, U.A.: S-ADDOPT: decentralized stochastic first-order optimization over directed graphs. IEEE Control Syst. Lett. **5**, 953–958 (2021)
16. Ren, C., Lyu, X., Ni, W., Tian, H., Song, W., Liu, R.P.: Distributed online optimization of fog computing for internet of things under finite device buffers. IEEE Internet Things J. **7**(6), 5434–5448 (2020)
17. Saadatniaki, F., Xin, R., Khan, U.A.: Decentralized optimization over time-varying directed graphs with row and column-stochastic matrices. IEEE Trans. Autom. Control **65**, 4769–4780 (2020)
18. Shi, W., Ling, Q., Wu, G., Yin, W.: EXTRA: an exact first-order algorithm for decentralized consensus optimization. SIAM J. Optim. **25**(2), 944–966 (2015)
19. Song, Z., Shi, L., Pu, S., Yan, M.: Optimal gradient tracking for decentralized optimization, pp. 1–48 (2021). http://arxiv.org/abs/2110.05282
20. Xi, C., Mai, V.S., Xin, R., Abed, E.H., Khan, U.A.: Linear convergence in optimization over directed graphs with row-stochastic matrices. IEEE Trans. Autom. Control **63**, 3558–3565 (2018)
21. Xin, R., Khan, U.A.: A linear algorithm for optimization over directed graphs with geometric convergence. IEEE Control Syst. Lett. **2**, 313–318 (2018)
22. Xin, R., Khan, U.A., Kar, S.: Variance-reduced decentralized stochastic optimization with accelerated convergence. IEEE Trans. Signal Process. **68**, 6255–6271 (2020)
23. Xin, R., Xi, C., Khan, U.A.: FROST–Fast row-stochastic optimization with uncoordinated step-sizes. In: Eurasip Journal on Advances in Signal Processing, vol. 2019, pp. 1–14. EURASIP Journal on Advances in Signal Processing (2019)

24. Zeng, J., Li, M., Liu, J., Wu, J.: Global coordinative optimization for energy management in distributed renewable energy generation system. In: Proceedings of the 29th Chinese Control Conference, pp. 1797–1801 (2010)
25. Zhang, J., You, K.: Fully asynchronous distributed optimization with linear convergence in directed networks **2**, 1–14 (2021). http://arxiv.org/abs/1901.08215

Differential Evolution Constrained Optimization for Peak Reduction

Min Wang(✉), Ting Wen, Xiaoyu Jiang, and Anan Zhang

Southwest Petroleum University, Chengdu 610500, China
wangmin80616@163.com

Abstract. Peak-to-average power ratio (PAPR) reduction is a major practical problem in Orthogonal Frequency Division Multiplexing (OFDM) systems. Many techniques, for example, use convex optimization to reduce the PAPR of OFDM signals. Existing methods either minimize the PAPR under the error vector magnitude (EVM) constraint or minimize the EVM under the PAPR constraint. Because of the complexities of these respective problems, researchers have proposed an interior point method with a computational complexity of $O(N^3)$. In this paper, we define a new problem that aims to find a feasible solution under the EVM and PAPR constraints, and propose a Differential Evolution Peak Reduction (DEPR) algorithm to deal with it. The motivation of the new problem lies in the fact that in applications, finding a feasible solution efficiently is more desirable than finding an optimal one exhaustively. The proposed DE algorithm minimizes the PAPR through a parallel direct search in the noise vector space. The computational complexity of the new algorithm is $O(N\log_2 N)$. Simulation results show that the number of computational steps of the new approach are 10% less than those of the convex optimization approaches without degrading the bit error rate (BER). Three empirical values were also obtained to ensure the efficiency and effectiveness of our algorithm.

Keywords: Convex optimization · Differential evolution · Error vector magnitude · Interior point method · Constrained optimization · Peak-to-average power ratio

1 Introduction

Orthogonal frequency division multiplexing (OFDM) is an attractive technique for achieving a high data transmission rate and is robust to frequency-selective fading channels [30]. It has been regarded as a core technology of various communication systems, such as asymmetric digital subscriber loop (ADSL), IEEE 802.11a/g, IEEE 802.16a and digital audio broadcast (DAB). However, OFDM's high peak-to-average power ratio (PAPR) has been a major obstacle in the implementation of power-efficient transmitters. High PAPR leads to at least two major

L. Fang et al. (Eds.): CICAI 2022, LNAI 13606, pp. 361–373, 2022.
https://doi.org/10.1007/978-3-031-20503-3_29

problems. First, it increases the complexity of the analog-to-digital and digital-to-analog converters. Second, it reduces the efficiency of the radio frequency power amplifier.

A number of approaches have been proposed to deal with the PAPR problem. These techniques include amplitude clipping [17,31], clipping and filtering [3,21,32], coding [10,22], tone reservation (TR) [2], tone injection (TI) [8,24], active constellation extension (ACE) [11,13], and multiple signal representation techniques such as partial transmit sequence (PTS) [9,12,25], selected mapping (SLM) [7,18,20] and interleaving [15,19]. The most simple and widely used method is to clip the signal to limit the PAPR below a threshold level, but this causes both in-band distortion and out-of-band radiation [4,29]. Furthermore, it requires many iterations to achieve the desired PAPR reduction. For this reason, in the recent literature [16,26,28], PAPR reduction has been cast as a convex optimization problem. The greatest advantage of convex optimization techniques is that a global optimal solution can be guaranteed. However it has been shown that these approaches are very poor in practice.

In this paper, we define a new problem that aims to find a feasible solution under the EVM and PAPR constraints and propose a differential evolution (DE) Peak Reduction algorithm to deal with it. The optimization objective of this new approach is to minimize run time by searching for a desirable PAPR reduction instead of the globally optimal solution. In most communication applications, finding a feasible solution efficiently is more desirable than finding an optimal one exhaustively. The PAPR constraint guarantees that the transmitter signal vector does not exceed a predetermined threshold, and the EVM constraint guarantees proper receiver operation that is specified by most modern communications standards.

The DEPR algorithm has four major steps to solve the optimal PAPR problem. First, we randomly select the initial parameter values uniformly from intervals along the noise space. Second, a mutation expands the search space. Third, the crossover acquires successful solutions from the previous generation. Finally, the lowest function value is selected for the next generation, and mutation, crossover, and selection continue until some stopping criterion is reached. The computational complexity of our approach is $O(N\log_2 N)$.

Simulations on the MATLAB platform were carried out to evaluate the performance of the proposed algorithm with data that used quadrature amplitude modulation(QAM). The simulations assumed that the data were 16-QAM modulated and the system contained N subcarriers. Simulation results show that the number of computational steps of the new approach are less than 10% of those required for the interior point method in [1,27]. The results also show that the algorithm has good efficiency and effectiveness without degrading the bit error rate (BER).

2 Problem Definition

In this section, we revisit two PAPR reduction problems. The first aims to minimize EVM subject to the PAPR constraint, while the second aims to minimize

PAPR subject to the EVM constraint. We also define a new problem that aims to find a feasible solution under both the EVM and PAPR constraints.

2.1 EVM Minimization Subject to the PAPR Constraint

PAPR is an important measure of the signal. In many cases, we would like to obtain signals that do not exceed the given PAPR threshold. However, we would also like to minimize the EVM measure. Some researchers have essentially addressed the following problem [16].

Problem 1. **ICF -optimization.**

Input: Original frequency-domain symbols X_k, PAPR upper bound $PA_{\max}$.
Output: The clipped frequency-domain symbols X_k'.
Constraints:

(i) $\frac{\max|x'(n)|^2}{E\left[|x'(n)|^2\right]} \leq PA_{\max}$;
(ii) $X_{outband}(m) = 0$.

Optimization objective: $\min EVM$.

Note that X_k and X_k' here are the original and clipped OFDM frequency-domain symbols, respectively.

The constraint is the desirable PAPR upper bound $PA_{\max}$ threshold.

$$\frac{\max|x'(n)|^2}{E\left[|x'(n)|^2\right]} \leq PA_{\max}. \tag{1}$$

In Equation (6), $x'(n)$ is composed of the optimized time domain transmitter symbols after IFFT. It can be expressed as follows:

$$x'(n) = IDFT(X'(k))_{LN}. \tag{2}$$

Here,

$$X'(k) = X_{inband} \cdot H_m, \tag{3}$$

where X_{inband} is the in-band components of $X(k)$, the operator $\cdot$ denotes the element-by-element product, and H_m denotes the optimal filter. In this paper, X_{outband} is set to zero.

Problem 1 aims to minimize the clipped signal EVM within the constraints of the PAPR. It modifies the classic ICF with a new convex optimization procedure. The optimized procedure dramatically decreases the number of required iterations. However, there is no need to obtain a globally optimal EVM by exhaustive search in practice. Furthermore, it only considers optimal filters. In further research on this technique, further consideration could be given to the overall process of optimization.

2.2 PAPR Minimization Subject to the EVM Constraint

There are two challenges in efficient OFDM transmission design. One is power amplifier (PA) non-linearity. The other is the OFDM's high PAPR. Alok et al. [1] attempted to find the global minimum PAPR of the transmission signal. At the same time, to acquire an acceptable BER after the clipping, the transmitted signal should also satisfy the EVM constraint. The maximum allowed EVM is empirically determined.

Problem 2. **PAPR minimization.**

Input: The original time-domain symbols $x(n)$ and EVM upper bound E_{max}.
Output: The optimized time-domain symbols $x'(n)$.
Constraints:

(i) $\sqrt{\frac{\sum_{k=0}^{N-1}|X(k)-X'(k)|^2}{\sum_{k=0}^{N-1}\left|X(k)\right|^2}} \leq E_{\max};$

(ii) $X'(k) \leq m_k, k = 1, 2, \cdots, N.$

Optimization objective: $\min PAPR$.

Problem 2 has two constraints. One is the EVM upper bound $E_{\max}$. The other is that out-of-band components must meet the spectral mask requirements. The individual carrier power constraints can be defined as:

$$X'(k) \leq m_k, k = 1, 2, \cdots, N, \tag{4}$$

where m is the spectral mask requirements and $X'(k)$ is the constellation that is actually transmitted.

Researchers have recognized that choosing the free carrier value to minimize PAPR is a convex optimization problem [6,14,23]. However, they have not considered the data carrier error. Alok et al. [1] defined the PAPR minimization under the EVM constraint problem that aims to obtain the globally minimum PAPR. At the same time, the optimized signal BER is significantly lower than that of the clipped signal because of the EVM constraint. The major challenge in this approach is to develop a fast and feasible technique for solving the problem.

3 Algorithm Design

This section presents the DEPR algorithm. The computational complexity of the algorithm is also analyzed.

3.1 DEPR Algorithm

We revised the DEPR algorithm to deal with our problem. The DEPR algorithm includes two main parts. One is to search for the optimal objective. The optimal objective selected for PAPR optimization is the noise vector. Our DEPR algorithm is employed to find the best clipped signal by a parallel search for the noise vector. The other is to handle the constraints. By meeting the various practical constraints, the algorithm attempts to find a satisfactory solution and hence replace the interior point method.

There are four major steps of our DEPR algorithm. In step 1, the initial vector population is chosen randomly and should cover the entire parameter space. In step 2, we generate new parameters by adding the weighted difference between two population vectors to a third vector. This operation is called mutation. In step 3, the mutated vector parameters are then mixed with the parameters of another predetermined vector, the target vector, to yield the so-called trial vector. This step is often referred to as crossover. In step 4, a selection operation is executed between the trial and target vectors. The lowest function value is selected for the next generation, and mutation, crossover and selection continue until some stopping criterion is reached.

Next, we elaborate on each step of our DEPR algorithm in more detail.

Step 1. Initialization operation

The population of DE consists of n-dimensional noise vectors. The initial generator noise vectors are generated randomly between the upper and lower limits of the noise values. Individuals in the first generation can be shown as:

$$E_{ij}^{(0)} = E_j^L + rand_j(0,1) \times (E_j^U - E_j^L), \quad i = 1, 2, \cdots, N, j = 1, 2, \cdots, D, \tag{5}$$

where E_j^L and E_j^U are the lower and upper bounds, respectively, and $rand_j(0,1)$ is a uniformly distributed random value such that $rand_j(0,1) \in [0,1]$. The upper and lower bounds of the noise vector $E(k)$ are chosen as (0,0.5).

Step 2. Mutation operation

Here, we adopt the basic variation: DE/rand/1/bin. Taking into account each target vector $\boldsymbol{E}_{i,t}$ at generation t, a mutant vector $\boldsymbol{V}_{i,t}$ is defined by:

$$V_{ij}^{(t+1)} = E_{r3j}^{(t)} + F \times (E_{r1j}^{(t)} - E_{r2j}^{(t)}), \tag{6}$$

where indexes $r1$, $r2$, and $r3$ represent mutually different integers that are different from i and randomly generated over $[1,Np]$. Parameter F is a real-valued mutation factor and $F \in [0,2]$. We use three random individuals $E_{r3j}^{(t)}$, $E_{r1j}^{(t)}$, and $E_{r2j}^{(t)}$ that maintain the population diversity and keep it from falling into local extreme points.

Step 3. Crossover operation

The crossover operator is designed to increase the diversity of the population. The target vector $\boldsymbol{E}_{i,t}$ is mixed with the mutant vector, using a binomial crossover operation, to form the trial vector. The crossover operation is:

$$U_{ij}^{(t+1)} = \begin{cases} V_{ij}^{(t+1)}, \; rand_j(0,1) \leq CR \; or \; j = j_{rand}, j = 1, 2, \cdots, D; \\ E_{ij}^{(t)}, \qquad\qquad\qquad rand_j(0,1) > CR, \end{cases} \tag{7}$$

where $i = 1, 2, \cdots, Np$, $j = 1, 2, \cdots, n$, index j_{rand} is a randomly chosen integer, $rand_j(0,1)$ is the jth output of a uniform random number generator, and

Table 1. Simulation time of our proposed method and the interior-point method with N=1,024 subcarriers.

	16QAM $EVM < 6\%$		64QAM $EVM < 6\%$	
	Interior point method	Proposed method	Interior point method	Proposed method
10	20.219	2.232	19.656	2.643
20	37.295	2.914	36.969	2.829
50	93	5.84	93.014	5.622
100	202.321	10.359	203.606	10.386
500	1098	47.138	1099	46.389
	Binary Phase Shift Keying (BPSK) $EVM < 6\%$		16QAM $EVM < 12\%$	
	Interior point method	Proposed method	Interior point method	Proposed method
10	19.505	1.856	19.534	1.851
20	37.803	2.687	37.22	2.984
50	93.322	5.59	98.088	5.81
100	208.11	10.103	208.035	10.18
500	1175	45.928	1174	46.305

crossover probability $(CR) \in [0, 1]$ is the crossover control parameter. The condition $j = j_{rand}$ is introduced to ensure that the trial vector $\boldsymbol{U}_{i,t}$ differs from its target vector $\boldsymbol{E}_{i,t}$ by at least one element.

Step 4. Selection operation

Selection operates by comparing fitnesses of individuals to generate the next generation population. After evaluating the target vector $\boldsymbol{E}_{i,t}$ and trial vector $\boldsymbol{U}_{i,t}$, the better one is preserved for the next generation:

$$\boldsymbol{E}_{i,t+1} = \begin{cases} \boldsymbol{U}_{i,t}, \, if \ f(U_{ij}^{(t+1)}) \leq f(E_{ij}^{(t)}); \\ \boldsymbol{E}_{i,t}. \qquad otherwise. \end{cases} \tag{8}$$

For a minimization optimization problem, individuals with lower fitness values are retained. The iterative procedure can be terminated when an acceptable solution or the maximum number of generations is reached.

3.2 Complexity Analysis

In this paper, we use computational complexity theory to measure the algorithm efficiency by counting the number of basic operations.

Problems 1 and 2 can be formulated as a convex optimal module that can be solved by standard interior-point methods. The computational complexity of the first iteration is:

$$O(N^3 + 2LN\log_2(LN)). \tag{9}$$

In the subsequent iterations, the computational complexity is:

$$O(N^3 + LN\log_2(LN)). \tag{10}$$

Let the maximum iteration number be G. Then the computational complexity of the proposed procedure for the interior point method is:

$$O(GN^3 + (G+1)LN\log_2(LN))) = O(GN^3). \tag{11}$$

We now analyze the time complexity of the DEPR algorithm. Let Np be the population size and G be the maximum number of iterations. First, we generate the initial population randomly within the noise space. The time complexity for initializing Np individuals is $O(Np)$.

Second, we generate the mutant vector, as described in Step 2. The mutant vector for each ith target vector from the current population, there other distinct parameter vectors are sampled randomly from the current population. These indices are randomly generated once for each mutant vector. Therefore, the time complexity for mutating Np individuals is $O(Np)$.

Third, the target vector $\boldsymbol{E}_{i,t}$ is mixed with the mutant vector to form the trial vector. Binomial crossover is performed on variables whenever a randomly generated number between 0 and 1 is less than or equal to the CR value. The time complexity for crossing between two population of size Np is $O(Np)$;

Finally, we must select the individual with lower fitness value. If the new trial vector yields an equal or lower value of the objective function, it replaces the corresponding target vector in the next generation; otherwise the target is retained in the population. At the same time, the number of basic operations of computer the value to the objective function is an N-point FFT. In our approach we set N=1,024, where N is the number of subcarriers. The time complexity for tournament selection is $O(Np * N\log_2 N)$. Therefore, the worst total time complexity is:

$$O(Np) + O(Np) + O(Np) + O(Np \times N\log_2 N) = O(Np \times N\log_2 N) \tag{12}$$

In this paper, we set $Np = 100$ to achieve a good optimization result. This is far less than the actual number of subcarriers. Let G is the maximum number of iterations, the number of the multiplications needed for the DEPR algorithm to find a sub-optimal solution is:

$$O(GN\log_2 N). \tag{13}$$

This is significantly lower than the existing interior point method.

4 Simulations and Analysis

Simulations were conducted on an Intel core 2 Duo CPU of 2.4 GHz and 2 GB RAM computer in the MATLAB 7.10.0.499 (R2010a) environment. The simulations assumed that the data were 16-QAM modulated and the system contained N = 1,024 subcarriers and had a 20-MHz bandwidth. We validated the performance of the parameter tuning, efficiency, effectiveness, and robustness of our DEPR method.

4.1 Efficiency of the DEPR Algorithm

An efficient algorithm ensures that a solution can be found in reasonable time. This efficiency can be assessed by estimating the computational complexity (time and space) of a method. It can also be estimated by simulation tests that compare two performance criteria.

The first performance criterion is its runtime compared with other exhaustive approaches. Table 1 lists the simulation times for our approach and the interior point method in [27] and [1]. These results show that the number of computational steps used for the new approach are 10% fewer than the interior-point method. This simulation provides a rough estimate of algorithm time complexity. In other words, it is appropriate to assess the newly defined problem.

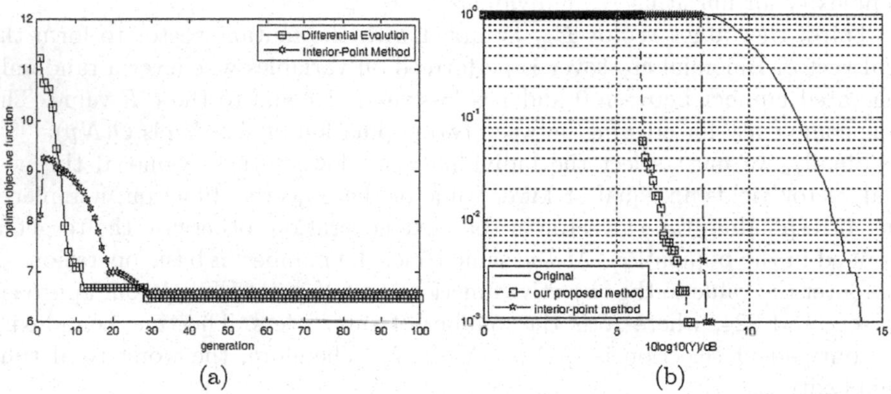

Fig. 1. (a) Performance comparison between our approach method and the interior-point method. (b) PAPR-reduction performance of original method, our proposed method, and the interior-point method (1,024 subcarriers, 16-QAM, L = 4).

The second performance criterion is the number of iterations compared with other exhaustive approaches. Figure 1(a) shows the evolution of the best solutions obtained when the interior-point method and our proposed DEPR algorithm are employed. From these results, we can compare the algorithm in terms of the convergence speed, and we can see that our approach needed only about 10 generations while the interior-point method needed almost 30 generations to obtain the final solution. It is clear that our approach is significantly faster than the interior-point method for finding optimum solutions.

4.2 Effectiveness of Our Proposed DEPR Algorithm

In this subsection, we compare the effectiveness of the algorithm. The effectiveness was estimated by measuring the BER and reduction of PAPR.

Figure 1(b) plots the CCDFs of the PAPR for the original symbols, modified symbols using the interior-point method [1,27], and modified symbols using our proposed method. The complementary cumulative distribution function (CCDF) of the PAPR is one of the most frequently used parameter for analyzing PAPR reduction. The desired PAPR was set to 6.3 dB at a probability of 10^{-2} using our proposed method. From the figure, it can be seen that our proposed method can reduce PAPR by a further 0.5 dB compared with the interior-point method for the same number of iterations. However, for our approach, the PAPR is reduced by about 6.3 dB at a probability of 10^{-2} after ten iterations. However, the interior-point method required 30 iterations to achieve the same level of PAPR reduction. Thus, our approach requires far fewer iterations to achieve the desired PAPR. Figure 1(b) shows that our method satisfies the PAPR constraints and could be used in an actual communication system.

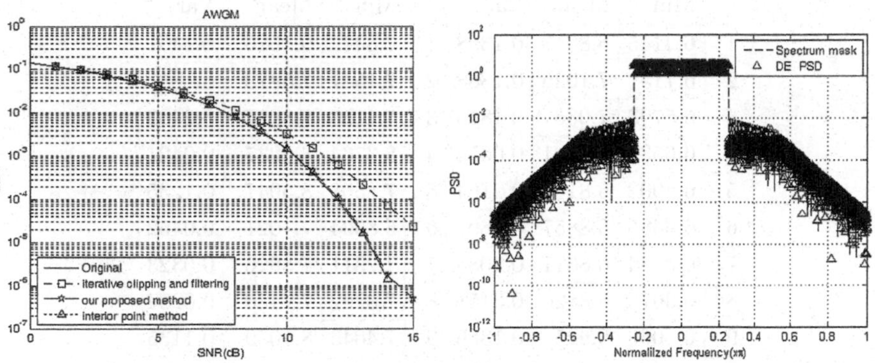

(a) (b)

Fig. 2. (a) BER comparison for original OFDM symbols: iterative clipping and filtering method, clipped OFDM symbols using interior point method, and our proposed method in an AWGN channel. (b) Power spectral density comparison between our approach and the spectral mask.

Figure 2(a) plots the BER curves of the original signal, the iterative clipping and filtering, the interior-point method, and our proposed method in an additive white Gaussian noise (AWGN) channel. The iterative clipping and filtering curves corresponding to modified symbols are located to the right of the original signal curve because the clipping procedure causes signal distortion. It is clear that the signal-to-noise ratio (SNR) of our proposed OFDM symbols are better than iterative clipping and filtering for a given BER because our proposed method meets the EVM constraint. For example, at a BER level of 10^{-5}, the iterative clipping and filtering method yields an SNR of about 1.1 dB less than that of our proposed method. Our proposed scheme and the interior point method have nearly the same BER performance because both of them satisfy the EVM constraint.

4.3 Stability of the DEPR Algorithm

A robust method finds the optimum in most cases, not in a few cases by random chance. Gradient-based methods such as the interior-point method are often very robust if proper initialization is used. The robustness of our DEPR algorithm may be estimated by repeatedly training and evaluating the type of optima (local/global). In this simulation, 20 independent runs were executed. In this case, the control parameters were set to $F = 0.5$ and $CR = 0.9$.

Table 2. Stability of our proposed algorithm and the interior-point method.

Our proposed method				Interior point method			
	Min	Mean	Var		Min	Mean	Var
1	6.4145	6.8617	0.1398	1	7.847	9.2526	0.1461
2	6.715	7.1025	0.1888	2	7.1547	7.6879	0.188
3	6.5793	7.035	0.1815	3	7.1155	8.8817	0.2463
4	6.6535	7.0001	0.0674	4	8.3764	8.6237	0.0421
5	6.4902	6.8137	0.2164	5	7.8558	8.5047	0.1235
6	6.6485	6.8857	0.0897	6	6.8594	7.1021	0.0462
7	6.3384	6.8611	0.2086	7	7.2311	7.8626	0.0323
8	6.3675	6.8126	0.315	8	7.3292	7.7436	0.0284
9	6.5066	6.925	0.1886	9	7.4046	8.1059	0.1135
10	6.1081	6.1081	0.3831	10	7.7981	8.4268	0.0512
11	6.4343	6.4343	0.2417	11	7.2276	10.7179	0.7817
12	6.4484	6.7851	0.1827	12	8.1171	8.6676	0.1553
13	6.5411	7.0315	0.1311	13	8.1316	9.2653	0.2547
14	6.6601	7.1033	0.1358	14	7.7967	7.7701	0.1073
15	6.2587	6.8774	0.2662	15	8.4539	8.5701	0.0174
16	6.5612	6.9604	0.2061	16	7.2951	7.8482	0.2217
17	6.5442	6.9613	0.2347	17	7.5368	7.7816	0.0397
18	6.1451	6.6771	0.2618	18	9.3532	11.1603	0.2687
19	6.7665	7.1049	0.1182	19	7.7482	8.2518	0.04474
20	6.712	7.0052	0.1698	20	7.965	8.3764	0.0447

Table 2 lists the minimum, average value, and average variance of the optimization objectives. From these results, we can see that the local or global solution can be found by the DEPR algorithm. It should be noted that because of the stochastic behavior of the DE, the solutions vary from one run to another. However, the variance is small, indicating that the data are stable. Our proposed method can be considered a robust method, that is, it finds the feasible solutions in most cases.

4.4 EVM

Figure 2(b) shows the power spectral density (PSD) curves of our proposed algorithm and the specified spectral mask in IEEE 802.16 systems. From these results, we can see that the spectrum of our proposed signal meets the spectrum mask exactly. The spectral mask is adopted from the 802.16 standard [5]. We can see in Fig. 2(b) that our proposed method can satisfy both the in-band EVM constraint and the out-of-band spectral constraint.

5 Conclusion

In this paper, we define a hardware-feasible PAPR optimization problem that aims to find a satisfactory solution with respect to PAPR and EVM constraints as quickly as possible, instead of exhaustively searching for the optimal solution. From the viewpoint of applications, we do not consider the optimization of certain objectives necessary. We proposed the DEPR algorithm to solve the new problem with high efficiency. It can directly search in parallel for the noise vector to further reduce the complexity of our algorithm. The time complexity of our approach is $O(N \log_2 N)$. Simulation results show that the number of computational steps required by the new approach is 10% less than the number needed by the interior-point method without degrading the BER. We believe that a real-time implementation is feasible given the capabilities of current-day CMOS technology.

References

1. Aggarwal, A., Meng, T.H.: Minimizing the peak-to-average power ratio of OFDM signals using convex optimization. IEEE Trans. Signal Process. **54**(8), 3099–3110 (2006)
2. Chen, J.C., Li, C.P.: Tone reservation using near-optimal peak reduction tone set selection algorithm for PAPR reduction in OFDM systems. IEEE Signal Process. Lett. **17**(11), 933–936 (2010)
3. Chen, L., Fang, Y., Wang, M.: PAPR reduction in optical OFDM systems using asymmetrically clipping and filtering technique. J. Comput. Inf. Syst. **8**(7), 2733–2740 (2012)
4. Cuteanu, V., Isar, A.: PAPR reduction of OFDM signals using hybrid clipping-companding scheme with sigmoid functions. In: International Conference on Applied Electronics, pp. 75–78 (2011)
5. Eklund, C., Marks, R.B., Ponnuswamy, S.: WirelessMAN: inside the IEEE 802.16 Standard for Wireless Metropolitan Area Networks. Standards Information Network (2006)
6. Gatherer, A., Polley, M.: Controlling clipping probability in DMT transmission. In: Conference Record of the Thirty-First Asilomar Conference on Signals, Systems & Computers, 1997, vol. 1, pp. 578–584. IEEE (1997)
7. Gay, M., Lampe, A., Breiling, M.: PAPR reduction in OFDM using selected mapping with adaptive clipping at the transmitter, and sparse reconstruction at the receiver. In: Proceedings of OFDM 2014; 18th International OFDM Workshop 2014 (InOWo 2014), pp. 1–8. VDE (2014)

8. Han, S.H., Cioffi, J.M., Lee, J.H.: Tone injection with hexagonal constellation for peak-to-average power ratio reduction in OFDM. IEEE Commun. Lett. **10**(9), 646–648 (2006)
9. Hou, J., Ge, J., Li, J.: Peak-to-average power ratio reduction of OFDM signals using pts scheme with low computational complexity. IEEE Trans. Broadcast. **57**(1), 143–148 (2011)
10. Joshi, A., Saini, D.S.: Performance analysis and PAPR reduction of coded OFDM (with RS-CC and turbo coding) system using modified SLM, PTS and DHT precoding in fading environments. WSEAS Trans. Commun. **12**(1), 14–28 (2013)
11. Kang, B.M., Ryu, H.G., Ryu, S.B.: A PAPR reduction method using new ace (active constellation extension) with higher level constellation. In: IEEE International Conference on Signal Processing and Communications, 2007. ICSPC 2007, pp. 724–727. IEEE (2007)
12. Kang, S.G., Kim, J.G., Joo, E.K.: A novel subblock partition scheme for partial transmit sequence OFDM. IEEE Trans. Broadcast. **45**(3), 333–338 (1999)
13. Krongold, B.S., Jones, D.L.: Par reduction in OFDM via active constellation extension. IEEE Trans. Broadcast. **49**(3), 258–268 (2003)
14. Krongold, B.S., Jones, D.L.: An active-set approach for OFDM par reduction via tone reservation. IEEE Trans. Signal Process. **52**(2), 495–509 (2004)
15. Latinovic, Z., Bar-Ness, Y.: SFBC MIMO-OFDM peak-to-average power ratio reduction by polyphase interleaving and inversion. IEEE Commun. Lett. **10**(4), 266–268 (2006)
16. Luo, Z.Q., Yu, W.: An introduction to convex optimization for communications and signal processing. IEEE J. Sel. Areas Commun. **24**(8), 1426–1438 (2006)
17. Mohani, S.P., Sarkar, S., Sutaone, M.S.: Adaptive amplitude clipping PAPR reduction technique using extended peak reduction tone set. Networking Commun. Eng. **5**(5), 256–259 (2013)
18. Naeiny, M.F., Marvasti, F.: Selected mapping algorithm for PAPR reduction of space-frequency coded OFDM systems without side information. IEEE Trans. Veh. Technol. **60**(3), 1211–1216 (2011)
19. Sakran, H.Y., Shokair, M.M., Elazm, A.A.: Combined interleaving and companding for PAPR reduction in OFDM systems. Progress in Electromagn. Res. C **6**, 67–78 (2009)
20. Shao, Y., Chi, N., Fan, J., Fang, W.: Generation of 16-QAM-OFDM signals using selected mapping method and its application in optical millimeter-wave access system. IEEE Photon. Technol. Lett. **24**(15), 1301–1303 (2012)
21. Sharma, C., Sharma, P.K., Tomar, S., Gupta, A.: A modified iterative amplitude clipping and filtering technique for PAPR reduction in OFDM systems. In: 2011 International Conference on Emerging Trends in Networks and Computer Communications (ETNCC), pp. 365–368. IEEE (2011)
22. Sharma, C., Tomar, S.K., Gupta, A.: PAPR reduction in OFDM system using adapting coding technique with pre distortion method. WSEAS Trans. Comput. **9**, 255–262 (2011)
23. Tellado, J., Cioffi, J.M.: Efficient algorithms for reducing par in multicarrier systems. In: 1998 IEEE International Symposium on Information Theory, 1998. Proceedings, p. 191. IEEE (1998)
24. Tuna, C., Jones, D.L.: Tone injection with aggressive clipping projection for OFDM PAPR reduction. In: 2010 IEEE International Conference on Acoustics Speech and Signal Processing (ICASSP), pp. 3278–3281. IEEE (2010)

25. Varahram, P., Al-Azzo, W.F., Ali, B.M.: A low complexity partial transmit sequence scheme by use of dummy signals for PAPR reduction in OFDM systems. IEEE Trans. Consumer Electr. **56**(4), 2416–2420 (2010)
26. Wang, C., Leung, S.H.: Par reduction in ofdm through convex programming. In: IEEE International Conference on Acoustics, Speech and Signal Processing, 2008. ICASSP 2008, pp. 3597–3600. IEEE (2008)
27. Wang, Y., Luo, Z.: Optimized iterative clipping and filtering for PAPR reduction of OFDM signals. IEEE Trans. Commun. **59**(1), 33–37 (2011)
28. Wang, Y.C., Yi, K.C.: Convex optimization method for quasi-constant peak-to-average power ratio of OFDM signals. IEEE Signal Process. Lett. **16**(6), 509–512 (2009)
29. Wang, Y., Chen, W., Tellambura, C.: Genetic algorithm based nearly optimal peak reduction tone set selection for adaptive amplitude clipping PAPR reduction. IEEE Trans. Broadcast. **58**(3), 462–471 (2012)
30. Wong, C.Y., Cheng, R.S., Lataief, K.B., Murch, R.D.: Multiuser OFDM with adaptive subcarrier, bit, and power allocation. IEEE J. Sel. Areas Commun. **17**(10), 1747–1758 (1999)
31. Xu, L., Chen, D.: ADC spectral testing allowing amplitude clipping. In: 2013 IEEE International Instrumentation and Measurement Technology Conference (I2MTC), pp. 1526–1529. IEEE (2013)
32. Zhong, X., Qi, J., Bao, J.: Using clipping and filtering algorithm to reduce PAPR of OFDM system. In: 2011 International Conference on Electronics, Communications and Control (ICECC), pp. 1763–1766. IEEE (2011)

Evolutionary Multitasking Optimization for Multiobjective Hyperspectral Band Selection

Pu Xiong, Xiangming Jiang(✉), Runyu Wang, Hao Li, Yue Wu, and Maoguo Gong

Xidian University, Xian 710043, China
omegajiangxm@gmail.com

Abstract. Hyperspectral remote sensing technology combines imaging technology and spectral technology, which greatly promotes the development of remote sensing science. However, the large amount of data, high redundancy and high data dimension of hyperspectral images will cause many problems such as "curse of dimensionality". Feature extraction and band selection are two main methods to reduce the dimensionality and retain information for practical application. Compared with the feature extraction, band selection aims to select a band subset to reduce dimensionality, which can maintain the physical meaning of the original band. However, many band selection methods usually face many problems such as high computational cost, local optimization, poor classification accuracy and so on. In this paper, considering the evolutionary multitasking optimization algorithm has the characteristics of processing multiple tasks at the same time to improve the search efficiency, band selection is modeled as a multitasking optimization problem, and the evolutionary multitasking optimization algorithm is used as search strategy to select the band subset. Using hyperspectral remote sensing dataset as the experiment, the result shows that compared with other methods, the band subsets obtained by the evolutionary multitasking optimization algorithm have excellent overall classification accuracy and average classification accuracy, which will contribute to practical application.

Keywords: Hyperspectral image · Band selection · Evolutionary algorithm · Multitask optimization

1 Introduction

Hyperspectral remote sensing relies on imaging spectrometer technology. Different from panchromatic remote sensing images or multispectral remote sensing images obtained by optical remote sensing, hyperspectral remote sensing images have the following main characteristics: high spectral resolution, continuous image imaging, and unified atlas. These three characteristics enable hyperspectral remote sensing images to provide characteristic information that

L. Fang et al. (Eds.): CICAI 2022, LNAI 13606, pp. 374–385, 2022.
https://doi.org/10.1007/978-3-031-20503-3_30

is incomparable to other images. However, at the same time, the large number of bands in hyperspectral remote sensing images also leads to the multiplication of the data volume of hyperspectral images, and the huge data volume brings challenges to the transmission and storage of hyperspectral data. In addition, a large amount of similar information is carried between adjacent bands, resulting in extremely high correlation between bands, resulting in serious information redundancy, which is not conducive to effective data processing. Therefore, while retaining effective information as much as possible, reducing data redundancy is a very important research direction in the field of hyperspectral remote sensing research.

Hyperspectral remote sensing images have hundreds of spectral bands, and the spectral resolution is generally about 10 nm. Although a large number of bands provide rich feature information for hyperspectral remote sensing, the huge amount of hyperspectral remote sensing data, serious information redundancy, high band correlation, and high data dimension increase the complexity of data storage, transmission and algorithm calculation burden. In addition, the curse of dimensionality is often faced in the process of high-dimensional data processing [1]. In order to solve the above problems, according to the characteristics of large amount of information and high redundancy of hyperspectral remote sensing image data, the hyperspectral band selection problem can be modeled as a multitasking optimization problem. Mining image information from multiple angles and improving the search efficiency of band subsets. Therefore, it is of great value and significance to study multitasking optimization methods for hyperspectral band selection.

This paper mainly uses the multitasking optimization theory to solve the problem of band selection of hyperspectral remote sensing image data, and proposes a multitasking optimization method for hyperspectral band selection. Multitasking optimization can simultaneously optimize multiple tasks and improve the timeliness of task processing through information transfer and knowledge sharing between different tasks. In this paper, the hyperspectral band selection problem is modeled as two multi-objective optimization tasks and the two tasks are optimized simultaneously. According to the multi-objective functions corresponding to different tasks, the non-dominated solutions corresponding to each multi-objective optimization task can be obtained, and suitable band combinations can be selected from them.

2 Background

In this section, we will introduce the relevant basic principles involved in this article.

2.1 Multi-objective Optimization

Different from the single-objective optimization problem that only considers one objective function, there are a lot of problems in reality that need to optimize multiple objectives at the same time, that is, *multi-objective optimization problems*(MOP). Since multiple objective functions usually conflict with each other, it is generally impossible to guarantee that there must be an optimal solution that can satisfy all objective functions at the same time and reach the extreme value. Therefore, unlike most single-objective optimization solutions, multi-objective optimization problems generally require the use of multi-objective optimization algorithms to find a non-dominated solution to solve the problem.

Without loss of generality, a multi-objective optimization problem can often be generalized as an objective minimization problem. As shown in Eqs. (1) and (2), a multi-objective minimization optimization problem can be expressed as follows:

$$\min F(x) = (f_1(x), f_2(x), \cdots f_n(x)) \tag{1}$$

$$s.t. x = (x_1, x_2, \cdots x_n) \in \Omega \tag{2}$$

$F(x)$ is a vector with n objective functions, x is a vector with n decision variables, corresponding to a solution to a multi-objective problem, and Ω represents a feasible decision space. Since multiple objective functions in multi-objective optimization conflict with each other, optimizing one objective alone may lead to the deterioration of other objective functions, making it impossible to guarantee that there must be at least one objective function that can simultaneously make all objective functions in the feasible decision space reach the optimal solution. Therefore, the multi-objective optimization problem is different from the single-objective optimization problem, and the essence of the solution is to find a compromise solution that can optimize all the objectives. The compromise solution can also be called a non-dominated solution, which generally needs to be determined by the domination relationship.

In the multi-objective optimization problem, the relationship between the optimal solutions is generally described by the dominance relationship. For the multi-objective minimization optimization problem, it is described without loss of generality. One solution x_u dominates the other solution x_v and needs to satisfy the formula at the same time The two conditions shown in (3) and (4):

$$\forall i = 1, 2, \cdots, m \quad f_i(x_u) <= f_i(x_v) \tag{3}$$

$$\exists j = 1, 2, \cdots, m \quad f_j(x_u) < f_j(x_v) \tag{4}$$

When the above conditions are met, it can be considered that x_u is non-dominated and x_v is dominated, that is, x_u dominates x_v, which is expressed as $x_u \succ x_v$, where the symbol "$\succ$" represents the dominance relationship. If there is a solution x^* such that in the feasible decision space there is no other solution $x \in \Omega$ such that $x \succ x^*$ holds, then x^* is It can be called a Pareto Optimal Solution. The set of all Pareto optimal solutions is called the *Pareto Optimal Set*(PS). In the Pareto optimal solution set, the set of target value vectors corresponding to the Pareto optimal solution constitutes the *Pareto Optimal Front*(PF). The solutions on the Pareto frontier have the characteristic that the decrease of any one objective function value will lead to the increase of one or more other objective function values. Therefore, solving a multi-objective optimization problem can be transformed into using a multi-objective optimization algorithm to find a set of solutions that are as close to the Pareto front as possible.

2.2 The Basic Theory of Multitasking Optimization

Traditional intelligent optimization algorithms usually only optimize and solve one problem. However, the information used by such single-task intelligent optimization algorithms based on population iteration is limited, which often leads to low algorithm search efficiency. Inspired by multitasking learning and transfer learning, the multitasking optimization method is based on the similarity between different tasks, by transferring the useful knowledge obtained in the process of solving an optimization task to another similar task. Accelerates the solution of another task, so as to achieve the purpose of optimizing multiple tasks at the same time and improving the efficiency of algorithm search. Compared with the previous single-task optimization algorithm, the evolutionary multitasking optimization algorithm can realize the optimization of multiple tasks in a population at the same time and improve the search efficiency through positive knowledge transfer and sharing between tasks. Therefore, multitasking optimization theory and related algorithms have become a popular research direction of intelligent optimization.

Inspired by multitasking learning and transfer learning, multitasking optimization refers to the process of solving multiple tasks at the same time, based on the similarity between tasks, transfer and share the useful knowledge obtained in the process of solving a certain task, thereby helping to solve another similar optimization task. In other words, during the search process, transferring the search progress of a task to another task can improve the search speed to a certain extent, thereby improving the problem of low efficiency of single-task optimization. Different from multi-objective optimization, which only focuses on the simultaneous optimization and solution of multiple objectives, multitasking optimization aims to solve multiple independent optimization tasks at the same time and use positive knowledge transfer and sharing between tasks to improve the efficiency of population search. Therefore, there is a clear distinction between multitasking optimization and multi-objective optimization [2].

Considering that the band selection method proposed in this paper is a hyperspectral band selection framework based on multitasking optimization, each task is a multi-objective optimization problem, so the following will focus on the *Multi-Objective Multifactorial Evolutionary*(MO-MFEA) algorithm for introduction [3]. Based on the MFEA algorithm, the MO-MFEA algorithm can solve multiple multi-objective optimization problems simultaneously in a multitasking environment. The MO-MFEA algorithm draws on the related concepts of skill factor and scalar fitness in the MFEA algorithm, and extends the concept of factor ranking in the MFEA algorithm, so that a population can solve multiple multi-objective optimization problems at the same time.

The MO-MFEA algorithm sorts individuals into the most suitable task according to their best factors. In the same task, the individual with large scalar fitness is in the non-dominant position, while the individual with small scalar fitness is in the dominant position, so that the Pareto frontier of each multi-objective optimization problem can be obtained. In this way, a population can be used to solve multiple multi-objective optimization problems at the same time.

In order to rank individuals, the MO-MFEA algorithm draws on the idea of the NSGA-II algorithm, evaluates the factor rank of individuals through non-dominated ranking and crowding degree distance, and selects a smaller ranking rank and a larger ranking rank in different tasks. Individuals with crowded distances are retained as elites, so that multiple multi-objective optimization problems can be solved simultaneously through a population in a multitasking environment.

3 Methodology

This section mainly introduces the multitasking modeling method for hyperspectral band selection. For hyperspectral band selection, it is usually required that the selected band combination has the characteristics of high information content and low redundancy, and the optimal number of bands should be achieved as much as possible. Combined with the evolutionary multitasking optimization algorithm, this paper proposes a hyperspectral band selection framework based on multitasking optimization.

3.1 A Framework for Hyperspectral Band Selection Based on Multitasking Optimization

The input data of the proposed hyperspectral band selection framework based on multitasking optimization is hyperspectral image data, and the hyperspectral image is modeled for the characteristics of large amount of information and high redundancy. The main modeling idea is to use the information entropy and the number of bands as two objective functions to establish a multi-objective optimization task, which is referred to as task 1 for short. Another multi-objective optimization task is established with the information divergence and the number of bands as two objective functions, which is referred to as task 2. The MO-MFEA algorithm is used as the band search strategy in the band selection to optimize the task one and the second task at the same time. According to different objective functions corresponding to each task, non-dominated solutions corresponding to different tasks are obtained to realize hyperspectral band selection and finally output the obtained band subsets.

Task 1 starts from the characteristic that hyperspectral images are rich in information. In order to reduce the loss of information as much as possible, it is generally necessary to select a band that carries a larger amount of information. *Information Entropy* (IE) [4]is derived from Shannon's information theory and can be used to quantify information. For hyperspectral images, the greater the information entropy of a certain band, the greater the amount of information carried by the band. Therefore, the information entropy can be used as an information measurement index to select the band with a larger amount of information [5,6]. Usually, each band in the hyperspectral image can be regarded as a random variable X, and the corresponding histogram is the probability distribution of the band, whose probability can be represented by $P(x)$ and satisfy $\sum_{x \in \Omega} P(x) = 1$. The information entropy of a certain band can be calculated according to the following formula (5).

$$H(x) = -\sum_{x \in \Omega} P(x) log P(x) \tag{5}$$

Therefore, taking the information entropy and the number of bands as two objective functions can constitute a multi-objective optimization task, that is, task one, which is represented by $F_1(x)$. The objective function of task 1 is expressed as formula (6), where $f_{11}(x)$ represents the number of bands, $f_{12}(x)$ represents the reciprocal of the information entropy of the corresponding band subset, the smaller the value of $f_{11}(x)$, the less the number of bands, the smaller the value of $f_{12}(x)$, the greater the information entropy of the selected band subset, and the higher the amount of information contained in the band subset.

$$\min F_1(x) = \begin{cases} f_{11}(x) = length(x) \\ f_{12}(x) = 1 / \sum_{i=1}^{f_{11}(x)} H(x_i) \end{cases} \tag{6}$$

The second task starts from the characteristics of hyperspectral images with large redundancy. In order to reduce the redundancy between the bands, it is usually necessary to select the least similar band combination as possible. *Information Divergence* (ID), also known as Relative Entropy, is mainly used to measure the asymmetry of the difference between two probability distributions. For hyperspectral images, the greater the information divergence between the two bands, the greater the difference between the two bands, and the lower the similarity. Therefore, the information divergence can be used as an information measure to select the band combination with low redundancy [6]. According to formula (7) and the probability distribution of the hyperspectral bands, the information divergence between the two bands can be calculated [7].

$$D(x_i, x_j) = \sum_{x \in \Omega} P_i(x) log \frac{P_i(x)}{P_j(x)} + \sum_{x \in \Omega} P_j(x) log \frac{P_j(x)}{P_i(x)} \tag{7}$$

Therefore, taking the information divergence and the number of bands as two objective functions can constitute a multi-objective optimization task, that is, the second task, which is represented by $F_2(x)$. The objective function of task 2 is expressed as formula (8), where $f_{21}(x)$ represents the number of bands, $f_{22}(x)$ represents the reciprocal of the information divergence of the corresponding band subset, the smaller the value of $f_{21}(x)$, the less the number of bands, The smaller the value of $f_{22}(x)$, the greater the information divergence of the selected band subsets, and the lower the redundancy of the band subsets.

$$\min F_2(x) = \begin{cases} f_{21}(x) = length(x) \\ f_{22}(x) = 1/\sum_{i=1}^{f_{21}(x)} \sum_{j=1}^{f_{21}(x)} D(x_i, x_j) \end{cases} \tag{8}$$

3.2 Hyperspectral Band Selection Based on Multitasking Optimization Algorithm

Based on the hyperspectral band selection framework of multitasking optimization, the MO-MFEA algorithm is used as the band search strategy in the band selection to optimize the first and second tasks at the same time. In order to improve the optimization efficiency in the optimization process, a hyperspectral band selection method based on a multitasking optimization algorithm is proposed as shown in Fig. 1. The main processes include population initialization, population evolution, and population selection and merging.

In the evolution process, the crossover operation adopts the multi-point crossover operator, and the mutation operation adopts the bit-flip mutation operator. The iteration number parameter is 50, the population number parameter is 100, the number of bands is set according to the number of bands in different hyperspectral datasets, the crossover probability is 0.5, the mutation probability is 0.05, and the random mating probability RMP parameter is 0.8.

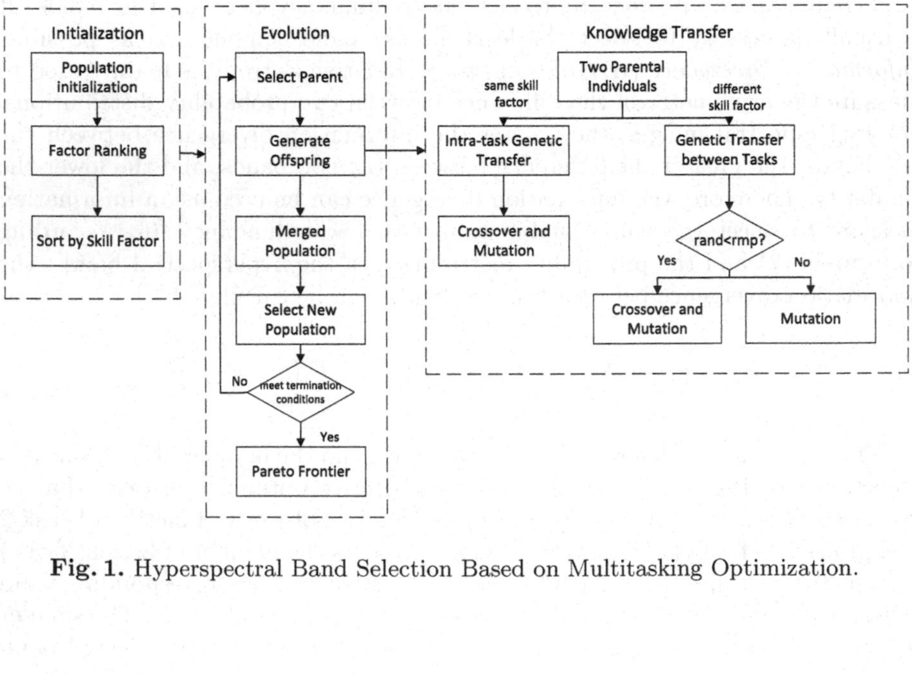

Fig. 1. Hyperspectral Band Selection Based on Multitasking Optimization.

4 Experimental Study

This section mainly evaluates and analyzes the band combinations obtained by the multitasking optimization method through experiments. Experiments are performed using classic hyperspectral remote sensing datasets from Indian Pines. The comparison methods include the ranking algorithm based on *Information Entropy*(IE) [8], the ranking algorithm based on Information Divergence (ID) [9], the ranking algorithm based on *Maximum Variance Principal Component Analysis* (MVPCA) [10] and the adaptive subspace-based algorithm proposed in recent years. Band selection algorithm (*Adaptive Subspace Partition Strategy*, ASPS) [11] and volume gradient-based fast algorithm (*Volume-Gradient-Based Band Selection*, VGBS) [12]. *The overall classification accuracy* (OA) and *Average Classification Accuracy* (AA) were used as evaluation criteria to evaluate the classification results in the experiments. Overall classification accuracy refers to the proportion of correctly classified individuals among all individuals, which can reflect the overall classification performance of a subset of bands. The average classification accuracy is the average of the classification accuracy of all ground object categories, which can reflect the classification performance of the band subset for each ground object category. Two types of classifiers are mainly used for classification, namely, *support vector machine* (SVM) [13] and *K-nearest neighborhood* (KNN) [14]. In all experiments, 10% of the sample data was randomly selected as the training set, while the rest of the data was used as the test set. To minimize the effects of random errors caused by the random selection of sample data, 10 replicate experiments for each method on each dataset will be performed and the average will be taken as the final result. Finally, in the process of repeating the experiment, the program execution time required by each band selection method was recorded and averaged.

4.1 Pavia University Dataset Experimental Results and Evaluation

The Pavia University dataset was acquired by the *Reflective Optics System Imaging Spectrometer optical sensor* (ROSIS) on the campus of the University of Pavia in northern Italy. The dataset consists of 115 bands, each with 610 $\times$ 340 pixels, with spectral resolutions ranging from 0.43 μm to 0.86 μm. After data preprocessing to remove 12 noise bands with low signal-to-noise ratio, the remaining 103 bands are used for experiments. There are 9 types of ground objects in this dataset, please refer to Table 1 for details.

The Pareto fronts obtained by the proposed multitasking optimization method on the Pavia University dataset are shown in Fig. 2, where (a) is the Pareto front of task one, and (b) is the Pareto front of task two, the horizontal axis represents the first objective function value of each multi-objective task, that is, the number of bands, and the vertical axis represents the second objective function value, that is, the reciprocal of information entropy or information divergence. In order to facilitate comparison with other methods and save calculation time, only individuals whose number of bands is less than 45 in the final population are retained and only individuals whose number of bands is not

Table 1. Pavia University data feature name and corresponding number of pixels

Heading level	Example	Font size and style
Category number	Feature name	Number of pixels
1	Asphalt	6631
2	Meadows	18649
3	Gravel	2099
4	Trees	3064
5	Painted metal sheets	1345
6	Bare soil	5029
7	Bitumen	1330
8	Self-blocking bricks	3682
9	Shadows	947

higher than 40 are considered for evaluation. In each interval, based on the scalar fitness of the individual on the corresponding task, the individual with the largest scalar fitness is selected as the representative in the interval for evaluation.

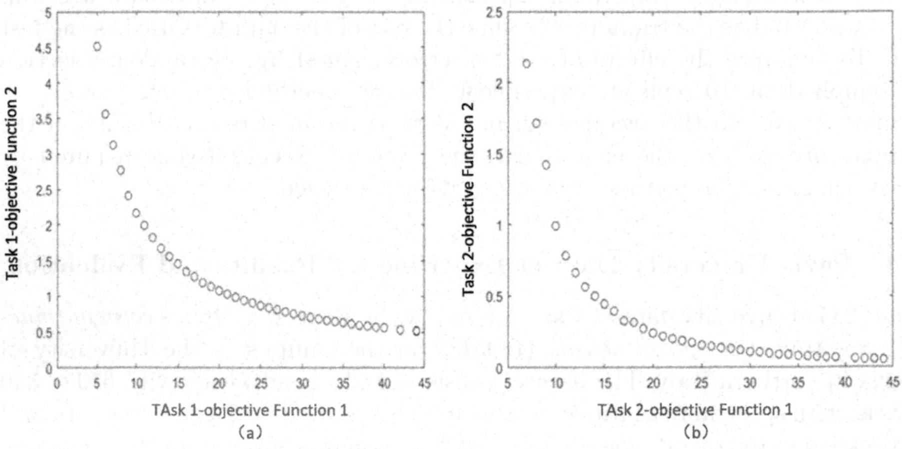

Fig. 2. Pareto frontier obtained by the algorithm on the Pavia University dataset.

The overall classification accuracy and average classification accuracy obtained by the SVM and KNN classifiers are shown in Fig. 3. The average classification accuracy is evaluated based on the number of bands of 15 and the number of bands of 30. Among them, (a) is the overall classification accuracy line graph obtained by the SVM classifier, (b) is the overall classification accuracy line graph obtained by the KNN classifier, (c) is the average classification accuracy bar graph obtained by the SVM classifier, and (d) is the average classification accuracy bar graph obtained by the KNN classifier. The evaluation results

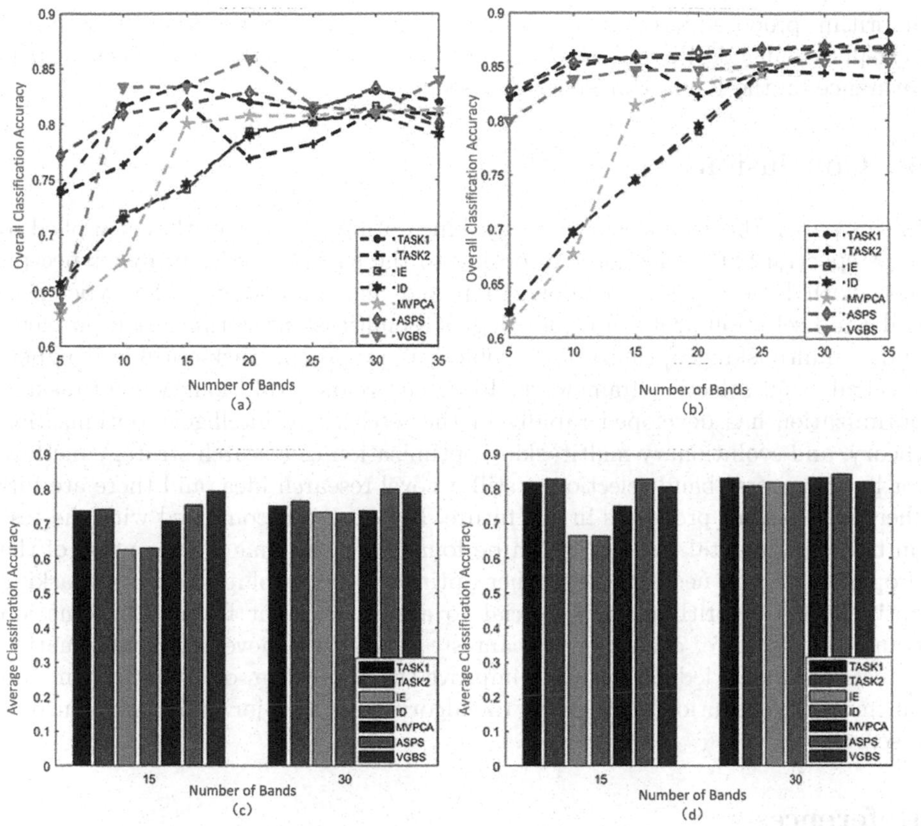

Fig. 3. Experimental results on the Pavia University dataset.

of band selection and subsets obtained by the multitasking optimization method correspond to TASK1 and TASK2 in the graph, and the experimental results of the comparison algorithms are presented in Fig. 3. For the overall classification accuracy evaluation, whether it is the SVM classifier or the KNN classifier has high overall classification accuracy when the number of bands is small, and when the number of bands increases, the overall classification accuracy is higher. The overall classification accuracy is similar to that of the comparison method, but relatively speaking, compared with the band combination obtained by VGBS, the performance of the band combination obtained by task one is slightly worse on the SVM classifier, but the band combination obtained by task one is more stable. Compared with other methods, the band subsets obtained by task one are significantly better than theirs. For the average classification accuracy, no matter when the number of bands is fixed at 15 or when the number of bands is 30, the average classification accuracy of the band subset obtained by multitasking optimization is much better than the three sorting methods of IE, ID and MVPCA. In terms of average classification accuracy, the performance of the algorithm proposed in this paper is very similar to that of the recently proposed best-performing ASPS and VGBS, indicating that the method has excellent performance on the Pavia University dataset.

5 Conclusion

In this paper, the evolutionary multitasking optimization algorithm is applied to hyperspectral band selection. According to the characteristics of hyperspectral data with large amount of information and high redundancy, the hyperspectral band selection problem is modeled as a multitasking optimization problem, and a multitasking optimization problem is proposed. Task-optimized hyperspectral band selection framework. In recent years, evolutionary multitasking optimization has developed rapidly in the research of intelligent optimization theory, and evolutionary multitasking optimization as a search strategy method for hyperspectral band selection is still a novel research idea, and there are further development prospects in the future. For example, combined with the test on the experimental dataset, it can be found that the time consumption of the proposed method needs to be further optimized. For evolutionary multitasking optimization algorithms, it is crucial to achieve efficient forward information transfer. Therefore, designing relevant strategies to achieve active information transfer and knowledge sharing to improve the parallelism of the algorithm and optimize the execution efficiency of the algorithm is a major challenge in current research.

References

1. Hammer, P.: Adaptive control processes: a guided tour (r. bellman) (1962)
2. Ong, Y.-S., Gupta, A.: Evolutionary multitasking: a computer science view of cognitive multitasking. Cogn. Comput. **8**(2), 125–142 (2016)
3. Gupta, A., Ong, Y.-S., Feng, L., Tan, K.C.: Multiobjective multifactorial optimization in evolutionary multitasking. IEEE Trans. Cybern. **47**(7), 1652–1665 (2016)
4. Shannon, C.E.: A mathematical theory of communication. ACM SIGMOBILE Mob. Comput. Commun. Rev. **5**(1), 3–55 (2001)
5. Gong, M., Zhang, M., Yuan, Y.: Unsupervised band selection based on evolutionary multiobjective optimization for hyperspectral images. IEEE Trans. Geosci. Remote Sens. **54**(1), 544–557 (2015)
6. Zhang, M., Gong, M., Chan, Y.: Hyperspectral band selection based on multi-objective optimization with high information and low redundancy. Appl. Soft Comput. **70**, 604–621 (2018)
7. MartÍnez-UsÓMartinez-Uso, A., Pla, F., Sotoca, J.M., García-Sevilla, P.: Clustering-based hyperspectral band selection using information measures. IEEE Trans. Geosci. Remote Sens. **45**(12), 4158–4171 (2007)
8. Datta, A., Ghosh, S., Ghosh, A.: Combination of clustering and ranking techniques for unsupervised band selection of hyperspectral images. IEEE J. Sel. Top. Appl. Earth Obs. Remote Sens. **8**(6), 2814–2823 (2015)
9. Chang, C.-I., Wang, S.: Constrained band selection for hyperspectral imagery. IEEE Trans. Geosci. Remote Sens. **44**(6), 1575–1585 (2006)

10. Chang, C.-I., Du, Q., Sun, T.-L., Althouse, M.L.: A joint band prioritization and band-decorrelation approach to band selection for hyperspectral image classification. IEEE Trans. Geosci. Remote Sens. **37**(6), 2631–2641 (1999)
11. Wang, Q., Li, Q., Li, X.: Hyperspectral band selection via adaptive subspace partition strategy. IEEE J. Sel. Top. Appl. Earth Obs. Remote Sens. **12**(12), 4940–4950 (2019)
12. Geng, X., Sun, K., Ji, L., Zhao, Y.: A fast volume-gradient-based band selection method for hyperspectral image. IEEE Trans. Geosci. Remote Sens. **52**(11), 7111–7119 (2014)
13. Cristianini, N., Shawe-Taylor, J., et al.: An Introduction to Support Vector Machines and Other Kernel-Based Learning Methods. Cambridge University Press, Cambridge (2000)
14. Cover, T., Hart, P.: Nearest neighbor pattern classification. IEEE Trans. Inf. Theory **13**(1), 21–27 (1967)

Robotics

Research on Control Sensor Allocation of Industrial Robots

Bin Yang, Zhouzhou Huang, and Wenyu Yang(✉)

College of Mechanical Science and Engineering, Huazhong University of Science and Technology, Wuhan, China
mewyang@hust.edu.cn

Abstract. To establish an efficient and scientific control sensor allocation process for industrial robots, a sensor allocation and optimization method based on dynamic parameters of robot joints was proposed. Based on the screw theory, the kinematics model of the robot was constructed, and the kinematics model was deduced according to the Lagrange equation, so as to determine the key dynamic parameters of the joints and the whole robot. According to the key dynamic parameters, the sensor types and allocation positions could be selected. The detection ability index, reliability, cost function, and the impact on robot motion were taken as the evaluation indexes of the sensor allocation problem. Then, considering all the actual constraints, the sensor allocation planning of the robot was made by combining the weight distribution method. Last, an example of the sensor allocation implementation process of the TA6R3 robot was demonstrated in detail.

Keywords: Industrial robot · Screw theory · Key dynamic parameters · Lagrange equation · Sensor allocation planning

1 Introduction

Industrial robots have the advantages of high repetition accuracy, good reliability, and strong applicability. They can replace the manual execution of repetitive production operations, especially in a harsh production environment. Therefore, industrial robots are widely used in automobiles, machinery, electronics, logistics, and other industries [1, 2]. In actual production, especially for important production environments, such as the assembly task of large Aerospace instruments and the processing task of ultra-precision instruments, the robot must ensure the accuracy of operation. So it needs to be guaranteed by a fast and accurate control system [3, 4]. The sensors are the data connection channel between robots and the control system. The operator and the control system need to observe and control the robot motion state according to the data information obtained by sensors [5, 6].

As an important part of the robot, the sensors arranged and equipped in the industrial robot are mainly divided into control sensors and external measurement sensors, namely internal sensors and external sensors [7]. The former is mainly used to measure and control the robot's own state, such as joint speed, rotation angle, output torque, etc. External

L. Fang et al. (Eds.): CICAI 2022, LNAI 13606, pp. 389–401, 2022.
https://doi.org/10.1007/978-3-031-20503-3_31

measurement sensors are related to specific operational tasks. The most common external measurement sensors include visual sensors, tactile sensors, etc [8]. Control sensors are important built-in detection equipment for industrial robots. The operation control command of the robot needs to start and stop according to the detection data of control sensors. Reasonable control sensor allocation can ensure good operation performance of the robot. Therefore, it is of great significance to study the allocation and optimization of control sensors.

In recent years, researchers have carried out a lot of related research on the allocation and optimization of industrial robot control sensors. Zhang et al. [9] selected three laser displacement sensors and arranged them at the feeding point to detect the position data. Then, a position compensation control system for the grinding robot was developed. Yang et al. [10] systematically analyzed the motion state monitoring and control principle of the series robot. They proposed optimal sensor allocation planning based on meta-action theory. Qiao et al. [11] aimed at monitoring the position and attitude health of industrial robots. Through error modeling and analysis technology, the sensor allocation strategy was determined, and a seven-dimensional measurement system was built. Zhang et al. [12] studied the optimal allocation of sensors in equipment, with the minimum cost and the harm of missed detection faults as the optimization objective. Yu et al. [13] studied the fault diagnosis methods of machine tools, robots, and other mechanical equipment. Then, a multi-objective optimal sensor allocation method according to the fault-measuring point mutual information matrix was proposed. It is very important to obtain accurate force information in some task scenarios, and the force sensor allocation problem is also the focus of current robot research [14, 15]. The existing sensor allocation methods based on target task parameter identification ensure the rationality of sensor allocation to a certain extent. However, there is still a lack of in-depth explanation of the underlying causes of sensor allocation problems. The actual constraints considered by many allocation methods are not comprehensive and perfect.

The paper constructs a general form kinematic and dynamic model of the industrial robot by using the screw theory [16] and the Lagrange equation [17]. Through kinematic and dynamic analysis, the key dynamic parameters of the robot are obtained. According to the key dynamic parameters, the corresponding types of control sensors and layout positions can be further determined and selected. Then, the sensor allocation problem is converted into a multi-objective programming problem by considering the constraints such as space constraints, cost, sensor characteristics, and robot motion stability. The remaining paper is organized as follows. The kinematics and dynamics modeling and analysis of the industrial robot are described in Sect. 2. Section 3 explains the proposed allocation method of robot control sensors in this paper. Section 4 presents an example of a TA6R robot to verify our method. The conclusion is drawn in Sect. 5.

2 Kinematics and Dynamics Analysis of Robot

2.1 Kinematic Model

According to Chasles theorem [18, 19], the arbitrary motion of a rigid body can be represented by a screw motion. The twist coordinate of the rigid body can be expressed as: $\boldsymbol{\xi} = [\boldsymbol{\omega}; \boldsymbol{\upsilon}]^{6\times1}$, where $\boldsymbol{\omega}$, $\boldsymbol{\nu}$ $\boldsymbol{R}^{3\times1}$ are the rotational and translation twist. Then, the twist of the rigid body can be expressed by:

$$\hat{\xi} = \begin{bmatrix} \hat{\omega} & \boldsymbol{\upsilon} \\ \mathbf{0} & 0 \end{bmatrix}^{4\times4} \in se(3) \tag{1}$$

The industrial robot can be simplified as a rigid motion system. The twist coordinate of an arbitrary joint of the robot can be written as:

$$\boldsymbol{\xi} = \left[j\boldsymbol{\upsilon} + i\boldsymbol{\omega}; i\boldsymbol{r} \times \boldsymbol{\omega}\right]^{\mathrm{T}} \tag{2}$$

where $\boldsymbol{r}$ is the coordinate of a point on the joint axis. For rotation joint: $I = 1, j = 0$. Otherwise: $I = 1, j = 0$.

Industrial robots are generally composed of several joints in series. Figure 1 shows the schematic diagram of two adjacent joints in the robot. $\boldsymbol{l}$ represents the unit direction vector of the joint motion axis. For m joint: $\boldsymbol{l}_m = j\upsilon_m + i\omega_m$. Based on Eq. (2), the twist of joint m can be expressed as a function of $\boldsymbol{l}_m$:

$$\boldsymbol{\xi}_m = [\mathbf{l}_m; i\boldsymbol{r}_m \times \mathbf{l}_m]^{\mathrm{T}} \tag{3}$$

In the screw kinematics analysis method, the joints of the robot are considered motion transfer operators. Assuming that $\boldsymbol{R}$ and $\boldsymbol{P}$ are position vector and attitude matrix, the motion transformation matrix of a joint can be expressed as an index matrix operator:

$$\mathbf{g}_m = \exp\left(\hat{\xi}_m\theta_m\right) = \begin{bmatrix} \mathbf{R} & \mathbf{P} \\ \mathbf{0}^{1\times3} & 1 \end{bmatrix} \tag{4}$$

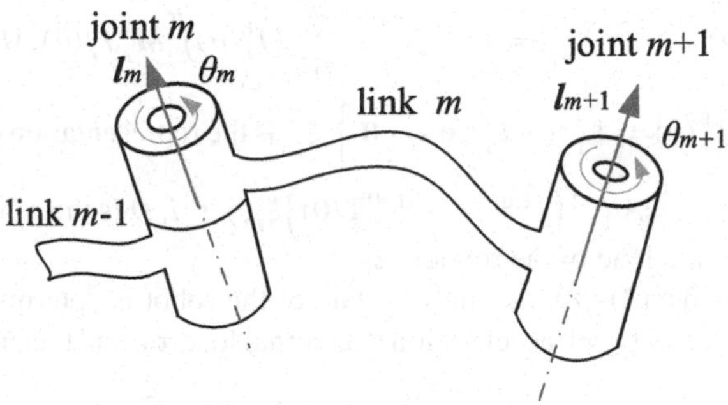

Fig. 1. Two adjacent joints of the robot

The twist exponential product model is used to describe the motion posture state of the robot. The kinematic model of the industrial robot is obtained as:

$$^{\mathrm{S}}_{\mathrm{T}}\boldsymbol{T}(\boldsymbol{\theta}) = \exp\left(\hat{\xi}_1\theta_1\right)\cdots\exp\left(\hat{\xi}_n\theta_n\right)^{\mathrm{S}}_{\mathrm{T}}\boldsymbol{T}(\mathbf{0}) \tag{5}$$

where $\boldsymbol{\xi}_i, \theta_i$ are twist coordinate and motion scale of joint i respectively. $^{\mathrm{S}}_{\mathrm{T}}\boldsymbol{T}(\mathbf{0})$ represents the initial pose of the robot end-effector.

The relationships between the joint velocity and the end effector velocity can be written as follows:

$$\begin{bmatrix} \boldsymbol{w} \\ \boldsymbol{v} \end{bmatrix} = {}_{\mathrm{T}}^{\mathrm{S}}\boldsymbol{J}^{s}(\theta)\dot{\theta} \tag{6}$$

where ${}_{\mathrm{T}}^{\mathrm{S}}\boldsymbol{J}^{s}(\theta) = \left[\boldsymbol{\xi}_1, \cdots, Ad\left(\exp\left(\hat{\xi}_1\theta_1\right)\cdots\exp\left(\hat{\xi}_{n-1}\theta_{n-1}\right)\right)\boldsymbol{\xi}_n\right]$ is the Jacobian matrix of the robot.

2.2 Dynamics Analysis

Let the load of joint k be: $\boldsymbol{\tau}_{k0} = \left[\boldsymbol{F}_{kT0}\ \boldsymbol{M}_{kT0}\right]^{\mathrm{T}}$. $\boldsymbol{G}_k(\theta) = \begin{bmatrix} \sum m_k\mathbf{g} \\ \sum(\mathbf{p}_k \times m_k\mathbf{g}) \end{bmatrix}$ represents the force vectors generated by joint gravity. The driving force of the joint can be calculated [20]:

$$\boldsymbol{\tau}_k = \begin{bmatrix} \boldsymbol{J}_{vk}m_k\dot{\boldsymbol{v}}_k \\ \boldsymbol{J}_{wk}\left(\boldsymbol{I}\dot{\boldsymbol{w}}_k + \boldsymbol{C}_k\left(\dot{\theta}_k, \theta_k\right)\right) \end{bmatrix} + \boldsymbol{G}_k(\theta) + \boldsymbol{\tau}_{k0} \tag{7}$$

where $\boldsymbol{J}$ represents the transformation matrix from the joint coordinate to the calculation coordinate system, and $\boldsymbol{C}$ is the coupling matrix of centrifugal force and Coriolis force. $\mathbf{c}_{ij} = \sum_{k=1}^{n}\frac{1}{2}\left(\frac{\partial\mathbf{D}_{ij}}{\partial\theta_k} + \frac{\partial\mathbf{D}_{ik}}{\partial\theta_j} - \frac{\partial\mathbf{D}_{jk}}{\partial\theta_i}\right)\theta_k$. According to the Lagrange equation, the dynamic model of the robot is obtained as follows:

$$\boldsymbol{\tau}=\mathbf{D}(\theta)\ddot{\theta} + \mathbf{C}\left(\dot{\theta}, \theta\right) + \boldsymbol{G}(\theta) + \boldsymbol{\tau}_0 \tag{8}$$

where $\boldsymbol{D}(\theta) = \sum_{i=1}^{n}\left(\boldsymbol{J}_i^{\mathrm{b}}(\theta)\right)^{\mathrm{T}}\boldsymbol{M}_i^{\mathrm{b}}\boldsymbol{J}_i^{\mathrm{b}}(\theta)$; $\boldsymbol{M}_i^{\mathrm{b}} = \begin{bmatrix} m_i\mathbf{E}_{3\times3} & 0 \\ 0 & {}^{Ci}\boldsymbol{I}_i \end{bmatrix}$; $\boldsymbol{J}_i^{\mathrm{b}}(\theta)=\left[\boldsymbol{\xi}_1^{+} \cdots \boldsymbol{\xi}_i^{+}\ \mathbf{0} \cdots \mathbf{0}\right]$, $\boldsymbol{\xi}_j^{+}$ is the representation of j joint in coordinate system $\{i\}$: $\boldsymbol{\xi}_j^{+}=Ad^{-1}\left(e^{\hat{\xi}_j\mathbf{q}_j}\cdots e^{\hat{\xi}_j\mathbf{q}_i}{}_i^0\mathbf{T}(0)\right)\boldsymbol{\xi}_j^{s}, j \le i$. $\boldsymbol{D}$ is the robot inertia matrix in joint space, and the load of the robot is $\boldsymbol{\tau}_0$.

According to Eqs. (2)–(8), the motion state of the robot is determined by the joint motion scale θ, velocity $[\boldsymbol{v}, \boldsymbol{w}]$, acceleration $\ddot{\theta}$, external load $\boldsymbol{\tau}_0$, and the inherent structural parameters.

Therefore, in order to ensure the monitoring, accuracy, and operability of industrial robots in performing production tasks, appropriate sensing equipment should be configured to monitor and control dynamic parameters such as motion scale, velocity, acceleration, and external load.

3 Control Sensor Allocation Planning

The control system monitors and controls the operation movement of the robot by controlling the detection data of the sensor. The allocation of the control sensor must first ensure the detectability and resolution of the motion data [13]. The corresponding sensors are configured according to the key dynamic parameters obtained from the kinematics and dynamics analysis of the robot. According to the actual industrial environment and demands, the constraints of cost, space, and other factors should also be considered.

3.1 Sensor Allocation Planning Function

Let $X = [x_1, x_2, \ldots, x_n]$ be the sensor allocation vector. x_j, M_j are the number and the type number of sensors at measuring point j. K_j is the type number of dynamic parameters to be detected, where $M_j \leq K_j$. The sensor detection ability (S_s), reliability (R_s), cost (C_s), and influence on robot motion (E_s) are used as the evaluation indexes of sensor allocation.

3.1.1 Evaluation of Detection Ability

Let $F = [f_1, f_2, \ldots, f_m]$ be the evaluation factor of the detection ability of m kinds of sensors. Then the sensor detection capability of class i dynamic parameters at the measuring point j can be described by the detection capability index s_{ij}:

$$s_{ij} = r_{ij} f_i \tag{9}$$

where r_{ij} is the number of type i sensor at measuring point j, and $x_j = \sum_{j \in N} r_{ij}$.

Based on Eq. (9), the evaluation index value of the overall sensor detection capability of the mechanical equipment is further calculated: $S_s = \sum_{i \in M} \sum_{j \in N} r_{ij} f_{ij}$.

3.1.2 Reliability

Assuming that $\boldsymbol{p} = [p_1, p_2, \ldots, p_m]$ is the prior failure probability of sensors. Then, the reliability of the whole machine is:

$$R_s = 1 - (\max_{\forall i} P_j) \tag{10}$$

where $P_j = 1 - \prod_{i \in M} (1 - p_{ij}^{r_{ij}})$ is the unreliability of the sensor at measuring point j.

3.1.3 Cost Function

Calculate the relative cost coefficient of sensor allocation according to the cost calculation formula:

$$C_s = 0.5 + [\sum_{j \in M} (c_j x_j)]/[2 \sum_{j \in N} (c_j q_j)] \tag{11}$$

where q_j is the upper limit of x_j. c_j is the cost factor of the sensor on measuring point j.

3.1.4 Impact on Equipment

The installation of sensors will inevitably increase the load of the robot. It will change the motion reliability of the system. So the effect should be minimized. Set m_{ij} as the influence factor of the sensor on the motion reliability, and the influence index of the configured sensor on the equipment can be calculated:

$$E_s = \sum_{i \in M} \sum_{j \in N} r_{ij} m_{ij} \tag{12}$$

According to the constraints of dynamic parameters, reliability, cost, and the influence of sensor on motion, the sensor allocation problem is transformed into a multi-objective nonlinear programming problem. The mathematical model of the general sensor allocation problem can be expressed as:

$$\begin{aligned} Z &= \{max\, S_s,\ max\, R_s, min\, C_s\, min\, E_s\} \\ &\text{s.t. } S_s;\, R_s;\, C_s;\, E_s \end{aligned} \tag{13}$$

In order to reduce the cost, the built-in sensors of specific industrial robots should be fully considered. At the same time, due to the different emphasis on actual demands, the importance of S_s, R_s, C_s, and E_s may be different. Assuming that the performance of the same type of sensor is the same. Set s_{ij} as the number of the i-th built-in sensor at measuring point j. Combined with the weight distribution method, the sensor allocation planning with built-in sensing equipment:

$$\min Z' = -k_1 S_s'' - k_2 R_s'' + k_3 C_s'' + k_4 E_s''$$

$$\text{s.t.} \begin{cases} C_s' = C_s,\, S_s' = \sum_{i \in M} \sum_{j \in N} (r_{ij} + s_{ij}) f_{ij} \\ P_j' = 1 - \prod_{i \in M} (1 - p_{ij}^{(r_{ij}+s_{ij})}) \\ R_s' = 1 - (\max_{\forall i} P_j') \\ E_s' = \sum_{i \in M} \sum_{j \in N} (r_{ij} + s_{ij}) m_{ij} \\ 0 \le \forall x_j \le q_j - s_j \end{cases} \tag{14}$$

where k is the assigned weight coefficient, and $\sum k = 1$. S_s'', R'', C_s'' and E_s'' are the four evaluation index function expressions after certain mathematical processing.

3.2 Control Sensor Allocation Process

In this section, the sensor allocation analysis includes the determination of key dynamic parameters and the solution process of the sensor allocation planning function. The specific process of implementing control sensor allocation by the proposed method is as follows:

1. The robot is analyzed, and the number and basic types of robot joints are determined.
2. The kinematics and dynamics models of the robot are constructed. Then the key dynamic parameters of joints and the whole robot can be determined. Namely: $\boldsymbol{\theta}$, [$\boldsymbol{v}$, $\boldsymbol{w}$], [$\dot{\mathbf{v}}$, $\dot{\mathbf{w}}$] and τ_0.
3. Determine the optional type and typeset of sensors. Combined with the spatial constraints of the robot, the rough boundary conditions of the sensor allocation of each joint are determined.
4. Firstly, without considering the existing built-in sensors, the allocation planning function is determined by the sensor requirements of the target robot: If the importance of S_s, R_s, C_s, and E_s are regarded as equal, select the planning function of Eq. (13). Otherwise, it is necessary to determine the weight of the four evaluation indexes, and then select Eq. (14).
5. Substituting the weight, accuracy, priori-failure rate P and cost factor of the sensor into Eq. (13) or Eq. (14) to solve the allocation planning function.
6. Then, considering the existing built-in sensors of the robot, the sensor allocation results are obtained from Eq. (14). If the robot body has no built-in sensor, directly perform Step 7.
7. Select the optimal solution from the feasible solution space and verify it. If it does not meet the requirements, return to Step 4 and improve the constraints.

4 Case Study

In this section, a TA6R robot, as shown in Fig. 2, is used to show the implementation process of the proposed joint sensor allocation method in detail. TA6R robot is a typical 6-DOF series industrial robot, Joints 1, 2, and 3 are the lumbar joint, shoulder joint, and elbow joint, respectively. And joints 4, 5, and 6 together form a wrist joint.

The TA6R3 robot contains 6 rotation joints. According to the method in Sect. 2, the kinematics and dynamics model of the robot can be obtained:

$$ {}_{\mathrm{T}}^{\mathrm{S}}\boldsymbol{T}(\boldsymbol{\theta}) = \exp\left(\hat{\xi}_1\theta_1\right)\cdots\exp\left(\hat{\xi}_6\theta_6\right){}_{\mathrm{T}}^{\mathrm{S}}\boldsymbol{T}(\mathbf{0}) \tag{15} $$

$$ \begin{cases} \tau_i = \begin{bmatrix} \boldsymbol{J}_{vi} m_i \dot{\boldsymbol{v}}_i \\ \boldsymbol{J}_{wi}\left(\boldsymbol{I}\dot{\boldsymbol{w}}_i + \boldsymbol{C}_i\left(\dot{\theta}_i, \theta_i\right)\right) \end{bmatrix} + \boldsymbol{G}_i(\theta) + \tau_{k0} \\ \tau = \boldsymbol{D}(\theta)\ddot{\theta} + \boldsymbol{C}\left(\dot{\theta}, \theta\right) + \boldsymbol{G}(\theta) + \tau_0 \end{cases} \tag{16} $$

According to Eqs. (15) and (16), the dynamic parameters of the TA6R3 robot are divided into motion parameters and mechanical parameters. For the convenience of analysis, this section only considers the detection of joint motion parameters, so the dynamic parameters to be monitored by the robot are mainly rotation rate and rotation angle. So that there is $1 \le \mathrm{M}_j, \mathrm{K}_j \le 2 (j \in \mathrm{N})$. Select the magnetoelectric speed sensor and displacement vibration sensor with the same inherent detection ability. The single weight is about 100 g and the accuracy is 5%. Consult the data of the sensor manufacturer to determine the prior failure rate P and cost factor c_j of these sensors. Set the sensor parameters according to the actual working conditions of the robot, as shown in Table 1.

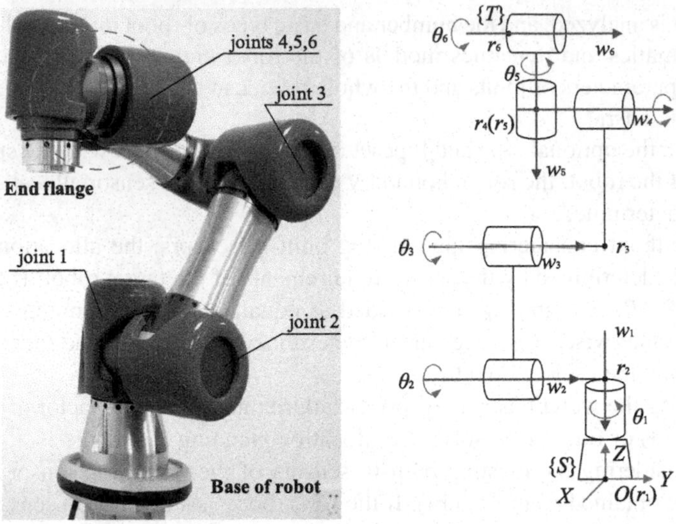

Fig. 2. TA6R3 robot

Table 1. Parameters of sensors

Joint	F	P	c_j	m_{ij}	Q
1	0.95	0.1	1.0	0.03	4
2	0.95	0.1	1.0	0.03	4
3	0.95	0.1	1.0	0.03	4
4	0.95	0.1	1.0	0.05	4
5	0.95	0.1	1.0	0.05	4
6	0.95	0.1	1.0	0.05	4

The sensor allocation is considered qualified if the success rate of parameters to be accurately detected exceeds 90%. If the built-in sensor of the robot is not considered, the above parameters are substituted into Eq. (13), and the multi-objective integer programming problem is solved by using the function in the toolbox of MATLAB. Firstly, the four objective functions are considered separately. The convergence curve of the objective function in single objective programming obtained by the NSGA-II algorithm [21], as shown in Fig. 3. Obviously, when it evolves to the 20th generation, all the four objective functions have converged.

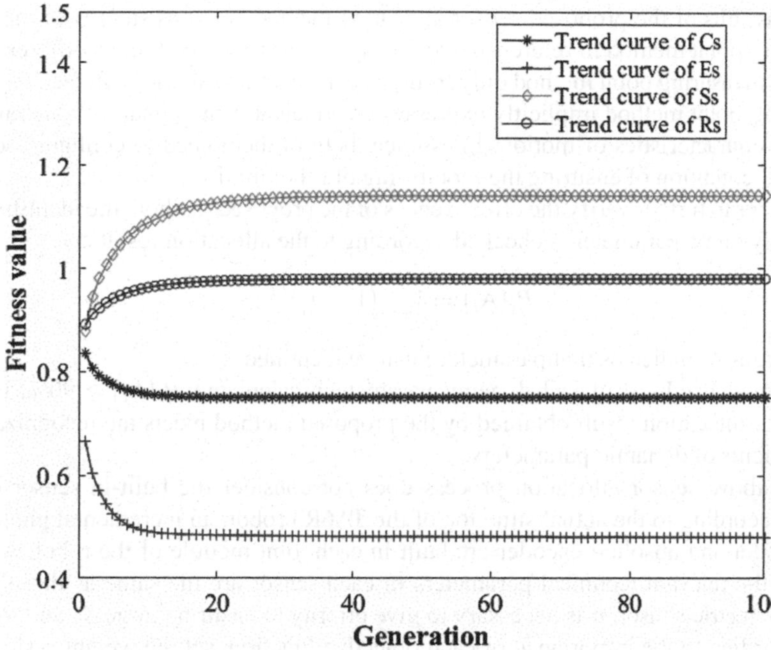

Fig. 3. The convergence curve of the objective function in single objective programming

In this example, the importance of the objective functions in the multi-objective planning of robot sensor allocation are the same. The feasible solutions to the sensor allocation planning problem can be obtained: $2 \leq x_j \leq 4$ ($i = 1, 2, \ldots, 6$).

Further considering the principle of cost reduction and simplest installation, the most appropriate optimal solution is selected: $X = [2\ 2\ 2\ 2\ 2\ 2]$ (where $r_{ij} = 1$). If the robot control system has a strong data processing ability or has low requirements for the response speed, the rotation angle (or rate) can be obtained through integral or differential calculation according to the rate data (or rotation angle). Then the optimal solution can be optimized as: $X = [1\ 1\ 1\ 1\ 1\ 1]$. The proposed method and the fault-measuring point-based sensor allocation method [13] are used to configure the sensor respectively. The feasible solutions are shown in Table 2.

Table 2. Feasible solutions to the sensor allocation problem

Key dynamic parameters	Feasible solutions	
	Proposed method	Fault-measuring point method
$[w, v]$	$4 \geq r_{ij} = x_j \geq 1$	$4 \geq r_{ij} = x_j \geq 1$
Θ	$4 \geq r_{ij} = x_j \geq 1$	$4 \geq r_{ij} = x_j \geq 1$
$[w, v] \cap \theta$	$2 \geq r_{ij} \geq 1 \cap 2 \geq x_j \geq 1$	$2 \geq r_{ij} \geq 1 \cap 2 \geq x_j \geq 1$

The results of the proposed sensor allocation method are consistent with the fault-measuring point method in reference. Obviously, the proposed method is different from the fault-measuring point method only from the perspective of analysis. In fact, the fault-measuring point method implicitly expresses the kinematic and dynamic constraints as the fault characteristics of motion. In essence, both of them need to configure sensors under the condition of ensuring the monitoring of robot motion.

In order to further verify the effectiveness of the proposed method, the identifiability of each dynamic parameter is checked according to the allocation results:

$$P_i\{\mathrm{A}_i\} = 1 - \left(1 - p_{ij}\right)^{r_{ij}} \tag{17}$$

where event A_i indicates that parameter i can be identified.

The recognizability of each dynamic parameter is calculated: $P_i\{\mathrm{A}_i\} \geq 90\%$. That is, the sensor allocation result obtained by the proposed method meets the recognizability requirements of dynamic parameters.

The above sensor allocation process does not consider the built-in sensor of the robot. According to the actual situation of the TA6R3 robot, an incremental photoelectric encoder and absolute encoder are built in each joint module of the robot. Assuming that the relevant technical parameters of each sensor are the same as those of the magnetoelectric sensor, it is necessary to give priority to ensuring large S_S and R_S.

According to the importance of each objective function, set the weight assigned to each evaluation index as: [0.35 0.35 0.15 0.15]. S_s, C_s, and E_s are all linear programming functions. It should be noted that the numerical magnitude of S_S is much larger than the other three objective functions. If the objective function is directly set to the reciprocal sum of four objective functions, the optimal solution of sensor allocation planning must be $\boldsymbol{X} = [4\ 4\ 4\ 4\ 4\ 4]$. Therefore, we need to deal with these objective functions first. If the arrangement limit of sensors at each measuring point is 1 state, the values of S_s, C_s, and E_s should be normalized first, in which the original function of C_S has been normalized. Incremental photoelectric encoder and absolute encoder are built in each joint module of the robot, that is, $s_{ij} = 1$. Combined with the prior failure probability of the sensor and the success rate index of dynamic parameter detection, S_S must meet the requirements, so only the other three objective functions need to be considered. According to the analysis results, the allocation planning function is re-summarized. The simulation results are shown in Fig. 4. The objective function has converged when it evolves to the 20th generation, and the optimal value of the comprehensive objective is -0.0859. Finally, the optimal solution obtained by the sensor allocation planning is $\{x_{ij} + r_{ij}\} = [2\ 2\ 2\ 2\ 2\ 2]$.

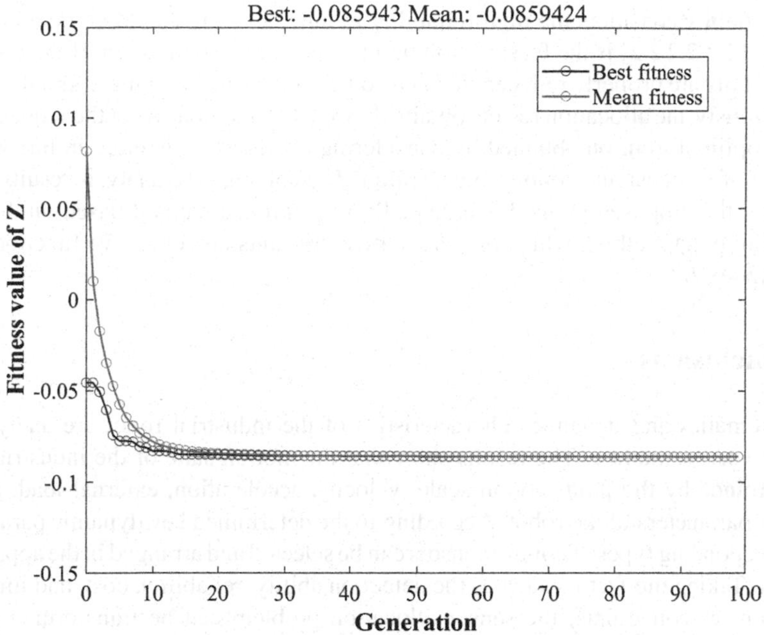

Fig. 4. Convergence curve of objective functions

The sensor allocation planning functions are constructed by the proposed method and the fault-measuring point method, respectively. The optimal solutions are shown in Table 3.

Table 3. Optimal solutions to sensor allocation problem

Sensor allocation planning	Optimal solutions	
	Proposed method	Fault-measuring point method
Total	$x_j + r_{ij} = 2$	$x_j + r_{ij} = 3$
X	$x_j = 0$	$x_j = 1$

The optimal sensor allocation scheme obtained by the proposed method is $\{x_{ij} + r_{ij}\} = [2\ 2\ 2\ 2\ 2\ 2]$ (where $r_j = 2$). That is, the built-in sensors of the robot have met the requirements. So there is no need to add other sensors. Therefore, in the future, only appropriate force sensors need to be configured at the end, and the motion of the TA6R3 robot can be controlled by the built-in control sensors and the end force sensors. The recognizability of each dynamic parameter is calculated: $P_i\{\mathrm{A}_i\} = 99\% > 90\%$.

The fault-measuring point method does not consider the influence on robot motion(E_s). The optimal allocation scheme of the sensor obtained by the fault-measuring point method is $\{x_{ij}+r_{ij}\} = [3\ 3\ 3\ 3\ 3\ 3]$ (where $r_j = 2$), and $\boldsymbol{X}' = [1\ 1\ 1\ 1\ 1\ 1]$, $P_i\{\mathrm{A}_i\} = 99\% > 90\%$. The optimal solution obtained by the fault-measuring point method mainly considers the high-reliability requirements. It should be noted that $\{x_{ij} + r_{ij}\} = [2\ 2\ 2\ 2\ 2\ 2]$ is the feasible solution for the objective function. If the reliability requirements are properly reduced, it can also be transformed into the optimal solution.

Obviously, the allocation results obtained by the two methods meet the requirements, and the optimal solution obtained by considering the index E_s is more in line with the principle of low cost and simplest installation. The comparative analysis results further show that the proposed method is essentially an improved method based on the fault-measuring point method, which increases the constraints and objective function of the impact index E_s.

5 Conclusions

The kinematics and dynamics characteristics of the industrial robot are analyzed by using the screw method. The results show that the motion state of the industrial robot is determined by the joint motion scale, velocity, acceleration, external load, and the inherent parameters of the robot. According to the determined key dynamic parameters, the corresponding types of control sensors can be selected and arranged in the appropriate position. Taking the factors such as the detection ability, reliability, cost, and impact on equipment as constraints, the sensor allocation problem can be transformed into an integer programming problem. In this paper, the allocation process of sensors is mainly divided into two steps: (1) Key dynamic parameters are obtained according to the results of kinematics and dynamics analysis. (2) The actual constraints are comprehensively considered to transform thc allocation problem into the problem of integer programming.

In the Case study, a TA6R robot is used to show the implementation process of the proposed joint sensor allocation method in detail, and the results show the practicability and effectiveness of our method. At the same time, compared with the fault-measuring point method, the comparative analysis results further show that the proposed method is essentially an improved method based on the fault-measuring point method.

References

1. Kevin, M.L., Frank, C.P.: Modern Robotics: Mechanics, Planning, and Control. Cambridge University Press, Cambridge (2017)
2. Meng, M., Zhou, C., Chen, L., et al.: A review of the research and development of industrial robots. J. Shanghai Jiaotong Univ. **50**(S1), 98–101 (2016)
3. Tian, L., Chen, H., Zhu, J., et al.: A study of optimal sensor placement based on the improved adaptive simulated annealing genetic algorithms. J. Vib. Eng. **25**(03), 238–243 (2012)
4. Chen, Y.: Fine Operation Control Method Based on Multi-sensor for Space Manipulator. Beijing University of Posts and Telecommunications, Beijing (2017)
5. Aazam, M., Zeadally, S., Harras, K.A.: Deploying fog computing in industrial internet of things and industry 40. IEEE Trans. Ind. Inform. **10**(14), 4674–4682 (2018)
6. Liu, S., Huang, T., Yen, J.: Comparison of sensor fusion methods for an SMA-based hexapod biomimetic robot. Robotic Auton. Syst. **58**(5), 737–744 (2010)
7. Lin, S., Chen, S., Li, C.: Welding Robot and Its Application. Machinery Industry Press, Beijing (2000)

8. Wang, W., Zhao, S., Sun, P., et al.: Crux techniques of multi-agent robot soccer system. J. Northeast Univ. **02**, 192–195 (2001)
9. Zhang, G., Li, Y., Zhang, H., et al.: Design of position compensation system for grinding robot based on displacement sensor. Sensors Microsyst. **40**(04), 74-76+79 (2021)
10. Yang, B., Zhang, G., Yu, H., et al.: Controlling the motion performance of mechanical products based on digital-twin. Comput. Integr. Manuf. Syst. **25**(06), 1591–1599 (2019)
11. Qiao, G., Schlenoff, C., Weiss, B.: Quick positional health assessment for industrial robot prognostics and health management (PHM). In: IEEE International Conference on Robotics and Automation (ICRA) (2017)
12. Zhang, L., Zhang, F.: Research on optimal sensor placement in equipment health management. Sensors Microsyst. **07**, 18–20 (2008)
13. Yu, B., Yang, S., Zhou, X.: Optimization of sensor allocation based on fault-measurement point mutual information. J. Zhejiang Univ. (Eng. Sci.) **46**(1), 156–162 (2012)
14. Li, Q., Yuan, H., Ma, X., et al.: Collision position detection for robotic end-effector using force/torque sensor. Comput. Integr. Manuf. Syst. **27**(01), 109–117 (2021)
15. Leng, Y., Chen, Z., He, X., et al.: Collision sensing using force/torque sensor. J. Sensors **3**, 1–10 (2016)
16. Dai, J.: Spinor Algebra, Lie Group and Lie Algebra. Higher Education Press, Beijing (2014)
17. Xiong, Y., Li, W., Chen, W., et al.: Robotics: Modeling, Control and Vision. Huazhong University of Science and Technology Press, Wuhan (2018)
18. He, R., Zhao, Y., Yang, S.: Kinematic-parameter identification for serial-robot calibration based on POE formula. IEEE Trans. Robot. **26**(3), 411–423 (2010)
19. Yang, B., Zhang, G., Ran, Y., et al.: Kinematic modeling and machining precision analysis of multi-axis CNC machine tools based on screw theory. Mech. Mach. Theory **140**, 538–552 (2019)
20. Craig, J.: Introduction to Robotics: Mechanics and Control. Pearson Education Inc, Upper Saddle River (2005)
21. Zhou, W., Xiao, B., Ran, Y., et al.: Kinematic accuracy mapping of NC machine tools based on meta-action units. J. Harbin Inst. Technol. **52**(1), 170–177 (2020)

Event-Based Obstacle Sensing and Avoidance for an UAV Through Deep Reinforcement Learning

Xinyu Hu, Zhihong Liu(✉), Xiangke Wang, Lingjie Yang, and Guanzheng Wang

National University of Defense Technology, Changsha 410073, China
zhliu@nudt.edu.cn

Abstract. Event-based cameras can provide asynchronous measurements of changes in per-pixel brightness at the microsecond level, thereby achieving a dramatically higher operation speed than conventional frame-based cameras. This is an appealing choice for unmanned aerial vehicles (UAVs) to realize high-speed obstacle sensing and avoidance. In this paper, we present a sense and avoid (SAA) method for UAVs based on event variational auto-encoder and deep reinforcement learning. Different from most of the existing solutions, the proposed method operates directly on every single event instead of accumulating them as an event frame during a short time. Besides, an avoidance control method based on deep reinforcement learning with continuous action space is proposed. Through simulation experiments based on AirSim, we show that the proposed method is qualified for real-time tasks and can achieve a higher success rate of obstacle avoidance than the baseline method. Furthermore, we open source our proposed method as well as the datasets.

Keywords: Event camera · UAV · Collision sensing and avoidance · Deep reinforcement and learning

1 Introduction

In recent years, UAVs have not only greatly facilitated human life in civil fields such as aerial photography, geological monitoring and agriculture, but also have played an increasingly important role in military tasks such as reconnaissance and communication [7]. As more and more UAVs are used, the risk of UAVs colliding with other objects increases significantly. A large number of studies have shown that the real-time sense and avoid of UAVs is extremely important [2,11]. Complex and highly dynamic search or navigation missions, such as searching for targets in forests or jungles obscured by tree canopies, pose urgent requirements

Supported by Science and Technology Innovation 2030-Key Project of “New Generation Artificial Intelligence” under Grant 2020AAA0108200 and National Natural Science Foundation of China under Grant 61906209 and 61973309.

L. Fang et al. (Eds.): CICAI 2022, LNAI 13606, pp. 402–413, 2022.
https://doi.org/10.1007/978-3-031-20503-3_32

for the UAV's high-speed sensing and avoidance capabilities. Therefore, sense and avoid with low latency is one of the key technologies for fast autonomous navigation of drones.

There are two main approaches for high-speed SAA in unknown environments: one is mapping and path planning, and the other is learning paradigm to achieve end-to-end obstacle avoidance. In [9], a depth camera is used to obtain Euclidean distance fields, combined with Rapidly-exploring Random Tree (RRT) to achieve high-speed obstacle avoidance for quadrotors. [12,23] obtain the mapping of sensing information from monocular cameras to UAV maneuvers by means of supervised learning and reinforcement learning. Wang et al. use 2D LIDAR information combined with a deep reinforcement learning algorithm to achieve collision avoidance of UAV clusters [22].

In order to obtain more precise state estimation, multiple sensors are used. In [19], binocular depth cameras, down-sight cameras, and IMU are used to compute motion primitives that enable obstacle avoidance in unstructured, GPS-rejected, crowded outdoor environments at 12 m/s. Loquercio et al. [10] use binocular cameras and depth cameras combined with imitation learning to achieve high-speed flight in the field.

Due to the limitations of micro-UAV payload and computing resources, vision sensors are widely used for aggressive maneuvering. However, conventional cameras have limitations in hardware, such as operating frequency and dynamic range. Thus, only can careful and precise decision and prediction in high latency be applied to achieve high-speed SAA. Furthermore, for situations where the environment light changes significantly, the mode of conventional cameras operating at frame rate leads to perceptual failure. By contrast, the emerging event cameras can overcome the disadvantages of high latency and low dynamic range of conventional cameras, which have attracted wide attention in recent years.

Event cameras have been favorably implemented on UAVs in the last decade. There are three main application areas. The first one is to achieve tracking of moving targets in the form of morphological filtering of events . The dualcopter can track the diameter line of the disk rotating rapidly (1600/s) in one-dimensional attitude, which is in form of event frame [3]. In [13,14], event streams are clustered into event frames to perform edge detection of single geometric target in aggressive environments. The second is to incorporate other sensors to estimate optical flow, ego-motion in high dynamic environment. A contrast maximization function wrt. event frames to estimate 3D rotational motion of the UAV [5]. Rebecq et al. [16] track features from event frames to achieve state estimation of itself and pose estimation of obstacles. Event streams are learned by self-supervised learning and supervised learning for deblurring, ego-motion estimation, and optical flow estimation [17,24] . The last is to assist the UAV to avoid obstacles. There are few studies using event camera to realize the obstacle avoidance of UAVs. Dealing with event frames, [4] combines the ego-motion estimation, morphological filtering algorithm and path planning algorithm to make it. It is challenging to adjust every function parameter, although only the rotation of the camera is considered. With the help of the depth camera, ego-motion estimation of translation is considered despite rotation part when dodging fast-moving objects [6]. Supervised learning paradigm is used to model the event frame deblurring, ego-motion estimation and optical flow estimation in [24], which contributes to obstacle avoidance. Vemprala et al. [1] firstly use reinforcement learning to achieve high-speed obstacle avoidance on an unmanned car with event frames as the inputs. In summary, although the event frame is widely used in different formats, it is not suitable for event streams in high frequency due to the sparsity of events. Besides, learning-based methods can avoid complex modeling and can realize SAA based on a data-driven manner, which is a promising approach for SAA using event cameras.

In this paper, we construct the event dataset using the event camera plugin and simulation environment in Airsim[1]. Then, a pre-trained feature-extractor network is implemented using variational auto-encoder (VAE), whose output is considered as the input states of deep reinforcement learning framework. And this algorithm extracts features from every event, which means it can adapt to variable control frequency. Finally, a control policy based on the proximal policy optimization algorithm (PPO) is designed to achieve high-speed SAA in the dense and static obstacle environment. The experiment results based on the AirSim simulator show that the proposed method works well in real time, and our avoidance policy achieves better obstacle avoidance than [21]. We open the source codes of our method along with the datasets[2].

The remainder of this paper is organized as follows. The problem description is given in Sect. 2. Section 3 introduces the preliminaries of our work. The overall approaches of the research are explained in Sect. 4. Next, a series of experiments are carried out in Sect. 5. Finally, conclusions are given in Sect. 5.

2 Problem Description

This paper explores the problem of high-speed SAA in high dynamic range, dense obstacle environments by using event cameras. A quadrotor equipped an event camera navigates in the dense obstacle environments without collisions with the obstacles. Figure 1 illustrates the problem of sense and avoid for the UAV that we are going to solve. The UAV is expected to reach the endpoint area without any collision.

3 Preliminaries

3.1 Working Principle of Event Camera

The working principle of dynamic vision sensor (DVS), which is a type of event cameras, is described as follows. Every pixel can generate "events" independently and asynchronously. Given a pixel location (x, y), the logarithmic brightness at

---

[1] https://github.com/microsoft/AirSim.

[2] Our code and environment can be found at https://gitcode.net/weixin_40807529/event-based_uav_saa.

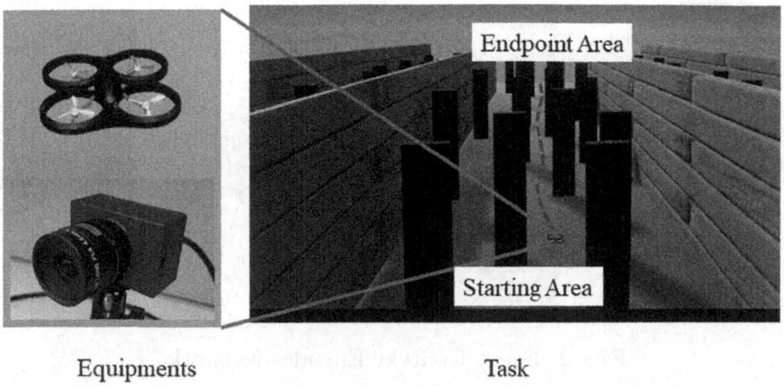

Fig. 1. Task setting of SAA.

that pixel $L(x, y, t)$ is measured. If the change of it exceeds a set threshold C, "event"(x, y, t, p) is output, where p ($p \in \{-1, 1\}$) is called polarity and stands for the increase/decrease of the logarithmic brightness. Therefore, formula (1) represents trigger conditions for events and formula (2) for event streams.

$$L(x, y, t) - L(x, y, t - \Delta t) \geq pC \tag{1}$$

$$E_\tau = \{e_k\}_{k=1}^{N} = \{(x_k, y_k, t_k, p_k)\}_{k=1}^{N} \tag{2}$$

3.2 Event Stream Processing

Given event data as an arbitrary long stream E_τ, the objective is to map it to a compressed vector representing the latent state of the observation through an encoder function $f_e(E_\tau)$. To overcome the disorder of data, PointNet [15] and EventNet [18] are referenced, which can obtain feature vectors of spatial and polarity information. Temporal information is extracted feature vectors through a Transformer network [20], which is summed up with spatial feature vectors. Finally, VAE is used to reduce the feature dimensions and create a smooth, continuous, and consistent representation. The whole sensing network, which is called "Event Context Encoder Network" (ECEN) [21] , is shown in Fig. 2. The input to the ECEN is an event stream of different length, which is extracted into a space-time feature vector of 1024 dimensions through three dense layers. Then, the space-time feature vector is encoded to a latent vector through two convolutional layers. The currently mentioned network layers are combined into "Event Context Network" (ECN). Besides, the decoder part of the ECEN also consists of three convolutional layers to reconstruct the event-constructed image. The self-supervised learning paradigm is applied thanks to the decoder part.

4 Approach

The problem is modeled as a partially observable Markov decision process (POMDP).Then, we introduce the policy network architecture.

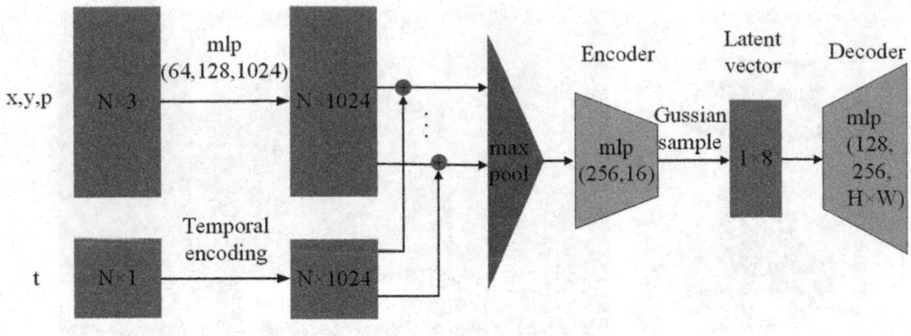

Fig. 2. Event Context Encoder Network.

4.1 Partially Observable Markov Decision Process

A partially observable Markov decision process is described as a six-element tuple(S, A, P, R, Ω, O). S refers to the state space, A refers to the action space of the agent, P refers to the state-transition model, R refers to the reward function, Ω refers to the observation space, and O refers to the observation probability distribution given the system state. In this problem modeling, we mainly focus on the observation space, action space, and the reward function.

The observation o_t consists of the latent vectors of event streams at times t, $t-1$, and $t-2$. The action space is a set of feasible velocities for the UAV. Considering the goal is to reach the target ahead without collision, we set the action space $a_t = [v_x, v_y]$,with $v_x \in [-1, 1]m/s$ and $v_y \in [0, 1]m/s$. The reward function consists of two parts. One is with respect to the goal $r_{g,t}$, the other is with respect to the collision $r_{c,t}$. The reward function is as follows:

$$r_t = r_{g,t} + r_{c,t} \tag{3}$$

When the UAV approaches or reaches its destination, it is rewarded by $r_{g,t}$. If the distance between the UAV and destination is less than $1m$, the task is finished and the UAV gets $r_{arrival}$ as the reward. Otherwise, if the distance between the UAV and the destination is getting smaller, it also gets a reward with parameter k_d. The details w.r.t. $r_{g,t}$ are shown in Eq. 4, where p_t stands for the position of UAV and g for its destination :

$$r_{g,t} = \begin{cases} r_{arrival}, if\ \|p_t - g\| \le 1 \\ k_d(\|p_{t-1} - g\| - \|p_t - g\|), otherwise \end{cases} \tag{4}$$

As for the collision-free requirement, it depends on the callback from Airsim. If the collision state is true, the reward gets a negative value, otherwise it is zero.

$$r_{c,t} = \begin{cases} r_{collision}, if\ collison\ state = true \\ 0, otherwise \end{cases} \tag{5}$$

4.2 Event-based Collision Sensing and Avoidance Algorithm

In order to construct the deep reinforcement learning method to make decisions, an efficient feature-extractor network should be designed. First, we collect a large number of event data in the simulation environment to train the ECEN, as described in the second session of the preliminaries. Thus we can obtain the temporal and spatial latent vectors of event streams. Second, the actor-critic network is attached to the ECN to implement PPO from stable baselines 3[3]. Finally, the pretrained weights of the ECN are fixed to output the latent vectors as the observation, and then the actor-critic network weights are updated based on PPO. The whole algorithm is shown in Algorithm 1. In detail, the input of the policy network corresponds with the observation space o_t, o_{t-1}, o_{t-2}. The output of the network is a two-element tensor (v_x, v_y). The end-to-end learning network is shown in Fig. 3.

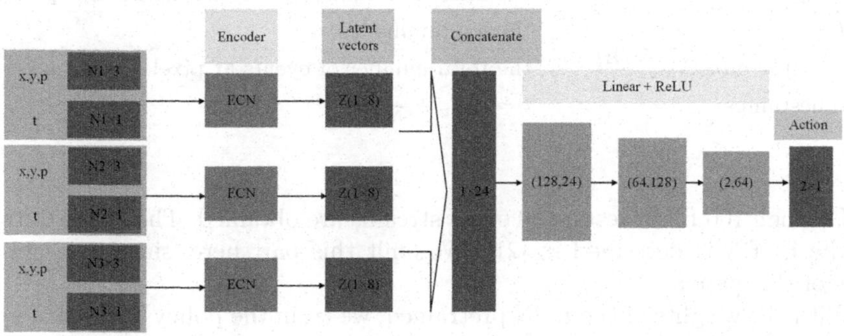

Fig. 3. Event-based collision sensing and avoidance learning network.

5 Simulation Experiments and Results

Training the policies of deep reinforcement learning in the real world usually takes a lot of time and resources. To deal with this, we build a simulation world in Unreal Engine for training and validation based on Airsim[1]. Unreal Engine[4] is a state-of-the-art engine and editor with photorealistic rendering, dynamic physics and effects, and much more on an open, extensible platform. Thus, the simulation we make is close to the real world. The experiment process and results are described in detail in this section.

5.1 Training

Experimental Setup. First, we apply RRT* [8] algorithm to control the drone collecting events in simulation world shown in Fig. 1, which is a long corridor

[3] https://github.com/DLR-RM/stable-baselines3.
[4] https://www.unrealengine.com/zh-CN/.

with a side length of over 100 m. The drone with an RGB camera is supposed to take off at the front end and reach the destination area while avoiding the dense obstacles.

AirSim simulates the event camera in Python, which computes the event stream from two consecutive RGB images. This function is now encapsulated an extension package named event-sim in AirSim. The resolution and the log luminance threshold (TOL) can be tuned to achieve a level of visual fidelity and performance of the event simulation. Details about the parameters of the event camera are shown in Table 1.

Table 1. Simulator Parameters.

Resolution	Width*Height(64*64)
TOL	0.1
N_{max}	the maximum number of events at a single pixel
L	the log luminance
$N_e(u) = min(N_{max}, \frac{\Delta L(u)}{TOL})$	the total number of events at pixel u
timestamps	$t = t_{prev} + \frac{\Delta T}{N_e(u)}$

Through 100 flight tests, 100 event streams are obtained. The detail of training the ECEN is described in [21]. We omit this part here, since it is not the focus of this paper.

With the weights of the ECN pretrained, we train the policy network according to Algorithm 1. The policy network starts to learn avoidance strategy and update the actor-critic network weights when the drone moves towards the goal area. If the drone crashes into the obstacles, the task is failed and it will be set to the original point.

Training Results. Training the policy network takes a lot of computing resources and time. Our training is carried out on a computer, whose CPU is Intel Core i9-12900k and GPU is NVIDIA RTX3090. We use the method in [21] as the baseline denoted as *discrete_baseline*, and the same hyperparameters and the simulation environment as [21] is used in the experiment. In order to improve the collision avoidance performance, we propose two modifications over the baseline. One is applying a new reward function as shown in Eq. 3 (denoted as *discrete_proposed*). The other is applying a continuous action space (denoted as *continuous_proposed*). The hyperparameters used in this paper are listed in Table 5. The training results of three methods mentioned above are shown in Fig. 4. The three methods are all trained for 500 episodes. The average episode reward of the *discrete_baseline* method reaches 50, which is only 25 percent of the desired one. Note that we use the same code and hyperparameters as [1]. The average episode reward of the *discrete_propose* method can converge to 135. The average episode reward of the *continuous_proposed* method converges to 190,

Algorithm 1 Event-based SAA learning algorithm for the UAV

Initialize policy network parameters θ_0
Define $MAX\_episode, clip\_range, n\_epochs, target\_kl$
for $i = 0$; $i < MAX\_episode$; $i++$ **do**
  interact with environment and process the event streams: $s_t = \sum_{i=0}^{2} f_e(E_{t-i})$
  collection$\{s_t, a_t, r_t\}$ with θ_k
  calculate the advantage function value : $A^{\theta_k}(s_t, a_t)$
  if i % $n\_step == 0$ **then**
    for $j = 0; j < n_{epochs}$; $j++$ **do**
      find θ to the optimal $J_{PPO}(\theta)$ via:
      $J^{\theta^k}_{PPO}(\theta) = J^{\theta^k}(\theta) - \beta KL(\theta, \theta^k)$
      $ratio = \frac{p_\theta(a_t|s_t)}{p_{\theta^k}(a_t|s_t)}$
      $ratio = \begin{cases} 1 - clip\_range & if ratio < (1 - clip\_range) \\ ratio & if\ (1 - clip\_range) < ratio < (1 + clip\_range) \\ 1 + clip\_range & if\ ratio > (1 + clip\_range) \end{cases}$
      $J^{\theta^k}(\theta) \approx \sum_{(s_t,a_t)} ratio \cdot A^{\theta^k}(a_t, s_t)$
      if $|KL(\theta, \theta^k)| > target\_kl$ **then**
        break
      //Update Policy
      $\theta_k = \theta$

which is the highest of the three. Because the environment includes the collision-free corridor, the initial position of the UAV is set randomly at the beginning of the collision-free corridor. If the UAV is set at that position many times during an episode, there may be a spike of the average episode reward shown in the very beginning of the *discrete_baseline*'s reward.

5.2 Validating Experiments

It is known that the learning network is trained offline and validated online. In this section, we provide the validating results for the feature extraction and the collision avoidance. In the end, we explore the end-to-end control policy based on the continuous-action model, which involves the observation of deep reinforcement learning.

Efficiency of the ECN. The ECN is trained by inputting a certain number of events in a time sequence. However, the lengths of event streams are varying in real-time flights. In order to prove it works in real time, we test the encoder of event streams in different lengths generated during real-time flight. Figure 6 shows the decoder's results of different event streams from in image form. The upper images are constructed by original events, and the lower ones show the results of the ECEN. The lengths of their event streams range from 1000 to 4000. The two rows of images are displayed in the viridis colormap, the value

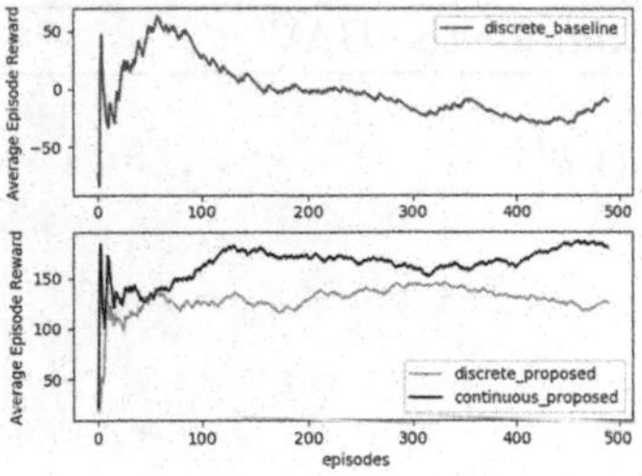

Fig. 4. The average episode rewards in training.

learning rate	0.0003
steps	2048
batch size	32
epochs	10
gamma	0.99
gae_lambda	0.95
clip_range	0.2

Fig. 5. Hyperparameters parameters of the continuous_proposed method.

of whose every pixel represents the product of polarity and firing time. In the viridis colormap, the value gradually increases as the color changes from blue to yellow. Whatever the event stream is dense or sparse, the rectangular shape of the obstacle can be reconstructed from the ECEN in Fig. 6, which means the sensing network is able to encode the location.

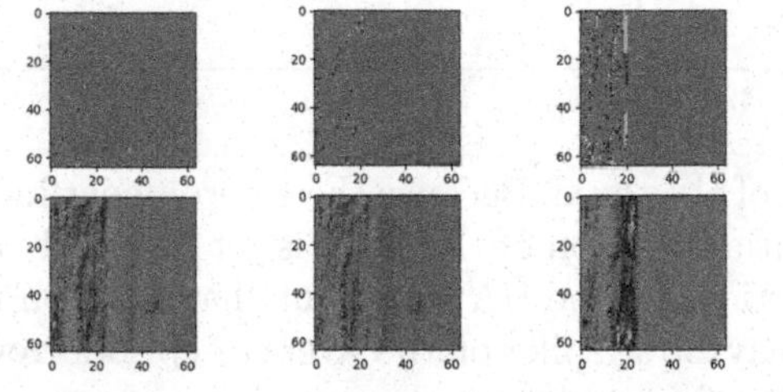

Fig. 6. The feature extractor of event streams in different length.

Success Rate of Obstacle Avoidance. In this section, we conduct ten trials for the three methods in the simulation environment. The initial position of the UAV is set randomly. If the UAV passes through the corridor without collision and reaches the goal, it is regarded as a successful trial. The paths recorded in ten trials are shown in Fig. 7. As depicted in Fig. 7, the success rate of obstacle avoidance can be counted, which is shown in Fig. 8(a). Figure 8(b) explains the mean and standard error length of the ten trails. Although the *discrete_baseline* method and the *discrete_proposed* method both only reach a success percentage of 10, the *discrete_proposed* method can get an average of 10 m longer than the *discrete_baseline* method. Especially, the *continuous_proposed method* can reach a 60 percent success rate and the longest goal with a mean value of 70 m. From these statistics, the new reward function and continuous action space can improve avoidance in the simulation scene. On the one hand, only when the agent goes through the whole long corridor does the reward function in [21] lead to obtaining a positive reward. This reward design is so sparse that the agent learns the desired policy slowly and is prevented from exploring the environment to accumulate rewards. On the other hand, the continuous action space is more suitable for smooth control in the avoidance task.

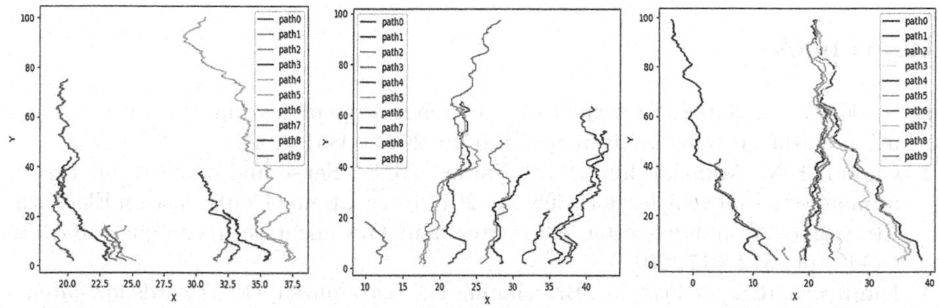

(a) 10 trails using discrete_baseline (b) 10 trails using discrete_proposed (c) 10 trails using continuous_proposed

Fig. 7. 10 trails for three model in simulation environment.

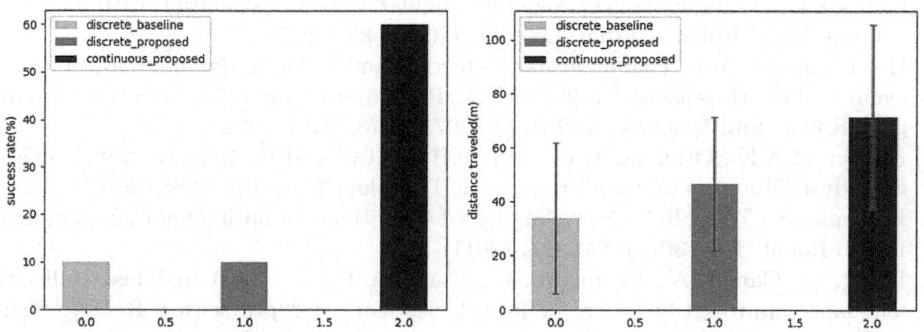

(a) the success percentage (b) the mean and standard error of travelled length

Fig. 8. The success percentage and length of ten avoidance trails.

6 Conclusion

This paper aims to explore the problem of collision avoidance for the UAV with an event camera. With this motivation, firstly we have modeled the problem with the help of partially observation Markov decision process and define the observation space, action space, and reward function. Secondly, the deep reinforcement learning algorithm based on PPO for collision sensing and avoidance has been proposed. We have validated the real-time efficiency of the sensing network and explored the probability of end-to-end control by using it as a front-end sensing.

Finally, the experiment results based on AirSim have demonstrated that our method outperforms the baseline method in collision avoidance success rate and the overall traveled length. In the future, we are going to fine-tune the network size to adapt to the actual event camera perception range and design new reward function to train a new model realizing avoiding dynamic obstacle. Moreover, long short-term memory (LSTM) will be applied to the dynamic obstacle sensing and avoidance algorithm.

References

1. Arakawa, R., Shiba, S.: Exploration of reinforcement learning for event camera using car-like robots. arXiv preprint arXiv:2004.00801 (2020)
2. Chand, B.N., Mahalakshmi, P.D., Naidu, V.P.S.: Sense and avoid technology in unmanned aerial vehicles: a review. In: 2017 International Conference on Electrical, Electronics, Communication, Computer, and Optimization Techniques (ICEECCOT), pp. 512–517 (2017)
3. Dimitrova, R.S., Gehrig, M., Brescianini, D., Scaramuzza, D.: Towards low-latency high-bandwidth control of quadrotors using event cameras. In: 2020 IEEE International Conference on Robotics and Automation (ICRA), pp. 4294–4300. IEEE (2020)
4. Falanga, D., Kleber, K., Scaramuzza, D.: Dynamic obstacle avoidance for quadrotors with event cameras. Sci. Robot. **5**(40), eaaz9712 (2020)
5. Gallego, G., Scaramuzza, D.: Accurate angular velocity estimation with an event camera. IEEE Robot. Autom. Lett. **2**(2), 632–639 (2017)
6. He, B., et al.: Fast-dynamic-vision: detection and tracking dynamic objects with event and depth sensing. In: 2021 IEEE/RSJ International Conference on Intelligent Robots and Systems (IROS), pp. 3071–3078. IEEE (2021)
7. Bahrin, M.A.K., Othman, M.F., Azli, N.H.N., Talib, M.F.: Industry 4.0: a review on industrial automation and robotic. J. Teknologi **78**(6–13), 9285 (2016)
8. Karaman, S., Frazzoli, E.: Sampling-based algorithms for optimal motion planning. Int. J. Robot. Res. **30**(7), 846–894 (2011)
9. Liang, L., Carrio, A., Sampedro, C., Campoy, P.: A robust and fast collision-avoidance approach for micro aerial vehicles using a depth sensor. Remote Sens. **13**(9), 1796 (2021)
10. Loquercio, A., Kaufmann, E., Ranftl, R., Müller, M., Koltun, V., Scaramuzza, D.: Learning high-speed flight in the wild (2021)
11. Lyu, Y., Kang, T., Pan, Q., Zhao, C., Hu, J.W.: UAV sense and avoidance: concepts, technologies, and systems. Sci. Sinica Inf. **49**(5), 520–537 (2019)
12. Michels, J., Saxena, A., Ng, A.Y.: High speed obstacle avoidance using monocular vision and reinforcement learning. In: Proceedings of the 22nd International Conference on Machine Learning, pp. 593–600 (2005)
13. Mueggler, E., Huber, B., Scaramuzza, D.: Event-based, 6-DOF pose tracking for high-speed maneuvers. In: 2014 IEEE/RSJ International Conference on Intelligent Robots and Systems, pp. 2761–2768. IEEE (2014)
14. Ni, Z., Bolopion, A., Agnus, J., Benosman, R., Régnier, S.: Asynchronous event-based visual shape tracking for stable haptic feedback in micro robotics. IEEE Trans. Rob. **28**(5), 1081–1089 (2012)

15. Qi, C.R., Su, H., Mo, K., Guibas, L.J.: PointNet: deep learning on point sets for 3D classification and segmentation. In: Proceedings of the IEEE Conference on Computer Vision and Pattern Recognition, pp. 652–660 (2017)
16. Rebecq, H., Horstschaefer, T., Scaramuzza, D.: Real-time visual-inertial odometry for event cameras using keyframe-based nonlinear optimization (2017)
17. Sanket, N.J., et al.: Evdodgenet: deep dynamic obstacle dodging with event cameras. In: 2020 IEEE International Conference on Robotics and Automation (ICRA), pp. 10651–10657. IEEE (2020)
18. Sekikawa, Y., Hara, K., Saito, H.: Eventnet: asynchronous recursive event processing. In: Proceedings of the IEEE/CVF Conference on Computer Vision and Pattern Recognition, pp. 3887–3896 (2019)
19. Spitzer, A., Yang, X., Yao, J., Dhawale, A., Michael, N.: Fast and agile vision-based flight with teleoperation and collision avoidance on a multirotor (2020)
20. Vaswani, A., et al.: Attention is all you need. In: Advances in Neural Information Processing Systems 30 (2017)
21. Vemprala, S., Mian, S., Kapoor, A.: Representation learning for event-based visuomotor policies. In: Advances in Neural Information Processing Systems 34 (2021)
22. Wang, G., Liu, Z., Xiao, K., Xu, Y., Yang, L., Wang, X.: Collision detection and avoidance for multi-UAV based on deep reinforcement learning. In: 2021 40th Chinese Control Conference (CCC), pp. 7783–7789. IEEE (2021)
23. Xie, L., Wang, S., Markham, A., Trigoni, N.: Towards monocular vision based obstacle avoidance through deep reinforcement learning (2017)
24. Zhu, A.Z., Yuan, L., Chaney, K., Daniilidis, K.: Ev-flownet: self-supervised optical flow estimation for event-based cameras. arXiv preprint arXiv:1802.06898 (2018)

Suspension Control of Maglev Train Based on Extended Kalman Filter and Linear Quadratic Optimization

Fengxing Li[1], Yougang Sun[1,2(✉)], Hao Xu[3], Guobin Lin[1,2], and Zhenyu He[1]

[1] Institute of Rail Transit, Tongji University, Shanghai 201804, China
{1851111,1989yoga,1753072}@tongji.edu.cn
[2] National Maglev Transportation Engineering R&D Center, Tongji University, Shanghai 201804, China
[3] China Railway Eryuan Engineering Group Co., Ltd., Chengdu 611830, China

Abstract. In recent years Maglev transport has received more and more attention because of its green, environmentally friendly and wide speed domain. The suspension control system is one of the core components of a Maglev train, and its open-loop instability, strong non-linearity and complex operating environment make the design of the control algorithm a great challenge. The suspension control system of Maglev train is plagued by noise and partial state unpredictability, and suspension stability may tend to deteriorate, so this paper proposes a suspension control method based on the extended Kalman filter algorithm to address the problem. Specifically, a mathematical model of the single-point suspension system is established firstly. Then the corresponding state observer is designed using the principle of the extended Kalman filter algorithm for the process, measurement noise and state unpredictability problems. Then the linear quadratic optimal control with feed-forward control and the extended Kalman filter are combined to propose a suspension controller suitable for the complex environment of Maglev trains. Finally, through numerical simulation, we have verified that the proposed method is able to achieve stable suspension and good dynamic performance of the system while overcoming the effects of process and measurement noise and estimating the velocity of the airgap, with an overshoot of the actual gap of approximately 0.4%, a rise time of 0.36 s, an adjustment time of 0.64 s to reach the 2% error band, and a tolerance band between the Kalman filter estimate and the actual value of only 0.13 mm.

Keywords: Maglev trains · Intelligent control · Extended Kalman filter · Suspension control · Process noise

1 Introduction

As a new green public transport mode, Maglev technology has shown great potential and competitiveness in urban rail transport and intercity high-speed transport with its

Supported in part by the National Key R&D Program of China(2016YFB1200601), in part by Science and Technology Research Project of CREEC (KYY2019031(19–22)) and in part by the National Natural Science Foundation of China (51905380).

L. Fang et al. (Eds.): CICAI 2022, LNAI 13606, pp. 414–425, 2022.
https://doi.org/10.1007/978-3-031-20503-3_33

advantages of high speed, low noise, low vibration, safety and environmental protection, becoming a future development direction of transport. The suspension control system of Maglev trains is one of the core technologies in train design. The EMS type Maglev system has inherent open-loop instability and non-linear characteristics, and it requires active adjustment of electromagnetic forces to make the system stable. However, due to the complex electromagnetic environment, the actual working environment is harsh and the suspension control system is subject to more external disturbances. The system will introduce two main types of external noise. One is measurement noise, which is generated by external electromagnetic disturbances or changes in temperature and humidity acting on the sensors. The other is process noise, which comes from disturbances in the power supply, the performance of the chopper, force disturbances brought about by the anti-roll beam of the bogie, etc. In addition, the velocity of the airgap is an important feedback state required for stable suspension, which is difficult to measure directly in engineering practice and is mainly obtained by integrating the measured value of the acceleration sensor, or differentiating the measured value of the airgap sensor. But the airgap differentiation amplifies noise and interference in the gap signal [1], integrators tend to cause problems such as integration saturation. Thus, specific observers or signal processing methods are needed to estimate or predict state changes before they can be used for suspension control.

In order to achieve stable suspension of Maglev trains, many domestic and foreign scholars have applied intelligent control algorithms in recent years. For example, Sun *et al.* [2] designed a sliding mode controller based on exponential convergence law and added radial basis function (RBF) neural network observer with minimum parameter learning method to enhance convergence speed and meet real-time control. Wang *et al.* [3] proposed a control strategy based on a linear quadratic optimal control method with a multi-electromagnet model for a high-speed Maglev train permanent magneto-electric levitation system, and verified that the control strategy has better performance under external force disturbance conditions. Wang *et al.* [4] introduced fuzzy control rules on PID control and compared the results of fuzzy PID control with a Maglev train-bridge system with equivalent linearized electromagnetic forces, which improved ride comfort and reduced the overall dynamic response of the bridge. Sun *et al.* [5] proposed a fuzzy sliding mode control law for ELS to enable the system to reach a new sliding mode surface at the stabilization point in finite time, slowing down the speed of the states crossing the surface and achieving global robustness. Xu *et al.* [6] proposed an adaptive robust control method for suspension control of Maglev vehicles with state constraints, transforming Maglev vehicles into interconnected uncertain system, and then proposed robust control to construct an adaptive law to model the system uncertainty and ensure the performance of the uncertain levitation system. Sun *et al.* [7] designed an extended state observer to ensure that the airgap is constrained within a certain range, and then designed a semi-supervised controller and output constrained controller to handle the unmeasurable airgap velocity and generalized external disturbances. Zhai *et al.* [8] estimated the airgap value of a faulty sensor by designing a state observer to improve the fault tolerance in case of sensor failure. Zhang *et al.* [9] proposed a disturbance observer-based sliding mode control strategy to compensate for system uncertainties and external disturbances. Yang *et al.* [10] used a disturbance observer-based control method for robust control by designing suitable disturbance compensation gains. The above research and analysis focus mainly on the theoretical application of the suspension control algorithm itself, which feeds the relevant states of the system directly into the controller for use, ignoring measurements, process noise and the actual situation of whether the states are measurable or not. The presence of these disturbances is likely to have an impact on the observer and controller performance, thus reducing levitation stability. Therefore, for these situations, suitable algorithms need to be selected, and filters and observers need to be designed to achieve stable suspension control.

In this paper, a control method for suspension systems of Maglev vehicles based on extended Kalman filter and linear quadratic optimum is proposed to address the problems raised above. The single-point levitation system is first modelled according to electromagnetic force, dynamics and electrical models. This is followed by the design of a state observer using the features of the Kalman filter algorithm for filtering and state observation, and then a control law combined with the linear quadratic optimum algorithm. In order to verify the effectiveness of the method, MATLAB/Simulink is applied for numerical simulation.

In conjunction with the content of this paper, the specific arrangements for what follows are as follows: in Sect. 2 the single-point suspension control system for Maglev trains is modelled and the control objectives are presented. In Sect. 3, the extended Kalman filter is applied to design the state observer and then combine the LQR control law with feedforward to design the suspension controller. In Sect. 4 the suspension control process of the system is simulated by MATLAB/Simulink, and the dynamic performance is analyzed and calculated. Finally, conclusions are given in Sect. 5.

2 Description of the Problem

2.1 Modelling Process

The EMS type Maglev train single point suspension system has a closed circuit of electromagnet and track, the mathematical model of the system is shown in Fig. 1 [11]. The system consists of an electromagnetic force model, a kinetic model and an electrical model.

Neglect the leakage flux and assume that the flux is uniformly distributed at the airgap, then the flux in the electromagnet circuit of the single point suspension system is

$$\phi = B \cdot A \tag{1}$$

where, B is the magnetic induction intensity of each part of the electromagnet circuit, A is the airgap cross-sectional area. Ignore the magnetic circuit term of the electromagnet core.

$$Ni = 2\delta \frac{B}{\mu_0} \tag{2}$$

where i is the solenoid coil current, δ is the airgap, N is the number of turns of the solenoid winding, μ_0 is the air magnetic permeability.

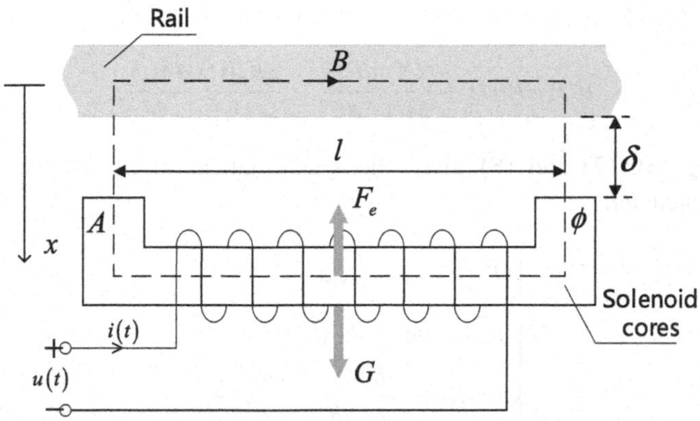

Fig. 1. Schematic diagram of the mathematical model structure of a single point suspension system

So, the magnetic field strength of the magnetic circuit can be obtained as

$$B = \frac{\mu_0 N i}{2\delta} \tag{3}$$

Substitute Eq. (3) into Eq. (1) yields [12]

$$\phi = \frac{\mu_0 N A i}{2\delta} \tag{4}$$

According to the Maxwell equation and the Biot-Savart theorem, the expression for the electromagnetic force can be obtained as

$$F_e(i,\delta) = \frac{\partial W}{\partial \delta} = \frac{\partial(\frac{1}{2}Li^2)}{\partial \delta} = \frac{\partial(\frac{1}{2}\frac{\mu_0 N^2 A}{2\delta}i^2)}{\partial \delta} = -\frac{\mu_0 N^2 A}{4}(\frac{i}{\delta})^2 \tag{5}$$

In this equation, μ_0, N, A all are constants, so let a constant $K_e = \mu_0 N^2 A/4$, then the above equation can be transformed to

$$F_e(i,\delta) = -K_e(\frac{i}{\delta})^2 \tag{6}$$

When the train is stably suspended at a certain height, assuming that the electromagnetic force is at the center of the electromagnet, the vertical direction of the electromagnet is subject to its own downward gravity and upward electromagnetic attraction. Ignoring external interference forces, when the two are equal to reach equilibrium, according to Newton's second law can be obtained from the single point suspension model in the vertical direction of the kinetic equation is.

$$m\ddot{\delta} = mg + F_e(i,\delta) \tag{7}$$

Assuming that the total resistance in the control circuit of the single point levitation system is R, and that the voltage applied to the solenoid coil is $u(t)$, and the excitation current is $i(t)$, the electrical equation for the levitation solenoid is

$$u(t)=Ri(t)+\frac{2K_e}{\delta(t)}\frac{di(t)}{dt}-\frac{2K_e i(t)}{[\delta(t)]^2}\frac{d\delta(t)}{dt} \tag{8}$$

Combining (6), (7) and (8) gives the mathematical model of the single point suspension system as

$$\begin{cases} F_e(i,\delta)=-K_e(\frac{i}{\delta})^2 \\ m\ddot{\delta}=mg+F_e(i,\delta) \\ u=Ri+\frac{2K_e}{\delta}\frac{di}{dt}-\frac{2K_e i}{\delta^2}\frac{d\delta}{dt} \end{cases} \tag{9}$$

2.2 Control Objectives

The mathematical model of the single point suspension system established in 2.1 is written as the state space equation shown in Eq. (10). The state variables selected are the airgap, the speed of airgap and the coil current, i.e. $X=\left[x_1\ x_2\ x_3\right]^T=\left[\delta\ \dot{\delta}\ i\right]^T$, the control quantity selected is the voltage across the solenoid u. Assuming that airgap sensor is used in the control system to measure the output $Y=x_1=\delta$, and the state space equation of the system is as follows.

$$\begin{cases} \dot{x}_1=x_2 \\ \dot{x}_2=g-\frac{K_e}{m}\frac{x_3^2}{x_1^2} \\ \dot{x}_3=\frac{x_2x_3}{x_1}-\frac{Rx_1x_3}{2K_e}+\frac{x_1}{2K_e}u \\ Y=x_1 \end{cases} \tag{10}$$

The control objectives of the single point suspension control system are.

(1) The airgap is brought precisely and quickly to the desired gap value required for operation.
(2) Negligible external disturbance forces in the levitation process.
(3) The control system is able to filter effectively after process and measurement noise is introduced into the system; and the Kalman filter outputs a usable estimate back to the controller.

3 Design and Analysis

3.1 Extended Kalman Filter Design

Since the single point suspension model is a non-linear system, the Extended Kalman Filter is applied. The Extended Kalman Filter is a non-linear extension of the conventional

Kalman Filter [13], where the core approach is to perform a Taylor series expansion of the system to obtain a linearized state space equation, followed by iterative operations using the conventional Kalman Filter algorithm [14].

In this section the EKF is designed. Firstly, for the state space equations in (10), discretize it with sampling period T and introduce process noise and measurement noise.

$$\begin{aligned} X_k &= f(X_{k-1}, u_k) + w_k \\ Y_k &= h(X_k) + v_k \end{aligned} \tag{11}$$

In this equation, w_k is the one-dimensional control noise, applied to the control output u_k. v_k is the one-dimensional airgap sensor noise. w_k and v_k are both normally distributed Gaussian noise, and are independent of each other. The covariance matrices of the two are Q_n, R_n. Let a posteriori estimate of the Kalman filter be $\hat{X}_k$, a priori estimate be $\hat{X}_k^-$, and a priori and a posteriori estimates of the error covariance matrix be P_k^-, P_k respectively. Then the estimation process of the extended Kalman filter is as follows [15, 16].

First calculate the priori estimate of the state quantity:

$$\hat{X}_k^- = f(X_{k-1}, u_k) \tag{12}$$

The Jacobian matrix $F_k = \frac{\partial f}{\partial X}\Big|_{X_k=\hat{X}_k^-}$ of f is then calculated and the a priori estimate of the error covariance matrix is

$$P_k^- = F_k P_k F_k^T + Q_n \tag{13}$$

Then calculate the Jacobian matrix $H_k = \frac{\partial h}{\partial X}\Big|_{X_k=\hat{X}_k^-}$ of h, followed by the Kalman gain matrix K_k:

$$K_k = P_k^- H_k^T \left(H_k P_k^- H_k^T + R_n\right)^{-1} \tag{14}$$

Then the optimal estimate of the state can be calculated as follows.

$$\hat{X}_k = \hat{X}_k^- + K_k(Y_k - \hat{X}_k^-) \tag{15}$$

Simultaneously update the posterior estimates of the error covariance matrix.

$$P_k = (I - K_k H_k) P_k^- \tag{16}$$

In this equation, I is the unit matrix. From the five recursive Eqs. (12) to (16). The EKF module is constructed by writing the MATLAB Function for the extended Kalman filter algorithm in MATLAB. The input quantity is X_0, P_0, Q_n, R_n, T, and the output quantity is $\hat{X}_k, P_k$.

3.2 Design and Analysis of an LQR Control Law with Feedforward

According to the primary approximation theorem, the stability of a single-point suspension system near the working equilibrium point can be determined by its corresponding linear system, so a linear system control method can be applied to the linearized model and then directly used for the original non-linear system. The widely used and simple linear LQR control (Linear Quadratic Regulator) is used in this paper.

LQR control is based on state feedback and its idea is to construct a state feedback control matrix K that multiplies the state variables. Then pass them into the control law [17].

Select performance indicator function:

$$J = \frac{1}{2}\int_0^{\infty}\left[X^T QX + u^T Ru\right]\mathrm{dt} \tag{17}$$

where Q, R are the positive definite matrices, Q is the state weight and R is the control weight, which are generally taken as a diagonal array. The purpose of this performance indicator is to design the system to minimize the combined tracking error and energy consumption. The state space and performance index of the closed-loop system at this point is obtained from $u = -KX$ as

$$\dot{X} = (A - BK)X = A_d X,\ J = \frac{1}{2}\int_0^{\infty} X^T\left(Q + K^T RK\right)X\,\mathrm{dt} \tag{18}$$

Use the minimax principle to find the minimum of a performance indicator function, then the problem can be transformed into finding the Riccati equation [18].

$$A^T P + PA + Q - PBR^{-1}B^T P = 0 \tag{19}$$

The state feedback matrix $K = R^{-1}B^T P$ is obtained by finding its positive definite solution P [19].

In order for the system to follow a given desired gap value, feed-forward control needs to be introduced to eliminate steady-state errors. Assuming that all state quantities of the system can be estimated by the observer, then following the input signal r can be achieved by using the feedforward matrix S after full state feedback by LQR. The new control law is

$$u = -KX + Sr \tag{20}$$

The feed-forward matrix expression is derived as

$$S = \left[C(BK - A)^{-1}B\right]^{-1} \tag{21}$$

3.3 Controller Design Based on Extended Kalman Filtering and LQR

The suspension control strategy in this paper is shown in Fig. 2. After linearizing the system at the operating equilibrium point to obtain its linear system, the LQR algorithm is applied to calculate its feedback control matrix, which is multiplied by the full state estimate observed by the extended Kalman filter, and together with feedforward control can achieve the system levitating at the desired gap value.

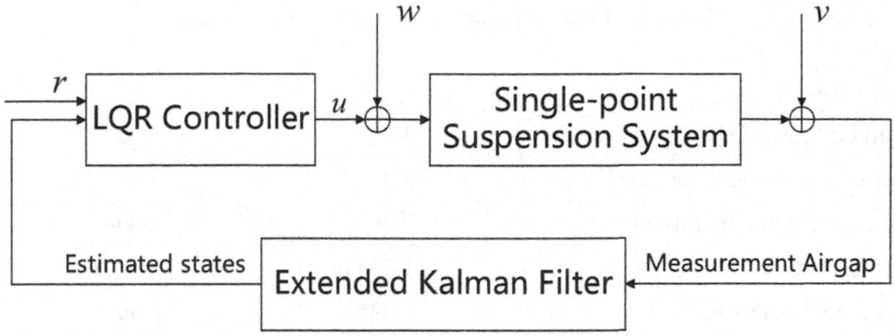

Fig. 2. Schematic diagram of a single-point suspension control strategy based on extended Kalman filter and LQR

The system is first expanded at the working equilibrium point $X_0 = \left[\delta_0\ 0\ i_0\right]^T$ using Taylor series to obtain an approximate linearized model as

$$\dot{X} = \begin{bmatrix} 0 & 1 & 0 \\ \frac{2g}{\delta_0} & 0 & \frac{-2}{\delta_0}\sqrt{\frac{K_e g}{m}} \\ 0 & \sqrt{\frac{mg}{K_e}} & \frac{-R\delta_0}{2K_e} \end{bmatrix} \begin{bmatrix} \delta \\ \dot{\delta} \\ i \end{bmatrix} + \begin{bmatrix} 0 \\ 0 \\ \frac{\delta_0}{2K_e} \end{bmatrix} u \tag{22}$$

At this point, choose the appropriate weighting matrix Q ($Q \in R^3$)and R ($R \in R^1$), solve the Riccati equation to obtain the feedback matrix K, use the extended Kalman filter to estimate the observed values of each state quantity of the system, and substitute into the feedback matrix to obtain the control law.

4 Numerical Simulation Results

In order to validate the suspension control system of the Maglev train combining the extended Kalman filter and the LQR algorithm, simulations are carried out in MATLAB/Simulink in this paper. The module is built in Simulink according to the mathematical model of the single-point levitation system established in Sect. 2, and the parameters are set in MATLAB with reference to the information related to the actual physical model.

The relevant parameters of the single point suspension system model are shown in Table 1.

Set the state weighting matrix and control weighting matrix as

$$Q = diag(\left[1\ 1\ 150\right]), R = 4$$

The state feedback matrix is calculated as

$$K = \left[k_1\ k_2\ k_3\right] = \left[-77447\ -1035\ 22\right]$$

The process noise and measurement noise in the system are set in the simulation to be Gaussian noise with a mean value of 0. Their covariances are

$$Q_n = 0.1V^2, R_n = 0.05\ \text{mm}^2$$

Table 1. Related parameters of simulation model

Physical quantity	value	Units
Solenoid coil resistance R	4	Ω
Number of turns of solenoid coil N	350	/
Suspension solenoid quality m	750	kg
Air permeability μ_0	1.2541e−6	H/m
Electromagnet pole area S	0.025	m^2
Gravitational acceleration g	9.81	m/s^2
Desired air gap δ_0	0.008	m
Initial air gap δ_i	0.016	m
Initial current i_i	44.29	A
Sampling period T	0.0001	s

Set the initial state posterior estimates and the initial posterior error covariance matrix of the extended Kalman filter to

$\hat{X}(0) = \left[\, \delta_i \; 0 \; i_i \,\right]^T, P(0) = 1e - 6.$

Three main sets of simulations were carried out to allow the air gap to follow a fixed gap value, a stepped wave and a sine wave to verify the effectiveness of the proposed method. The simulation of group (1) were performed for ablation experiments. As shown in Fig. 3, the blue line shows the control results of combining the EKF and LQR and the green one shows those of not using the EKF and only using the LQR, which introduces the measured air gap and its differentiation directly into the controller. It is clear that blue line is able to track fixed values more accurately, whereas without the EKF the jitter of air gap is very pronounced, tracking less well. Thus, the good filtering and estimation effect of the EKF is demonstrated. By calculating its dynamic characteristics, the actual air gap overshoot is approximately 0.4%, the rise time to reach 90% of steady state is 0.36 s and the adjustment time to reach the 2% error band is 0.64 s. The system exhibits good dynamic characteristics. Whereas it can be seen from the enlarged graphs that the

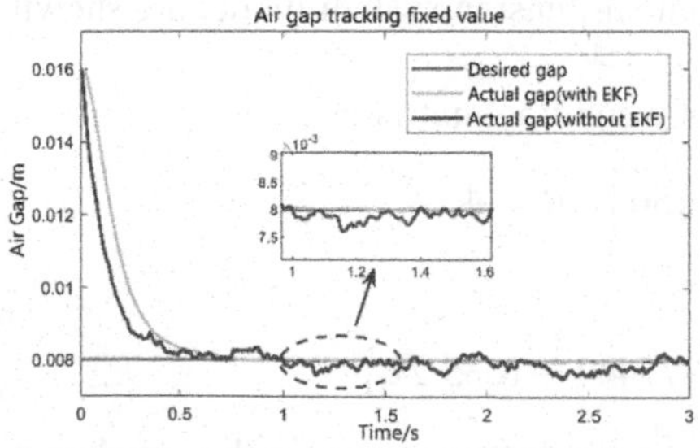

Fig. 3. Fixed suspension gap response

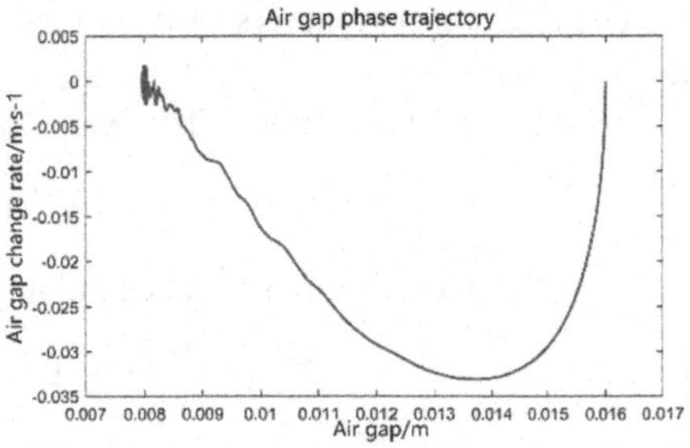

Fig. 4. Air gap phase trajectory

air gap has slight fluctuations when reaching steady state due to the presence of process and measurement noise. It can also be seen from the phase trajectory curve in Fig. 4 that the air gap is smoother before converging to the desired gap and fluctuates around the steady state value.

As can be seen from the comparison graph of the three air gaps in Fig. 5, the sensor measures the gap with very spurious white noise, while the Kalman filter estimate fluctuates less and is close to the actual gap, which reflects its filtering effect. In order to analyze the accuracy of the estimation, the consistency curve shown in Fig. 6 was drawn with the actual gap as the horizontal axis and the estimated gap as the vertical axis. The upper deviation of the curve relative to the contour is calculated to be 0.06 mm and the lower deviation is 0.07 mm, so the tolerance band of the Kalman filter estimate relative to the true value is 0.13 mm, which is a good agreement between the two, indicating that the Kalman filter has a very obvious observation and filtering effect in the presence of noise.

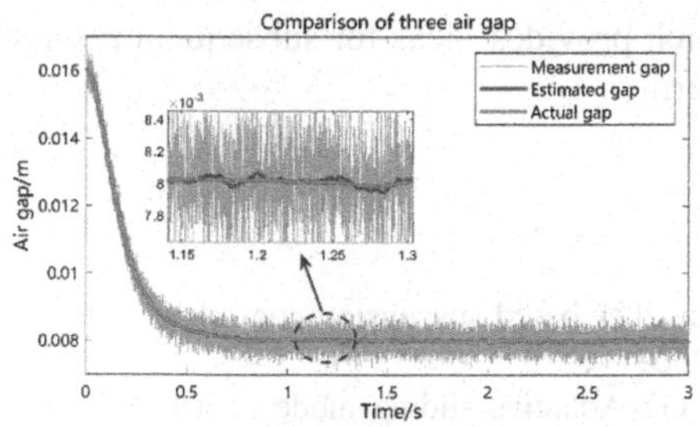

Fig. 5. Comparison of three air gap

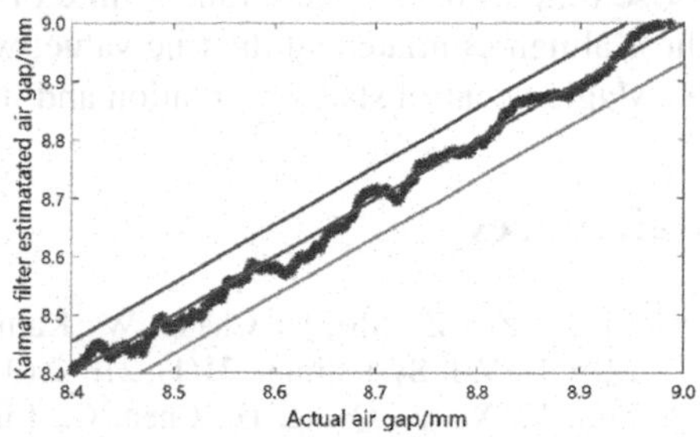

Fig. 6. Air gap consistency curve

The simulation results for groups (2) and (3) are shown in Fig. 7 and Fig. 8 and it can be seen that the tracking dynamic characteristics of the system are good with almost no overshoot. In Fig. 8, it can be observed that there is some lag in the actual gap relative to the desired gap, which may be due to the existence of lag in the filter. The desired gap is continuously transformed and the controller does not introduce the first and second

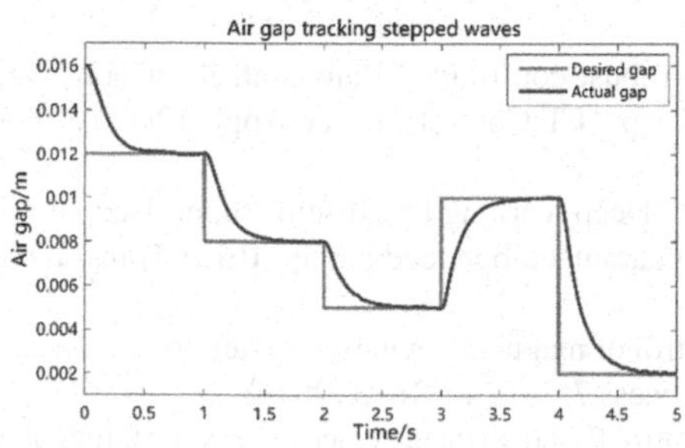

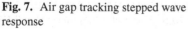

Fig. 7. Air gap tracking stepped wave response

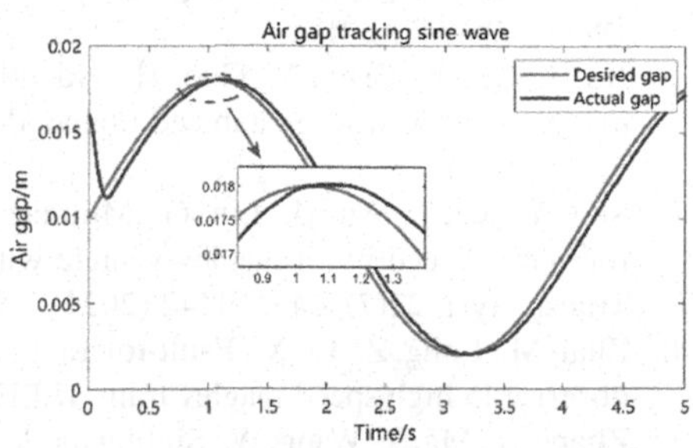

Fig. 8. Air gap tracking sine wave response

order derivatives of the desired gap, which can be minimized by adjusting the control parameters to reduce the lag time.

5 Conclusion

In this paper, a suspension control method based on the extended Kalman filter and the LQR algorithm is proposed to address the problem that previous suspension control algorithm studies often ignore the system process noise and sensor measurement noise, and that the air gap speed cannot be measured directly. By establishing a non-linear single-point suspension model, the extended Kalman filter is designed as a state observer, and the estimated state quantities are used in conjunction with a linear quadratic optimal control method with feedforward to achieve effective noise filtering and optimal state estimation. It is demonstrated through simulation that the control method can quickly and stably track a given fixed gap, step and sine wave with a system overshoot of 0.4%, a rise time of 0.36 s and a tuning time of 0.64 s, with a small tolerance band between the Kalman estimate and the true value, which provides ideas for subsequent research on Maglev control state observation and filtering.

References

1. Li, X., Zhi, Z., She, L., Chang, W.: Kalman filter based suspension control algorithm for MAGLEV. J. Syst. Simul. **21**(1), 216–219 (2009)
2. Sun, Y., Xu, J., Qiang, H., Chen, C., Lin, G.: Adaptive sliding mode control of maglev system based on RBF neural network minimum parameter learning method. Measurement **141**, 217–226 (2019)
3. Wang, Z., Long, Z., Li, X.: Levitation control of permanent magnet electromagnetic hybrid suspension maglev train. Proc. Inst. Mech. Eng. Part I J. Syst. Control Eng. **232**(3), 315–323 (2018)
4. Wang, B., Zhang, Y., Xia, C., Li, Y., Gong, J.: Dynamic analysis of high-speed maglev train–bridge system with fuzzy proportional–integral–derivative control. J. Low Frequency Noise Vibr. Active Control **41**(1), 374–386 (2022)
5. Sun, Y., Li, W., Xu, J., Qiang, H., Chen, C.: Nonlinear dynamic modeling and fuzzy sliding-mode controlling of electromagnetic levitation system of low-speed maglev train. J. Vibroeng. **19**(1), 328–342 (2017)
6. Xu, J., Du, Y., Chen, Y., Guo, H.: Adaptive robust constrained state control for non-linear maglev vehicle with guaranteed bounded airgap. IET Control Theory Appl. **12**(11), 1573–1583 (2018)
7. Sun, Y., Xu, J., Wu, H., Lin, G., Mumtaz, S.: Deep learning based semi-supervised control for vertical security of maglev vehicle with guaranteed bounded airgap. IEEE Trans. Intell. Transp. Syst. **22**(7), 4431–4442 (2021)
8. Zhai, M., Long, Z., Li, X.: Fault-tolerant control of magnetic levitation system based on state observer in high speed maglev train. IEEE Access **7**, 31624–31633 (2019)
9. Zhang, S., Ma, S., Wang, W.: Sliding mode control based on disturbance observer for magnetic levitation positioning stage. J. Electr. Eng. Technol. **13**(5), 2116–2124 (2018)
10. Yang, J., Zolotas, A., Chen, W., Michail, K., Li, S.: Robust control of nonlinear MAGLEV suspension system with mismatched uncertainties via DOBC approach. ISA Trans. **50**(3), 389–396 (2011)

11. Shi, J., Wei, Q., Zhao, Y.: Dynamic simulation of maglev with two degree on flexible guideway. J. Syst. Simul. **19**(3), 519–523 (2007)
12. Sun, Y., Xie, J., Xu, J., Qiang, H.: Simulation platform design and control research for magnetic suspension systems of low-speed maglev train. J. Mach. Des. **35**(05), 14–19 (2018)
13. Chui, C., Chen, G.: Kalman Filtering with Real-Time Applications, 4th edn. Springer, Berlin (2009). https://doi.org/10.1007/978-3-662-02508-6
14. Lv, G., Qin, P., Miao, Q., Liu, M., Jiao, P.: Research of extended Kalman filter based on multi-innovation theory. J. Chin. Comput. Syst. **37**(03), 576–580 (2016)
15. Li, W., Jia, Y., Du, J.: Distributed consensus extended Kalman filter: a variance-constrained approach. IET Control Theory Appl. **11**(3), 382–389 (2017)
16. Li, W., Sun, J., Jia, Y., Du, J., Fu, X.: Variance-constrained state estimation for nonlinear complex networks with uncertain coupling strength. Dig. Signal Process. **67**, 107–115 (2017)
17. Zhou, Y.: Research on LQR control of improved genetic algorithm for ring inverted pendulum system (03), 31–32+35 (2020)
18. Liu, A., Yang, Z., Yin, H., Wang, Y.: LQR active magnetic bearing control based on genetic algorithm optimization. Machine Tool Hydraulics **48**(14), 157–162 (2020)
19. Lin, F., Ma, H., Lu, Y.: ESO_LQR composite control method for airborne three-axis PTZ. J. Shenyang Aerospace Univ. **38**(01), 47–53 (2021)

EO-SLAM: Evolutionary Object Slam in Perceptual Constrained Scene

Chen Jiahao[1,2] and Li Xiuzhi[1,2](✉)

[1] Faculty of Information Technology, Beijing University of Technology, Beijing, China
xiuzhi.lee@163.com

[2] Beijing Key Laboratory of Computational Intelligence and Intelligent System, Beijing, China

Abstract. Object-level SLAM systems can capture semantic instances in the environment to help systems better understand intelligence tasks. However, due to limitation of the sensor's viewing angle, classical object-level SLAM systems fail to parameterize instantiated objects under certain conditions. In this work, we propose a evolutionary object SLAM system which is termed EO-SLAM. In the absence of sufficient angular information, it parameterizes the object's observation as the cylinder model and allows the quality improvement based on multi-frame accumulated observations to upgrade the dual quadric model. Additionally, data association schemes that combine 2D tracing with 3D model fitting are explored, enabling the system to operate accurately and robustly in both short and long term. The evaluation results indicate that EO-SLAM achieves the best tradeoff between speed and accuracy.

Keywords: Object SLAM · Dual quadric · Cylinder

1 Introduction

The simultaneous localization and mapping (SLAM) based on mobile robots utilizes the geometric information of the environment to assist the robot to fulfill autonomous task [4,13,17]. However, it is difficult to realize environmental understanding related tasks due to the lack of high-level semantic information. Semantic SLAM [1,18,21] is a promising technique for addressing such problems and receives much attention in robotic community.

As a branch of the semantic SLAM system, object SLAM plays a fundamental role, which capitalizes semantic object instances as features to estimate the ego-motion of the robot, recover the 3D pose and volume of the object, and suppress dynamic data distortions, which provides semantic guidance for the Intelligent task. We're also interested in Object SLAM due to its robustness against missing correspondences in hostile scenarios.

In contrast to other semantic slam systems that assign category attributes to each map point, the object slam system instantiates whole objects as landmarks. It is intuitively more robust against strong disturbances. Object SLAM capitalizes positon, pose and spatial attributes of which are expressed as the simplified

L. Fang et al. (Eds.): CICAI 2022, LNAI 13606, pp. 426–438, 2022.
https://doi.org/10.1007/978-3-031-20503-3_34

mathematical model. Cube-SLAM [24] generates cuboid object candidates from the upper edge of the current anchors and the vanishing point sampling. However, the model relies on line extraction in the appearance of the object and the application scenario is limited. Besides, SLOAM [8,10] is proposed based on lidar to generate tree diameter estimations as cylinder models. QuadricSLAM [5,14] expresses objects as dual quadrics, which have more explicit mathematical and physical meaning. Liao et al. [9] establish the symmetrical assumption used an RGBD camera to initialize quadrics with a single frame and extend it to a mono SLAM system by adding space and scale constraints in SO-SLAM [3].

Combining the two forms of cube and quadric, EAO-SLAM [23] uses different characterization forms for objects with different prior shapes. Objects in the regular form are represented by the cube model, while irregular objects are represented by dual quadrics. Ok et al. [15] propose a hierarchical representation method to solve the dilemma that 3D object models fail to be generated in time due to the lacking of appropriate perspectives.

Data association and object modeling are core components of the difference between object SLAM and classic SLAM. Bowman et al. [1] first use a probability-based method to model the data association process (PDA). Although achieving high correlation accuracy, the PDA algorithm with the high computational cost is hard to support real-time performance. Under the limitation of the complexity of object number, the Hungarian tracking algorithm for object SLAM [7] is proposed by Li et al. to associate the projected boundary box from the 3D cube model with the detected anchor in the current frame. Besides, bag of words in SLAM system [16] is formulated as a maximum weighted bipartite matching problem. However, artificially designed feature descriptors are difficult to ensure their robustness.

In addition, another creative idea is to analyze the statistical distribution of features and utilize whether the observed points comply with the distribution of map points as a guide for object tracking and merging. The theory is first proposed by Mu [12]. More recently, Wu et al. [23] propose EAO-SLAM, which integrates parametric and nonparametric methods, indicating superior correlation performance.

In this paper, we will focus on two key links to facing challenging tasks: (1) The existing data association methods have poor robustness in dealing with complex environments containing multiple object instances; (2) The object pose and volume estimation are inaccurate due to the limited viewing angle, especially in the ground robot system.

In this work, we propose the EO-SLAM, an RGBD evolutionary object SLAM system to address the combination of short-term and long-term data association, and pose estimation special to the limited observation. The main contributions of our paper are:

(1) A combination of short-term and long-term data association schemes is proposed to achieve long-term and stable tracking of observation objects.

(2) A scalable cylinder model formulating parameterization of objects is proposed. The degenerate model was built when observability is insufficient, while the complete model was built when observations are informative.

The framework of the system is shown in Fig. 1.

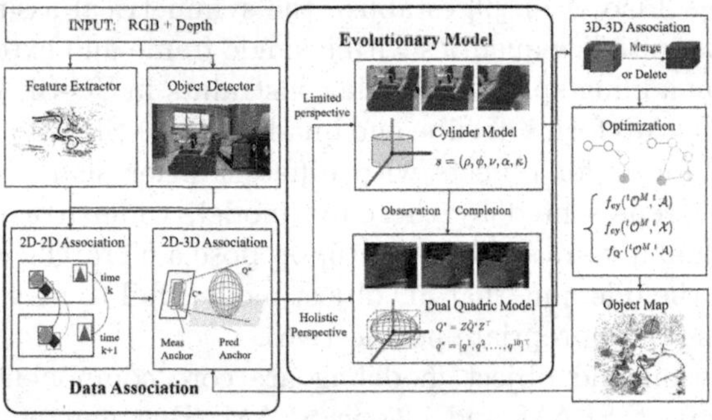

Fig. 1. The Framework diagram of EO-SLAM system.

2 Mathematics of Quadric and Cylinder Model

2.1 Notation

The system employs an RGBD camera as main sensor, and $\mathbf{P} = \mathbf{K}[\mathbf{I}_{3\times3}\ \mathbf{0}_{3\times3}]\mathbf{T}$ is the camera projection matrix that contains intrinsic $\mathbf{K}$ and extrinsic camera parameters ($\mathbf{T}$).

Given the current timestamp t, for N observable object sequences $\mathcal{O}^W = \{O_j^W\}_{j=1}^N$ in world coordinates $\{W\}$, the i-th observation anchor ${}^tA_i = [{}^tb_i, {}^tC_i, {}^tP_i]^\top$ is expressed, corresponding to the spatial property of anchor ${}^tb_i = [x_{lt}, y_{lt}, x_{rb}, y_{rb}]^\top$ (upper left and lower right corner coordinates of anchor), the category of the object ${}^tC_i$, and the probability of detection ${}^tP_i$. Setting I_t valid observation within the current frame, observation sequences are marked as ${}^t\mathcal{A} = \{{}^tA_i\}_{i=1}^{I_t}$.

If observed objects O_j at time t have corresponding instances on the map $\{M\}$ it is recorded as ${}^tO_j^M = [({}^t\theta_j^M)^\top, ({}^tP_j^M)^\top, ({}^tV_j^M)^\top]^\top$, where the orientation is ${}^t\theta_j^M$, the position is ${}^tP_j^M$, and the volume is ${}^tV_j^M$. Make ${}^t\mathcal{O}^M = \{{}^tO_j^M\}_{j=1}^{J_t}$ represent the sequence of object features of size J_t associated with the current observation. At the same time, in the current coordinate system $\{F\}$, there is a sequence of feature points ${}^t\mathcal{P}^F = \{{}^tp_k^F\}_{k=1}^{K_t}$ composed of K_t feature points ${}^tp_k^C$. Pose ${}^tT_F^W = \begin{bmatrix} {}^tR_F^W & {}^tt_F^W \\ \mathbf{0} & 1 \end{bmatrix}$ at time t in $\{W\}$ can be inferred from the sequence of feature points ${}^t\mathcal{P}^F$ and object observation ${}^t\mathcal{O}^M$.

2.2 Dual Quadric Representation

Object SLAM system pursues to establish a mathematical model to describe objects in a parametric manner. These parameters are regarded as landmarks to participate in map optimization and assist the robot in ego-motion. Dual-quadric, proposed by Nicholson et al. [14], has a more complete mathematical representation than other models in expressing the spatial attributes of objects. Thus, EO-SLAM utilizes the dual quadric model to describe the observed object instance.

Due to the limitation of the length of paper, the dual quadric theory is no longer introduced in detail. If the reader is interested in the relevant work, refer to the relevant paper.

2.3 Cylinder Representation

In the application of robots, constructing the parameterized quadratic surface is not always successful. On the one hand, it is difficult for some objects to observe changes in the pitch angle under the mobile robot's planar trajectory, which results in generating an unobservable problem. On the other hand, cameras, especially RGBD cameras, have a small Field of View (FOV), and cannot obtain all of the information about large objects from a single frame, which is often discarded in traditional algorithms. Inspired by the literatures [9,15], the hierarchical object construction method is proposed in EO-SLAM. The cylinder model is modeled when achieving no complete information about the object and will be reconstructed as a dual quadric model until sufficient observation angles are obtained.

When initializing the quadric model, several situations may occur result in failing to build a self-enclosing model: (1) If the eigenvalues do not satisfy three positive values and one negative number, the degraded phenomena will occur; (2) The object detection anchor edge coincides with the image boundary, i.e., only part of the object is within the FOV of camera.

In these cases, EO-SLAM will parameterize a cylinder [11] $s = (\rho, \phi, \nu, \alpha, \kappa)$ instead of dual quadric. To facilitate visualization and representation, the center points of the upper cylinder boundary and the lower cylinder boundary are marked as o_U, o_B, which is combined with the previous cylinder model to form an extended parameter $s_e = (s, o_u, o_b)$; The following Table explains the parameters in detail:

- ρ: The closest distance from the point on the cylinder to the origin;
- ϕ: The angle between the projection on the xy plane and the x axis;
- ν: Angle between $\mathbf{n}$ and z axis;
- α: The angle between $\mathbf{a}$ and $\mathbf{n}$ about the partial derivative of ν;
- $1/\kappa$: The radius of cylinder;
- λ_u: The distance from the upper boundary on the cylinder to the origin projected along the normal;
- λ_b: The distance from the lower boundary of a cylinder to the origin projected along the normal;

Given $(\rho, \phi, \nu, \alpha, \kappa)$, the cylinder normal vector $\mathbf{n}$ and the axis of the cylinder $\mathbf{a}$ are:

$$\begin{aligned} \mathbf{n} &= (\cos\phi \sin\nu, \sin\phi \sin\nu, \cos\nu) \\ \mathbf{a} &= \mathbf{n}^{\nu} \cos\alpha + \bar{\mathbf{n}}^{\phi} \sin\alpha \end{aligned} \tag{1}$$

Here: $\mathbf{n}^{\nu} = (\cos\phi\cos\nu, \sin\phi\cos\nu, -\sin\nu)$ is the partial derivative of $\mathbf{n}$ with respect to v, and $\bar{\mathbf{n}}^{\phi} = \frac{n^{\phi}}{\sin\nu} = (-\sin\phi, \cos\phi, 0)$ is the partial derivative of $\mathbf{n}$ with respect to ϕ.

Different from quadric, the construction of cylinder is implemented by split-fit method. For each observation in ${}^{t}\mathcal{A}$, point clouds within each anchor are obtained first. After down-sampling, point clouds ${}^{t}\mathrm{SC}$ are obtained by eliminating exterior points using euclidean spatial clustering and plane fitting algorithms. Then, the algorithm uses ${}^{t}\mathrm{SC}$ to fit the cylinder model and gets the initial value of the cylinder model ${}^{t}s^{0}$ and ${}^{t}s_{e}^{0}$. Where,

$$o_u = (\rho + 1/\kappa)\mathbf{n} + \lambda_u \mathbf{a}; \quad o_b = (\rho + 1/\kappa)\mathbf{n} + \lambda_b \mathbf{a} \tag{2}$$

However, the segmented model is rough and susceptible to outliers, so it can be iteratively optimized by non-linear optimization methods.

Consider the singularity dilemma when the curvature κ decreases, EO-SLAM attempts to approximate the point-to-cylindrical distance, which has the same zero set and the same derivative at the zero set. Expressed as follows:

$$\hat{d}_s(\mathbf{s}, \mathbf{p}) = \frac{\kappa}{2}(|\mathbf{p}|^2 - 2\rho\langle \mathbf{p}, \mathbf{n}\rangle - \langle \mathbf{p}, \mathbf{a}\rangle^2 + \rho^2) + \rho - \langle \mathbf{p}, \mathbf{n}\rangle \tag{3}$$

The optimized cylinder model ${}^{t}s_f$ can be obtained by optimizing the least squares formula and $\arg\min \sum_l \hat{d}_s^2(\mathbf{s}, \mathbf{p}_l)$ is found by using the Gauss-Newton method. Considering the tradeoff between accuracy and efficiency, iterations are bounded by 3. Finally, all point clouds are traversed, all outliers that do not meet the criteria are rejected, and the upper and lower edges are updated according to the new model to form the extended parameter ${}^{t}s_{e}^{f}$.

For cylinder objects, each observation is checked to determine whether a quadric model conforming to the constraints of the ellipsoid can be constructed using the new observations. Once the quadric model is built, the cylinder model will be permanently destructed.

To generate geometric errors projection method, we discretize the upper and lower boundaries of the cylinder, which means replacing the continuous boundaries with a discretized set of 3D points. The position of an anchor at the current time t is based on the maximum width and length. A three-dimensional Bresenham algorithm [2] is applied to avoid the floating-point problem.

Using the symmetry of circles, we only need to scan one-eighth of a circle. First, assume that all points are on a plane with a height of c and that the center of the circle is (a, b, c), corresponding to the center of the upper boundary o^{u} and the center of the lower boundary o^{b}, respectively, with a radius of $1/\kappa$. The 2D Bresenham algorithm is used to generate point sequences ${}^{S}\mathcal{X}_u = \{{}^{S}x_i^{u} \in \mathbb{R}^{3\times1}\}_{i=1}^{n_u}$, ${}^{S}\mathcal{X}_b = \{{}^{S}x_j^{b} \in \mathbb{R}^{3\times1}\}_{j=1}^{n_b}$ for the upper and lower boundaries. Compared with the real coordinate system $\{R\}$, they differ by a rotation matrix R_R^S. Assuming that coordinate system $\{S\}$ can be represented as unit matrix $R_S^W = \mathbf{I}$, and the corresponding true coordinate system $\{R\}$ is represented as $R_W^R = [\mathbf{n}, \mathbf{n} \times \mathbf{a}, \mathbf{a}]$ by a combination of vectors in a cylinder, then the rotation matrix $R_S^R = R_W^R (R_W^S)^\top = R_W^R$ is utilized, so the point set is transformed to obtain the real point sequence in the world coordinate system.

$$^R\mathcal{X}_u = \{R_S^{RS} x_i^u\}_{i=1}^{n_u};\ ^R\mathcal{X}_B = \{R_S^{RS} x_j^b\}_{j=1}^{n_b} \tag{4}$$

Let $X_u \in \mathbb{R}^{4 \times n_u}, X_b \in \mathbb{R}^{4 \times n_b}$ represents each element in $^R\mathcal{X}_u, ^R\mathcal{X}_b$ and stack it horizontally. Projecting the above point sequence into the image with the size of $w \times h$, the set of point sequences $\mathcal{Y}_u, \mathcal{Y}_b$ combined by Y_u, Y_b is obtained:

$$Y_u = \mathbf{P} X_u;\ Y_b = \mathbf{P} X_b \tag{5}$$

Using the above information, the prediction anchor $^t\hat{b}_i$ can be inferred:

$$\begin{aligned} ^t\hat{b}_i = \{&\max_x(\min_x(\mathcal{Y}_u, \mathcal{Y}_b), 0), \max_y(\min_y(\mathcal{Y}_u, \mathcal{Y}_b), 0), \\ &\min_x(\max_x(\mathcal{Y}_u, \mathcal{Y}_b), w), \min_y(\max_y(\mathcal{Y}_u, \mathcal{Y}_b), h)\} \end{aligned} \tag{6}$$

By analogy with a quadric, we can construct a similar geometric error term:

$$f_{\mathbf{cy}}(^t\mathcal{O}^M, ^t\mathcal{A}) = \sum_i^{I_t} ||^t b_i - ^t\hat{b}_i||^2 \tag{7}$$

3 Data Association

3.1 2D to 2D Data Association

2D data association establishes correspondence of observed object in adjacent frames. It is necessary to determine whether the current observation is a "false" observation, which may result from the false detection, ghosting, or miss detection of the object by the detector. Only observations with more than 3 frames of continuous stability can be considered "true" observations. This "true" observation will be stored and projected to the current moment for each frame of observation correction.

In short-term tracking, using the MOT tracking scheme, only a small amount of computational resources is consumed to achieve high accuracy, thus gaining community attention. The 2D data association of the EO-SLAM adopts the ByteTrack algorithm [25]. Considering that most of the detection results for obscured objects are low-scoring boxes, ByteTrack separates high and low scoring boxes and is simple to track obscured objects from low scoring boxes, which is robust to obscuring.

However, 2D-based data association is unstable in long-term tracking due to the lack of sufficient identifiers, so it is necessary to introduce 2D-3D and 3D-3D data association algorithms.

3.2 3D to 2D Data Association

By associating the 2D anchor obtained with the existing 3D model on the map, the model parameters are optimized to locate ego-motion more accurately during long-term tracking. In addition, when the observation is missing, the 3D model can be utilized to predict the current moment instead of the observation, which makes the tracking more continuous and robust. Among them, 2D-3D data association plays a very important role.

Depending on the model, there are two methods for 2D-3D data association. The first strategy is to infer 2D features (feature points, edges, and so on) into 3D space and fit the model on the map by minimizing distance (As in Formula 7). The other strategy is to project models within the FOV to the current frame, and associates all observations with the hungarian or bipartite matching algorithms [6]. Both types of association have both its pros and cons. The former is more robust while the latter is more accurate.

3.3 3D to 3D Data Association

In 3D-3D data association, the algorithm focuses on dealing with the problem of overlapping and eliminating multiple models. Depending on the distance between the centroids and the volume of any two objects on the map, the instances of objects are defined that have three states: separation, overlap, and dependency.

Since data associations do not completely eliminate mismatches, there may be multiple object instances on a map corresponding to one object in the world, which are dependencies or overlaps between echo other. However, under normal conditions, there are many overlaps and dependencies in the real world, such as books in cabinets and keyboards on tables. It is not effective to exclude these disturbances by simply judging whether two object models are attached or overlapping.

In addition, object SLAM implies the assumption that all objects are static, which, in many cases, is not meet. Some objects disappear from the view as a result of human movement. Ideally, this object instance should also be cleared from the map.

Taken together, a series of rules are specified:

- In the object detector pre-training category, the object that can accommodate other objects are marked C_S, otherwise C_N;
- No overlap or dependence is allowed for all objects classified as C_s;
- For all objects classified as C_n, overlap is not allowed, but interdependence is allowed.
- A object which volume is σ, and category is C_S, allow several objects which are not to exceed σ, and are classified as C_N to attach or overlap;
- From time t_i to t_{i+15}, if object instances on the map are located in the camera's view range and the detector is not able to obtain the corresponding observation data, the object will be moved and the object instance on the map will be destroyed.

4 Experiments and Evaluation

4.1 Backgrounds

Several apples-to-apples comparisons are conducted between classic object slam systems and the proposed system in this paper on several public datasets and real robot datasets to verify the comparison of trajectory accuracy, evolutionary model performance, and data association effect.

The systems involved in the comparison include QuadricSLAM [14], EAO-SLAM [23], DwB [9], EO-SLAM, and the dataset includes the multiple sequences of the TUM-RGBD dataset [20] which contains rich object information, the 'office' sequence of CoRBS dataset [22] and OpenLORIS-Scene dataset [19]. The ground truth object IDs are then manually labeled for those semantic measurements. Since not all codes are not fully open source, we have completed some of them as presented in the paper. Although consistency with the original is not guaranteed, consistency is maintained in all sub-sections to ensure fairness of comparison. Unless specifically mentioned, parameters used for testing are uniform.

All the experiments in this paper are carried out on a laptop with an Intel Core i7-11800H CPU and one GeForce RTX 3060 GPU.

4.2 Comparison of Trajectory

In these experiments, we compare the Time Consumption (TC) and Absolute Trajectory Error (ATE) of the estimated camera poses against other slam systems in Table 1 and Table 2.

Table 1. Track accuracy of multiple SLAM systems on different datasets.

Sequence	Motion Mode	QuadricSLAM	EAO-SLAM	DwB	EO-SLAM
TUM/fr1/desk	Handhold	0.0166	0.0159	**0.0150**	0.0152
TUM/fr1/desk2	Handhold	0.0245	0.0244	**0.0239**	0.0244
TUM/fr2/desk	Handhold	0.0327	0.0260	**0.0258**	0.0266
TUM/fr3/l_o_h	Handhold	0.0171	0.0164	**0.0158**	0.0163
TUM/fr3/teddy	Handhold	0.0263	0.0279	0.0263	**0.0260**
CoRBS/office	Handhold	0.0318	0.0270	**0.0266**	0.0273
TUM/pioneer_slam	Robot	0.0632	0.0566	0.0548	**0.0512**
OpenLORIS-Scene/office1-1	Robot	2.497	2.247	**2.189**	2.283
OpenLORIS-Scene/home1-3	Robot	x	2.4440	2.1825	**2.0924**

The quantitative results of the experiments on different sequences can be found in Table 1 and time-consuming comparison is in Tab 2. These results demonstrate an improvement in trajectory quality over the QuadricSLAM and EAO-SLAM but display a slight decrease in performance when compared to

Table 2. Time consumption (unit: second) of multiple SLAM systems on different datasets.

Sequence	QuadricSLAM	EAO-SLAM	DwB	EO-SLAM
TUM/fr1/desk	**16.213**	19.012	23.452	19.938
TUM/fr1/desk2	**18.109**	20.168	27.512	19.135
TUM/fr2/desk	**75.034**	95.547	109.701	82.980
TUM/fr3/l_o_h	**73.244**	101.109	114.822	93.915
TUM/fr3/teddy	74.464	82.477	89.751	**73.484**
CoRBS/office	**13.391**	20.747	19.821	15.960
TUM/pioneer_slam	40.657	56.724	54.744	**40.589**
OpenLORIS-Scene/office1-1	**12.135**	15.553	18.373	14.425
OpenLORIS-Scene/home1-3	x	49.8013	57.2624	**46.0331**

DwB. Furthermore, time consumption for EO-SLAM is significantly better than EAO-SLAM and DWB. This is due to EAO-SLAM needs to extract spatial straight lines, and DWB has a large number of point cloud processes, so time-consuming and memory consumption are high. EO-SLAM guarantees accuracy while consuming only a small amount of computing resources and has better performance.

In addition, we believe that the accuracy of the trajectory is also related to the mode of motion. In a HandleSLAM system, the gap of localization accuracy for each object slam system is not very large, but in the sequence of the robot motion, DWB and EO-SLAM is superior to other sequences. This is because in the process of robot motion, due to the limited angle of view and lack of observation of pitch angle, an effective object instance cannot be formed. The DWB algorithm uses the principle of symmetry to estimate unobserved regions, while EO-SLAM explores a cylindrical model to fit. Despite accuracy, EO-SLAM is weaker than DWB, but it reduces direct operations on a large number of point clouds and operational burden.

4.3 Evolutionary Model

Unlike previous SLAM systems, EO-SLAM explores a evolutionary model in which a cylinder model will be constructed instead of a dual quadric model when the system observation is insufficient or when the anchor intersects the edge of the image, as shown in Fig. 2. QuadricSLAM is no longer shown because he failed to generate the correct quadric.

It can be seen from the figure that among all slam systems, only DWB and EO-SLAM are available in all SLAM systems to generate object instances in a map when the observation frame or the observation angle is not sufficient. Only EO-SLAM can generate an object instance in a map when the anchor intersects the edge of the image.

4.4 Data Association

The data association means to find the correspondences between the object IDs assigned to semantic measurements in ground truth dataset and the IDs assigned, as shown in Fig. 3. At the same time, considering that data association only concerns the class and probability of objects, we ignore the shape characteristics of all associated objects in this chapter.

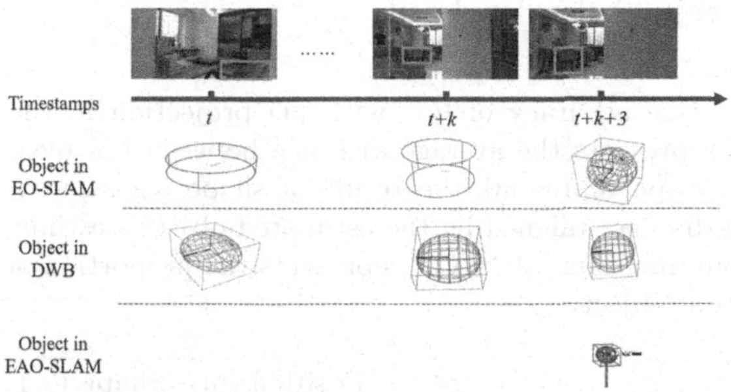

Fig. 2. Generation process of object instances in different slam systems.

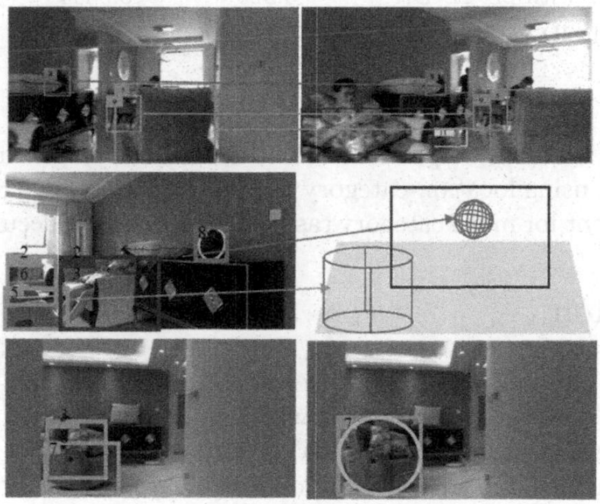

Fig. 3. Data association processes in different dimensions. The first row of images is 2D-2D data association, the second row is 2D-3D data association, and the last row is 3D-3D data association. The 3D space is mapped onto the image for visualization. It can be seen that overlapping objects are corrected for the same observation (projected as a yellow quadratic curve). (Color figure online)

Different from MOT, which treats multiple object tracking as a weighted binary graph problem, long-term tracking in three-dimensional space is more complex. An object that can be tracked successfully in a short-term may not be tracked successfully in the map. Based on this, we will use the objects in the map instead of the 2D tracker for correlation statistics only at the end of the slam system scan to participate in the correlation comparison. Association comparison is divided into two parts: category accuracy and location matching. When the measurement category of an object is consistent with the true value category of an object, the category matching is considered successful. When the distance between the measured centroid and the true centroid of the object is less than half the maximum radius of the object, the location matching is considered successful. See Table 3 below for details.

Table 3. The average accuracy of IoU with 2D projection method and 3D fitting method. Position represents the average distance between the measured objects and the real objects. Shape represents the result of shape assessment and is measured using the Jaccard distance aligned by the estimated object's estimated bounding box and the actual bounding box. MES/GT represents the proportion of observed objects occupying the actual object.

SLAM System	Position (m)	Shape (%)	MES/GT (%)
EAO-SLAM	**0.075**	0.4463	0.903
EO-SLAM with PDA	0.121	0.4711	0.817
EO-SLAM only with 2D-2D DA	0.325	0.5095	0.333
EO-SLAM with 3D-2D and 2D-2D DA	0.109	**0.4450**	**0.924**

In addition, we also compared two methods of tracking using only location information and using location-category joint information. The results show that category judgment for multi category tasks can improve the accuracy of tracking.

5 Conclusion

In this paper, we propose an object-level SLAM system named EO-SLAM. The system utilizes a evolutionary model, which builds a scalable regular cylinder model when the viewing angle is limited or the observability is insufficient, and builds a dual quadric model when the observability is sufficient. We derive the projection of the object model in the current frame by mathematical derivation and associate it with the observation. A long-short term data association scheme is proposed to achieve stable object tracking. We demonstrate the reliability of EO-SLAM by comparing it with several classical algorithms on multiple datasets.

References

1. Bowman, S.L., Atanasov, N., Daniilidis, K., Pappas, G.J.: Probabilistic data association for semantic slam. In: 2017 IEEE International Conference on Robotics and Automation (ICRA), pp. 1722–1729. IEEE (2017)

2. Bresenham, J.E.: Algorithm for computer control of a digital plotter. IBM Syst. J. **4**(1), 25–30 (1965)
3. Chen, S.W., et al.: SLOAM: semantic lidar odometry and mapping for forest inventory. IEEE Robot. Autom. Lett. **5**(2), 612–619 (2020)
4. Forster, C., Pizzoli, M., Scaramuzza, D.: SVO: fast semi-direct monocular visual odometry. In: 2014 IEEE International Conference on Robotics and Automation (ICRA), pp. 15–22. IEEE (2014)
5. Hosseinzadeh, M., Latif, Y., Pham, T., Suenderhauf, N., Reid, I.: Structure aware SLAM using quadrics and planes. In: Jawahar, C.V., Li, H., Mori, G., Schindler, K. (eds.) ACCV 2018. LNCS, vol. 11363, pp. 410–426. Springer, Cham (2019). https://doi.org/10.1007/978-3-030-20893-6_26
6. Kuhn, H.W.: The hungarian method for the assignment problem. Naval Res. Log. Quart. **2**(1–2), 83–97 (1955)
7. Li, J., Meger, D., Dudek, G.: Semantic mapping for view-invariant relocalization. In: 2019 International Conference on Robotics and Automation (ICRA), pp. 7108–7115. IEEE (2019)
8. Liao, Z., Hu, Y., Zhang, J., Qi, X., Zhang, X., Wang, W.: So-slam: semantic object slam with scale proportional and symmetrical texture constraints. IEEE Robot. Autom. Lett. **7**(2), 4008–4015 (2022)
9. Liao, Z., et al.: Object-oriented slam using quadrics and symmetry properties for indoor environments. arXiv preprint arXiv:2004.05303 (2020)
10. Liu, X., et al.: Large-scale autonomous flight with real-time semantic slam under dense forest canopy. IEEE Robot. Autom. Lett. **7**(2), 5512–5519 (2022)
11. Lukács, G., Marshall, A., Martin, R.: Geometric least-squares fitting of spheres, cylinders, cones and tori. RECCAD, Deliverable Doc. **2**, 671–675 (1997)
12. Mu, B., Liu, S.Y., Paull, L., Leonard, J., How, J.P.: Slam with objects using a nonparametric pose graph. In: 2016 IEEE/RSJ International Conference on Intelligent Robots and Systems (IROS), pp. 4602–4609. IEEE (2016)
13. Mur-Artal, R., Tardós, J.D.: Orb-slam2: an open-source slam system for monocular, stereo, and RGB-D cameras. IEEE Trans. Rob. **33**(5), 1255–1262 (2017)
14. Nicholson, L., Milford, M., Sünderhauf, N.: Quadricslam: dual quadrics from object detections as landmarks in object-oriented slam. IEEE Robot. Autom. Lett. **4**(1), 1–8 (2018)
15. Ok, K., Liu, K., Roy, N.: Hierarchical object map estimation for efficient and robust navigation. In: 2021 IEEE International Conference on Robotics and Automation (ICRA), pp. 1132–1139. IEEE (2021)
16. Qian, Z., Patath, K., Fu, J., Xiao, J.: Semantic slam with autonomous object-level data association. In: 2021 IEEE International Conference on Robotics and Automation (ICRA), pp. 11203–11209. IEEE (2021)
17. Qin, T., Li, P., Shen, S.: Vins-mono: a robust and versatile monocular visual-inertial state estimator. IEEE Trans. Rob. **34**(4), 1004–1020 (2018)
18. Runz, M., Buffier, M., Agapito, L.: Maskfusion: real-time recognition, tracking and reconstruction of multiple moving objects. In: 2018 IEEE International Symposium on Mixed and Augmented Reality (ISMAR), pp. 10–20. IEEE (2018)
19. Shi, X., et al.: Are we ready for service robots? the openloris-scene datasets for lifelong slam. In: 2020 IEEE International Conference on Robotics and Automation (ICRA), pp. 3139–3145. IEEE (2020)
20. Sturm, J., Engelhard, N., Endres, F., Burgard, W., Cremers, D.: A benchmark for the evaluation of RGB-D slam systems. In: 2012 IEEE/RSJ International Conference on Intelligent Robots and Systems, pp. 573–580. IEEE (2012)

21. Sünderhauf, N., Pham, T.T., Latif, Y., Milford, M., Reid, I.: Meaningful maps with object-oriented semantic mapping. In: 2017 IEEE/RSJ International Conference on Intelligent Robots and Systems (IROS), pp. 5079–5085. IEEE (2017)
22. Wasenmüller, O., Meyer, M., Stricker, D.: CORBS: Comprehensive RGB-D benchmark for slam using kinect v2. In: 2016 IEEE Winter Conference on Applications of Computer Vision (WACV), pp. 1–7. IEEE (2016)
23. Wu, Y., Zhang, Y., Zhu, D., Feng, Y., Coleman, S., Kerr, D.: EAO-slam: Monocular semi-dense object slam based on ensemble data association. In: 2020 IEEE/RSJ International Conference on Intelligent Robots and Systems (IROS), pp. 4966–4973. IEEE (2020)
24. Yang, S., Scherer, S.: Cubeslam: monocular 3-d object slam. IEEE Trans. Rob. **35**(4), 925–938 (2019)
25. Zhang, Y., et al.: Bytetrack: multi-object tracking by associating every detection box. arXiv preprint arXiv:2110.06864 (2021)

ADRC Based Multi-task Priority Tracking Control for Collaborative Robots

Kun Fan[1], Yanhong Liu[1(✉)], Kuan Zhang[1], Guibin Bian[1,2], and Hongnian Yu[1,3]

[1] School of Electrical Engineering, Zhengzhou University, Zhengzhou 450001, China
liuyh@zzu.edu.cn

[2] Institute of Automation Chinese, Academy of Sciences, Beijing 1000000, China

[3] School of Engineering and the Built Environment, Edinburgh Napier University, Edinburgh EH14 1DJ, UK

Abstract. When collaborative robots perform multiple tracking tasks at the same time, the dynamics of each task will interact with each other. In addition, the uncertainty of robot model and external perturbation also limit the performance of the system. In order to tackle above problems, a task priority control framework combining ADRC and null space projection is proposed. First, by using the null-space projection, the new state variables which can realize the decoupling of task space inertia are obtained and the task space dynamics is inertially decoupled correspondingly. Then, by establishing the relationship between the new task state variables and the original task state variables, the task space dynamics represented by the original task state variables is got. Thirdly, based on the task space dynamics of each task, a ADRC controller is constructed to improve the tracking performance and disturbance attenuation property simultaneously. Simulation experiments on the UR5 robot verify the effectiveness of the proposed controller.

Keywords: Multi-task · Tracking control · Null space projection · Inertial decoupling of task space dynamics · ADRC

1 Introduction

With more and more applications of collaborative robots, the tasks that robots need to perform become more and more complex. In many scenarios, robots need to perform multiple tasks with different priorities at the same time. For example, when the robot drags the plate to carry objects, it should always keep the plate in the proper pose, which means that keeping the posture of the plate is more important than the movement of the plate. At present, the control schemes dealing with multi-task priority can be divided into two types: soft priority control based on numerical optimization and strict task priority control based on null-space projection. The former designs control actions for each task individually, and then uses task weights to determine the order of priority. In these cases, the multi-task Priority control problem is transformed into an optimization problem with equality and/or inequality constraints [1, 2]. However, the soft priority scheme will inevitably introduce the coupling between tasks, which will cause the tasks to compete

L. Fang et al. (Eds.): CICAI 2022, LNAI 13606, pp. 439–450, 2022.
https://doi.org/10.1007/978-3-031-20503-3_35

with each other in some way, resulting in the degradation of multi-task performance of the system. In the strict task priority control based on null-space projection, the Jacobian matrix of the low-priority task is projected into the null space of the high-priority task, so as to ensure the precise priority relationship of tasks at all levels.

On the basis of the strict priority scheme based on null space projection, the operating space formula (OSF) proposed by Khatib [3] is the basis for multi-task control at the torque level. In reference [3], a Jacobian pseudo-inverse matrix with dynamic consistency was proposed to achieve inertial decoupling of space dynamics of primary and secondary tasks of robots. Literature [4] decouple robot dynamics into Cartesian space dynamics for end-effector and zero-space dynamics for secondary tasks, and achieve compliant control in zero-space dynamics to deal with compliant contact between the robot body and the environment without affecting the performance of the main task. In literature [5], the method of [3] was extended to any number of priority task dynamics, and a new set of Jacobian matrices were designed by using the zero-space projection technology and a new set of velocity state vectors of the task space were obtained, and then the dynamic equations of the task space with inertial decoupling at each level were obtained. Based on this, a compliant controller is designed. Nakanishi et al. [6] simplified the representation method of the null space Jacobian matrix on the basis of [5] to reduce the amount of calculation, and realized the adjustment controller of force, position and direction by designing the reference acceleration in the new coordinate system of the controller. In the new reference speed coordinate system, Karami et al. [7] assumed the existence of a new state design reference trajectory, and then realized the design of compliant trajectory tracking controller. Although the decoupling of inertial matrix of cooperative robot multi-task is achieved in the above literature, only position error vector can be used in the multi-task controller due to the mismatching between the introduced new task state and the original task state, and the transient performance of this control scheme in trajectory tracking task cannot be further improved. To tackle the problem of mismatched task states, Dietrich et al. [8] proposed a new state vector for task decoupling by adding the hypothesis of the target task, and found the lower triangular mapping relationship between the new decoupling state vector and the original task state vector. Finally, the optimal closed-loop structure of each task was obtained by designing a compliant controller based on the calculated torque. But the closed-loop dynamics of each level of task are affected by higher priority acceleration and velocity errors. It should be noted that modeling error and external interference of robot system are not considered in the above literatures.

Active Disturbance Rejection Control (ADRC) is widely used in engineering because it can effectively suppress uncertainty and external interference only by using part of the known information of the system. An ADRC is considered in the trajectory tracking of the biped robot system [9] and the training of robotic enhanced limb rehabilitation [10]. In [11], ADRC strategy is adopted for serial manipulator based on input shaping. Although ADRC technology has been widely used in practical engineering, it has not been applied to multi-task priority control of robots.

This paper proposes a task priority control framework combining ADRC and null space projecting to solve the multi-priority task control problem of the collaborative robot with consideration of external disturbance and model uncertainty. The organization of this paper is as follows: Sect. 2 introduces the problem description of multi-task priority control and related task space inertial decoupling. Section 3 introduces the design of a multi-task priority controller based on ADRC. Section 4 verifies the effectiveness of the designed controller through simulation analysis. Section 5 summarizes the main work of this paper.

2 Problem Formulation

2.1 Robot Dynamics

The dynamics model of a collaborative robot with n degrees of freedom (Dof) is described as

$$M(q)\ddot{q} + C(q, \dot{q})\dot{q} + G(q) + F_f(q, \dot{q}) = \tau + \tau_{ext}. \tag{1}$$

where $q, \dot{q}, \ddot{q} \in R^n$ denote the joint configuration, velocity and acceleration, respectively. $M(q) \in R^{n\times n}$ is the symmetric positive definite inertia matrix, $C(q, \dot{q})\dot{q} \in R^n$ is the torque vector of Coriolis force and centrifugal force, $G(q) \in R^n$ is the gravity torque vector, $F_f(q, \dot{q}) \in R^n$ is the joint friction torque vector, $\tau \in R^n$ is the control input, $\tau_{ext} \in R^n$ represents the generalized external forces.

However, it is difficult to obtain accurate model information in practical robot system. Consider the modeling error as follows

$$\begin{aligned} M(q) &= M_o(q) - \Delta M(q) \\ C(q, \dot{q}) &= C_o(q, \dot{q}) - \Delta C(q, \dot{q}) . \\ G(q) &= G_o(q) - \Delta G(q) \end{aligned} \tag{2}$$

where $M_o(q)$, $C_o(q, \dot{q})$, $G_o(q)$ are the normal model of the robot dynamics, $\Delta M(q)$, $\Delta C(q, \dot{q})$, $\Delta G(q)$ are the model error of robot dynamics. In addition to model error, there is friction nonlinearity in manipulator dynamics $F_f(q, \dot{q})$ and external disturbance τ_{ext}. For the convenience of control, the modeling error, friction nonlinear term and external disturbance term are classified as the lumped disturbance term of the robot system, record as τ_d. Then the system of (1) with uncertainties can be described as

$$M_o(q)\ddot{q} + C_o(q, \dot{q})\dot{q} + G_o(q) = \tau + \tau_d. \tag{3}$$

$$\tau_d = \tau_{ext} - F_f(q, \dot{q}) + \Delta M(q)\ddot{q} + \Delta C(q, \dot{q})\dot{q} + \Delta G(q). \tag{4}$$

Assumption 1: The lumped disturbance term of robot system τ_d and its rate of change $\dot{\tau}_d$ are bounded, which satisfies $\|\tau_d\| < \alpha$, $\|\dot{\tau}_d\| < \beta$ for all t > 0 where α, β are positive constants.

2.2 Multi-task Priority and Task Object

Consider the multi-task priority with includes $r > 1$ hierarchy levels, and the task hierarchy is ranked in descending order of priority from highest to lowest by indexes $i = 1$ to $i = r$. Each task is defined as follows:

$$x_i = f_i(q),\ i = 1, 2, \ldots, r. \tag{5}$$

$$\dot{x}_i = J_i(q)\dot{q},\ i = 1, 2, \ldots, r. \tag{6}$$

where $x_i \in R^{m_i}$ is task-space position of the *i-th* task, which obtained by its forward kinematics $f_i(q) : R^n \to R^{m_i}$. $\dot{x}_i \in R^{m_i}$ is task-space velocity of the *i-th* task, which obtained by its Jacobian matrix $J_i(q) = \partial f_i(q)/\partial q \in R^{m_i \times n}$.

Assumption 2: All tasks can be structurally realized simultaneously, and the sum of dimensions of all tasks is the same as the number of joints of the collaborative robot, i.e.:$\sum_{i=1}^{i=r} m_i = n$.

Assumption 3: The robot will not appear singular configuration in the considered workspace, which means that the Jacobian of each task is full row rank.

The augmented Jacobian is defined as follows:

$$J_i^{aug}(q) = \left[J_1^T(q) \ldots J_i^T(q)\right]^T. \tag{7}$$

and the augmented task-space velocity vector is defined follows:

$$\dot{x}_i^{aug} = \left[\dot{x}_1^T \ldots \dot{x}_i^T\right]^T = J_i^{aug}(q)\dot{q}. \tag{8}$$

The object of each task is tracking a desire trajectory, which include continuous task-space positions, velocity, and accelerations are denoted by $x_{i,des}(t)$, $\dot{x}_{i,des}(t)$, $\ddot{x}_{i,des}(t) \in R^{m_i}$ for $i = 1, \ldots, r$ with time t.

2.3 Dynamic-Consistency Task-Space

To achieve a strict priority structure between multi task, null space projectors is utilized as follows:

$$N_i(q) = \begin{cases} I \text{ for } i = 1 \\ I - J_{i-1}^{\text{aug}}(q)^T J_{i-1}^{\text{aug}}(q)^{M+,T} \text{ for } i = 2 \ldots r \end{cases}. \tag{9}$$

where, $I \in R^{n \times n}$ is identity matrix, $J_{i-1}^{\text{aug}}(q)^{M+}$ is the dynamically consistent pseudoinversion of $J_{i-1}^{\text{aug}}(q)$, which specific expression as follows:

$$J_{i-1}^{\text{aug}}(q)^{M+} = M(q)^{-1} J_{i-1}^{\text{aug}}(q)^T \left(J_{i-1}^{\text{aug}}(q) M(q)^{-1} J_{i-1}^{\text{aug}}(q)^T\right)^{-1}. \tag{10}$$

If (7) is directly substituted into robot dynamics (3), the inertia matrix obtained has the problem of coupling between different tasks, which is not conducive to controller design. In order to solve this problem, dynamic decoupling of Jacobian matrix of tasks at all levels is required. The most direct scheme is shown as follows:

$$\overline{J}_i(q) = J_i(q)N_i(q)^T. \tag{11}$$

Then, the new task-space velocity vector $\dot{\overline{x}}$ is defined as follows

$$\dot{\overline{x}} = \begin{bmatrix} \dot{\overline{x}}_1 \\ \vdots \\ \dot{\overline{x}}_r \end{bmatrix} = \begin{bmatrix} \overline{J}_1(q) \\ \vdots \\ \overline{J}_r(q) \end{bmatrix} \dot{q} = \overline{J}\dot{q}. \tag{12}$$

where $\dot{\overline{x}} \in R^n$ is the complete velocity vector of the new task space, and $\overline{J} \in R^{n\times n}$ is Jacobian matrix of the new task space.

2.4 Inertial Decoupling of Task Space Dynamics

Substitute (12) into (3) to obtain the task space dynamics of inertial decoupling, as follows:

$$\Lambda(q)\left(\ddot{\overline{x}} - \dot{\overline{J}}\dot{q}\right) = \overline{J}(q)^{-T}(\tau + \tau_d - C_o(q, \dot{q})\dot{q} - G_o(q)). \tag{13}$$

where $\Lambda(q)$ is the block diagonal inertia matrix, details as follows:

$$\Lambda(q) = daig(\Lambda_1(q), \ldots, \Lambda_r(q)). \tag{14}$$

where $\Lambda_i(q) = \left(\overline{J}_i(q)M(q)^{-1}\overline{J}_i(q)^T\right)^{-1} \in R^{m_i\times m_i}$ for $i = 1, \ldots, r$.

However, the task space velocity vector $\dot{\overline{x}}$ in (13) does not match the original task velocity vector $\dot{x}$. This is not conducive to designing high-performance controllers. Literature [8] found the relationship between the two velocity vectors:

$$\dot{\overline{x}} = L(q)\dot{x}. \tag{15}$$

where $J = J_r^{aug}(q)$ is the complete Jacobian of original task space, $L(q) = \overline{J}J^{-1} \in R^{n\times n}$ is a lower triangular block matrix. It has the following properties

$$L_{i,j} = 0, \forall i < j; . \tag{16}$$

$$L_{i,j} = I, \forall i = j; . \tag{17}$$

where $L_{i,j} \in R^{m_i\times m_j}$ denoting the (i, j)-block of L. In more detail, the relationship between two velocity vectors at a particular priority can be obtained, as follows:

$$\dot{\overline{x}} = \dot{x}_i + \sum_{j=1}^{j=i-1} L_{i,j}(q)\dot{x}_j. \tag{18}$$

Differentiating (18) leads to

$$\ddot{\overline{x}}_i = \ddot{x}_i + \sum_{j=1}^{j=i-1}\left(L_{i,j}(q)\ddot{x}_j + \dot{L}_{i,j}(q, \dot{q})\dot{x}_j\right). \tag{19}$$

Substituting (19) into (13), we can obtain the dynamics equation under the original task state as follows:

$$\begin{bmatrix} \Lambda_1(q) & \dots & 0 \\ \vdots & \ddots & 0 \\ 0 & 0 & \Lambda_r(q) \end{bmatrix} \begin{bmatrix} \ddot{x}_1 - \dot{\overline{J}}_1 \dot{q} \\ \vdots \\ \ddot{x}_r - \dot{\overline{J}}_r \dot{q} \end{bmatrix} = \begin{bmatrix} \overline{J}_1(q)^{-T} \\ \vdots \\ \overline{J}_r(q)^{-T} \end{bmatrix} (\tau + \tau_d - C_o(q, \dot{q})\dot{q} - G_o(q)) - \begin{bmatrix} \Lambda_1(q) & \dots & 0 \\ \vdots & \ddots & 0 \\ 0 & 0 & \Lambda_r(q) \end{bmatrix} \begin{bmatrix} 0 \\ \vdots \\ \sum_{j=1}^{j=r-1} \left(L_{i,j}(q)\ddot{x}_j + \dot{L}_{i,j}(q, \dot{q})\dot{x}_j \right) \end{bmatrix}. \tag{20}$$

Equation (20) indicates that the dynamics of low task priority will be affected by the velocity and acceleration levels of high priority.

2.5 Task Priority Controller Design

The task priority controller is designed as follows:

$$\tau = C_o(q, \dot{q})\dot{q} + G_o(q) + \sum_{i=1}^{i=r} N_i J_i^T F_{i,ctrl}. \tag{21}$$

Substitute (21) into (20) to obtain the closed loop dynamics of each task, as follows:

$$\Lambda_i(q)\left(\ddot{x}_i - \dot{\overline{J}}_i \dot{q}\right) = F_{i,ctrl} + F_{i,d},\ i = 1, 2, \dots, r. \tag{22}$$

$$F_{i,d} = \overline{J}_1(q)^{-T}\tau_d - \Lambda_i(q)\sum_{j=1}^{j=i-1} \left(L_{i,j}(q)\ddot{x}_j + \dot{L}_{i,j}(q, \dot{q})\dot{x}_j \right). \tag{23}$$

where $F_{i,d}$ is the uncertain lumped disturbance of the *i-th* task. Due to the noise in the acceleration measurement of each task, all the acceleration and velocity related items with high priority will be regarded as disturbances.

Remark1: Since τ_d, $\dot{\tau}_d$, $\ddot{x}_j$ and $\dot{x}_j$ are bounded, so the uncertain lumped disturbance $F_{i,d}$ and $\dot{F}_{i,d}$ of the *i-th* task are bounded.

3 ADRC Control Law

In this section, the controller is designed according to the ADRC control paradigm for each level of task dynamics. Firstly, an extended state observer (ESO) is designed to estimate the task velocity states $\dot{x}_i$ and uncertain lumped disturbances $F_{i,d}$ at each level of task, and the stability and convergence of the observer were analyzed. Then, the estimated uncertain lumped disturbances are used to design controllers for each level of tasks based on computational torque control. Block diagram of multi-task priority control based on ADRC is shown in Fig. 1.

Define the extended state vector of the *i-th* task as

$$z_i = \begin{bmatrix} z_{i1} \\ z_{i2} \\ z_{i3} \end{bmatrix} = \begin{bmatrix} x_i \\ \dot{x}_i \\ \Lambda_i(q)^{-1} F_{i,d} \end{bmatrix} \in R^{3m_i}. \tag{24}$$

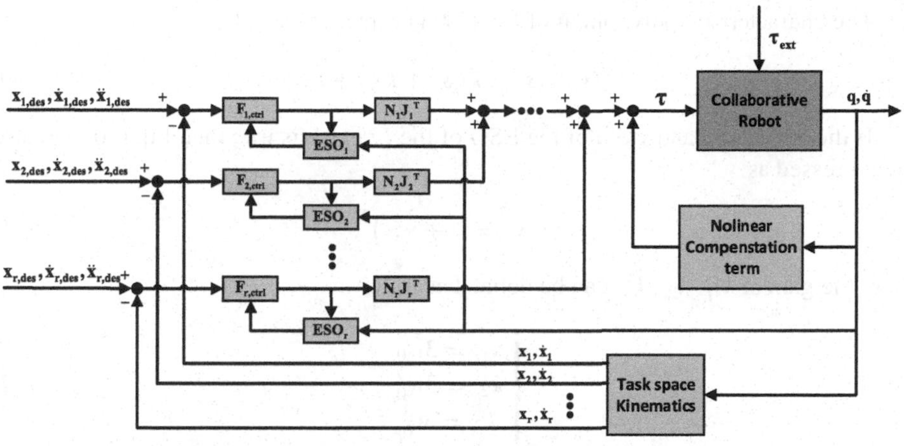

Fig. 1. Block diagram of multi-task priority control based on ADRC

Then system dynamic of the *i-th* task (22) can be expressed as follows,

$$\begin{cases} \dot{z}_{i_1} = z_{i_2} \\ \dot{z}_{i_2} = z_{i_3} + \Lambda_i(q)^{-1} F_{i,ctrl} + \dot{J}_i \dot{q} \ . \\ \dot{z}_{i_3} = \Lambda_i(q)^{-1} \dot{F}_{i,d} + \dot{\Lambda}_i(q)^{-1} F_{i,d} \end{cases} \tag{25}$$

The estimated vector of z_i is represented as follows

$$\hat{z}_i = \begin{bmatrix} \hat{z}_{i_1} \\ \hat{z}_{i_2} \\ \hat{z}_{i_3} \end{bmatrix} = \begin{bmatrix} \hat{x}_i \\ \hat{\dot{x}}_i \\ \Lambda_i(q)^{-1} \hat{F}_{i,d} \end{bmatrix} \in R^{3m_i}. \tag{26}$$

The dynamics of the designed extended state observer are as follows:

$$\begin{cases} \dot{\hat{z}}_{i_1} = \hat{z}_{i_2} - k_{1_i}(\hat{z}_{i_1} - z_{i_1}) \\ \dot{\hat{z}}_{i_2} = \hat{z}_{i_3} + \Lambda_i(q)^{-1} F_{i,ctrl} + \dot{J}_i \dot{q} - k_{2_i}(\hat{z}_{i_1} - z_{i_1}) . \\ \dot{\hat{z}}_{i_3} = -k_{3_i}(\hat{z}_{i_1} - z_{i_1}) \end{cases} \tag{27}$$

where k_{1_i}, k_{2_i}, k_{3_i} are positive definite symmetric matrices of the observe gains of the *i-th* task.

Subtracting (27) from (25), the observers error dynamic can be represented as

$$\begin{cases} \dot{\tilde{z}}_{i_1} = \tilde{z}_{i_2} - k_{1_i} \tilde{z}_{i_1} \\ \dot{\tilde{z}}_{i_2} = \tilde{z}_{i_3} - k_{2_i} \tilde{z}_{i_1} \ . \\ \dot{\tilde{z}}_{i_3} = \dot{F}_{i,d} - k_{3_i} \tilde{z}_{i_1} \end{cases} \tag{28}$$

where $\tilde{z}_{i_1} = z_{i_1} - \hat{z}_{i_1}, \tilde{z}_{i_2} = z_{i_2} - \hat{z}_{i_2}, \tilde{z}_{i_3} = z_{i_3} - \hat{z}_{i_3}$,, In terms of $\tilde{z}_{i_1}$,, the dynamic of (28) can be expressed as

$$\dddot{\tilde{z}}_{i_1} + k_{2_i} \ddot{\tilde{z}}_{i_1} + k_{1_i} \dot{\tilde{z}}_{i_1} + k_{3_i} \tilde{z}_{i_1} = \Lambda_i(q)^{-1} \dot{F}_{i,d}. \tag{29}$$

The characteristic polynomial of Eq. (29) is expressed as follows

$$\lambda(s) = s^3 + k_{1_i}s^2 + k_{2_i}s + k_{3_i}. \tag{30}$$

If the expected bandwidth of the ESO of the *i-th* task is w_{0_i}, then Eq. (30) can also be expressed as

$$\lambda(s) = \left(s + w_{0_i}\right)^3. \tag{31}$$

then, the gain of k_{1_i}, k_{2_i}, k_{3_i} can be defined as follows

$$\begin{cases} k_{1_i} = 3w_{0_i} \\ k_{2_i} = 3w_{0_i}^2 \\ k_{3_i} = w_{0_i}^3 \end{cases}. \tag{32}$$

Remark2: The lumped disturbance estimation error $\tilde{z}_{i_3}$asymptotically converges to the origin, i.e., $\lim_{t\to\infty}\left(F_{i,d} - \Lambda_i(q)^{-1}\hat{z}_{i_3}\right) = 0$..

The gain selection above can ensure that the eigenvalues of the system dynamic Eq. (29) fall on the left half plane, thus ensuring that ESO is asymptotically stable.

Based on the known model information, the task-space sliding model controller of the *i-th* task is designed as follows:

$$F_{i,ctrl} = \Lambda_i(q)\left(\ddot{x}_{i,des} + K_{d_i}\dot{e}_i + K_{p_i}e_i - \dot{J}_i\dot{q}\right) - \hat{z}_{i_3}. \tag{33}$$

where $e_i = x_{i,des} - x_i$, $\dot{e}_i = \dot{x}_{i,des} - z_{i_2}$ are position error and velocity error in the *i-th* task, K_{p_i} is positive definite symmetric matrices, K_i are positive definite symmetric matrices of the controller gains of the *i-th* task, $K_{s_i} > \Lambda_i(q)^{-1}(F_{i,d} - \hat{z}_{i_3})$is the switching gain.

Substitute (34) into (22) to obtain the closed-loop dynamics of the *i-th* task,

$$\ddot{e}_i + K_{d_i}\dot{e}_i + K_{p_i}e_i + \tilde{z}_{i_3} = 0. \tag{34}$$

Remark3: Because the lumped disturbance estimation error $\tilde{z}_{i_3}$. asymptotically converges to the origin, so the closed-loop dynamics of the *i-th* task show the following dynamics $\ddot{e}_i + K_{d_i}\dot{e}_i + K_{p_i}e_i = 0$.

Theorem1: For the collaborative robot system (1) that need to realize multiple tracking tasks simultaneously, if the controller is designed as (21), (33) based on the ESO (27), then each task state of the collaborative robot system will asymptotically converge to the desired trajectory of each task space.

4 Simulations

In this section, in order to verify the effectiveness of the proposed controller, a collaborative UR5 simulation platform with 6 Dof is builted in Matlab, and three priority tasks are defined to verify the proposed scheme. The orientation of the robot end-effector is the first priority task, the Z-axis trajectory tracking of the end-effector in the world coordinate system is the second priority task, and the X and Y axis trajectory tracking is the third priority task.

The initial configuration of UR5 robot is given by $q_0 = [-44.2, -23.89, 64.629, -220.7, -45.79, 0]^T$. As the first priority task, the desired direction of the end-effector is $R_d = roty(90)$, As the second priority task, the desired trajectory of the end-effector in the z-axis of the world coordinate system is $x_{2,des} = 0.1 + 0.1\sin(t)$, $\dot{x}_{2,des} = 0.1\cos(t)$, $\ddot{x}_{2,des} = -0.1\sin(t)$. As the third priority stage, the desired trajectory of the end-effector in the X and Y directions of the world coordinate system is $x_{3,des} = \left[0.1\sin(t) + 0.55\ 0.1\cos(t) - 0.4\right]^T$, $\dot{x}_{3,des} = \left[0.1\cos(t)\ -0.1\sin(t)\right]^T$, $\ddot{x}_{3,des} = \left[-0.1\sin(t)\ -0.1\cos(t)\right]^T$. In order to verify the anti-disturbance capability, a 10N disturbing force was applied along the Z-axis direction of the world coordinate system on the end-effector during 5 s of the simulation, and 1N and 5N torques were applied around the Y-axis and Z-axis directions. The controller gain selection of each tasks at each level is shown in Table 1.

Table 1. Table captions should be placed above the tables.

Task level	Task description	controller gain selection
1	orientation of end-effector	$K_{p_1} = 400I_{3\times3}$, $K_{d_1} = 40I_{3\times3}$, $w_{0_1} = 200$
2	end-effector in Z-axis	$K_{p_2} = 100$, $K_{d_2} = 20$, $w_{0_2} = 300$
3	end-effector in X,Y-axis	$K_{p_3} = 100I_{2\times2}$, $K_{d_3} = 20I_{2\times2}$, $w_{0_3} = 500$

In order to verify the performance of the proposed controller. Compared with the priority-considering computational torque control (PcCTC) and the priority-non-considering computational torque control (PncCTC).

The simulation results are shown in Figs. 2 and 4. The first 5 s of simulation is used to illustrate the convergence of different controller. It can be seen from the simulation results that PncCTC can not guarantee the fast tracking performance of the most important tasks, and the dynamics of each task affect each other. Figures 2 and 3 depicts PncCTC has the worst trace performance among high-priority tasks. PcCTC ensures the dynamic performance of high-priority tasks, but the dynamic performance of low-priority tasks is affected by the dynamic performance of high-priority tasks. Figure 2 shows that the end-effector direction can be quickly tracked. As can be seen from Fig. 4, the end-effector tracking in the X,Y-axis is affected by high task priority, and the convergence speed is the slowest. The proposed controller The controller proposed in this paper can achieve the same tracking effect as PcCTC in the first priority task as shown in Fig. 2, and better tracking effect can be achieved in the second and third priority tasks by compensating the interference from the high priority task as shown in Figs. 3and 4.

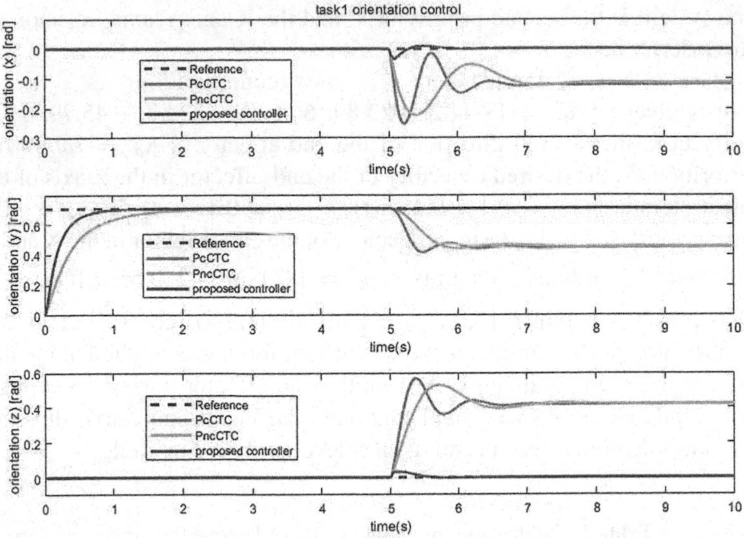

Fig. 2. Simulation #1 For the task with the first priority, the convergence speed and anti-disturbance ability of the proposed controller are compared with PcCTC and PncCTC

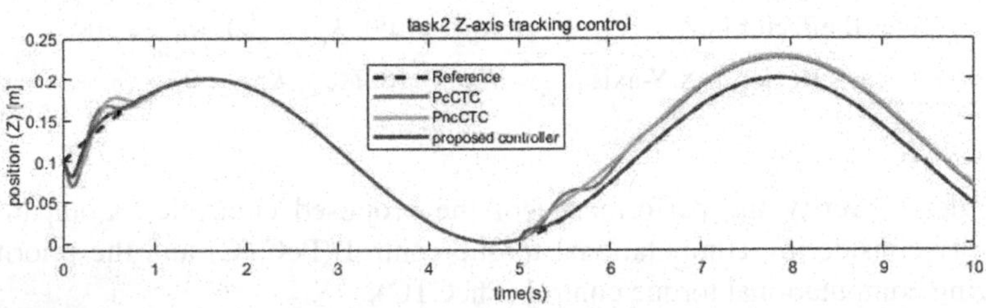

Fig. 3. Simulation #1 For the task with the second priority, the convergence speed and anti-disturbance ability of the proposed controller are compared with PcCTC and PncCTC

The last 5 s of the simulation is used to illustrate the anti-disturbance capability of different controller. It can be seen from the simulation results that the proposed controller can still track the desire trajectories of each tasks even with uncertain disturbance. The Fig. 2 ~ 4 shows the dynamic performance of each task of the robot after receiving uncertain disturbance, the proposed controller can realize the task states convergence quickly within 1 s of uncertain disturbance. The model-based controller PcCTC and PncCTC cannot converge when disturbed.

ITAE indexes of different methods for each task are shown in Table 2. It can be seen from the table that the proposed controller is optimal in both convergence performance and disturbance rejection performance.

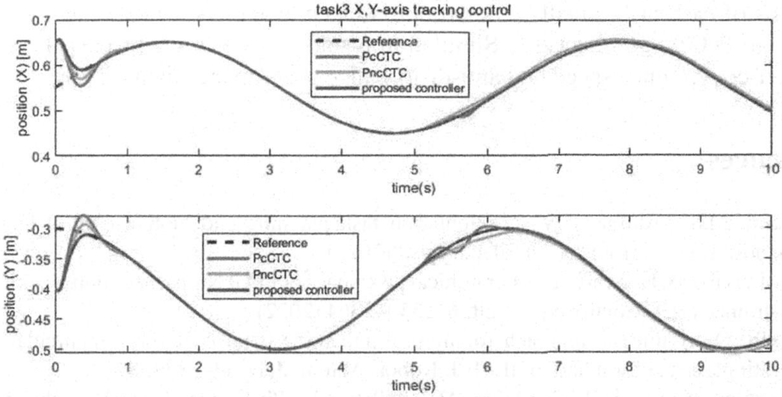

Fig. 4. Simulation #1 For the task with the third priority, the convergence speed and anti-disturbance ability of the proposed controller are compared with PcCTC and PncCTC

Table 2. ITAE index of different controllers for each task.

Task level	The first 5 s of simulation		The last 5 s of simulation	
	Controller	Ieae index	Controller	Ieae index
1	PcCTC	0.0342	PcCTC	18.7804
	PncCTC	0.1473	PncCTC	18.5272
	Proposed controller	0.0342	Proposed controller	0.0871
2	PcCTC	0.0045	PcCTC	0.8314
	PncCTC	0.0080	PncCTC	0.9260
	Proposed controller	0.0021	Proposed controller	0.0273
3	PcCTC	0.0114	PcCTC	0.0598
	PncCTC	0.0081	PncCTC	0.3103
	Proposed controller	0.0039	Proposed controller	0.0003367

5 Conclusion

In this paper, a task priority control framework combining ADRC and null space projection is proposed to solve the multi-task priority control problem of collaborative robot considering disturbance and model uncertainty. By using the null-space projection, the new state variables which can realize the decoupling of task space inertia are obtained, and the task space dynamics of inertia decoupling is obtained. By establishing the relationship between the new task state variables and the original task state variables, the task space dynamics represented by the original task state variables is obtained. Then, the ADRC controller is designed and its stability is analyzed according to the task space dynamics of each task. Finally, UR5 robot is taken as the controlled object and compared with PcCTC and PncCTC. Simulation results show that the proposed controller has faster convergence speed and anti-disturbance performance than CTC.

References

1. Pereira, M.S., Adorno, B.V.: Manipulation task planning and motion control using task relaxations. J. Control, Autom. Electr. Syst. (2022)
2. Lee, J., Bakolas, E., Sentis, L.: Hierarchical task-space optimal covariance control with chance constraints. IEEE Control Syst. Lett. **6**, 2359–2364 (2022)
3. Khatib, O.: A unified approach for motion and force control of robot manipulators: the operational space formulation. IEEE J. Robot. Autom. **3**(1), 43–53 (1987)
4. Sadeghian, H., Villani, L., Keshmiri, M., Siciliano, B.: Task-space control of robot manipulators with null-space compliance. IEEE Trans. Rob. **30**(2), 493–506 (2013)
5. Ott, C., Dietrich, A., Albu-Schäffer, A.: Prioritized multi-task compliance control of redundant manipulators. Automatica **53**, 416–423 (2015)
6. Karami, A., Sadeghian, H., Keshmiri, M., Oriolo, G.: Force, orientation and position control in redundant manipulators in prioritized scheme with null space compliance. Control. Eng. Pract. **85**, 23–33 (2019)
7. Karami, A., Sadeghian, H., Keshmiri, M., Oriolo, G.: Hierarchical tracking task control in redundant manipulators with compliance control in the null-space. Mechatronics **55**, 171–179 (2018)
8. Dietrich, A., Ott, C.: Hierarchical impedance-based tracking control of kinematically redundant robots. IEEE Trans. Rob. **36**(1), 204–221 (2019)
9. Martínez-Fonseca, N., Castañeda, L.Á., Uranga, A., Luviano-Juárez, A., Chairez, I.: Robust disturbance rejection control of a biped robotic system using high-order extended state observer. ISA Trans. **62**, 276–286 (2016)
10. Madoński, R., Kordasz, M., Sauer, P.: Application of a disturbance-rejection controller for robotic-enhanced limb rehabilitation trainings. ISA Trans. **53**(4), 899–908 (2014)
11. Li, W.P., Luo, B., Huang, H.: Active vibration control of flexible joint manipulator using input shaping and adaptive parameter auto disturbance rejection controller. J. Sound Vib. **363**, 97–125 (2016)

Model Predictive Tracking Control for USV with Model Error Learning

Siyu Chen, Huiping Li(✉), and Fei Li

School of Marine Science and Technology, Northwestern Polytechnical University, Xi'an 710072, China

lihuiping@nwpu.edu.cn

Abstract. This paper is concerned with the learning-based model predictive control (MPC) for the trajectory tracking of unmanned surface vehicle (USV). The accuracy of system model has a significant influence on the control performance of MPC. However, the complex hydrodynamics and the complicated structure of USV make it difficult to capture the accurate system model. Therefore, we present a learning approach to model the residual dynamics of USV by using Gaussian process regression. The learned model is employed to compensate the nominal model for MPC. Simulation studies are carried out to verify the effectiveness of the proposed method.

Keywords: Model predictive control · Trajectory tracking · Gaussian process regression · Unmanned surface vehicles

1 Introduction

Unmanned surface vehicle (USV) plays an important role in hydrological observation, seabed mapping, ocean reconnaissance, maritime rescue and other tasks. The related control technologies have received increasing attention from researchers, government and the military. However, the realization of high-quality tracking control for USV under the constraints of actuator and environment is a challenging issue. In particular, the accurate control of USV at high speed is not only of great significance to the safe operation, but also relates to the efficient task completion. Accordingly, the accurate tracking control requires higher precision of the model. Due to the hydrodynamic effect and the complexity of hull surface structure, the dynamics of USV is highly nonlinear and time-variant [8] To deal with this issue, most existing literatures [1] designed the controller based on the simplified model, where the input force and torque are assumed to be linear. However, such simplification holds only if USV runs with the medium and low speed, and the rudder angle is small. For the higher speed case, hydrodynamics changes dramatically. In addition, area immersed in water and resistance coefficient will be different from the formal case. Such issue becomes more critical for the small hard clipper ship of non-drainage type, which will lead to the stern sinking and bow lifting [3]. Therefore, there has an

L. Fang et al. (Eds.): CICAI 2022, LNAI 13606, pp. 451–461, 2022.
https://doi.org/10.1007/978-3-031-20503-3_36

urgent demand to model the USV system by considering the nonlinearity of the hydrodynamic force.

In order to establish the nonlinear dynamic model of USV, many literatures have focused on the identification methods. Traditional parametric modeling methods, such as least squares, extended Kalman filter, least squares support vector machine regression approximate the real model with finite orders [11]. However, there exists a contradiction between the complexity and feasibility of the model, since complex model will increase the difficulty of controller design and calculation. Differently, non-parametric modeling method such as Support vector regression (SVR) [18], local weighted learning (LWL) [2] and recursive neural network (RNN) [12] can obtain accurate prediction without the structural information. As one of the typical non-parametric modeling method, Gaussian process regression (GPR) is widely employed to construct the system model of mechanical arm [4], racing car [9], etc. GPR is able to optimize the hyperparameters by maximizing a marginal likelihood function while settling the overfitting issue. In addition, GPR has the advantages of requiring less identification data which can be noisy and even include outliers [17]. Therefore, in this paper, we utilize GPR to model the system dynamics of USV.

To realize the tracking control of USV, controller designed by the methods such as sliding mode control (SMC) [16], PID control and nonlinear control [7] have shown their effectiveness. However, these methods are not able to cope with the constraints in USV, which may result in the performance degradation. As one of the most advanced control technologies, model predictive control (MPC) is widely studied because of its constraints handling ability and the optimization-based property. It provides a feasible scheme to the tracking control problems of USV. Note that the control performance of MPC relies on the accurate model Therefore, the combination of MPC and GPR is a hot research area in the control field [10]. For instance, in the field of marine systems, the multi-output Gaussian process regression was used to identify the dynamic model of container ships and shown better performance than RNN [15]. Xue et al. employed the Gaussian process regression to identify the handling model of ship and fit the real data [19]. Chen et al. proposed a non-parametric modeling method for ship based on similarity regression of sparse Gaussian process, which solved the problem of the application of GP in big data [5]. Dong et al. proposed Gaussian process based model predictive control (GPMPC) to improve the tracking accuracy of MPC, which learned the sailing heading change from the sampled data and used the learned model as the prediction model to form a heading controller [6].

Although complete ship models have been established in all the above studies, it is difficult to realize real-time application in MPC, because a large number of sampled data are required to cover the state space to ensure the generalization ability. In other way, with the known of partial model, only learning the residual dynamics can simplify the learning progress, which needs fewer training points. In order to enable the GPMPC controller achieving relatively accurate trajectory tracking of USV with efficient computation, this paper proposes a learning-based MPC method. On the basis of the known simple nominal model, Gaussian process regression is employed to learn the residual dynamics which is difficult to identify. Comparison simulations are performed to verify the effectiveness and advantages of the proposed algorithm.

2 USV Modeling

In this section, two different mathematical models of the USV are presented. The complete model is assumed to be a real-world system and the simplified model refers to as the nominal model.

2.1 Complete Model

Usually, the manoeuvring mathematical modeling group (MMG) model is regarded as the complete model of USV [14]. The model parameters can be computed by empirical regression formula method if the structural parameters of USV are known. Research results have shown that this model fits the experimental data well [13]. Therefore, in this paper, we assume the MMG model as the complete model of USV.

The USV dynamics can be described by 3-DOF model as in Fig. 1. The Earth-fixed coordinate system is denoted by $O-XY$ and the body-fixed coordinate system is represented as $o-xy$. u, v, and r are surge velocity, sway velocity, and yaw rate, respectively. δ is the rudder angle and ψ is the heading angle.

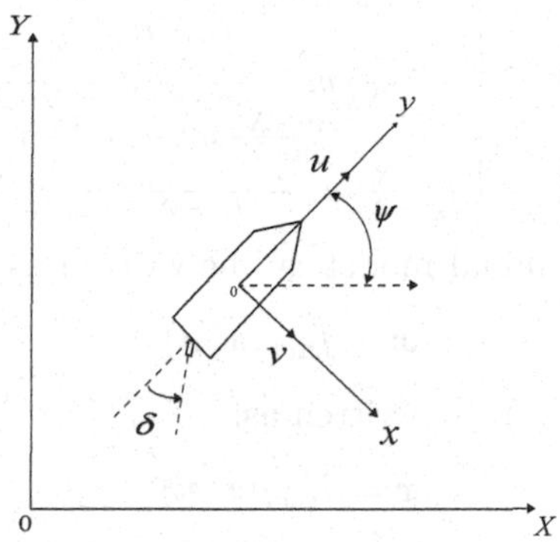

Fig. 1. Underactuated surface vessel.

The MMG model of USV can be written as:

$$\begin{array}{l} m(\dot{u} - y_G\dot{r} - vr - x_G r^2) = X_\Sigma \\ m(\dot{v} + x_G\dot{r} + ur - y_G r^2) = Y_\Sigma \\ I_{zz}\dot{r} + m[x_G(\dot{v} + ur) - y_G(\dot{u} - vr)] = N_\Sigma \end{array}, \tag{1}$$

$$\begin{array}{l} X_\Sigma = X_{ctrl} + X_{\dot{u}}\dot{u} + X_u u + X_{|u|u}|u|u + X_{vr}vr + X_{rr}r^2 \\ Y_\Sigma = Y_{ctrl} + Y_{\dot{v}}\dot{v} + Y_{\dot{r}}\dot{r} + Y_v v + Y_r r + Y_{|v|v}|v|v + Y_{|r|r}|r|r \\ N_\Sigma = N_{ctrl} + N_{\dot{v}}\dot{v} + N_{\dot{r}}\dot{r} + N_v v + N_r r + N_{|v|v}|v|v + N_{|r|r}|r|r \end{array}. \tag{2}$$

Equation (1) is the motion of equation of USV. m is the mass and $Y_{\dot{v}}$, $X_{\dot{u}}$, $N_{\dot{r}}$ represent added mass. x_G, y_G respectively represent the location of the USV's center of gravity in x, y direction. I_{zz} is the moments of inertia in yaw motion. Equation (2) is the expression of the resultant force on the USV. X_{ctrl} and N_{ctrl} represent forward thrust and yaw moment respectively, which are function of propeller speed and rudder angle. In this paper, we set $Y_{ctrl} = 0$ for simplicity. Parameters like Y_v, X_u represent the linear hydrodynamic coefficient and parameters like $X_{|u|u}$, $Y_{|v|v}$ represent the nonlinear hydrodynamic coefficient. These parameters can be computed by empirical regression formula if all structural parameters of the USV are known.

2.2 Nominal Model

Unfortunately, not all the structural parameters of USV can be available. In this case, the model can be obtained by system identification. But it is unrealistic to conduct the comprehensive and accurate system identification for every USV because of the high experimental cost. Therefore, the nonlinear hydrodynamic term of the dynamics is usually ignored in motion control studies, and x_G, y_G are assumed to be 0. Parameters of the simplified nominal model can be obtained through the maneuverability test of small rudder angle. The resulting 6-dimesional nominal model is as:

$$\dot{\boldsymbol{x}} = \begin{bmatrix} \dot{x} \\ \dot{y} \\ \dot{\psi} \\ \dot{u} \\ \dot{v} \\ \dot{r} \end{bmatrix} = \begin{bmatrix} u\cos\psi - v\sin\psi \\ u\sin\psi + v\cos\psi \\ \dot{\psi} = r \\ \frac{m-Y_{\dot{v}}}{m-X_{\dot{u}}}vr - \frac{-X_u}{m-X_{\dot{u}}}u + \frac{1}{m-X_{\dot{u}}}X \\ -\frac{m-X_{\dot{u}}}{m-Y_{\dot{v}}}ur - \frac{-Y_v}{m-Y_{\dot{v}}}v \\ \frac{Y_{\dot{v}}-X_{\dot{u}}}{I_{zz}-N_{\dot{r}}}uv - \frac{-N_r}{I_{zz}-N_{\dot{r}}}r + \frac{1}{I_{zz}-N_{\dot{r}}}N \end{bmatrix}. \tag{3}$$

For simplicity, the nominal model can be written as:

$$\dot{\boldsymbol{x}} = f_{nor}(\boldsymbol{x}, \boldsymbol{u}). \tag{4}$$

The complete model can be written as:

$$\dot{\boldsymbol{x}} = f_{real}(\boldsymbol{x}, \boldsymbol{u}). \tag{5}$$

Compared with the nominal model, the complete model can be expressed by integration of the nominal model and a residual term:

$$f_{real}(\boldsymbol{x}, \boldsymbol{u}) = f_{nor}(\boldsymbol{x}, \boldsymbol{u}) + g(\boldsymbol{x}, \boldsymbol{u}), \tag{6}$$

where $g(\boldsymbol{x}, \boldsymbol{u})$ represents the residual dynamics of the system.

3 MPC with Model Learning

In this section, the framework of Gaussian process regression and how to construct training sets are introduced. Combined with the MPC theory, the Gaussian process based MPC is proposed. The nominal model in MPC is compensated by the residual model learned by Gaussian process.

3.1 Gaussian Process Based Residual Model

A set of training data for Gaussian process regression is presented by

$$D = \left([x_i]_{i=1}^{n}, [y_i]_{i=1}^{N}) \right), \tag{7}$$

where x_t is input vector, y_t is output vector. Their relationship is given by

$$y_i = f(x_i) + \omega, \omega \sim N(0, \delta_n^2) \tag{8}$$

In the Gaussian process regression [10], the joint distribution of y and $f(x_*)$ is

$$\begin{bmatrix} y \\ f(x_*) \end{bmatrix} \sim N\left(0, \begin{bmatrix} \kappa(X,X) + \delta_n^2 I & \kappa(X,x_*) \\ \kappa(x_*,X) & \kappa(x_*,x_*) \end{bmatrix}\right). \tag{9}$$

The common used kernel function is radial basis function (RBF), which is expressed as:

$$\kappa(x_i, x_j) = \delta_f^2 \exp\left(-\frac{1}{2}(x_i - x_j)^T \boldsymbol{l}^{-2}(x_i - x_j)\right), \tag{10}$$

where $\boldsymbol{l}$ is the diagonal length proportional matrix, δ_f is the data variance and δ_n is the priori noise variance. δ_f and $\boldsymbol{l}$ are the kernel's hyperparameters, which can be computed by using the conjugate gradient (CG) algorithm to maximize the following marginal likelihood function:

$$-\log p(y|X,\theta) = \frac{1}{2}y^T(\kappa(X,X)+\delta_n^2 I)^{-1}y + \frac{1}{2}\log\left|(\kappa(X,X)+\delta_n^2 I)^{-1}\right| + \frac{n}{2}\log 2\pi. \tag{11}$$

Then, the predictive mean and variance of $f(x_*)$ can be written as:

$$f(x_*) = \kappa(X, x_*)(\kappa(X,X) + \delta_n^2 I)^{-1} y, \tag{12}$$

$$Var[f(x_*)] = \kappa(x_*, x_*) - \kappa(X, x_*)(\kappa(X,X) + \delta_n^2 I)^{-1}\kappa(X, x_*). \tag{13}$$

According to practical requirements, it's necessary to discretize the system model. We use an explicit Runge-Kutta method of 4th order to integrate $\dot{\boldsymbol{x}}$ given initial state $\boldsymbol{x}_k$, input $\boldsymbol{u}_k$ and integration step T:

$$\boldsymbol{x}_{k+1} = f_{RK4}(\boldsymbol{x}_k, \boldsymbol{u}_k). \tag{14}$$

So that the formulation (7) can be rewritten as:

$$f_{cor}(\boldsymbol{x}_k, \boldsymbol{u}_k) = f_{nor}(\boldsymbol{x}_k, \boldsymbol{u}_k) + g(\boldsymbol{x}_k, \boldsymbol{u}_k). \tag{15}$$

The correction model can be rewritten as:

$$f_{cor}(\boldsymbol{x}_k, \boldsymbol{u}_k) = f_{nor}(\boldsymbol{x}_k, \boldsymbol{u}_k) + \boldsymbol{B}_d \mu(g(\boldsymbol{z}_k)), \tag{16}$$

$$\boldsymbol{z}_k = \boldsymbol{B}_z \left[\boldsymbol{x}_k^T \ \boldsymbol{u}_k^T\right]^T, \tag{17}$$

$$g(\boldsymbol{z}_k) \sim gp(\mu(\boldsymbol{z}), \kappa(\boldsymbol{z}, \boldsymbol{z}_*)), \tag{18}$$

where $\boldsymbol{B}_d$ determines the state modified by GPR. $\boldsymbol{B}_z$ is the selection matrix and determines the feature vector. In this paper, the error of $\boldsymbol{u}_k$, $\boldsymbol{v}_k$ and $\boldsymbol{r}_k$ are taken as the independent regression variables.

Regressor variable can be written as:

$$\boldsymbol{a}_{ek} = \boldsymbol{\mu}(\boldsymbol{z}_k) = \begin{bmatrix} \mu_{uk}(\boldsymbol{z}_k) \\ \mu_{vk}(\boldsymbol{z}_k) \\ \mu_{rk}(\boldsymbol{z}_k) \end{bmatrix}, \tag{19}$$

$$\boldsymbol{a}_{ek} = \begin{bmatrix} \frac{\boldsymbol{u}_{k+1} - \hat{\boldsymbol{u}}_{k+1}}{T} \\ \frac{\boldsymbol{v}_{k+1} - \hat{\boldsymbol{v}}_{k+1}}{T} \\ \frac{\boldsymbol{r}_{k+1} - \hat{\boldsymbol{r}}_{k+1}}{T} \end{bmatrix}, \tag{20}$$

where $\boldsymbol{u}_{k+1}$, $\boldsymbol{v}_{k+1}$, $\boldsymbol{r}_{k+1}$ are state variables predicted according to the nominal model, $\hat{\boldsymbol{u}}_{k+1}$, $\hat{\boldsymbol{v}}_{k+1}$, $\hat{\boldsymbol{r}}_{k+1}$ are the real state at time $k+1$. T is the sampling time. $\boldsymbol{a}_{ek}$ is the acceleration error, which captures the unmodeled dynamics of the USV system. In this paper, $\boldsymbol{z}_k$ and $\boldsymbol{a}_{ek}$ constitute the training set of the GPR:

$$D = ([\boldsymbol{z}_i]_{i=1}^{n}, [\boldsymbol{a}_{ei}]_{i=1}^{N}). \tag{21}$$

Given the training set D and test samples $\boldsymbol{z}_*$, the residual dynamics is predicted by the GPR model.

3.2 Gaussian Process Based MPC

Assuming the model is accurate, the nonlinear model predictive control (NMPC) stabilizes a system subject to its dynamics $\dot{\boldsymbol{x}} = f(\boldsymbol{x}, \boldsymbol{u})$ along state and input trajectories $\boldsymbol{x}^*(t)$, $\boldsymbol{u}^*(t)$, by minimizing a cost function as in:

$$\min_{u} \int l(\boldsymbol{x}, \boldsymbol{u}), \tag{22}$$

Subject to

$$\dot{\boldsymbol{x}} = f_{dyn}(\boldsymbol{x}, \boldsymbol{u}), \tag{23}$$

$$\boldsymbol{x}(t_0) = \boldsymbol{x}_{init}, \tag{24}$$

$$\boldsymbol{h}(\boldsymbol{x}, \boldsymbol{u}) \leq 0, \tag{25}$$

where $\boldsymbol{x}(t_0)$ is the initial state and $\boldsymbol{h}(\boldsymbol{x}, \boldsymbol{u}) \leq 0$ represents constraints.

If just partial dynamics is available, we model the residual dynamics by GPR. On the basis of the correction model, the Gaussian process based MPC can be represented as follows:

$$\min_{u} \boldsymbol{x}_N^T \boldsymbol{P} \boldsymbol{x}_N + \sum_{k=0}^{N} (\boldsymbol{x}_k^T \boldsymbol{Q} \boldsymbol{x}_k + \boldsymbol{u}_k^T \boldsymbol{R} \boldsymbol{u}_k), \tag{26}$$

Subject to

$$f_{cor}(\boldsymbol{x}_k, \boldsymbol{u}_k) = f_{nor}(\boldsymbol{x}_k, \boldsymbol{u}_k) + B_d g(\boldsymbol{x}_k, \boldsymbol{u}_k), \tag{27}$$

$$\boldsymbol{z}_k = \boldsymbol{B}_z[\boldsymbol{x}_k^T, \boldsymbol{u}_k^T]^T, \tag{28}$$

$$g(\boldsymbol{x}_k, \boldsymbol{u}_k) = \mu(\boldsymbol{z}_k), \tag{29}$$

$$\boldsymbol{x}_0 = \boldsymbol{x}_{init}, \tag{30}$$

$$u_{\min} \leq \boldsymbol{u}_k \leq u_{\max}, \tag{31}$$

where $f_{cor}(\boldsymbol{x}_k, \boldsymbol{u}_k)$ is the correction model.

4 Simulation

In this section, we choose ship model KVLCC2_L7 as the simulation objects. According to empirical regression formula, the hydrodynamic coefficients of the complete model and nominal model can be calculated from the structural parameters in Table 1. The control of circle and lemniscate trajectory tracking are simulated by the proposed GP-MPC and the traditional NMPC.

Table 1. Principal parameter of the KVLCC2_L7 model.

Parameter	Value	Parameter	Value
Length, L/m	7.12	Propeller diameter, D_p/m	0.216
Breadth, B/m	1.27	Rudder span length, H_R/m	0.345
Draught, T/m	0.46	Rudder area, A_R/m^2	0.0539
Block coefficient, C_b	0.81	Displacement volume, V/m^3	3.27

4.1 Performance Evaluation

In order to capture the real dynamics of USV and build the training set, we collect the data from complete model through maneuverability simulation firstly. Then the acceleration error can be computed by Eq. (20), which represents the dynamics residuals. The inputs and states of the system, as well as the dynamics residuals, constitute the original dataset.

Because the complexity of GPR depends on the number of training points, we subsampled the dataset by using k-means algorithm. Before performing simulations, we determine the dataset scale of GPR by analyzing the relationship between the number of training points with the computation time, and the relationship between the number of training points with the optimization performance. The results are shown in Fig. 2

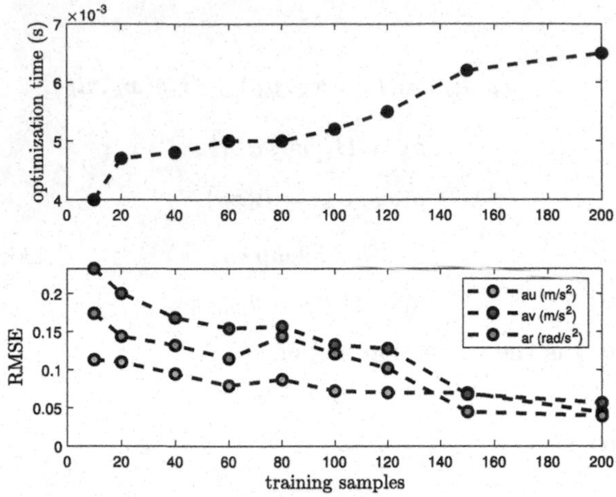

Fig. 2. Trade-off between number of training samples, GP performance and optimization time.

We can see that the optimization performance is almost stable when the number of training points exceeds 150. Thus, in order to achieve the desired performance of calculation time and prediction error, we set the dataset scale as 150 in this simulation.

In the simulations, the parameters of the GP-MPC and NMPC are: $Q = diag(10, 10, 1, 1, 1)$, $R = diag(0.1, 0.1)$, $N = 15$. The initial state is $[0, -3, 0, 0, 0, 0]$. The simulation results are shown in Fig. 3 and Fig. 4.

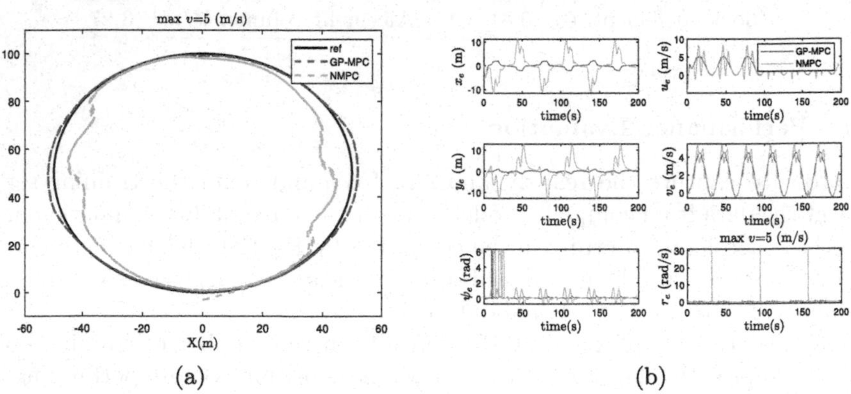

Fig. 3. Circle trajectory tracking and state error.

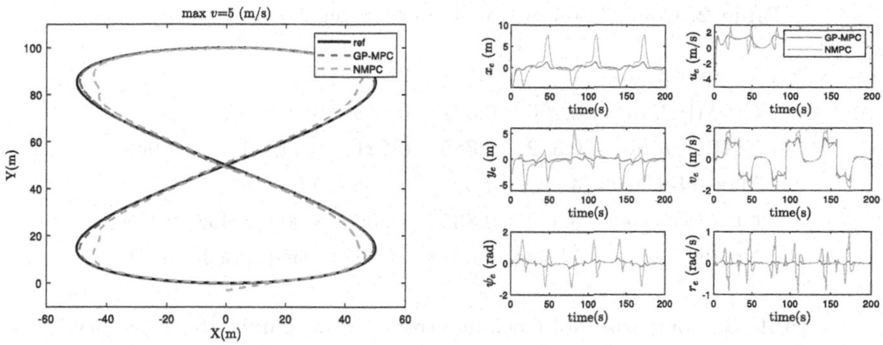

(a) (b)

Fig. 4. Lemniscate trajectory tracking and state error.

The trajectory and the states error in Fig. 3, Fig. 4 show that GP-MPC has obvious effects on improving the accuracy of trajectory tracking and reducing the error compared with NMPC.

4.2 Performance Comparison

Since the adjustment time and overshoot of the GP-MPC and the NMPC are different, the position tracking error of the last lap is used to express the control accuracy. The simulation results are shown in Fig. 5, Table 2 and Table 3.

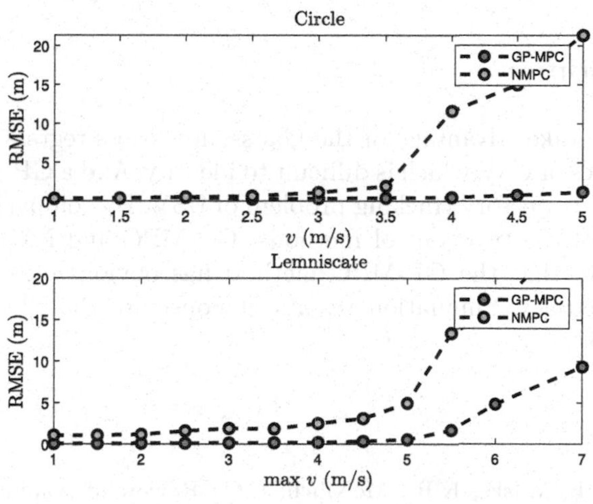

Fig. 5. Position tracking error.

Table 2. Comparison of tracking errors on the circle trajectory.

Max ref-u, m/s	1	1.5	2	2.5	3	3.5
GP-MPC, m	0.296	0.371	0.488	0.264	0.344	0.489
NMPC, m	0.332	0.385	0.521	0.650	1.113	1.967
Max ref-u, m/s	4	4.5	5	5.5	6	6.5
GP-MPC, m	0.492	0.835	1.206	1.009	3.438	9.159
NMPC, m	11.624	15.218	21.359	crash	crash	crash

Table 3. Comparison of tracking errors on the lemniscate trajectory.

Max ref-u, m/s	1	1.5	2	2.5	3	3.5
GP-MPC, m	0.051	0.071	0.097	0.124	0.149	0.171
NMPC, m	1.068	1.071	1.172	1.163	1.902	1.846
Max ref-u, m/s	4	4.5	5	5.5	6	7
GP-MPC, m	0.204	0.343	0.548	1.682	4.827	9.358
NMPC, m	2.564	3.187	4.951	13.386	16.072	27.513

It can be seen that when the speed increases to a certain extent, the nominal model will seriously mismatch with the real system, which leads to the consequences that the traditional NMPC cannot keep up with the reference trajectory and the tracking state cannot converge. Whereas, the GP-MPC can still achieve good tracking accuracy at high speed due to the correction of the residual dynamics.

5 Conclusion

In this work, we take advantage of the Gaussian process regression to learn the residual dynamics of USV which is difficult to identify. And a GP-MPC method is proposed for the trajectory tracking problem of USV. By comparing the tracking position errors of the two control methods, GP-MPC and NMPC, at different speeds, we verify that the GP-MPC method has obvious effect on improving the tracking accuracy. Simulation results demonstrate the effectiveness of the proposed method.

References

1. Ashrafiuon, H., Muske, K.R., McNinch, L.C.: Review of nonlinear tracking and setpoint control approaches for autonomous underactuated marine vehicles. In: Proceedings of the 2010 American Control Conference, pp. 5203–5211. IEEE (2010)
2. Bai, W., Ren, J., Li, T.: Modified genetic optimization-based locally weighted learning identification modeling of ship maneuvering with full scale trial. Futur. Gener. Comput. Syst. **93**, 1036–1045 (2019)

3. Bonci, M., De Jong, P., Van Walree, F., Renilson, M., Huijsmans, R.: The steering and course keeping qualities of high-speed craft and the inception of dynamic instabilities in the following sea. Ocean Eng. **194**, 106636 (2019)
4. Carron, A., Arcari, E., Wermelinger, M., Hewing, L., Hutter, M., Zeilinger, M.N.: Data-driven model predictive control for trajectory tracking with a robotic arm. IEEE Robot. Autom. Lett. **4**(4), 3758–3765 (2019)
5. Chen, G., Wang, W., Xue, Y.: Identification of ship dynamics model based on sparse gaussian process regression with similarity. Symmetry **13**(10), 1956 (2021)
6. Dong, Y., Wu, N., Qi, J., Chen, X., Hua, C.: Predictive course control and guidance of autonomous unmanned sailboat based on efficient sampled gaussian process. J. Marine Sci. Eng. **9**(12), 1420 (2021)
7. Dong, Z., Wan, L., Li, Y., Liu, T., Zhang, G.: Trajectory tracking control of underactuated USV based on modified backstepping approach. Int. J. Naval Archit. Ocean Eng. **7**(5), 817–832 (2015)
8. Han, J., Xiong, J., He, Y., Gu, F., Li, D.: Nonlinear modeling for a water-jet propulsion USV: an experimental study. IEEE Trans. Industr. Electron. **64**(4), 3348–3358 (2016)
9. Hewing, L., Liniger, A., Zeilinger, M.N.: Cautious NMPC with gaussian process dynamics for autonomous miniature race cars. In: 2018 European Control Conference (ECC), pp. 1341–1348. IEEE (2018)
10. Li, F., Li, H., He, Y.: Adaptive stochastic model predictive control of linear systems using gaussian process regression. IET Control Theory Appl. **15**(5), 683–693 (2021)
11. Luo, W., Moreira, L., Soares, C.G.: Manoeuvring simulation of catamaran by using implicit models based on support vector machines. Ocean Eng. **82**, 150–159 (2014)
12. Moreira, L., Soares, C.G.: Dynamic model of manoeuvrability using recursive neural networks. Ocean Eng. **30**(13), 1669–1697 (2003)
13. Mu, D., Wang, G., Fan, Y., Zhao, Y.: Modeling and identification of podded propulsor unmanned surface vehicle. ICIC Express Lett. Part B: Appl. **8**(2), 245–253 (2017)
14. Ogawa, A., Kasai, H.: On the mathematical model of manoeuvring motion of ships. Int. Shipbuild. Prog. **25**(292), 306–319 (1978)
15. Ramirez, W.A., Leong, Z.Q., Nguyen, H., Jayasinghe, S.G.: Non-parametric dynamic system identification of ships using multi-output gaussian processes. Ocean Eng. **166**, 26–36 (2018)
16. Shi, Y., Shen, C., Fang, H., Li, H.: Advanced control in marine mechatronic systems: a survey. IEEE/ASME Trans. Mechatron. **22**(3), 1121–1131 (2017)
17. Torrente, G., Kaufmann, E., Föhn, P., Scaramuzza, D.: Data-driven MPC for quadrotors. IEEE Robot. Autom. Lett. **6**(2), 3769–3776 (2021)
18. Wang, X.G., Zou, Z.J., Yu, L., Cai, W.: System identification modeling of ship manoeuvring motion in 4 degrees of freedom based on support vector machines. China Ocean Eng. **29**(4), 519–534 (2015)
19. Xue, Y., Liu, Y., Ji, C., Xue, G., Huang, S.: System identification of ship dynamic model based on gaussian process regression with input noise. Ocean Eng. **216**, 107862 (2020)

Other AI Related Topics

A Hybrid Pattern Knowledge Graph-Based API Recommendation Approach

Guan Wang, Weidong Wang(✉), and Dian Li

Faculty of Information Technology, Beijing University of Technology, Beijing, China
wangweidong@bjut.edu.cn, iamlidian@emails.bjut.edu.cn

Abstract. There are a large number of application program interfaces (APIs) on the Internet. Due to frequently updating APIs, programmers are prompted to frequently consult API documents in the process of software development. However, the traditional query approaches of the documents have certain limitations. For example, the programmers need to know the API name as a prerequisite and are often unable to obtain the expected search results because of the difference in understanding between the description of the problem and the description of the documents. Only these "known-unknown" information can be queried, and it is difficult to query the "unknown-unknown" information. To address the limitations, we establish the knowledge graph of software source code combined with knowledge which is derived from the documents, Github project code warehouse, Stack overflow platform, and local project code warehouse. Moreover, we propose a hybrid pattern knowledge graph-based API recommendation approach for programmers to complete the query task of unknown-unknown information. Finally, we constructed large-scale real experiments. Evaluation results prove that the proposed approach significantly outperforms the state-of-the-art approach.

Keywords: Knowledge graph · API · Machine learning

1 Introduction

In the field of modern software intelligent development, it is a more efficient means of accurately calling Application programming interfaces, shortly APIs, indispensable programming tasks. However, the rapid updating of those interfaces and their description documents on the Internet [15–17], as well as the increasing complexity of software projects and their related documents defined by the projects [14], bring great inconveniences in querying and understanding APIs for programmers [6].

Actually, the traditional API query approach may not meet the needs of programmers. It has two limitations. First, programmers need to know the name of the query interface in advance [19]. In the scenario, the query needs to be carried out through the query description. So it is often unable to match because

L. Fang et al. (Eds.): CICAI 2022, LNAI 13606, pp. 465–476, 2022.
https://doi.org/10.1007/978-3-031-20503-3_37

of the lexical gap between the natural language description of the query information and the content of the target API documents. Second, the contents of the documents could be simple and incomplete. Some function parameters or class function descriptions are missing, which could result in ambiguity for programmers. At the same time, the traditional documents more or less rely on text narration, to some extent, such long texts make it difficult for programmers to obtain all useful information.

For instance, the programmers need to know the name as the prerequisite when querying its detailed information, and the description of the function is often unable to obtain the expected search results because of the difference in understanding between the description of the problem and the description of the documents. Only these "known-unknown" information can be queried, and it is difficult to query the "unknown-unknown" information.

To address these limitations above, this paper proposes a hybrid pattern knowledge graph-based intelligent APIs recommendation approach, which combines knowledge graph established by APIs-related information, which is derived from the Internet program documents, Github project code warehouse, Stack overflow platform, and local project code warehouse. The classification of query statements and the hybrid pattern approach based on rule template and text understanding ensure the search speed of query and improve the accuracy of query results, which provides a time guarantee for programmers in the search process of development.

In order to evaluate the proposed approach, we constructed a large-scale dataset by enriching the public dataset BIKER [9]. Compared with transitional approaches, the MRRs and MAPs are improved by at least 25% and 31%, respectively. And our approach outperforms BIKER by at least 20% inaccuracy. In addition, the proposed approach greatly improves the efficiency of a query.

2 Related Work

In the aspect of knowledge graph query approaches based on rule template, Berant et al. [1] integrated the traditional semantic parsing [20] of directly mapping into KGQA as a solution. They solves the problem that the training semantic parser needs the logical form data manually labeled as supervision and vocabulary table coverage, and then they proposed the famous dataset as the baseline.

Yahya et al. [2] implement an Agendail system, which is combined with imitation learning in the training process of the semantic parser to ensure accuracy and improve the running speed. At the same time, They added text description information in relation-triples to form a quadruple (S, P, O, X) form using the description information to achieve word sense disambiguation. However, this approach modifies the tuple structure and is difficult to combine with the existing approaches.

Fader et al. [3] proposed an OQA system that translates problem statements into standard query statements through predefined templates and then uses standard query statements to translate them into similar entities or relationships that exist in the knowledge graph. This approach makes the query content close connected with the knowledge map. However, the template of the approach needs to be constructed manually in advance, and it cannot adaptively update along with the knowledge graph and the new problem set.

Lopez et al. [4] proposed an Aqualog Q&A system using the triple matching approach to retrieve similar triples from RDF data by calculating similarity, which has good accuracy for simple problem queries. However, for the complex problems, the triplets cannot be directly used to capture semantics. The system procedure uses the approach of query template, using a pipeline of template matching to try to match all templates. This approach has high interpretability and fully corresponds to simple problem queries. But the semantic parsing ability of complex statements is weak.

In the aspect of the machine learning-based KGQA approach, Bordes et al. [5] proposed the knowledge graph embedding approach, namely TransE. The improved TransE encodes the question-answer path and the surrounding subgraph embedding to vectors. Weston et al. [5] proposed the MemNN framework, taking the memory network as the core, and embedded all the contents of the knowledge map into the vector space to determine the relationship between the system output and the natural language query statement. Yavuz et al. [7], based on the AGENDAIL system, used BiLSTM to express the contents around entities by vectors to predict the types of entities and ranked them with entity similarity, which improved the ranking results. Bao et al. [8] used CNN to encode the problem statement and the semantic query graph and implemented the similarity analysis. The optimal query graph was selected as the intermediate logic form and converted to the query statement. This approach can better deal with complex and multi-hop queries, but the search space is redundant.

In summary, the approaches based on rule templates have strong interpretability, but weak adaptability and low query accuracy for complex problems. Most machine learning-based approaches have improved and optimized some capabilities based on the rule template approaches. Simultaneously, some machine learning-based approach have improved the ability of complex problem queries compared with the rule template approaches. However, they need more time for training and query. And for some simple problem queries, the accuracy of the machine learning-based approach is less or even lower than the rule template approaches.

Therefore, we consider a hybrid approach to achieve both the high time efficiency and the accuracy in dealing with simple and complex queries.

3 A Hybrid APIs Query Approach via Knowledge Graph

The framework of the Hybrid APIs query approach via knowledge graph mainly consists of two parts: (1) knowledge graph construction; (2) hybrid APIs query and visualization.

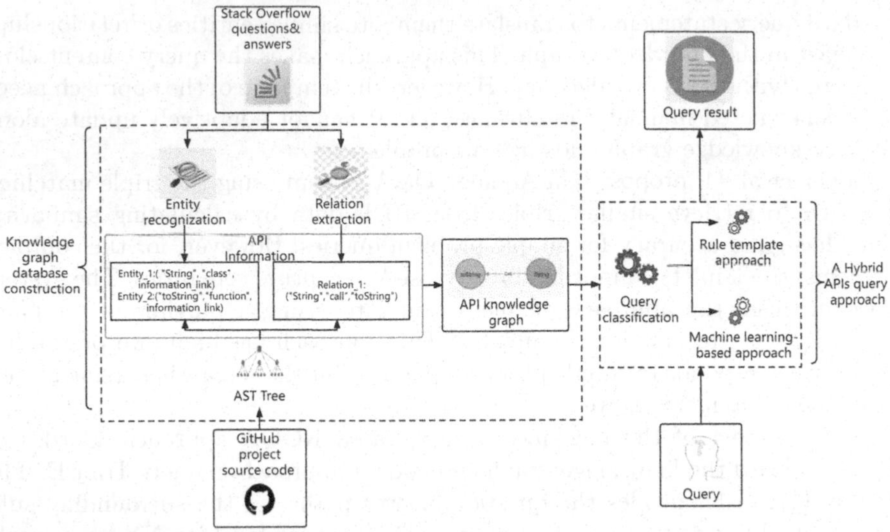

Fig. 1. The framework of the Hybrid APIs query approach via knowledge graph.

3.1 Knowledge Graph Database Construction

First, we acquire the basic APIs related information through their documents. Then, we extract the categories using the tags of HTML pages. Subsequently, we obtain 10k software projects from GitHub according to their stars and label those projects with related information. At the same time, the questions and discussions for the projects are stored in the form of question-answer pairs.

Second, we compile the source codes to obtain their abstract syntax trees (AST). According to the structure of AST, we obtain the relationships between API entities, and then these relationships are stored in the form of triples.

Then, we achieve knowledge extraction from unstructured information in the API description documents and the Q&A pairs of GitHub projects. First, we run named entity recognition, because the text information has an obvious temporal relationship. We use long short-term memory, LSTM [10], to label the text. To improve accuracy, we add a conditional random field, CRF [11], to assist us to obtain the correct sequence of labels. Thereafter, based on the Chinese vocabulary database and the extracted entity vocabulary, we use segmentation and part-of-speech tagging to process the text. Second, the dependency parsing tree is generated by the dependency parsing of part-of-speech tagging.

By using semantic role labeling to process dependency the parsing tree, we obtain the semantic role labels of words. We extract triples of relationships from the semantic role labeling result according to predefined rules, e.g., in positive relational structures, we abstract (agent, predicate verb, and patient), as the supplement to our knowledge.

Finally, we build a knowledge map based on extracted entities and relation triples. We extract 5 types of entities and 9 types of relationships. We construct the knowledge graph of APIs based on the Neo4j database. Texts such as the function description, parameter descriptions, and sample code of classes or functions are stored in the form of hyperlinks as the properties of entities.

3.2 A Hybrid APIs Query Approach

In this section, we firstly classify query statements as regular queries or irregular queries through Rules 1 and 2. If the query is judged like a regular query, then we select template rule, and vice versa, we select machine learning processing.

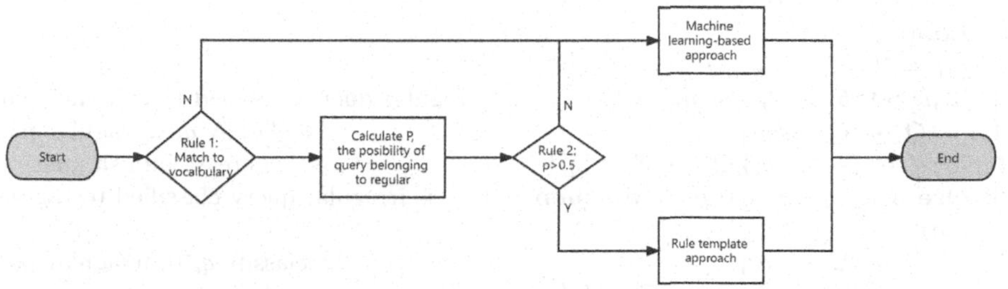

Fig. 2. The framework of the Hybrid APIs query approach via knowledge graph.

First, we randomly collected descriptions from 100 query issues among such corresponding answers from the top 10000 JDK queries for APIs in Java SE and C++ standard library on the stack overflow platform as the training dataset Q. Then, the template rule and machine learning processing are used to query Q, respectively. Each test query q_i records the rule and template t_{gi} and machine learning query time t_{mi}, and the accuracy of top k query results $prec_{gi}$, $prec_{mi}$. n this paper, k is initialized as 3. We labeled each query statement q_i as (a_i, b_i), where

$$a_i = \frac{prec_{gi} - prec_{mi}}{prec_{gi} + prec_{mi}}, b_i = \frac{t_{gi} - t_{mi}}{t_{gi} + t_{mi}} \tag{1}$$

Next, we classify the training set Q. The query with maximum ai, which satisfies $b_i > 0$, is viewed as the query q_1. The query with maximum bi that satisfies $a_i > 0$, is considered as the query q_2. The query with the $maximum(\sqrt{a_i^2 + b_i^2})$ which satisfying $a_i > 0, b_i > 0$is recognized as the query q_3. Take set $\{q_1, q_2, q_3\}$ as the initialized set of the regular query set. Similarly, The query with minimum ai which satisfies $b_i < 0$ is identified as the query q_1'. The query with minimum bi which satisfies $a_i < 0$ as the query q_2', and the query with the $maximum(\sqrt{a_i^2 + b_i^2})$ that satisfies $a_i < 0andb_i < 0$ is configured as the query q_3'. Take set $\{q_1', q_2', q_3'\}$ as the initial set of the irregular query set. The maximum likelihood estimation (MLE) for a gaussian distribution Susan of $\{ q_1, q_2, q_3\}$ N_1 and the MLE for a gaussian distribution Susan of (q_1', q_2', q_3') N_2 are both calculated.

Algorithm 1. Query Classification

Input: question set Q
Output: requgalr question set Q1 ,non-regular question set Q2
Initialize:
$q_1 \leftarrow q_n, a_n = \arg\max(a_i)\ , b_n > 0, q_n \in Q$
$q_2 \leftarrow q_n, b_n = \arg\max(b_i)\ , a_n > 0, q_n \in Q$
$q_3 \leftarrow q_n, b_n = \arg\max\left(\sqrt{a^2+b^2}\right)\ , a_n > 0, b_n > 0, q_n \in Q$
$q_1' \leftarrow q_n, a_n = \arg\min(a_i)\ , b_n < 0, q_n \in Q$
$q_2' \leftarrow q_n, b_n = \arg\min(b_i)\ , a_n < 0, q_n \in Q$
$q_3' \leftarrow q_n, b_n = \arg\max\left(\sqrt{a^2+b^2}\right)\ , a_n < 0, b_n < 0, q_n \in Q$
calculate the MLE of gaussian distribution of$q_1, q_2, q_3 N_1$,the MLE of gaussian distribution of$q_1', q_2', q_3' N_1$.

Loop:
  $p_{1i} = N_1(a_i, b_i), q_i \in Q$
  if $a_i > 0, b_i > 0, p_{1i} < p_{2i}$ **then** ▷ q_i is regular query classified to irragular part
    $Q_1 \leftarrow Q_1 \cup \{q_i\}$ ▷ classify q_i to ragular part
    $q_i' \leftarrow (-a_i, -b_i)$,$Q_2' \leftarrow Q_2' \cup \{q_i'\}$ ▷ create punish query q_i'
  else if $a_i < 0, b_i < 0, p_{1i} > p_{2i}$ **then** ▷ q_i is irregular query classified to ragular part
    $Q_2 \leftarrow Q_2 \cup \{q_i\}$ ▷ classify q_i to irragular part
    $q_i' \leftarrow (-a_i, -b_i)$,$Q_1' \leftarrow Q_1' \cup \{q_i'\}$ ▷ create punish query q_i'
  else if $p_{1i} > p_{2i}$ **then** ▷ other queries classified by $p_{1i} and p_{2i}$
    $Q_1 \leftarrow Q_1 \cup \{q_i\}$
  else
    $Q_2 \leftarrow Q_2 \cup \{q_i\}$
  calculate the MLE of gaussian distribution of $Q_1 \cup Q_1' N_1'$,the MLE of gaussian distribution of $Q_2 \cup Q_1' N_1'$. $N_1 \leftarrow N_1'$,$N_2 \leftarrow N_2'$
  $Q_1' \leftarrow \{\}$,$Q_2' \leftarrow \{\}$ ▷ clear punish queries sets
End Loop
Retrun regular question set Q_1,non-reqular question set Q_2;

For each training phase, Both probability $p_1 i$ and $p_2 i$ of each real query q_i belonging to the regular query set Q_1 and the irregular query set Q_2 are calculated. For the query issue q_i satisfying $a_i > 0, b_i < 0$, or $a_i < 0, b_i > 0$, q_i is viewed as N_1 if $p_1 i > p_2 i$ and q_i is classified as N_2 if $p_1 i < p_2 i$. For q_i satisfying $a_i > 0, b_i > 0$, if $p_1 i > p_2 i$, it belongs to Q_1; if $p_2 i < p_1 i$, a punish false $q_i'(-q_1, -q_2)$ is calculated, which is classified as Q_2', and q_i is classified as Q_1. For$a_i < 0, b_i < 0$, vice versa. The new maximum likelihood MLE for a gaussian distribution of $\{Q1 \cup Q1'\}$, $N1'$ and the new MLE for a gaussian distribution of $\{Q_2 \cup Q_2'\}$, $N2'$ are calculated. Repeat the above training phase untilN_1, N_2 don't change.

For query issue q_i belonging to Q_1, the scores for two query set $y_i = (1, 0)$ are labeled. For query q_i belonging to Q_2, the scores to two query sets $y_i = (0, 1)$ are also labeled. The word vector sequence of q_i as the input and the score y_i as the output are used for LSTM training. The trained LSTM takes the word vector sequence x of a new query statement as input and the output is $y(y_1, y_2)$. y_1, y_2

is the scores of the query statement belonging to the regular query and irregular query, respectively. the probability p that the query belongs to the regular query is calculated. where

$$p = \frac{e^{y_1}}{e^{y_1} + e^{y_2}} \tag{2}$$

The following rules are defined to classify a query:
Rule 1: Query statements contain words in the entity or relational vocabulary.
Rule 2: The probability of belonging to irregular query $p > 0.5$.

The query statement satisfying all the above rules belongs to the irregular query, otherwise, it belongs to the irregular query issue.

For regular problems, we use rule templates-based approach to process. We segment the query sentences with Chinese vocabulary and run dependency parsing. According to the rules, the specific dependency parsing relationship related to the words in the entity table is selected and transformed into query statements of Cypher. For the scenario where the query statement does not obtain the query result, the entity mentioned in the statement and its two-hop nodes are viewed as the candidate entity set E. We choose the entity which has the maximum text similarity with query statement as the answer.

For irregular problems, we usermachine learning-based approach to process. First, the entities and relationships in the knowledge graph are transformed into two matrices M_1, M_2. Each row in a matrix represents a vector representation of an entity or relationship. Then, the TransE technology [12] is introduced to train the matrix $M_1 and M_2$. Through training, the triple (h, l, t) relationship in the knowledge map can be expressed as $h+l = t$, in other words, the relationship triple can be expressed by the word vectors of the entity and relation. Next, we use GRU [13], a gated recurrent unit, to train the queries above. For the trained GRU model, the output of each query statement is the vector $(y_1, y_2, ..., y_n)$, which is a vector representation of the entity node, where n is the dimension of the entity vector. Thus, the entity node and relationship in the query statement and knowledge map are mapped to the same space. By calculation $sim(y_i, e_i)$, the nodes that have $max(sim(y_i, e_i))$ is selected as the query results, Where $sim(y_i, e_i)$ is the cosine similarity between query statement sequence vector, and knowledge graph entity node word vector e_i.

4 Experimental Evaluation

4.1 Comparisons with Traditional Approaches

In the section, we answer the question of how effective the hybrid query approach proposed is compared with the rule template query and the machine learning query in performance individually. First, we use the public dataset released by BIKER [9] as our evaluation dataset. The machine learning query uses query data from BIKER dataset when training. Therefore, we remove the query issues used in training and random sample 20% queries as the test datasets. In addition, we use 100 queries based on the hot query issues in the stack overflow platform as a supplement to queries. For evaluation metrics in this paper, we select MRR and MAP to evaluate three query approaches. Then, we also use the Wilcoxon signed-rank test [18] to check whether the hybrid approach is statistically significant compared to two separate query approaches. If the corresponding Wilcoxon signed-rank test result p is less than 0.05, we consider that one approach performs significantly better than the other one at the confidence level of 95%. The experimental results are as follows:

Table 1. Comparison with separate approach.

Approach	BIKER's data set		Our data set	
	MRR	MAP	MRR	MAP
Hybrid Pattern Recommendation	0.624	0.587	0.476	0.443
Rule and Template-based	0.392	0.375	0.311	0.283
Machine Learning-based	0.467	0.411	0.382	0.337
Improvement.Rule and Template-based	0.592	0.565	0.531	0.565
Improvement.Machine Learning-based	0.336	0.428	0.246	0.315

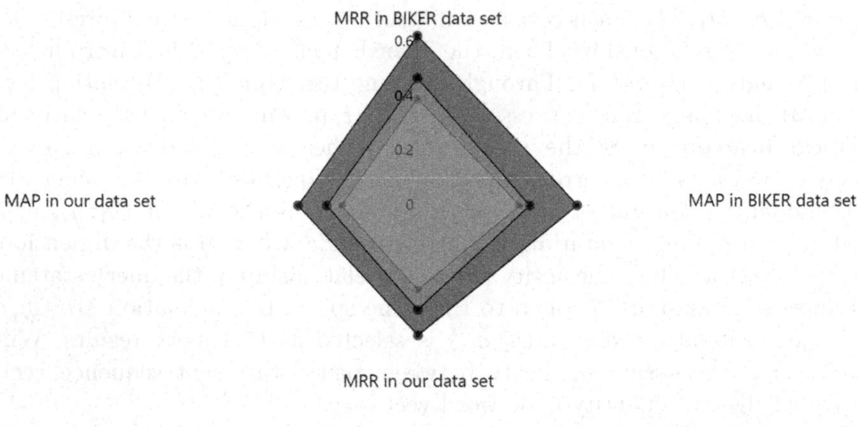

Fig. 3. The framework of the Hybrid APIs query approach via knowledge graph.

We can clearly see that the performance of the hybrid approach has been significantly improved compared with the two other query approaches. For the former, the MRR and map are enhanced by 25% and 31%, respectively, which is enough to prove that the query classification in the hybrid approach proposed is effective and necessary.

4.2 Comparisons with a Hybrid Baseline Approach

We compared the performance with the existing APIs recommendation baseline BIKER [9] to verify whether our hybrid approach is accurate and effective. We used the dataset in rq1 and eliminated the answers without JDK to ensure fairness. In terms of evaluation metrics, we still chose MAP and MRR. The results are as follows:

Table 2. Comparison with baseline.

Approach	BIKER's data set		Our data set	
	MRR	MAP	MRR	MAP
Hybrid Pattern Recommendation	0.624	0.587	0.476	0.443
BIKER	0.522	0.458	0.328	0.271
cre Improvement	0.195	0.282	0.451	0.635

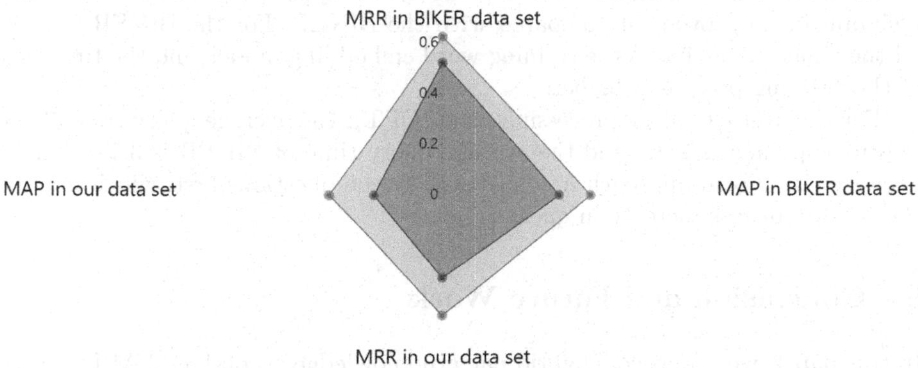

Fig. 4. The framework of the Hybrid APIs query approach via knowledge graph.

For BIKER's dataset, our hybrid approach has a 20% improvement in MRR and a 28% improvement in MAP. For our dataset, MRR and MAP are increased by 45% and 63%, respectively. And all the results of the Wilcoxon signed-rank test below 0.01. It can be found that the improvement of our dataset is more obvious, which may be because although the answer to the query is about GDK, the query statements and calculation involve APIs outside the training set. In summary, the hybrid APIs query approach has certain improvements compared with BIKER.

4.3 Overheads

In the section, we answer the question of how time efficient the hybrid APIs query approach is. To verify the time efficiency of our approach, we compare the model construction time and query time with the existing baseline. For the comparison of training time, we retrain the model of the hybrid approach and record the training time of the model. For the query time, we randomly sampled 100 queries from the test set in rq1, and test the hybrid approach and baseline approach, then record the processing time of each query. The average and variance of time are calculated as the indicators of query efficiency and stability. The results are as follows:

Table 3. Comparison efficient with baseline.

Approach	MTT	QPT
Hybrid Pattern Recommendation	18 min	2.9 s/ query
BIKER	30 min	3.1 s/ query

In the model training time (MTT), the hybrid approach is 18 min, and it has a significant improvement compared with the BIKER. For the BIKER, almost all the time cost is due to the training word embedding model, and the time cost of the training process is higher.

For the average query processing time (QPT), the average query time of the hybrid approach is 2.9 s, and the average query time of BIKER is 3.1 s. This is because the hybrid approach needs to classify the questions first, which may be the reason for less increase in query time.

5 Conclusion and Future Work

In this paper, we proposed a hybrid pattern knowledge graph-based APIs recommendation approach to address the limitation issues in query accuracy and efficiency. By classifying the query statements into regular and non-regular queries, the hybrid approach proposed in this paper improves query accuracy and reduces query time. In the experiment, we enriched the BIKER's dataset, and then proved the effectiveness of the hybrid query approach, comparing with two independent recommendation approaches.

For the future work, we will develop an IDE plug-in so that developers can easily query APIs-related information in programming tasks. At the same time, we consider further extending the types of programming languages supported by our hybrid pattern recommendation approach.

Acknowledgement. This work was supported in part by the Beijing Municipal Science and Technology Project under Grant KM201910005031, and in part by the Education and Teaching Research Fund of the Beijing University of Technology under Grant ER2020B011.

References

1. Berant, J., Liang, P.: Imitation learning of agenda-based semantic parsers. Trans. Assoc. Comput. Linguist. **3**, 545–558 (2015)

2. Yahya, M., Berberich, K., Elbassuoni, S., Weikum, G.: Robust question answering over the web of linked data. In: Proceedings of the 22nd ACM international conference on Information and Knowledge Management, pp. 1107–1116. Association for Computing Machinery, California, San Francisco, USA (2013)
3. Fader, A., Zettlemoyer, L., Etzioni, O.: Open question answering over curated and extracted knowledge bases. In: Proceedings of the 20th ACM SIGKDD International Conference on Knowledge Discovery and Data Mining, pp. 1156–1165. Association for Computing Machinery, East Lansing, MI, USA (2014)
4. Lopez, V., Motta, E.: Ontology-driven question answering in AquaLog. In: Meziane, F., Métais, E. (eds.) NLDB 2004. LNCS, vol. 3136, pp. 89–102. Springer, Heidelberg (2004). https://doi.org/10.1007/978-3-540-27779-8_8
5. Bordes, A., Weston, J., Usunier, N.: Open question answering with weakly supervised embedding models. In: Calders, T., Esposito, F., Hüllermeier, E., Meo, R. (eds.) ECML PKDD 2014. LNCS (LNAI), vol. 8724, pp. 165–180. Springer, Heidelberg (2014). https://doi.org/10.1007/978-3-662-44848-9_11
6. Wang, X., Wu, H., Hsu, C.H.: Mashup-oriented API recommendation via random walk on knowledge graph. IEEE Access **7**, 7651–7662 (2019)
7. Yavuz, S., Gur, I., Su, Y., Srivatsa, M., Yan, X.: Improving semantic parsing via answer type inference. In: Proceedings of the 2016 Conference on Empirical Methods in Natural Language Processing, pp. 149–159. Association for Computational Linguistics, Austin, Texas (2016)
8. Bao, J., Duan, N., Zhou, M., Zhao, T.: Knowledge-based question answering as machine translation. In: Proceedings of the 52nd Annual Meeting of the Association for Computational Linguistics (Volume 1: Long Papers), pp. 967–976. Association for Computational Linguistics, Baltimore, Maryland (2014)
9. Huang, Q., Xia, X., Xing, Z., Lo, D., Wang, X.: API method recommendation without worrying about the task-API knowledge gap. In: 2018 33rd IEEE/ACM International Conference on Automated Software Engineering (ASE), pp. 293–304. IEEE, Montpellier, France (2018)
10. Shi, X., Chen, Z., Wang, H., Yeung, D.Y., Wong, W.K., Woo, W.C.: Convolutional LSTM network: a machine learning approach for precipitation nowcasting. In: Advances in Neural Information Processing Systems, pp. 293–304. Curran Associates, San Francisco, US (2015)
11. Le, T.A., Arkhipov, M.Y., Burtsev, M.S.: Application of a hybrid Bi-LSTM-CRF model to the task of Russian named entity recognition. In: Filchenkov, A., Pivovarova, L., Žižka, J. (eds.) AINL 2017. CCIS, vol. 789, pp. 91–103. Springer, Cham (2018). https://doi.org/10.1007/978-3-319-71746-3_8
12. Bordes, A., Usunier, N., Weston, J., Yakhnenko, O.: Translating embeddings for modeling multi-relational data. In: Advances in Neural Information Processing Systems 26, pp. 2787–2795. Curran Associates, Red Hook (2013)
13. Cho, K., van Merrienboer, B., Gülçehre, Ç., Bougares, F., Schwenk, H., Bengio, Y.: Learning phrase representations using RNN encoder-decoder for statistical machine translation. CoRR arXiv:1406.1078 (2014)
14. Bui, N.D.Q., Yu, Y., Jiang, L.: SAR: learning cross-language API mappings with little knowledge. In: Proceedings of the 2019 27th ACM Joint Meeting on European Software Engineering Conference and Symposium on the Foundations of Software Engineering, pp. 796–806. ACM, Tallinn Estonia (2019)
15. Li, H., et al.: Improving API caveats accessibility by mining API caveats knowledge graph. In: 2018 IEEE International Conference on Software Maintenance and Evolution (ICSME), pp. 183–193. IEEE, Madrid (2018)

16. Zhang, J., Jiang, H., Ren, Z., Zhang, T., Huang, Z.: Enriching API documentation with code samples and usage scenarios from crowd knowledge. IEEE Trans. Softw. Eng. **47**(6), 1299–1314 (2021)
17. Kwapong, B., Fletcher, K.: A knowledge graph based framework for web API recommendation. In: 2019 IEEE World Congress on Services, pp. 115–120. IEEE, Milan, Italy (2019)
18. Wen, M., Wu, R., Cheung, S.C.: Locus: locating bugs from software changes. In: Proceedings of the 31st IEEE/ACM International Conference on Automated Software Engineering, pp. 262–273. ACM, Singapore Singapore (2016)
19. McMillan, C., Grechanik, M., Poshyvanyk, D., Xie, Q., Fu, C.: Portfolio: finding relevant functions and their usage. In: Proceedings of the 33rd International Conference on Software Engineering, pp. 111–120. IEEE, Minneapolis, MN, USA (2011)
20. Berant, J., Chou, A., Frostig, R., Liang, P.: Semantic parsing on freebase from question-answer pairs. In: Proceedings of the 2013 Conference on Empirical Methods in Natural Language Processing, pp. 1533–1544. Association for Computational Linguistics, Stroudsburg, PA, United States (2013)

Intelligence Quotient Scores Prediction in rs-fMRI via Graph Convolutional Regression Network

Hao Zhang[1], Ran Song[1,4(✉)], Dawei Wang[2,4], Liping Wang[3], and Wei Zhang[1,4]

[1] School of Control Science and Engineering, Shandong University, Jinan, China
ransong@sdu.edu.cn
[2] Department of Radiology, Qilu Hospital of Shandong University, Jinan, China
[3] School of Information Science and Engineering, Shandong Normal University, Jinan, China
[4] Institute of Brain and Brain-Inspired Science, Shandong University, Jinan, China

Abstract. The intelligence quotient (IQ) scores prediction in resting-state functional magnetic resonance imaging (rs-fMRI) imagery is an essential biomarker in understanding autism spectrum disorder (ASD)' mechanisms and in diagnosing and treating the disease. However, existing intelligence quotient prediction methods often produce unsatisfactory results due to the complexity of brain functional connections and topology variations. Besides, the important brain regions which contribute most to IQ predictions are often neglected for priority extraction. In this paper, we propose a novel Graph Convolutional Regression Network for IQ prediction that consists of an attention branch and a global branch, which can effectively capture the topological information of the brain network. The attention branch can learn the brain regions' importance based on a self-attention mechanism and the global branch can learn representative features of each brain region in the brain by multilayer GCN layers. The proposed method is thoroughly validated using ASD subjects and neurotypical (NT) subjects for full-scale intelligence and verbal intelligence quotient prediction. The experimental results demonstrate that the proposed method generally outperforms other state-of-the-art algorithms for various metrics.

Keywords: rs-fMRI · Intelligence quotient prediction · GCN · Attention

1 Introduction

Autism spectrum disorder is a condition related to brain development that impacts how a person perceives and socializes with others, causing problems in social interaction and communication. Measuring a person's IQ can help diagnose this disorder and decoding how intelligence is engrained in the human brain is vital in the understanding of ASD [1,5].

L. Fang et al. (Eds.): CICAI 2022, LNAI 13606, pp. 477–488, 2022.
https://doi.org/10.1007/978-3-031-20503-3_38

As a common non-invasive imaging technique, resting-state functional magnetic resonance imaging (rs-fMRI) can capture the interactions between different brain regions and may find diagnostic biomarkers [28]. Based on rs-fMRI data, the changes in brain structural and functional connectivity of subjects with diseases can be shown. Functional connectivity networks are important fingerprints that may provide insights into individual variability in brain activation patterns in IQ prediction [29].

In the last two decades, we have witnessed the rapid development of clinical score regression using rs-fMRI images [9,18,30]. However, the intelligence quotient prediction in ASD is a less explored topic. Research about investigating how neural correlates of intelligence scores are altered by ASD is almost absent. On the one hand, most techniques require elaborate design, which depends on heavily on domain knowledge [14,23,24]. On the other hand, the complexity of brain functional connections and topology variations may lead to suboptimal performance of IQ prediction. As a result, it is desirable to design a fully automated IQ prediction method.

There are many handcrafted IQ prediction methods that have been proposed. For example, Park et al. [17] proposed to use a linear model to predict IQ scores using the results of the connectivity analysis, which is obtained by group independent component analysis (ICA) and weighted degree values in brain networks. Dryburgh et al. [5] proposed to predict intelligence scores from brain functional connectomes using a CPM protocol [22], including (i) connectional feature selection, feature summarization, (iii) model building, and (iv) model evaluation. However, these existing handcrafted methods generally treat feature selection and regression score as two separate steps. Besides, it is challenging to extract high-level topological features from the brain connection network accurately, which may lead to suboptimal performance of IQ prediction.

In recent years, the method based on deep learning has demonstrated excellent performance in medical image analysis [21,26,27,31]. In particular, Graph neural networks (GNN) provide an efficient way to learn depth graph structures for non-Euclidean data (such as functional brain networks), and it is rapidly becoming an advanced method that can improve performance in a variety of neural science tasks [15]. The nodes of the graph can be defined as the brain regions of interest (ROIs), and the edges of the graph can be defined as functional connections between these ROIs. GNNs can combine brain ROI regions with graph representation by explicitly capturing the topological information of brain networks, which are suitable for analyzing the relationship between brain regions. For example, Hanik et al. [7] designed a regression GNN model for predicting IQ scores from brain connectivity in rs-fMRI images. Huang et al. [8] proposed a spatio-temporal directed acyclic graph learning method to predict IQ scores using rs-fMRI images. However, they neglected to prioritize the extraction of important brain regions which contribute most to IQ prediction. Recent findings have shown that some ROIs are more important than others in the prediction of IQ scores [19].

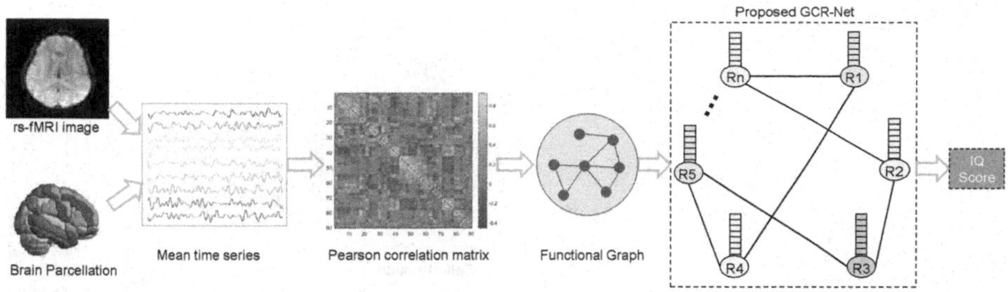

Fig. 1. The pipeline of the proposed framework. The rs-fMRI image is parcellated by an atlas and transferred to a functional graph. Then the constructed graph is input to the proposed GNN to predict the IQ score.

To this end, we propose a novel Graph Convolutional Regression Network (GCR-Net) for IQ prediction with rs-fMRI in an end-to-end fashion, which consists of an attention branch and a global branch. The attention branch can learn the brain regions' importance based on a self-attention mechanism, and then encode it in the embedding vector properly. The global branch can learn representative features of each brain region in the brain by utilizing multilayer GCN layers to aggregate the features of its neighboring vertices. Combined with the branch and global branch, brain region features can be fully learned and accurately realize the predictions of the intelligence quotient.

2 Methodology

2.1 Pipeline

Figure 1 illustrates the pipeline of the proposed framework. For each subject, we construct the brain graph based on the rs-fMRI image of the subject. First, the brain of the subject depicted by the rs-fMRI image is parcellated into 116 ROIs with Automated Anatomical Labeling (AAL) atlas [25], then the mean time series are extracted from each brain ROI. Next, the Pearson correlation [2] is conducted to calculate the pairwise associations between the time series of brain parcels, representing the functional connectivity network $\mathbf{A} \in \mathbb{R}^{n \times n}$ (i.e., correlation matrix):

$$Q\left(r_i, r_j\right) = \frac{\operatorname{Cov}\left(v_i, v_j\right)}{\sigma_{v_i}, \sigma_{v_j}}. \tag{1}$$

where $\operatorname{Cov}(v_i, v_j)$ denotes the cross-covariance between v_i (ROI_i) and v_j (ROI_j), and σ_v is the standard deviation of v. n is the number of the ROIs in the brain. Therefore, the brain graph $G = \{\mathbf{V}, \mathbf{A}\}$ is constructed where each node corresponds to an ROI, i.e. $\mathbf{V} = \{v_1, \ldots, v_n\}$. The constructed graph is input to the proposed GCR-Net to predict the IQ scores.

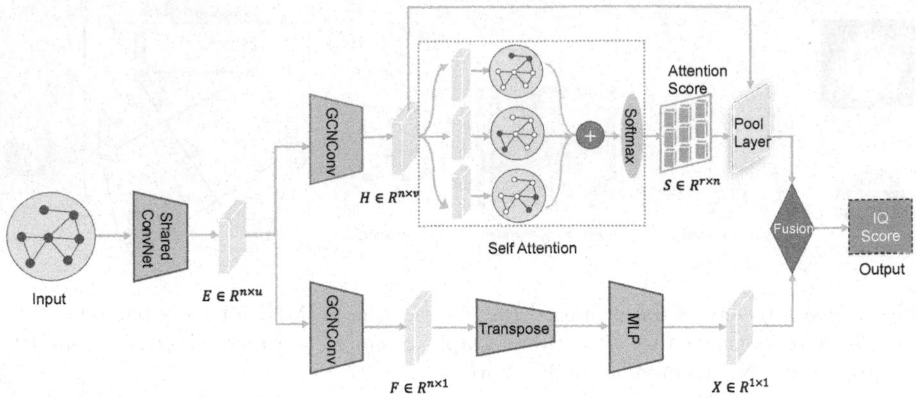

Fig. 2. The framework of the proposed GCR-Net consists of an attention branch and a global branch.

2.2 Proposed Network

Some recent studies have shown that some ROIs are more strongly associated with suffering from ASD [19]. Therefore, it is essential to focus on these important brain areas. To this end, we propose a new framework, GCR-Net, to extract the discriminative node features and selects the important nodes in the brain graph. As shown in Fig. 2, the proposed GCR-Net consists of an attention branch and a global branch. The attention branch can learn the brain regions important in the brain based on a self-attention mechanism. The global branch can learn representative features and output which brain regions contribute most to IQ prediction.

First of all, the shared ConvNet, consisting of a graph convolution (GC) layer and a dropout layer, takes the constructed graph G as input and produces the corresponding embeddings $\mathbf{E} \in \mathbb{R}^{n \times u}$. u represents the number of channels of a node feature. Here GC layer can be considered as a Laplacian smoothing operator for node features and can smooth each node's features over the graph's topology. Specifically, the l-th ($l \in \{1, 2\}$) GC layer with ReLU activation can be defined referring to [12] as:

$$\begin{aligned} \mathbf{X}^{(l)} &= \mathrm{ReLU}(\mathrm{GC}(\mathbf{X}, \mathbf{A}) \\ &= \mathrm{ReLU}\left(\mathbf{D}^{-\frac{1}{2}}\mathbf{A}\mathbf{D}^{-\frac{1}{2}}\mathbf{X}^{(l-1)}\mathbf{W}^{(i)}\right), \end{aligned} \tag{2}$$

where $\mathbf{D} = \mathrm{diag}\left(\sum_j \mathbf{A}(1, j), \sum_j \mathbf{A}(2, j), \ldots, \sum_j \mathbf{A}(n, j)\right)$ is the degree matrix of $\mathbf{A}$ and $\mathbf{W}^{(l)}$ denotes a trainable weight matrix of the l-th layer. $\mathbf{X}^{(l)} = \left[x_1^{(l)}, x_2^{(l)}, \ldots, x_n^{(l)}\right]^\top$ denotes the node representation of the l-th GC layer. As initialization we choose $\mathbf{A}^{(0)}$ to be equal to identity matrix $\mathbf{I}$. The output $\mathbf{E} = \mathbf{X_1}$ of shared ConvNet is then input into the two branches attention branch and global branch respectively.

In the attention branch, we firstly get a set of representations $\mathbf{H} \in \mathbb{R}^{n \times v}$ for nodes in the graph by a GC layer, where v is the number of channels of a node feature. Our goal is to obtain the important representation of nodes, while $\mathbf{H}$ can not provide the importance of nodes, which is variable in size. In other words, its size is still determined by the number of N nodes, so next, we use the self-attention mechanism to learn the importance of nodes and encode them into a unified graph representation, with the size invariant:

$$S = \text{softmax}\left(\mathbf{W}_{s2} \tanh\left(\mathbf{W}_{s1}\mathbf{H}^{\top}\right)\right), \tag{3}$$

where $\mathbf{W}_{s1} \in \mathbb{R}^{d \times v}$ and $\mathbf{W}_{s2} \in \mathbb{R}^{r \times d}$ denote two trainable weight matrices generated by two fully connected layers. The purpose of introducing the matrix $\mathbf{W}_{s1}$ is to linearly transform the node representation from v-dimensional space to d-dimensional space, and then the tanh function is used to increase the nonlinear ability of the neural network. $\mathbf{W}_{s2}$ can be regarded as r views to infer the importance of each node in the graph. Then the softmax function is applied to derive standardized importance of each node in the graph, which means in each view the summation of all the node importance is 1. Thus, the attention score $\mathbf{S} \in \mathbb{R}^{r \times n}$ is obtained and then sent to a pool module to obtain the prediction score $\hat{\mathbf{X}} \in \mathbb{R}^{1 \times 1}$ of this branch.

Specifically, the pooling module includes matrix multiplication and maximum pooling. We first compute the graph representation $e \in \mathbb{R}^{r \times v}$ by multiplying $\mathbf{S}$ with $\mathbf{H}$:

$$e = \mathbf{SH}. \tag{4}$$

Then we use max pooling to obtain the value with the largest weight to represent the IQ score.

In the global branch, to get the global information of the graph, we first utilize a GC layer with one output channel to generate a global feature $\mathbf{F} \in \mathbb{R}^{n \times 1}$:

$$\mathbf{F} = \text{GC}(\mathbf{E}, \mathbf{A}), \tag{5}$$

where n is the number of nodes in the graph, 1 is the feature dimension of each node. Next, to represent the feature dimension of the graph with n, an additional transpose operation is required for feature dimension transformation. Taking the transposed features as the current global feature, we then use a multi-layer perceptron (MLP) layer with one output channel to get the prediction score $\mathbf{X} \in \mathbb{R}^{1 \times 1}$, which is the same size as the output of the attention branch. To sum up, the operation is expressed below:

$$\mathbf{X} = \mathbf{W}_{s3}\mathbf{F}^{\top}, \tag{6}$$

where $\mathbf{W}_{s3}$ denote the trainable weight parameters generated by MLP layer. Then, $\mathbf{X}$ and $\hat{\mathbf{X}}$ is fused by an element-wise mean operation to obtain the final prediction score.

3 Experimental Results

3.1 Experimental Settings

Our GCR-Net was implemented on a PyTorch framework with a single GPU (RTX 3090). Adaptive moment estimation (Adam) [11] was employed for network optimization. The initial learning rate was set to 0.001, with a weight decay of 0.0005. A poly learning rate policy [32] with a power 0.9 was used, and the maximum epoch was 100. We adopt the mean square error (MSE) loss function for constraining the proposed GCR-Net. In addition, u, v, and r are experimentally set to 64, 2, and 4 respectively.

Dataset: A publicly available multi-site rs-fMRI dataset, Autism Brain Imaging Data Exchange (ABIDE)[1], was used in our work. The data was preprocessed using four different preprocessing pipelines, including Connectome Computation System (CCS), Configurable Pipeline for the Analysis of Connectomes (CPAC), Data Processing Assistant for Resting-State fMRI (DPARSF), Neuroimaging Analysis Kit (NIAK). Due to the possible biases caused by different sites, we used a randomly sampled sub-dataset, which consists of 226 NT subjects (with mean age $= (15 \pm 3.6)$) and 202 ASD subjects (with mean age $= (15.4 \pm 3.8)$). The same sets were also used by [5,7]. FIQ and VIQ scores in the NT cohort have means 111.573 ± 12.056 and 112.787 ± 12.018, whereas FIQ and VIQ scores in the ASD cohort have means 106.102 ± 15.045 and 103.005 ± 16.874, respectively.

Evaluation Metrics: To verify the generality and robustness of the proposed method, 3-fold stratified cross-validation was used for the evaluation on the prediction FIQ and VIQ in NT and ASD subjects respectively, where each fold is used once in turn for the test while the remaining 2 folds form the training set. The results of the regression are averaged over the 3 times of tests. To facilitate better observation and objective evaluation of the IQ prediction method, we adopted two metrics including the mean absolute error (MAE) and the root mean squared error (RMSE):

$$\text{MAE} = \frac{1}{m}\sum_{i=1}^{m} |y_i - \hat{y}_i|, \tag{7}$$

$$\text{RMSE} = \sqrt{\frac{1}{m}\sum_{i=1}^{m} (y_i - \hat{y}_i)^2}, \tag{8}$$

where $\widehat{y}$ and y represent the predicted results and the ground-truth labels, respectively. m is the number of samples. They are always positive, and a lower value indicates better performance. MSE is highly biased toward higher values, while RMSE is better in terms of reflecting performance when dealing with large error values.

[1] http://preprocessed-connectomes-project.org/abide.

3.2 Performance Evaluation

In this section, we compare our proposed method with five different methods, including three state-of-the-art general duty GNN-based methods (PNA [3], k-GNN [16], GraphSAGE [6]) and two other IQ score regression methods (CPM [22], and RegGNN [7]). Note that PNA is implemented in two versions: sum aggregation and identity scaling only (denoted by PNA-S); various aggregation (sum, mean, var, and max) and scaling (identity, amplification, and attenuation) methods (denoted by PNA-V), which is detailed in [3]. The codes of these comparison methods are available online[2,3,4,5,6] and the open-source code of these methods were used to train the model.

Regression Results on the ASD Subjects. The qualitative results in ASD subjects are presented in Table 1. Our method achieves the best regression performance in terms of all metrics for the prediction of both FIQ score and VIQ score, which demonstrates the advantage of GCR-Net with both attention branch and global branch to build optimal topological representations of the brain graph. In particular, the regression performance of the task-related method, CPM, is significantly lower than ours, which shows that due to the inherent complexity of the rs-fMRI data, it is difficult to extract representative features from them through hand engineering.

Table 1. Comparison of regression methods on the ASD subjects. The ↓ denotes that the lower, the better.

Method	FIQ		VIQ	
	MAE ↓	RSME ↓	MAE ↓	RMSE ↓
CPM [22]	12.533	15.965	14.171	18.834
PNA-S [3]	17.162	22.023	19.955	25.848
PNA-V [3]	15.671	20.085	22.518	28.804
k-GNN [16]	17.264	22.034	18.146	23.047
GraphSAGE [6]	13.451	16.873	13.504	17.853
RegGNN [7]	12.564	15.624	13.090	17.250
Ours	**12.349**	**15.389**	**13.003**	**17.120**

[2] https://github.com/esfinn/cpm_tutorial.
[3] https://github.com/lukecavabarrett/pna.
[4] https://github.com/basiralab/RegGNN.
[5] https://github.com/chrsmrrs/k-gnn.
[6] https://github.com/williamleif/GraphSAGE.

Table 2. Comparison of regression methods on the NT subjects. The ↓ denotes that the lower, the better.

Method	FIQ		VIQ	
	MAE ↓	RSME ↓	MAE ↓	RMSE ↓
PNA-S [3]	17.217	21.008	12.838	16.130
PNA-V [3]	20.109	25.113	14.695	18.903
CPM [22]	9.672	12.440	9.517	**12.049**
k-GNN [16]	14.490	17.919	14.900	19.585
GraphSAGE [6]	10.152	12.563	11.681	14.594
RegGNN [7]	9.768	12.270	10.195	13.044
Ours	**9.489**	**12.019**	**9.054**	12.714

Regression Results on the NT Subjects. In this section, we evaluate the proposed GCR-Net on NT subjects. The qualitative results are presented in Table 2. The GCR-Net achieves 9.489, 12.019 best MAE, and RMSE values for the prediction of FIQ and 9.054 best MAE value for the prediction of VIQ, which indicates that the high-level topological information of the brain connectivity extracted by GCR-Net is useful for regressing IQ scores. We observed that although the machine learning method CPM outperformed other GNN-based methods, our GCR-Net outperformed CPM for all tasks according to MAE and RMSE, with the exception for the VIQ (RMSE) task. Such results demonstrate the ability of the GCR-Net to extract high-level topological features from the brain connection network.

3.3 Biomarker Detection

To better understand which brain regions contribute most to the FIQ score and FIQ score regression on both NT and ASD subjects, we investigate the model weights output by global branch, which are important for delivering discriminative features. We assume that the more accurate the classifier, the more reliable the biomarker will be found. We choose the best model for interpretability analysis. The five brain regions most important for regression FIQ and FIQ on ASD subjects and NT subjects are shown in Fig. 3 and Fig. 4 respectively.

For regression FIQ scores on ASD subjects, the top five most weighted brain regions are precuneus_L (PCUN.L), paracentral_lobule_R (PCL.R), cerebellum_7b_L (CRBL7b.L), cerebellum_10_L (CRBL10.L), and vermis_1_2 (Vermis12). For regression VIQ scores on ASD subjects, the top five most weighted brain regions are insula_L (INS.L), insula_R (INS.R), putamen_L (PUT.L), hippocampus_L (HIP.L), and fusiform_R (FFG.R). The precuneus is involved in many high-level cognitive functions, such as episodic memory, self-related information processing, and various aspects of consciousness. The function of the hippocampus is responsible for most recent memory in humans. These brain regions emphasized in this study are related to the neurological functions of

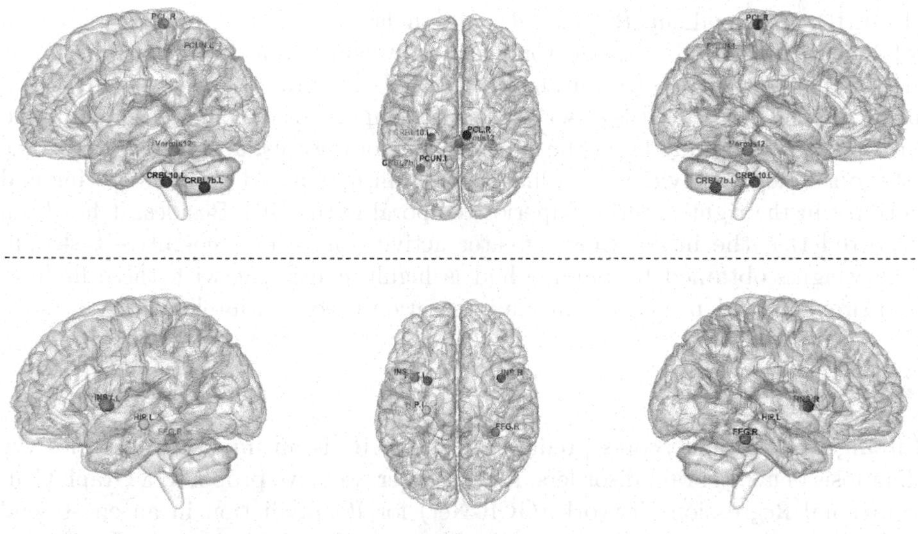

Fig. 3. The 5 brain regions most important for regression FIQ score (row 1) and VIQ score (row 2) on ASD subjects according to the output of the learned weight by GCR-Net.

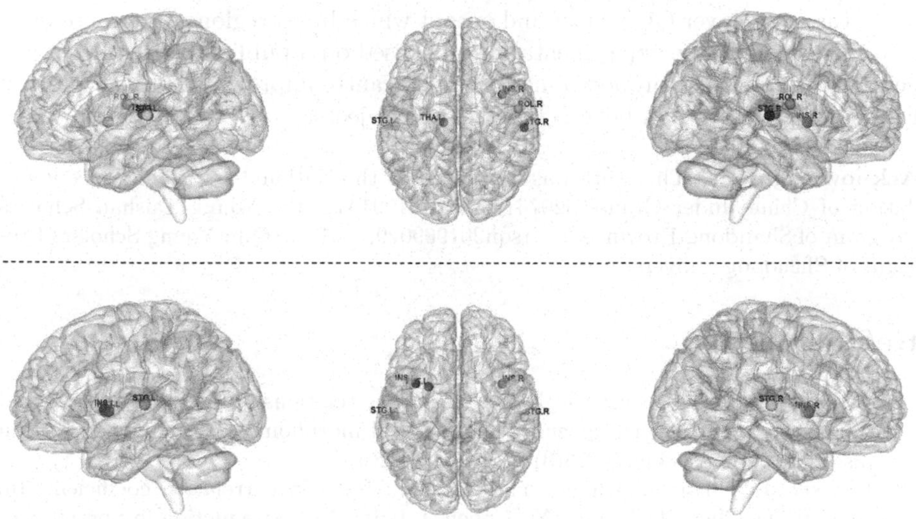

Fig. 4. The 5 brain regions most important for regression FIQ score (row 1) and VIQ score (row 2) on NT subjects according to the output of the learned weight by GCR-Net.

social communication, perception, and execution, which are highly consistent to those related to IQ scores [13,20].

For regression FIQ scores on NT subjects, the top five most weighted brain regions are insula_R (INS.R), temporal_sup_L (STG.L), rolandic_oper_R

(ROL.R), temporal_sup_R (STG.R) and thalamus_L (THA.L). For regression VIQ scores on NT subjects, the top five most weighted brain regions are insula_R (INS.R), insula_L (INS.L), putamen_L (PUT.L), temporal_sup_L(STG.L), temporal_sup_R (STG.R). It is observed that the important regions for IQ prediction in NT subjects mainly lie in the superior temporal gyrus, and insula areas. The previous study has evidenced a link between insight-based problem solving and activity in the right anterior superior-temporal gyrus [10]. Besides, it has been reported that the insula shows greater activity in various cognitive tasks [4]. The weights obtained by our method is highly consistent with these findings and provides evidence on the insula integration theory of intelligence.

4 Conclusion

Intelligence quotient scores prediction in rs-fMRI is an important step for the diagnosis of many brain disorders. In this paper, we have proposed a Graph Convolutional Regression Network (GCR-Net) for IQ prediction in an end-to-end fashion, which consists of an attention branch and a global branch. The former can learn the brain regions important in the brain based on a self-attention mechanism and pays more attention to regions that are important for IQ prediction. The latter can learn representative features of each brain region in the brain by using the multi-layer GCN layer and output which brain regions contribute most to IQ prediction. The experimental results based on a publicly available dataset demonstrate that the proposed method significantly improves the FIQ score and FIQ score prediction on both NT and ASD subjects.

Acknowledgment. This work was supported by the National Natural Science Foundation of China under Grants 62076148 and 61991411, the Young Taishan Scholars Program of Shandong Province No. tsqn201909029, and the Qilu Young Scholars Program of Shandong University.

References

1. Bedford, S.A., et al.: Large-scale analyses of the relationship between sex, age and intelligence quotient heterogeneity and cortical morphometry in autism spectrum disorder. Mol. Psychiatry **25**(3), 614–628 (2020)
2. Benesty, J., Chen, J., Huang, Y., Cohen, I.: Pearson correlation coefficient. In: Benesty, J., Chen, J., Huang, Y., Cohen, I. (eds.) Noise Reduction in Speech Processing, vol. 2, pp. 1–4. Springer, Heidelberg (2009). https://doi.org/10.1007/978-3-642-00296-0_5
3. Corso, G., Cavalleri, L., Beaini, D., Liò, P., Veličković, P.: Principal neighbourhood aggregation for graph nets. Adv. Neural. Inf. Process. Syst. **33**, 13260–13271 (2020)
4. Critchley, H.D., et al.: The functional neuroanatomy of social behaviour: changes in cerebral blood flow when people with autistic disorder process facial expressions. Brain **123**(11), 2203–2212 (2000)
5. Dryburgh, E., McKenna, S., Rekik, I.: Predicting full-scale and verbal intelligence scores from functional connectomic data in individuals with autism spectrum disorder. Brain Imaging Behav. **14**(5), 1769–1778 (2020)

6. Hamilton, W., Ying, Z., Leskovec, J.: Inductive representation learning on large graphs. In: Advances in Neural Information Processing Systems, vol. 30 (2017)
7. Hanik, M., Demirtaş, M.A., Gharsallaoui, M.A., Rekik, I.: Predicting cognitive scores with graph neural networks through sample selection learning. Brain Imaging Behav. 1–16 (2021)
8. Huang, S.G., Xia, J., Xu, L., Qiu, A.: Spatio-temporal directed acyclic graph learning with attention mechanisms on brain functional time series and connectivity. Med. Image Anal. **77**, 102370 (2022)
9. Huang, Z., et al.: Parkinson's disease classification and clinical score regression via united embedding and sparse learning from longitudinal data. IEEE Trans. Neural Netw. Learn. Syst. (2021)
10. Jung-Beeman, M., et al.: Neural activity when people solve verbal problems with insight. PLoS Biol. **2**(4), e97 (2004)
11. Kingma, D.P., Ba, J.: Adam: a method for stochastic optimization. arXiv preprint arXiv:1412.6980 (2014)
12. Kipf, T.N., Welling, M.: Semi-supervised classification with graph convolutional networks. arXiv preprint arXiv:1609.02907 (2016)
13. Li, D., Karnath, H.O., Xu, X.: Candidate biomarkers in children with autism spectrum disorder: a review of MRI studies. Neurosci. Bull. **33**(2), 219–237 (2017)
14. Li, T., et al.: Pot-GAN: pose transform GAN for person image synthesis. IEEE Trans. Image Process. **30**, 7677–7688 (2021)
15. Li, X., et al.: BrainGNN: interpretable brain graph neural network for fMRI analysis. Med. Image Anal. **74**, 102233 (2021)
16. Morris, C., et al.: Weisfeiler and leman go neural: higher-order graph neural networks. In: Proceedings of the AAAI Conference on Artificial Intelligence, vol. 33, pp. 4602–4609 (2019)
17. Park, B.Y., Hong, J., Lee, S.H., Park, H.: Functional connectivity of child and adolescent attention deficit hyperactivity disorder patients: correlation with IQ. Front. Hum. Neurosci. **10**, 565 (2016)
18. Peraza, L.R., et al.: fMRI resting state networks and their association with cognitive fluctuations in dementia with Lewy bodies. NeuroImage Clin. **4**, 558–565 (2014)
19. Plitt, M., Barnes, K.A., Wallace, G.L., Kenworthy, L., Martin, A.: Resting-state functional connectivity predicts longitudinal change in autistic traits and adaptive functioning in autism. Proc. Natl. Acad. Sci. **112**(48), E6699–E6706 (2015)
20. Press, C., Weiskopf, N., Kilner, J.M.: Dissociable roles of human inferior frontal gyrus during action execution and observation. Neuroimage **60**(3), 1671–1677 (2012)
21. Ronneberger, O., Fischer, P., Brox, T.: U-Net: convolutional networks for biomedical image segmentation. In: Navab, N., Hornegger, J., Wells, W.M., Frangi, A.F. (eds.) MICCAI 2015. LNCS, vol. 9351, pp. 234–241. Springer, Cham (2015). https://doi.org/10.1007/978-3-319-24574-4_28
22. Shen, X., et al.: Using connectome-based predictive modeling to predict individual behavior from brain connectivity. Nat. Protocols **12**(3), 506–518 (2017)
23. Song, R., Zhang, W., Zhao, Y., Liu, Y.: Unsupervised multi-view CNN for salient view selection and 3D interest point detection. Int. J. Comput. Vision **130**(5), 1210–1227 (2022)
24. Song, R., Zhang, W., Zhao, Y., Liu, Y., Rosin, P.L.: Mesh saliency: an independent perceptual measure or a derivative of image saliency? In: Proceedings of the IEEE/CVF Conference on Computer Vision and Pattern Recognition, pp. 8853–8862 (2021)

25. Tzourio-Mazoyer, N., et al.: Automated anatomical labeling of activations in SPM using a macroscopic anatomical parcellation of the MNI MRI single-subject brain. Neuroimage **15**(1), 273–289 (2002)
26. Xia, L., et al.: A nested parallel multiscale convolution for cerebrovascular segmentation. Med. Phys. **48**(12), 7971–7983 (2021)
27. Xia, L., et al.: 3D vessel-like structure segmentation in medical images by an edge-reinforced network. Med. Image Anal. 102581 (2022)
28. Xiao, L., et al.: Multi-hypergraph learning-based brain functional connectivity analysis in fMRI data. IEEE Trans. Med. Imaging **39**(5), 1746–1758 (2019)
29. Yao, D., et al.: A mutual multi-scale triplet graph convolutional network for classification of brain disorders using functional or structural connectivity. IEEE Trans. Med. Imaging **40**(4), 1279–1289 (2021)
30. Yoshida, K., et al.: Prediction of clinical depression scores and detection of changes in whole-brain using resting-state functional MRI data with partial least squares regression. PLoS ONE **12**(7), e0179638 (2017)
31. Zhang, H., et al.: Cerebrovascular segmentation in MRA via reverse edge attention network. In: Martel, A.L., et al. (eds.) MICCAI 2020. LNCS, vol. 12266, pp. 66–75. Springer, Cham (2020). https://doi.org/10.1007/978-3-030-59725-2_7
32. Zhao, H., Shi, J., Qi, X., Wang, X., Jia, J.: Pyramid scene parsing network. In: Proceedings of the IEEE Conference on Computer Vision and Pattern Recognition, pp. 2881–2890 (2017)

A Novel In-Sensor Computing Architecture Based on Single Photon Avalanche Diode and Dynamic Memristor

Jiyuan Zheng[1(✉)], Shaoliang Yu[2], Jiamin Wu[1,3], Yuyan Wang[1], Chenchen Deng[1], and Zhu Lin[1]

[1] Beijing National Research Center for Information Science and Technology, Tsinghua University, Beijing 100084, China
zhengjiyuan@tsinghua.edu.cn
[2] Research Center for Intelligent Optoelectronic Computing, Zhejiang Lab, Hang Zhou 311121, China
[3] Department of Automation, Tsinghua University, Beijing 100084, China

Abstract. In this paper, A novel In-Sensor computing architecture is proposed and the fundamental device is demonstrated by experiment. Single photon avalanche diode integrated with dynamic memristor functions as the basic neuron and single photon level visible light is proved to be capable to manipulate the weight of optoelectronic neurons, indicating that the novel architecture is highly energy efficient and sensitive.

Keywords: Opto-electronic devices · Artificial intelligence · Solid-state materials

1 Introduction

The escalating development of artificial intelligence (AI) greatly challenges the available computing hardware resources like CPU, and GPU in their performance of computing power and energy efficiency. With Moore's Law meeting with its upper bond limit, it has become an international common sense that the computing hardware's architecture needs to be reformed. Many successful examples like Google TPU, intel Loihi, and IBM True north have attracted great attention. AI algebra has inspired the design of many computing hardware.

Enabled by a massive interconnections of neurons, human retina and visual cortex are capable to sense and manipulate complex image information. In retina, light sensitive neurons preprocess the image information by passing the signal to the next stage of neurons. Preprocessing includes image quality enhancement, image feature extraction, and object recognition. The information is processed parallelly and energy efficiently.

Supported by the Ministry of Science and Technology of China under contract Nos. 2021ZD0109900 and 2021ZD0109903 and National Natural Science Foundation of China No. 62175126.

L. Fang et al. (Eds.): CICAI 2022, LNAI 13606, pp. 489–500, 2022.
https://doi.org/10.1007/978-3-031-20503-3_39

The commonly used AI imaging hardware system, however, don't function like human visual system, especially in the signal readout stage. Signals In-Sensors like charge-coupled device (CCD) arrays and complementary metal oxide semiconductor (CMOS) arrays are read out row by row or column by column, then they are transited to memory and processing units putting in series with the sensors through bus lines.

Most imaging processing applications generate large amount of redundant data and the useless data generated by sensors could reach more than 90% [1]. Inspired by the human vision system and by shifting some processing tasks to sensors, researchers allow in situ computing and reducing data movement. Aiming to imitate the retina's preprocessing function, early vision chips could only achieve low-level processing, such as image filtering and edge detection [2]. Gradually, the goal for AI vision chips shifted to high-level processing, including recognition and classification. Moreover, the development of programmable vision chips were proposed around 2006, aiming to flexibly deal with various processing scenes through software control [3]. In 2021, Liao et al. [4] summarized the principle of the biological retina and discussed developments in neuromorphic vision sensors in emerging devices. Wan et al. [5] provided an overview of the technology of electronic, optical, and hybrid optoelectronic computing for neuromorphic sensory computing.

2 Related Work

Photon imaging sensors are widely applied in microscopy, astronomy, biomedical imaging, quantum computing and other high technology domains. Different from conventional imaging sensors, photon imaging sensor has ultra-high sensitivity that can detect single-photon level weak signal. It can even be used in photon number counting, two photon microscopy, biamedical fluorescence lifetime measurement, areospace radiation detection, photon time of flight analysis. In photon imaging applications, it is not abnormal that the target is small and the photon pace is unpredictable, high density pixels and high frame refresh rate are necessary to meet the requirement. However, in conventional Von-Neumann architecture, imaging devices (e.g. CCD or CMOS) and processing devices are seperated, and imaging data is transited from imaging part to processing part through bus line. The data transport occupies tramendous bandwidth and generate severe heat, so the frame rate is seriously dragged especially when the pixel scale is large. Moreover, there is lots of redundant data transporting in bus line, which further brings down the computing speed and energy efficiency. It is commonly accepted that new imaging architecture needs to be developed to fulfill the imaging development. Human retina is an in-sensor computing system that envolves sensing, memorizing, and computing capabilities, which is highly energy efficient and fast in processing image signals. The first artificial vision computing chip was proposed and demonstrated by Mead et al. in 1990s. Some simple Filtering and edge detection functions were realized. In 2006, re-programmable vision computing chips were demonstrated, which is capable to deal with different image processing missions through software control. After 2012, with AlexNet's success, vision computing chips meets with the cusp. Typically, there are two main stream in vision chip technology, In-Sensor computing and near sensor computing [2,3,6].

(1) Architectures with computing inside sensing units. In this type of architecture, by placing the photodetector directly into the analog memory and computing unit, researchers form a processing element (PE) [3,7,8]. The PEs are then developed to possess in situ sensing and to deal with the analog signals obtained by the sensors. The advantage of this type of architecture is highly parallel processing speed. However, due to the analog memory and computing unit taking up a large volume, the PEs are much larger than the sensor, which leads to a low pixel fill factor and limits the image resolution.

(2) Architectures with computing near the sensing units. Due to the low fill factor issue, most vision chips cannot incorporate in situ sensing and computing architecture. Instead, the pixel array and processing circuits are separated physically while still being connected in parallel on a chip [3,6], which makes independent design possible according to the system's requirements.

For In-Sensor computing system, each pixel is a processing unit put vertically onto a sensor. The image processing is highly parallel and fast. However, conventional analog processor is much larger than the sensor in size, which reduces the filling factor of the pixel array and brings down the imaging resolution. Small size analog processor is highly decired for In-Sensor computing. Since the low filling factor for In-Sensor computing limits its applications, near sensor computing was proposed as a compromise. Sensors and processors are put as close as possible but still seperated. The data transition speed and energy efficiency can be improved at some level but is still barely satisfied regarding the application development. To address this issue, new architecture envolving envolve material system and device structure needs to be taken into considerations.

The current vision chip only has a neuron scale of 10^2–10^3, which is much smaller than those of the retina and cortex (10^{10}).

3 Methodology

Here, we propose a novel In-Sensor computing architecture. Each pixel in the sensor array is a series connection of single photon avalanche diode (SPAD) with dynamic memristor (ARS). SPAD is a highly sensitive device that is capable to detect single-photon level weak signal [9–22], while ARS is a device capable to memorize the stimuli introduced into it [23–41]. In such an architecture, extremely weak image signal is converted into electric stimuli by the SPAD and introduced into the ARS, of which the resistance has a critical change. The change of ARS's resistance could influence the output current's magnitude, which can be regarded as computing results. Therefore, the sensor array combines functions of sensing and computing (Fig. 1).

The working principle is: 1, when there is no photon illumination, the voltage drops mainly across SPAD, the voltage over ARS is nearly 0, and the conductive

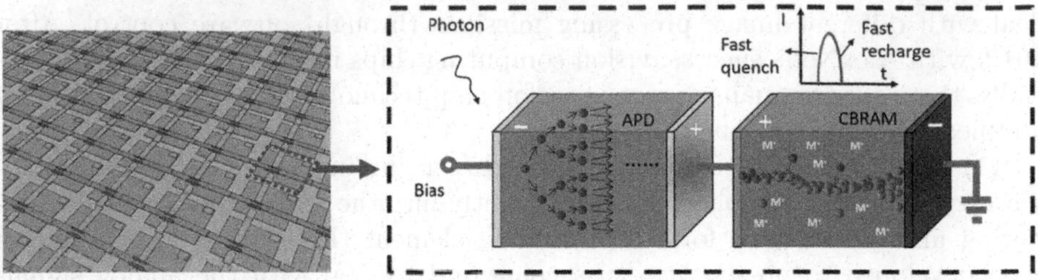

Fig. 1. Novel in-sensor computing architecture proposed in this work. Optical neuron is a series connection between SPAD and ARS.

bridge in ARS dissolves and the ARS is in its off state (high resistance). 2, When there is light illumination, avalanche happens in the SPAD, generating a voltage pulse across the ARS, which drives the conductive ions inside the ARS to drift and deposit on the electrode and form a conductive bridge. When the deposition speed is faster than the dissolving speed, the resistance of the ARS drops significantly, which reduces the R-C time constant. With a fast recharge process, the SPAD takes back the voltage share across ARS. The conductive bridge dissolves and the ARS turns back to off state, during which process, the series system generates a current pulse, of which the peak value is inversely proportional to the minimum value of the ARS's resistance R_{min}. 3, R_{min} is a dynamic parameter which is determined by the competition between forming and dissolving process of conductive bridge. R_{min} can be tuned by the light pulse repetition rate. When the repetition rate is high, the interval time between avalanche event is short, therefore the dissolving time is reduced, which makes the conductive bridge to be stronger and R_{min} smaller. On the contrary, when the repetition rate is low, the interval time between avalanche events is long, the dissolving time of conductive bridge is enlarged, which makes the conductive bridge weak and R_{min} larger. So, with a higher repetition rate light pulse (brighter light signal) to train the optical neurons, the ARS is more active and the resistance is low, which means when it is used to sense light, a higher peak electric pulse will be generated. With a lower repetition rate light pulse (weaker light signal) to train the optical neurons, the ARS is less active and the resistance is larger, which means when it is used to sense light, a lower peak electric pulse will be generated. Thus, the optical neuron can be used to memorize the feature of the light pulses, which is a basic In-Sensor computing element.

Similar to human retina, an array is designed by the optical neurons. As is shown in Fig. 2a, at checkpoint t, a predesigned light signal P_{min}(t) is introduced into the array and set the responsivity of the optical neuron at row i and column j to be R_{min}(t). As the resistance of the ARS could be temporarily kept, it can be regarded as unchanged if the inference process is fast enough. As is shown in Fig. 2b, all the neurons are parallelly connected, and the inference process generates a current pulse, $\mathrm{I(t)} = \sum_{(i,j=1)}^{m,n} S_{ij} R_{ij}(t)$. With an iteration of training and inference processes (the time interval is δ), a series sets of current outputs

$I(t + x\delta)$, (x = 1, 2 ... l) will be generated. As is shown in Fig. 2c, the generation of these current is building the first layer in neuro network, combined with a following electric network, a deep neuro network can be formed. By using this architecture, image signal can be pre-treated parallelly with high speed, which significantly compresses the signal and excludes redundant image signals. The number of nodes in hidden layers is greatly reduced.

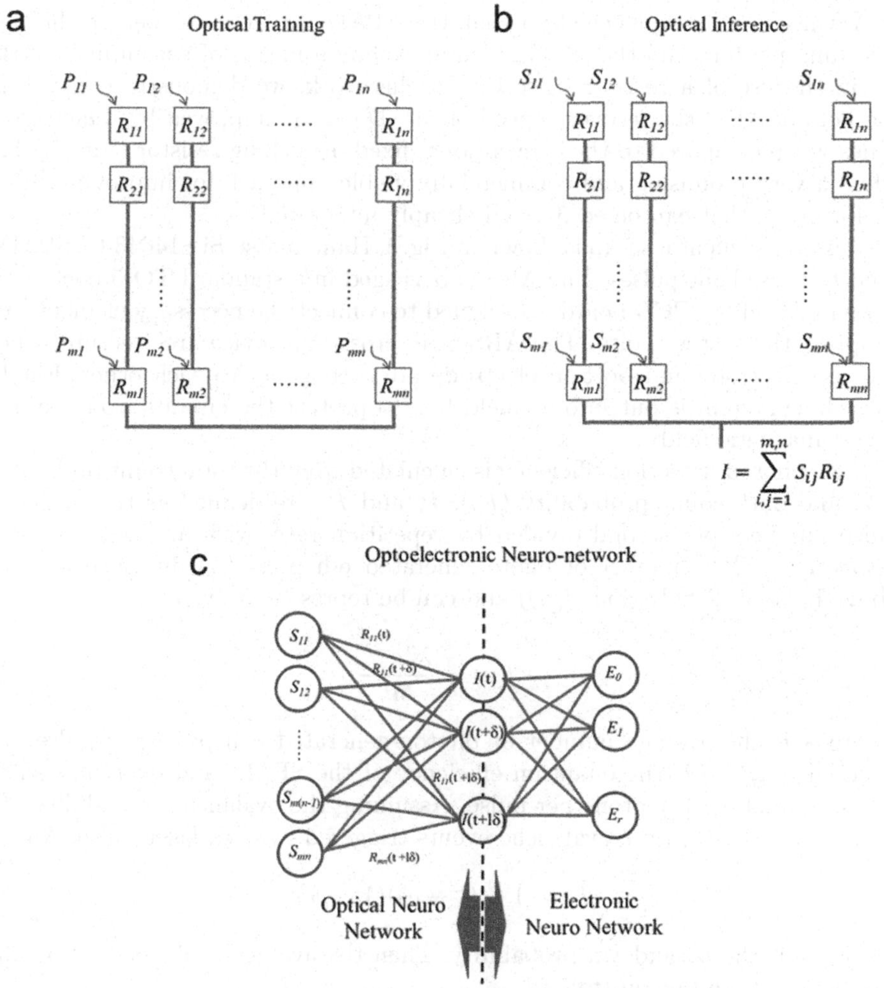

Fig. 2. Single photon sensitive in-sensor computing architecture. a) Training method. b) inference method. c) A merged optoelectronic deep neuro network.

4 Experiments

In recent research, we have primarily investigated the performance of the basic element composing a series connection of SPAD with ARS [9,39]. This work focuses on the single photon detection performance.

SPAD performance is generally unsatisfactory when detecting high repetition rate single photons (>10 MHz), especially when the optical sensing area is large (>100 µm) to ensure high sensitivity. The avalanche currents generated in a SPAD needs to be quenched so that the SPAD can reset to detect the next incoming photon, and the simplest, most common means of accomplishing this is via the use of a resistor in series. In this work, we demonstrate that with the help of ARS, the recovery speed of SPAD can be improved by a factor of 8 times when compared to the conventional fixed quenching resistor (Fig. 2). This offers a very promising and commercially viable approach for improving SPAD performance that can be easily, and cheaply integrated.

The experimental setup is shown in Fig. 3, Hamamatsu Si S14643-02 SPAD is used to sense light pulses. The ARS is packaged in a standard TO-5 package by Au wire bonding. PCB board is designed to connect the necessary elements and suppress the system noise. The ARS is in cross-bar device architecture, where the top electrode and bottom electrode are orthogonal to each other. Finally, the whole system is put into a shield box to protect the circuits from external electromagnetic field.

The photon detection efficiency is calculated from the total count probability (P_t) and dark count probability (P_d). P_t and P_d are defined as the avalanche pulse numbers per second divided by repetition rate, with and without light, respectively. The number of photo-generated e-h pairs (n) during each pulse obeys Poisson distribution ($f(n)$) and can be represented by [11]

$$f(n) = \frac{\lambda^n e^{-\lambda}}{n!} \tag{1}$$

where λ is the average number of photon-generated e–h pairs per pulse, and is equal to $\eta\overline{n}$. η is the quantum efficiency of the SPAD, and *overlinen* is the average number of photons per pulse. Assuming the avalanche probability (P_a) is the same between the avalanche events triggered by each laser pulse. And

$$P_a = 1 - (1 - p_d)(1 - p_b)^n \tag{2}$$

where p_b is the breakdown probability. Then the average avalanche probability per pulse, P_t, can be written as

$$P_t = \sum\nolimits_{n=0}^{\infty} P_a f(n) = 1 - (1 - p_d)^{-\overline{n}\eta P_b} \tag{3}$$

Therefore, the SPDE of the SPAD, which is equal with ηp_b, can be expressed as

$$SPDE = \frac{1}{n} ln(\frac{1 - P_d}{1 - P_t}) \tag{4}$$

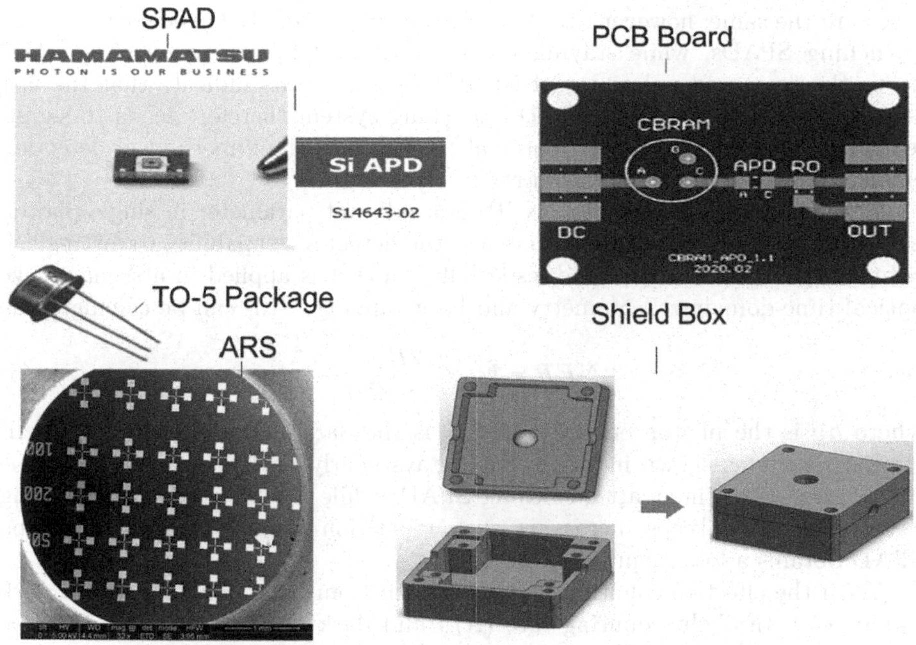

Fig. 3. Experimental setup of the basic element.

The single photon detection performance is plotted in Fig. 4. A threshold is set to get the effective counting from the avalanche response curve saved from the oscilloscope. The counting is controlled by a timer, when laser pulse comes, the timer is triggered, when the timer is running and avalanche pulse exceeds the threshold, the timer is stopped and the pulse is regarded as an effective counting. Then, the timer is reset only when the next laser pulse comes. This counting mechanism is widely used in commercial photon counting technology. Single photon data gotten for conventional quenching method is analyzed and plotted in Fig. 4. When the threshold is taken to be lower than 5 mV, the counting rate under light illumination and dark are both unreasonably high, and thus the single photon detection efficiency (SPDE) doesn't make sense. A threshold larger than 5 mV will give a reasonable counting performance. So, in this paper, the threshold for conventional quenching is taken to be 5 mV. When the threshold is larger than 5 mV, the counting rate and SPDE stay at reasonable range.

Single photon data gotten for conventional quenching method is plotted and compared in Fig. 4. When the threshold is larger than 30 mV, the counting rate doesn't change too much when varying the threshold. So, in this paper, the threshold for conventional quenching is taken to be 30 mV.

As shown in Fig. 4, the count rate of conventional quenching system saturates when the light repetition rate increases from 1 MHz to 50 MHz. In contrast, the ARS quenching response continues increasing linearly. The SPDE is then calculated and plotted in Fig. 4. At a low repetition rate, the SPDE for the two cases are the same: however the efficiencies drop off rapidly in the conventional-quenching SPADs, while staying nearly constant for smart-quenching SPAD up to the measured pulse rate of 50 MHz. These results indicate that the fast response of the ARS based SPAD quenching system therefore leads to significant advantages over the conventional passive quenching method in detection efficiency at high input repetition rates.

The noise equivalent power (NEP) is a relevant parameter in single photon detection. It is a figure of merit to reflect the detector's capability to distinguish photon signal from background, especially when it is applied in a scenario like optical time-domain reflectometry and laser ranging. NEP can be calculated as

$$NEP = hv \times \frac{(2R_{dc})^{1/2}}{SPDE} \tag{5}$$

where hv is the photon energy, and R_{dc} is the dark count rate. From experimental results as shown in Fig. 4, NEP stays nearly unchanged at a low value $(0.2 \times 10^20\frac{W}{Hz^{1/2}})$ in smart-quenching SPAD, while for conventional-quenching SPAD, the NEP keeps increasing with repetition rate. So, smart-quenching SPAD obtains a lower equivalent noise.

With the effective counting number gotten from light signal (C_L) and dark signal (C_d), the light counting rate (R_L) and dark counting rate (R_d) can be calculated by, $R_L = C_L/T_i$ and $R_L = C_L/T_i$, where T_i is the total time duration for the saved data. Then, SPDE can be calculated from Eq. 4.

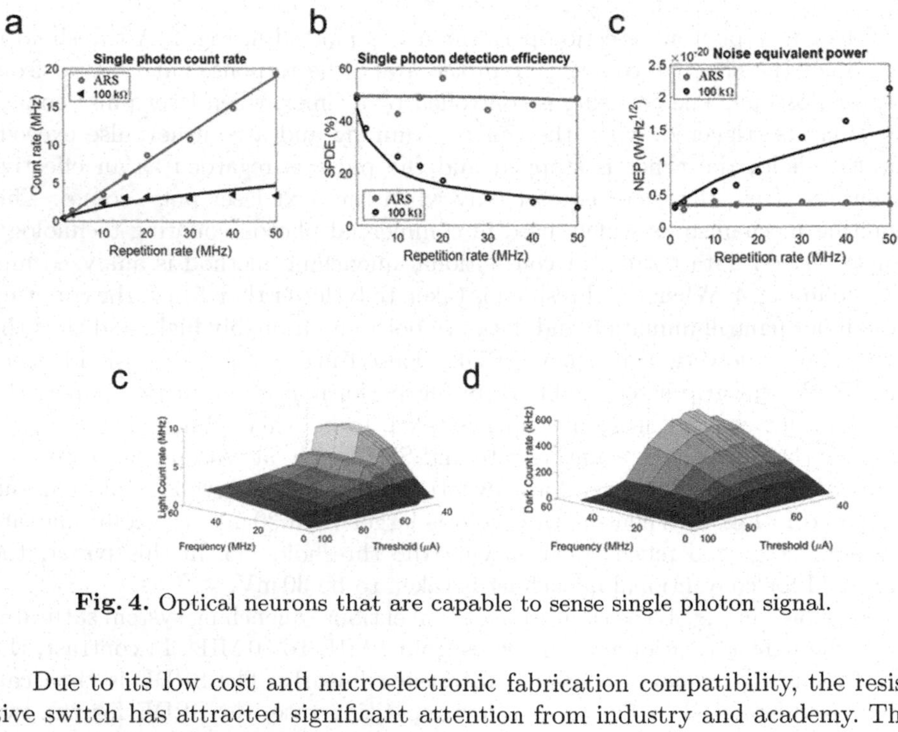

Fig. 4. Optical neurons that are capable to sense single photon signal.

Due to its low cost and microelectronic fabrication compatibility, the resistive switch has attracted significant attention from industry and academy. The resistive switch has shown great potential in neuromorphic networks, computing, and memory.

The resistive switch is a type of device that the resistance could be tuned electronically. The switching mechanism can be categorized into Phase change, Magneto-switch, Ferro-switch, Mott transition, and Filamentary switch. The resistive switch has been successfully deployed in performing logic operations, reroute signals, self-reconfigure system, rectifier, and field-programmable gate array (FPGA)-like functionality. Beyond, multi-level internal state resistive switch has shown great potential in realizing synapse emulation, which is a critical part of neuromorphic computing and network. During the past two decades, the resistive switch has demonstrated its significant advantages in compatibility with CMOS technology, cheap in cost, and stability. In this paper, the resistive switch is explored a step further into the optics field. The optical neuron in this work is demonstrated based on a quenching resistor made by resistive switch.

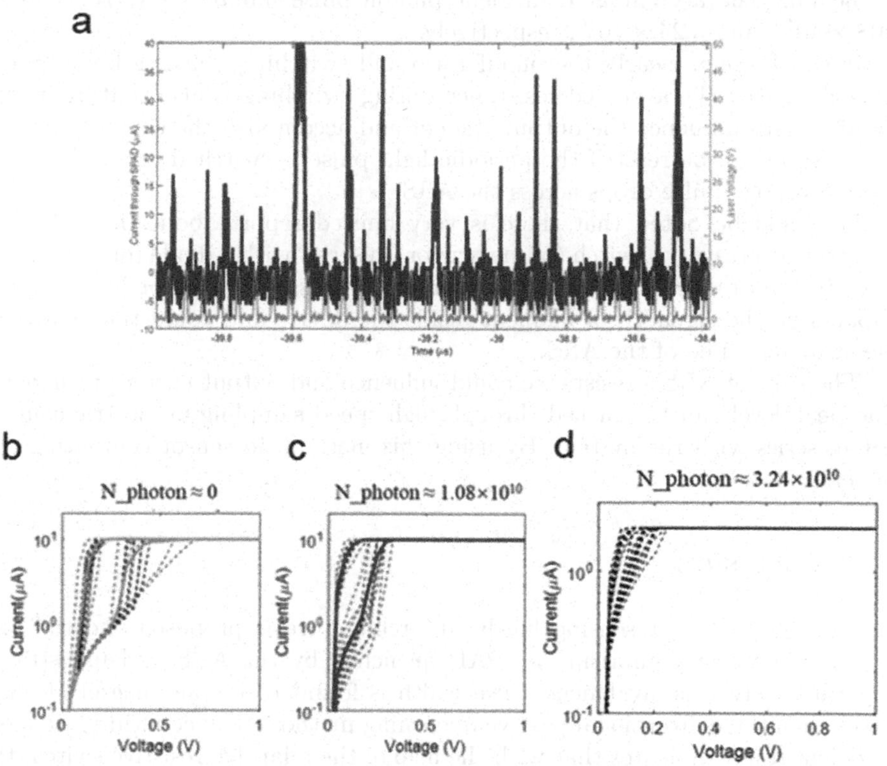

Fig. 5. Single photon detection performance of the optical neuron. The average resistance of the ARS is manipulated by photon pulse numbers.

In this paper, the resistive switch is further explored in smart sensing. It is demonstrated that this device could smartly break the R-C limit of passive quenching SPAD and yield a steep and reversible avalanche pulse, which is critical to achieving a low cost, compact and fast SPAD.

Here, a 5 nm Al_2O_3 dielectric layer sandwiched by Ag and Pt electrode ARS is used as the resistive switch to quench commercial APD. The avalanche pulse is found to be about ten times steeper, and a high repetition rate response has been significantly improved in ARS based SPAD system than in the conventional fixed resistor based SPAD system. The ARS is found to be volatile, and the detection is continuous (Fig. 5a).

Finally, as is shown in Fig. 5, the optical memorization capability of the ARS is tested. The average resistance of the ARS shifts when the incident photon number accumulates. In the experiment, 6 MHz repetition rate single photon pulse is used to test the memorizing capability of the ARS quenching system. The resistances of the ARS are tested off-site at time point of 0 s, 30 min and 1 h 30 min. The accumulated incident photon pulse numbers (N_{photon}) are 0, 1.08×10^{10}, and 3.24×10^{10}, respectively.

It can be seen clearly the on-off ratio and switching voltages both have a critical change, the averaged resistance during switching is also shifting, which could in turn influence the output current and accomplish the computing task. The system can be reset if the periodic light pulse is switched off or there is a reverse voltage pulse drops across the ARS.

It should be noted that there is very limited report about the switching behavior of a resistive switch (or memristor) in series with a diode (or capacitor), wherein the drastic charging and recharging process of capacitor impose great impacts on the strength of filament inside the device and could thus influence the switching time of the ARS.

The shift of ARS's resistance could influence and output current peak level. The peak level can be tracked through high speed sampling or electric counter put in series with the system. By using this method, in sensor computing can be realized.

5 Conclusion

In summary, a smart sensing hardware architecture is proposed and the fundamental element comprising a SPAD quenched by the ARS is demonstrated experimentally. The avalanche pulse width is found to be one magnitude narrower than the conventional passive quenching method. The counting rate measurement result indicates that with the help of the adaptive resistive switch, the response is significantly improved than the conventional method in a high repetition rate. The ARS is verified to be volatile, which is critical for SPAD to have a continuous counting capability. Moreover, the ARS is verified to have a temporary memorization capability of which the averaged resistance can be tunned by a periodic light stimuli. This SPAD and ARS series system is a fundamental element to support the in-sensor computing architecture.

References

1. Chai, Y.: In-sensor computing for machine vision. Nature **579**(7797), 32–33 (2020)
2. Zhou, F., Chai, Y.: Near-sensor and in-sensor computing. Nat Electron. **3**(11), 664–671 (2020)
3. Liu, L., Wu, N.: Artificial intelligent vision chip. Micro/Nano Electron. Intell. Manuf. **1**, 12–19 (2019)
4. Liao, F., Zhou, F., Chai, Y.: Neuromorphic vision sensors: principle, progress and perspectives. J Semicond. **42**(1), 013105 (2021)
5. Wan, T., Ma, S., Liao, F., Fan, L., Chai, Y.: Neuromorphic sensory computing. Sci. China Inf. Sci. **65**, 141401 (2022)
6. Wu, N.: Neuromorphic vision chips. Sci. China Inf. Sci. **61**, 060421 (2018)
7. Komuro, T., Kagami, S., Ishikawa, M.: A dynamically reconfigurable SIMD processor for a vision chip. IEEE J. Solid-State Circuits **39**(1), 265–8 (2004)
8. Jendernalik, W., Blakiewicz, G., Jakusz, J., Szczepanski, S., Piotrowski, R.: An analog sub-miliwatt CMOS image sensor with pixel-level convolution processing. IEEE Trans. Circuits Syst. I Regul. Pap. **60**(2), 279–289 (2013)
9. Zheng, J., et al.: Dynamic quenching of a single-photon avalanche photodetector using an adaptive resistive switch. Nat. Commun. **13**, 1517 (2022)
10. Pan, W., Zheng, J., Wang, L., Luo, Y.: A future perspective on the in-sensor computing. Engineering (2022). https://doi.org/10.1016/j.eng.2022.01.009
11. Cova, S., Ghioni, M., Lacaita, A., Samori, C., Zappa, F.: Avalanche photodiodes and quenching circuits for single-photon detection. Appl. Opt. **35**, 1956–1976 (1996)
12. Tachella, J., et al.: Real-time 3D reconstruction from single-photon lidar data using plug-and-play point cloud denoisers. Nat. Commun. **10** (2019). https://doi.org/10.1038/s41467-019-12943-7
13. Kollorz, E., Penne, J., Hornegger, J.: Gesture recognition with a time-of-flight camera. Int. J. Intel. Syst. Technol. Appl. **5**, 334–343 (2008)
14. Cui, Y., Schuon, S., Chan, D., Thrun, S., Theobalt, C.: Proceedings of IEEE Conference on Computer Vision and Pattern Recognition (CVPR, 2010), pp. 1173–1180, San Francisco, CA (2010)
15. Craddock, A.N., et al.: Quantum interference between photons from an atomic ensemble and a remote atomic ion. Phys. Rev. Lett. **123** (2019). https://doi.org/10.1103/PhysRevLett.123.213601
16. Meda, A., et al.: Quantifying backflash radiation to prevent zero-error attacks in quantum key distribution. Light Sci. Appl. **6** (2017). https://doi.org/10.1038/lsa.2016.261
17. Wengerowsky, S., Joshi, S.K., Steinlechner, F., Hubel, H., Ursin, R.: An entanglement-based wavelength-multiplexed quantum communication network. Nature **564**, 225 (2018). https://doi.org/10.1038/s41586-018-0766-y
18. Tenne, R., et al.: Super-resolution enhancement by quantum image scanning microscopy. Nat. Photon. **13**, 116 (2019). https://doi.org/10.1038/s41566-018-0324-z
19. Zhang, J., Itzler, M.A., Zbinden, H., Pan, J.W.: Advances in InGaAs/InP single-photon detector systems for quantum communication. Light Sci. Appl. **4** (2015). https://doi.org/10.1038/lsa.2015.59
20. Marano, D., et al.: Silicon photomultipliers electrical model extensive analytical analysis. IEEE Trans. Nucl. Sci. **61**, 23–34 (2014). https://doi.org/10.1109/TNS.2013.2283231

21. Bronzi, D., et al.: Fast sensing and quenching of CMOS SPADs for minimal afterpulsing effects. IEEE Photon. Tech. Lett. **25**, 776–779 (2013). https://doi.org/10.1109/LPT.2013.2251621
22. Tisa, S., Guerrieri, F., Zappa, F.: Variable-load quenching circuit for single-photon avalanche diodes. Opt. Express **16**, 2232–2244 (2008). https://doi.org/10.1364/OE.16.002232
23. Wong, H.S.P., et al.: Metal-oxide RRAM. Proc. IEEE **100**, 1951–1970 (2012). https://doi.org/10.1109/JPROC.2012.2190369
24. Fan, L.L., et al.: Growth and phase transition characteristics of pure M-phase VO2 epitaxial film prepared by oxide molecular beam epitaxy. Appl. Phys. Lett. **103** (2013). https://doi.org/10.1063/1.4823511
25. Wang, Z., et al.: Memristors with diffusive dynamics as synaptic emulators for neuromorphic computing. Nat. Mater. **16**, 101–108 (2017). https://doi.org/10.1038/NMAT4756
26. Berggren, K., et al.: Roadmap on emerging hardware and technology for machine learning. Nanotechnology **32** (2021). https://doi.org/10.1088/1361-6528/aba70f
27. Zhang, W., et al.: Neuro-inspired computing chips. Nat. Electron. **3**, 371–382 (2020). https://doi.org/10.1038/s41928-020-0435-7
28. Li, C., et al.: Long short-term memory networks in memristor crossbar arrays. Nat. Mach. Intell. **1**, 49–57 (2019). https://doi.org/10.1038/s42256-018-0001-4
29. Li, C., et al.: Analogue signal and image processing with large memristor crossbars. Nat. Electron. **1**, 52–59 (2018). https://doi.org/10.1038/s41928-017-0002-z
30. Ielmini, D., Wong, H.S.P.: In-memory computing with resistive switching devices. Nat. Electron. **1**, 333–343 (2018). https://doi.org/10.1038/s41928-018-0092-2
31. Jerry, M., et al.: 2017 IEEE International Electron Devices Meeting (2017)
32. Yang, J.J., et al.: High switching endurance in TaOx memristive devices. Appl. Phys. Lett. **97** (2010). https://doi.org/10.1063/1.3524521
33. Intel: Intel and Micron Produce Breakthrough Memory Technology (2015). https://newsroom.intel.com/news-releases/intel-and-micron-produce-breakthrough-memory-technology/gs.laqacz
34. Menzel, S., von Witzleben, M., Havel, V., Boettger, U.: The ultimate switching speed limit of redox-based resistive switching devices. Faraday Discuss. **213**, 197–213 (2019). https://doi.org/10.1039/c8fd00117k
35. Midya, R., et al.: Anatomy of Ag/Hafnia-based selectors with 1010 nonlinearity. Adv. Mater. **29** (2017). https://doi.org/10.1002/adma.201604457
36. Yu, S.M., Wong, H.S.P.: Compact modeling of conducting-bridge random-access memory (CBRAM). IEEE Trans. Electron Devices **58**, 1352–1360 (2011). https://doi.org/10.1109/TED.2011.2116120
37. Shukla, N., Ghosh, R.K., Grisafe, B., Datta, S.: 2017 IEEE International Electron Devices Meeting (2017)
38. Wang, W., et al.: Volatile resistive switching memory based on Ag Ion Drift/Diffusion Part I: numerical modeling. IEEE Trans. Electron. Devices **66**, 3795–3801 (2019). https://doi.org/10.1109/TED.2019.2928890
39. Zheng, J., et al.: Quenching of single photon avalanche photodiodes with dynamic resistive switches, vol. 11721 SI (SPIE, 2021)
40. Lanza, M., et al.: Recommended methods to study resistive switching devices. Adv. Electron. Mater. **5** (2019). https://doi.org/10.1002/aelm.201800143
41. Lin, Q., et al.: Dual-layer selector with excellent performance for cross-point memory applications. IEEE Electron. Device Lett. **39**, 496–499 (2018). https://doi.org/10.1109/LED.2018.2808465

Lane Change Decision-Making of Autonomous Driving Based on Interpretable Soft Actor-Critic Algorithm with Safety Awareness

Di Yu(✉), Kang Tian, Yuhui Liu, and Manchen Xu

Beijing Information Science and Technology University, Beijing 100192, China
yudizlg@aliyun.com

Abstract. Safe and efficient lane change behavior is indispensable and significant for autonomous driving. A new lane change decision-making scheme is proposed for autonomous driving based on Soft Actor-Critic (SAC) algorithm. Combined the kinematics information with visual image information of ego vehicle, a multi-data fusion convolutional neural network structure is constructed to improve the perception ability of the algorithm with continuous action space. Moreover, the interpretability of the algorithm is enhanced through feature refinement of surrounding vehicles and key road environment information with attention mechanism. In order to ensure driving safety and driving efficiency, the reward function is constructed comprehensively considering the driving risk assessment and competitive consciousness. In static scenario and stochastic dynamically one, the experiments are conducted in the CARLA simulator and the results demonstrate the scheme proposed is effective and feasible.

Keywords: Lane change · Autonomous driving · Attention mechanism · SAC · Safety awareness

1 Introduction

1.1 A Subsection Sample

Highly smart integration of artificial intelligence, internet and big data technology, intelligent transportation system is the fundamental way to relieve traffic congestion and reduce traffic accidents with the core of modern intelligent vehicles [1]. With respect to autonomous driving application, safe and effective decision making is crucial according to the surrounding environmental information and road rules in a complex traffic environment. Being an indispensable driving behavior for overtaking and navigation purposes, lane change decision making is important and has attracted great attention.

The methods for lane change decision making can be mainly classified into rule-based and end-to-end ones. Owing to the requirements of considering many

L. Fang et al. (Eds.): CICAI 2022, LNAI 13606, pp. 501–512, 2022.
https://doi.org/10.1007/978-3-031-20503-3_40

types of traffic rules and complex environmental factors, the rule-based decision making method is greatly limited and not effective to more traffic scenarios [2]. However, the end-to-end learning methods realizes the mapping from perception to decision-making without modeling, based on deep reinforcement learning (DRL) algorithm with the powerful perception and learning capabilities. And many excellent advances have demonstrated successful lane change behaviors with DRL [3–6]. The lane change decision making of trucks were realized with deep Q learning (DQN) algorithm with the equivalent performance to common reference models on highways [3]. Modeled as a partially observable Markov decision process (MDP) and a motion planning problem, the lane change problem was tackled by applying DQN algorithm [4]. In dense traffic scenarios, proximal policy optimization (PPO) algorithm was applied to learn continuous control strategy to achieve better lane changing and lane merging behaviors than model predictive control [5]. More interesting, based on imitation learning and reinforcement learning, lane change decision-making method was proposed to learn macro decisions from expert demonstration data and apply deep deterministic policy gradient algorithm to obtain optimization strategies [6].

Attention mechanism makes people selectively focus on important information within vision field while ignoring other visible information in cognitive science, and has been introduced into the lane change [7–10] and other tasks [11–13] of autonomous driving, because it not only reduces the calculation amount of model training, but also improves the interpretability of the algorithm to a certain extent. Combined with temporal and spatial attention, a hierarchical DRL algorithm was proposed to learn lane change behaviors in dense traffic [7]. In the highway scenarios, different self-attention mechanisms were embedded in deep reinforcement learning algorithm to further extract features of the BEV image and vector inputs to make successful lane change decision making [8]. With the aid of gated recurrent unit neural network and long short-term memory (LSTM), lane change intentions were thoroughly recognized based on the vehicle trajectory data [9] and construction of utility functions [10], respectively. To obtain robust autonomous driving policies, a new scheme was proposed with latent attention augmentation with scene dynamics approximation [11]. And Seq2Seq model with LSTM was constructed with soft attention module [12]to enhance the trajectory prediction ability of autonomous vehicles in complex highway scenarios.

Safety is a crucial problem for end-to-end learning method of autonomous driving, and researchers usually solve it by setting safety constraints and verifications [13–15] or combining model theory [16–18]. The risk assessment method was proposed based on probability model which applied DRL algorithms to find the optimal strategy with the minimum expected risk [13]. The lane change safety problem was tackled by limiting the action space to a safe action subspace based on PPO [14]. Moreover, a safety verification method was developed to avoid dangerous actions and prevent unsafe conditions with DRL algorithm [15]. On the basis of DDQN algorithm, regret decision model was proposed to evaluate whether the decision instructions were safe so as to ensure the lane change safety [16]. Through introducing safety intervention module, a DRL method was developed based on security PPO to make vehicles avoid dangerous behaviors in complex interactive environment [17]. And a hierarchical reinforcement learning method was given considering longitudinal reactions with social preference estimation to make safe overtaking [18].

Recent achievements have accomplished successful lane change tasks, yet the decision making is few investigated with both safety awareness and attention mechanism. Inspired by [7,9,13], a new end-to-end scheme is proposed for lane change decision-making of autonomous driving based on SAC algorithm with continuous action space. The main contributions include:

(1) The perception ability of the algorithm is improved combined with multi-data fusion method. Moreover, attention mechanism is introduced to achieve feature refinement and help the ego vehicle pay attention to surrounding vehicles and key road information so that the interpretability of the algorithm is strengthened.

(2) The reward function is constructed with safety awareness and competition awareness so as to ensure not only driving safety with the minimum driving risk but also driving efficiency.

(3) Compare different lane change decision making methods in static and stochastic dynamic scenarios and obtain the better one.

The paper is organized into five sections. The related fundamentals are summarized in Sect. 2. And the lane change decision making scheme proposed is illustrated in Sect. 3 as well as its detailed descriptions. Then the experiment results are analyzed in Sect. 4. Finally, the conclusion is presented in Sect. 5.

2 Fundamentals

2.1 Reinforcement Learning

Reinforcement learning is suitable for solving sequential decision-making problems, where the agent learns the optimal strategy by constantly interacting with the environment and improves its own strategy in a continuous trial-and-error and feedback learning. MDP is often used to denote the reinforcement learning process, which can be defined as a five-tuple $<S, A, P, r, \gamma>$, and the state space S represents a finite set of states, action space A represents a limited set of actions that can be taken, P indicates the probability by which the current state will shift to a subsequent state after taking an action in a certain state, reward r represents an immediate reward for taking an action based on the current state, $\gamma \in [0, 1]$ is a discount factor that represents the trade-off between immediate rewards and long-term returns.

The goal of reinforcement learning is to learn the optimal strategy

$$\pi^* = \arg\max_{\pi} E_{(s_t,a_t)\sim\rho_\pi}(\sum_{k=0}^{\infty} \gamma^k r_k) \tag{1}$$

where s_t represents the observed state information of the current environment obtained by the agent at time t, a_t is the action generated according to the current strategy π, ρ_π is the probability distribution that the sequence state s_t and action a_t follow when the agent interacts with the environment according to the policy.

2.2 The SAC Algorithm

The SAC algorithm is an off-policy one that overcomes the sample complexity problem of on-policy algorithm and solves the problem of fragile convergence by introducing entropy regularization terms. Entropy is used to describe the uncertainty of a random variable or event, and the greater the degree of uncertainty is, the greater the entropy value is. The training goal of the SAC algorithm is to maximize both the expected return and the entropy of the strategy, so it enhances the exploration ability of the algorithm and tends to prefer a strategy with strong randomness and higher returns.

The diagram of the principle of SAC algorithm is shown in Fig. 1. It is seen that the overall structure of the SAC algorithm consists of three parts, namely the actor network, the critic network 1 and the critic network 2. The critic network 1 and the critic network 2 have the same structure, and both have a pair of online networks and target networks with the same neural network structure, while the actor network has only one neural network structure. The goal is to learn optimal strategy

$$\pi^* = \arg\max_{\pi} \sum_{t} E_{(s_t,a_t)\sim\rho_\pi}[r(s_t,a_t)+\alpha\mathcal{H}(\pi(\cdot|s_t))] \tag{2}$$

$$\mathcal{H}(\pi(\cdot|s_t)) = -\sum_{t=0}^{T} \pi(\cdot|s_t)\log\pi(\cdot|s_t) \tag{3}$$

where $\mathcal{H}(\pi(\cdot|s_t))$ represents the entropy value of an action under the state s_t, and the coefficient α is used to balance the relationship between the exploration ability and desired return of the algorithm.

In the SAC algorithm, the input and the output of the actor network are the current state s_t and action a_t, respectively. The inputs of the online critic network are both the current state s_t and the action a_t, and the outputs are the state value functions Q_1 and Q_2, respectively, and their minimum value is taken as $Q(s_t, a_t)$ used to calculate the loss function. The inputs of the target critic network are both the state s_{t+1} and the action a_{t+1}, and their minimum output is used to calculate the target state value y_t according to the following equation,

$$y_t = r_{t+1}(s_t,a_t) + \gamma(\min_{i=1,2}Q_{\theta'_i}(s_{t+1},a_{t+1}) - \alpha\log(\pi_\phi(a_{t+1}|s_{t+1}))) \tag{4}$$

The loss function of the critic network is

$$J_Q(\theta_i) = E_{(s_t,a_t)\sim\mathcal{D}}[\frac{1}{2}(Q_{\theta_i}(s_t,a_t) - y_t)^2], i = 1,2 \tag{5}$$

The online critic network parameters θ_i and actor network parameters $\phi_i, i = 1, 2$ are updated by the stochastic gradient descent method. The target critic network parameters $\theta'_i, i = 1, 2$ are soft updated.

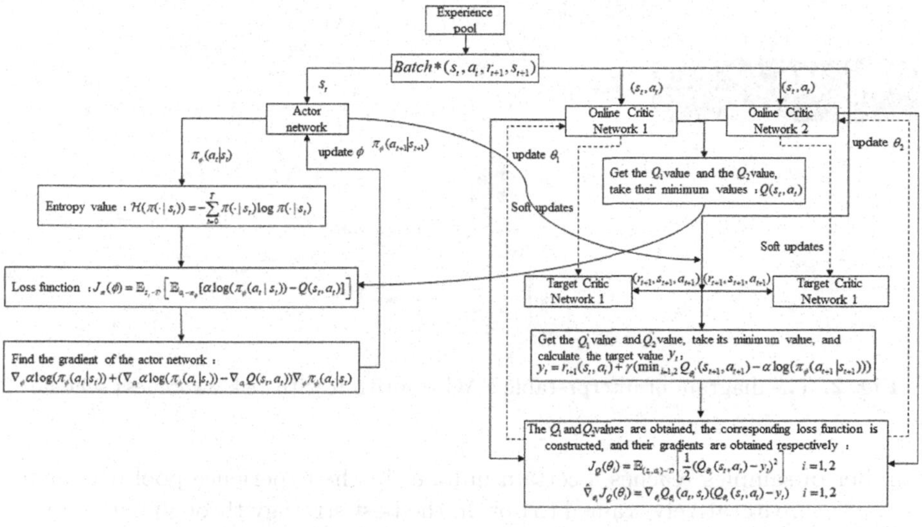

Fig. 1. The diagram of the principle of SAC algorithm.

3 Interpretable SAC with Safety Awareness

The diagram of the scheme proposed is shown in Fig. 2 of the lane change decision making, which consists of the server CARLA part and the client one. CARLA is a simulation platform used to simulate the real world, which helps researchers develop and verify various algorithms related to autonomous driving systems. The researchers operate the client to control the vehicles through decision making algorithms. The environment information or operation instructions can be transferred through Python API interface between the server and the client. The decision making module is the core of the interpretable SAC algorithm with safety awareness, which consists of input module, SAC module, reward function module, experience pool module and output one.

The input module includes the kinematics information of the ego vehicles and the front view image information. The SAC algorithm module receives the data and outputs decision making actions through extracting features and updating neural networks parameters. After receiving the actions, the output module sends it to CARLA through the python API interface, and the internal vehicle controller executes the operation actions. At the same time, CARLA simulator returns the new environment information to the input module as the interaction feedback, and the reward function module gives the corresponding rewards or punishments according to the driving state of the current ego vehicles. When the

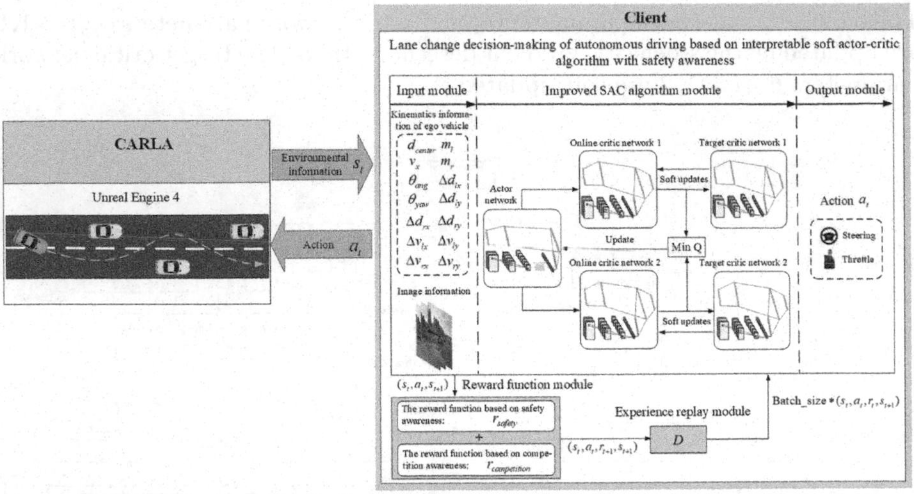

Fig. 2. The diagram of interpretable SAC algorithm with risk awareness scheme.

number of samples reaches a certain number in the experience pool, the neural networks are iteratively trained to obtain the best strategy through batch updating of randomly collected samples. The reward functions are comprehensively considered with safety and competitive awareness to improve driving safety and task execution efficiency in the reward function module. Then the critic network and actor network are detailed illustrated as well as the state space and the action space and the reward function as follows.

3.1 The Critic Network and the Actor Network

The diagram of the actor network is shown in Fig. 3. To improve the perception ability of the SAC algorithm, multi-data fusion is applied on the kinematics information and visual image information of ego vehicle so as to enhance the decision making effectiveness. For the low dimensional kinematic information of ego vehicles, the full connection layer is used to obtain the corresponding feature vectors. For high-dimensional visual image information, two convolution layers and two pooling layers are designed for local perception and feature dimensionality reduction, and the resulting feature map is flattened. Among them, a compact general attention module is embedded to focus on the environment information closely related to the lane changing task in the visual image so as to refine the features and improve the algorithm performance. Then the two feature streams are fused through the concatenated operation, and the action is output through the full connection layer. Except that the inputs increase the decision action information and no attention module is embedded in them, the other structures of the critic networks are similar to the actor network.

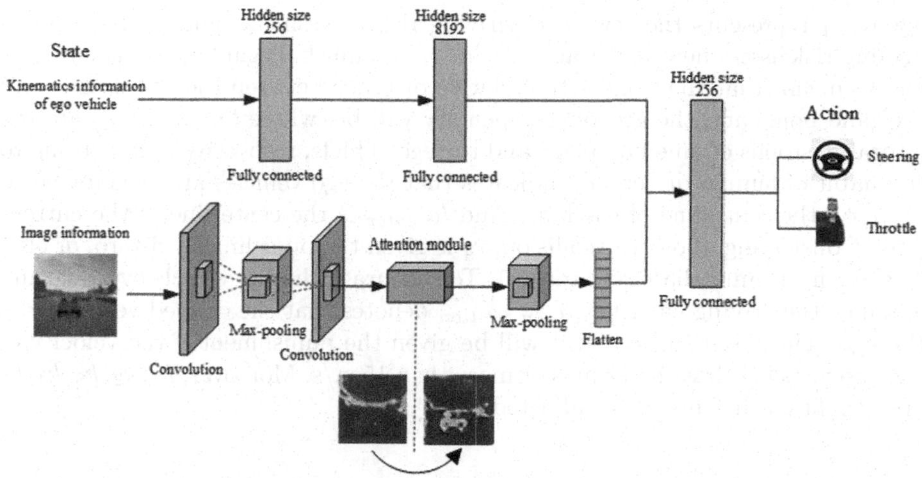

Fig. 3. The diagram of the structure of the actor network.

3.2 The State Space and the Action Space

The state space contains kinematic data information and visual image information of the ego vehicle. The kinematics information of the ego vehicle includes its speed and yaw angle, as well as the relative distance and the relative speed between the ego vehicle and the surrounding vehicles, and so on. In order to realize the humanoid lane change operation of autonomous vehicles in complex scenarios, the continuous lateral steering action and the longitudinal throttle action are considered in the action space. At the same time, in order to prevent the ego vehicle from learning too conservative behaviors, the braking action is not considered here.

3.3 The Reward Function

Combined with the safety reward functions[13] and comprehensively considering the improvement of driving efficiency, the following reward functions are constructed including safety awareness part and competition awareness one, as follows below.

$$r = r_{safety} + r_{competition} \tag{6}$$

$$\begin{aligned} r_{safety} &= k_1 \cdot r_1 + k_2 \cdot r_2 + k_3 \cdot r_3 + k_4 \cdot r_4 \\ &= k_1 \cdot (2 - \varepsilon_t) - k_2 \cdot e^{-\frac{(la_{ld} - la_{cv})^2}{2\sigma^2}} + k_3 \cdot e^{-\frac{(la_{center} - la_{cv})^2}{2\sigma^2}} + k_4 \cdot r_4 \end{aligned} \tag{7}$$

$$r_{competition} = \begin{cases} v - 5, & v < 5\,\mathrm{m/s} \\ 0.2v - 1, & 5\,\mathrm{m/s} \le v < 10\,\mathrm{m/s} \\ 3 - -0.2v, & 10\,\mathrm{m/s} \le v \le 15\,\mathrm{m/s} \\ -10, & v > 15\,\mathrm{m/s} \end{cases} \tag{8}$$

where r_1 represents the reward of driving risk assessment, and ε_t denotes the driving risk assessment result at any time t. To punish illegal lane change behavior, r_2 means that the smaller the relative distance between the ego vehicle and the lane boundary, the greater the penalty will be, where la_{ld} and la_{cv} are the lateral positions of lane boundary and the ego vehicle, respectively. According to the habit of human drivers, r_3 indicates that the ego vehicles are encouraged to drive on the centerline of the lane. And la_{center} is the centerline of the current lane. Considering about the collision, r_4 is given the punishment of zero, or else, it is given the immediate rewards of 1. To encourage the ego vehicle overtake and enhance the driving efficiency, $r_{competition}$ denotes that the desired velocity v of the ego vehicle is 10 m/s, and it will be given the punishment if the velocity of the ego vehicle is less than 5 m/s or more than 15 m/s. Moreover, k_1, k_2, k_3, k_4, k_5 are weight coefficients to be adjusted.

4 Experiments

Under the static scenario (SS) and the stochastic dynamic scenario (SDS), different experiments are conducted in CARLA simulator. In detail, these experiments are performed on Ubuntu20.04 operating system and Pytorch framework. The system environment includes R7-5800 CPU, RTX3060 GPU and 16 GB RAM, as well as Python3.7 interpreter. And the scenarios setting are almost similar to the reference [13]. The road has the length of 420 m under these two scenarios. For SS, 10–26 stationary vehicles are randomly displayed on the straight road, which act as the static obstacles for the ego vehicle in the training and evaluation stages, moreover, their positions are specified according to Gaussian probability sampling method. For SDS, based on the initial settings of SS, the surrounding vehicles adopt automatic driving mode to drive forward safely without collision, and the speed limit is 10 m/s. Under these two scenarios, the ego vehicle is expected to autonomously drive safely without collisions and accomplish the driving task fast.

The experiments conducted include SAC, SAC+CNN, SAC+CNN+SE and SAC+CNN+CBAM. And SAC+CNN denotes the SAC algorithm considering the kinematic information and image information of the ego vehicle, SAC+CNN +SE and SAC+CNN+CBAM denote the SAC algorithm with multi-data fusion and attention mechanisms SE [19] and CBAM [20], respectively. Moreover, comparison results are shown in Table 1 of these four methods according to the average reward and the distance accomplished under SS and SDS, where the comparison of the average reward can be further clearly demonstrated in Fig. 4. It can be seen that these four methods can all achieve lane change driving task and arrive at the destination under two scenarios. Moreover, SAC+CNN+CBAM method obtains the better performance than other three ones, meanwhile, the simple SAC method embodies slower driving performance with lower perception capability even though the ego vehicle can accomplish the driving tasks. It should be paid more attention on the algorithm performance of SAC+CNN+SE method, which is even lower than SAC+CNN method. Thus it is clear that

SAC+CNN+CBAM method is most suitable for the ego vehicle to make the safe and effective lane change decision making under these two scenarios because of the stronger perception and feature refinement capabilities, consistent with the statement of the interpretable SAC algorithm with safety awareness.

Table 1. Comparison of the different methods under SS and SDS.

Scenarios	Algorithms	Average reward	Distance (m)
SS	SAC	1399.66	416.73
	SAC+CNN	1891.84	420.11
	SAC+CNN+SE	1724.50	420.20
	SAC+CNN+CBAM	1968.26	419.95
SSD	SAC	1328.42	420.01
	SAC+CNN+SE	1664.10	420.28
	SAC+CNN+CBAM	1556.95	420.10
	SAC+CNN	1939.57	420.04

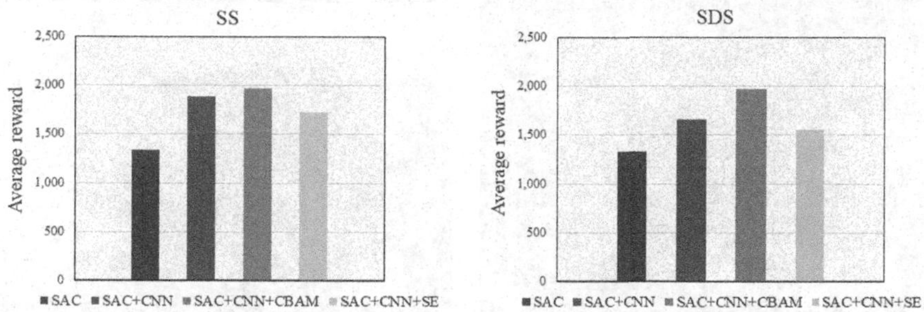

Fig. 4. The diagram of the average reward of different methods under SS and SDS.

The examples of visualization results are shown in Fig. 5. The top heat maps are for the static scenario, and the below ones are for the stochastic dynamic scenario, where the heat maps of each line are the front-view image obtained from CARLA, the input feature maps and the output feature maps of the attention modules, respectively. Particularly, the upper left heat maps are obtained from CBAM, and the upper right ones are obtained from SE, so are for the below left and below right ones. It is seen that the attention modules can also further emphasize the key road information such as ahead vehicles and surrounding environment so as to make the ego vehicle avoid collisions effectively and accomplish the driving task smoothly.

Compared with SE which only has channel-wise attention module, CBAM can refine more detailed critical feature information owing to have sequential spatial and channel-wise ones, such as the lane lines on the road are emphasized as shown in the below left heat maps while these information is invisible in the right ones. Therefore, it further verifies above-mentioned illustration.

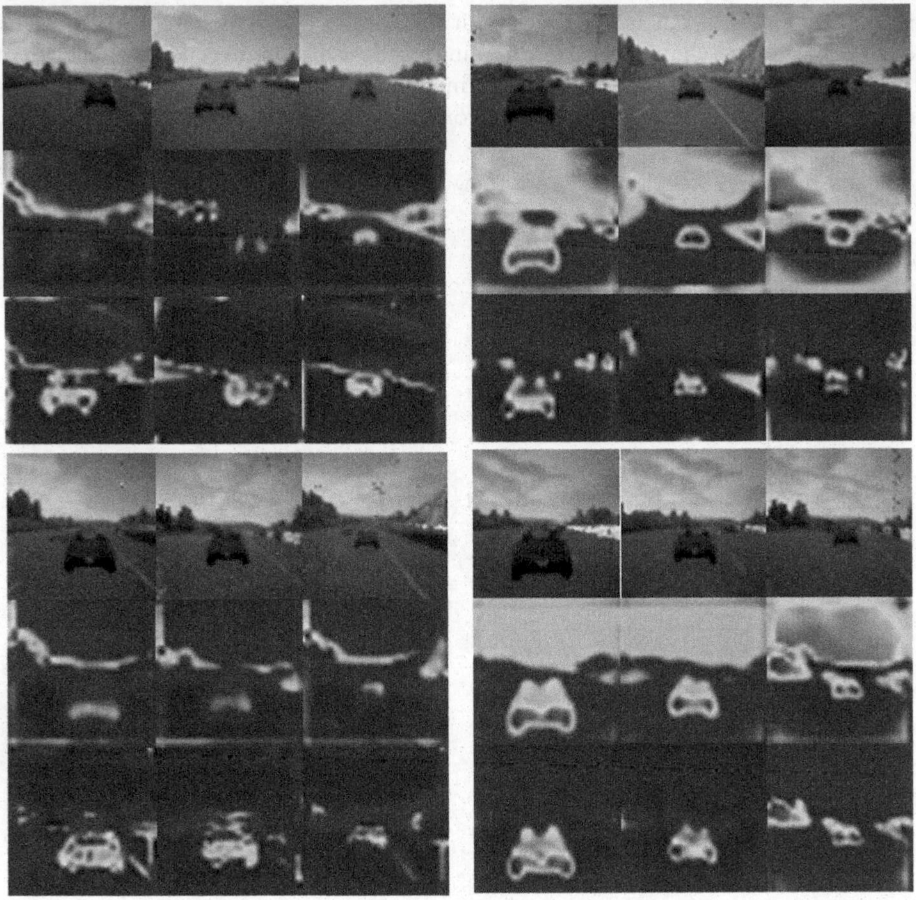

Fig. 5. The examples of visualization results.

5 Conclusion

Based on the SAC algorithm, a new end-to-end scheme is proposed for the lane change decision making of autonomous driving. From the interpretability and safety consciousness perspective, attention mechanism is applied to refine the features of multi-data fusion perception information so that the ego vehicle can pay more attention on key surrounding vehicles and environment clearly. Moreover, according to the risk assessment method, the comprehensive reward function is designed considering safety and competition awareness so as to encourage the ego vehicle to achieve the driving task smoothly and effectively without collisions. At last, the experiments further verify the validity of the interpretable SAC algorithm with safety awareness proposed in the paper.

References

1. Biondi, F., Alvarez, I., Jeong, K.A.: Human-vehicle cooperation in automated driving: a multidisciplinary review and appraisal. Int. J. Hum.-Comput. Interact. **35**(11), 932–946 (2019)
2. Qin, Y.Q., Wang, J.R., Xie, J.M., et al.: Lane change model for weaving area based on dynamic safety spacing. J. Saf. Environ. 1–9 (2022). https://doi.org/10.13637/j.issn.1009-6094.2022.0007
3. Cao, X., Wan, H.Y., Lin, Y.F., et al.: High-value prioritized experience replay for off-policy reinforcement learning. In: Proceedings of the 31st IEEE International Conference on Tools with Artificial Intelligence, pp. 1510–1514. IEEE, USA (2019)
4. Jiang, S., Chen, J., Shen, M.: An interactive lane change decision making model with deep reinforcement learning. In: Proceedings of the 7th IEEE International Conference on Control and Mechatronics and Automation, pp. 370–376. IEEE, Netherlands (2019)
5. Saxena, D.M., Bae, S., Nakhaei, A., et al.: Driving in dense traffic with model-free reinforcement learning. In: Proceedings of the 2020 IEEE International Conference on Robotics and Automation, pp. 5385–5392. IEEE, France (2020)
6. Song, X., Sheng, X., Cao, H., et al.: Lane-change behavior decision-making intelligent vehicle based on limitation learning and reinforcement learning. Automot. Eng. **43**(01), 59–67 (2021)
7. Chen, Y., Dong, C., Palanisamy, P., et al.: Attention-based hierarchical deep reinforcement learning for lane change behaviors in autonomous driving. In: Proceedings of 2019 IEEE/RSJ International Conference on Intelligent Robots and Systems, pp. 3697–3703. IEEE, China (2019)
8. Wang, J., Zhang, Q., Zhao, D.: Highway lane change decision-making via attention-based deep reinforcement learning. J. Autom. Sci. **9**(3), 567–569 (2022)
9. Zhao, J.D., Zhao, J.M., Qu, Y.C., et al.: Vehicle lane change intention recognition driven by trajectory data. J. Transp. Syst. Eng. Inf. Technol. 1–13 (2022). https://kns.cnki.net/kcms/detail/11.4520.U.20220518.1133.004.html
10. Song, X.L., Zeng, Y.B., Cao, H.T., et al.: Lane change intention recognition method based on an LSTM network. China J. Highway Transp. **34**(11), 236–245 (2021)
11. Cheng, R., Agia, C., Shkurti, F., et al.: Latent attention augmentation for robust autonomous driving policies. In: Proceedings of 2021 IEEE/RSJ International Conference on Intelligent Robots and Systems, pp. 130–136. IEEE, Czech Republic (2021)
12. Han, H., Xie, T.: Lane change trajectory prediction of vehicles in highway interweaving area using Seq2Seq-attention network. China J. Highway Transp. **33**(6), 106–118 (2020)
13. Li, G.F., Yang, Y.F., Li, S., et al.: Decision making of autonomous vehicles in lane change scenarios: deep reinforcement learning approaches with risk awareness. Transp. Res. Part C-Emerg. Technol. **134**, 1–18 (2022)
14. Krasowski, H., Wang, X., Althoff, M.: Safe reinforcement learning for autonomous lane changing using set-based prediction. In: Proceedings of the 23rd IEEE International Conference on Intelligent Transportation Systems, pp. 1–7. IEEE, Greece (2020)
15. Kamran, D., Engelgeh, T., Busch, M., et al.: Minimizing safety interference for safe and comfortable automated driving with distributional reinforcement learning. In: Proceedings of 2021 IEEE/RSJ International Conference on Intelligent Robots and Systems, pp. 1236–1243. IEEE, Czech Republic (2021)

16. Chen, D., Jiang, L.S., Wang, Y., et al.: Autonomous driving using safe reinforcement learning by incorporating a regret-based human lane-changing decision model. In: Proceedings of the 2020 American Control Conference, pp. 4355–4361. IEEE, USA (2020)
17. Ye, F., Cheng, X., Wang, P., et al.: Automated lane change strategy using proximal policy optimization-based deep reinforcement learning. In: Proceedings of the 31st IEEE Intelligent Vehicles Symposium, pp. 1746–1752. IEEE, USA (2020)
18. Lu, H., Lu, C., Yu, Y., et al.: Autonomous overtaking for intelligent vehicles considering social preference based on hierarchical reinforcement learning. Auton. Innov. **5**, 195–208 (2022)
19. Hu, J., Shen, L., Sun, G.: Squeeze-and-excitation networks. In: Proceedings of the IEEE Conference on Computer Vision and Pattern Recognition, pp. 7132–7141. IEEE, Salt Lake City, UT, USA (2018)
20. Woo, S., Park, J., Lee, J.Y., et al.: CBAM: convolutional block attention module. In: Proceedings of the 15th European Conference on Computer Vision, pp. 3–19. IEEE, Germany (2018)

Demo

GLRNet: Gas Leak Recognition via Temporal Difference in Infrared Video

Erqi Huang, Linsen Chen, Tao Lv, and Xun Cao(✉)

School of Electronic Science and Engineering, Nanjing University, Nanjing, China
{erqihuang,linsen_chen,lvtao}@smail.nju.edu.cn, caoxun@nju.edu.cn

Abstract. Gas Leak Recognition (GLR) is a task where existing techniques face significant challenges, such as the need for infrared cameras, various industrial scenes, and validated data. In this work, We demonstrate Gas Leak Recognition Network (GLRNet), a network with temporal difference inputs and temporal shifting operations. GLRNet integrating module is designed by analogy to human perception, which is a physical constraint in the feature representation. The synergy of our proposed GLRNet and infrared camera we have developed is an emerging comprehensive system that achieves state-of-the-art (SOTA) results in real-world data captured in several chemical industrial parks across the country. Our demo video is at https://youtu.be/glt4DMeNXDU.

Keywords: Gas leak recognition · Temporal difference · Infrared video

1 Introduction

Gas Leak Recognition has been thought of as a key factor in warding off accidents of various kinds [2,3] and ensuring the safety of people's lives and property [7,9]. Recent works by researchers have established that analysis of optical gas imaging (OGI) [10,11] and deep learning [5,13,18] are most popular methods for GLR. The technology of OGI to recognize GLR provides the quantification of detect limits, while deep learning methods predict sequences of labels by learning feature representations from numerous training samples. Neither analysis of OGI nor deep learning recognize gas leaks efficiently. The isolation of problem-solving existed in the process of OGI [8,19], where data acquisition and pattern recognition are separate. Furthermore, in these deep learning methods [1,12,16], feature representations are extremely lacking in physical constraints when designing the network.

In this work inspired by physical constraints [6,14,15], we demonstrate GLRNet: a temporal difference and shifting-based convolutional neural networks (CNNs). GLRNet begins by dividing the infrared video into multiple clips along the temporal dimension. Next, our system uses the information fusion approach [17] that builds on a fusion block to link multimodal inputs to extract features of its temporal motion and spatial shape. The feature representation is then

L. Fang et al. (Eds.): CICAI 2022, LNAI 13606, pp. 515–520, 2022.
https://doi.org/10.1007/978-3-031-20503-3_41

generated from a set of temporal shifting blocks, a high efficiency and performance CNN modules that enable GLRNet to model temporal features without extra computation. This step accurately maps high-level physical information to low-level visual features in both the temporal and spatial dimensions. Finally, the results of GLRNet are obtained by the fully connected layer (FC), a linear mapping that matches the number of recognition classes. GLRNet consists of several blocks which are trained end-to-end for their specific functions.

This demonstration shows a physics-based methodology for GLR that achieves the SOTA results in a real-world dataset captured in a chemical industrial park. GLRNet is a model according to perceptual process that demonstrates: (1) the use of temporal difference as input; (2) an approach that interacts with the temporal and the spatial information; (3) temporal shifting block implements similar functionality to 3D CNNs [4].

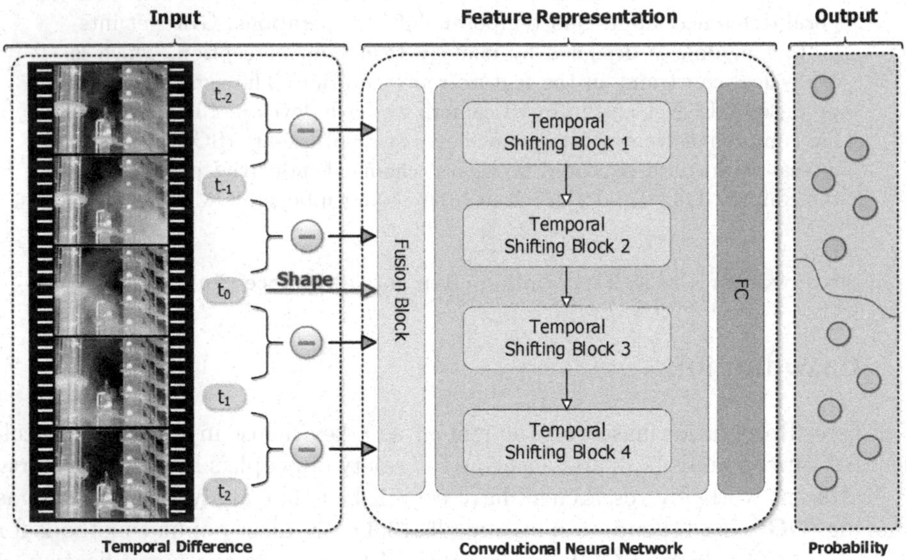

Fig. 1. Overview of Gas Leak Recognition Network (GLRNet).

2 Gas Leak Recognition Network

As shown in Fig. 1, GLRNet is a method with the temporal differences and spatial characteristics of infrared video for recognition. The inputs of fusion block consist of spatial shapes and temporal differences, from which can obtain multi-modality features simultaneously. Multiple temporal shifting blocks then extract features hierarchically for representation after data fusion. At the end of the feature representation, the fully connected layer figures out the result of GLR. Detailed information on the block is described below.

Fusion Block: The purpose of the fusion block is to aggregate the information from the infrared video about spatial shapes and temporal differences. Firstly, the differences in the infrared video are concatenated as a whole along the channel dimension. It is then added as a residual to a convolutional feature of a single infrared frame. In this way, the fusion block can have good aggregation of the spatio-temporal information.

Temporal Shifting Block: The requirements of GLR are not only the high efficiency in computation but also the ample correlation information of temporal dimension between infrared video frames. We use a high-performance temporal shift block with a large field of perception to extract spatio-temporal features. It moves part of the channels by -1, another by $+1$ along the temporal dimension, and the rest un-shifted. Therefore, the temporal shifting block has an extra-large temporal perceptual field for highly complex feature representation.

In conclusion, GLRNet, an end-to-end designed network, applies the temporal difference and shifting to extract and represent feature widely. The Network to capture both morphological and motion information is achieved in a unified framework when the features represent by spatio-temporal fusion and temporal shifting. With the above design, GLRNet shows excellent performance on the GLR task.

3 Experimental Evaluation

3.1 Experimental Setup

There are 30,874 video sequences sampled in the real-world dataset obtained from chemical industrial parks, and randomly divided into training and testing sets in a 3:1 ratio to ensure rational experimental settings. In the experiments, each video sequence of the dataset is scaled to 224×224 with a batch size of 128. The initial learning rate is 0.01, which is reduced to $1/10$ when the performance of the testing set is saturated. The training epoch is 60. The training and testing servers are equipped with 2 Nvidia GeForce RTX 3090. To appraise accurately, the evaluation criteria for GLR selected Accuracy and Recall. The recall is the number of positive classes predicted as a positive class out of all labels.

When the cost of False Negative is greater than that of False Positive, we should select our best model using Recall, which is the ratio of correctly predicted outcomes to all predictions. It is also known as sensitivity or specificity. If a gas leak tested positive (Actual Positive) is predicted as negative, the cost associated with a False Negative can become extraordinarily high since explosions are involved.

3.2 Comparative Evaluation

We compare GLRNet with three classical algorithms: (1) VideoGasNet [12] considers the size of the gas leak to recognize by 3D CNNs; (2) ShuffleNet [16] is

Table 1. Comparisons of prediction performance with existing methods.

Methods	VideoGasNet [12]	ShuffleNet [16]	I3D [1]	GLRNet
Accuracy ↑	81.90%	95.31%	97.74%	**98.52%**
Recall ↑	67.42%	96.62%	97.31%	**98.78%**

a lightweight approach, well suited to edge computing, and plays a crucial role in various fields; (3) I3D [1] is a typical 3D CNN-based approach introduced in multiple CV tasks. These results are all evaluated under the same experimental conditions. Table 1 shows the performance of each algorithm on the gas leakage dataset. GLRNet achieves SOTA performance on all metrics, outperforming existing methods such as VideoGasNet, ShuffleNet, and I3D. The superiority of GLRNet in accuracy allows it has a great chance to recognize gas leaks in real-world scenarios (Fig. 2).

Fig. 2. Demonstration of gas leak recognition on infrared video taken at a chemical industrial park. See supplement for full video. (Digital zoom Recommended.)

4 Demonstration and Conclusion

This research presents an innovative and integral approach to gas leak recognition. The approach comprising CNNs has achieved video-level recognition tasks in practical. The trained CNNs are encapsulated into the infrared camera and utilized for the entire recognition system. Besides, the verification of GLR on the gas leak validates by a scientific method, which cross-validate and joint extrapolate between multi-scale information. Finally, as data accumulates, the performance of the approach will be further improved by advanced deep learning methods.

References

1. Carreira, J., Zisserman, A.: Quo vadis, action recognition? A new model and the kinetics dataset. In: proceedings of the IEEE Conference on Computer Vision and Pattern Recognition, pp. 6299–6308 (2017)
2. Eckerman, I.: The Bhopal saga: causes and consequences of the world's largest industrial disaster. Universities Press (2005)
3. Gålfalk, M., Olofsson, G., Crill, P., Bastviken, D.: Making methane visible. Nat. Clim. Change **6**(4), 426–430 (2016)
4. Ji, S., Xu, W., Yang, M., Yu, K.: 3D convolutional neural networks for human action recognition. IEEE Trans. Pattern Anal. Mach. Intell. **35**(1), 221–231 (2012)
5. Kopbayev, A., Khan, F., Yang, M., Halim, S.Z.: Gas leakage detection using spatial and temporal neural network model. Process Saf. Environ. Prot. **160**, 968–975 (2022)
6. Lin, J., Gan, C., Han, S.: TSM: temporal shift module for efficient video understanding. In: Proceedings of the IEEE/CVF International Conference on Computer Vision, pp. 7083–7093 (2019)
7. Meribout, M.: Gas leak-detection and measurement systems: prospects and future trends. IEEE Trans. Instrum. Meas. **70**, 1–13 (2021)
8. Olbrycht, R., Kałuża, M.: Optical gas imaging with uncooled thermal imaging camera-impact of warm filters and elevated background temperature. IEEE Trans. Industr. Electron. **67**(11), 9824–9832 (2019)
9. Ravikumar, A.P., Brandt, A.R.: Designing better methane mitigation policies: the challenge of distributed small sources in the natural gas sector. Environ. Res. Lett. **12**(4), 044023 (2017)
10. Ravikumar, A.P., et al.: Repeated leak detection and repair surveys reduce methane emissions over scale of years. Environ. Res. Lett. **15**(3), 034029 (2020)
11. Ravikumar, A.P., Wang, J., Brandt, A.R.: Are optical gas imaging technologies effective for methane leak detection? Environ. Sci. Technol. **51**(1), 718–724 (2017)
12. Wang, J., Ji, J., Ravikumar, A.P., Savarese, S., Brandt, A.R.: VideogasNet: deep learning for natural gas methane leak classification using an infrared camera. Energy **238**, 121516 (2022)
13. Wang, J., et al.: Machine vision for natural gas methane emissions detection using an infrared camera. Appl. Energy **257**, 113998 (2020)
14. Wang, L., Tong, Z., Ji, B., Wu, G.: TDN: temporal difference networks for efficient action recognition. In: Proceedings of the IEEE/CVF Conference on Computer Vision and Pattern Recognition, pp. 1895–1904 (2021)
15. Wang, L., et al.: Temporal segment networks: towards good practices for deep action recognition. In: Leibe, B., Matas, J., Sebe, N., Welling, M. (eds.) ECCV 2016. LNCS, vol. 9912, pp. 20–36. Springer, Cham (2016). https://doi.org/10.1007/978-3-319-46484-8_2
16. Zhang, X., Zhou, X., Lin, M., Sun, J.: ShuffleNet: an extremely efficient convolutional neural network for mobile devices. In: Proceedings of the IEEE Conference on Computer Vision and Pattern Recognition, pp. 6848–6856 (2018)
17. Zhou, K., Chen, L., Cao, X.: Improving multispectral pedestrian detection by addressing modality imbalance problems. In: Vedaldi, A., Bischof, H., Brox, T., Frahm, J.-M. (eds.) ECCV 2020. LNCS, vol. 12363, pp. 787–803. Springer, Cham (2020). https://doi.org/10.1007/978-3-030-58523-5_46

18. Zhou, K., Wang, Y., Lv, T., Linsen, C., Qiu, S., Cao, X.: Explore spatio-temporal aggregation for insubstantial object detection: benchmark dataset and baseline. In: Proceedings of the IEEE Conference on Computer Vision and Pattern Recognition, p. 1 (2022)
19. Zimmerle, D., Vaughn, T., Bell, C., Bennett, K., Deshmukh, P., Thoma, E.: Detection limits of optical gas imaging for natural gas leak detection in realistic controlled conditions. Environ. Sci. Technol. **54**(18), 11506–11514 (2020)

ATC-WSA: Working State Analysis for Air Traffic Controllers

Bo Liu[1], Xuanqian Wang[1], Jingjin Dong[2], Di Li[2], and Feng Lu[1](✉)

[1] State Key Laboratory of VR Technology and Systems, Beihang University, Beijing, China
lufeng@buaa.edu.cn
[2] North China Air Traffic Management Bureau CAAC, Beijing, China

Abstract. Air traffic controllers (ATCs) are required to focus on flight information, make instant decisions and give instructions to pilots with high attention and responsibility. Human factors related to aviation risks should be monitored, such as fatigue, distraction, and so on. However, existing methods have two major problems: 1) Wearable or invasive devices may interfere with ATCs' work; 2) Appropriate state indicator for ATCs is still not clear. Therefore, we propose a working state analysis solution, called ATC-WSA, and solve the above questions by 1) Computer vision and speech techniques without contact; 2) Specific models and indexes optimized by collected real ATCs' data, including video, audio, annotation, and questionnaire. Three layers' architecture is designed for AI detection, state analysis, and high-level indexes calculation. Overall, our demo can monitor and analyze the working state of ATCs and detect abnormal states in time. Key parts of this demo have already been applied to North China Air Traffic Control Center (Beijing) and the control tower of Beijing Capital International Airport.

Keywords: Air traffic controller · State analysis · Keyword spotting

1 Introduction

For safe and orderly air traffic flow, air traffic controllers (ATCs) need to monitor and process a variety of flight-related information attentively and give instructions to pilots by radio. Based on the survey of human factors in aviation accidents, the majority of errors are made by ATCs when they are distracted, tired and asleep [10]. Therefore, working state monitoring and analysis are of vital importance in this key profession.

Most of the existing state assessment methods for ATCs are conducted by questionnaires [2], simulation [7] or physiological measurements like EEG, fNIR, and heart rate [9] before and after work. However, these methods could not reflect the real-time state of ATCs and may bring extra interference when ATCs are on duty. Additionally, how to find appropriate indexes to describe the state remains unclear.

L. Fang et al. (Eds.): CICAI 2022, LNAI 13606, pp. 521–525, 2022.
https://doi.org/10.1007/978-3-031-20503-3_42

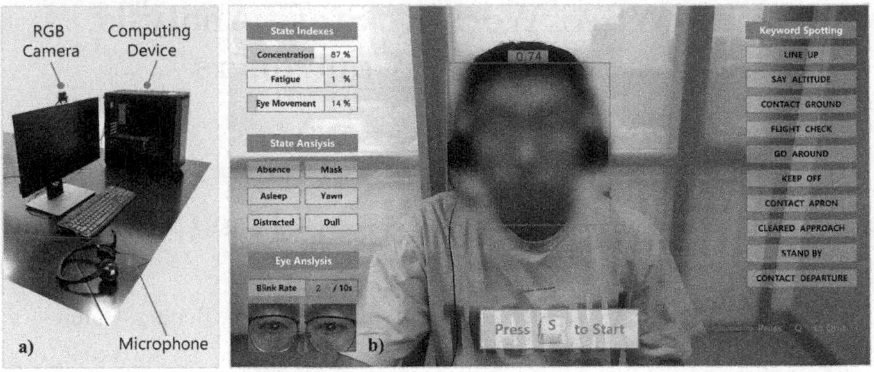

Fig. 1. Demonstration overview. a) Setup. b) User interface.

In this paper, we build a real-time demo called ATC-WSA which could detect abnormal working states and spot spoken keywords. It has two advantages: 1) **Contactless monitoring**. With only an ordinary RGB camera and microphone, the video and audio streams are analyzed in real-time; 2) **Adaptation to ATC environments**. Models and indexes are optimized by collecting real ATC datasets, including video, audio, annotation, and questionnaire. The key parts of this demo have already been verified and applied to control centers and towers.

2 Setup and Data Collection

2.1 Setup

The demo system is running on a Windows 10 computing device with a GeForce GTX 1080 graphics card. The speed achieves 32 FPS. An RGB camera, headset microphone, and display screen are connected to the computing device via USB and HDMI cables, as shown in Fig. 1.

2.2 Data Collection

To train and refine this demo, we collect a large-scale dataset in real ATC working environments. The dataset contains:

- **Video:** At the North China Air Traffic Control Center (Beijing), we deployed two cameras in two seats, and collected 3TB videos for two months. Videos cover 74 ATCs aged from 25 to 40.
- **Audio:** At the control tower of Beijing Capital International Airport, we collected 780 h of audio data with a microphone in front of the ATCs.
- **Annotation:** Representative data were selected, and video data are labeled with blink state, face landmarks and head poses, while audio data are labeled with keywords.
- **Questionnaire:** For subjective work state assessment, 19 questions are designed for ATCs including working experience, sleep time, workload, and fatigue level. Finally, 68 ATCs filled out the questionnaire after work.

3 Architecture

ATC-WSA is an integrated and practical demo system that could be used for ATCs. The architecture can be divided into 3 layers as shown in Fig. 2:

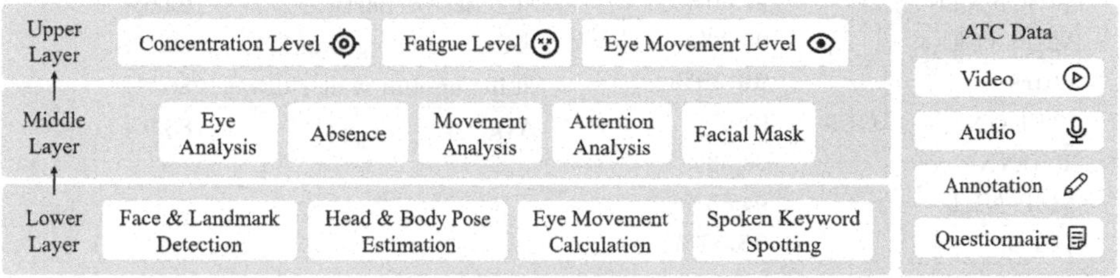

Fig. 2. Architecture of the ATC-WSA

1) Lower layer: This layer provides basic AI detection abilities, like face detection, pose estimation, and spoken keyword spotting. We use MTCNN [13] for fast and accurate face detection. The detected faces will then be processed by a two-stream network inspired by PFLD [4], which predicts face landmarks and head poses simultaneously from the MobileNetV3 [5] backbone. Models are trained and optimized on ATC data in Sect. 2.2.

2) Middle layer: Based on the lower layer, the middle layer analyzes some common states of ATCs like eye state, absence, movement, attention, and facial mask.

3) Upper layer: To give an overall and high-level view of ATCs' state, we analyze ATCs' data and suggestions, and finally propose three state indexes in upper layer which are concentration level, fatigue level, and eye movement level. These indexes are calculated based on the output of the lower and middle layer and normalized into a percentage.

Algorithms are implemented using Pytorch [11] 1.11 and OpenCV [1] 4.5 by C++. Optimized on real ATC data, our system is robust to various head poses, camera positions, ambient noise, and light conditions.

4 Experiments

4.1 Blink Detection and Keyword Spotting

For blink detection, we compare our method with some popular eye state detection methods in Table 1 on two datasets. Closed Eyes in the Wild (CEW) has 1192 faces with both eyes closed and 1231 with eyes open. We also construct the Closed Eye in ATC (CEA) dataset with 2803 images covering challenging conditions like low light, various head poses, and glasses reflection. Our method achieves the best performance on both datasets.

We also test our keyword spotting module on 526 ATC audio clips, which are manually segmented by 5 keywords. Table 2 reports the keyword spotting accuracy.

Table 1. Blink detection accuracy

Method	Dataset	Acc
P-FDCN [6]	CEW	94.90%
TSCNN [12]		98.51%
WBCNNTL [8]		97.40%
MSP-NET [3]		95.36%
Ours		**99.29%**
P-FDCN	CEA	94.67%
TSCNN		96.53%
MSP-NET		94.85%
Ours		**98.57%**

Table 2. Keyword spotting accuracy

Keywords	Acc
"Contact apron"	91.3%
"Contact ground"	84.2%
"Contact departure"	92.6%
"Keep off the ground"	89.6%
"Line up and wait runway"	86.7%
Avg	88.9%

4.2 Control Task of Upper Layer Indexes

To test the upper layer indexes, five volunteers are invited to finish three tasks: 1) Control the concentration level over 70% for 10 s; 2) Control the fatigue level over 60% for 3 s; 3) Control the eye movement level over 25% for 5 s. Each task is repeated 5 times by each volunteer. The success rate is shown in Table 3.

Table 3. Success rate on each task

	1) Concentration	2) Fatigue	3) Eye movement
User 1	100%	100%	80%
User 2	100%	100%	100%
User 3	100%	80%	100%
User 4	100%	100%	60%
User 5	100%	100%	100%
Avg	100%	96%	88%

5 Attendees Experience

To make attendees feel involved, our demo is designed with four tasks: 1) Trigger any 5 keywords. 2) Trigger any 3 states. 3) Control blink rate above 10 times/10 s. 4) Control upper layer index to satisfy: Concentration > 70%, Fatigue > 60%, Eye Movement > 40%. Finally, "Congratulations" will be shown on the screen if all tasks are finished. On average, it takes about 1 min and 45 s to experience the whole demo.

6 Conclusion

We propose a working state analysis demo for ATCs, called ATC-WSA, by using only a common camera and microphone. Based on computer vision and speech recognition technology, we could make multi-level state analysis without contact. AI models are optimized by our collected ATC data. Key parts of the demo have been deployed in real air traffic control environments.

References

1. Bradski, G.: The OpenCV Library. Dr. Dobb's J. Softw. Tools (2000)
2. Chang, Y.H., Yang, H.H., Hsu, W.J.: Effects of work shifts on fatigue levels of air traffic controllers. J. Air Transp. Manag. **76**, 1–9 (2019)
3. Gu, W.H., Zhu, Y., Chen, X.D., He, L.F., Zheng, B.B.: Hierarchical cnn-based real-time fatigue detection system by visual-based technologies using msp model. IET Image Proc. **12**(12), 2319–2329 (2018)
4. Guo, X., et al.: Pfld: a practical facial landmark detector. arXiv preprint arXiv:1902.10859 (2019)
5. Howard, A., et al.: Searching for mobilenetv3. In: Proceedings of the IEEE/CVF International Conference on Computer Vision, pp. 1314–1324 (2019)
6. Huang, R., Wang, Y., Guo, L.: P-fdcn based eye state analysis for fatigue detection. In: 2018 IEEE 18th International Conference on Communication Technology (ICCT), pp. 1174–1178. IEEE (2018)
7. Karikawa, D., Aoyama, H., Takahashi, M., Furuta, K., Ishibashi, A., Kitamura, M.: Analysis of the performance characteristics of controllers' strategies in en route air traffic control tasks. Cogn. Technol. Work **16**(3), 389–403 (2014)
8. Liu, Z.T., Jiang, C.S., Li, S.H., Wu, M., Cao, W.H., Hao, M.: Eye state detection based on weight binarization convolution neural network and transfer learning. Appl. Soft Comput. **109**, 107565 (2021)
9. Pagnotta, M., Jacobs, D.M., de Frutos, P.L., Rodriguez, R., Ibáñez-Gijón, J., Travieso, D.: Task difficulty and physiological measures of mental workload in air traffic control: a scoping review. Ergonomics (just-accepted), pp. 1–54 (2021)
10. Pape, A.M., Wiegmann, D.A., Shappell, S.A.: Air traffic control (atc) related accidents and incidents: a human factors analysis (2001)
11. Paszke, A., et al.: Pytorch: An imperative style, high-performance deep learning library. Advances in neural information processing systems 32 (2019)
12. Sanyal, R., Chakrabarty, K.: Two stream deep convolutional neural network for eye state recognition and blink detection. In: 2019 3rd International Conference on Electronics, Materials Engineering & Nano-Technology (IEMENTech), pp. 1–8. IEEE (2019)
13. Zhang, K., Zhang, Z., Li, Z., Qiao, Y.: Joint face detection and alignment using multitask cascaded convolutional networks. IEEE Signal Process. Lett. **23**(10), 1499–1503 (2016)

3D-Producer: A Hybrid and User-Friendly 3D Reconstruction System

Jingwen Chen[1], Yiheng Zhang[1], Zhongwei Zhang[2], Yingwei Pan[1(✉)], and Ting Yao[1]

[1] JD AI Research, Beijing, China
chenjingwen.sysu@gmail.com, yihengzhang.chn@gmail.com, panyw.ustc@gmail.com, tingyao.ustc@gmail.com
[2] Dalian University of Technology, Dalian, China
dlut0zzw@gmail.com

Abstract. In this paper, we introduce a hybrid and user-friendly 3D reconstruction system for objects, which seamlessly integrates a deep neural network model for 3D shape reconstruction and a multi-band image blending algorithm for realistic texturing into a single encapsulated application, dubbed as 3D-Producer. Compared to expensive laser scanning devices, our 3D-Producer offers an economical and practical solution, where a textured 3D model for a target object can be easily created by general users with mobile devices, a physical turntable, and our application. An illustration video can be found here (https://github.com/winnechan/3D-Producer/blob/main/3D-Producer-Demo-480P.mp4).

Keywords: 3D Reconstruction system · Deep learning

1 Introduction

Recently, deep learning has demonstrated its effectiveness in a series of 2D/3D vision tasks [4,6,10,11,14,18–21,23,28,29]. In between, 3D reconstruction techniques [8,16,17,22] has attracted increasing attention in both academia and industry due to the emerging trend of virtual reality and autonomous driving. Compared to expensive and professional laser scanning devices, image-based 3D modeling is a cheap and practical solution for general users to produce 3D models of objects by taking photos with their mobile devices (e.g., mobile phones or digital cameras).

In this paper, we develop a hybrid and user-friendly 3D reconstruction system named 3D-Producer, which seamlessly combines deep neural network model NeuS [27] for accurate 3D shape reconstruction and multi-band image blending algorithm [1,2] for realistic texturing together to generate fancy 3D models. Compared with existing softwares (e.g., openMVS [3], Meshroom [9], COLMAP

L. Fang et al. (Eds.): CICAI 2022, LNAI 13606, pp. 526–531, 2022.
https://doi.org/10.1007/978-3-031-20503-3_43

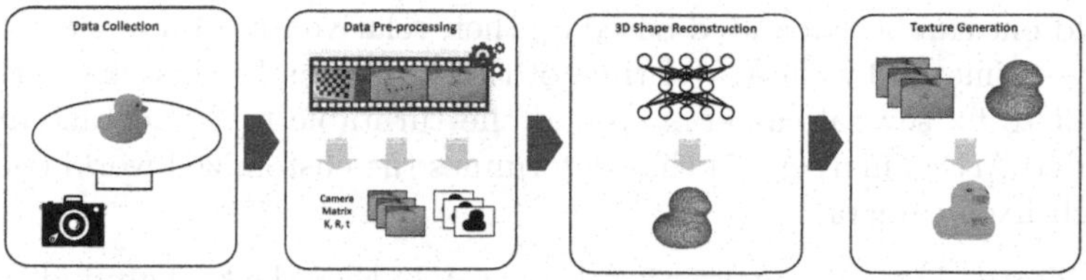

Fig. 1. Our 3D-Producer system consists of four modules: 1) data collection module, 2) data pre-processing module, 3) 3D shape reconstruction module, and 4) texture generation module.

[24,25]), our 3D-Producer utilizes deep neural network to model the 3D shape as implicit surface representation, alleviating the difficulty of correspondence point matching in conventional multi-view stereo and increasing the resolution of the 3D shape compared to those deep learning based methods [13,26] with explicit volumetric representation. Additionally, we propose to leverage a turntable coupled with mobile devices for standard and convenient data collection.

2 System Architecture

Our 3D-Productor aims to achieve accurate and visually pleasing 3D reconstruction results. As shown in Fig. 1, our 3D-Productor consists of four modules: 1) data collection module, 2) data pre-processing module, 3) 3D shape reconstruction module, and 4) texture generation module. In the following part of this section, we will describe each module in details sequentially.

2.1 Data Collection Module

In the data collection module, users need to take some images/videos for later stages by following several simple guidance.

Chessboard Data. To build up the correspondence between 3D points on the surface of the target object and 2D pixels on the rendered image for reconstruction, the intrinsic parameters and distortion coefficients of the camera need to be determined (the procedure is well-known as camera calibration). Following the typical algorithm [30], the users are asked to take a video above a provided chessboard from multiple angles with each camera.

ArUco Marker Data. Besides the intrinsic parameters and distortion coefficients, the extrinsic parameters of the camera (i.e., camera pose) should also be estimated for the complete rendering from 3D points to 2D pixels, which are represented as the projection matrix from world coordinate to camera coordinate. To reduce the freedom in the projection matrix estimation, we fix the origin of the world coordinate to the center of a turntable and the z-axis is always perpendicular to the turntable plane. This way, a new pose of the camera relative to the old one can be calculated based on their relative rotation angle, which can be easily estimated by the travel time of the turntable. In this phase, the users are asked to fix several cameras around the turntable first and take an image of provided ArUco markers [7] that determines the customized world coordinate with each fixed camera.

Object Data. Then, the marker is taken away to keep the background clean and the target object is placed on the turntable. When the turntable is running, the fixed cameras are utilized to capture videos for the target object from different angles/viewpoints.

2.2 Data Pre-processing Module

Camera Intrinsic Parameters and Initial Pose Estimation. We exploit OpenCV[1] to pre-process the chessboard and ArUco marker data collected in Sect. 2.1 for each video to derive their corresponding intrinsic parameters and initial poses.

Pose Calculation for New Viewpoint. Once the initial pose P for a camera is determined, the new pose P' for a new viewpoint after the turntable rotates by an angle of θ can be estimated by $P' = RP$. The transformation matrix R is computed as

$$R = I + sin(\theta)N + (1 - cos(\theta)N^2),$$

where I is the identity matrix and N is the antisymmetric matrix of z-axis vector (i.e., (0, 0, 1)). If the travel time of one round and the video frame rate are known, the pose for a sampled frame (i.e., a sampled viewpoint) can be easily calculated.

Object Mask Generation. To achieve more accurate 3D modeling results, our 3D-Producer additionally integrates PaddleSeg [5,12] into the system for one-click object mask generation. This step is optional depending on the users.

2.3 3D Shape Reconstruction Module

Our 3D-Productor adopts Neus [27] for 3D shape reconstruction of the target object by taking the pre-processed data from Sect. 2.2. Specifically, Neus proposes to represent the surface of an object as the zero-level set of a signed distance function, which achieves state-of-the-art 3D reconstruction performances on several benchmarks. The training objective of NeuS is to minimize the difference between the rendered 2D image and the ground truth of a specific viewpoint. When the training finishes, we apply marching cubes algorithm [15] to convert the implicit 3D representation to explicit polygon mesh.

[1] https://github.com/opencv/opencv.

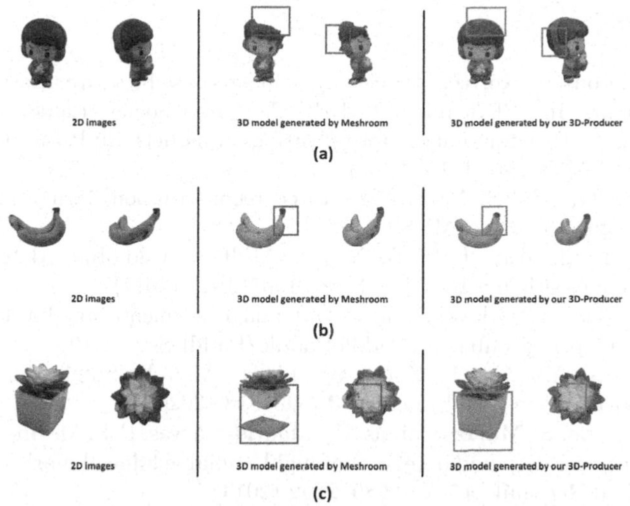

Fig. 2. Examples of textured 3D models generated by our 3D-Producer system and Meshroom. The three columns in each example are the pre-processed 2D images sampled from the videos taken by the users, the models generated by Meshroom and our 3D-Producer, respectively. Zoom in for better view.

2.4 Texture Generation Module

The reconstructed 3D model from Sect. 2.3 is without any texture. To improve the realism of the 3D model, we further integrates a multi-band image blending algorithm [1,2] into the system to generate the texture for the 3D model. Particularly, we reuse the stable and outstanding implementation of the texturing algorithm from Meshroom[2].

3 Results

In this part, we show several textured 3D models generated by our 3D-Producer system and compare with those produced by Meshroom which is based on Structure-from-Motion [24] in Fig. 2, where the three columns illustrate the pre-processed 2D images sampled from the videos taken by the users, the models generated by Meshroom and our 3D-Producer, respectively. As shown, our 3D-Producer generates more visually pleasing results by adopting deep model NeuS for accurate 3D shape reconstruction while Meshroom outputs 3D models with less accurate shapes due to the failure of correspondence point matching, which is the key factor in Structure-from-Motion. Specifically, for the example on the first row (i.e., Fig. 2(a)), some parts of the toy's head are missing in the model generated by Meshroom while the corresponding parts in the model created by our 3D-Producer are complete. Similar results are observed in the example on the third row (i.e., Fig. 2(c)).

2 https://github.com/alicevision/meshroom.

References

1. Allène, C., Pons, J., Keriven, R.: Seamless image-based texture atlases using multi-band blending. In: ICPR, pp. 1–4. IEEE Computer Society (2008)
2. Baumberg, A.: Blending images for texturing 3d models. In: Rosin, P.L., Marshall, A.D. (eds.) BMVC, pp. 1–10 (2002)
3. Cernea, D.: OpenMVS: Multi-view stereo reconstruction library (2020). http://cdcseacave.github.io/openMVS
4. Chen, X., Ma, H., Wan, J., Li, B., Xia, T.: Multi-view 3d object detection network for autonomous driving. In: CVPR, pp. 1907–1915 (2017)
5. Contributors, P.: Paddleseg, end-to-end image segmentation kit based on paddlepaddle. http://github.com/PaddlePaddle/PaddleSeg (2019)
6. Deng, J., Pan, Y., Yao, T., Zhou, W., Li, H., Mei, T.: Single shot video object detector. IEEE Trans. Multimedia **23**, 846–858 (2020)
7. Garrido-Jurado, S., Muñoz-Salinas, R., Madrid-Cuevas, F.J., Marín-Jiménez, M.J.: Automatic generation and detection of highly reliable fiducial markers under occlusion. Pattern Recognit. **47**(6), 2280–2292 (2014)
8. Gkioxari, G., Johnson, J., Malik, J.: Mesh R-CNN. In: ICCV. IEEE (2019)
9. Griwodz, C., Gasparini, S., Calvet, L., Gurdjos, P., Castan, F., Maujean, B., Lillo, G.D., Lanthony, Y.: Alicevision Meshroom: an open-source 3D reconstruction pipeline. In: Proceedings of the 12th ACM Multimedia Systems Conference - MMSys 2021. ACM Press (2021)
10. He, K., Zhang, X., Ren, S., Sun, J.: Deep residual learning for image recognition. In: CVPR, pp. 770–778 (2016)
11. Li, Y., Yao, T., Pan, Y., Mei, T.: Contextual transformer networks for visual recognition. IEEE Trans. Pattern Anal. Mach. Intell. (2022)
12. Liu, Y., Chu, L., Chen, G., Wu, Z., Chen, Z., Lai, B., Hao, Y.: Paddleseg: a high-efficient development toolkit for image segmentation (2021)
13. Lombardi, S., Simon, T., Saragih, J.M., Schwartz, G., Lehrmann, A.M., Sheikh, Y.: Neural volumes: learning dynamic renderable volumes from images. ACM Trans. Graph. **38**(4), 65:1–65:14 (2019)
14. Long, F., Qiu, Z., Pan, Y., Yao, T., Luo, J., Mei, T.: Stand-alone inter-frame attention in video models. In: Proceedings of the IEEE/CVF Conference on Computer Vision and Pattern Recognition, pp. 3192–3201 (2022)
15. Lorensen, W.E., Cline, H.E.: Marching cubes: a high resolution 3d surface construction algorithm. In: Stone, M.C. (ed.) SIGGRAPH. ACM (1987)
16. Mescheder, L.M., Oechsle, M., Niemeyer, M., Nowozin, S., Geiger, A.: Occupancy networks: learning 3d reconstruction in function space. In: CVPR, pp. 4460–4470 (2019)
17. Mildenhall, B., Srinivasan, P.P., Tancik, M., Barron, J.T., Ramamoorthi, R., Ng, R.: NeRF: representing scenes as neural radiance fields for view synthesis. In: Vedaldi, A., Bischof, H., Brox, T., Frahm, J.-M. (eds.) ECCV 2020. LNCS, vol. 12346, pp. 405–421. Springer, Cham (2020). https://doi.org/10.1007/978-3-030-58452-8_24
18. Pan, Y., Chen, Y., Bao, Q., Zhang, N., Yao, T., Liu, J., Mei, T.: Smart director: an event-driven directing system for live broadcasting. TOMM **17**(4), 1–18 (2021)
19. Pan, Y., Li, Y., Yao, T., Mei, T., Li, H., Rui, Y.: Learning deep intrinsic video representation by exploring temporal coherence and graph structure. In: IJCAI, pp. 3832–3838 (2016)

20. Pan, Y., Yao, T., Li, Y., Mei, T.: X-linear attention networks for image captioning. In: CVPR (2020)
21. Qi, C.R., Su, H., Mo, K., Guibas, L.J.: Pointnet: Deep learning on point sets for 3d classification and segmentation. In: CVPR, pp. 652–660 (2017)
22. Rematas, K., Liu, A., Srinivasan, P.P., Barron, J.T., Tagliasacchi, A., Funkhouser, T.A., Ferrari, V.: Urban radiance fields. CoRR abs/2111.14643 (2021)
23. Ren, S., He, K., Girshick, R., Sun, J.: Faster r-cnn: towards real-time object detection with region proposal networks. IEEE Trans. Pattern Anal. Mach. Intell. **39**(6), 1137–1149 (2016)
24. Schönberger, J.L., Frahm, J.M.: Structure-from-motion revisited. In: CVPR (2016)
25. Schönberger, J.L., Zheng, E., Pollefeys, M., Frahm, J.M.: Pixelwise view selection for unstructured multi-view stereo. In: ECCV (2016)
26. Sitzmann, V., Thies, J., Heide, F., Nießner, M., Wetzstein, G., Zollhöfer, M.: Deepvoxels: Learning persistent 3d feature embeddings. In: CVPR, pp. 2437–2446 (2019)
27. Wang, P., Liu, L., Liu, Y., Theobalt, C., Komura, T., Wang, W.: Neus: Learning neural implicit surfaces by volume rendering for multi-view reconstruction. In: NeurIPS, pp. 27171–27183 (2021)
28. Yao, T., Li, Y., Pan, Y., Wang, Y., Zhang, X.P., Mei, T.: Dual vision transformer. arXiv preprint arXiv:2207.04976 (2022)
29. Yao, T., Pan, Y., Li, Y., Ngo, C.W., Mei, T.: Wave-vit: unifying wavelet and transformers for visual representation learning. In: Proceedings of the European Conference on Computer Vision (ECCV) (2022)
30. Zhang, Z.: A flexible new technique for camera calibration. IEEE Trans. Pattern Anal. Mach. Intell. **22**(11), 1330–1334 (2000)

Contactless Cardiogram Reconstruction Based on the Wavelet Transform via Continuous-Wave Radar

Shuqin Dong, Changzhan Gu(✉), and Xiaokang Yang

MoE Key Lab of Artificial Intelligence, Shanghai Jiao Tong University, Shanghai, China
changzhan@sjtu.edu.cn

Abstract. Cardiogram is an important representation of the human's heart function. However, the most of the widely used cardiograms need contact sensor like electrocardiogram (ECG) and phonocardiogram (PCG), which is uncomfortable for daily healthcare. In this paper, we present a contactless radar-based cardiogram reconstruction technique. This technique includes linear motion recreation process and continuous wavelet transform. At the first step of the proposed technique, the linear signal demodulation algorithm is employed to obtain the accurate reconstruction of the phase shift caused by chest-wall motion detected by continuous-wave radar. Secondly, wavelet transform is used for cardiogram extraction from recovered chest-wall motion. To validate the accuracy of the recreated cardiogram, the error of RR interval with respect to ECG was also computed. With a custom-designed 24 GHz continuous-wave radar, the practical experiment was carried out. The results show that the RR intervals estimation reached 99.3% accuracy with respect to the simultaneously detected ECG signal, which shows the proposed technique has the great contactless cardiogram reconstruction performance.

Keywords: Continous-wave radar · Cardiogram reconstruction · Linear signal demodulation algorithm

1 Introduction

As a common diagnostic basis for heart disease, cardiograms has gained a huge popularity in clinical diagnosis [1,2]. However, they usually rely on contact electrodes to collect heart electrical activity signals, such as electrocardiogram (ECG) and seismocardiogram (SCG). The recent study has documented a "RF-SCG" and "Doppler cardiogram" (DCG) detected remotely by radar. They both describe the combined atrial and ventricular motions conducted to the skin [4,5], and provides the similar waveform characteristic with respect to SCG signal.

The operation principle of radar-based vital sign detection is that the transmitted wave can be scattered by human chest and the received signal would contain the chest-wall displacement information as shown in Fig. 1 [3]. Obviously,

L. Fang et al. (Eds.): CICAI 2022, LNAI 13606, pp. 532–536, 2022.
https://doi.org/10.1007/978-3-031-20503-3_44

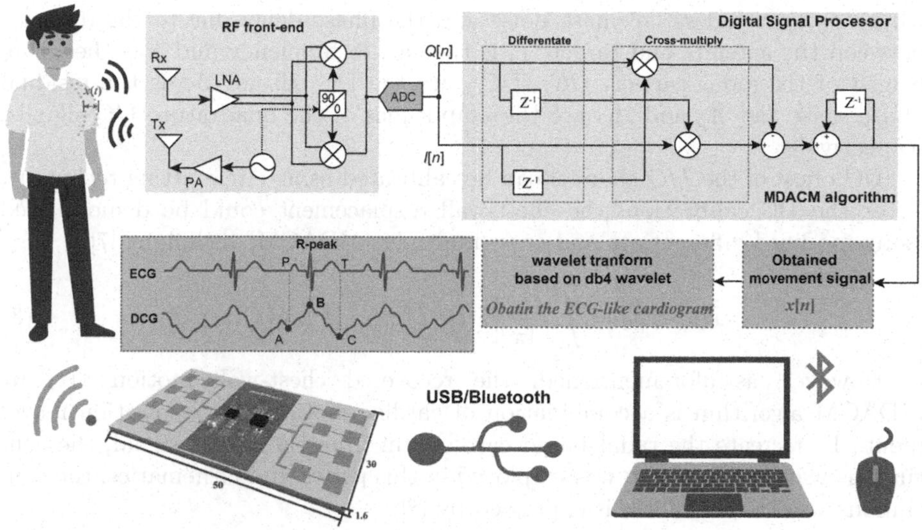

Fig. 1. Contactless cardiogram reconstruction based on the wavelet transform via continuous-wave radar.

the obtained signal is a mixed signal containing both respiration and heartbeat motions. The aforementioned measurement of RF-SCG and DCG requires subject hold his/her breath to obtain accurate waveform. For normal situation, the RF-SCG and DCG waveform would be easily distorted by the large interference due to respiration.

In this work, we present a contactless radar-based cardiogram reconstruction technique. This technique includes linear motion recreation process and continuous wavelet transform. The experimental results show that the extracted cardiogram aligned well with the ECG signal, and the RR intervals estimated by radar-based cardiogram reached 99.3% accuracy with respect to the simultaneously detected ECG signal, which indicates the proposed technique can be used for further contactless heart rate variability (HRV) measurement [6].

2 Theory

According to the principle of continuous-wave radar, the baseband signals containing the information of target motion $x(t)$ can be expressed as:

$$I(t) = A_I \cdot \cos\left[\frac{4\pi x(t)}{\lambda} + \phi_0 + \Delta\phi\right] + DC_I \quad (1)$$

$$Q(t) = A_Q \cdot \sin\left[\frac{4\pi x(t)}{\lambda} + \phi_0 + \Delta\phi\right] + DC_Q \quad (2)$$

where $\phi_0(t)$ includes the phase noise and the phase delay due to the distance between the antenna and target, f_c is the carrier frequency and λ is the wavelength of the radar carrier, DC_I/DC_Q are the DC offsets, $\Delta\phi$ is the residual phase noise and A_I and A_Q are the amplitudes of the quadrature I/Q signals, respectively.

DC offest of the I/Q channel can be calibrated using circle fitting technique. After the DC calibration, the chest-wall displacement could be demodulated using modified differentiate and cross-multiply (MDACM) technique [7]:

$$x(t) + d_0 = \int_0^t \frac{\lambda}{4\pi}\left[I(t)\cdot Q'(t) - I'(t)\cdot Q(t)\right] \tag{3}$$

However, as aforementioned, the recovered chest-wall motion $x(t)$ by MDACM algorithm is a combination of cardiac motion and respiration movement. To recreate the radar-based cardiogram from the mixed signal, the continuous wavelet transform was employed in this paper. In mathematics, the continuous wavelet transform is expressed by [8]:

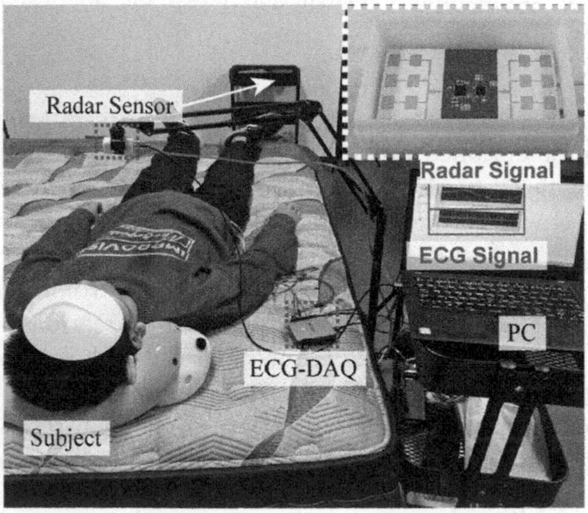

Fig. 2. The experimental setup. The insert is the picture of custom-designed radar system.

$$CWT(a,b) = \frac{1}{|\sqrt{a}|}\int_{-\infty}^{+\infty} x(t)\Psi^*\left(\frac{t-b}{a}dt\right) \tag{4}$$

where $x(t)$ is the signal being processed, $\Psi^*(t)$ is the complex conjugate of the continuous mother wavelet $\Psi(t)$, and a and b are the scale and translational value, respectively. $\Psi\left(\frac{t-b}{a}dt\right)$ is the daughter wavelet which is simply the translated and scaled versions of the mother wavelet. In this work, the *db*4 wavelets [9] have been selected for extracting characteristic waves.

3 Experiment

Figure 2 show the experimental setup. The experiment was performed in a regular laboratory environment on a 24-year-old man who weighted 60kg and 182 tall. The subject was measured in supine posture on the bed with their chest facing upside. The radar sensor was attached to a fixed construction and was placed around 40-cm above the chest of the subject. During the measurement, the subject was breathing normally. For comparison, a ECG device manufactured based on chip AD8232 was employed to collect the synchronized ECG data. Besides, a multi-sensor data synchronized acquisition system based on NI DAQ is designed to sample these signals with the same clock.

The experimental results are shown in Fig. 3. It is seen that the chest-wall displacement reconstructed by MDACM algorithm is not flat, because it is a combination signal of respiration and heart motion. Figure 3(b) shows the cardiogram detection results by ECG and contactless ECG-like cardiogram from radar. Obviously, it is seen that the extracted ECG-like cardiogram has well correspondence with ECG signal, and the peaks on extracted waveform occur at the same moment with R-peak of ECG. Figure 3(c) shows The root mean square error (RMSE) and the normalized RMSE (NRMSE) of RR intervals are 5.5 ms/0.69%.

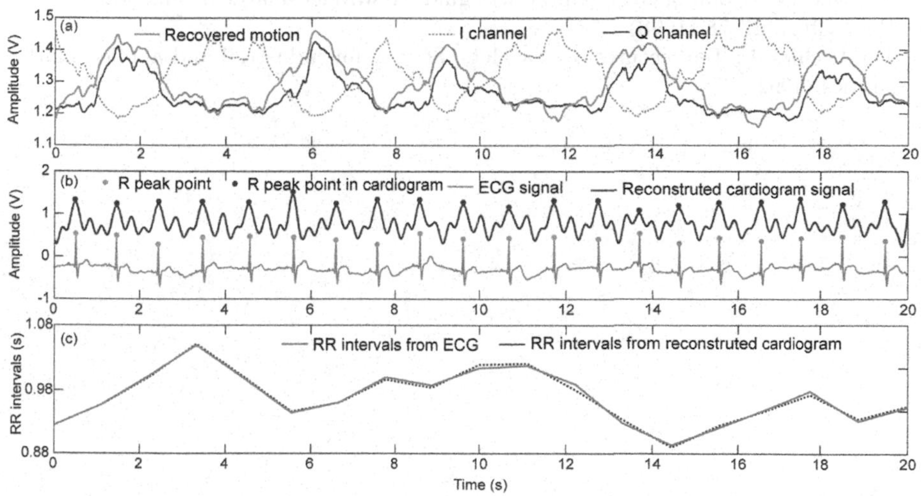

Fig. 3. The experimental results. (a) shows the baseband signals and the recovered chest-wall motion via MDACM algorithm. (b) shows the recreated cardiogram signal via wavelet transform and the simultaneously detected ECG signal. (c) The RR intervals estimation results by ECG signal and reconstructed cardiogram.

Acknowledgement. This work was supported in part by the National Science Foundation of China under Grant 62171277, and in part by the Shanghai Municipal Science and Technology Major Project under Grant 2021SHZDZX0102.

References

1. Paiva, R.P., et al.: Beat-to-beat systolic time-interval measurement from heart sounds and ECG. Physiol. Meas. **33**(2), 177–194 (2012)
2. Inan, O.T., et al.: Ballistocardiography and seismocardiography: a review of recent advances. IEEE J. Biomed. Health Inform. **19**(4), 1414–1427 (2015)
3. Zhang, Y., Qi, F., Lv, H., Liang, F., Wang, J.: Bioradar technology: recent research and advancements. IEEE Microw. Mag. **20**(8), 58–73 (2019)
4. Ha, U., Assana, S., Adib, F.: Contactless seismocardiography via deep learning radars. In: Proceedings of the 26th Annual International Conference on Mobile Computing and Networking, pp. 1–14 (2020)
5. Dong, S., et al.: Doppler cardiogram: a remote detection of human heart activities. IEEE Trans. Microw. Theory Techn. **68**(3), 1132–1141 (2020)
6. Xhyheri, B., Manfrini, O., Mazzolini, M., et al.: Heart rate variability today. Prog. Cardiovasc. Dis. **55**(3), 321–331 (2012)
7. Xu, W., Li, Y., Gu, C., et al.: Large displacement motion interferometry with modified differentiate and cross-multiply technique. IEEE Trans. Microw. Theory Techn. **69**(11), 4879–4890 (2021)
8. Torrence, C., Compo, G.P.: A practical guide to wavelet analysis. Bull. Am. Meteor. Soc. **79**(1), 61–78 (1998)
9. Daubechies, I.: Ten lectures on wavelets. Society for industrial and applied mathematics (1992)

Intelligent Data Extraction System for RNFL Examination Reports

Chunjun Hua[1], Yiqiao Shi[1], Menghan Hu[1](✉), and Yue Wu[2](✉)

[1] East China Normal University, Shanghai, China
{51205904049,71215904075}@stu.ecnu.edu.cn,
mhhu@ce.ecnu.edu.cn

[2] Ninth People's Hospital Affiliated to Shanghai Jiaotong University School of Medicine, Shanghai, China
eyedrwuyue@gmail.com

Abstract. Glaucoma is the collective term for a group of diseases that cause damage to the optic nerve. Retina nerve fiber layer (RNFL) thickness is an indicative reference for evaluating the progression of glaucoma. In this demo paper, we proposed an intelligent data extraction system for RNFL examination report, which can extract the RNFL thickness data from the report photo. The system consists of two procedures viz. target area segmentation and structure data extraction. This system can reduce the amount of data that needs to be entered manually, thus reducing the manual workload in electronic health records (EHRs) system. The demo video of the proposed system is available at: https://doi.org/10.6084/m9.figshare.20098865.v1.

Keywords: Glaucoma · Computer vision · Image processing · Electronic health records

1 Introduction

With the rapid development of information technology, electronic health records (EHRs) are increasingly used in the management of various diseases [8,17,18]. Glaucoma, as a lifelong disease, can also benefit from EHR system, as studies have shown that this system can greatly improve the management of glaucoma [5, 9,13,15]. More importantly, with strong capabilities of data collection, EHR system also makes it possible to apply data-driven deep learning to glaucoma research [2,11,20]. In addition, the previous application of deep learning and image processing in glaucoma treatment [21,23] can also benefit more patients and doctors with the help of EHR system.

Retina nerve fiber layer (RNFL) thickness is an indicative reference for evaluating the progression of glaucoma [10,19], so RNFL data is one of the common data types in EHR systems. Typically, RNFL data is obtained through regular RNFL examination. Therefore, for patient-facing EHR systems, RNFL data is often presented as photos of RNFL examination reports.

L. Fang et al. (Eds.): CICAI 2022, LNAI 13606, pp. 537–542, 2022.
https://doi.org/10.1007/978-3-031-20503-3_45

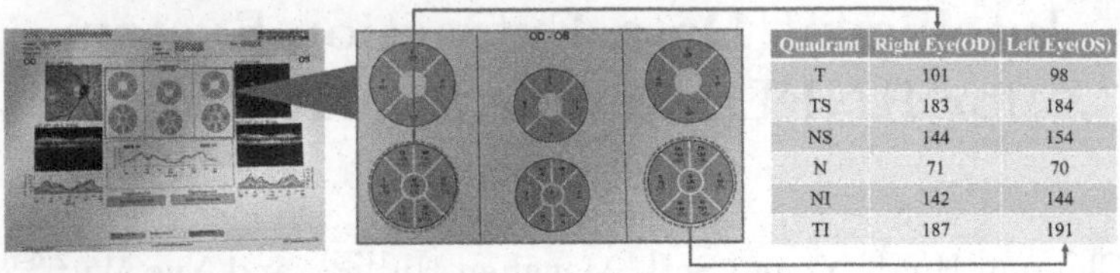

Fig. 1. One example of RNFL examination report and the data that needs to be extracted.

Figure 1 shows what data should be extracted from RNFL examination report. For each eye there are six quadrants of data that indicate the thickness of the layer, which add up to total 12 groups of data for each report. This is a challenge for the data entry personnel of EHR systems, especially when the upload volume is so large that the amount of data to be entered increases exponentially. Hence, the development of intelligent extraction system for such reports is of considerable importance to the reduction of manual workload.

2 Methods

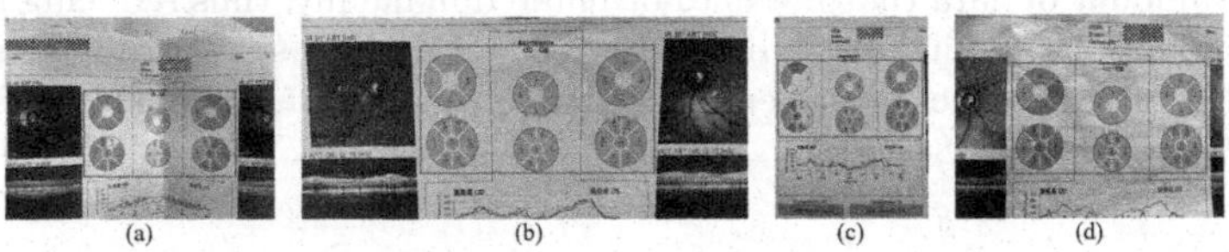

Fig. 2. Actual photos uploaded by users: (a)–(d) show four example images uploaded by users, and the region within red box contains the target data. (Color figure online)

The RNFL report is generally one kind of template document, on which structured data recognition and extraction can be easily carried out to obtain the needed data. In practice, however, the photos uploaded by users are often incomplete, with the considerable portion of report missing, as demonstrated in Fig. 2. Although incomplete, these images all contain the area where the target 12 sets of data are located. We can consider this area as the template document and apply structured data extraction to it so that the target data can be obtained. Therefore, the whole extraction method can be divided into two steps: target area segmentation and structure data extraction.

Figure 3 shows the pipeline of the proposed system. The target area recognition and segmentation is carried out based on YOLOv3 [14]. Combining the essence of ResNet [6], DenseNet [7] and FPN [12], YOLOv3 is a mature and efficient solution for object detection. Thanks to its efficient network structure and training strategy, YOLOv3, in our case, reached an average 99.4% recognition accuracy with only 100 labeled RNFL report images as training set.

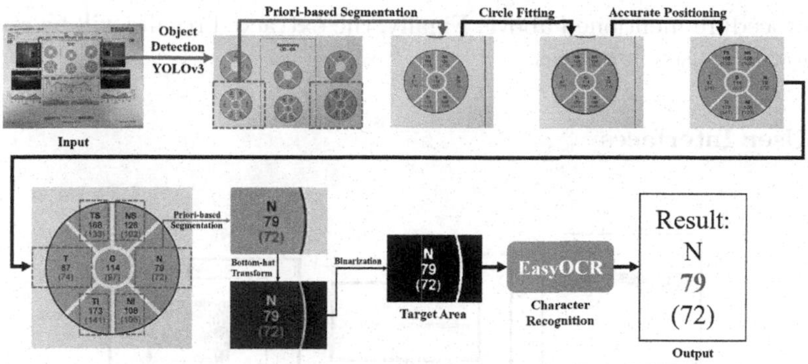

Fig. 3. Pipeline of the proposed system

As for structured data extraction, it consists of a series of image processing and optical character recognition (OCR) procedure. First, the segmented area will be re-segmented based on location priors to get approximate areas containing the two circles corresponding to left and right eyes. Second, Hough circle detection [22] is carried out to get the precise position of the circle area. Then the areas containing the six quadrants of data are obtained through another prior-based segmentation. After that, each area is binarized and sent to OCR module-EasyOCR [1]. As an open-source Python library based on deep learning, EasyOCR draws on the CRNN network put forward by An et al. [16], which is composed of three main modules: feature extraction (ResNet [6]), sequence labeling (LSTM [3]) and decoding (CTC [4]). Finally, the numbers in the images are recognized as the final output.

3 Demonstration Setup

3.1 System Design

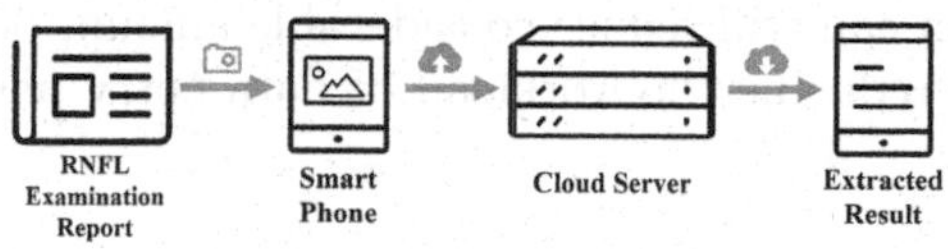

Fig. 4. Workflow of the developed system

Figure 4 shows the workflow of the whole demonstration system. In our demonstration, the user needs to prepare the photo of the RNFL report. Then the developed program is used to do image processing and data communication based on front and back-end architecture. The photo of the RNFL report is first uploaded to the cloud sever. The server will then carry out image processing and

OCR procedure mentioned above. Finally, the extracted results will be returned and shown on user interfaces.

3.2 User Interfaces

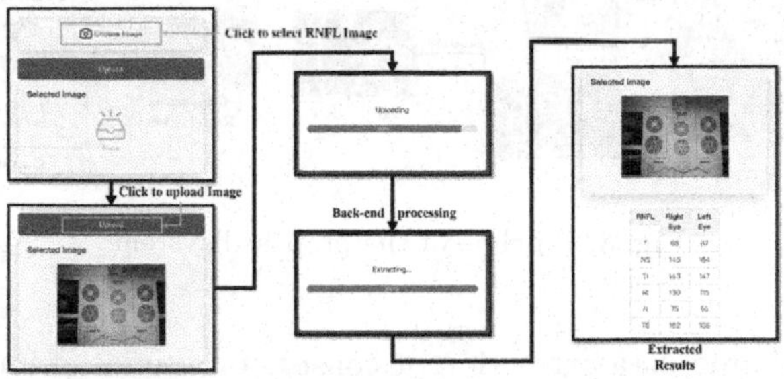

Fig. 5. User Interfaces of the demonstration program

Figure 5 shows the specially designed user interfaces of our demonstration program, which are simple and clear, with high accessibility. The app offers a variety of options that users can either upload the image from photo library of their smart phones or take the photo of the report directly on site. With the selected photo displayed on the screen, users can easily verify the consistency of the result if necessary.

4 Visitors Experience

The system turned out to be trouble-free and user-oriented according to the feedback from the users. We tested the proposed system on 77 user-uploaded RNFL report photos from our partner's EHR system. It achieved an average recognition accuracy of 72.40% in our validation test. Although not completely accurate, this system can still reduce considerable amount of data that needs to be entered, thereby reducing the workload of data entry personnel to a certain extent.

5 Conclusion

In this paper, the intelligent data extraction system is proposed for retina nerve fiber layer examination report. With user-friendly interface, the demo program can recognize and extract the needed data from the uploaded photo of RNFL reports. Implemented in electronic health records systems, the proposed extraction system can reduce the manual workload for data entry.

References

1. Easyocr homepage. https://www.jaided.ai/easyocr/
2. Baxter, S.L., Marks, C., Kuo, T.T., Ohno-Machado, L., Weinreb, R.N.: Machine learning-based predictive modeling of surgical intervention in glaucoma using systemic data from electronic health records. Am. J. Ophthalmol. **208**, 30–40 (2019). https://doi.org/10.1016/j.ajo.2019.07.005
3. Cheng, J., Dong, L., Lapata, M.: Long short-term memory-networks for machine reading, pp. 551–561, January 2016. https://doi.org/10.18653/v1/D16-1053
4. Graves, A., Fernández, S., Gomez, F., Schmidhuber, J.: Connectionist temporal classification: Labelling unsegmented sequence data with recurrent neural networks. In: ICML 2006, pp. 369–376. Association for Computing Machinery, New York (2006). https://doi.org/10.1145/1143844.1143891
5. Hamid, M.S., Valicevic, A., Brenneman, B., Niziol, L.M., Stein, J.D., Newman-Casey, P.A.: Text parsing-based identification of patients with poor glaucoma medication adherence in the electronic health record. Am. J. Ophthalmol. **222**, 54–59 (2021). https://doi.org/10.1016/j.ajo.2020.09.008
6. He, K., Zhang, X., Ren, S., Sun, J.: Deep residual learning for image recognition. In: 2016 IEEE Conference on Computer Vision and Pattern Recognition (CVPR), pp. 770–778 (2016). https://doi.org/10.1109/CVPR.2016.90
7. Huang, G., Liu, Z., Van Der Maaten, L., Weinberger, K.Q.: Densely connected convolutional networks. In: 2017 IEEE Conference on Computer Vision and Pattern Recognition (CVPR), pp. 2261–2269 (2017). https://doi.org/10.1109/CVPR.2017.243
8. Jhaveri, R., John, J., Rosenman, M.: Electronic health record network research in infectious diseases. Clin. Ther. **43**(10), 1668–1681 (2021). https://doi.org/10.1016/j.clinthera.2021.09.002
9. Lee, E.B., Hu, W., Singh, K., Wang, S.Y.: The association among blood pressure, blood pressure medications, and glaucoma in a nationwide electronic health records database. Ophthalmology **129**(3), 276–284 (2022)
10. Leung, C.K., et al.: Retinal nerve fiber layer imaging with spectral-domain optical coherence tomography: analysis of the retinal nerve fiber layer map for glaucoma detection. Ophthalmology **117**(9), 1684–1691 (2010). https://doi.org/10.1016/j.ophtha.2010.01.026
11. Li, F., et al.: A deep-learning system predicts glaucoma incidence and progression using retinal photographs. J. Clin. Investigation **132**(11), June 2022. https://doi.org/10.1172/JCI157968
12. Lin, T.Y., Dollár, P., Girshick, R., He, K., Hariharan, B., Belongie, S.: Feature pyramid networks for object detection. In: 2017 IEEE Conference on Computer Vision and Pattern Recognition (CVPR), pp. 936–944 (2017). https://doi.org/10.1109/CVPR.2017.106
13. Pandit, R.R., Boland, M.V.: The impact of an electronic health record transition on a glaucoma subspecialty practice. Ophthalmology **120**(4), 753–760 (2013). https://doi.org/10.1016/j.ophtha.2012.10.002
14. Redmon, J., Farhadi, A.: Yolov3: an incremental improvement. arXiv e-prints (2018)
15. Robbins, C.C., et al.: An initiative to improve follow-up of patients with glaucoma. Ophthalmol. Sci. **1**(4), 100059 (2021). https://doi.org/10.1016/j.xops.2021.100059

16. Shi, B., Bai, X., Yao, C.: An end-to-end trainable neural network for image-based sequence recognition and its application to scene text recognition. IEEE Trans. Pattern Anal. Mach. Intell. **39**(11), 2298–2304 (2017). https://doi.org/10.1109/tpami.2016.2646371
17. Singer, E., et al.: Characterizing sleep disorders in an autism-specific collection of electronic health records. Sleep Med. **92**, 88–95 (2022). https://doi.org/10.1016/j.sleep.2022.03.009
18. Taxter, A.J., Natter, M.D.: Using the electronic health record to enhance care in pediatric rheumatology. Rheumatic Disease Clinics North Am. **48**(1), 245–258 (2022). https://doi.org/10.1016/j.rdc.2021.08.004
19. Tuulonen, A., Airaksinen, P.J.: Initial glaucomatous optic disk and retinal nerve fiber layer abnormalities and their progression. Am. J. Ophthalmol. **111**(4), 485–490 (1991). https://doi.org/10.1016/S0002-9394(14)72385-2
20. Wang, S.Y., Tseng, B., Hernandez-Boussard, T.: Deep learning approaches for predicting glaucoma progression using electronic health records and natural language processing. Ophthalmol. Sci. **2**(2), 100127 (2022). https://doi.org/10.1016/j.xops.2022.100127
21. Yao, C., Tang, J., Hu, M., Wu, Y., Guo, W., Li, Q., Zhang, X.P.: Claw u-net: a unet variant network with deep feature concatenation for scleral blood vessel segmentation. Lecture Notes in Computer Science, pp. 67–78 (2021). https://doi.org/10.1007/978-3-030-93049-3_6
22. Yuen, H., Princen, J., Illingworth, J., Kittler, J.: Comparative study of hough transform methods for circle finding. Image Vis. Comput. **8**(1), 71–77 (1990). https://doi.org/10.1016/0262-8856(90)90059-E
23. Zhang, L., Wu, Y., Hu, M., Guo, W.: Automatic image analysis of episcleral hemangioma applied to the prognosis prediction of trabeculotomy in sturge-weber syndrome. Displays **71**, 102118 (2022). https://doi.org/10.1016/j.displa.2021.102118

NeRFingXR: An Interactive XR Tool Based on NeRFs

Shunli Luo[1,4], Hanxing Li[3,4], Haijing Cheng[4], Shi Pan[2], and Shuangpeng Sun[2](✉)

[1] Liaoning Technical University, Fuxin, China
[2] BNRist, Tsinghua University, Beijing, China
pengcheng786@gmail.com
[3] Northwest A&F University, Yangling, China
[4] Hangzhou Yilan Technology, Hangzhou, China

Abstract. With the rise of the metaverse, XR technologies have gained wide attention. However, traditional XR technologies require a high precision explicit 3D scene appearance and geometric model representation to achieve a more realistic visual fusion effect, which will make XR technologies require high computational power and memory capacity. Recent studies have shown that it is feasible to implicitly encode 3D scene appearance and geometric models by position-based MLPs, of which NeRF is a prominent representative. In this demo, we propose an XR tool based on NeRF that enables convenient and interactive creation of the XR environments. Specifically, we first train the NeRF model of XR content using Instant-NGP to achieve an efficient implicit 3D representation of XR content. Second, we contribute a depth-awareness scene understanding approach that automatically adapts different plane surfaces for XR content placement and more realistic real-virtual occlusion effects. Finally, we propose a multi-nerf joint rendering method to achieve natural XR content occlusion from each other. This demo shows the final result of our interactive XR tool.

Keywords: Mixed reality · Neural radiance fields

1 Introduction

The underlying logic of traditional XR technology is usually to use 3D technology, image analysis, tracking technology, real-time rendering technology and other graphic image processing technology, to fuse the real scene captured by the camera and the projection of virtual objects, to create the illusion of virtual objects really exist in the scene. However, the result of the fusion is not a complete projection of the real three-dimensional scene, there will always be some discrepancies between the picture and the real scene. For example, when inserting a 3D virtual object into a 2D image, if the orientation and size of the virtual object are not set correctly, it will make the inserted virtual object look abrupt on the fused image. A natural idea is to reconstruct the real 3D scene in

L. Fang et al. (Eds.): CICAI 2022, LNAI 13606, pp. 543–547, 2022.
https://doi.org/10.1007/978-3-031-20503-3_46

real time, then place the virtual objects in the 3D scene, and finally render the fused image. However, such an operation will bring huge computing and memory consumption, and it is difficult to achieve the real-time requirement when the resolution requirement is high.

Compared with the traditional XR technology, which represents 3D scenes explicitly, NeRF uses an MLP to represent the mathematical functions of parametric 3D scene appearance and geometric models, which greatly reduces the memory required for storing 3D scenes. Moreover, after NeRF [1] was proposed, a large number of research teams have been attracted to further explore this implicit 3D representation method. Among them, Object-NeRF [2] proposes a two-branch structure using different MLPs for representing the scene and object 3D model respectively. The 3D model of the object branch can be used for editing operations such as copying and rotation. For the original NeRF representation is a static scene, NeRF-Editing [3] proposes a deformable 3D scene representation, which can be combined with the traditional mesh matching method to learn the transformation relationship between the spatial points of the deformed and original static scene, thus realizing the deformable NeRF editing effects. The most recent advance of NeRF is Instant-NGP [4], which uses multi-resolution hash coding as input coding to increase the training and inference speed of NeRF by several orders of magnitude, realizing real-time rendering and greatly expanding the application scope of implicit 3D representations.

Our interactive XR tool is based on both depth-based 3D scene understanding and NeRFs. In which we use multi-resolution hash coding as input coding to accelerate the training and inference speed of NeRF and finally achieves realistic rendering under interactive frame rates. Second, an efficient depth plane detection method based on curvature residual analysis was proposed to realize 3D-scene aware rational placement of any XR contents. Finally, we organize the NeRF models corresponding to XR contents into a unified virtual scene based on the relative position relationship of multiple XR contents in the real scene, and achieve real-time joint rendering and natural mutual occlusion of multiple XR contents.

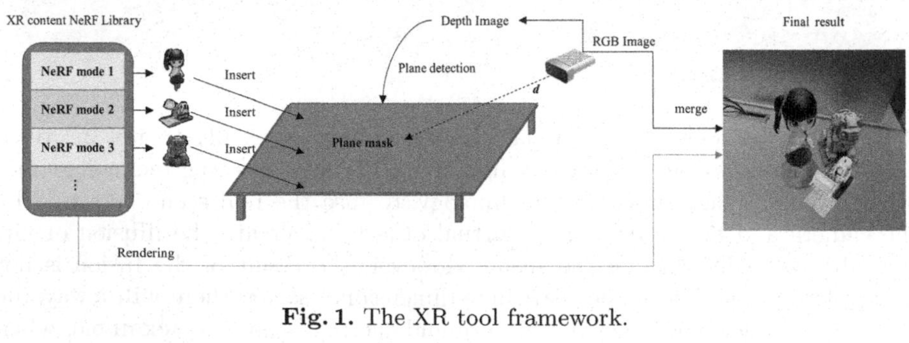

Fig. 1. The XR tool framework.

2 Demo Setup

The implementation of this interactive XR tool consists of three main steps: 1. Implicit 3D reconstruction based on NeRFs; 2. Detecting the planes from the 3D scene for inserting XR contents; 3. Render the XR content inserted in the scene and blend it with the real scene RGB image, as illustrated in Fig. 1.

To obtain an implicit 3D model of XR content, we first need to get the RGB image sequence of XR content, then use colmap preprocessing to generate the camera parameters for training the nerf, and finally, use multi-resolution hash coding as the input code to train the nerf model and save it.

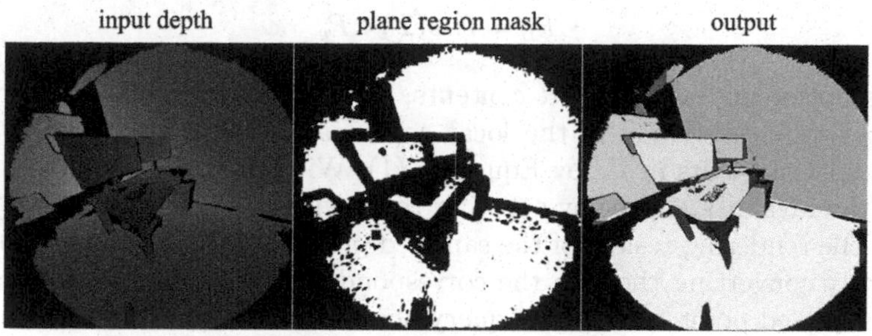

Fig. 2. Plane detection results.

In order to detect planes accurately in real time, we propose a new plane detection method for depth images based on curvature residual analysis. First, we use two mean filters with different kernel sizes to filter the depth image out from Kinect camera. Then, we calculate the curvature residual image which is the pixel value differences of these two smoothed images. The regions with small curvature residual can be seen as potential planar areas. With these areas we perform components connection analysis and select candidate regions, then convert pixels of each region to 3D points with camera intrinsic parameters. Finally, for each candidate point cloud region, principal component analysis (PCA) is implemented to get plane parameters and fitting mean square errors. Those candidate planes with small errors are identified as final planes.

Normally, we set the camera's pose relative to the nerf model, and then query in the nerf model to get the rendering result. However, in our XR tool, multiple nerf models are involved, and if each XR content is rendered once separately, this will greatly increase the burden of the system. Therefore, we propose a method for joint rendering of multiple nerf models.

First, we take the center P_{mean} of the horizontal plane with the largest area in the plane detection result as the origin of V_w, and the direction of the plane normal to the real kinect camera coordinate system as the Y-axis direction of V_w. And, it is assumed that the real kinect camera is placed horizontally, so the X-axis direction of V_w can be consistent with the X-axis direction of the real kinect camera coordinate system C_w. According to the Y-axis direction and X-axis direction of V_w, we can get the Z-axis direction of V_w. Also, we can get the transformation of C_w to V_w as T_{Cw2Vw} and set the virtual camera V_c pose as T_{Cw2Vw}.

Second, suppose we have XR content, whose local coordinate system is L, and the Y-axis axes of L are the same as the Y-axis axes of V_w. When we insert XR content into the point P under C_w, the transformation of L to V_w is T. The initial rotation matrix of T is the unit array, and the initial translation transformation is the coordinate difference between P and P_{mean} under C_w. Since P_{mean} is a known quantity and P can be obtained directly through the depth map, the coordinates P_L in L of any point P_v in V_w can be expressed as

$$P_L = Inv(T) * P_v \tag{1}$$

When inserting multiple XR contents, we can also establish the correspondence between the points in the local coordinate system of each XR content and the spatial points in V_w by Equation (1). With the correspondence, we can sample the rays in V_w space by emitting them with the virtual camera V_c as the center. The rendering results of the sampled points in V_w space can be obtained by directly converting them to the corresponding XR content local coordinate system sampled point coordinates query according to Equation (1), thus realizing the joint rendering of multiple XR contents. Moreover, when multiple XR contents exist in occlusion, it is possible to decide which XR content belongs to the true foreground based on which XR content intersection point the ray is closest to, thus realizing a true occlusion relationship.

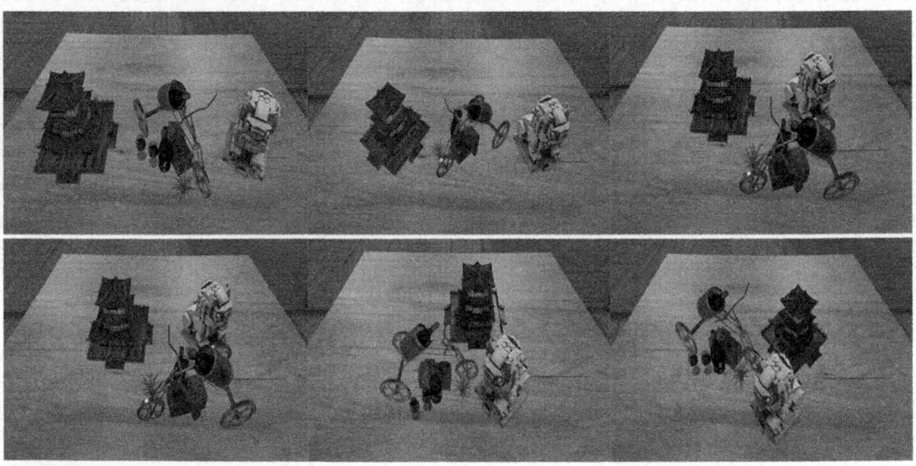

Fig. 3. XR tool results.

3 Results

The results of our proposed plane detection method are shown in Fig. 2. The left side of the figure is the input depth image, the middle is the detected planar areas which can be used for XR content placement, and the right side is the final plane mask. It is evident from the output image that our proposed method in this demo accurately detects all critical planes in the scene in real-time with minimal memory and CPU consumption. The final effects of the XR tool is shown in Fig. 3.

Acknowledgement. We thank Prof. Borong Lin and Tao Yu from Tsinghua University for their insightful comments and discussions.

References

1. Mildenhall, B., Srinivasan, P.P., Tancik, M., Barron, J.T., Ramamoorthi, R., Ng, R.: NeRF: representing scenes as neural radiance fields for view synthesis. In: Vedaldi, A., Bischof, H., Brox, T., Frahm, J.-M. (eds.) ECCV 2020. LNCS, vol. 12346, pp. 405–421. Springer, Cham (2020). https://doi.org/10.1007/978-3-030-58452-8_24
2. Bangbang Y., et al.: Learning object-compositional neural radiance field for editable scene rendering. In: Proceedings of the IEEE/CVF International Conference on Computer Vision (ICCV), pp. 13779–13788. IEEE, Montreal, Canada (2021)
3. Yu-Jie, Y., Yang-Tian, S., Yu-Kun, L., Yuewen, M., Rongfei, J., Lin, G.: NeRF-editing: geometry editing of neural radiance fields. In: Proceedings of the IEEE/CVF Conference on Computer Vision and Pattern Recognition (CVPR), pp. 18353–18364. IEEE, New Orleans, Louisiana (2022)
4. Thomas, M., Alex, E., Christoph, S., Alexander, K.: Instant neural graphics primitives with a multiresolution hash encoding. ACM Trans. Graph. **41**(4), 102:1–102:15 (2022). https://doi.org/10.1145/3528223.3530127

A Brain-Controlled Mahjong Game with Artificial Intelligence Augmentation

Xiaodi Wu[1], Yu Qi[2(✉)], Xinyun Zhu[1], Kedi Xu[1], Junming Zhu[3], Jianmin Zhang[3], and Yueming Wang[1(✉)]

[1] Qiushi Academy for Advanced Studies, Zhejiang University, Hangzhou, China
ymingwang@zju.edu.cn

[2] MOE Frontier Science Center for Brain Science and Brain-machine Integration, and the Affiliated Mental Health Center & Hangzhou Seventh People's Hospital, Zhejiang University School of Medicine, Hangzhou, China
qiyu@zju.edu.cn

[3] The Second Affiliated Hospital of Zhejiang University School of Medicine, Hangzhou, China

Abstract. Brain-computer interface (BCI) provides a direct pathway from the brain to external devices, which have demonstrated potential in rehabilitation and human-computer interaction. One limitation of current BCI lies in the low precision and stability in control, thus the user scenario and the interaction process play an important role in BCI applications. We propose to leverage the ability of artificial intelligence (AI) to improve the easy-of-use of BCIs. We designed a brain-controlled mahjong game with AI augmentation. As the user plays the game, an AI system continuously monitors the game and suggests several optimal candidate options, and the user directly selects from the candidates rather than from all the mahjong tiles in hand. This design enables the mahjong game with easy control, thus improving the user experience. Our system is evaluated by a human subject with an invasive BCI system. Results demonstrate that the subject can fluently control the system to play the mahjong game.

Keywords: Brain-computer interface · System design · AI augmentation

1 Introduction

Brain-computer interface (BCI) establishes a pathway from the brain to external devices such as computers and neuroprosthetics [1–4]. Especially, BCI provides a novel way of human-computer interaction, which gives rise to new applications such as BCI typewriters [5,6] and BCI games [7,8].

The challenge of current BCIs lies in the limited bandwidth and low precision in control. In continuous control tasks, the state-of-the-art non-invasive BCIs could mind-control a robotic arm to continuously track and follow a computer cursor [9]. Invasive

Demo video: https://bcidemo.github.io/BCIMahjongDemo/.

Supplementary Information The online version contains supplementary material available at https://doi.org/10.1007/978-3-031-20503-3_47.

L. Fang et al. (Eds.): CICAI 2022, LNAI 13606, pp. 548–553, 2022.
https://doi.org/10.1007/978-3-031-20503-3_47

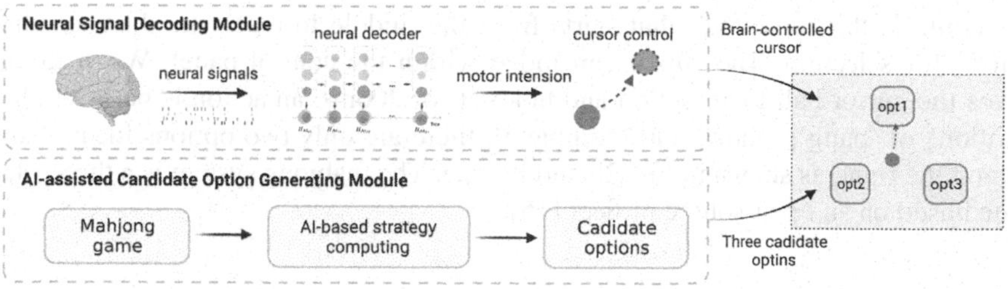

Fig. 1. Framework of the AI-based BCI mahjong system.

BCIs demonstrated higher bandwidth compared with non-invasive ones, while the control performance is far behind natural motor controls. Meanwhile, the nonstationary property of neural signals leads to unstable BCI control [10]. With the limitations of BCI systems, the system design and the user interaction process play an important role in leveraging the performance of BCIs. To improve the easy-of-use of BCI systems, efforts have been made. Several studies design special typewriter interfaces to facilitate BCI-based interactions [5, 11]. The Berlin Brain-Computer Interface presents the novel mental typewriter Hex-o-Spell which uses an appealing visualization based on hexagons, and improves the bandwidth of BCI systems [5]. Some approaches design games specifically tailored to the characteristics of direct brain-to-computer interaction [8, 12, 13]. Besides, a few studies involve AI language model to improve performance of BCI communication system [14].

Inspired by studies mentioned above, this study proposes an artificial intelligence augmentation approach to improve the human-computer interaction with a BCI system. We propose a BCI mahjong game with artificial intelligence assistance to improve the easy-of-use of BCI system. As the user plays the game, an AI system helps suggest optimal candidate options. Instead of directly choosing a tile from the tiles in hand, which usually requires delicate operations, the user can choose from the candidate options. In this way, the difficulty of operation can be highly reduced. A user can play the mahjong game with easy control, such that improves the user experience.

2 The BCI Mahjong System

The framework of the proposed BCI mahjong system is illustrated in Fig. 1. The system consists of two modules, a neural signal decoding module and an AI-based candidate option generating module. The neural decoding module decodes neural signals to control a computer cursor. The AI-based option generating module computes three optimal candidate options based on the current game situation. A user can control the cursor ball to choose an option from the candidates.

2.1 User Interface

The user interface is illustrated in Fig. 2a. The major part is a control panel where three candidate options generated by the AI system are placed at the top, left and right. The user controls the cursor ball that starts from the middle to reach the three candidate options for selection. The cursor can move within the control panel. When the user moves the cursor ball to an option and holds it for 200ms, an action is performed. For situations of 'pung', 'chow' and 'complete', there are only two options located at the left and the right, as shown in Fig. 2b and Fig. 2c. The mahjong system is a three-player game based on an open-source project [15].

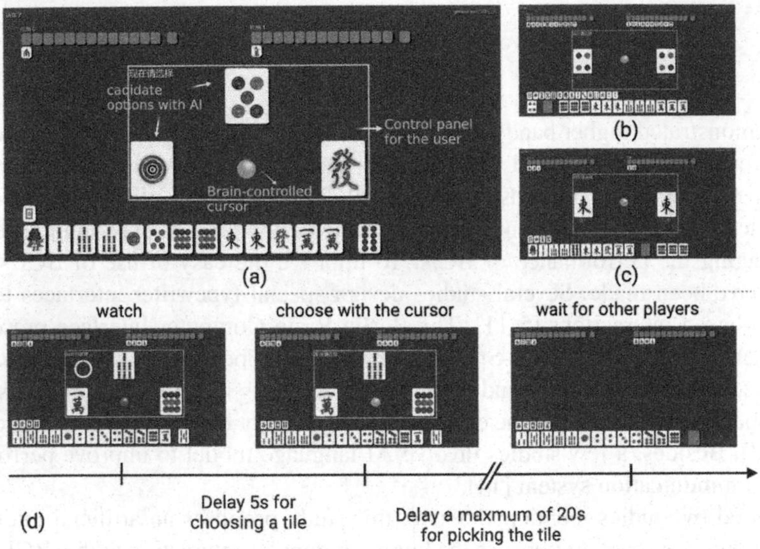

Fig. 2. User interface of the mahjong game.

The play process is illustrated in Fig. 2d. Firstly, the user watches and thinks for five seconds, during which the cursor ball does not move. Then the user controls the cursor to select an option. If the user fails to select an option within 20 s, the system will randomly select one option from the three candidates.

2.2 The AI Augmentation Algorithm

The goal of mahjong game is to start from current hand to reach ready hand in minimum steps. In the optimal case, we start from current hand and search all paths to find the fastest one. However, the high randomness and a lot of asymmetric information of mahjong bring a vast search space and sampling space to standard search algorithms. In order to obtain feasible solutions in acceptable time, we adopt the shanten-based heuristic evaluation algorithm, which uses the function based on the shanten number as the evaluation function.

We have some definitions as follows. The shanten number is the minimum number of tiles required by current hand before ready hand. The useful tile is the tile that can be drawn at next step that will decrease the shanten number. The improved tile is the tile

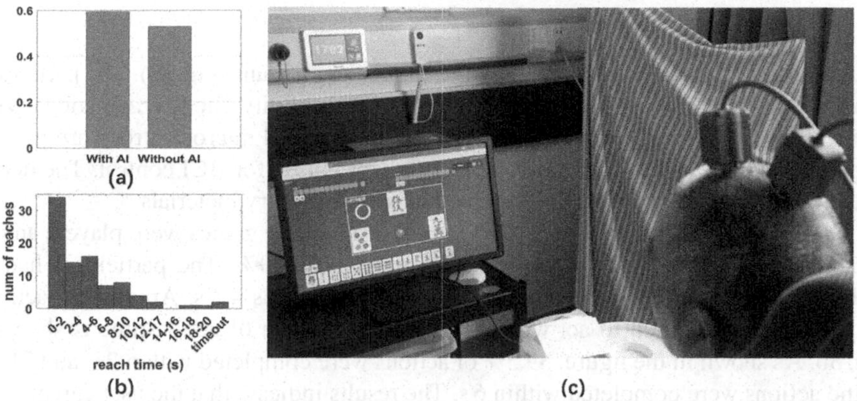

Fig. 3. Performance of the AI-augmented mahjong game.

that can be drawn at next step that will not decrease the shanten number but increase the number of useful tiles. When the player needs to play a tile, we use the evaluation function to evaluate all the actions as follows: $Score_0 = 9 - n_{shanten}, Score_1 =$ number of useful tiles, $Score_2 =$ number of improved tiles. Specifically, the constant 9 guarantees a minimum benchmark 1 of $Score_0$. Every time the player needs to play a tile, we compare the score array $\{Score_n\}$ for all the possible actions in lexicographical order, then the options with the top three score array are presented as candidates for the users to choose from. The pseudocode of the mahjong AI approach is in Algorithm S1.

2.3 The Neural Decoding Algorithm

The neural decoding algorithm is a dynamic ensemble model (DyEnsemble) [16,17]. The DyEnsemble approach is a Bayesian filter with an assembled measurement function, which can cope with nonstationary neural signals. The model settings and model calibration process is similar to [16]. After model calibration, the user can control a cursor ball with the brain signals.

3 Results

3.1 Evaluation of the AI Augmentation

We first evaluate the AI augmentation performance. In this experiment, a total of six participants were involved, and each participant played 10–21 mahjong games both with and without the AI assistance. The results are shown in Fig. 3a. Without the AI assistance, there are 44 winnings out of 83 games, with a winning rate of 53.0%. With the AI assistance, there are a total of 47 winnings out of 79 games, with a winning rate of 59.5%. The AI assistance effectively improves the winning rate by 12.3%.

3.2 Online Performance

The online experiment is carried out with a human participant (Fig. 3c). The participant is a 74-year-old male who suffers from movement disability due to traumatic cervical spine injury at C4 level. Two 96-channel Utah intracortical microelectrode arrays were implanted into the participant's left primary motor cortex for BCI control. The details of the neural signal recording are specified in supplementary materials.

On the experiment day, a total of 5 rounds of mahjong games were played, among which two rounds were won, with a winning rate of 40%. The participant had 89 mahjong playing actions, and the average action time was 4.4 s. Among the actions, only two of them failed to act within 20 s. The histogram of action time is shown in Fig. 3b. As shown in the figure, 39.5% of actions were completed within 2 s, and 71.9% of the actions were completed within 6 s. The results indicate that the user can play the mahjong game fluently with the BCI control. The way of improving the user experience of BCI systems using AI assistance is valuable and extendable to other BCI systems.

Acknowledgment. This work was partly supported by the grants from National Key R&D Program of China (2018YFA0701400), Key R&D Program of Zhejiang (2022C03011), Natural Science Foundation of China (61906166, 61925603), the Lingang Laboratory (LG-QS-202202-04), and the Starry Night Science Fund of Zhejiang University Shanghai Institute for Advanced Study (SN-ZJU-SIAS-002).

References

1. Hochberg, L.R., et al.: Neuronal ensemble control of prosthetic devices by a human with tetraplegia. Nature **442**(7099), 164 (2006)
2. Bouton, C.E., et al.: Restoring cortical control of functional movement in a human with quadriplegia. Nature **533**(7602), 247–250 (2016)
3. Pandarinath, C., et al.: High performance communication by people with paralysis using an intracortical brain-computer interface. Elife **6**, e18554 (2017)
4. Willett, F.R., et al.: Hand knob area of premotor cortex represents the whole body in a compositional way. Cell **181**(2), 396–409 (2020)
5. Blankertz, B., et al.: The berlin brain-computer interface presents the novel mental typewriter hex-o-spell (2006)
6. Yu, H., Qi, Y., Wang, H., Pan, G.: Secure typing via BCI system with encrypted feedback. In: 2021 43rd Annual International Conference of the IEEE Engineering in Medicine and Biology Society (EMBC), pp. 4969–4973. IEEE (2021)
7. Bos, D.O., Reuderink, B.: Brainbasher: a BCI game. In: Extended Abstracts of the International Conference on Fun and Games, pp. 36–39. Eindhoven University of Technology Eindhoven, The Netherlands (2008)
8. Yoh, M.S., Kwon, J., Kim, S.: Neurowander: a BCI game in the form of interactive fairy tale. In: Proceedings of the 12th ACM International Conference Adjunct Papers on Ubiquitous Computing-Adjunct, pp. 389–390 (2010)
9. Edelman, B.J., et al.:: Noninvasive neuroimaging enhances continuous neural tracking for robotic device control. Sci. Robot. **4**(31), eaaw6844 (2019)
10. Sussillo, D., et al.: Making brain-machine interfaces robust to future neural variability. Nat. Commun. **7**(1), 1–13 (2016)

11. Müller, K.R., Blankertz, B.: Toward noninvasive brain-computer interfaces. IEEE Sig. Process. Mag. **23**(5), 126–128 (2006)
12. Wang, Q., Sourina, O., Nguyen, M.K.: EEG-based "serious" games design for medical applications. In: 2010 International Conference on Cyberworlds, pp. 270–276. IEEE (2010)
13. Kerous, B., Skola, F., Liarokapis, F.: EEG-based BCI and video games: a progress report. Virtual Reality **22**(2), 119–135 (2018)
14. Willett, F.R., Avansino, D.T., Hochberg, L.R., Henderson, J.M., Shenoy, K.V.: High-performance brain-to-text communication via handwriting. Nature **593**(7858), 249–254 (2021)
15. Gleitzman, B.: Mahjong (2013). [source code]. https://github.com/gleitz/mahjong.git
16. Qi, Y., et al.: Dynamic ensemble Bayesian filter for robust control of a human brain-machine interface (2022). https://doi.org/10.1109/TBME.2022.3182588
17. Qi, Y., Liu, B., Wang, Y., Pan, G.: Dynamic ensemble modeling approach to nonstationary neural decoding in brain-computer interfaces. arXiv preprint arXiv:1911.00714 (2019)

VAFA: A Visually-Aware Food Analysis System for Socially-Engaged Diet Management

Hang Wu, Xi Chen, Xuelong Li, Haokai Ma, Yuze Zheng, Xiangxian Li, Xiangxu Meng, and Lei Meng(✉)

School of Software, Shandong University, Jinan, China
wuhang@mail.sdu.edu.cn, lmeng@sdu.edu.cn

Abstract. In this demo, we demonstrate a visually-aware food analysis (VAFA) system for socially-engaged diet management. VAFA is capable of receiving multimedia inputs, such as the images of food with/without a description to record a user's daily diet. A set of AI algorithms for food classification, ingredient identification, and nutritional analysis are provided with this information to produce a nutrition report for the user. Furthermore, by profiling users' eating habits, VAFA can recommend individualized recipes and detect social communities that share similar dietary appetites for them. With the support of state-of-the-art AI algorithms and a large-scale Chinese food dataset that includes 300K users, 400K recipes, and over 10M user-recipe interactions, VAFA has won several awards in China's national artificial intelligence competitions.

Keywords: Diet management · Classification · Recommendation · Community discovery

1 System Overview

With the development of the Internet and AI technology, and the increasing concerns about more balanced nutritional health, systems for healthy diet management [6,11,12] are emerging all over the world. Nevertheless, the existing research and databases, such as food logs [2], are generally based on the western diet data, which leads to insufficient large-scale datasets and analysis results for Chinese users and recipes. Therefore, this paper proposes a visually-aware food analysis (VAFA) system for socially-engaged diet management.

As shown in Fig. 1, VAFA enables users to record their dietary intake through multimedia inputs and produce a nutrition report using a set of artificial intelligence algorithms that are trained by a self-collected large-scale dataset containing over 300K users, 400K recipes, and 10M user-recipe interactions. As a result, VAFA achieves a cutting-edge performance of 87.3% in food classification. More importantly, VAFA can analyze users' eating behaviors for personalized recommendations, and it empowers users with community discovery via the calculation of diet similarity and builds a social-based relationship network named the social cloud. Such functions may enhance the system adherence of the users.

L. Fang et al. (Eds.): CICAI 2022, LNAI 13606, pp. 554–558, 2022.
https://doi.org/10.1007/978-3-031-20503-3_48

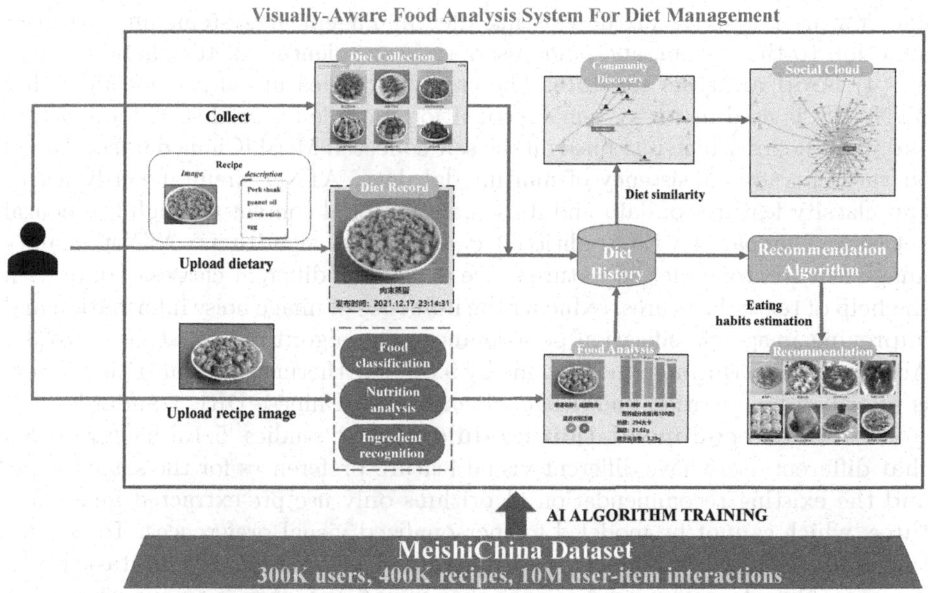

Fig. 1. Framework of VAFA system for diet management.

2 Demo Setups

The VAFA system for Chinese diet management is implemented on Windows10 OS with six 2.8Ghz CPUs.

(1) **Front end**: The Vue framework [1] which focuses only on the view layer and uses a bottom-up incremental development design is applied to build the system's user interfaces and interact with the back-end using Axios. It will check the user's input to ensure the correctness of the input data sent to the back-end, and then send operation requests to access the back-end services based on the user's operation, and upon receiving the returned result processed by the back-end, it will render the data and present it to the user immediately.

(2) **Back end**: The Flask framework [4] and PyTorch are used to build the backend and implement state-of-the-art AI algorithms. It provides APIs to the front-end for various system services and is responsible for receiving user requests sent by the front-end. It will call the corresponding functions to process the requests according to the type of requests, and finally, return the processing results to the front-end. Moreover, the PostgreSQL is installed in this part to access and manage the system's data more efficiently.

(3) **Social cloud**: Users with similar eating habits are detected by GHF-ART [10] as social communities, and then the community relationships are visualized in the form of a social cloud network using ECharts [7], a JavaScript-based data visualization chart library, so that users can easily dig out the recipes they may be interested in and other users similar to them based on this network. The purpose of this module is aimed to solve the problems of lack of social elements

and low user activity in the existing diet management system, improve user adhesion to the system, and guide users to form scientific dietary habits.

(4) **Food analysis module**: The cross-modal feature alignment algorithm ATNet [8] is used in our system's recognition module for predicting food classes and ingredients, which is trained on the self-collected MeishiChina dataset. Based on the semantic consistency of multimodal data, ATNet simultaneously learns and classify features of data and its semantic modal content through the neural network. Compared with traditional classification algorithms, ATNet enables image features to guide the feature space division of different classes of data with the help of textual features, reducing the influence of image noise information and improving image classification performance. The algorithm has achieved 87.3% Accuracy, outperforming the baseline by 9.6%. Furthermore, the nutrition report is informed by nutrition knowledge which follows Chinese DRIs standards.

(5) **Food recommendation module**: Recent studies [5,13] have revealed that different users have different visual feature preferences for the same recipe, and the existing recommendation algorithms only use pre-extracted image features, which cannot be modeled for personalized visual preferences. To address this problem, we used PiNet [9], a dual-gated heterogeneous multitask learning algorithm, to recommend recipes that may be of interest to our users. It is trained by more than 10 million user-item interactions, which has achieved the Top-10 precision of 8.11%, outperforming the existing methods by 6.29%. Furthermore, according to users' feedback, our recommendation technology is able to achieve better performance than traditional methods in practical application scenarios, recommending recipes more accurately for users.

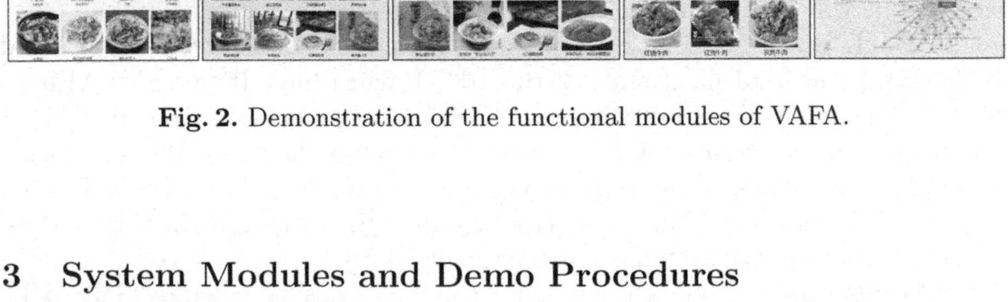

Fig. 2. Demonstration of the functional modules of VAFA.

3 System Modules and Demo Procedures

VAFA contains five primary modules: diet record module, diet collection module, diet recognition module, diet recommendation module and social cloud module. The diet record module enables users to upload daily diets; the diet collection module allows users to collect recipes they are interested in; the diet recognition module analyzes food images uploaded by users and generates nutrition reports; the diet recommendation module extracts users' individual preferences for the personalized recommendation; the social cloud module enhances recipe sharing through the visualization of users' dietary relationships. As shown in Fig. 2, the demo procedures of VAFA will be demonstrated as follows:

i) **Illustration of the homepage of system website**: The homepage includes various information such as healthy diets, nutrition knowledge and the top recommended recipes of the week. In addition, this page serves as the initial page of VAFA, where users can access other modules.
ii) **Interpretation of the user's profile page**: The user's profile page includes the user's diet history, such as diet record, diet collection, and the recommended recipes. The diet record displays information about recipes that users uploaded. As for the diet collection, it shows the recipes that users collected, which are sorted in reverse chronological order allowing users to access their favorite recipes at any time. In addition, the diet recommendation is able to precisely recommend recipes based on users' collaborative information [3] and visual feature.
iii) **Account of the detailed recipe description**: The recipe page describes the specific information of the recipe. Users are able to view the recipe's detail page including the name, the preparation, the directions, and other relevant information about the recipe when they click on any recipe image on any page of the system. From this page, users can learn how this recipe is cooked and discover recipes they may be interested in.
iv) **Formulation of the nutrition report**: The nutrition report is automatically generated by an intelligent food analysis algorithm from the uploaded image, which contains the recipe's name, the top five ingredients with the highest probability, and the nutrition facts. In particular, the nutritional knowledge is obtained from Web APIs provided on the internet, which follows the Chinese dietary reference intakes standards.
v) **Explanation of the social cloud**: The social cloud page presents a social network in the form of points and lines based on users' dietary relationships. With a click on a user's avatar or a recipe in the social network, VAFA will redirect to the user's personal page or the recipe's detailed description. If users have similar preferences, such as having favorited the same recipe, these user nodes will be connected with this recipe.

4 Conclusion

In this demo paper, we demonstrate a novel food analysis system based on leading-edge AI algorithms for socially-engaged diet management. The existing research are generally based on the western food datasets, which are not adapted to the demands of Chinese users and are poor in intelligence. To address these issues, VAFA deploys frontier AI algorithms such as food classification, ingredient identification, diet recommendation, and nutrition analysis, with the support of self-collected large-scale Chinese food datasets to analyze the food images uploaded by users and provide them with nutrition reports. By estimating users' diet habits, it can more precisely recommend personalized recipes. More significantly, it enables users with community discovery via calculating diet similarity. Note that the VAFA system's demo video is shared at: https://b23.tv/Vy5uIB8.

Acknowledgments. This work is supported in part by the Excellent Youth Scholars Program of Shandong Province (Grant no. 2022HWYQ-048) and the Oversea Innovation Team Project of the "20 Regulations for New Universities" funding program of Jinan (Grant no. 2021GXRC073)

References

1. Vue. https://vuejs.org/. Accessed 8 Oct 2021
2. Aizawa, K., Ogawa, M.: FoodLog: multimedia tool for healthcare applications. IEEE Multimedia **22**(2), 4–8 (2015)
3. Chen, C.M., Wang, C.J., Tsai, M.F., Yang, Y.H.: Collaborative similarity embedding for recommender systems. In: The World Wide Web Conference, pp. 2637–2643 (2019)
4. Copperwaite, M., Leifer, C.: Learning Flask Framework. Packt Publishing Ltd., Birmingham (2015)
5. Elsweiler, D., Trattner, C., Harvey, M.: Exploiting food choice biases for healthier recipe recommendation. In: Proceedings of the 40th International ACM SIGIR Conference on Research and Development in Information Retrieval, pp. 575–584 (2017)
6. Kerr, D.A., et al.: The connecting health and technology study: a 6-month randomized controlled trial to improve nutrition behaviours using a mobile food record and text messaging support in young adults. Int. J. Behav. Nutr. Phys. Activity **13**(1), 1–14 (2016)
7. Li, D., et al.: ECharts: a declarative framework for rapid construction of web-based visualization. Vis. Informat. **2**, 136–146 (2018)
8. Meng, L., et al.: Learning using privileged information for food recognition. In: MM, pp. 557–565 (2019)
9. Meng, L., Feng, F., He, X., Gao, X., Chua, T.: Heterogeneous fusion of semantic and collaborative information for visually-aware food recommendation. In: MM, pp. 3460–3468 (2020)
10. Meng, L., Tan, A.H.: Community discovery in social networks via heterogeneous link association and fusion. In: Proceedings of the International Conference on Data Mining, pp. 803–811. SIAM (2014)
11. Merler, M., Wu, H., Uceda-Sosa, R., Nguyen, Q.B., Smith, J.R.: Snap, eat, repeat: a food recognition engine for dietary logging. In: Proceedings of the 2nd International Workshop on Multimedia Assisted Dietary Management, pp. 31–40 (2016)
12. Ming, Z.-Y., Chen, J., Cao, Yu., Forde, C., Ngo, C.-W., Chua, T.S.: Food photo recognition for dietary tracking: system and experiment. In: Schoeffmann, K., et al. (eds.) MMM 2018. LNCS, vol. 10705, pp. 129–141. Springer, Cham (2018). https://doi.org/10.1007/978-3-319-73600-6_12
13. Yang, L., Cui, Y., Zhang, F., Pollak, J.P., Belongie, S.J., Estrin, D.: Plateclick: bootstrapping food preferences through an adaptive visual interface. In: Proceedings of the 24th ACM International Conference on Information and Knowledge Management, CIKM 2015, pp. 183–192. ACM (2015)

XIVA: An Intelligent Voice Assistant with Scalable Capabilities for Educational Metaverse

Jun Lin[1,2](✉), Yonghui Xu[1], Wei Guo[1], Lizhen Cui[1], and Chunyan Miao[2]

[1] Shandong University, Jinan, China
junlin@ntu.edu.sg, {guowei,clz}@sdu.edu.cn
[2] Nanyang Technological University, Singapore, Singapore
ascymiao@ntu.edu.sg

Abstract. We introduce XIVA, an intelligent voice assistant to support Chinese voice interaction for the future educational metaverse system. Unlike existing commercial voice assistants within a closed or limited ecosystem, XIVA focuses on providing open programming interfaces to third-party developers, to supporting them to develop new voice commands and functions. XIVA can be customized and run on different operating systems and different devices, such as a raspberry Pi, a personal computer, a VR headset or other intelligent devices developed by metaverse practitioners. In this demo, we present XIVA's capabilities of basic voice interaction and various capabilities extended by third-party for smart classroom operation control.

Keywords: Voice assistant · Education · Metaverse

1 Introduction

1.1 Background

The word "Metaverse" has appeared frequently in the public eye since 2021, and now it has become one of the hottest "international words". Metaverse, which integrates emerging information technologies such as virtual reality, augmented reality, mixed reality, extended reality, 5G, cloud computing, artificial intelligence, digital twin and blockchain etc., is considered to be a new generation of network interconnection technology in the future after the Internet and the Internet of Things. By integrating the above-mentioned various new-generation information technologies, Metaverse produces a new type of internet application and social form integrating virtual and real. It brings immersive experience to users by expanding reality technology, and generates a mirror image of the real world based on digital twin technology. It uses blockchain technology to build a trusted virtual digital economic system, so as to closely integrate the virtual world with the real world in the economic system, social system, and credit system, which allows each user to produce digital contents and contribute to the construction of the virtual world with open interfaces [1].

L. Fang et al. (Eds.): CICAI 2022, LNAI 13606, pp. 559–563, 2022.
https://doi.org/10.1007/978-3-031-20503-3_49

As the carrier of the fusion of the virtual world and the real world, Metaverse has brought great opportunities for changes in life, social interaction, entertainment, working environment and other scenarios for all domains, including the education. Especially in the post-COVID-19 era and the time after China's implementation of the "double reduction policy" for education, which requires to effectively reduce the heavy homework burden and after-school training burden of students in the stage of Chinese compulsory education. The educational metaverse can build a smart learning space with the physical meta entity, including teachers, students, schools and other education entities as the core, as well as various education resource ecology, social communication, learning evaluation, analysis and certification as the key links, to form a future education mode with virtual reality symbiosis and cross-border fusion exploration [2, 3].

The changing speed of scientific and technological has surpassed the evolution speed of the traditional education mode. The life cycle of new knowledge and technology from birth, diffusion to extinction has become shorter, as well as the life cycle of new information has been shortened too. All educational subjects and traditional educational paradigms are facing new challenges. In the past, learning was for creation, but now the process of learning itself is creation. The educational metaverse can help break the time and space limitations to realize the digital upgrading of the traditional educational model, effectively reduce the cost of learning and teaching, improve the imbalance of educational resources, promote the balance of the distribution of high-quality teaching resources, and finally make lifelong learning, interdisciplinary learning, fair learning and human-machine mutual learning possible.

1.2 Motivation

Educational metaverse can bring a more customized and personalized learning space and experience, and provide a learning and growth plan that adapts to each student's psychological characteristics and thinking habits, so as to make up for the learning gap in the actual educational environment. The combination of virtual and real digital learning scenarios and highly interactive learning methods created by educational metaverse will greatly stimulate students' enthusiasm for acquiring knowledge. At the same time, the technological development of virtual characters in "Metaverse" may promote the generation of virtual teachers or learning assistants, so as to improve the quality of accompanying learning and make up for the emotional lack of online learning interaction. Highly simulated human-computer interactive learning and accompanying learning may become an important direction of educational reform. In fact, the current modern "teacher-centered teaching method" as the core has only a history of about 300 years. Before that, the education methods, forms and scenes were diversified. It is foreseeable that educational metaverse will change the way, form and scene of education in the future.

One of the development directions of educational metaverse in the future is to cultivate a new generation of talents with immersive "listening, speaking, reading, and writing" literacy skills. At the same time, giving virtual teachers or learning assistants in the metaverse the basic abilities of "listening, speaking, reading, and writing" is also the primary task of building the future educational metaverse. From this perspective, this paper proposes an architecture design of intelligent voice assistant [4, 5] with knowledge and function expansion capabilities. Combined with our practical experience of the demo development, this paper also provides a design idea and direction for the interactive system between virtual people and real people in the future research and development of educational metaverse.

2 XIVA: Xiaohui Chinese Intelligent Voice Assistant

Xiaohui Chinese Intelligent Voice Assistant (XIVA) project aims to build a virtual voice assistant that will assist educators/learners to improve their teaching/learning efficiency in an educational metaverse in the future. Comparing with existing commercial voice assistants within a closed or limited ecosystem [6], XIVA focuses on providing open interfaces to third-party developers to supporting them to develop new functionalities easily.

2.1 System Architecture

The system architecture of XIVA is designed as Fig. 1.

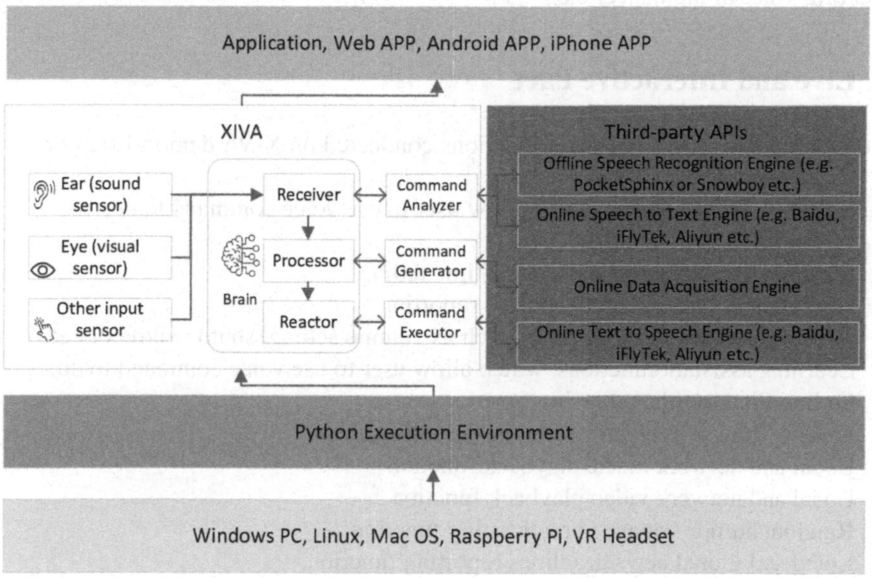

Fig. 1. Xiaohui Chinese intelligent voice assistant (XIVA) architecture.

2.2 Main Features

The XIVA demo v1.0 is developed using Python language that can run in the Python execution environment on a variety of operating systems and different hardware platforms, including but not limited to Windows, Linux, Mac OS, Raspberry Pi, or even a VR device.

XIVA's core system refers to the architecture of human-like information perception and process system, which is composed of: 1) external perception module, including sound, vision and other input sensors; 2) brain process module, which can receive the voice or other commands from external sensors, and then analyze commands. Finally it generates and send some new command or feedback to reactor to execute. During the process, brain module may call 3) third-party APIs that run some AI algorithms or engines in cloud servers. Those third-party APIs include some online or offline speech recognition and vocal processing module, personalized configuration processing module, data acquisition module, knowledge retrieval and learning processing module and new capability expansion interface module, etc. As those third-party modules mostly run in a cloud server, from the perspective of establishing a trusted computing and communicating environment, we use the blockchain technology mentioned in the paper [7] to record and save the activity data for each call and return.

XIVA provides sample code including standard Python function structure and Web API framework, which allow third-parties to develop new voice command parsing and execution code. XIVA operators also can customize the characteristics of the virtual assistant according to the operation region, country, institution, local customs and personal characteristics, etc., such as reporting local weather, news, traffic condition, and selecting voice of announcer etc.

3 Live and Interactive Part

In the demo[1], we will show main functions conducted on XIVA demo v1.0 as below:

1) System basic functions, which allow user to use voice command to do:
- Introduction to Xiaohui
- System cache cleaning and system time telling
- Operating system status/IP address reporting
- Execution of system commands such as volume setting, shutdown/reboot etc.

2) Learning assistant functions, which allow user to use voice command to do:
- Online Chinese idiom parsing function
- Online Chinese-English word translation function
- Local and network music playback function
- Local and network video playback function
- Random humor sentence broadcasting function
- Local and global news headlines reporting function

3) Teaching assistant functions, which allow user to use voice command to do:
- Personalized to-do reminder function and city weather forecast function
- Conference room availability checking and reservation functions
- Smarter classroom slideshow playing and control functions

[1] https://www.hzzlink.com/cicai22demo/CICAI22-XIVA-DemoVideo.mp4.

4 Conclusion

XIVA is an open Chinese intelligent voice assistant designed to support future human-machine interactions in the educational metaverse. With XIVA, programmers can develop new voice command and functions integrating into the virtual assistant. For future work, we would support continuous development and test environments and provide more functions in XIVA.

Acknowledgments. This research is supported, in part, by (1) Joint SDU-NTU Centre for Artificial Intelligence Research (C-FAIR), Shandong University, China; (2) The Joint NTU-UBC Research Centre of Excellence in Active Living for the Elderly (LILY), Nanyang Technological University, Singapore; (3) China-Singapore International Joint Research Institute (CSI-JRI), Guangzhou, China; (4) Alibaba-NTU Singapore Joint Research Institute (JRI), Nanyang Technological University, Singapore.

References

1. Lik-Hang L., et al.: All One needs to know about metaverse: a complete survey on technological singularity, virtual ecosystem, and research agenda (2021). https://doi.org/10.13140/RG.2.2.11200.05124/8
2. Yuyang W., Lik-Hang L., Tristan B., Pan H.: Re-shaping Post-COVID-19 teaching and learning: a blueprint of virtual-physical blended classrooms in the metaverse era. https://arxiv.org/abs/2203.09228. Accessed 20 Jun 2022
3. Bokyung K., Nara H., Eunji K., Yeonjeong P., Soyoung J.: Educational applications of metaverse: possibilities and limitations. J. Educ. Eval. Health Prof. 18(32). Published Online 13 Dec 2021. https://doi.org/10.3352/jeehp.2021.18.32 (2021)
4. Polyakov E.V., Mazhanov M.S., Rolich A.Y., Voskov L.S., Kachalova M.V., Polyakov S.V.: Investigation and development of the intelligent voice assistant for the internet of things using machine learning. In: 2018 Moscow Workshop on Electronic and Networking Technologies (MWENT), pp. 1–5. IEEE, Moscow (2018)
5. Sweeney, Miriam, E., Davis, E.: Alexa, are you listening? An exploration of smart voice assistant use and privacy in libraries. Inf. Technol. Libraries (Online); Chicago, 39(4), 1–21 (2020)
6. López, G., Quesada, L., Guerrero, L.A.: Alexa vs. Siri vs. Cortana vs. Google Assistant: A Comparison of Speech-Based Natural User Interfaces. In: Nunes, I. (eds) Advances in Human Factors and Systems Interaction. AHFE 2017. Advances in Intelligent Systems and Computing, vol 592. Springer, Cham. 241–250 (2018). https://doi.org/10.1007/978-3-319-60366-7_23
7. Jun L., Bei L., Lizhen C. Chunyan M.: Practices of Using Blockchain Technology in e-Learning. In: 5th International Conference on Crowd Science and Engineering (ICCSE 2021), pp. 55–60. ACM, Jinan (2021). https://doi.org/10.1145/3503181.3503191

Weld Defect Detection and Recognition System Based on Static Point Cloud Technology

Changzhi Zhou[1(✉)], Siming Liu[1], Fei Huang[1], Qian Huang[2(✉)], Anbang Yan[2], and Guocui Luo[2]

[1] Shanghai Shipbuilding Technology Research Institute, Shanghai, China
ndt@csscstri.com

[2] School of Computer and Information, Hohai University, Nanjing, China
huangqian@hhu.edu.cn

Abstract. At present, welding technology has been widely used in the industrial field, and high-quality welding is very important in the entire industrial process. Therefore, the detection and identification of welding quality are of great significance to the development of the industrial industry. Weld defects are mainly divided into internal defects and external defects. External defects can be seen directly with the naked eye, but internal defects cannot be seen directly. Non-destructive testing is required. Currently, the main non-destructive testing is manual evaluation, which is relatively subjective and limited by factors such as the technical level of technicians. This application is aimed at welding defects in the industrial process, based on the welding image generated by X-ray to detect the weld quality, use the MGLNS-Retinex algorithm and the improved region growth algorithm to detect the weld defect area, and innovatively pass the defect point cloud. For defect classification and identification, the defect point cloud can reduce the amount of data and improve the operation speed while achieving a high classification accuracy.

Keywords: Weld defects · Defects detection · Defect identification · Pointnet++

1 Introduction

X-ray weld defect detection and identification [1] is a series of detections using the difference in the absorption of X-ray energy between the weld defect area and the non-weld defect area. [2–4] In terms of detection, researchers use methods [5–8] such as edge detection, mathematical morphological image processing [6], and multiple thresholding [7]. In terms of identification, researchers mainly use vector machines and other methods. In practical applications, there are problems such as low image contrast, noise interference, and poor brightness, resulting in low detection results and poor real-time performance.

In this application, combined with actual production, given the characteristics of X-ray images and the shortcomings of common segmentation algorithms,

L. Fang et al. (Eds.): CICAI 2022, LNAI 13606, pp. 564–569, 2022.
https://doi.org/10.1007/978-3-031-20503-3_50

the image enhancement based on the multi-granularity local noise suppression Retinex algorithm [18] is firstly performed on the weld image, and the MGLNS-Retinex threshold is used to perform a second image enhancement on the weld. The value processing [19], through the statistical waveform analysis of the gray value of the line, extracts the weld area. In the intercepted weld area, the weld area is first enhanced to improve the uneven gray distribution of the weld and highlight the texture of the defect. Then, an improved area growth algorithm is proposed to detect weld defects, and good results are obtained. Check the effect. Combining the defects in weld defects and the characteristics of the point cloud [15], this system proposes the concept of the defect point cloud. The defect point cloud is generated by the weld defect, which reduces the amount of sample data while retaining the characteristics of the defect, reducing the interference of the defect background. Finally, the defect point cloud is used to classify the defects in the weld defect through the PointNet++ network [11–13,16,17].

2 Design and Implementation

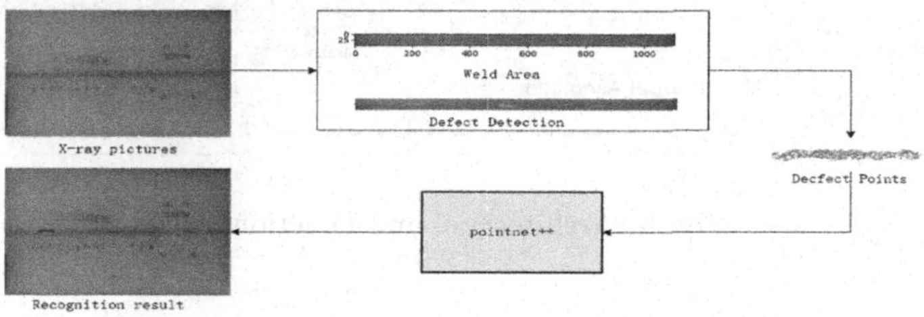

Fig. 1. System Composition

As shown in Fig. 1, considering the digital detection, automation, and intelligent requirements of welding quality management in actual industrial manufacturing, the weld defect detection and identification system mainly includes defect detection and defect identification. As shown in Fig. 2, there are five types of welding defects in the current industry standard: bubbles, cracks, slags, noFusion, and noPenetration. The five kinds of defects are further divided into circular defects (bubbles), irregular defects (cracks), and elongated defects (slags, noFusion, noPenetration) according to their shape. The system divides the weld area through the detection module, locates and detects each defect of the weld, and then generates a defect point cloud. The recognition module can classify five kinds of defects through the PointNet++ [14] network.

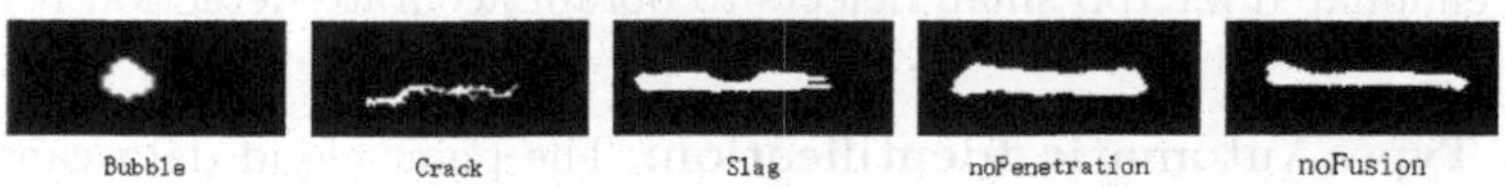

Fig. 2. Defects Type

2.1 Architecture and Functions

Figures 3 and 4 shows the architecture of the system in this paper. The main functions are defect detection and defect recognition. The training data set of the recognition model is manually generated by professional reviewers using annotation software. The system supports the iterative optimization of the recognition model.

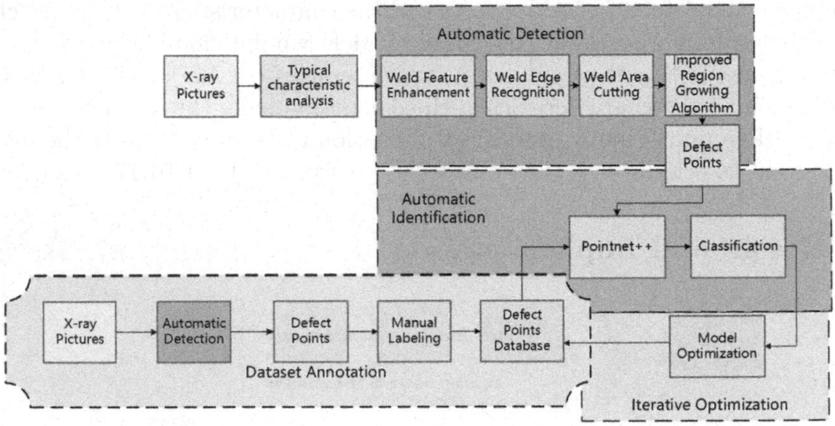

Fig. 3. Architecture and Functions

Defect Automatic Detection: Defect detection firstly divides the weld area from the original image, and uses the improved area growing algorithm to complete the detection of specific defects in the cut area.

Cutting: The image is processed by the Multi Granularity Local Noise Suppression Retinex (MGLNS-Retinex) algorithm, and then the threshold is iterated to obtain the MGLNS-Retinex threshold [9] [10], which is used to threshold the image to distinguish the weld area. as the picture shows.

Improved Region Growth Detection Defects: Improvements are made to the problem that a single scale in the region growing algorithm cannot cover all defects well. The algorithm sets different scales to determine the regional growth seed points, uses the seed points to formulate different expansion schemes for defect similarity expansion, and takes the union to obtain a perfect coverage defect map. At the same time, based on the actual industrial requirements for welding, noise is eliminated for too small defects to obtain accurate detection results.

Defect Type Automatic Identification: The point cloud data can remove the interference of unnecessary pixels around the defect, close to the real characteristics of the object, and the point data is stored in an array, which is simple in

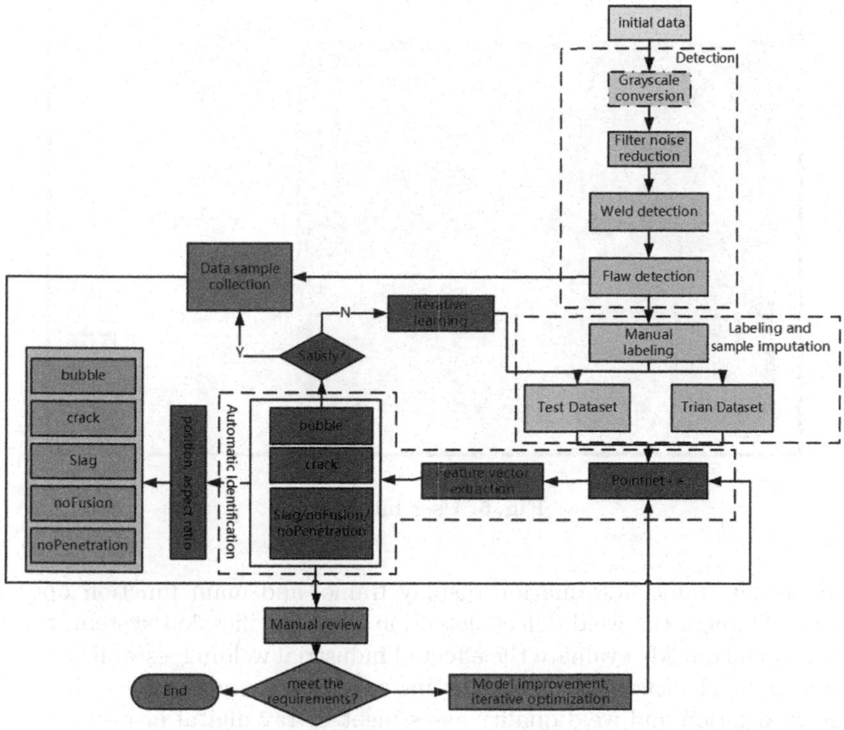

Fig. 4. Detailed Structure

expression, small in data scale, and can well support defect classification. Therefore, the defect point cloud is used as the dataset. The system uses the pointnet++ network for feature learning to complete defect identification. Compared with the image data scale, the defect point cloud recognition has low computational complexity and high recognition efficiency, which meets the recognition requirements of industrial inspection.

Datasets: Select defects detected in X-ray weld images to generate point cloud datasets. The number of point cloud sets with large differences is unified, and the coordinates of the points are centralized and normalized. Each defect data sample includes all pixel points of a single defect, and each pixel point stores abscissa and ordinate position and grayscale information.

Pointnet++ classification network [14] [15]: Put the marked data set into the improved pointnet++ network for weld defect data, and iteratively train the classification model. Use the trained model to identify flawed point clouds.

2.2 User Interface

As shown in Fig. 5, according to the detection and recognition function requirements, the software interface design mainly includes an image display frame,

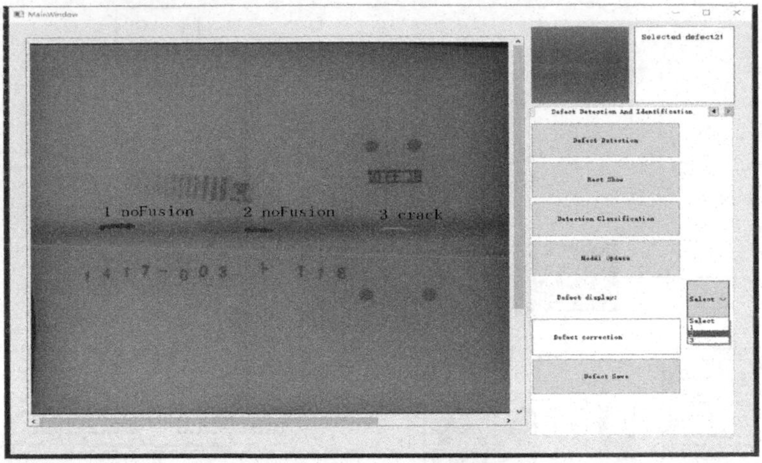

Fig. 5. User Interface

detail display frame, information display frame, and main function operation buttons. Through the weld defect detection and identification system, the staff can easily and quickly evaluate the effect of industrial welding, establish the basis for automatic identification of engineering applications, and achieve the goals of image recognition and weld quality assessment by ray digital inspection.

2.3 Performance

In the actual test, the software can run stably, and the defect detection effect is good. In the existing data set, the defect detection success rate reaches 90%, and the defect classification accuracy reaches 83%. The classification model is learnable, and with the expansion of the data set, the model classification accuracy can be further improved. The system has been completed and will soon be used in actual industrial inspection.

Acknowledgment. This work is partly supported by the National Key Research and Development Program of China under Grant No. 2018YFC0407905, the Postgraduate Research & Practice Innovation Program of Jiangsu Province under Grant No.SJCX2_0161, and the Key Research and Development Program of Jiangsu Province under Grant No. BE2016904, the Fundamental Research Funds of China for the Central Universities under Grant No. B200202188, the Jiangsu Water Conservancy Science and Technology Project under Grant No.2018057.

References

1. Chen, H., Li, J., Zhang, X.: Application of visual servoing to an X-ray based welding inspection robot. In: CONFERENCE 2005, International Conference on Control and Automation, vol. 2, pp. 977–982. IEEE (2005). https://doi.org/10.1109/ICCA.2005.1528263

2. Tabary, J., Hugonnard, P., Mathy, F.: INDBAD: a realistic multi-purpose and scalable X-ray simulation tool for NDT applications. Conference **1**, 1–10 (2007)
3. Zhu, W., Ma, W., Su, Y.: Low-dose real-time X-ray imaging with nontoxic double perovskite scintillators. Journal **9**(1), 1–10 (2020)
4. Lashkia, V.: Defect detection in X-ray images using fuzzy reasoning. Image Vis. Comput. J. **19**(5), 261–269 (2001)
5. Gang, W., Liao, T.W.: Automatic identification of different types of welding defects in radio-graphic images. Journal **35**(8), 519–528 (2002)
6. Mahmoudi, A., Regragui, F.: Fast segmentation method for defects detection in radiographic images of welds. In: CONFERENCE 2009, IEEE/ACS International Conference on Computer Systems and Applications, pp. 857–860. https://doi.org/10.1109/AICCSA.2009.5069430
7. Anand, R.S., Kumar, P.: Flaw detection in radiographic weld images using morphological approach. Journal **39**(1), 29–33 (2006)
8. Felisberto, M.K., Lopes, H.S., Centeno, T.M.: Flaw detection in radiographic weld images using morphological approach. Journal **102**(3), 238–249 (2006)
9. Gayer, A., Saya, A., Shiloh, A.: Automatic recognition of welding defects in real-time radiography. Journal **23**(3), 131–136 (1990)
10. Hou, W., Zhang, D., Wei, Y.: Review on computer aided weld defect detection from radiography images. Journal **10**(5), 1878 (2020)
11. Sheshappanavar, S V., Kambhamettu, C.: A novel local geometry capture in Pointnet++ for 3D classification. In: CONFERENCE 2020, CVF Conference on Computer Vision and Pattern Recognition Workshops (CVPRW). IEEE (2020). https://doi.org/10.1109/CVPRW50498.2020.00139
12. Seo, H., Joo, S.: characteristic analysis of data preprocessing for 3D point cloud classification based on a deep neural network. Journal **41**(1), 19–24 (2021)
13. Chen, Y., Liu, G., Xu, Y.: PointNet++ network architecture with individual point level and global features on centroid for ALS point cloud classification. Journal **13**(3), 472 (2021)
14. Qi, C R., Li, Y., Hao, S.: PointNet++: Deep Hierarchical Feature Learning on Point Sets in a Metric Space. In: (eds.) CONFERENCE 2017
15. Qi, C R., Hao, S., Mo, K.: PointNet: deep learning on point sets for 3D classification and segmentation. Journal (2016)
16. Garcia-Garcia, A., Gomez-Donoso, F., Garcia-Rodriguez, J.: PointNet: a 3D convolutional neural network for real-time object class recognition. In: CONFERENCE 2016, 2016 International Joint Conference on Neural Networks (IJCNN). IEEE (2016). https://doi.org/10.1109/IJCNN.2016.7727386
17. Zheng, D H., Xu, J., Chen, R X.: Generation method of normal vector from disordered point cloud. In: CONFERENCE 2009, 2009 Joint Urban Remote Sensing Event. IEEE (2009). https://doi.org/10.1109/URS.2009.5137560
18. Jiang, B., Zhong, M.: Improved histogram equalization algorithm in the image enhancement. Journal **44**(6), 702–706 (2014)
19. Setty, S., Srinath, N K., Hanumantharaju, M C.: An improved approach for contrast enhancement of spinal cord images based on multiscale retinex algorithm. Journal arXiv:1408.2997 (2016)

3D Human Pose Estimation Based on Multi-feature Extraction

Senlin Ge[1], Huan Yu[2], Yuanming Zhang[2], Huitao Shi[2], and Hao Gao[1(✉)]

[1] College Automation, College Artificial Intelligence, Nanjing University of Posts and Telecommunications, Nanjing, China
tsgaohao@gmail.com

[2] Shanghai Institute of Aerospace Electronic Technology, Shanghai, China

Abstract. As an important computer vision task, 3D human pose estimation has received widespread attention and many applications have been derived from it. Most previous methods address this task by using a 3D pictorial structure model which is inefficient due to the huge state space. We propose a novel approach to solve this problem. Our key idea is to learn confidence weights of each joint from the input image through a simple neutral network. We also extract the confidence matrix of heatmaps which reflects its feature quality in order to enhance the feature quality in occluded views. Our approach is end-to-end differentiable which can improve the efficiency and robustness. We evaluate the approach on two public datasets including Human3.6M and Occlusion-Person which achieves significant performance gains compare with the state-of-the-art.

Keywords: 3D human pose estimation · Multi-view · End-to-end

1 Introduction

Recovering 3D human pose and motion from multiple views has attracted a great deal of attention over the last decades in the field of computer vision, which has a variety of applications such as activity recognition, sports broadcasting [1] and retail analysis [2]. The ultimate goal is to estimate 3D locations of the body joints in a world coordinate system from multiple cameras. While remarkable advances have been made in reconstruction of a human body, there are few works that address a more challenging setting where the joints are occluded in some views.

The methodology for multi-view 3D pose estimation in many existing studies includes two steps. In the first step, it tries to estimate the 2D poses in each camera view independently, for example, by Convolutional Neural Networks (CNN) [3,4]. Then in the second step, it recovers 3D pose by aggregating the 2D poses from all of the views. One typical method is to use the 3D Pictorial Structures model (3DPS), which directly estimate the 3D pose by exploring an ample state space of all possible human keypoints in the 3D space [5,6]. However, this method has large quantization errors because of the huge state space needed to explore.

L. Fang et al. (Eds.): CICAI 2022, LNAI 13606, pp. 570–581, 2022.
https://doi.org/10.1007/978-3-031-20503-3_51

On the other hand, the mainstream two-stage approach does not make full use of the information in multiple camera views and the relationship between views. They simply use the traditional triangulation or 3DPSM to restore the 3D human skeleton [7] which makes the pose estimation performance is not well in some challenging scenarios.

In this paper, we propose to solve the problem in a different way by making the best of information which extracted from the multi-view images and intermediate features. The method is orthogonal to the previous efforts. The motivation behind our method is that a joint occluded in some views may be visible in other views. So it is generally helpful to evaluate a weight matrix for each joint of each view. To that end, we present a novel approach for the 3D human pose estimation. Figure 1 shows the pipeline.

We first obtain more accurate 2D poses by jointly estimating them from multiple views using a CNN based approach. At the same time, we calculate the confidence of each joint under each view through a simple neural network. If a joint is occluded in one view, its feature is also likely corrupted. In this case, we hope to give a small confidence to the joint so that the high-quality joint in the visible views is dominant which will improve the performance in the subsequent triangulation process.

Second, when we get the initial heatmap from the 2D pose estimation, using a learning weight network to extract the confidence matrix of heatmaps under each view. Then the heatmap from each view is weighted and fused by use of the confidence matrix to obtain the final heatmaps. We apply the SoftMax operator to get the 2D positions of keypoints. Finally, the 2D positions of keypoints with the joints' confidence calculated in the first step are passed to the triangulation module that outputs the 3D pose.

We evaluate our approach on two public datasets including Human3.6M [8] and Occlusion-Person Dataset [9]. It has achieved excellent results on the two datasets. Furthermore, we compare our method to a number of standard multiview 3D pose estimation methods to give more detailed insights. Our method is end-to-end differentiable which improves the efficiency and robustness on 3D pose estimation.

2 Related Work

In this section, we briefly review the related works that utilize the techniques of this paper.

2.1 Single-view 2D Human Pose Estimation

The goal of 2D human pose estimation is to localize human anatomical keypoints or parts in one RGB image. With the introduction of "DeepPose" by Toshev et al. [10], many existing deep learning-based methods have achieved amazing results [11–13].

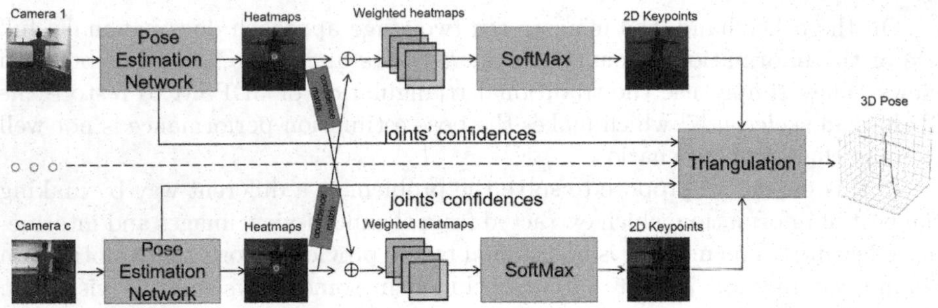

Fig. 1. The framework of our approach. It takes multi-view images as input and outputs the heatmaps and joints' confidences through the Pose Estimation Network. Then the heatmaps are weighted and fused by use of the confidence matrix to obtain the weighted heatmaps. The 2D positions of keypoints are inferred from the weighted heatmaps. Finally the 2D positions together with the joints' confidence are fed into the triangulation module to get the 3D pose position.

For 2D human pose estimation, current state-of-the-art methods can be typically categorized into two classes: top-down method and bottom-up method. In general, top-down methods [14–18] first detects people and then have estimated the pose of each person independently on each detected region. The bottom-up [19,20] methods jointly label part detection candidates and associated them to individual people by a matching algorithm. In our work, we choose the top-down method because of their higher accuracy. We adopt the SimpleBaseline [17] as the 2D human pose estimation backbone network.

2.2 Multi-view 3D Human Pose Estimation

Different from estimating from a single image, the goal of multi-view 3D human pose estimation is to get the ground-truth annotations for the monocular 3D human pose estimation. Most previous efforts can be divided into two categories. The first class is analytical methods [17,21–23] which explicitly model the relationship between a 2D and 3D pose according to the camera geometry. They first model human body by simple primitives and then optimize the model parameters through the use of multi-view images until the body model can be explained by the image features. The advantage of the analytical methods is that it can deal with the occlusion problem well because of the inherent structure prior embedded in human body model. However, due to the need to optimize all model parameters at the same time, the entire state space is huge, resulting in heavy computation in inference.

The second class is predictive methods which often flow a two-step framework by use of the powerful neural networks. They first detect the 2D human pose from all the camera views and then recover the 3D human pose by the use of triangulation or 3D Pictorial Structures model(3DPS). In [7], a recursive pictorial structure model was proposed to speed up the inference process. Recent work [25] has proposed to use 1 D convolution to jointly address the cross-view fusion and 3D pose reconstruction based on plane sweep stereo. [24] proposed a volumetric triangulation method to project the feature maps produced by the 2D human pose backbone into 3D volumes, which were then used to predict 3D poses. The shortcoming of this method is that not making full use of the information of images and feature maps which lead to the poor effect in the face of occlusion. On the contrary, our approach is efficient on multi-view 3D human pose estimation which benefits from the weight extraction network.

3 Method

In this section, we present our proposed approach for multi-view 3D human pose estimation. We assume that we have synchronized video streams from multiple cameras with known parameters which capture performance of a single person in the scene. The goal is to detect and predict human body poses in 3D given images captured from all views.

The overview of our approach is shown in Fig. 1. It first estimates 2D pose heatmaps and produces the joint confidence for each view. Then the heatmaps from all camera views are fused through the confidence matrix which extracted by a learning weight network. Finally, input the 2D positions of the keypoints and joints' confidence into the algebraic triangulation module to produce the 3D human pose.

3.1 2D Pose Detector and Joint Confidence

The 2D pose detector backbone h_p with learnable weight θ_p consists of a ResNet-152 network, followed by a series of transposed convolutions and a 1 * 1 kernel convolutional neutral network that outputs the heatmaps:

$$H_{c,j} = h_p(I_c; \theta_p), c = 1, 2, ..., C \quad (1)$$

where I_c denotes the image in the cth view, $H_{c,j}$ denotes the heatmap of the jth keypoint in the cth view.

In addition to output the heatmaps, we propose a simple network to extract the joint confidence for each view. The network structure as shown in Fig. 2. Starting from the images with known camera parameters, we apply some convolutional layers to extract features. Then the features are down-sampled by max pooling and feed into three fully connected layers that outputs the joint confidence:

$$\omega_{c,j} = h_\omega \left\{ I_c; \theta_\omega \right\} \quad (2)$$

where h_ω denotes the joint confidence learnable module. θ_ω denotes the confidence learnable weight. $\omega_{c,j}$ denotes the confidence of the jth keypoint in the cth view.

It is worth noting that the joint confidence network is jointly trained end-to-end with the 2D pose detector backbone.

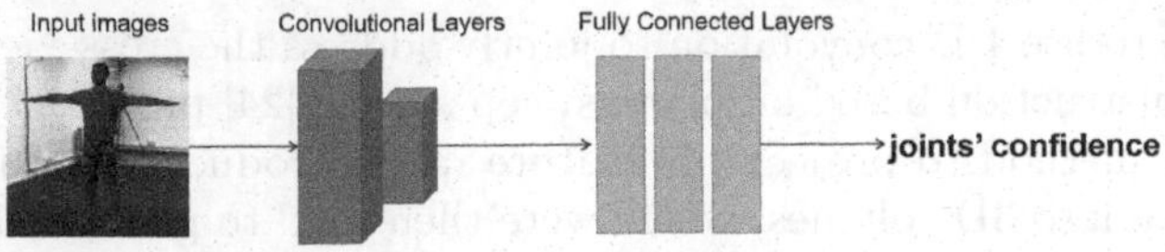

Fig. 2. The structure of the joints' confidence network.

3.2 Confidence Matrix for Heatmap Fusion

Our 2D pose detector takes multi-view images as input, generate initial pose heatmaps respectively for each, and then using a learning weight network to extract the confidence matrix of heatmaps which reflect the heatmap quality in each view. Finally the heatmaps are weighted and fused by use of the confidence matrix to obtain the final heatmaps. The core of this stage is to find the corresponding keypoints between all views.

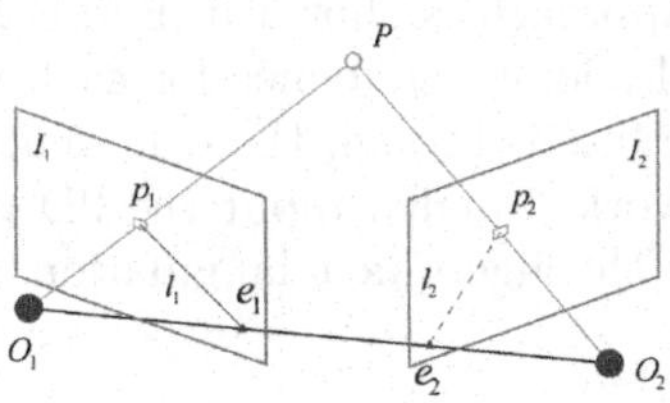

Fig. 3. Illustration of the epipolar geometry. For an image point p_1 in one view, we can find its corresponding point p_2 in another view and the corresponding point lies on the epipolar line.

We assume that there is a point P in 3D space as shown in Fig. 3. Note that we use homogeneous coordinate to represent a point. According to the equal up to a scale, the corresponding points p_1, p_2 in the two cameras are:

$$p_1 = KP \tag{3}$$

$$p_2 = K(RP + t) \tag{4}$$

where K is camera intrinsics, R and t are the rotation matrix and translation vector between two camera views.

From Fig. 3, we can find that the 3D point P, the image points p_1, p_2, and the camera centers O_1 and O_2 on the same plane which calls the epipolar plane. The l_1, l_2 which epipolar plane intersects with the two image planes calls epipolar line. In particular, we have

$$l_2 = Fp_1 \tag{5}$$

$$l_1 = F^T p_2 \tag{6}$$

where F is fundamental matrix which can be derived from K, R and t.

For each point p_1 in the first view, the epipolar geometry helps us to ensure the corresponding point p_2 in another view lie on the epipolar line l_2. However, since we do not know the depth of the point P, it is difficult to determine the exact location of point p_2. We decided to use the sparsity of heatmap to solve this problem. The heatmap has a small number of large responses near the joint location and a large number of zeros at other locations. So we select the largest response on the epipolar line as the corresponding point. Then we have:

$$\hat{H}_{c,j} = \lambda_c H_{c,j} + \sum_{u=1}^{N} \lambda_u max H_{u,j'} \tag{7}$$

where $\hat{H}$ denotes the weighted heatmap. λ is the weight of heatmap extracted from a learning weight network and N is the number of camera views.

To keep the gradients to flow back to heatmaps, we use soft-argmax to calculate the 2D positions of the joints:

$$x_{c,j} = e^{\hat{H}_{c,j}} / (\int_{q \in \Omega} e^{\hat{H}_{c,j}(q)}) \tag{8}$$

where $\hat{H}_{c,j}$ denotes the weighted heatmap of the jth keypoint in the cth view and Ω is the domain of the heatmap.

3.3 3D Pose Reconstruction

Given the estimated 2D poses and heatmaps from all views, we can reconstruct the 3D pose in several ways. The 3DPS model is one of them, but the large quantization errors result from exploring the huge state space may largely degrade the reconstruction. In order to fully integrate the information extracted from the multiview images and heatmaps, we make use of the point triangulation method with the joint confidence learned form neutral network for efficient inference.

The point triangulation is an efficient 3D pose estimation method with strong theoretical supports which reduces the finding of the 3D coordinates of a joint y_j to solving the overdetermined system of equations on homogeneous 3D coordinate vector of the joint $\widetilde{y}_j$:

$$A_j \widetilde{y}_j = 0 \tag{9}$$

where A_j is a matrix concatenating the homogeneous 3D vectors of all views for the jth keypoint.

However, traditional triangulation method can not solve the occlusion problem in views because it treats the joint in different views evenly without considering the joint may be occluded in some views, leading to unnecessary degradation of the final triangulation result. To deal with the problem, we add joint confidence which generated by a learnable module when triangulating and we have:

$$(\omega_j \circ A_j) \widetilde{y}_j = 0 \tag{10}$$

where ω_j is the joint confidence matrix which is in the same size of and $\circ$ denotes the Hadamard product.

We use Singular Value Decomposition of the matrix $E = U\sum V^T$ to solve the equation above. We set $\widetilde{y}$ as the last column of V. Then we can get the final 3D coordinates of a joint y_i by dividing the homogeneous 3D coordinate vector $\widetilde{y}$ by its fourth coordinate:

$$y = \frac{\widetilde{y}}{(\widetilde{y})_4} \tag{11}$$

3.4 Loss Function

For our method, the gradients pass from the output prediction to the input images which makes the method trainable end to end. The loss function for training the network contains two parts, the 3D mean square error (MSE) loss and the 2D joint smooth loss.

The 3D MSE loss between the estimated heatmaps and ground truth heatmaps is defined as:

$$L_{mse} = \sum_{c=1}^{C} \|H_k - H_{gt,k}\|_2 \tag{12}$$

The 2D joint smooth loss between the estimated 2D keypoints coordinates and the ground truth 2D keypoints coordinates:

$$L_{2d} = \sum_{c=1}^{C} \|y_c - y_{gt,c}\|_1 \tag{13}$$

The total loss of our method is defined as:

$$L = L_{mse} + L_{2d} \tag{14}$$

4 Experiments

4.1 Datasets and Metrics

We conduct experiments on two standard datasets for multi-view 3D human pose estimation.

Human3.6M: Human3.6M is currently the largest multi-view 3D human pose estimation dataset. It provides synchronized images captured by four cameras which includes around 3.6 million images. Following the standard evaluation protocol used in the literature, subjects S1, S5, S6, S7, S8 are used for training and S9, S11 for testing [26–28]. To avoid over-fitting to the background, We also use the MPII dataset [29] to augment the training data.

Occlusion-Person: The Occlusion-Person dataset adopt UnrealCV to render multi-view images from 3D models. In particular, thirteen human models of different clothes are put into nine different scenes such as bedrooms, offices and living room. The scenes is captured by eight cameras. The dataset consists of thirty-six thousand frames and propose use objects such as sofas and desks to occlude some human body joints.

Metrics: The 2D pose estimation is measured by Percentage of Correct Keypoints (PCK) which measures the percentage of the estimated joints whose distance from the ground-truth joints is smaller than t times of the head length. Following the previous work, we set t to be 0.5 and head length to be 2.5% of the human bounding box width for all benchmarks.

The 3D pose estimation accuracy is measured by Mean Per Joint Position Error (MPJPE) between the estimated 3D pose and the ground-truth:

$$MPJPE = \frac{1}{M}\sum_{i=1}^{M} \|p_i^3 - \bar{p}_i^3\|_2 \tag{15}$$

where $y = [p_1^3, ..., p_M^3]$ denotes the ground-truth 3D pose, $\bar{y} = [\bar{p}_1^3, ..., \bar{p}_M^3]$ denotes the estimated 3D pose and M is the number of joints in a pose.

4.2 Results on Human3.6M

2D Pose Estimation Results. The 2D pose estimation results are shown in Table 1. We compare our approach to two baselines. The first is NoFuse which estimates 2D pose independently for each view without joints' confidence and weighted heatmaps. The second is HeuristicFuse which uses a fixed confidence for each heatmap according to Eq.(7). The patameter λ is set to be 0.5 by cross-validation. From the table, we can see that the performance of our approach is better than the two baselines. The average improvement is 10.6%. This demonstrates that our approach can effectively refine the 2D pose detection.

Table 1. The 2D pose estimation accuracy (PCK) of the baseline methods and our approach on the Human3.6M dataset

Methods	Root	Belly	Nose	Head	Hip	Knee	Wrist	Mean
NoFuse	95.8	77.1	86.4	86.2	79.3	81.5	70.1	82.3
HeuristicFuse	96.0	79.3	88.4	86.8	83.1	84.5	75.2	84.7
Ours	**96.5**	**94.9**	**96.3**	**96.4**	**96.0**	**92.5**	**85.9**	**94.1**

3D Pose Estimation Results. Table 2 shows the 3D pose estimation errors of the baselines and our approach. We also compare our approach with the RANSAC baseline which is the standard method for solving robust estimation problems. We can see from the table that our approach outperforms the other three baselines by 3.39, 0.89, 2.21mm respectively in average. Considering these baselines are already very strong, the improvement is significant. The results demonstrate that adding the confidence matrix of heatmaps and joints' confidence can significant improve the performance and model's robustness.

In addition, we also compare our approach with existing state-of-the-art methods for multi-view 3D pose estimation. The results are presented in Table 3.

From the table we can find that our approach surpasses the state-of-the-arts in average. The performance of our approach is 27.3mm with improvement of 3.6mm comparing with the second best method. The improvement is significant considering that the error of the state-of-the-art is already very small. We also show some 2D and 3D pose estimation results in Fig. 4.

Table 2. The 3D pose estimation error (mm) of the baseline methods and our approach on the Human3.6M dataset.

Methods	MPJPE,mm
NoFuse	30.7
HeuristicFuse	28.2
RANSAC	29.5
Ours	**27.3**

Table 3. The 3D pose estimation error (mm) of the state-of-the-arts and our approach on the Human3.6M dataset

Methods	MPJPE,mm
Pavlakos et al. [30]	56.9
Tome et al. [31]	52.8
Qiu et al. [7]	31.1
Gordon et al. [32]	30.9
Ours	**27.3**

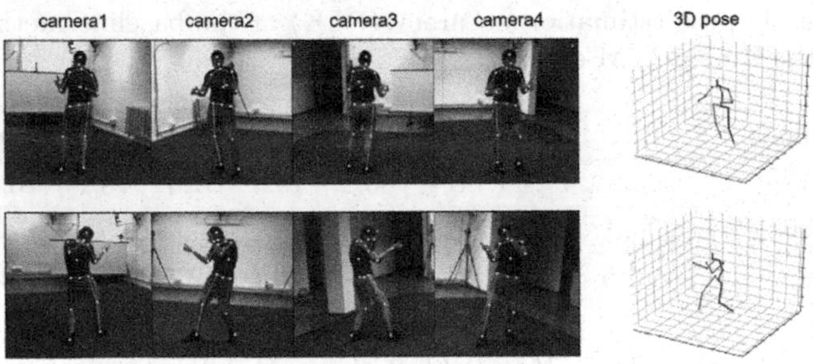

Fig. 4. Examples of 2D and 3D pose estimation on Human3.6M dataset.

4.3 Results on Occlusion-Person

2D Pose Estimation Results. Table 4 shows the results on the Occlusion-Person dataset. From the table we can find that our approach is significantly improved compared with NoFuse. This is reasonable because the features of the

Table 4. The 2D pose estimation accuracy (PCK) of the baseline methods and our approach on the Occlusion-Person dataset

Methods	Hip	Knee	Ankle	Shlder	Elbow	Wrist	Mean
NoFuse	63.4	21.5	17.0	29.5	14.6	12.4	30.9
HeuristicFuse	76.9	59.0	73.4	63.5	49.0	54.8	65.0
Ours	**97.7**	**94.4**	**91.0**	**97.9**	**91.0**	**93.1**	**94.2**

occluded joints are severely corrupted and the results demonstrate the advantage of our approach for dealing with occlusion.

3D Pose Estimation Results. The results of 3D pose estimation error (mm) are presented in Table 5. The result of NoFuse is 41.64mm which is a large error. The performance of our model is 9.67mm which means our approach can better handle occlusions than other baselines. Since there are very few works have report results on this new dataset, we only compare our approach to the three baselines. Figure 5 shows some examples of 2D and 3D pose estimation on Occlusion-Person dataset.

Table 5. The 3D pose estimation error (mm) of the baseline methods and our approach on the Occlusion-Person dataset.

Methods	MPJPE,mm
NoFuse	41.6
HeuristicFuse	13.4
RANSAC	12.4
Ours	**9.7**

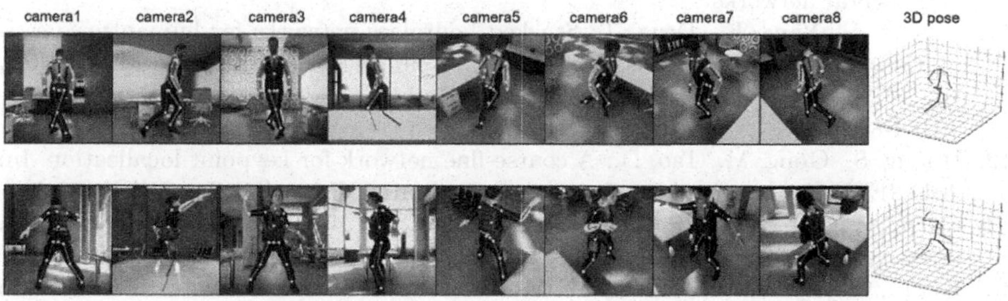

Fig. 5. Examples of 2D and 3D pose estimation on Occlusion-Person dataset.

5 Conclusion

In this paper, we present a novel approach for multi-view human pose estimation. Different from previous methods, we propose to extract weights of views and heatmaps to reflect their quality. The experimental results on the two datasets validate that the approach is efficient and robust to occlusion.

References

1. Bridgeman, L., Volino. M., Guillemaut, J.Y., et al.: Multi-person 3D pose estimation and tracking in sports. In: 2019 IEEE/CVF Conference on Computer Vision and Pattern Recognition Workshops (CVPRW). IEEE (2019)
2. Tu, H., Wang, C., Zeng, W.: VoxelPose: towards multi-camera 3D human pose estimation in wild environment. In: Vedaldi, A., Bischof, H., Brox, T., Frahm, J.-M. (eds.) ECCV 2020. LNCS, vol. 12346, pp. 197–212. Springer, Cham (2020). https://doi.org/10.1007/978-3-030-58452-8_12
3. Zhe, C., Simon, T., Wei, S.E., et al.: Realtime multi-person 2D pose estimation using part affinity fields. IEEE (2017)
4. He, K., Gkioxari, G., Dollár, P., et al.: Mask R-CNN. IEEE Trans. Pattern Anal. Mach. Intell. (2017)
5. Joo, H., Simon, T., Li, X., et al.: Panoptic studio: a massively multiview system for social interaction capture. IEEE Trans. Pattern Anal. Mach. Intell. 99 (2016)
6. Belagiannis, V., Sikandar, A., et al.: 3D pictorial structures revisited: multiple human pose estimation. IEEE Trans. Pattern Anal. Mach. Intell. **38**, 1929–1942 (2016)
7. Qiu, H., Wang, C., Wang, J., et al.: Cross view fusion for 3D human pose estimation. University of Science and Technology of China; Microsoft Research Asia; TuSimple; Microsoft Research (2019)
8. Ionescu, C., Papava, D., Olaru, V., et al.: Human3.6M: large scale datasets and predictive methods for 3D human sensing in natural environments. IEEE Trans. Pattern Anal. Mach. Intell. **36**(7), 1325–1339 (2014)
9. Zhang, Z., Wang, C., Qiu, W., et al.: AdaFuse: adaptive multiview fusion for accurate human pose estimation in the wild. arXiv e-prints (2020)
10. Oberweger, M., Wohlhart, P., Lepetit, V.: DeepPose: human pose estimation via deep neural networks
11. Newell, A., Yang, K., Deng, J.: Stacked hourglass networks for human pose estimation. In: Leibe, B., Matas, J., Sebe, N., Welling, M. (eds.) ECCV 2016. LNCS, vol. 9912, pp. 483–499. Springer, Cham (2016). https://doi.org/10.1007/978-3-319-46484-8_29
12. Huang, S., Gong, M., Tao, D.: A coarse-fine network for keypoint localization. In: 2017 IEEE International Conference on Computer Vision (ICCV). IEEE (2017)
13. Carreira, J., Agrawal, P., Fragkiadaki, K., et al.: Human pose estimation with iterative error feedback. IEEE (2015)
14. Ke, L., Chang, M.C., Qi, H., et al.: Multi-scale structure-aware network for human pose estimation (2018)
15. Fang, H.S., Xie, S., Tai, Y.W., et al.: RMPE: Regional Multi-person Pose Estimation. IEEE (2017)
16. Chen, Y., Wang, Z., Peng, Y., et al.: Cascaded pyramid network for multi-person pose estimation

17. Xiao, B., Wu, H., Wei, Y.: Simple baselines for human pose estimation and tracking. arXiv e-prints (2018)
18. Li, J., Wang, C., Zhu, H., et al.: CrowdPose: efficient crowded scenes pose estimation and a new benchmark (2018)
19. Pishchulin, L., Insafutdinov, E., Tang, S., et al.: DeepCut: joint subset partition and labeling for multi person pose estimation. IEEE (2016)
20. Insafutdinov, E., Pishchulin, L., Andres, B., et al.: DeeperCut: a deeper, stronger, and faster multi-person pose estimation model. arXiv e-prints (2016)
21. Amin S, Andriluka M, Rohrbach M, et al. Multi-view Pictorial Structures for 3D Human Pose Estimation[C]// British Machine Vision Conference 2013. 2013
22. Wang, C., Wang, Y., Lin, Z., et al.: Robust estimation of 3D human poses from a single image. arXiv e-prints (2014)
23. Ramakrishna, V., Kanade, T., Sheikh, Y.: Reconstructing 3D human pose from 2D image landmarks. In: Fitzgibbon, A., Lazebnik, S., Perona, P., Sato, Y., Schmid, C. (eds.) ECCV 2012. LNCS, vol. 7575, pp. 573–586. Springer, Heidelberg (2012). https://doi.org/10.1007/978-3-642-33765-9_41
24. Iskakov, K., Burkov, E., Lempitsky, V., et al.: Learnable triangulation of human pose. In: 2019 IEEE/CVF International Conference on Computer Vision (ICCV). IEEE (2020)
25. Lin, J., Lee, G.H.: Multi-view multi-person 3D pose estimation with plane sweep stereo (2021)
26. Chen, C.H., Ramanan, D.: 3D human pose estimation = 2D pose estimation + matching. IEEE (2017)
27. Pavlakos, G., Zhou, X., Derpanis, K.G., et al.: Coarse-to-fine volumetric prediction for single-image 3D human pose. In: IEEE Conference on Computer Vision & Pattern Recognition. IEEE (2017)
28. Martinez, J., Hossain, R., Romero, J., et al.: A simple yet effective baseline for 3D human pose estimation. IEEE Computer Society (2017)
29. Andriluka, M., Pishchulin, L., Gehler, P., et al.: Human pose estimation: new benchmark and state of the art analysis. In: Computer Vision and Pattern Recognition (CVPR). IEEE (2014)
30. Pavlakos, G., Zhou, X., Derpanis, K.G., et al.: Harvesting multiple views for marker-less 3D human pose annotations. IEEE (2017)
31. Tome, D., Toso, M., Agapito, L., et al.: Rethinking pose in 3D: multi-stage refinement and recovery for markerless motion capture. IEEE (2018)
32. Gordon, B., Raab, S., Azov, G., et al.: FLEX: parameter-free multi-view 3D human motion reconstruction (2021)

Visual Localization Through Virtual Views

Zhenbo Song[1,2], Xi Sun[2(✉)], Zhou Xue[2], Dong Xie[1], and Chao Wen[2]

[1] Nanjing University of Science and Technology, Nanjing, China
[2] ByteDance, Beijing, China
sunxi.ustc@bytedance.com

Abstract. This paper addresses the problem of camera localization, i.e. 6 DoF pose estimation, with respect to a given 3D reconstruction. Current methods often use a coarse-to-fine image registration framework, which integrates image retrieval and visual keypoint matching. However, the localization accuracy is restricted by the limited invariance of feature descriptors. For example, when the query image has been acquired at the illumination (day/night) not consistent with the model image time, or from a position not covered by the model images, retrieval and feature matching may fail, leading to false pose estimation. In this paper, we propose to increase the diversity of model images, namely new viewpoints and new visual appearances, by synthesizing novel images with neural rendering methods. Specifically, we build the 3D model on Neural Radiance Fields (NeRF), and use appearance embeddings to encode variation of illuminations. Then we propose an efficient strategy to interpolate appearance embeddings and place virtual cameras in the scene to generate virtual model images. In order to facilitate the model image management, the appearance embeddings are associated with image acquisition conditions, such as daytime, season, and weather. Query image pose is estimated through similar conditional virtual views using the conventional hierarchical localization framework. We demonstrate the approach by conducting single smartphone image localization in a large-scale 3D urban model, showing the improvement in the accuracy of pose estimation.

Keywords: Visual localization · Pose estimation · View synthesis · Structure from motion

1 Introduction

In recent years, visual place recognition [5,14] is a popular problem in computer vision and has applications in various domains (*e.g.* shared and collaborative augmented reality). Currently, most 6-DoF visual localization methods

This work was completed when Zhenbo Song was an intern at ByteDance. This work was supported in part by the Jiangsu Funding Program for Excellent Postdoctoral Talent under Grant 2022ZB268.

L. Fang et al. (Eds.): CICAI 2022, LNAI 13606, pp. 582–587, 2022.
https://doi.org/10.1007/978-3-031-20503-3_52

are model-based [8,15]. For these approaches, a 3D model of the environment is first reconstructed beforehand by SfM (Structure-from-Motion) methods [16]. Then the query image is matched directly to the reconstructed point cloud, and estimates the absolute camera pose in the 3D model coordinate system. For large-scale use cases, an image retrieval stage is often added before 2D-3D matching to narrow down the search space. After searching a set of reference frames in the large database of model images, the query image is only matched with 3D points visible by reference images. Afterwards, the camera pose is estimated with 2D-3D matches using PnP solvers [6].

Throughout the pipeline from reconstruction to localization, invariant visual features always play an important role. Robust image-level indexable representation and keypoint-level local descriptors are two main topics for improving the application performance. Accordingly, deep learning-based approaches have widely employed in the two-step localization pipeline, such as NetVLAD [1], SuperPoint [3], ASLFeat [9], *etc.* However, it is still a challenge for visual matching when handling large changes in appearance, namely illumination and structural changes. A typical phenomenon can be summarized from previous works, that matching across large changes is easier when both the query image and the model image are acquired at the similar condition, and from approximately the same viewpoint [17]. Collecting such diverse model images is costly and time-consuming. Therefore, one possible solution [12,17] is to simulate novel views from existing model images. The model image database can be expanded, covering more viewpoint and illumination changes. The query image is matched not only in the real model images but also through these virtual views. Although there are many existing approaches working toward this idea [7,13], they can only synthesize with limited-range transformations or heavily rely on precise 3D models.

To generate high-quality virtual views controllably, we propose to use the neural radiance field method [11] for its great performance in view synthesis. Specifically, the sparse point reconstructed by SfM is utilized as depth supervision to accelerate the training of NeRF [2]. Besides, to encode variation of illuminations, we use appearance embeddings from the NeRF-W [10]. In this way, we can easily control both the synthesis viewpoint and illumination. We demonstrate the whole pipeline of our visual localization system in this work. To summarize, the major contributions are two-fold.

- We propose a efficient and controlablle novel view synthesis method, to enrich the model image database used for visual localization.
- We present a conditional selection visual localization pipeline based on virtual views, to boost the pose estimation accuracy.

2 3D Scene Representation

The 3D map is reconstructed from a large set of individual image collections using SfM method *Colmap* [16]. Then the recovered 3D point cloud is used for

NeRF training as sparse depth supervision. When generating novel images, viewpoints are evenly put at additional positions not covered by the original images. Novel appearances are obtained by uniform interpolation on the appearance embeddings. The final localization database is conditional real and virtual combined image sets, where all images are aligned to the same 3D model. One image set is associated with one specific weather and day-and-night illuminations. For efficient storage, each image is presented as a global feature vector, as well as a set of keypoints with corresponding feature descriptors and 3D coordinates. This reconstruction workflow is illustrated in Fig. 1.

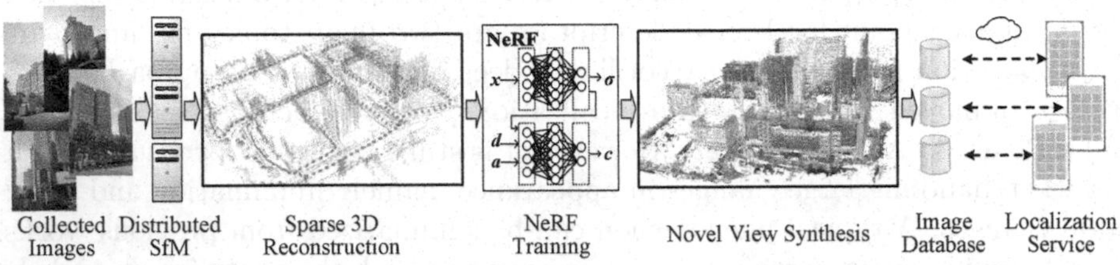

Fig. 1. The system diagram for the proposed method.

2.1 Sparse Reconstruction

In this paper, we use high-end cell phones on the market for high-quality image acquisition. Collected images are tagged with rough GPS labels, so that we can split the images into a consistent size of subsets according to geo-information. We empirically choose adequate images for each submap, to cover enough geographical area. By running the SfM parallelly, sub-maps are built simultaneously under the bounded computational complexity. After accomplishing sub-maps, all views in the same sub-map are bonded as a camera rigid for later merging maps. Towards robust merging, we also add images to each sub-map from neighboring sub-maps, making each sub-map have shared views with its neighbors. The SfM applied a two-step merging strategy. Cameras are first aligned with shared views to get a coarse global map. Then, feature matching is conducted at sub-map levels. Several rounds of global bundle adjustment (BA) are lastly employed for better reconstructing a global consistent 3D map. Besides, to get more matching between day-and-night images, image keypoints and descriptors are obtained using deep neural network, *e.g.* SuperPoint [3] and ASLFeat [9].

2.2 NeRF Reconstruction

After sparse reconstruction, the camera poses are leveraged to train the NeRF model. Since SfM also produces the sparse 3D points, we employ the sparse depth as a supervisory signal following the DS-NeRF [2]. Specifically, depth recovered by SfM is applied to construct the depth uncertainty and the ray distribution loss, supporting fast and stable training of NeRF. Besides, we also integrate appearance embeddings proposed by NeRF-W [10]. The appearance embeddings are dependent on images, which means each image has an independent embedding vector. We relate model images with their time stamp and weather condition forming a dictionary of conditional embeddings. However, image collection cannot fully cover all illumination/weather conditions. To simulate new conditions not included in the image database, we apply embedding interpolation and expand the dictionary.

2.3 View Synthesis

As aforementioned, novel views include new viewpoints and new visual appearances. Here, we describe the procedure from two aspects, *i.e.* viewpoint generating and appearance-aware view rendering. We first align constructed scene with the geographic map, and identify the reachable area on the road plane. New views are placed on these areas. At each viewpoint, 4 horizontal rotations are sampled to cover 360° sweep. The vertical rotation for each view is randomly generate from −30° to 30°. For appearance rendering, we use embedding dictionary to synthesize a set images at each view. We also remove novel views that is too close to the real views to avoid repetition. New image feature and keypoint descriptors are generated in advance by NetVLAD [1] and ASLFeat [9].

3 Visual Localization

We follow the hierarchical framework to formulate the visual localization system. As the application is running on smartphones, rough locations, weather conditions, and illuminations can be obtained through internet data. Subsequently, a subset of database images is chosen for matching with the query image. Then a more accurate initial location can be gotten though virtual views retrieval. However, synthesized views may have artifacts, so the keypoint matching could fail from time to time. To enhance robustness for point-level matching, we also include keypoints from real images which are close to the retrieved views. The L_2 distance of descriptors are calculated, and keypoints passing the ratio test are selected. A standard P3P-RANSAC [4] is applied to compute the camera pose.

Table 1. Comparison of camera pose accuracy.

Accuracy	<(0.25 m, 2°)		<(0.50 m, 5°)		<(5.00 m, 10°)	
Condition	Day	Night	Day	Night	Day	Night
Colmap [16]	0.5123	0.1301	0.7134	0.1517	0.7334	0.1898
w/o virtual views	0.5312	0.2055	0.7292	0.2372	0.7999	0.4504
w/ virtual views	0.5788	0.3570	0.7424	0.4539	0.9092	0.7196

4 Experiment and Conclusion

We validate the proposed approach with iPhone 12 Pro Max. The target scene is an enterprise campus, covering over 10 buildings and 3 blocks. Table 1 presents the localization results. It shows that a significant improvement achieved with respect to localization accuracy. This paper demonstrate that localization accuracy can be boosted through adequate database views, even from virtual views.

References

1. Arandjelovic, R., Gronat, P., Torii, A., Pajdla, T., Sivic, J.: Netvlad: Cnn architecture for weakly supervised place recognition. In: Proceedings of the IEEE Conference on Computer Vision and Pattern Recognition, pp. 5297–5307 (2016)
2. Deng, K., Liu, A., Zhu, J.Y., Ramanan, D.: Depth-supervised nerf: fewer views and faster training for free. arXiv preprint arXiv:2107.02791 (2021)
3. DeTone, D., Malisiewicz, T., Rabinovich, A.: Superpoint: self-supervised interest point detection and description. In: Proceedings of the IEEE Conference on Computer Vision and Pattern Recognition Workshops, pp. 224–236 (2018)
4. Fischler, M.A., Bolles, R.C.: Random sample consensus: a paradigm for model fitting with applications to image analysis and automated cartography. Commun. ACM **24**(6), 381–395 (1981)
5. Hausler, S., Garg, S., Xu, M., Milford, M., Fischer, T.: Patch-netvlad: multi-scale fusion of locally-global descriptors for place recognition. In: Proceedings of the IEEE/CVF Conference on Computer Vision and Pattern Recognition, pp. 14141–14152 (2021)
6. Horaud, R., Conio, B., Leboulleux, O., Lacolle, B.: An analytic solution for the perspective 4-point problem. Comput. Vis. Graph. Image Process. **47**(1), 33–44 (1989)
7. Irschara, A., Zach, C., Frahm, J.M., Bischof, H.: From structure-from-motion point clouds to fast location recognition. In: 2009 IEEE Computer Society Conference on Computer Vision and Pattern Recognition (CVPR 2009), 20–25 June 2009, Miami, Florida, USA (2009)
8. Liu, L., Li, H., Dai, Y.: Efficient global 2d–3d matching for camera localization in a large-scale 3d map. In: 2017 IEEE International Conference on Computer Vision (ICCV) (2017)
9. Luo, Z., Zhou, L., Bai, X., Chen, H., Zhang, J., Yao, Y., Li, S., Fang, T., Quan, L.: Aslfeat: learning local features of accurate shape and localization. In: Proceedings of the IEEE/CVF Conference on Computer Vision and Pattern Recognition, pp. 6589–6598 (2020)
10. Martin-Brualla, R., Radwan, N., Sajjadi, M.S., Barron, J.T., Dosovitskiy, A., Duckworth, D.: Nerf in the wild: neural radiance fields for unconstrained photo collections. In: Proceedings of the IEEE/CVF Conference on Computer Vision and Pattern Recognition, pp. 7210–7219 (2021)
11. Mildenhall, B., Srinivasan, P.P., Tancik, M., Barron, J.T., Ramamoorthi, R., Ng, R.: NeRF: representing scenes as neural radiance fields for view synthesis. In: Vedaldi, A., Bischof, H., Brox, T., Frahm, J.-M. (eds.) ECCV 2020. LNCS, vol. 12346, pp. 405–421. Springer, Cham (2020). https://doi.org/10.1007/978-3-030-58452-8_24

12. Purkait, P., Zhao, C., Zach, C.: Spp-net: deep absolute pose regression with synthetic views. arXiv preprint arXiv:1712.03452 (2017)
13. Rolin, P., Berger, M.O., Sur, F.: View synthesis for pose computation. Machine Vision and Applications (2–3) (2019)
14. Sarlin, P.E., Cadena, C., Siegwart, R., Dymczyk, M.: From coarse to fine: robust hierarchical localization at large scale. In: Proceedings of the IEEE/CVF Conference on Computer Vision and Pattern Recognition, pp. 12716–12725 (2019)
15. Sattler, T., Leibe, B., Kobbelt, L.: Improving image-based localization by active correspondence search. In: Fitzgibbon, A., Lazebnik, S., Perona, P., Sato, Y., Schmid, C. (eds.) ECCV 2012. LNCS, vol. 7572, pp. 752–765. Springer, Heidelberg (2012). https://doi.org/10.1007/978-3-642-33718-5_54
16. Schonberger, J.L., Frahm, J.M.: Structure-from-motion revisited. In: Proceedings of the IEEE Conference on Computer Vision and Pattern Recognition, pp. 4104–4113 (2016)
17. Torii, A., Arandjelovic, R., Sivic, J., Okutomi, M., Pajdla, T.: 24/7 place recognition by view synthesis. In: Proceedings of the IEEE conference on computer vision and pattern recognition. pp. 1808–1817 (2015)

A Synchronized Multi-view System for Real-Time 3D Hand Pose Estimation

Zhipeng Yu and Yangang Wang(✉)

Laboratory of Measurement and Control of Complex Systems of Engineering, Ministry of Education, Southeast University, Nanjing, China

yangangwang@seu.edu.cn

Abstract. In this demo, we present a high-speed and high-quality synchronous multi-view system, which has extremely low latency (10^{-8}s) benefiting the high-fidelity performance of 3D hand pose estimation. Our synchronous multi-view system adopts the unified trigger signal generated by the newly designed synchronous signal generator to control all cameras to work at the same time. It can not only be used for real-time hand pose estimation but also for other moving targets, e.g., body, cloth, general objects, etc. With the help of the developed multi-view system, we have achieved real-time 3D hand pose estimation results.

Keywords: Synchronization · Multi-view system · Hand pose estimation

1 Brief Description

Hand pose estimation has long been the focus of the computer version [4] [18], AR, and computer graphics. Many recent works can estimate the hand joints with high accuracy [6] [5]. Some works estimate the 3D position of hand joints from an RGB/depth camera [8] [9] [10] [11] [12], but it's difficult to estimate the depth of hand joints, the others use the multi-view system to estimate the 3D position [13] [14].

However, camera synchronization and the high capture speed for hand pose estimation were not given much attention, which can improve pose estimation algorithms' understanding of spatiotemporal synchronization. We present a combined hardware and software solution to jointly estimate hands. This multi-view hand pose estimation system can be a platform for data acquisition, experimental operation, and real-time hand pose estimation. It mainly has two characteristics: high-speed and high-synchronization data acquisition.

This work was supported in part by the National Natural Science Foundation of China (No. 62076061), Natural Science Foundation of Jiangsu Province (No. BK20220127) and the "Zhishan Young Scholar" Program of Southeast University (No. 2242021R41083).

All the authors from Southeast University are affiliated with the Key Laboratory of Measurement and Control of Complex Systems of Engineering, Ministry of Education, Nanjing, China.

L. Fang et al. (Eds.): CICAI 2022, LNAI 13606, pp. 588–593, 2022.
https://doi.org/10.1007/978-3-031-20503-3_53

Fig. 1. hand pose estimation demo. With the synchronized multi-view system, we have achieved real-time 3D hand pose estimation. Our system can run at 50fps with 4-cameras. The left part shows our newly designed synchronous signal generator, the middle is a demo hardware configuration of our multi-view system, and the right is the 3D hand pose estimation result.

2 Demo Setup

Our demo system consists of hardware and software in two parts. The former is described in Sect. 2.1 and the latter is described in Sect. 2.2. When the system works, the hardware capture system will control the camera parameters and cap the synchronized sequence. Then it transfers data to the demo software running on the acquisition computer, where we use the pose estimation method we described in Sect. 2.3 to estimate hands.

2.1 Hardware Setup

The hardware of the demo system mainly contains three parts: a synchronous signal generator, industrial cameras, and an acquisition computer. When working, the synchronous signal is sent from the main board of the hardware synchronizer, which is used to ensure the synchronization of the long sequences. After being copied and gained from the other synchronizer from the slave boards, the synchronous signal reaches each camera respectively and controls the camera to take a picture at the same time. Then the collected data is transmitted to the acquisition computer through the usb3.0 line, which can be previewed in real time on the display control computer.

To take into account the sequence Resolution and shooting speed, we finally select the DaHeng camera as the system's camera, whose resolution is 2048*1536 and shooting speed can reach 110 fps. high resolution and high shoot speed bring huge data transmission pressure. The data transmission speed for only one camera can reach 300MB/s. To make one acquisition computer do carry more cameras, we choose the PCIe expansion card to extend the usb3.0 ports instead of making full use of the ports on the computer motherboard, that is because many ports on the computer motherboard share bandwidth with each other.

2.2 Software Setup

The software of the demo system is also composed by three parts: control software, acquisition software, and synchronous signal generator software. **Control software** is directly used by the operator, which can control all the cameras' capture parameters, preview the camera images and show hands pose estimation result. The control software can correspond to many acquisition software and help the operator shield the bottom layer's differences, such as the different cameras or different computer. **Acquisition software** is the software running on the acquisition computer and directly controlling all the cameras. The use of the multithreaded method to ensure all the cameras can store the sequence timely, helping to share the pressure of data transmission and calculation. **Synchronous signal generator software** is running on the synchronous signal generator, which allows the user to set the frequency and times of the synchronized signal.

2.3 Hand Pose Estimation

After getting the synchronous and high fps sequences, we use the Hand-3D-Studio [1] to get the 3D hand global pose. We run the SRHandnet [2] on the acquisition software with an independent thread to get the hands 2D key points $K_i \in R^{3\times21}$, where i is the camera index. Then the 2d key points for each camera view will be transferred to the control software through the local area network. After receiving the 2d key points, the control software will use the existing 3d hands to match the 2d key points and update the 3d hands' location if matching successfully. For the 2d key points not being matched with 3d hands, we use the matching matrix optimization method [3] to assemble a new 3d hand. Finally, we draw the hands in 3D space and the results are shown in the demo video.

3 Experimental Results

In this section, we qualitatively and quantitatively tested our real-time hand pose estimation system.

To visualize the pose estimation results of our hands, we conducted three-dimensional real-time pose estimation of hands with different gestures in the laboratory. The experimental equipment includes a synchronous signal generator, four industrial cameras, and an acquisition computer. The pose estimation results are qualitatively displayed as Fig. 2.

In our experiment, the acquisition computer used is configured as Intel I5 11400 and NVIDIA Quadro P4000. With four cameras and multi-hands, our method can reconstruct the 3d hands pose at 50 fps.

Our method also has some limitations. If the hands are too far from the cameras or there is occlusion, the accuracy of estimation will decrease.

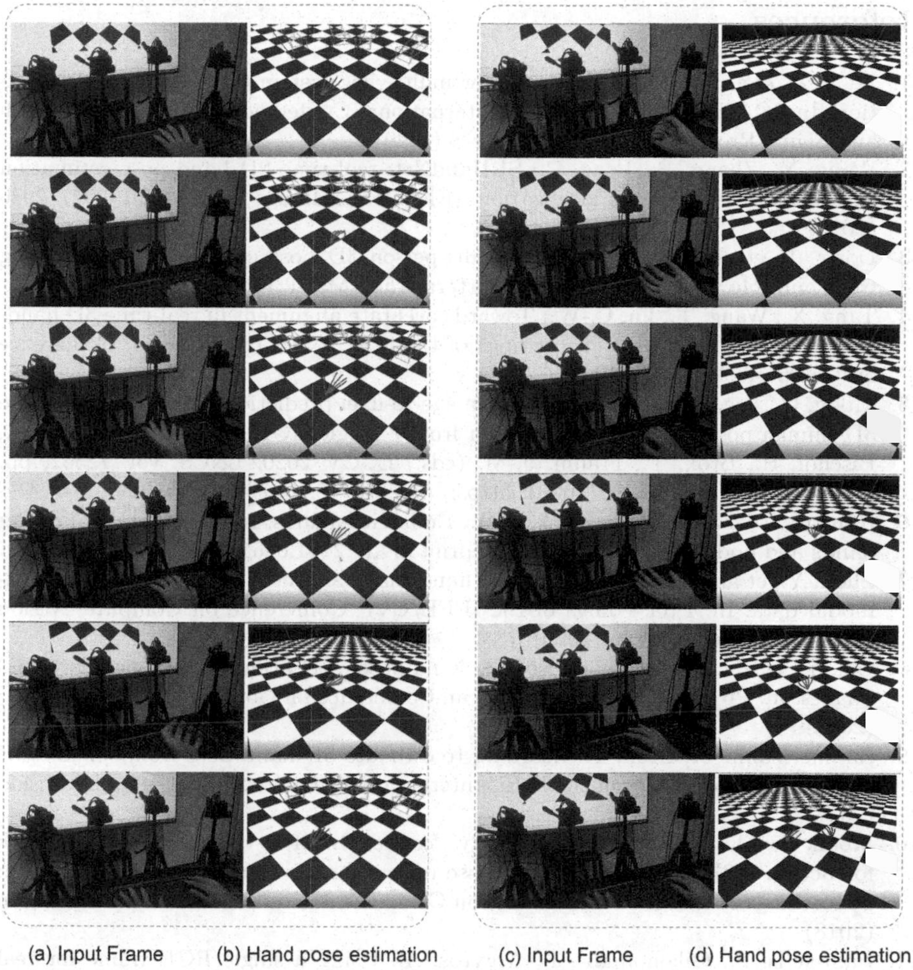

Fig. 2. Hand pose estimation results. With our high-speed and high-quality synchronous multi-view system, the hands pose estimation can be a real-time and high-quality demo. The left is our multi-view system, and the right is our hand pose estimation result.

4 Conclusion

In this demo, we present a high-speed and high-quality multi-view system, and a real-time pose estimation method for hands. The former can ensure the synchronization of the sequences and the latter reconstruct hands in real-time. When the system works, our hand pose estimation system can not only reconstruct hands in real-time but also acquire the data of different scenarios, which can contribute to the pose estimation method research.

References

1. Zhao, Z., et al.: Hand-3D-studio: a new multi-view system for 3D hand reconstruction. In: ICASSP 2020–2020 IEEE International Conference on Acoustics, Speech and Signal Processing (ICASSP). IEEE (2020)
2. Wang, Y., Zhang, B., Peng, C.: SRHandNet: real-time 2D hand pose estimation with simultaneous region localization. IEEE Trans. Image Process. **29**, 2977–2986 (2019)
3. Dong, J., et al.: Fast and robust multi-person 3D pose estimation and tracking from multiple views. IEEE Trans. Pattern Anal. Mach. Intell. (2021)
4. Tang, X., Wang, T., Fu, C.-W.: Towards accurate alignment in real-time 3D hand-mesh reconstruction. In: Proceedings of the IEEE/CVF International Conference on Computer Vision (2021)
5. Moon, G., Lee, K.M.: I2L-MeshNet: image-to-lixel prediction network for accurate 3D human pose and mesh estimation from a single RGB image. In: Vedaldi, A., Bischof, H., Brox, T., Frahm, J.-M. (eds.) ECCV 2020. LNCS, vol. 12352, pp. 752–768. Springer, Cham (2020). https://doi.org/10.1007/978-3-030-58571-6_44
6. Romero, J., Tzionas, D., Black, M.J.: Embodied hands: modeling and capturing hands and bodies together. arXiv preprint arXiv:2201.02610 (2022)
7. Zhou, Y., et al.: Monocular real-time hand shape and motion capture using multi-modal data. In: Proceedings of the IEEE/CVF Conference on Computer Vision and Pattern Recognition (2020)
8. Cai, Y., et al.: Weakly-supervised 3D hand pose estimation from monocular RGB images. In: Proceedings of the European Conference on Computer Vision (ECCV) (2018)
9. Zimmermann, C., Brox, T.: Learning to estimate 3D hand pose from single RGB images. In: Proceedings of the IEEE International Conference on Computer Vision (2017)
10. Moon, G., Chang, J.Y., Lee, K.M.: V2v-PoseNet: voxel-to-voxel prediction network for accurate 3D hand and human pose estimation from a single depth map. In: Proceedings of the IEEE Conference on Computer Vision and Pattern Recognition (2018)
11. Panteleris, P., Oikonomidis, I., Argyros, A.: Using a single RGB frame for real time 3D hand pose estimation in the wild. In: 2018 IEEE Winter Conference on Applications of Computer Vision (WACV). IEEE (2018)
12. Baek, S., Kim, K.I., Kim, T.-K.: Weakly-supervised domain adaptation via GAN and mesh model for estimating 3D hand poses interacting objects. In: Proceedings of the IEEE/CVF Conference on Computer Vision and Pattern Recognition (2020)
13. Ren, P., et al.: Mining multi-view information: a strong self-supervised framework for depth-based 3D hand pose and mesh estimation. In: Proceedings of the IEEE/CVF Conference on Computer Vision and Pattern Recognition (2022)
14. Corona, E., et al.: LISA: learning implicit shape and appearance of hands. In: Proceedings of the IEEE/CVF Conference on Computer Vision and Pattern Recognition (2022)
15. He, Y., et al.: Epipolar transformers. In: Proceedings of the IEEE/CVF Conference on Computer Vision and Pattern Recognition (2020)
16. Huang, L., Tan, J., Liu, J., Yuan, J.: Hand-Transformer: non-autoregressive structured modeling for 3D hand pose estimation. In: Vedaldi, A., Bischof, H., Brox, T., Frahm, J.-M. (eds.) ECCV 2020. LNCS, vol. 12370, pp. 17–33. Springer, Cham (2020). https://doi.org/10.1007/978-3-030-58595-2_2

17. Malik, J., et al.: Deephps: end-to-end estimation of 3D hand pose and shape by learning from synthetic depth. In: 2018 International Conference on 3D Vision (3DV). IEEE (2018)
18. Spurr, A., et al.: Cross-modal deep variational hand pose estimation. In: Proceedings of the IEEE Conference on Computer Vision and Pattern Recognition (2018)
19. Wang, Y., Rao, R., Zou, C.: Personalized hand modeling from multiple postures with multi-view color images. Comput. Graph. Forum. 39(7) (2020)

Monocular Real-Time Human Geometry Reconstruction

Qiao Feng[1], Yebin Liu[2], Yu-Kun Lai[3], Jingyu Yang[1], and Kun Li[1](✉)

[1] Tianjin University, Tianjin, China
fengqiao@tju.edu.cn, yjy@tju.edu.cn, lik@tju.edu.cn
[2] Tsinghua University, Tianjin, China
liuyebin@mail.tsinghua.edu.cn
[3] Cardiff University, Cardiff, UK
laiy4@cardiff.ac.uk

Abstract. Monocular human reconstruction is very popular in computer vision and graphics, which can be a backbone for various applications, such as 3DTV and Holographic Telepresence systems. In this demo, we present the first 30+FPS monocular real-time human geometry reconstruction system, which achieves high-quality results and real-time speed at the same time. Our system reconstructs a 3D human mesh for each frame with Fourier Occupancy Field (FOF), which is a very efficient representation for monocular reconstruction. We adapt it to support the perspective camera setup and build a concise reconstruction framework. HDRis and ray tracing rendering are used to improve the results on the real captured images. Furthermore, the newly designed framework can ignore background automatically, eliminating the need for segmentation, which reduces the usage of GPU and enables more post-processing. Our system can be deployed on PCs with high-performance GPUs and common USB web cameras, which is a consumer-accessible solution for human geometry reconstruction.

Keywords: 3D reconstruction · Real-time system · Human reconstruction · Geometry processing

1 Introduction

Several works, such as PIFu [1], PIFuHD [2], and ICON [3], have been proposed to reconstruct high-fidelity human geometry from a single image. Typically, this problem is addressed with implicit neural representation. The inference of these methods however are very time-consuming. Li *et al.* [4] design an efficient sampling scheme to accelerate the inference of implicit neural representation which enables 15 fps processing time with a 256^3 spatial resolution occupancy voxel grid. However, most graphics applications require the 3D objects in the form of mesh. To be compatible with this, the marching cube algorithm is used to extract the mesh from the voxel grid, which further reduces the frame rate. Thus, existing methods have difficulties achieving high-quality results and real-time speed at the same time.

L. Fang et al. (Eds.): CICAI 2022, LNAI 13606, pp. 594–598, 2022.
https://doi.org/10.1007/978-3-031-20503-3_54

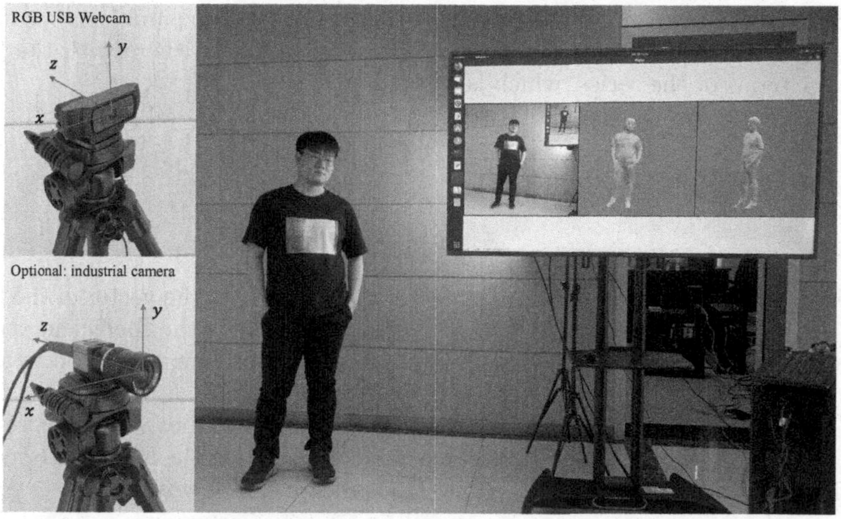

Fig. 1. Our monocular real-time human geometry reconstruction demo can produce high-quality meshes at 30FPS. Industrial camera can be used instead to further improve the performance.

The purpose of this demo is to demonstrate monocular real-time human geometry reconstruction, using an incremental version of the Fourier occupancy field (FOF) proposed in [6], which can be a backbone for various computer vision and graphics applications. The FOF is a representation that makes it possible to reconstruct a 3D object with a 2D CNN framework directly.

Compared to the original FOF, this demo adapts it to support the perspective camera setup and makes it eliminate the need for segmentation which results in a more compact system. Moreover, we optimize the processing of training data and further improve the performance.

2 Brief Description

Here, we briefly describe the FOF representation adapted from [6], which is the basis of this demo. Given a 3D object, it can be normalized into a $[-1, 1]^3$ cube and be represented with a 3D occupancy field $F : [-1, 1]^3 \mapsto \{0, 0.5, 1\}$, where the occupancy function $F(x, y, z)$ is defined as

$$F(x,y,z) = \begin{cases} 1, & (x,y,z) \text{ is inside the object,} \\ 0.5, & (x,y,z) \text{ is on the surface of the object,} \\ 0, & (x,y,z) \text{ is outside the object.} \end{cases} \quad (1)$$

Such an occupancy field can be considered as a composition of 1D occupancy functions along the lines orthogonal to the xy-plane. Each of them passing through a particular point $(x^*, y^*) \in [-1, 1]^2$ on the xy-plane, which can be written as $f(z) = F(x^*, y^*, z) : [-1, 1] \mapsto \{0, 0.5, 1\}$. We expand such a family of 1D occupancy functions to Fourier series, and represent them with the first $2N+1$ terms of the series, which is written as:

$$f(z) = \frac{a_0}{2} + \sum_{n=1}^{\infty} \left(a_n \cos(nz) + b_n \sin(nz)\right), z \in [-1, 1], \tag{2}$$

$$\hat{f}(z) = \boldsymbol{b}^\top(z)\boldsymbol{c}, \tag{3}$$

where $\boldsymbol{b}(z) = [1/2, \cos(z), \sin(z), \ldots, \cos(Nz), \sin(Nz)]^\top$ is the vector of the first $2N+1$ basis functions , and $\boldsymbol{c} = [a_0, a_1, b_1, \ldots, a_N, b_N]^\top$ is the coefficient vector which provide a more compact representation of the approximated 1D occupancy function $\hat{f}(z)$.

This process can be extended to the entire xy-plane so that we can represent the original 3D occupancy field with a more efficient and compact 2D vector field orthogonal to the view direction. This process can be written as:

$$\hat{F}(x, y, z) = \boldsymbol{b}^\top(z)\boldsymbol{C}(x, y), \tag{4}$$

where $C(x, y) : [-1, 1]^2 \mapsto \mathbb{R}^{2 \times N+1}$ is the 2D vector field used to represent the 3D object, namely Fourier Occupancy Field (FOF). Equation 4 also shows how to recover an approximated occupancy field from a FOF. In the practical implementation, $C(x, y)$ is discretized and represented as a multi-channel image where each pixel is a coefficient vector corresponding to a line orthogonal to the xy-plane.

Thus, a 3D human geometry can be represented with a multi-channel 2D map and N is chosen as 31 in our implementation. Such a structure is compatible with the existing CNN frameworks and can be estimated with a image-to-image neural network, which is very efficient and have significant advantages over implicit neural representations. Once the FOF is estimated, the approximated occupancy field can be recovered according to Eq. 4 and the mesh can be extracted with the marching cube algorithm [5].

3 Demo Setup

Our system consists of a PC with a high-performance GPU, a monitor, and an RGB camera. A single RTX-3090 GPU is used for reconstruction and rendering. For the convenience of common customers, a USB webcam (Logitech C920 PRO) is used, which may be the bottleneck of the system and reduce the frame rate slightly. An industrial camera (FLIR BFS-U3-13Y3C-C) can be used instead to improve the performance. The PC used in our system has an Intel Core i9-10900X CPU and 64GB memory, which is far more than enough for our demo. A monitor is used to preview the reconstructed results in real-time and can be adapted to different situations. Figure 1 shows the overview of our system.

4 Practical Implementation

Our demo system is implemented in python. OpenCV is used to get images from the camera, and TensorRT is used to provide efficient neural network inference. The marching cube algorithm [5] is implemented as a PyTorch Cuda extension to be compatible with the pipeline. Pytorch3D is used as the renderer to provide a real-time preview of the reconstructed meshes. All processing above is implemented in a single loop but not in multiple processes. Because the GPU is already fully utilized. Python multiprocessing may result in additional performance loss.

As described in Sec. 2, we use an image-to-image network based on HR-Net-W32-V2 [7] as the backbone of our demo system. Equation 4 is integrated with the network and transformed into a single TensorRT engine. The network is trained to have the ability to ignore background automatically. Therefore, the captured images are fed into the engine directly, eliminating the need for segmentation. The resolution of the input image is 512^2, and the resolutions of the reconstructed FOF and voxel grid are 256^2 and 256^3 respectively. As the input images are captured with a perspective camera, the meshes extracted from the voxel grids are in NDC space. Thus, the vertices of the meshes should be transformed into the world space before rendering, which is the inverse process of perspective projection. This process is implemented with PyTorch and integrated into the rendering module.

Our backbone is trained on THuman2.0 [8] which contains 526 human models with various clothing in different poses. We render 64 views high-fidelity images for each model in randomly selected HDRi environment texture with ray tracing rendering. All the HDRis we used are collected from Poly Haven[1]. The HDRi not only serves as a panoramic background image but also provides ambient lighting. A perspective camera is used for rendering, the parameters of which are the same as the USB webcam used in our system. Figure 2 shows some results.

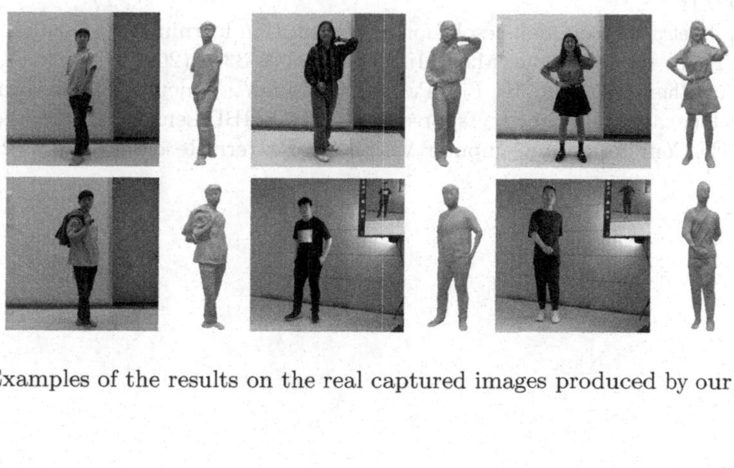

Fig. 2. Examples of the results on the real captured images produced by our system.

[1] https://polyhaven.com/.

5 Conclusion

In this demo, we present a monocular real-time human geometry reconstruction, which can produce meshes from a single RGB camera at 30+FPS and can be a backbone for various computer vision and graphics applications. In the future, we will implement it with CPP and Cuda to further improve the performance and release a library for free non-commercial exploitation.

Acknowledgment. This work was supported in part by the National Natural Science Foundation of China (62171317 and 62122058).

References

1. Saito, S., Huang, Z., Natsume, R., Morishima, S., Ka-nazawa, A., Li, H.: PIFu: Pixel-aligned implicit function for high-resolution clothed human digitization. In: Proceedings of 2019 IEEE/CVF International Conference on Computer Vision (ICCV) (2019)
2. Saito, S., Simon, T., Saragih, J., Joo, H.: PIFuHD: multi-level pixel aligned implicit function for high-resolution 3D human digitization. In: Proceedings of the IEEE Conference on Computer Vision and Pattern Recognition CVPR (2020)
3. Xiu, Y., Yang, J., Tzionas, D., Black, M.J.: ICON: implicit clothed humans obtained from normals. In: Proceedings of the IEEE/CVF Conference on Computer Vision and Pattern Recognition (CVPR) (2022)
4. Li, R., Xiu, Y., Saito, S., Huang, Z., Olszewski, K., Li, H.: Monocular real-time volumetric performance capture. In: Proceedings of European Conference on Computer Vision (ECCV) (2020)
5. Lorensen, W.E., Cline, H.E.: Marching Cubes: a high resolution 3D surface construction algorithm. ACM SIGGRAPH Comput. Graphics **21**(4), 163–169 (1987)
6. Feng, Q., Liu, Yebin., Lai, Y.-K., Yang, J., Li, K.: FOF: learning fourier occupancy field for monocular real-time human reconstruction. https://arxiv.org/abs/2206.02194
7. Wang, J., et al.: Deep high-resolution representation learning for visual recognition. IEEE Trans. Pattern Anal. Mach. Intell. **43**, 3349–3364 (2019)
8. Tao, Yu., Zheng, Z., Guo, K., Liu, P., Dai, Q., Liu, Y.: Function4D: real-time human volumetric capture from very sparse consumer RGBD sensors. In: Proceedings of the IEEE Conference on Computer Vision and Pattern Recognition (CVPR) (2021)

EasyPainter: Customizing Your Own Paintings

Yuqi Zhang, Qian Zheng(✉), and Gang Pan

Zhejiang University, Zhejiang, China
{yq_zhang,qianzheng,gpan}@zju.edu.cn

Abstract. With the advance in deep learning techniques, increasing attention has been paid to studying the creation capacity of artificial intelligence (AI) models. AI-based painting image generation not only streamlines tedious tasks of professional users but also enlightens a broad range of casual users for artistic creation. This paper focuses on customizing painting images with limited inputs, *i.e.*, a reference painting image as the exemplar and a conditional input that users could easily specify to reflect their ideas. We show the challenges of directly applying existing solutions to the problem of painting customizing and provide a solution to address them. We also develop the EasyPainter, a user-friendly interface system to better understand AI's creative and exploratory potential in customizing painting images. Our demo video is available here.

Keywords: Painting customizing · Deep learning · Image translation

1 Introduction

Recent rapid advances in machine learning have led to an acceleration of interest in artificial intelligence (AI) research, fostering the exploration of possible applications of AI in various fields. 'AI and Art' has attracted tremendous attention. Among existing research, visual art analysis is considered a mysterious and distant discipline in the public domain [2]. On the other hand, researchers pay increasing attention to studying AI creation capacity on images with the advance of Generative Adversarial Networks (GANs) [4]. Exemplar-based image translation, which takes inputs of an exemplar and a conditional input, directly illustrates the translation results and intuitively reflects the creative capacity of an AI model.

Existing advances in exemplar-based image translation (*e.g.*, [16,18,21]) have achieved promising results on real images. However, these methods are trained by real images and fail for artistic works such as painting images. That is, the domain gap between real images and stylized ones makes the prior learned from real images less reliable when testing on stylized ones, which leads to poor translation results (detailed in Sec. 3). Collecting all painting styles to train an AI model to address the domain gap is almost impossible.

L. Fang et al. (Eds.): CICAI 2022, LNAI 13606, pp. 599–604, 2022.
https://doi.org/10.1007/978-3-031-20503-3_55

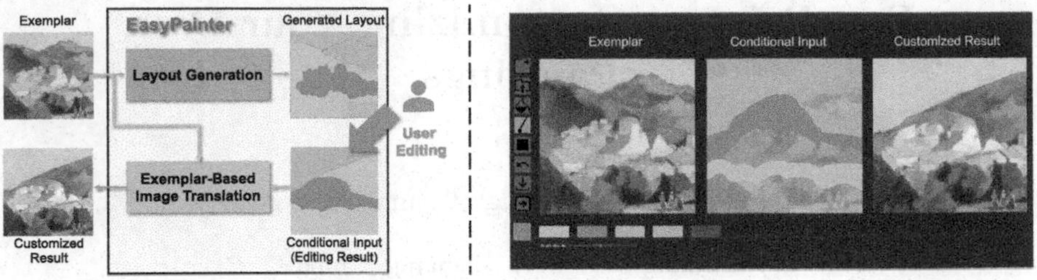

Fig. 1. Overview of EasyPainter (left) and its user interface (right). EasyPainter takes the input of an exemplar, automatically generates the layout for user editing, and outputs the customized painting images. The layout generation method is implemented by Deeplab [3,8]. Note that the conditional input could also be obtained from other images via layout generation methods.

Fig. 2. Application scenarios: conditional inputs generated from other images (left) and progressively edited by users (right). Conditional inputs or generated layouts are placed at left corners of corresponding images.

Motivated by the insights that 1) coarse features (*e.g.*, sharp edges) are more consistent among images with different styles, and 2) coordinate transport will not distort the style, this paper proposes a novel method to compute the correlation matrix, *i.e.*, it imposes dominant features for coordinate transport. We show that our method facilitates the generation of high-quality customized painting images. We further develop the system of EasyPainter, allowing both professional and casual users to customize their painting images. This work is beneficial to understanding the creative and exploratory potential of AI in the context of art.

2 Proposed System and Applications

Figure. 1 illustrates the system of the proposed EasyPainter. EasyPainter allows professional users such as artists to create thumbnail sketches for streamlining tedious tasks or quick experiments. Moreover, it also enlightens a wide range of casual users, such as children, to create customized drawings through simple

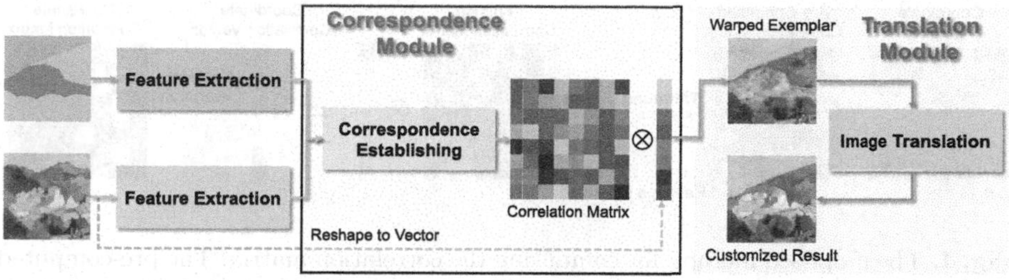

Fig. 3. Framework of state-of-the-art exemplar-based image translation methods [16, 18, 21] and our method. It consists of a correspondence module and a translation module. The correspondence module takes inputs of a conditional input and an exemplar, and extracts features of them by two feature extraction networks. The correspondence between these two inputs, represented by a correlation matrix, is established in the feature space by measuring the cosine similarity or doing optimal transport [14, 16]. The output of the correspondence module is the warped exemplar (or warped feature), calculated by multiplying the reshaped vector of the exemplar (or reshaped vector of its feature) and the correlation matrix. Note that we only show the warped exemplar for clarity. The translation module takes the input of the warped exemplar(or warped feature) and outputs a customized result.

editing. The user-friendly interface provides a simple interaction with the automatically generated layout and timely demonstrates the customized results for quick feedback.

EasyPainter is developed to deal with painting images with any style and content. We show two application scenarios of EasyPainter: 1) conditional inputs generated from other images, and 2) conditional inputs progressively edited by users, in Fig. 2.

3 Methodology

Image-to-image translation methods learn the mapping between different image domains. A category of methods (*e.g.*, [7, 9, 10, 22]) resort to a conditional generative adversarial network and directly synthesize output images of the target domain given user inputs. Among all image-to-image translation methods, exemplar-based image translation methods [1, 6, 11–13, 15, 17] could control the output style based on the exemplar and achieved increasing attention in recent years. As our target is to deal with painting images with various styles, we adopt the framework of state-of-the-art exemplar-based image translation methods [16, 18, 21] in EasyPainter.

Figure 3 shows the overview of our framework, which is the same as those in [16, 18, 21]. As can be observed, the key idea of this framework is how to establish the correspondence between the exemplar and the conditional input, based on which the exemplar (or feature map) is warped and guides the generation of translation results.

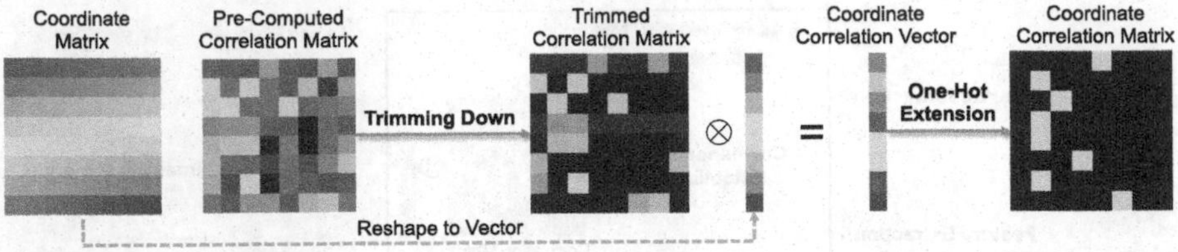

Fig. 4. The proposed method for computing the correlation matrix. The pre-computed correlation matrix could be obtained by several ways (*e.g.*, [16,18,21]). In our implementation, we calculate it based on UNITE [16]. After obtaining the pre-computed correlation matrix, we trim it down and retain the dominant elements (*i.e.*, the top-K largest elements) in each row by setting the rest elements to be zero. We then multiply the trimmed correlation matrix and the reshaped coordinate vector (indicates the pixel positions of the exemplar) to obtain the coordinate correlation vector. This vector is extended to our coordinate correlation matrix, where each row is a one-hot vector (element value is set to be 1). The position of the one-value element is determined by the value of the coordinate correlation vector at the corresponding dimension.

Fig. 5. For each subfigure, from left to right: visual quality comparison of warped exemplars (W.E.), and customized results from UNITE [16] and EasyPainter. Conditional inputs or generated layouts are placed at top left corners of corresponding images.

Existing solutions [5,16,18,19,21], trained with real images, could not generalize well to painting images with various unseen styles and content for two reasons. First, these methods conduct dense feature matching over a large search space. The domain gap caused by unseen painting image styles makes the search significantly more complex, so they are prone to establishing unreliable correspondences, including false and many-to-one matching. Second, these methods transport and fuse features (or pixel values) at different locations to generate the warped features (or warped exemplar), which brings blurry customized results and could destroy the style of the exemplar.

To this end, this paper proposes a novel method to compute the correlation matrix for correspondence. As shown in Fig. 4, it imposes dominant features from a pre-computed correlation matrix to eliminate the impact of unreliable matching. That is, by trimming down the pre-computed correlation matrix and retaining dominant elements in each row, we force the feature of the conditional input at each position only associating to that of exemplar at most K positions, which facilitates discarding unreliable matching and maintaining reliable one. We further use the reshaped coordinate vector rather than the feature (or exemplar) vector for transport. Different from the transport of features (or pixel values) that produce interpolated features (or pixel values), the transport of coordinates is essential to interpolate coordinates and will not introduce artifacts from fused features (or values). Hence, it facilitates the preservation of the style features.

The implementation and training details of our method are same as UNITE [16], except for the computation of the correlation matrix (see Fig. 4). The K is set to be 10. We train the model on ADE20K dataset [20]. Figure 5 shows the comparison of warped exemplar and the final results with UNITE [16]. As can be observed, our method provides more sharp and accurate style guidance and achieves better customized results than UNITE [16].

Acknowledgments. This work is supported by Natural Science Foundation of China (No. 61925603).

References

1. Bansal, A., Sheikh, Y., Ramanan, D.: Shapes and context: in-the-wild image synthesis and manipulation. In: Proceedings of Computer Vision and Pattern Recognition, pp. 2317–2326 (2019)
2. Cetinic, E., She, J.: Understanding and creating art with AI: review and outlook. ACM Trans. Multimedia Comput. Commun. Appl. (TOMM) **18**(2), 1–22 (2022)
3. Chen, L.C., Papandreou, G., Kokkinos, I., Murphy, K., Yuille, A.L.: DeepLab: semantic image segmentation with deep convolutional nets, atrous convolution, and fully connected CRFs. IEEE Trans. Pattern Anal. Mach. Intell. **40**(4), 834–848 (2017)
4. Goodfellow, I., et al.: Generative adversarial nets. In: Advances in Neural Information Processing Systems **27** (2014)
5. He, M., Chen, D., Liao, J., Sander, P.V., Yuan, L.: Deep exemplar-based colorization. ACM Trans. Graph. (TOG) **37**(4), 1–16 (2018)
6. Huang, X., Liu, M.Y., Belongie, S., Kautz, J.: Multimodal unsupervised image-to-image translation. In: Proceedings of European Conference on Computer Vision, pp. 172–189 (2018)
7. Isola, P., Zhu, J.Y., Zhou, T., Efros, A.A.: Image-to-image translation with conditional adversarial networks. In: Proceedings of Computer Vision and Pattern Recognition, pp. 1125–1134 (2017)
8. kazuto1011: Deeplab with pytorch. https://github.com/kazuto1011/deeplab-pytorch
9. Kim, H., Jhoo, H.Y., Park, E., Yoo, S.: Tag2Pix: line art colorization using text tag with secat and changing loss. In: Proceedings of Computer Vision and Pattern Recognition, pp. 9056–9065 (2019)
10. Kim, J., Kim, M., Kang, H., Lee, K.H.: U-GAT-IT: unsupervised generative attentional networks with adaptive layer-instance normalization for image-to-image translation. In: International Conference on Learning Representations (2019)
11. Ma, L., Jia, X., Georgoulis, S., Tuytelaars, T., Van Gool, L.: Exemplar guided unsupervised image-to-image translation with semantic consistency. In: International Conference on Learning Representations (2018)

12. Qi, X., Chen, Q., Jia, J., Koltun, V.: Semi-parametric image synthesis. In: Proceedings of Computer Vision and Pattern Recognition, pp. 8808–8816 (2018)
13. Rozière, B., Riviere, M., Teytaud, O., Rapin, J., LeCun, Y., Couprie, C.: Inspirational adversarial image generation. IEEE Trans. Image Process. **30**, 4036–4045 (2021)
14. Villani, C.: Optimal transport: old and new, vol. 338. Springer (2009). 10.1007/978-3-540-71050-9
15. Wang, M., et al.: Example-guided style-consistent image synthesis from semantic labeling. In: Proceedings of Computer Vision and Pattern Recognition, pp. 1495–1504 (2019)
16. Zhan, F., et al.: Unbalanced feature transport for exemplar-based image translation. In: Proceedings of Computer Vision and Pattern Recognition, pp. 15028–15038 (2021)
17. Zhang, B., et al.: Deep exemplar-based video colorization. In: Proceedings of Computer Vision and Pattern Recognition, pp. 8052–8061 (2019)
18. Zhang, P., Zhang, B., Chen, D., Yuan, L., Wen, F.: Cross-domain correspondence learning for exemplar-based image translation. In: Proceedings of Computer Vision and Pattern Recognition, pp. 5143–5153 (2020)
19. Zhang, X., Zhang, X., Xiao, Z.: Deep photographic style transfer guided by semantic correspondence. Multimedia Tools Appl. **78**(24), 34649–34672 (2019). https://doi.org/10.1007/s11042-019-08099-7
20. Zhou, B., Zhao, H., Puig, X., Fidler, S., Barriuso, A., Torralba, A.: Scene parsing through ADE20K dataset. In: Proceedings of Computer Vision and Pattern Recognition, pp. 633–641 (2017)
21. Zhou, X., et al.: CoCosNet v2: full-resolution correspondence learning for image translation. In: Proceedings of Computer Vision and Pattern Recognition, pp. 11465–11475 (2021)
22. Zhu, J.Y., Park, T., Isola, P., Efros, A.A.: Unpaired image-to-image translation using cycle-consistent adversarial networks. In: Proceedings of International Conference on Computer Vision, pp. 2223–2232 (2017)

Gesture Interaction for Gaming Control Based on an Interferometric Radar

Yuchen Li, Jingyun Lu, Changzhan Gu(✉), and Xiaokang Yang

MoE Key Lab of Artificial Intelligence, Shanghai Jiao Tong University, Shanghai, China
changzhan@sjtu.edu.cn

Abstract. Non-contact hand gesture recognition emerged as a human-computer interactive method to facilitate human life. Compared with vision-based solutions, millimeter-wave radars consume much less power and the adverse effects of ambient environment interference such as lighting conditions can be minimized. Moreover, it can penetrate obstacles for detection and maintains a high motion detection accuracy. In this paper, a custom-designed 5.8 GHz interferometric radar system is presented and a novel real-time gesture recognition technique with a linear phase demodulation algorithm is proposed for gaming control. The experimets result shows that this system could recover the subject's motion with micrometer accuracy. To better demonstrate the performance of the proposed gesture sensing technique, a Super Mario Bro Demo is built and it shows that one can pass the mission easily with the defined hand gesture command.

Keywords: Hand gesture recongnition · Human-computer interaction · Interferometric radar

1 Introduction

Nowadays, human-computer interaction (HCI) has increasingly raised human interests due to the prevalence of the Internet of Things. HCI enables people to realize the information transfer between humans and systems through various methods. Comparing to using contact devices such as orientation sensors, accelerometer, gyroscope, etc., non-contact HCI shows its superiority of simple control and high flexibility [1] and is gradually introduced into the civilian market.

Unlike optical gesture recognition techniques, the radar sensor is insensitive to ambient light conditions and has the advantage of maintaining robustness even in the presence of occlusion. In recent years, radar-based gesture sensing approaches have attracted great attention from both academia and industry [2–10].

In this paper, a novel real-time hand gesture recognition technique based on the customized Doppler interferometric radar is proposed. This technique can not only obtain the hand's frequency, but also the detail information such as velocity and amplitude.

L. Fang et al. (Eds.): CICAI 2022, LNAI 13606, pp. 605–609, 2022.
https://doi.org/10.1007/978-3-031-20503-3_56

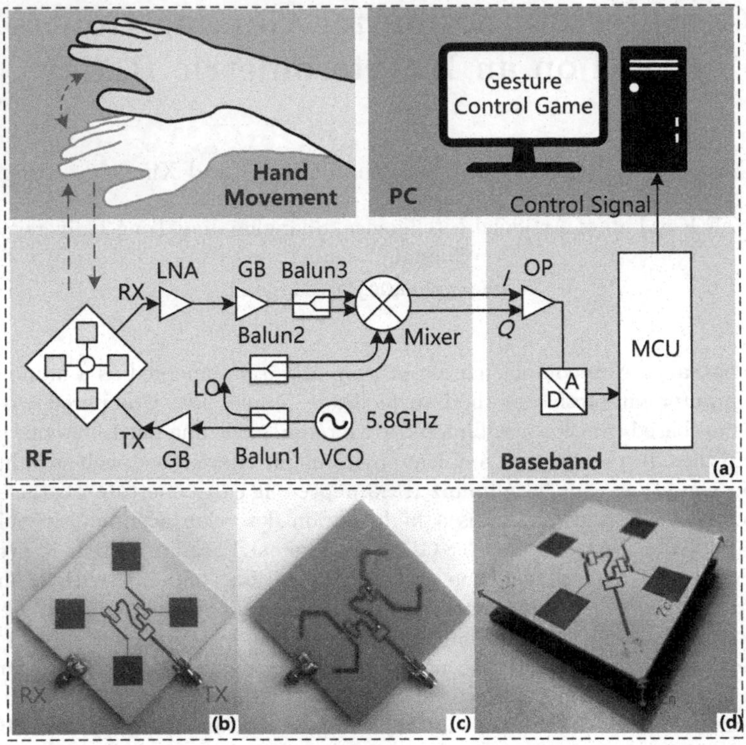

Fig. 1. Block diagram of the 5.8 GHz radar system (a), the top view (b), bottom view (c) of the implemented antenna, and the photograph of the radar system (d).

2 System Design

The system used for gesture recognition is a continuous wave radar which is tailored to work at a single frequency of 5.8 GHz. Figure 1 demonstrates the block diagram of the RF and baseband circuit. A voltage-controlled oscillator is utilized to generate the RF signal. The signal is divided into two channels by a balun. One part of the signal firstly passes through a power amplifier and is transmitted by the TX antenna, the other serves as a local oscillation signal for down-conversion. After being backscattered, the received signal is amplified by a low noise amplifier. A quadrature mixer down-converts the signal to output a pair of differential in-phase and a pair of differential quadrature-phase baseband signals which contain the information of the motion of the hand. After being differentially amplified by the intermediate-frequency amplifier, the baseband signals are delivered to the MCU for analog-to-digital conversion and real-time signal processing. The radar is custom-designed to operate at monostatic mode by a common-aperture antenna, which means TX and RX channels share the same antenna aperture. The photograph of the radar system is shown in Fig. 1(b)-(d).

The common-aperture architecture could not only obtain the highest SNR when the hand is directly above the radar but also miniaturize the system's dimension.

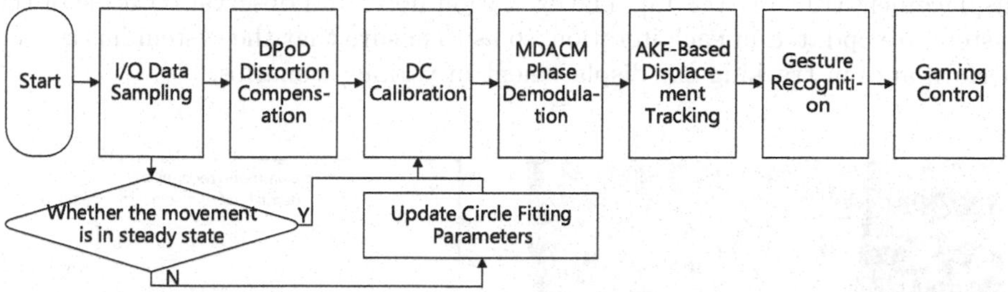

Fig. 2. Flow chart of the proposed radar-based gesture gaming control technique.

3 Principles and Signal Processing

The work flow of the radar-based gesture gaming control is shown in Fig. 2. A single-tone electromagnetic wave is transmitted from the TX antenna and illuminates the hand. Then some portion of the wave get modulated by the hand's displacement and is redirected back toward the radar. The signal is captured by RX antenna, went through directly down-converted, and split into two baseband signals (I/Q), which could be described as follows:

$$I(t) = A_I \cdot \cos\left[\frac{4\pi}{\lambda}\left(x\left(t\right) + d_0\right) + \Delta\theta(t)\right] + DC_I \tag{1}$$

$$Q(t) = A_Q \cdot \sin\left[\frac{4\pi}{\lambda}\left(x\left(t\right) + d_0\right) + \Delta\theta(t)\right] + DC_Q \tag{2}$$

where $x(t)$ is the hand's displacement, d_0 is the initial distance between the hand and the radar, $\Delta\theta(t)$ is the residual phase, DC_I and DC_Q are the unwanted DC offsets due to the environmental stationary scattering and circuit coupling[].

Since the system adopts AC coupling, the DPoD compensation is employed to restore accurate displacement information [11]. In addition, in practical application scenarios, circuit noise and channel errors will cause I/Q channel mismatch, namely DC_I and DC_Q. DC offest of the I/Q channel can be calibrated using circle fitting technique. When the motion is kept within a steady state, the constellation diagram of I/Q is relatively stable. When the motion position changes greatly, the parameters of the circle fitting need to be updated.

After the DPoD compensation and DC calibration, the hand's displacement could be demodulated using modified differentiate and cross-multiply (MDACM) technique [12]:

$$x(t) + d_0 = \int_0^t \frac{\lambda}{4\pi}\left[I(t) \cdot Q'(t) - I'(t) \cdot Q(t)\right] \tag{3}$$

However, the displacement information directly demodulated has a certain degree of distortion due to the stationary cluster, thermal noise and flicker noise from the circuit. A Sage-Husa adaptive Kalman filter algorithm is employed for displacement tracking [13,14]. The excitation noise and observation noise of the system are updated in each iteration, so as to ensure that the system has a good performance in tracking the displacement in various scenarios.

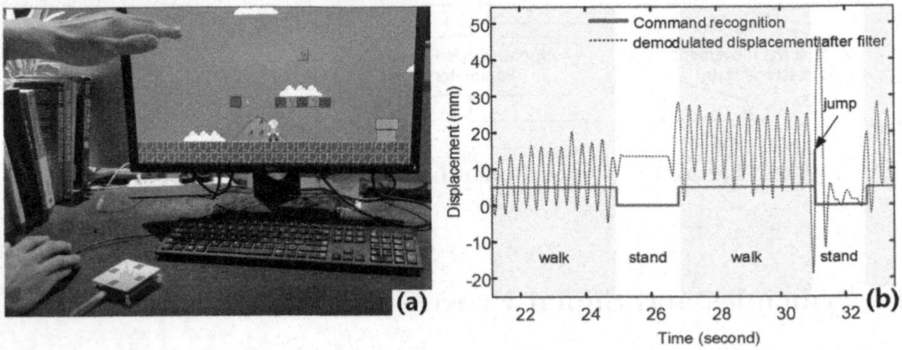

Fig. 3. Setup of the gaming control demonstration on super Mario Bros (a) and the command recognition results with different amplitudes.

4 Experiments and Demonstration

The digital processing algorithm is written on the MCU integrated on the chip, and users can use the radar to control the classic game Super Mario Bros, as shown in Fig. 3(a). The distance between the hand and the radar is 30 cm. The continuous motion pattern with small amplitude displacement and the triggered motion pattern with large amplitude displacement are identified by discriminating the displacement amplitude and velocity of the target. With the relationship between the hand displacement and command is determined, an experiment to evaluate the method's accuracy is carried as shown in Fig. 3(b). In this experiment, ten people are asked to do the three motions randomly, and by comparing the computer printed results, it turns out that the method can achieve 95% accuracy.

5 Conclusion

In this paper, a novel hand gesture recognition scheme for gaming control is proposed. A 5.8 GHz radar system is custom-designed and an demonstration experiment is carried out. The result shows that this system could recover the subject's motion with micrometer accuracy. Kinds of longitudinal gestures with different amplitude are recognized and it turns out that the proposed method achieved 95% precision. This method shows a novel way to help recognize the hand gesture more precisely and could be expanded to other frequencies easily.

Acknowledgement. This work was supported in part by the National Science Foundation of China under Grant 62171277, and in part by the Shanghai Municipal Science and Technology Major Project under Grant 2021SHZDZX0102.

References

1. Gu, C., Wang, J., Lien, J.: Motion sensing using radar: gesture Interaction and Beyond. IEEE Microwave Mag. **20**(8), 44–57 (2019)
2. Kim, Y., Toomajian, B.: Hand gesture recognition using micro-Doppler signatures with convolutional neural network. IEEE Access **4**, 7125–7130 (2016)
3. Fan, T., et al.: Wireless hand gesture recognition based on continuous-wave Doppler radar sensors. IEEE Trans. Microw. Theory Tech. **64**(11), 4012–4020 (2016)
4. Wang, F.-K., Tang, M.-C., Chiu, Y.-C., Horng, T.-S.: Gesture sensing using retransmitted wireless communication signals based on Doppler radar technology. IEEE Trans. Microw. Theory Tech. **63**(12), 4592–4602 (2015)
5. Wang, Z., Yu, Z., Lou, X., Guo, B., Chen, L.: Gesture-radar: a dual Doppler radar based system for robust recognition and quantitative profiling of human gestures. IEEE Trans. Hum. Mach. Syst. **51**(1), 32–43 (2021)
6. Skaria, S., Al-Hourani, A., Lech, M., Evans, R.J.: Hand-gesture recognition using two-antenna Doppler radar with deep convolutional neural networks. IEEE Sens. J. **19**(8), 3041–3048 (2019)
7. Wang, S., Song, J., Lien, J., Poupyrev, I., Hilliges, O.: Interacting with Soli: exploring fine-grained dynamic gesture recognition in the radio-frequency spectrum. In: Proceedings 29th Annual Symposium User Interface Software and Technology, pp. 851–860 (2016)
8. Wei, T., Zhang, X.: 2015. mTrack: high-precision passive tracking using millimeter wave radios. In: Proceedings of the 21st Annual International Conference on Mobile Computing and Networking, ACM, 117–129 (2015)
9. Sun, Y., Fei, T., Gao, S., Pohl, N.: Automatic radar-based gesture detection and classification via a region-based deep convolutional neural network. In: ICASSP 2019–2019 IEEE International Conference on Acoustics, Speech and Signal Processing (ICASSP), pp. 4300–4304 (2019)
10. Google Project Soli. https://atap.google.com/soli
11. Gu, C., Peng, Z., Li, C.: High-precision motion detection using low-complexity doppler radar with digital post-distortion technique. IEEE Trans. Microw. Theory. Tech. **64**(3), 961–971 (2016)
12. Xu, W., Li, Y., Gu, C., Mao, J.-F.: Large displacement motion interferometry with modified differentiate and cross-multiply technique. IEEE Trans. Microw. Theory Tech. **69**(11), 4879–4890 (2021)
13. Sinopoli, B., Schenato, L., Franceschetti, M., Poolla, K., Jordan, M.I., Sastry, S.S.: Kalman filtering with intermittent observations. IEEE Trans. Autom. Control **49**(9), 1453–1464 (2004)
14. Cattivelli, F.S., Sayed, A.H.: Diffusion strategies for distributed Kalman filtering and smoothing. IEEE Trans. Autom. Control **55**(9), 2069–2084 (2010)

Accelerating Allen Brain Institute's Large-Scale Computational Model of Mice Primary Visual Cortex

Zefan Wang[1], Kuiyu Wang[1], and Xiaolin Hu[1,2,3](✉)

[1] Department of Computer Science and Technology, Institute for Artificial Intelligence, State Key Laboratory of Intelligent Technology and Systems, BNRist, Tsinghua University, Beijing, China
{wang-zf20,wang-ky21}@mails.tsinghua.edu.cn, xlhu@tsinghua.edu.cn
[2] IDG/McGovern Institute for Brain Research, Tsinghua University, Beijing, China
[3] Chinese Institute for Brain Research (CIBR), Beijing, China

Abstract. Efficient simulation of large-scale biological neural networks is important for studying the working mechanisms of the brain. Allen Brain Institute proposed a large scale computational model of mice primary visual cortex (V1) which is so far the most detailed model of mice V1. The original model is composed of two parts: lateral geniculate nucleus (LGN) model and V1 spiking neural network (SNN) model. The original model has low computational efficiency and does not support GPU and multithreading acceleration. In this work we we propose several techniques for accelerating the original model. We refactored the original LGN model based on PyTorch and V1 SNN model based on NEST-Simulator, and enabled GPU acceleration and multithreading and multiprocess acceleration. Our LGN model achieved 60 times acceleration in computing speed. The building time of V1 SNN model was accelerated by 5.7 times. When using multiple threads and process, our V1 SNN model achieved 17.8 times acceleration on clusters compared with the original model. Our refactored model is helpful for computational research about the mice V1.

Keywords: Spiking neural network · Primary visual cortex · Computational acceleration

1 Introduction

Computational simulation of large-scale biological neural networks is one of the focuses in brain-like intelligence research. Two basic requirements for these simulations are large-scale and high efficiency. Neuroscience research has accumulated a large number of quantitative data about the structure of cortical neural

This work was supported by the National Key Research and Development Program of China (No. 2021ZD0200301) and by the National Natural Science Foundation of China (Nos. U19B2034, 62061136001, 61836014).

L. Fang et al. (Eds.): CICAI 2022, LNAI 13606, pp. 610–614, 2022.
https://doi.org/10.1007/978-3-031-20503-3_57

network, which makes it possible to build large-scale biological neural network models. Many previous works aim to reconstruct and simulate cortex neural network based on biological data [1,6,7,9–12].

Large-scale computational models of mice primary visual cortex (V1) are proposed by Allen Brain Institute. Through extensive literature research on experimental results of neuroscience, a model of layer four in V1 [2] and a complete model of V1 [3] are established. The complete model of V1 is based on Brain Modeling Toolkit(BMTK) [5], containing about 230 thousands of neurons and having similar dynamic properties compared with *in vivo* experiment data. It is so far the most detailed model of mice V1. The model is composed of two parts: lateral geniculate nucleus (LGN) model and V1 SNN model. The model of Allen Institute has the following disadvantages. First, the computation of LGN model part is very slow. Second, the building time for the V1 SNN model is too long to be parallelized, for it usually takes multiple times to build the network when using multiple threads or processes compared with using only one thread and one process. Third, the V1 SNN part of the model is based on BMTK, but BMTK doesn't support multithreading acceleration.

In this work, we propose several techniques to accelerate Allen Brain Institute's large-scale computational model of mice primary visual cortex(V1) [3]. The original model is refactored to enable multiple GPUs acceleration of the LGN model and to shorten the construction time of the V1 SNN model. Our V1 SNN model also supports multi-threaded acceleration based on OpenMP [4]. Our model improves the computing speed of LGN by about 180 times and improves the building time of the V1 SNN model by about 5.4 times. We tested the parallel acceleration performance of V1 SNN on multiple computing nodes and achieved a maximum acceleration ratio of about 17.8 times.

2 Acceleration Method

2.1 LGN Acceleration Method

In the Allen Brain's model, an LGN cell is modeled as a spatial-temporal separable kernel. The spatial kernel of a LGN cell is a Gaussian filter with different spatial size and location. The temporal kernel is a curve-shaped linear filter. The output firing-rate of the LGN cell is equal to the matrix product of the combined 3D spatial-temporal kernel and the input video. The original LGN model is implemented by the filternet module in BMTK, which has a low computational efficiency. Noticing that most of computation of the LGN model is matrix product, we re-implement the LGN model based on PyTorch [8], which has a lot of optimizations for matrix multiplication and supports GPU acceleration.

The LGN model is composed of 17400 LGN cells with different spatial position, spatial kernel parameters and temporal kernel parameters. To implement GPU acceleration of LGN model based on PyTorch, we model the spatial and temporal kernel of each LGN cell as a PyTorch tensor object. After LGN cells are created in memory in batch mode, they are loaded into the memory of one GPU or splitted and loaded into the memory of multiple GPUs. When multiple

GPUs are used, LGN computing tasks are allocated to each GPU in chronological order, which can effectively utilize the performance of multiple GPUs.

2.2 V1 SNN Acceleration Method

The V1 SNN model is composed of ˜230k cortex neurons modeled with Generalized Leaky Integrate and Fire(GLIF) units [14]. The SNN model is simulated with Runge-Kutta method using 0.25 ms as step size. The simulation can be and have already be implemented in NEST framework as highly paralleled process. However, the V1 SNN model provided along with BMTK is unpractical to be paralleled in MPI [13] or OpenMP interface. The reason is that building hundred thousands of neurons inevitably induces high overhead. While building neurons in unparalleled computation is still acceptable, the design of NEST framework requires duplicate building in every process and duplicate input cells in every threads. As a result, the model cannot be paralleled in reasonable time in BMTK code.

To optimize the building stage of SNN model, we re-ordered neurons in the model to build them in vectorized function calls to NEST backend. This speed-up not only reduces the building time for single-thread simulation of the V1 SNN model, but also makes paralleled running practical, because it reduces the time for building model when using multiple processes.

3 Experiments and Results

3.1 LGN Speed-up Based on GPU Acceleration

Compared with the original LGN model, our accelerated LGN model had a significant improvement in computing speed. Given the same video as input, it took about 30 min for the original LGN model to finish the computation when given this input. For our LGN model, when using one GPU (NVIDIA 2080ti), it took 15.3 s to build LGN cells and 90.9 s for calculation, and the total time was 116.2 s. When using 8 GPUs, it took 53.6 s to build LGN cells and 30.8 s for calculation, and the total time was 84.4 s. If we only considered the calculation time, our LGN model achieved about 60x acceleration. The result of the experiments is shown in Table 1.

To ensure that the output error of our model is within a reasonable range compared with the original model, we calculated the R-squared coefficient of original output and our model's output. The averaged R-squared coefficient is 0.9925, which means that our model's output is very close to the original model's output.

3.2 V1 Speed-up Based on Parallel Computation

We tested the building and simulation time of V1 SNN model provided by Allen Brain with our speed-up on cluster. Each node in the cluster adopted Intel (R)

Table 1. LGN model acceleration

Name	Device	Computing Time (s)
Original model	CPU	~1800
Our model	1 GPU	90.9
Our model	8 GPU	30.8

Xeon (R) CPU e5-2680 V4 chip. Each node had 28 cores and was configured with 132G memory. When using single process and single thread, the original model took 351 s to finish building, while our accelerated model took only 62 s to finish building, with a 5.7 times acceleration. The parallel computation experiments' result is shown in Table 2. We changed the number of processes with MPI interface, and number of threads per process with OpenMP interface. The cores used in each simulation was the product of the amount of threads and processes. The result shows that parallel computation stably increases building time and reduces simulation time of the model. We achieved about 9 times acceleration on single machine comparing to non-parallel simulation, and 17.8 times acceleration on clusters.

Table 2. V1 SNN model parallel computing acceleration

Threads (per process)	Processes	Building time (s)	Simulation time (s)
13	16	453.76	36.11
13	8	400.79	40.76
13	4	319.69	51.39
23	1	228.69	87.56
12	1	88.65	111.46
1	1	62.72	638.79

We compared our model's output and the original model's output. When given the same input LGN spike train and background spike train, our model got exactly the same output spike train as the original model, which means our model is totally the same as the original model.

4 Conclusion

In this work we accelerated Allen Brain Institute's mice V1 model, and our model enabled multiple GPUs acceleration for LGN model, shortened the building time for V1 SNN model and enabled multithreading acceleration based on OpenMP and multiprocess acceleration based on MPI. Our model achieved efficient simulation of mouse V1 model, and provided a basis for subsequent computational simulation research of mouse V1 functional characteristics.

References

1. Antolík, J., Monier, C., Frégnac, Y., Davison, A.P.: A comprehensive data-driven model of cat primary visual cortex. BioRxiv p. 416156 (2019)
2. Arkhipov, A., et al.: Visual physiology of the layer 4 cortical circuit in silico. PLoS Comput. Biol. **14**(11), e1006535 (2018)
3. Billeh, Y.N., et al.: Systematic integration of structural and functional data into multi-scale models of mouse primary visual cortex. Neuron **106**(3), 388–403 (2020)
4. Dagum, L., Menon, R.: OpenMP: an industry standard API for shared-memory programming. IEEE Comput. Sci. Eng. **5**(1), 46–55 (1998)
5. Dai, K., et al.: Brain Modeling Toolkit: an open source software suite for multiscale modeling of brain circuits. PLoS Comput. Biol. **16**(11), e1008386 (2020)
6. Markram, H., et al.: Reconstruction and simulation of neocortical microcircuitry. Cell **163**(2), 456–492 (2015)
7. Motta, A., et al.: Dense connectomic reconstruction in layer 4 of the somatosensory cortex. Science **366**(6469), eaay3134 (2019)
8. Paszke, A., et al.: PyTorch: an imperative style, high-performance deep learning library. In: Advances in Neural Information Processing Systems **32** (2019)
9. Potjans, T.C., Diesmann, M.: The cell-type specific cortical microcircuit: relating structure and activity in a full-scale spiking network model. Cereb. Cortex **24**(3), 785–806 (2014)
10. Ramaswamy, S., et al.: The neocortical microcircuit collaboration portal: a resource for rat somatosensory cortex. Front. Neural Circuits **9**, 44 (2015)
11. Reimann, M.W., Anastassiou, C.A., Perin, R., Hill, S.L., Markram, H., Koch, C.: A biophysically detailed model of neocortical local field potentials predicts the critical role of active membrane currents. Neuron **79**(2), 375–390 (2013)
12. Schmidt, M., Bakker, R., Shen, K., Bezgin, G., Diesmann, M., van Albada, S.J.: A multi-scale layer-resolved spiking network model of resting-state dynamics in macaque visual cortical areas. PLoS Comput. Biol. **14**(10), e1006359 (2018)
13. Snir, M., Gropp, W., Otto, S., Huss-Lederman, S., Dongarra, J., Walker, D.: MPI-the Complete Reference: the MPI core, vol. 1. MIT press, Cambridge (1998)
14. Teeter, C., et al.: Generalized leaky integrate-and-fire models classify multiple neuron types. Nat. Commun. **9**(1), 709 (2018). https://doi.org/10.1038/s41467-017-02717-4,https://doi.org/10.1038/s41467-017-02717-4

Artistic Portrait Applet, Robot, and Printer

Jingjie Zhu[1,2], Lingna Dai[2], Chenghao Xia[1], Chenyang Jiang[1,2], Weiyu Weng[1,2], Yiyuan Zhang[2], Jinglin Zhou[1], Fei Gao[1,2(✉)], Peng Li[2,3], Mingrui Zhu[4], and Nannan Wang[4]

[1] School of Computer Science and Technology, Hangzhou Dianzi University, Hangzhou 310012, China
gaofei@hdu.edu.cn

[2] AiSketcher Technology, Hangzhou 311200, China

[3] Advanced Institute of Information Technology (AIIT), Peking University, Hangzhou 311200, China

[4] ISN State Key Laboratory, Xidian University, Xi'an 710071, China

Abstract. In this paper, we present a series of demonstrations for artistic portrait drawing generation (APDG), including an applet, a robot, and a printer. To this end, we develop novel APDG algorithms which can translate a facial photo into high quality portraits of five artistic styles, i.e. the line-drawing, pen-drawing, pencil-drawing, abstract-drawing, and cartoon. Besides, we provide a number of templates to post-process the generated portraits. By simply pressing several buttons, users can obtain artistic portraits in the applet and order the robot/printer to draw/print them on beautiful postcards. The whole procedure only consumes about 2 min. Our demonstrations are easy to use and have achieved excellent user experience in various exhibitions.

Keywords: Artistic portrait drawing generation · Applet · Drawing robot · Printer · Raspberry Pi

1 Introduction

Artistic portraits drawing generation (APDG) has attracted huge attention in recent years. Inspired by the great success of deep learning, especially Generative Adversarial Networks (GANs) [8], researchers have proposed a large number of APDG methods for generating portraits of multiple styles, such as line-drawings [6,21–23], pencil sketches [2,14,24,25,27], pen-drawings [15,20,23], manga/cartoon [3,16,18], oil-paintings [11,18,26]. Most of these methods model the mapping from a facial photo to an artistic portrait as an image-to-image translation task [9,13,17] or a neural style transfer (NST) task [10]. These algorithms typically produce high-quality portraits for facial photos without extreme variations in pose, expression, and occlusion. However, it is still inconvenient for normal people to try such functions. Only few APDG demonstrations have been developed [19], including applications, websites, and robots [1,6]. It is almost

L. Fang et al. (Eds.): CICAI 2022, LNAI 13606, pp. 615–620, 2022.
https://doi.org/10.1007/978-3-031-20503-3_58

impossible for universal people to try these. It is significant to develop various formats of demonstrations to boost the experience of users.

In this work, we develop a series of demonstrations for artistic portrait drawing generation, including an applet, a robot, and a printer. Specially, we propose novel APDG algorithms for five artistic styles, including the line-drawing, pen-drawing, pencil-drawing, abstract drawing, and cartoon. Besides, we construct the software and hardware of all these demonstrations. An user can obtain multiple styles of artistic portraits by simply scanning a QR code and uploading a facial photo. Our demonstrations have served over 50,000 people in various exhibitions, and achieved excellent user experience.

2 Demonstrations

2.1 Overview

Figure 1 overviews the pipeline of our demonstrations. Our whole system includes an applet, a robot, a printer, and a series of algorithms on a server. We provide a specific QR code for each demonstration. First, an user can upload a facial photo and select an artistic style, after he/she scan the QR code. Second, the uploaded photo is processed and translated to the selected style of portrait on a remote server. Third, he/she can choose a template of a postcard, a wallpaper, or a head portrait. Finally, he/she can download the final version of the portrait in the applet, or order the robot to draw it on paper, or make the printer to print it, by simply pressing a button.

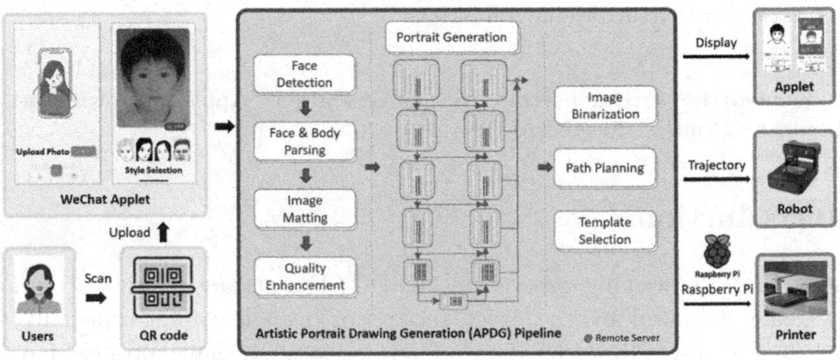

Fig. 1. Overview of our artistic portrait demonstrations.

2.2 Algorithms

On the remote server, after an image is uploaded, a series of algorithms are used for pre-processing, artistic portrait generation, and post-processing [6]. In the implementation, we pre-process the uploaded image by face detection, face alignment, face/body parsing, image matting, and quality enhancement. Afterwards, we select the corresponding APDG algorithm to translate the processed image to an artistic portrait. Finally, we process the obtained portrait by selectively using image binarization, fusion, or enhancement. Besides, we translate a portrait to the drawing trajectories of robot via morphological processing approaches [6].

We extend our previous GENRE method [12] for generating different types of portrait drawings. Specially, we use Pix2Pix [9] as the base network for generating pencil-drawings, and train it on the FS2k dataset [5]. For the other four styles, we use U^2Net [15] as the base network, due to its excellent performance in matting and generating pen-drawings. We train the modified network by combining the pixel-wise reconstruciton loss [9], boundary loss [12], and style loss [7]. Here, the boundary loss makes the generated portraits presenting accurate lines [4]. We use the same network architecture and loss functions for all the four artistic styles. To train the network, we draw about 1,200 line-drawings and abstract-drawings, and 300 pen-drawings and cartoons, respectively. Afterwards, we separately train the APDG network for each style by using the corresponding paired data.

2.3 Applet and Interface

We developed a WeChat applet so that universal people can try our algorithms easily. Besides, this applet is also used as the interface of the drawing robot and the printer. The interfaces of this applet is as shown in Fig. 2. By using a mobile phone, an user can upload an image, captured by the camera or selected from the album. Once selected a style, he/she can obtain the corresponding portrait in about 5 s. Afterwards, he/she can select a prefer template and download it. If a user scans the QR code of a robot or printer, he/she would inter the same applet but have a `draw` or `print` button in the final interface.

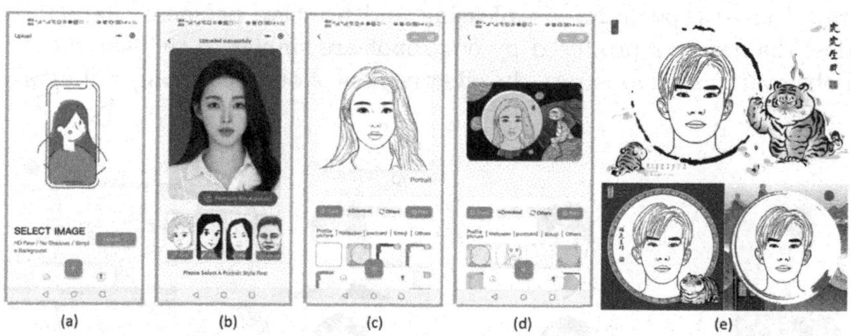

Fig. 2. The interface and operating pipeline of our APDG applet. (a)-(d) are interfaces about image uploading, style selection, template selection, and the final result; (e) some available postcard templates.

2.4 Robot

We further developed a portrait drawing robot. We provide each robot a specific QR code. Once an user scanned it, he/she would inter the applet and connect to the robot. The operation pipeline is as shown in Fig. 1. The generated artistic portrait would be converted to a trajectory sequence to control the drawing process of the robot. For users' convenience, we add assistant functions, such as voice prompts, video tips, paper sensor, and human sensor. The voice prompts and video tips tell a user what to do next and the state of the robot. The paper sensor detects whether there is a postcard or paper on the painting plate. The human sensor detects whether there is people nearby. The whole operation process costs about 2 min in general.

2.5 Printer

Finally, we developed a smart printer for printing artistic portraits. Users can generate artistic portraits and print them though the printer, by simply operating in the applet. We use a Raspberry Pi to connect the applet and a normal printer. The Raspberry Pi runs a TCP client program connected to the back end of the applet. This program is mainly responsible for monitoring the printer status, accepting print requests from the back end, and pre-processing the portraits, and transferring these formats to the printer. Here, in the pre-processing stage, the portrait is automatically rotated, resized, and then converted to `pdf` and `raw` data format, so that a printer can accept. We implement the message transmission mainly based on the CUPS project of Apple[1].

3 Results

We collect a number of facial photos from Web, and generate five styles of artistic portraits. Besides, we use the robot to draw each generated line-drawing on a postcard. Figure 3 illustrates results of our demonstrations. Obviously, the generated artistic portraits are pleasing and preserve personal characteristics. Besides, the results reproduced by our robot are similar to the generated ones. Such observations demonstrate the effectiveness of our APDG algorithm and our systems.

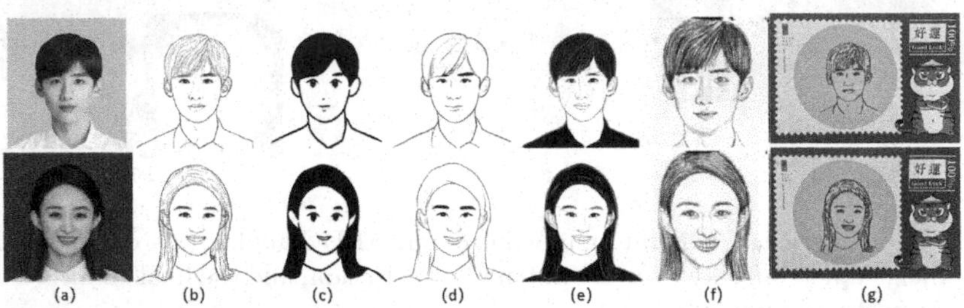

Fig. 3. Generated artistic portraits. (a) Input photo; (b)-(f) the generated line-drawing, cartoon, abstract drawing, pen-drawing, and pencil-drawing; (g) the line-drawing drawn by our robot on a postcard.

[1] https://github.com/apple/cups.

Acknowledgement. This work was supported by the National Natural Science Foundation of China under Grant No. 61971172, and the Hangzhou Science and Technology Development Program under Grant No. 20200401B20.

References

1. Asadi, E., Li, B., Chen, I.M.: Pictobot: a cooperative painting robot for interior finishing of industrial developments. IEEE Robot. Autom. Mag. **25**(2), 82–94 (2018)
2. Cao, B., Wang, N., Li, J., Hu, Q., Gao, X.: Face photo-sketch synthesis via full-scale identity supervision. Pattern Recogn. **124**, 108446 (2022)
3. Cao, N., Yan, X., Shi, Y., Chen, C.: AI-sketcher: a deep generative model for producing high-quality sketches. In: Proceedings of the AAAI Conference on Artificial Intelligence. vol. 33, 2564–2571 (2019)
4. Cheng, T., Wang, X., Huang, L., Liu, W.: Boundary-preserving mask R-CNN. In: Vedaldi, A., Bischof, H., Brox, T., Frahm, J.-M. (eds.) ECCV 2020. LNCS, vol. 12359, pp. 660–676. Springer, Cham (2020). https://doi.org/10.1007/978-3-030-58568-6_39
5. Deng-Ping, F., Ziling, H., Peng, Z., Hong, L., Xuebin, Q., Luc, V.G.: Deep facial synthesis: A new challenge. arXiv (2021)
6. Gao, F., Zhu, J., Yu, Z., Li, P., Wang, T.: Making robots draw a vivid portrait in two minutes. In: 2020 IEEE/RSJ International Conference on Intelligent Robots and Systems (IROS), pp. 9585–9591. IEEE (2020)
7. Gatys, L., Ecker, A., Bethge, M.: A neural algorithm of artistic style. J. Vis. **16**(12), 326–326 (2016)
8. Gui, J., Sun, Z., Wen, Y., Tao, D., Ye, J.: A review on generative adversarial networks: algorithms, theory, and applications. IEEE Trans. Knowl. Data Eng. 1–1 (2021)
9. Isola, P., Zhu, J.Y., Zhou, T., Efros, A.A.: Image-to-image translation with conditional adversarial networks. In: 2017 IEEE Conference on Computer Vision and Pattern Recognition (CVPR), pp. 1125–1134 (2017)
10. Jing, Y., Yang, Y., Feng, Z., Ye, J., Yu, Y., Song, M.: Neural style transfer: a review. IEEE Trans. Visual. Comput. Graphics **26**(11), 3365–3385 (2019)
11. Kim, J., Kim, M., Kang, H., Lee, K.H.: U-GAT-IT: unsupervised generative attentional networks with adaptive layer-instance normalization for image-to-image translation. In: International Conference on Learning Representations (2019)
12. Li, X., Gao, F., Huang, F.: High-quality face sketch synthesis via geometric normalization and regularization. In: 2021 IEEE International Conference on Multimedia and Expo (ICME), pp. 1–6. IEEE (2021)
13. Pang, Y., Lin, J., Qin, T., Chen, Z.: Image-to-image translation: methods and applications. IEEE Trans. Multimedia (2021)
14. Qi, X., Sun, M., Li, Q., Shan, C.: Biphasic face photo-sketch synthesis via semantic-driven generative adversarial network with graph representation learning. arXiv preprint arXiv:2201.01592 (2022)
15. Qin, X., Zhang, Z., Huang, C., Dehghan, M., Zaiane, O.R., Jagersand, M.: U2-Net: going deeper with nested U-structure for salient object detection. Pattern Recogn. **106**, 107404 (2020)
16. Su, H., Niu, J., Liu, X., Li, Q., Cui, J., Wan, J.: MangaGan: unpaired photo-to-manga translation based on the methodology of manga drawing. In: Proceedings of the AAAI Conference on Artificial Intelligence. vol. 35, pp. 2611–2619 (2021)

17. Wang, T.C., Liu, M.Y., Zhu, J.Y., Tao, A., Kautz, J., Catanzaro, B.: High-resolution image synthesis and semantic manipulation with conditional GANs, pp. 8798–8807 (2018)
18. Yang, S., Jiang, L., Liu, Z., Loy, C.C.: Pastiche master: exemplar-based high-resolution portrait style transfer. In: Proceedings of the IEEE/CVF Conference on Computer Vision and Pattern Recognition, pp. 7693–7702 (2022)
19. Yi, R., Liu, Y.J.: Multi-style artistic portrait drawing generation. In: 2021 IEEE International Conference on Multimedia and Expo Workshops (ICMEW), pp. 1–2. IEEE (2021)
20. Yi, R., Liu, Y.J., Lai, Y.K., Rosin, P.L.: APDrawingGAN: generating artistic portrait drawings from face photos with hierarchical GANs. In: IEEE Conference on Computer Vision and Pattern Recognition (CVPR 2019), pp. 10743–10752 (2019)
21. Yi, R., Liu, Y.J., Lai, Y.K., Rosin, P.L.: Unpaired portrait drawing generation via asymmetric cycle mapping. In: IEEE Conference on Computer Vision and Pattern Recognition (CVPR 2020), pp. 8214–8222 (2020)
22. Yi, R., Liu, Y.J., Lai, Y.K., Rosin, P.L.: Quality metric guided portrait line drawing generation from unpaired training data. IEEE Trans. Pattern Anal. Mach. Intell. (2022). https://doi.org/10.1109/TPAMI.2022.3147570
23. Yi, R., Xia, M., Liu, Y.J., Lai, Y.K., Rosin, P.L.: Line drawings for face portraits from photos using global and local structure based GANs. IEEE Trans. Pattern Anal. Mach. Intell. **43**(10), 3462–3475 (2020)
24. Yu, J., et al.: Toward realistic face photo-sketch synthesis via composition-aided GANs. IEEE Trans. Cybern. **51**(9), 4350–4362 (2020)
25. Zhang, M., Wang, N., Li, Y., Gao, X.: Bionic face sketch generator. IEEE Trans. Cybern. **50**(6), 2701–2714 (2019)
26. Zhou, X., et al.: CoCosNet v2: full-resolution correspondence learning for image translation. In: Proceedings of the IEEE/CVF Conference on Computer Vision and Pattern Recognition, pp. 11465–11475 (2021)
27. Zhu, M., Li, J., Wang, N., Gao, X.: A deep collaborative framework for face photo-sketch synthesis. IEEE Trans. Neural Netw. Learn. Syst. **30**(10), 3096–3108 (2019)

Sim-to-Real Hierarchical Planning and Control System for Six-Legged Robot

Yue Gao[1(✉)], Yangqing Fu[2], and Ming Sun[3]

[1] MoE Key Lab of Artificial Intelligence and AI Institute, Shanghai Jiao Tong University, Shanghai, China
yuegao@sjtu.edu.cn

[2] Department of Computer Science and Engineering, Shanghai Jiao Tong University, Shanghai, China
frank79110@sjtu.edu.cn

[3] Department of Automation, Shanghai Jiao Tong University, Shanghai, China
mingsun@sjtu.edu.cn

Abstract. Legged robots have attracted much attention both from industry and academia. Despite the recent progresses in robotics, planning and control in complex environments are still great challenges for legged robots. Generally, constructing the planning and control system for legged robots requires complex designing and parameter tuning. To reduce resource consumption and potential risk in this process, a sim-to-real hierarchical planning and control system is proposed in this paper. The proposed hierarchical system can improve the data efficiency and reduce the training cost utilizing the reinforcement-learning-based framework. Several experiments are conducted to demonstrate the feasibility and effectiveness of the proposed method.

Keywords: Sim-to-real · Robot control system · Hierarchical system

1 Introduction

Legged robot is a challenging research topic in the field of robotics. It has potential applications in military and civilian fields, such as military reconnaissance, intelligent manufacturing monitoring and disabled assistance [1–4]. Compared with tracked and wheeled robots, multi-legged robots have advantages in adapting to complex unstructured environments [5–7]. However, the planning and control system for the multi-legged robot is more complex due to its "multi-input, multi-output and multi-end-effector" property [8]. Compared with bipedal and quadruped robots, six-legged robots have better stability, bearing capacity and adaptability [9].

This work is supported by the National Natural Science Foundation of China (Grant No. 61903247), and Shanghai Municipal Science and Technology Major Project (Grant No. 2021SHZDZX0102).

L. Fang et al. (Eds.): CICAI 2022, LNAI 13606, pp. 621–625, 2022.
https://doi.org/10.1007/978-3-031-20503-3_59

In recent years, the field of multi-legged robot has made great progresses. Particularly when utilizing simulation environment, reinforcement learning methods have shown potentials in planning and control for unstructured environments [10]. For instance, with an end-to-end approach, a policy network can be trained to control the torques of the robot with different height maps as input [11,12]. In addition, sim-to-real framework facilitates the planner to learn the boundary of the motor skills of the robots [13].

To improve the adaptability and the intelligence of six-legged robots in diverse environments, a hierarchical planning and control system is proposed in this paper. In this two-layer framework, the bottom layer performs local motion planning and balance control, while upper layer adaptively learns the capabilities of the robot and generates global trajectory considering the capability embeddings of the robot. In the training process of the proposed framework, a Sim-to-Real pipeline is employed for efficiency and safety, where the robot is modeled in the simulator to train the policy for the planning and control system.

2 Methods

2.1 Sim-to-Real Training Framework

To simulate the kinematics and dynamics motion of the robots, the model of the six-legged robot is created in the simulator. In this process, the identified parameters of the real robot directly influence the accuracy of the simulated model. As shown in Fig. 1, the physical model reflects the topological motion characteristics of the six-legged robots, and the control policy is optimized in the simulation environment by utilizing reinforcement learning methods. Since a gap between reality and simulation still exists, data from the real-world data are utilized during training. The learned control policy network can be deployed on a real robot and evaluated in indoor and outdoor environments.

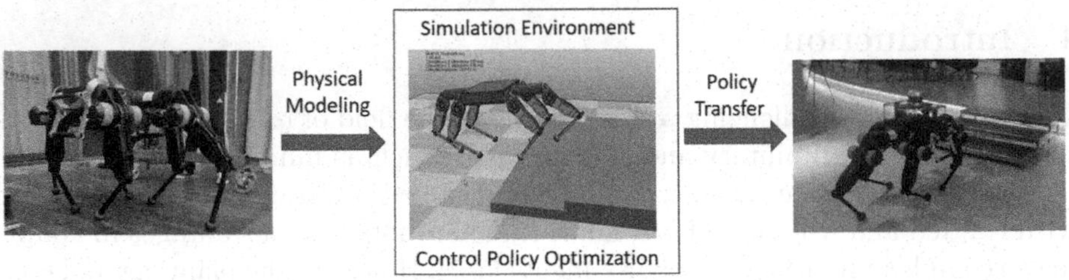

Fig. 1. Sim-to-real legged-robot policy learning process. Policy learned in the simulation environment can be transferred to a real robot.

2.2 Hierarchical Planning and Control System

To improve the accuracy and the robustness of the control policy for the six-legged robot, a hierarchical architecture is presented to integrate planning and control modules. As shown in Fig. 2, the hierarchical adaptive planning framework is designed, which has the advantages of efficient and accurate computation for global planning. This hierarchical structure contains a bottom and an upper layer. The bottom layer is responsible for the balance and local motion control of the six-legged robot, and the upper layer learns the capabilities of the bottom controller and provides the guiding trajectory by considering the surrounding height map and the target location. In the training process, the control module consists of an active compliance control method based on the whole-body dynamics model and an impedance control method for a single leg based on the foot-ground contact stiffness identification model.

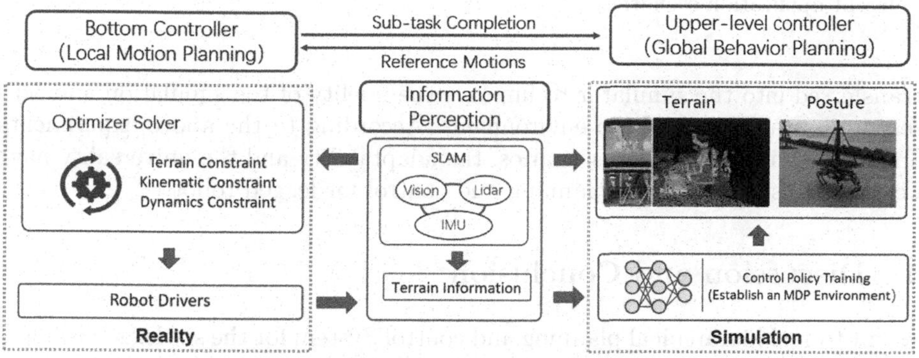

Fig. 2. The overall framework of proposed hierarchical planning and control system. The bottom-level controller is designed to control the robot's motion in real time according to the instructions from upper-level policy learned in the simulation environment.

3 Experiments

Based on the proposed intelligent system, a variety of six-legged robots have been developed and applied to different scenarios, as shown in Fig. 3. The mechanical structure of the high-load six-legged robot is designed based on the bionics of an octopus. The mechanism and motion topology of the robot are complex, hence the design of the controller is a complicated task. With the support of the proposed hierarchical system, the robot can be modeled and trained in the simulator, and then the system with the learned policy can be deployed to the real robot for real-time execution. In terms of the guide-dog robot, the employment of the sim-to-real framework can avoid the potential risks caused by unexpected behaviors in policy training and physical human-robot interaction.

For skiing and skating robots, the uncertainty and complexity of the environment is the major challenge for the planning and control system of the robots. Hence in the training process, the environment characteristics are parameterized to generate diverse training scenarios for better adaptability of the system. On the other hand, the data of the robot and the environment is collected and

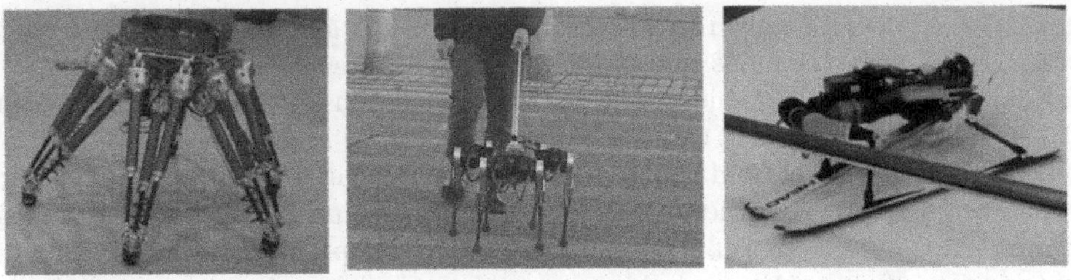

(a) High-load robot (b) Guide-dog robot (c) Skiing robot

Fig. 3. The experiments of several six-legged robots based on the sim-to-real planning and control system. The effectiveness of the proposed system has been verified in different application scenarios.

transferred into the simulator to improve the fidelity of the simulation and represent the variability of the environment. According to the above experiments and applications in various scenarios, the adaptability and the universality have been demonstrated on the planning and control for legged robots.

4 Discussion and Conclusion

A sim-to-real hierarchical planning and control System for the six-legged robot is proposed in this paper. The implementation and experiments show the feasibility and effectiveness of this framework in practical scenarios. In the future, more work can be done to increase the efficiency of each module. For instance, the efficiency of data utilization during training and the accuracy of model construction can affect the performance of this system.

References

1. Usha, M.N., Priyadharshini, S., Shree, K.R., Devi, P.S., Sangeetha, G.: Military reconnaissance robot. Int. J. Adv. Eng. Res. Sci. **4**(2), 237036 (2017)
2. Zhang, Z., Chen, S.: Real-time seam penetration identification in arc welding based on fusion of sound, voltage and spectrum signals. J. Intell. Manuf. **28**(1), 207–218 (2017)
3. Xiao, A., Tong, W., Yang, L., Zeng, J., Li, Z., Sreenath, K.: Robotic guide dog: leading a human with leash-guided hybrid physical interaction. In: IEEE International Conference on Robotics and Automation, pp. 11470–11476 (2021)
4. Al-dabbagh, A.H., Ronsse, R.: A review of terrain detection systems for applications in locomotion assistance. Robot. Auton. Syst. **133**, 103628 (2020)
5. Bartsch, S., Manz, M., Kampmann, P., Dettmann, A., Hanff, H., et al.: Development and control of the multi-legged robot mantis. In: International Symposium on Robotics, pp. 1–8 (2016)
6. Wellhausen, L., Hutter, M.: Rough terrain navigation for legged robots using reachability planning and template learning. In: IEEE/RSJ International Conference on Intelligent Robots and Systems, pp. 6914–6921 (2021)

7. Tennakoon, E., Peynot, T., Roberts, J., Kottege, N.: Probe-before-step walking strategy for multi-legged robots on terrain with risk of collapse. In: IEEE International Conference on Robotics and Automation, pp. 5530–5536 (2020)
8. Fang, L., Gao, F.: Type design and behavior control for six legged robots. Chin. J. Mech. Eng. **31**(1), 1–12 (2018)
9. De Santos, P.G., Cobano, J.A., Garcia, E., Estremera, J., Armada, M.A.: A six-legged robot-based system for humanitarian demining missions. Mechatronics **17**(8), 417–430 (2007)
10. Nie, B., Gao, Y., Mei, Y., Gao, F.: Capability iteration network for robot path planning. Int. J. Robot. Autom. **37**(3), 266–272 (2022)
11. Qin, B., Gao, Y., Bai, Y.: Sim-to-real: six-legged robot control with deep reinforcement learning and curriculum learning. In: International Conference on Robotics and Automation Engineering, pp. 1–5 (2019)
12. Erden, M.S., Leblebicioğlu, K.: Free gait generation with reinforcement learning for a six-legged robot. Robot. Auton. Syst. **56**(3), 199–212 (2008)
13. Hwangbo, J., et al.: Learning agile and dynamic motor skills for legged robots. Sci. Robot. **4**(26), eaau5872 (2019)

Author Index

Bai, Suli II-194
Bao, Qian II-559
Bian, Guibin III-439

Cai, Zebin II-33
Cao, Cong II-413
Cao, Jiale I-343
Cao, Jiantong II-84
Cao, Wenming I-42
Cao, Xun III-515
Chai, Jin I-280
Che, Keqin III-348
Chen, Bo II-339
Chen, C. L. Philip I-280
Chen, Enhong II-218, III-119, III-248
Chen, Hao I-560, III-180
Chen, Haoming II-339
Chen, Hongsheng I-584
Chen, Jingwen III-526
Chen, Jingyao II-291
Chen, Lei I-142
Chen, Linsen III-515
Chen, Liuqing II-33, II-303
Chen, Ming III-119
Chen, Siyu III-451
Chen, Wenbai I-330
Chen, Wensheng II-339
Chen, Xi III-554
Chen, Xiaoju III-167
Chen, Yaozong I-42
Chen, Yisheng I-280
Chen, Yunnong II-303
Chen, Zhengyan II-315
Chen, Zhengyin III-39
Chen, Zhi I-551
Chen, Zihao II-242
Chen, Zitan III-155
Cheng, Haijing III-543
Cheng, Mingyue III-248
Choi, Daewon I-526
Chu, Fuchen I-343
Cong, Xin I-179
Cui, Huili I-129
Cui, Jiashuai I-229
Cui, Lizhen II-438, III-559
Cui, Xiaosong III-26
Cui, Xiufen I-356
Cui, Zeyu II-496
Cui, Zhenchao I-538
Cui, Zhenhua I-80

Dai, Lingna III-615
Deng, Chenchen III-489
Deng, Jinrui I-658
Deng, Pingye II-291
Deng, Yue II-46
Ding, Hangqi I-293
Ding, Huiming I-490, II-59
Ding, Qing I-106
Ding, Shuxue II-375, III-167
Ding, Xilun I-538
Dong, Jing I-404
Dong, Jingjin III-521
Dong, Shuqin III-532
Dou, Hongkun II-46
Du, Junping III-236
Du, Qiuyu II-256
Du, Yichun II-33
Duan, Huiyu II-588

Fan, Kun III-439
Fan, Lei II-588
Fan, Lu I-490
Fang, Yanyan I-80
Feng, Qiao III-594
Fu, Yangqing III-621
Fu, Ying I-368, I-466

Gao, Fei III-615
Gao, Hao III-570
Gao, Hongxia I-584
Gao, Jiayi I-154
Gao, Shenghua I-80
Gao, Yue III-621
Gao, Zhuangzhuang I-622
Ge, Senlin III-570
Ge, Yuhao III-180
Gong, Maoguo I-293, III-374

Gu, Changzhan III-532, III-605
Guan, Bochen I-242
Guan, Weinan I-404
Guan, Zeli III-236
Guo, Jixiang II-256
Guo, Jun I-154
Guo, Ming I-356
Guo, Peini II-315
Guo, Wei II-206, III-559
Guo, Yaowei III-55
Guo, Zhenpeng III-223
Guo, Zhiqiang III-272

Han, Xiao II-206
Han, Yahong II-268, II-280
He, Mingrui II-33
He, Wei II-171
He, Weidong II-159
He, Yaqi III-71
He, Zhenyu III-414
Hou, Jingyi II-171
Hou, Muzhou II-413
Hou, Zeng-Guang II-291
Hounye, Alphonse Houssou II-413
Hu, Bo I-80
Hu, Fenyu II-496
Hu, Guangfeng II-182
Hu, Jian-Fang I-305, I-514
Hu, Menghan III-537
Hu, Xiaolin III-610
Hu, Xinyu III-402
Hu, ZhongZhi II-508
Hua, Chunjun III-537
Huang, Andi I-526
Huang, Chengyue II-230
Huang, Erqi III-515
Huang, Fei III-564
Huang, Hai II-3
Huang, Hanyun I-647, II-194
Huang, Meiyu III-55
Huang, Pingxuan I-80
Huang, Qian I-317, III-564
Huang, Ruqi I-3
Huang, Shasha I-584
Huang, Wei II-159
Huang, Xingming III-81
Huang, Xinshengzi I-67
Huang, Ye II-159
Huang, Yuhang II-559
Huang, Zhangjin I-622
Huang, Zhengliang II-450
Huang, Zhengxing II-362
Huang, Zhenya III-119
Huang, Zhouzhou III-389

Ji, Chaonan I-634
Ji, Cheng III-208
Ji, Mengqi I-3
Ji, Tongyu III-337
Ji, Xiangyang I-392
Ji, Yapeng II-15
Jia, Kunyang II-95
Jiahao, Chen III-426
Jiang, Chenyang III-615
Jiang, JinCen II-508
Jiang, Junzhe III-248
Jiang, Shangjun II-485
Jiang, Xiangming III-374
Jiang, Xiaoyu III-361
Jiang, Xuhao I-106
Jiang, Yuqi III-167
Jiao, Wenpin III-39
Jie, Biao II-626
Jin, Junwei II-462
Jin, Rui III-144

Kong, Ming II-362
Kong, Shuyi I-647, II-194
Kuang, Kun II-95, II-362, III-284

Lai, Jundong II-59
Lai, Yu-Kun I-129, III-594
Li, Ang III-236
Li, Chang I-317
Li, Di III-521
Li, Dian III-465
Li, Fan I-229
Li, Fang I-441
Li, Fei III-451
Li, Fengxing III-414
Li, Gang I-453
Li, Gao I-54
Li, Genghui III-325
Li, Hanxing III-543
Li, Hao I-293, I-526, III-374
Li, Heng I-647
Li, Hongjue II-46
Li, Huiping III-451
Li, Jing I-80
Li, Kun I-129, I-216, III-594

Li, Lei III-195
Li, Linhui II-601
Li, Manyi I-597
Li, Mengqi I-502
Li, Mengyao I-106
Li, Mingchen III-94
Li, Peng III-615
Li, Qingyan III-325
Li, Shanshan II-206
Li, Teng II-350
Li, Weiming II-256
Li, Weiyi I-242
Li, Xiangxian I-416, I-572, I-597, III-554
Li, Xin II-425, III-272
Li, Xixi II-182
Li, Xuelong III-554
Li, Yanting II-462
Li, Yidi II-315
Li, Yuchen III-605
Li, Yujie II-375, III-167
Li, Zejian I-30, II-520
Li, Zhenini III-81
Li, Zhenyu III-310
Li, Zhi III-248
Li, Zi-Xin III-71
Lian, Jing II-601
Liang, Guoheng I-584
Liang, Hailun II-230
Liang, Kongming I-154
Liang, Shufan II-256
Liang, Xingwei III-298, III-310
Liang, Zhiwei III-195
Liao, Jun II-520
Liao, Yihui I-490
Lin, Guobin III-414
Lin, Haodong II-532
Lin, Jun III-559
Lin, Qiuzhen III-132
Lin, Rui II-112
Lin, Siyou I-634
Lin, Zhu III-489
Ling, Xiao I-356
Ling, Zhenhua III-272
Liu, Bo III-521
Liu, Chen I-191
Liu, Chongliang II-182
Liu, Daowei II-147
Liu, Gaosheng I-268
Liu, Haohan II-3
Liu, Hong II-315, II-327
Liu, Jian II-268
Liu, Jiawei II-171
Liu, Jing II-425
Liu, Jinxing I-416
Liu, Kun II-559, III-39
Liu, Li II-520
Liu, Lijuan III-272
Liu, Mengya I-429
Liu, Ning II-438
Liu, Qi II-159, III-119, III-248
Liu, Qiang II-496
Liu, Qiaoqiao I-204
Liu, Qiong I-67
Liu, Shaohua II-46
Liu, Siming III-564
Liu, Wu II-559
Liu, Xudong II-71
Liu, Xuejiao III-55
Liu, Yan II-112
Liu, Yanhong III-439
Liu, Yanli I-242
Liu, Yanzhe III-26
Liu, Yebin I-634, III-594
Liu, Yifei III-284
Liu, Yuhui III-501
Liu, Zhaoci III-272
Liu, Zhenyi I-634
Liu, Zhigui II-450
Liu, Zhihao I-478
Liu, Zhihong III-402
Liu, Zhijie II-171
Liu, Zuozhu III-180
Lu, Feng III-521
Lu, Jingyun III-605
Lu, Ming I-356, II-545
Lu, Wei II-136
Lu, Weiming II-95
Lu, Weiwen II-399
Lu, Yunhua III-223
Luan, Yingxin II-350
Luo, Guocui III-564
Luo, Shunli III-543
Luo, Yong III-260
Lv, Tao III-515

Ma, Chao II-532
Ma, Haokai I-416, I-572, III-554
Ma, Lijia III-132
Ma, Lizhuang I-191, II-59
Ma, Nan II-387

Ma, Teng II-438
Ma, Wenping I-293
Ma, Zhanyu I-154
Mei, Kangdi III-272
Meng, Lei I-416, I-572, I-597, III-155, III-554
Meng, Xiangxu I-416, I-572, I-597, III-155, III-554
Miao, Chunyan III-559
Miao, Qiguang I-293
Miao, Yisheng II-206
Miao, Zhuang I-317
Miao, Ziling II-327
Min, Xiongkuo II-136, II-588
Ming, Li II-182
Muntean, Gabriel-Miro II-242

Nie, Yuzhou II-230
Niu, Shicheng I-584

Ou, Shifeng III-15

Pan, Binbin II-339
Pan, Gang III-599
Pan, Shi III-543
Pan, Wenjie I-255
Pan, Yingwei III-526
Pang, Yanwei I-343
Pang, Yunhe III-15
Pei, Zongxiang II-577
Peng, Bo I-404
Peng, Chenglei I-478
Philip Chen, C. L. II-462

Qi, Jing I-538
Qi, Yu III-548
Qi, Zhuang I-416, I-572, III-155
Qian, Wen I-502
Qiao, Jianbo II-438
Qiao, Ying II-147
Qin, Haibo I-142
Quan, Shuxue I-242

Ren, Xiaoyu II-588
Renzong, Lian II-613

Shan, Xiangxuan I-154
Shang, Kejun II-182
Shao, Wei II-577
Shao, Zengyang III-132
Shao, Zhuang I-343
Shen, Ao III-107
Shen, Liquan I-106
Shen, Wei II-588
Shen, Weinan I-280
Shi, Fangyu II-588
Shi, Huitao III-570
Shi, Yiqiao III-537
Song, Limei I-429
Song, Ran II-350, III-477
Song, Zhenbo III-582
Su, Ri II-413
Su, Yuxin II-327
Sun, Bingcai I-658
Sun, Chang II-124
Sun, Kailai I-166
Sun, Liang II-577, III-107
Sun, Lingyun I-30, II-33, II-303, II-520
Sun, Ming III-621
Sun, Shuangpeng III-543
Sun, Wei II-136
Sun, Weilin I-597
Sun, Xi III-582
Sun, Yougang III-414
Suo, Jinli I-634

Tan, Benying II-375, III-167
Tan, Chaolei I-305, I-514
Tan, Jianrong II-124
Tan, Tieniu I-404, II-496
Tan, Xin I-191
Tang, Hongan III-223
Tang, Huimei III-132
Tao, Dacheng I-18
Tao, Xiaofeng II-242
Tian, Jingqi II-545
Tian, Kang III-501
Tu, Geng III-310

Wan, Fang I-392
Wan, Jiaqiang III-223
Wan, Xiaohong II-399
Wan, Xing II-3
Wang, Chao III-337
Wang, Chen II-291
Wang, Cheng II-387
Wang, Chengdi II-256
Wang, Chengjia I-658
Wang, Cong II-218
Wang, Dahui II-399

Wang, Dawei III-477
Wang, Guan III-465
Wang, Guanzheng III-402
Wang, Haosong II-71
Wang, Huimin I-538
Wang, Jing II-450
Wang, Jun III-298
Wang, Junkui III-81
Wang, Kuiyu III-610
Wang, Le II-124
Wang, Lei III-195
Wang, Liang II-496
Wang, Liejun I-229
Wang, Linlin II-588
Wang, Liping III-477
Wang, Lulu I-317
Wang, Manman II-375
Wang, Meiling II-577
Wang, Min III-361
Wang, Nannan III-615
Wang, Qianlong III-298
Wang, Qiuli I-191
Wang, Ran III-155
Wang, Runyu III-374
Wang, Sanpeng II-112
Wang, Shuai II-462
Wang, Shuaiping I-441
Wang, Shunfei I-356
Wang, Shunli I-441
Wang, Sidong II-399
Wang, Suming III-337
Wang, Tao II-136
Wang, Wei I-404
Wang, Weidong III-465
Wang, Xiangke III-402
Wang, Xiaohao I-330
Wang, Xichao III-15
Wang, Xinwei I-166
Wang, XiTing II-508
Wang, Xu II-242
Wang, Xuanqian III-521
Wang, Xueqi I-658
Wang, Xueqian I-453
Wang, Yangang III-588
Wang, Yibo I-502
Wang, Yueming III-548
Wang, Yu-Long III-71
Wang, Yuqing I-572, I-597, III-155
Wang, Yuyan III-489
Wang, Zefan III-610
Wang, Zeke II-95
Wang, Zhengdong II-626
Wang, Zhengyong I-106
Wang, Zhenkun III-325
Wang, Zhibo I-356, II-545
Wang, Zhihao I-453
Wang, Zili II-124
Wei, Jiahao III-180
Wen, Chao II-473, III-582
Wen, Ting III-361
Wen, Yaqing III-3
Wen, Yonggang III-260
Wen, Zhiyuan III-298
Weng, Weiyu III-615
Wu, Fei II-95, II-362, III-284
Wu, Hang III-554
Wu, Hanxiao I-255
Wu, Jiamin III-489
Wu, Jingyu I-30, II-520
Wu, Jinlin II-496
Wu, Jinze III-119
Wu, Kaifang II-425
Wu, Le II-425
Wu, Shu II-496
Wu, Xiaodi III-548
Wu, Yiquan III-284
Wu, Yue I-293, III-374, III-537
Wu, Zhichao I-429
Wu, Zhixuan II-387
Wu, Zhonghao II-532

Xia, Chenghao III-615
Xiang, Xueshuang III-55
Xiang, Zhonghao I-429
Xiao, Jiatong I-429
Xiao, Junjin I-416
Xiao, Shuhong II-303
Xiao, Tong II-159
Xiao, Wenpeng I-93
Xie, Dong III-582
Xie, Shengli III-81
Xie, Yu II-473
Xie, Zaipeng III-208
Xie, Zhifeng I-490, II-59
Xing, Congying I-54
Xiong, Huimin III-180
Xiong, Pu III-374
Xiuzhi, Li III-426
Xu, Changlin III-3
Xu, Cheng I-93

Xu, Feng I-356, II-545
Xu, Hao III-414
Xu, Hongteng II-230
Xu, Kaiyue II-147
Xu, Kedi III-548
Xu, Long II-242
Xu, Manchen III-501
Xu, Mengting III-107
Xu, Ningcun II-291
Xu, Qinwen I-242
Xu, Ruifeng III-298, III-310
Xu, Tong II-218
Xu, Xinhang II-473
Xu, Xiuyuan II-256
Xu, Xuemiao I-93
Xu, Yangyang I-117
Xu, Yanmin II-59
Xu, Yonghui II-438, III-559
Xu, Yufei I-18
Xuan, Haibiao I-216
Xue, Zhe III-236
Xue, Zhou III-582

Yan, Anbang III-564
Yang, Bin III-389
Yang, Chun I-142
Yang, Hongxia II-95
Yang, Jiangchao II-95
Yang, Jing I-129
Yang, Jingyu I-268, III-594
Yang, Lei I-647, II-194
Yang, Lingjie III-402
Yang, Shaofu III-348
Yang, Tan II-84
Yang, Weipeng I-584
Yang, Wenyu III-389
Yang, Xiao II-438
Yang, Xiaokang II-136, II-532, II-588, III-532, III-605
Yang, Yang II-626
Yang, Yonghui II-425
Yang, You I-67
Yang, Yu I-392
Yang, Zhiji III-94
Yao, Ting III-526
Yao, Yongqiang II-387
Ye, Qixiang I-392
Yi, Hui II-485
Yi, Lin II-613
Yi, Zhang II-256
Yin, Aihua II-399
Yin, Xucheng I-142
Yin, Yu III-119
Yin, Zijin I-154
Yin, Zikai II-218
Yong, Wang II-613
You, Hongzhi II-399
Yu, Bin II-473
Yu, Di III-501
Yu, Hongnian III-439
Yu, Huan III-570
Yu, Liangwei I-106
Yu, Shaoliang III-489
Yu, Tao I-634
Yu, Yemin II-95
Yu, Zhipeng III-588
Yuan, Haitao II-438
Yuan, Tongtong I-380
Yuan, Ziang II-84
Yuankai, Wu II-613
Yue, Huanjing I-268
Yue, Shiyu II-303
Yue, Zhuang II-450

Zeng, Dan II-559
Zeng, Huanqiang I-255
Zeng, Yuhang I-368
Zhai, Guangtao II-136, II-588
Zhan, Yibing I-551, I-609, III-144
Zhang, Anan III-361
Zhang, Bo I-330
Zhang, Chen II-473
Zhang, Chutian II-291
Zhang, Daoqiang II-577, III-107
Zhang, Dengming II-303
Zhang, Gengchen I-368
Zhang, Hao III-477
Zhang, Huaidong I-93
Zhang, Jianmin III-548
Zhang, Jin II-387
Zhang, Jing I-18
Zhang, Jinsong I-216
Zhang, Kuan III-439
Zhang, Kun II-425
Zhang, Lefei I-54, I-117
Zhang, Lihua I-441, I-560
Zhang, Na II-171
Zhang, Pu II-291
Zhang, Qi I-204
Zhang, Qiming I-18

Zhang, Qingzhu I-380
Zhang, Rong III-144
Zhang, Ruyuan II-399
Zhang, Shengyuan I-30, II-520
Zhang, Shi-Xue I-142
Zhang, Shuyou II-124
Zhang, Tong I-280
Zhang, Wei II-350, III-477
Zhang, Xiaoqian II-450
Zhang, Xin II-159
Zhang, Xinliang II-601
Zhang, Xinwen III-260
Zhang, Xuming I-502
Zhang, Ye III-26
Zhang, Yifan II-601
Zhang, Yiheng III-526
Zhang, Ying I-658
Zhang, Yingkai I-466
Zhang, Yiyuan III-615
Zhang, Yuanming III-570
Zhang, Yufeng III-208
Zhang, Yuqi III-599
Zhang, Yuru II-339
Zhang, Yutao III-132
Zhang, Zhenyu II-532
Zhang, Zhixiang II-626
Zhang, Zhongwei III-526
Zhang, Zichen III-260
Zhang, Zicheng II-136
Zhao, Chunjiang I-330
Zhao, Guo I-609
Zhao, Haoli III-81
Zhao, Qianchuan I-166
Zhao, Shuaitao II-559
Zhao, Tianqi II-362
Zhao, Xiaochen I-634
Zhao, Yan I-560, II-601
Zhao, Yaping I-3
Zhao, Zeyu I-478
Zhen, Yankun II-33
Zhen, Zonglei II-399
Zheng, Chu II-112
Zheng, Chunyan III-284
Zheng, Haitian I-3
Zheng, Jiyuan III-489
Zheng, Qian III-599
Zheng, Wei-Shi I-305, I-514
Zheng, Yuze I-597, III-554
Zheng, Zerong I-634
Zheng, Zhi II-218
Zhong, Guoqiang I-526
Zhou, Changzhi III-564
Zhou, Fei II-280
Zhou, Guoxu II-15
Zhou, Guoyu I-538
Zhou, Jie II-626
Zhou, Jinglin III-615
Zhou, Siying III-284
Zhou, Tingting II-33, II-303
Zhou, Yangxi III-236
Zhu, Jianqing I-255
Zhu, Jingjie III-615
Zhu, Junming III-548
Zhu, Mingrui III-615
Zhu, Qiang II-362
Zhu, Suguo I-609
Zhu, Wenhan II-136
Zhu, Xiaobin I-255
Zhu, Xiaowei I-317
Zhu, Xinyun III-548
Zhu, Yucheng II-588
Zi, Lingling I-179
Zou, Hang I-204
Zou, Jianhong I-166
Zou, Li III-26
Zou, Longhao II-242
Zuo, Xingquan II-3
Zuo, Yingli II-577